T. E. Timell

Compression Wood in Gymnosperms

Volume 1

Bibliography, Historical Background, Determination, Structure, Chemistry, Topochemistry, Physical Properties, Origin, and Formation of Compression Wood

With 341 Figures

Springer-Verlag
Berlin Heidelberg New York Tokyo

T. E. TIMELL
State University of New York
College of Environmental Science and Forestry
Syracuse, New York 13210, U.S.A.

ISBN 3-540-15715-8 (in 3 Volumes) Springer-Verlag Berlin Heidelberg New York Tokyo
ISBN 0-387-15715-8 (in 3 Volumes) Springer-Verlag New York Heidelberg Berlin Tokyo

Library of Congress Cataloging in Publication Data. Timell, T. E. (Tore E.), 1921– . Compression wood in gymnosperms. Contents: v. 1. Bibliography, historical background, determination, structure, chemistry, topochemistry, physical properties, origin, and formation of compression wood –– v. 2. Occurrence of stem, branch, and root compression woods... –– v. 3. Ecology of compression wood formation, silviculture, and compression wood... 1. Compression wood––Collected works. 2. Gymnosperms––Collected works. I. Title. QK647.T5 585 85-25035
ISBN 0-387-15715-8 (U.S.:set)

This work is subject to copyright. All rights are reserved, whether the whole or part of the material is concerned, specifically those of translation, reprinting, re-use of illustrations, broadcasting, reproduction by photocopying machine or similar means, and storage in data banks.

Under § 54 of the German Copyright Law where copies are made for other than private use, a fee is payable to "Verwertungsgesellschaft Wort", Munich.

© Springer-Verlag Berlin Heidelberg 1986
Printed in Germany

The use of registered names, trademarks, etc. in this publication does not imply, even in the absence of a specific statement, that such names are exempt from the relevant protective laws and regulations and therefore free for general use.

Product Liability: The publisher can give no guarantee for information about drug dosage and application thereof contained in this book. In every individual case the respective user must check its accuracy by consulting other pharmaceutical literature.

Typesetting, printing and bookbinding: Konrad Triltsch, Graphischer Betrieb, D-8700 Würzburg
2131/3130-543210

To Anna

Preface

> *Felix qui potuit rerum cognoscere causas.*
>
> *Publius Vergilius Maro*

Forests have long been one of mankind's most important natural resources and not the least because forests are, if properly managed, renewable. They serve us in many different ways, but above all in providing us with wood, one of the most remarkable and useful of all natural materials. Reaction wood, compression wood in gymnosperms and tension wood in the arboreal angiosperms, serves the function of making it possible for trees to perform movements. In the ancient Ginkgo and in the conifers, the ability to form compression wood is of vital importance to each and every tree. Compression wood plays a crucial role in the regulation of tree form in these gymnosperms, and their arborescent habit probably depends on their ability to develop this tissue. Few forest and plantation trees are devoid of compression wood in their stem, and all of them have it in their branches. Unfortunately, what is necessary and beneficial for the tree in this case is harmful to mankind, for compression wood is a very serious defect in both sawtimber and pulpwood.

It is now almost 20 years since the last complete survey of compression wood was published, namely Arthur H. Westing's excellent review of this subject. My major objective in writing the present book has been to bring together in one single work everything that is currently known about compression wood. I have had two aims, namely, first, to summarize the entire field in its different facets and to prepare a book that would be comprehensive in both scope and bibliography. My second aim has been to trace the historical development of our present knowledge, reaching back to its beginnings. This seemed especially important in view of the fact that so much of the early research on compression wood has been forgotten or neglected by later investigators. In this context, I discovered, for example, that compression wood had been scientifically described for the first time in 1860 by K. G. Sanio. Many other instances could be mentioned, all of them testifying to the truth of Santayana's dictum that those who cannot remember the past are condemned to repeat it. One consequence of this historic approach has been that I have on occasions included older illustrations where I could have used more recent material of higher quality, my intention being to give credit to early investigators.

To the best of my ability I have written this book sine ira et studio, always trying to arrive at the truth, without bias or prejudice. Usually, I have let the facts speak for themselves, but when I have felt some comment or correction to be called for, I have not hesitated to take a stand, for this work was never intended to be an uncritical compilation of data. I realize that on occasion I might well have offended some fellow scientists. All I can do is to assure them that my criticism was never intended to be ad hominem. It was a greater scientist than any of us who said 2300 years ago "Amicus Plato, sed magis amica veritas."

The subject matter of this book encompasses many different disciplines, including botany, chemistry, cytology, ecology, entomology, forestry, genetics, mycology, paleobotany, physics, plant physiology, pulping technology, wood anatomy, wood technology, and zoology. Experts in these fields will undoubtedly find mistakes and omissions in this book. It might, indeed, seem presumptuous to attempt a work of this encyclopedic nature in an age of ever increasing specialization in the sciences. At first, I spent several years attempting to prepare myself for the writing of this book. Like a greater author before me, however, I eventually found that it is too late to look for instruments when the work calls for execution, and that whatever abilities I had brought to my task, with those I must finally perform it. My only qualification for writing this book has really been that I was willing to read and learn, and all that I can hope for now is that this single talent has been reasonably well employed.

If it is true, as it has been said, that a man will turn over half a library to make one book, then I have probably done so. A deliberate attempt has been made to account for all literature directly and most of that indirectly concerned with compression wood. Every effort has also been exerted to include results of recent research. The literature search was completed on October 15 of 1984, and the entire manuscript was delivered to the publisher's offices on October 18.

When one is discussing a subject with so many tracks as this, it is not surprising that the same station is passed more than once. Notwithstanding my attempts to avoid repetitions, there is some overlap between Chapters 10 and 14 regarding the incidence of compression wood in various conifer species and varieties. I am also aware of the fact that the relation between the angle of inclination and the extent of compression wood formation is considered in three places, namely in Chapters 10, 12, and 14. I have tried to alleviate such deficiencies by a liberal use of cross references between the different chapters. Tension wood has been mentioned when necessary, but comparisons have been restricted since this kind of reaction wood will form the subject of another book, now in preparation.

I have used the English measuring system or the older dimensions in the metric system throughout this book, occasionally converting some of the English into corresponding metric units. The SI system I have avoided entirely, and notably when expressing the density of wood, thus rendering the metric density numerically equal to the specific gravity. My reasons for not using the SI unit Pascal for pressure are the same as those expressed last year by my friend, the late Martin H. Zimmermann, when he wrote in the introduction to his last book

Xylem Structure and the Ascent of Sap: "The SI pressures are very abstract units. They may be useful for engineers, but certainly not for biologists. In the range we use them, we even have to switch back and forth between MPa and kPa, because of their awkward size. Thus I see nothing but disadvantages in using SI units in dealing with pressures in plants. I do not mind being considered either old-fashioned or reactionary for the advantage of being practical; units are supposed to be our servants, we are not their slaves!"

The organization of the subject matter of this book is largely straightforward. Chapters 11, 12, and 13, however, are an exception in this respect, and here I cannot hope to satisfy those who are perhaps not inclined to be pleased, since I have not been able to satisfy myself. I wanted to describe the classical experiments on the formation of compression wood at some length and in chronological order. Since the mechanism by which a tree initiates differentiation of compression wood still eludes us, it was thought best to confine this subject to a separate chapter (11), terminated by a critical summary. In Chapter 12 compression wood formation is treated as a gravitropic phenomenon together with a review of the effect of gravity on plants other than trees. Chapter 13 is devoted to physiological aspects and particularly to the effect of auxin. The related subjects of apical control in trees and the formation of normal earlywood and latewood are also discussed here.

The numerous ecological factors responsible for the occurrence of compression wood in trees have never been reviewed before. I have tried to account for most of the agencies that cause tree stems and branches to become displaced and a leader to be injured or destroyed. Both phenomena are associated with formation of often massive amounts of compression wood, when the stem and branches are returned to their original location, and the leader is replaced by one or several laterals. The factors involved here are of the most diverse kind, such as wind, snow, ice, hail, light, mammals, insects, and fungi. Despite the length of this chapter, I know only too well that it must be incomplete, and that much has probably been inadvertently omitted. The one insect that is able to induce directly the formation of compression wood, namely the balsam woolly aphid, is discussed in more detail in Chapter 20.

When I began my research on compression wood 20 years ago, I had high hopes that at some time in the near future the ultimate cause of compression wood initiation would become known. This happiness of getting to know the causes of things, to use Vergilius's phrase, we have, alas, not yet attained. Formation of compression wood is obviously a gravitropic phenomenon, yet it is also evident that gravity alone cannot be the primary cause. We have no information on how the unknown stimulus is perceived by the tree, nor do we know whether auxin plays any role in its transmittance. If this book should stimulate further research on this problem, it would have served at least one useful purpose. Compression wood, as pointed out by Edmund Sinnott, offers a very favorable material for morphogenetic research, at the same time as it is a problem of organic form in one of its simplest, yet also most puzzling manifestations.

I have spent many years and probably the best part of my life in preparing this book. Now that the work has been completed, I cannot but look upon it with some pleasure, not because it is at last finished and certainly not because of its merits, which are highly uncertain and perhaps even nonexistent, but because I enjoyed every minute of the time that I managed to devote to this book. Walking in the woods observing trees, working at the electron microscope, searching the literature in various libraries, reading, and writing, all were equally enjoyable. During the last few months, I even came to like the typing of the 8100 references that form the backbone of this book.

I am indebted to an unusually large number of persons and organizations for all the assistance that they have given me while I was preparing this book, not to mention my family and my publisher. The acknowledgments are so numerous that they had to be listed separately. I only wish to point out how indebted I also am to all those investigators and writers who have contributed to the subject matter of this work, which really is their book as much as it is mine. What Samuel Johnson said to Boswell on July 28, 1763 is no less true today than it was then: "Human experience, which is constantly contradicting theory, is the great test of truth. A system, built upon the discoveries of a great many minds, is always of more strength, than what is produced by the mere workings of one mind, which, of itself, can do little. There is not so poor a book in the world that would not be a prodigious effort were it wrought out entirely by a single mind, without the aid of prior investigators."

Syracuse, New York
December, 1985

T. E. TIMELL

Acknowledgments

*Es würde wenig von mir übrig bleiben
wenn ich alles abgeben müsste was
ich andern verdanke.*

J. W. Goethe

A full acknowledgment of all the assistance that I have received from so many individuals, organizations, and institutions while preparing this book would be impossible within the present confines. Only the most obvious instances of the help and services that made this work possible can be included here.

The resources and facilities of many libraries were indispensable for the preparation of this book. I am first and foremost indebted to the fine Moon Memorial Library at my own college. I wish to express my gratitude to the late T. J. Hoverter and to the present director of the Moon Library, D. F. Webster, and their faculty and staff. Particularly, I want to thank J. J. Petraites for his tireless efforts in providing me with literature from other libraries. I have also made full use of the resources of other, nearby libraries, such as the various science libraries at Syracuse University and, especially, the Mann Library at Cornell University. Francois Mergen and J. A. Miller kindly made available to me the facilities of the unique library of the School of Forestry and Environmental Studies at Yale University during two visits to New Haven.

In the course of my first sabbatic year, which was spent at the Institut für allgemeine Botanik at the ETH in Zürich, Albert Frey-Wyssling generously allowed me to use his departmental library, at that time still containing all the literature on reaction wood once collected by his predecessor Paul Jaccard. During my second sabbatic year, Horst Schulz and K. A. Sorg made it possible for me to work in the library of the Institut für Holzforschung in München. Wolfgang Knigge was most helpful in connection with a visit to Göttingen. I am also indebted to several other libraries, namely those at the Forest Products Laboratory in Madison, the Institute of Paper Chemistry in Appleton, the College of Forest Resources in Seattle, the Pulp and Paper Research Institute of Canada in Pointe Claire, and the Swedish Forest Products Research Laboratory in Stockholm.

Mrs. Marie Marchetti typed the entire text of this book not just once but a total of four times, beginning with my original, hand-written manuscript. I could never have prepared this book without her help, and I owe her a very special gratitude, not due to anybody else outside my own family. I am indebted to David Evers and Richard Correll for assistance with practically all line drawings. Photographic copies of these drawings, of all chemical formulas, and of many of the other illustrations were obtained through the courtesy of G. A. Snyder and his staff at this college, for which I am very grateful.

Exactly one quarter of the illustrations in this book are published here for the first time, most of them having been prepared for this purpose. I am indebted to W. A. Côté, Jr., D. Fengel, and K. Mühlethaler for making available to me the facilities of their electron microscopy laboratories in Syracuse, München, and Zürich, respectively. A. C. Day, M. Stoll, and M. V. Parthasarathy kindly taught me the necessary techniques. A. C. Day and J. J. McKeon advised me concerning the preparation of prints. I am indebted to Wilfred A. Côté, Jr. and Arnold C. Day for providing me with many light micrographs, some of them published here for the first time. A. C. Day and A. Rezanowich (Pulp and Paper Research Institute of Canada) generously assisted me in obtaining most of the scanning electron micrographs.

Many of the photographs of trees and tree parts in this book were obtained in the Heiberg Memorial Forest at Tully, New York, one of the several forest properties of this college. I wish to express my appreciation to J. H. Engelken for advice and for information regarding the history of the various conifer plantations.

While preparing this book, I approached literally hundreds of scientists, asking for original prints of already published illustrations. It is a remarkable fact that, with only one single exception, all of them most generously made available to me the requested material. Similarly, all publishers except one kindly granted me permission to reproduce previously published tables and figures without any charge. I am most thankful to both authors and publishers, but I cannot acknowledge here the sources of all previously published material. Such acknowledgments are made in the caption of the tables and in the legend of the figures by reference to the bibliography. I would be remiss, however, if I did not mention how indebted I am to the United States Department of Agriculture Forest Service for their prompt and courteous assistance. The ability of the USDA to make available all its published material, no matter how old, greatly facilitated the writing of this book. I have also incurred a deep obligation to the Canadian Forestry Service and to the Forestry Commission of Great Britain.

Several friends and colleagues kindly provided me with specimens of compression wood from various conifer species which I could not collect myself. Such samples were obtained from C. L. Brown (*Cedrus deodara*), D. Churchill and G. Scurfield (*Agathis robusta* and *Araucaria bidwillii*), H. A. Core (*Ginkgo biloba*), H. Harada and T. Koshijima (*Chamaecyparis obtusa, Cryptomeria japonica,* and *Sciadopitys verticillata*), R. Krahmer (*Libocedrus decurrens*), R. W. Kennedy (*Chamaecyparis lawsoniana, C. nootkatensis, Cupressus arizonica,* and *C. macrocarpa*), and G. Scaramuzzi (*Cupressus sempervirens*). Fred Shaw (Ginkgo Gem Shop, Vantage WA) generously gave me a specimen of petrified Douglas-fir wood containing compression wood. M. Bariska, finally, helped me in identifying a spruce tree growing in a park in Zürich. I wish to express my sincere gratitude to all these persons for their valuable help.

I am indebted to several scientists for providing me with translations into English of articles written in languages not accessible to me. Arthur H. Westing kindly gave me a translation of Kononchuk's (1888) Russian report on compression wood, and George Batki translated Zherebov's (1946) article from the Russian. Renata Marton assisted me with various communications in Polish. Ladislav J. Kucera was most helpful in translating from the Czech Douda's (1948) paper on compression wood formation. Jurasek's (1964) two papers on biodegradation of normal and compression woods were translated from the Slovak by M. Mahdalik. I am greatly indebted to K. Shimada for translating from the Japanese several of Onaka's (1935, 1937, 1940) important papers. M. Yumoto kindly provided me with a translation from the Japanese of an article on spiral thickenings in compression wood tracheids by Takaoka and Ishida (1974). Several colleagues helped me with the names of the two kinds of reaction woods, in their language. I wish to express my appreciation for this assistance to Øystein Ellefsen (Norwegian), Pentti Hakkila (Finnish), and Vladimir Nečesaný (Czech, Polish, Russian, and Slovak).

I am sincerely thankful to Emanuel Fritz for providing me with detailed information about the Flatiron Tree in the Humboldt Redwoods State Park in California. Robert W. Kennedy kindly brought my attention to the unreported presence of spiral compression wood in a black spruce tree, first observed by L. A. Josza and M. L. Parker in 1972. I am very grateful to W. O. Wilen for showing me the unique grove of large Norway spruce trees at Ware, Massachusetts, where the undamaged leader seems to have lost all control over the lower branches.

I am deeply indebted to Lore Kremser for her kind interest in this work. She generously gave me a copy of the remarkable book published in 1942 by her father, Franz Hartmann, *Das statische Wuchsgesetz bei Nadel- und Laubbäumen,* together with several of the original illustrations, and also decribed for me her father's research, in which she took an active part for many years as a professional forester.

Over the years, I have had the good fortune of receiving many manuscripts which at that time had not yet been published. I wish to express my appreciation for this assistance to R. R. Archer, J. D. Boyd, D. A. I. Goring, T. Koshijima, L. J. Kucera, H. H. Nimz, H. Schulz, N. Terashima, A. H. Westing, B. F. Wilson, and N. Yoshizawa. I am particularly indebted to Gerhard Casperson and to Masahide Yumoto for making available to me a large number of their unpublished manuscripts together with the original illustrations. H. H. Bosshard kindly obtained for me in Zürich Engler's (1918) and Jaccard's (1919) books on reaction wood. I am thankful to Walter Liese for providing me with Ascherson's biography of Karl Gustav Sanio. Several of the portraits in this book were obtained through the courtesy of K. A. Sorg. My brother-in-law Rolf Öfverholm kindly made available to me older books on Swedish silviculture from his extensive library.

During most of the years when this book was in preparation, I was fortunate in enjoying the important and steadfast support of R. E. Pentoney, formerly vice president of this college. I am thankful to J. D. Mabie for the innumerable times he has helped me during the last 20 years. In latter years, I have also received assistance from D. F. Behrend and J. W. Geis. For generous financial support, without which this book could not have been written, I am deeply indebted to my own college and to the USDA Cooperative State Research Service through the McIntire-Stennis program.

My friend and colleague R. Milton Silverstein for many years let me share his profound knowledge of the American language. For this, as well as for his constant, kind interest in this book I will always remain thankful. I have received much encouragement from Philip R. Larson to whom I owe a special gratitude for his friendly interest in this work and for his valuable, professional advice. I am also indebted to J. Maddern Harris for drawing my attention to recent research on compression wood. In the course of the last year, I have benefited from frequent correspondence and one visit with Reginald A. Balch who has generously shared with me experiences from his long life. His pioneering research on the balsam woolly aphid is only one facet of his wide-ranging interests in forest ecology. Arthur H. Westing provided me with the second of his two reviews of compression wood, at that time still in manuscript, and also made available to me his extensive bibliography. Later, he kindly read the manuscript of an early, brief version of this book. I wish to express my appreciation to him for this and also for his several helpful suggestions regarding the organization of the subject matter.

This book was initiated when the author was the L. T. Murray Distinguished Visiting Lecturer at the College of Forest Resources, University of Washington, Seattle. I am very grateful to my old friend and former colleague Kyösti V. Sarkanen, who not only provided me with a unique opportunity to teach a graduate course in wood science but also thereby set in motion the series of events that ultimately resulted in this book.

I am deeply indebted to my publisher, Springer-Verlag. First of all, I wish to express my sincere gratitude to Konrad F. Springer for his understanding and patience in the course of many years and for his hospitality during my visits to Heidelberg. I am very grateful to Dieter Czeschlik, my editor, for encouraging me to complete this work, for his wise counsel, for his willingness to offer assistance at all times, and for his kind interest in this endeavor. For the prompt publication and high quality production of this book I am indebted to J. von dem Bussche, D. Czeschlik, Th. Deigmöller, E. Schuhmacher, L. Teppert, and many others at Springer-Verlag in Heidelberg. I also wish to thank

Mark Licker at Springer-Verlag New York for much help. No author of a scientific book ever had a better publisher.

Last, but emphatically not least, I wish to express my deep gratitude to my family for their patience while I was working on this book, which often must have seemed destined never to be completed. I am particularly thankful to my long-suffering wife, to whom this book is most appropriately dedicated.

Syracuse, New York   T. E. TIMELL
December, 1985

Contents

Detailed contents appear at the beginning of each chapter

Volume 1

1	Introduction	1
2	Historical Background	29
3	Taxonomy, Designation, General Characteristics, and Determination of Compression Wood	45
4	The Structure of Compression Wood	81
5	Chemical Properties of Compression Wood	289
6	Distribution of Chemical Constituents	409
7	Physical Properties of Compression Wood	469
8	Origin and Evolution of Compression Wood	597
9	Formation of Compression Wood	623

Volume 2

10	Incidence and Occurrence of Compression Wood	707
11	Fundamental Factors Causing Formation of Compression Wood	983
12	Gravitropism and Compression Wood	1105
13	Physiology of Compression Wood Formation	1183
14	Inheritance of Compression Wood	1263

Volume 3

15	Ecology of Compression Wood Formation	1339
16	Silviculture and Compression Wood	1673
17	Mechanism of Compression Wood Action	1745
18	Compression Wood in Lumber, Plywood, and Board Manufacture	1799
19	Compression Wood in Pulp and Paper Manufacture	1831
20	Compression Wood Induced in Firs by the Balsam Woolly Aphid (*Adelges piceae*)	1907
21	Opposite Wood	1969

Gymnosperm Species
 Scientific Name, Author, Common Name, and Number
 of Times Cited 1999

Subject Index 2003

Abstract, Volume 1

The aim of this interdisciplinary book in three volumes is to present a critical review of what is currently known about compression wood, the reaction wood of the gymnosperms. With its 8100 references, the work is intended to be encyclopedic in nature. One quarter of the 932 illustrations and many of the 197 tables have not been published before.

The present, first volume deals with the basic properties and formation of compression wood. It begins with an outline of the bibliography and the early investigations. Taxonomy, designation, general characteristics, and determination are discussed next, followed by a description of the structure of the axial tracheids, the ray tracheids, the ray parenchyma, and the epithelial cells of compression wood in stem, branches, and roots. Following a survey of the general chemical composition of compression wood, each constituent is considered, namely cellulose, laricinan, galactan, galactoglucomannan, xylan, arabinogalactan, lignin, and the various extractives. A brief account is given of the microbial degradation of compression wood. The topochemistry of compression wood is introduced with a comparison between juvenile and mature woods, sapwood and heartwood, earlywood and latewood, and tracheids and ray cells. The distribution of the polysaccharides and lignin within the longitudinal tracheids is described and compared with that in normal wood.

The first of the physical properties to be considered is the density and its variation in normal, compression, branch, and knot woods. This is followed by a discussion of the moisture content, permeability, shrinkage, and swelling of compression wood. Its various strength characteristics are compared with those of normal wood. Chemical and anatomical factors determining these properties are discussed.

The origin and evolution of compression wood are discussed in the context of the fossil record and the nature of the compression wood present in the ancient Ginkgo and in primitive conifers and angiosperms. The dormant and active cambial zones are the subjects first dealt with in the formation of compression wood. Next is a description of the differentiating xylem, with special attention directed toward the origin of the helical cavities and ribs. This is followed by a discussion of the rate of formation of compression wood and its seasonal variation. The volume is concluded with a description of the transition between normal and compression woods.

Chapter 1 Introduction

CONTENTS

1.1 Compression Wood and Tension Wood 1
1.2 Bibliography 7
References 21

1.1 Compression Wood and Tension Wood

When a herbaceous plant is brought out of its natural, equilibrium position in space, a longitudinal growth promotion is initiated on its lower side, a growth promotion that is mediated by auxin and possibly also by other growth substances. In woody plants the response is different, since in this case the often considerable resistance of the already existing stem or lateral has to be overcome. Increased axial growth on the under side would not be sufficient in a tree. In a leaning stem, for example, radial growth is promoted on either the lower or upper side. As a result the stem develops an elliptic cross section with the long axis perpendicular to the plane of inclination and is now better able to support its own weight. This eccentric, radial growth is incapable of righting the stem, and such a righting is achieved, not by increased longitudinal growth on the under side, but by formation of special tissues that have the ability to force the tree back into its original orientation. Collectively, these tissues are referred to as *reaction wood*, as they develop when a tree reacts to a change in its environment.

In most gymnosperms, radial growth increases on the lower side of an inclined shoot or stem, and reaction wood also develops on this side. The type of reaction wood formed in gymnosperms is referred to as *compression wood*. It has the ability to expand along the grain while being formed, thus exerting an axial pressure that slowly bends the stem back to its original, usually vertical orientation. Branches that have been bent down develop compression wood on their under side and are thus able to resume their previous, predetermined orientation in space. All movements of orientation in the gymnosperms are effected in this way. Any stem or branch that has become displaced in any direction is restored to its original position with the aid of appropriately located compression wood.

In the arboreal, dicotyledonous angiosperms, radial growth in a leaning stem increases on the upper side although growth promotion on the lower side is not rare. Reaction wood, here referred to as *tension wood*, is normally formed on the upper side. It has the ability to contract in situ along the grain, thereby exerting a tensile stress that serves the purpose of righting the tree. Expressed simply, it can be said that compression wood pushes a leaning stem up, whereas tension wood pulls it up. The chemical composition, anatomy, and physical properties of compression and tension woods to a considerable extent reflect this difference in function between the two tissues.

Compression wood, which is the subject of this book, is formed in all Ginkgoales, Coniferales, and Taxales, that is, in the great majority of the Gymnospermae, but in neither the Cycadales, which produce very little xylem, nor in the Gnetales, which instead form tension wood. It is present in both shrubs and trees. The equilibrium orientation of a stem is in the vertical direction, that is, parallel to the plumb line. A difference of as little as a few degrees from the vertical is sufficient to trigger the formation of compression wood. Wind is a ubiquitous environmental factor, exerting an almost constant influence on trees, and for this reason alone very few coniferous trees are entirely free of compression wood. Branches are in equilibrium at an angle to the plumb line and are also fixed in the horizontal plane. Any displacement from this predetermined position causes formation of compression wood anywhere in the branch. Generally, branches are displaced downward by their own weight or by agents such as snow and ice, and for this reason compression wood in branches is usually located on their under side. Subterranean roots seldom, if ever, produce any compression wood.

The terminal leader at the top of a conifer stem can be damaged or destroyed by a variety of ecological agencies, such as wind, snow, ice, frost, mammals, insects, or fungi. When this happens, one or several of the adjacent laterals begins to bend upward and eventually assumes a vertical orientation, taking the place of the former leader. The upward curving is effected with the aid of compression wood, which is formed on the under side of the branch. After the latter has assumed its new role as a leader, the curvature at its base remains, and compression wood continues to develop on its lower side as long as the tree remains alive. Very large amounts of compression wood are produced in this way. Because the factors that can cause the loss of a leader in a conifer are both numerous and common and because each tree can have its leader replaced many times by not only one but by several laterals, this process probably causes formation of compression wood to an extent comparable to that brought about by displacement of a stem.

Ability to form compression wood and thus to retain or recover an orientation ensuring maximum growth and a leader capable of exerting apical control over the laterals below is of utmost survival value to any conifer tree. A conifer would not remain healthy for very long without this ability. Compression wood is probably as old as the gymnospermous habit itself, and it has been traced in fossils as far back as the Upper Jurassic. It is also formed by *Ginkgo biloba*, probably the world's oldest, still living tree species, which dates back to the Upper Devonian, some 280 million years ago. Compression wood is formed in xylem tissues consisting largely of tracheids, and it is therefore not surprising to find that it occurs not only in most of the gymnosperms, but also in certain primitive angiosperms with a high proportion of axial tracheids in their wood.

Compression wood was originally considered to be caused by compressive stresses in stems or branches and it was given its name for this reason. This view

1.1 Compression Wood and Tension Wood

was rejected long ago, although it has been revived in recent years. Because compression wood is always produced on the lower side of inclined stems and ceases to form when the stem has become parallel with the gravity vector, it was recognized some 80 years ago that gravity must be involved, and that formation of compression wood is, accordingly, a gravitropic phenomenon. It is now recognized that gravity probably is not the primary cause, for compression wood is produced also in the absence of any gravistimulus or in direct opposition to it. It has been shown, for example, that compression wood can form on the side of branches, causing them to move in a strictly horizontal plane, and it has long been known that if branches of conifers are bent upward, compression wood develops on the upper side, thereby forcing them down again. Compression wood can also occur in perfectly vertical stems, branches, and exposed roots. The view most widely accepted today is that the basic function of compression wood is to make it possible for a tree to restore to its original position by the shortest possible route any part of its stem or branches that has been brought out of its equilibrium location in space.

What the initial stimulus is and how it is received is still largely unknown. The stimulus is possibly transmitted with the aid of an auxin, presumably indoleacetic acid, which is capable of inducing formation of compression wood. In inclined, herbaceous plants a lateral transport of auxin occurs from the upper to the lower side where axial growth is promoted. Although it has been claimed that leaning conifers contain more auxin on their lower side than on the upper, conditions are very complex in trees, and it is far from certain that any difference in auxin concentration between the upper and lower sides of the stem is responsible for the formation of compression wood in mature trees, or even that such a difference exists at all. It has been suggested that there may be a difference in sensitivity to auxin between the upper and lower sides of an inclined bole, but nothing is known concerning the nature of this difference or what causes it.

A tissue almost identical with compression wood is produced in *Abies* trees attacked by the balsam woolly aphid when this insect injects its saliva into their phloem. The nature of the substance or substances in the saliva that stimulate the tree to form this compression wood is still unknown. Similarly, trees infected with *Diplodia pinea* form compression wood immediately in advance of the fungal hyphae. Again, what causes this remains to be determined.

Compression wood develops most frequently and rapidly in vigorous, fast-growing trees. Both eccentric growth and formation of compression wood reach their highest intensity in vigorous branches that have escaped the apical control of the leader and are bending upward. Radial growth is usually increased on the compression wood side of a stem or branch, but exceptions to this rule are not unknown, and there is actually no causal relationship between wide growth rings and compression wood. There is a continuous transition from normal wood via mild and moderate compression woods to fully developed, severe, or pronounced compression wood. Which tissue develops depends on the intensity and duration of the stimulus.

Compression wood is very common and probably more widespread than is generally appreciated. There is hardly a forest or plantation tree that does not have at least some compression wood in its stem, and branches always contain substantial amounts. Both virgin spruce forests and plantations of southern pines and *Pinus radiata* have been found to contain on the average 15% compression wood by volume. It is usual to consider compression wood as an "abnormal" tissue in contrast to "normal" wood, but from a physiological point of view both are, of course, equivalent. What is referred to as "normal" wood is only more common than compression wood, which develops under more unusual external conditions. All leaning, crooked, curved, or sinuous stems of conifers contain variable amounts of compression wood. This tissue is often also present in vertical, straight stems that had suffered some displacement at an earlier date but had been able to right and straighten themselves by forming appropriately located compression wood. Both the tendency of trees to develop stem curvatures and their ability to form compression wood have been found to be strongly inherited.

Compression wood is readily recognized thanks to its reddish-brown color, which is reflected in names such as "redwood", "Rotholz", and "bois rouge". It is an extremely hard and brittle wood, sometimes with a density twice that of corresponding normal wood. Anatomically, compression wood differs strikingly from normal wood. Earlywood and latewood in the accepted meaning of these terms do not exist, as almost the entire growth ring in severe compression wood is uniform in structure. Ray parenchyma cells are largely the same as in normal wood, but the longitudinal tracheids are entirely different. Compared to normal wood, the tracheids in compression wood are shorter and more distorted and have a thicker wall. A very characteristic feature is their rounded outline and the intercellular spaces that often surround them. The primary wall is the same as in normal wood, but S_1, the outer layer of the secondary wall, is much thicker. The inner portion of the S_2 layer is deeply fissured in most compression woods. Helically oriented cavities and ribs follow the flat orientation of the cellulose microfibrils in this layer, usually forming an angle of 30–50° to the tracheid axis. The helical cavities are a very characteristic feature of compression wood, and are responsible for some of the physical properties of this tissue. Fully developed compression wood lacks an S_3 layer.

Chemically, compression wood differs from normal wood in containing more lignin, galactan, and 1,3-glucan but less cellulose and galactoglucomannan. The distribution of lignin in the tracheids is different from that in normal wood and is characterized by a high concentration of lignin in the outer portion of the S_2 layer. More than 90% of the lignin is located within the tracheid wall. The ray cells have the same, high lignin content as in normal wood. The lignin differs from that in normal wood in several structural details, the most important of which is the higher content of unmethylated phenylpropane units. As a result, compression wood lignin is more highly condensed than that in normal wood. The galactan and 1,3-glucan present in compression wood occur in only trace amounts in normal wood.

In the green (natural, water-saturated) state compression wood has a higher compressive strength than normal wood and is also much more elastic. When air-dried, however, it is inferior to normal wood in almost all strength properties, particularly on an equal weight basis. Its greatest drawback is its exceed-

1.1 Compression Wood and Tension Wood

ingly high longitudinal shrinkage, which can attain 5−10%. Radial and tangential shrinkage, in contrast, are less than in normal wood. Compression wood is also a very hard and brash tissue. The high lignin content and large microfibril angle are probably responsible for its high compressive strength. The abnormally high longitudinal shrinkage is mostly a result of the large microfibril angle in the S_2 layer of the tracheids. The longitudinal expansion that occurs in developing compression wood has been attributed to several causes. The most likely explanation is that the expansion takes place when the tracheids are lignified.

Many of the properties of compression wood, while admirably suited to the function of this tissue in the living tree, are extremely undesirable in both pulpwood and lumber, and compression wood has always been considered to be a serious defect. Mechanical pulp cannot be made from compression wood, and chemical pulps prepared by the sulfite method have poor strength properties. For kraft pulps, the presence of compression wood, albeit undesirable, is less serious. Pulps containing compression wood fibers improve less on beating than normal pulps. When stems containing compression wood are sawn in either felling or conversion, saws often become bound because compressive stresses exerted by the compression wood are set free. The hard, dense compression wood is difficult to work and nail. When lumber contains both normal and compression wood, the high longitudinal shrinkage of the compression wood causes severe warping, distortion, and cross checking. Conifer knots are always hard and dense, a fact partly due to their high content of compression wood.

The wood present on the upper side of a branch or an inclined stem opposite to the compression wood is generally referred to as *opposite wood*. It has the same chemical composition as normal wood but differs in its structure from the latter, especially in the presence of a thick and highly lignified S_3 layer in the latewood tracheids. In many respects normal wood can be regarded as intermediate in structure and properties between opposite wood and compression wood.

The corresponding reaction tissue of the arborescent angiosperms, tension wood, has many characteristics directly opposite to those of compression wood. Whereas the latter is of universal occurrence in all conifers, tension wood is found almost exclusively in the arboreal dicotyledons, and among these it is not present in all genera or species. Unlike compression wood, tension wood exhibits much variability, not only with respect to occurrence, but also in its location and incidence in the stem and in its anatomy, ultrastructure, and physical properties.

Branches and leaning stems of angiosperms usually show eccentric growth toward the upper side, but growth promotion on the lower side is far from uncommon, as already mentioned. Tension wood is normally formed on the upper side of the stem or branch, regardless of the direction of the increased radial growth, but at times this tissue has also been observed on the lower side or all around a cross section. Tension wood is more difficult to detect than compression wood. Large, solid areas of tension wood are more rare than such areas of compression wood. Where they do occur, they are either lighter or darker in color than the normal wood. Tension wood can occur across an entire growth ring, but, in contrast to compression wood, it tends to form only in the outer portion of each increment. Compared to normal wood, it is largely the fibers that are modified in tension wood although the vessels are fewer and smaller. Sometimes, all fibers are modified over a sizable area but usually tension wood fibers are scattered among normal cells.

The ultrastructure of the fibers in tension wood is far more variable than that of the tracheids in compression wood. Their most outstanding characteristic is the presence of a thick, innermost cell wall layer which consists of axially oriented, highly crystalline cellulose microfibrils. The location of this so-called *gelatinous layer* or *G-layer*, which is not found in normal wood fibers, is variable, but most commonly it replaces the S_3 layer, next to the lumen. Gelatinous fibers are formed not only in tension wood but also in the secondary phloem of both angiosperms and gymnosperms. A number of hardwoods, for example *Tilia*, develop a tension wood without any gelatinous fibers, and in others, such as *Fraxinus*, these fibers are rare. The G-layer is almost invariably unlignified, and sometimes the lignin content of the other cell-wall layers is reduced. In other cases, however, the S_1 and S_2 layers have been found to be as highly lignified as in normal fibers. Due to the presence of the gelatinous layer, S_2 is often much reduced in size, sometimes to the extent that S_1 exceeds it in thickness, a fact that has a substantial influence on the physical properties of tension wood. Because of the presence of the unlignified, cellulosic G-layer, tension wood has a higher cellulose and a lower lignin content than corresponding normal wood. A galactan with an exceedingly complex structure occurs in tension wood from some genera, for example *Betula*, *Eucalyptus*, and *Fagus*, while it apparently is not formed in genera such as *Acer*, *Populus*, and *Ulmus*. The structure of the cellulose and lignin is the same in normal and tension woods. The cytology and physiology of tension wood formation have attracted considerable attention in later years. Like compression wood, tension wood serves the function in a tree of restoring a displaced part to its original position, for example in righting an inclined stem or maintaining a branch at its proper branch angle. Both phenomena would appear to be gravitropic in nature, but gravity is probably not the ultimate cause in either case. In direct contrast to compression wood, tension wood develops under conditions of auxin deficiency.

As could be expected from its high cellulose content, tension wood is stronger in tension than normal wood but it is inferior in most other strength properties and especially in compression. Tension wood shrinks considerably more longitudinally than normal wood. The reason for this has long been a moot question. One explanation is that the thin S_2 layer no longer dominates S_1, which has horizontally oriented microfibrils. The gelatinous layer, where the microfibrils are oriented parallel with the fiber axis, probably has no effect on the axial shrinkage.

Tension wood gives an excellent mechanical pulp and high yields of chemical pulps. The strength properties of the latter, however, are poor, and the pulps respond poorly to beating. Stems containing tension wood tend to contain numerous compression failures which result in weak sulfite pulps. Tension wood is difficult to saw because the gelatinous fibers are torn out instead of being cut, resulting in a woolly surface, and regions with compression wood readily col-

1.2 Bibliography

lapse during seasoning. Lumber containing tension wood commonly warps or splits due to the high longitudinal shrinkage of this tissue. There can be little doubt that tension wood is a serious defect in many hardwoods, but it is probably less harmful than compression wood.

1.2 Bibliography

Compression wood has been the subject of several extensive reviews and a large number of shorter articles. It has also been reviewed, discussed briefly, or at least mentioned in most books dealing with wood science and technology, and it is usually also referred to in books on pulp and paper technology. It has always been considered in books concerned with wood defects, since it is one of the most serious of wood degrades. Some books on tree physiology, forestry, and plant growth in general discuss compression wood at considerable length. The occurrence of compression wood in conifers must have been known for many centuries, although it does not seem to be mentioned in the Greek and Roman literature (Chap. 2). Carl von Linné (Linnaeus) referred to it in 1732. The structure of compression wood was first described in 1860 by the German botanist Sanio (Chap. 2.2). Forty years later, Robert Hartig published the results of his extensive investigations on the causes, occurrence, formation, and properties of compression wood in three papers (Chaps. 2.2 and 2.3). Many scientists have contributed to our knowledge of compression wood since this time.

There are at least 10 reviews of compression wood that can be considered as major, not only because of their size, but also because they have had a considerable impact on later contributions to this subject. Articles that must be regarded as reports of research will not be included here. Others constitute borderline cases. The first five of the ten reviews, namely those by Hartig (1901), Mork (1928), Trendelenburg (1932), Pillow and Luxford (1937), and Onaka (1949) all contain a wealth of original data. The following five by Spurr and Hyvärinen (1954), Low (1964), White (1965), Westing (1965, 1968), and Wilson and Archer (1977) are essentially reviews, although the last two also contain unpublished material.

Shortly before his death in 1901, Robert Hartig (Fig. 1.1) summarized some of his recent research, including that dealing with compression wood, in a little book entitled *Holzuntersuchungen. Altes und Neues* (Fig. 1.2). Only half of the 99 pages in this book are devoted to compression wood, but they contain a surprising wealth of information, largely from Hartig's own investigations, on the causes, physiology, occurrence, structure, and chemistry of compression wood. Unlike most later reviews, this book also includes a discussion of opposite wood. It is illustrated with many excellent line drawings and with photographs of trees and wood. The emphasis is on the factors causing formation of compression wood, its occurrence in trees as mediated by various ecological agencies, and its anatomy and ultrastructure. Hartig's book has obviously had a large influence on later investigators. Even today it remains one of the best illustrated of all major reviews of compression wood, and little in it has become

Fig. 1.1. Robert Hartig (1839–1901). Professor of Botany, Forstliche Hochschule, Eberswalde (1866–1878). Professor of Forest Botany, Universität München (1878–1901). (Courtesy of K. A. Sorg)

outdated. It has, of course, long been out of print, and is now available only in a few libraries.

Mork's (1928) long article *Om tennar* (*On Compression Wood*) is a highly perceptive review of compression wood. It also contains results of his own studies on the occurrence of compression wood and its relationship to ring width and to earlywood and latewood. Particularly noteworthy are his many original observations on the structure of compression wood and opposite wood. After a critical review of all earlier contributions, Mork presents a thoughtful discussion of possible causes of compression wood formation. The theory he proposes here anticipates some of the views later held by Hartmann and Münch. Mork devotes considerable attention to the peculiar, so-called spiral compression wood, which is still unexplained. He has universally been credited with being the first to report this remarkable phenomenon, but it had actually been described 10 years earlier (Chap. 10.3.8.3). The review concludes with a brief survey of the role of compression wood as a defect in lumber and pulpwood. Mork's article would undoubtedly have exerted a greater influence on later investigators if it had been written in a language more commonly read than Norwegian. That it nevertheless became fairly well known in the English-speaking world is probably to be attributed to Berg, who a few years later (1931) wrote a review of it for the Journal of Forestry, where he summarized both Mork's ideas and his major results.

Fig. 1.2. The title page of Hartig's book *Holzuntersuchungen. Altes und Neues,* Verlag von Julius Springer, Berlin 1901

Fig. 1.3. Reinhard Trendelenburg (1907–1941). Director of Research (1935–1940) and Professor (1940–1941), Technische Hochschule, München

Fig. 1.4. Maxon Young Pillow (1899–1980). Supervisory Technologist, Forest Products Laboratory, US Forest Service, Madison, Wisconsin (1924–1967). (Courtesy of US Forest Service)

A different type of review was published in 1932 by Trendelenburg (Fig. 1.3). Causes, occurrence, structure, and chemistry of compression wood are briefly considered, but most of the paper is devoted to its physical properties, such as density, shrinkage and swelling, and especially the various strength properties. The survey of these properties is extremely comprehensive and authoritative and includes practically all information on this subject available at the time. The data, which are presented in numerous tables and figures, are critically evaluated throughout. Trendelenburg also devotes attention to the relation between the mechanical properties of compression wood and its function in the living tree.

The review by Pillow (Fig. 1.4) and Luxford from 1937 can rightly be considered a classic. Partly based on an earlier treatise by Pillow (1933), this review covers all aspects of compression wood, surveys earlier contributions, and includes much new information. It is well illustrated with many excellent photographs. Topics discussed include structure, chemical, physical, and mechanical properties of compression wood, its occurrence in different portions of tree stems, ecological factors, effects of silvicultural practices on formation of compression wood, and the influence of compression wood on lumber. Theoretical and practical aspects are treated with equal authority in this survey. Although nearly 50 years have now passed since its publication, this review remains one of the most cited contributions in the literature on compression wood.

The treatise by Onaka (1949) (Fig. 1.5) on compression and tension wood represents a remarkable achievement. It is a review at the same time as it sum-

Fig. 1.5. Fumihiko Onaka (1904–1950). Professor, Kyoto University, Kyoto (1937–1950). (Courtesy of H. Harada)

marizes the author's own contributions, and like Hartig's book from 1901 it was completed shortly before the author's death. Originally published in Japanese in 1949, this article was translated into English by the Canadian Government in 1956. It begins with an interesting discussion of the causes and physiology of compression wood formation, largely based on earlier work by German investigators but also on Onaka's own results. The epinasty concept is discussed in great detail. All possible cases of the occurrence of compression wood in stems, branches, and roots are considered, and this is followed by a brief survey of the ecological factors involved. The subsequent sections on the structure and the chemical, physical, and mechanical properties of compression wood contain an enormous wealth of data, some of them published here for the first time. The treatment of tension wood is as thorough as that of compression wood. Onaka's survey of reaction wood is still highly relevant in all but a few respects. One can only regret that the English translation was never issued in printed form, and that it lacks all of the original illustrations and some of the references.

The review by Spurr and Hyvärinen appeared in 1954, 2 years before Onaka's article became available in English. This is a review in the strict sense of the word, with no original data, few personal opinions, and a fairly complete bibliography. It is actually devoted only to the physiology and ecology of compression wood formation. Most of the relevant information on these subjects is presented together with a brief but interesting discussion.

Low, in 1964, published a review of compression wood which is comprehensive and well balanced. Unlike the earlier survey by Spurr and Hyvärinen (1954), it deals with all aspects of its topic and gives no preference to any particular area. A forester by profession, Low very likely prepared this fine review in connection with his extensive investigations on the occurrence of compression wood in *Pinus sylvestris* in Scotland. The vast material is well organized, and all major contributions are briefly discussed, including the complex physiology of compression wood formation. This is also the first review since that by Pillow and Luxford (1937) to consider the effects of environmental factors and silvicultural practices on the formation of compression wood.

One year later, in 1965, there appeared an entirely different review of reaction wood by White, portions of which were later incorporated into the monograph on the structure of wood by Jane et al. (1971). This article has the advantage of containing numerous illustrations, but it is somewhat uneven in its treatment of the subject. The anatomy of compression wood is discussed at some length, but occurrence and properties of compression wood are only mentioned briefly, perhaps because the review, at least according to its title, is supposed to be concerned only with the anatomy of reaction tissues. Possible causes of reaction wood formation are critically evaluated. There is also a section dealing with reaction phloem. One weakness of this otherwise interesting article is the omission of most of the contributions from Germany, nor is there any reference to the important results obtained by Onaka.

The most extensive review of compression wood currently available is that by Westing, published in 1965 with an updating addition in 1968. This summary, a monument of scholarship and thoroughness, is unique first of all in its size, namely a total of 123 pages with 575 references in the first and 165 in the second part. No such encyclopedic treatment of compression wood had ever been attempted before. It is also unique in its exhaustive treatment of the physiological aspects of compression wood formation, a subject touched upon very lightly in similar articles. Finally, this superb review is unique in its detailed discussion of the early German literature on compression wood. The labors involved in the preparation of this work must have been Herculean, but the result amply justifies them.

The organization of Westing's review is somewhat unusual. The effect of compression wood on wood utilization is part of the introduction, and the anatomy and chemistry are discussed in a section devoted to the physiology of compression wood formation. Subjects treated relatively briefly are the structural, chemical, and physical characteristics of compression wood and ecological factors involved in its formation. The central part consists of the sections concerned with the physiology, functional significance, and mechanism of action of compression wood. The treatment of these subjects is admirable. The physiology of compression wood formation, and especially the gravitational and hormonal aspects, are presented in great detail and with numerous references to related phenomena in herbaceous plants. The review of the mechanism of compression wood action is in itself an original contribution to this subject. The list of references is the most complete ever prepared and a deliberate effort was obviously made to include all information published about compression wood per se[1]. As Westing (1965) himself points out, all articles dealing with compression wood are not included. It should perhaps also be added that, according to the author's own estimate, only about one third of the publications cited are directly concerned with compression wood. Westing's reviews have been frequently cited ever since they were first published, and they must have contributed to the renewed interest in compression wood that has been apparent during the last 20 years.

As its title indicates, the excellent review by Wilson and Archer (1977), is largely restricted to the induction and mechanical action of compression wood, both areas to which the two authors have made many contributions in later years. The anatomy of compression and tension woods is briefly outlined, but there is no mention of their chemical or physical properties. Induction of reaction wood is discussed in detail, including the role of various growth substances and their distribution in trees. The mechanism by which reaction wood effects its righting action is treated in depth, together with the nature of growth stresses and strains in trees. Recent reviews of reaction wood by Wilson and Archer (1979) and Wilson (1981) also deal with the stresses and strains associated with the formation of this wood.

There are many other reviews of compression wood, less extensive than the 10 now discussed but still comprehensive enough to deserve special mention. A much-cited article was published in 1949 by Dadswell and Wardrop under the direct title: *What is Reaction Wood?* It deals largely with Australian species, but contains much pertinent general information, including a section on the formation of compression and tension woods. Rendle, in 1956, published a brief but well-illustrated review of compression wood, part of which was in the same year incorporated into a booklet on reaction wood issued by the Princes Risborough Laboratory in England (Anonymous 1956). Both focus on the occurrence and detection of compression wood in standing timber and in lumber. An updated booklet was published in 1972. In this newer version, compression wood is allotted only half the space devoted to tension wood (Anonymous 1972).

Probably the most overlooked review of compression wood is that written by Jones (1956–1957) for a Welsh journal. It is not a long paper but it contains the more important facts. Knigge's survey of reaction wood from 1958 is more comprehensive. Anatomy, physical properties, mechanical characteristics, and the role of compression wood in wood utilization are given special emphasis in this fine review. An important and well-illustrated summary of the anatomy,

---

[1] Anybody who wishes to use Westing's references should be cautioned that, for unknown reasons, the author has systematically omitted the indefinite and definite articles from the beginning of all titles, and not only in the English but also in the German publications.

chemistry, and pulping characteristics of reaction wood was prepared by Dadswell et al. (1958) for the conference on *Fundamentals of Papermaking Fibres*, held at Cambridge, England in 1957. It is a relatively brief survey, but it contains much information, especially on the problems encountered in pulping of compression wood.

Core et al. (1961) reviewed the characteristics of compression wood in some North American conifers. This paper is largely concerned with the structure of compression wood and is accompanied by many excellent illustrations. In the same year, Correns (1961) and Croon (1961) published two reviews of reaction wood. Unfortunately, Croon's article, which emphasizes structure and chemistry, is written in Swedish. As a result, a simultaneous, but more limited, summary by Hale et al. (1961) has been cited more frequently in the literature. A review of reaction wood, prepared by Dadswell in 1960 but not published until 1963, contains much of the relevant information on this subject and also includes a discussion. A more specialized survey is that by du Toit (1964), who summarized factors that could possibly cause formation of compression wood. It is an interesting article, albeit not inclusive. It is also somewhat surprising to find that factors such as light and bark pressure are each allotted as much space as that devoted to gravity.

The year 1965 saw the publication of several review articles on compression wood, in addition to those by Westing and White. They all differ considerably from one another. Casperson (1965a) contributed a thorough review of the anatomy of reaction wood with over 140 references and many illustrations. Especially valuable in this article are the references to the older literature on the structure of compression wood. The review by Côté and Day (1965) on the same subject is very brief but contains a large number of photomicrographs and electron micrographs. Wardrop's (1965) companion review on the formation and function of reaction wood is an interesting account of the physiological aspects of this subject.

The stimulating review on the effect of gravity on the formation of reaction wood by Robards (1969) has a brief summary of the structure of compression and tension woods. While the author first claims that gravity probably is the primary factor responsible for formation of reaction wood, his subsequent discussion makes it clear that the problem is more complex than this. In 1973 Scurfield published an article where he attempted to show that lignification might be the force which makes it possible for inclined tree stems to right themselves, in the case of gymnosperms by causing the compression wood to expand on the under side of the bole. In this connection he also presented a brief review of compression and tension woods, including their anatomy and function. An informative and more conventional review is that reported in the same year by von Pechmann (1973) at the fifth IUFRO meeting in South Africa. The topics emphasized here are the anatomy, physical properties, and occurrence of compression wood.

There are many brief reviews of compression wood besides those mentioned. Grossenbacher in 1915 published an excellent, exhaustive discussion of radial growth in trees, where he devoted much attention to eccentric growth in conifers and the associated formation of compression wood. Surveys of compression wood have also been contributed by Pillow in 1930, Tikka in 1935, and Clarke in 1940.

Ewald König was a self-taught German scientist with an intense interest in wood and wood utilization, who spent almost his entire life writing on these subjects (Mombächer 1971). In his many articles and several books, König often discussed compression wood, always correctly, yet in terms understandable by a layman. A complete list of all König's publications referring to compression wood cannot be included here. Some of the better-known ones appeared in 1942, 1946, and 1952. Compression wood is also discussed in his book *Sortierung und Pflege des Holzes* from 1956. His 1957 book on wood defects (*Fehler des Holzes*) contains informative summaries of both eccentric growth and reaction wood.

The book by Boas (1947) on commercial timbers of Australia has a review of compression wood with several illustrations. Under the title *Abnormal Wood*, Jane (1952) discussed the anatomy of reaction wood. A similar summary, and one of the few in French, is that by Campredon (1953). The review of compression wood by Lukić-Simonović (1955) contains much information on the physical properties of this wood in *Picea omorica*. Excellent summaries of available information on compression wood were published by Ollinmaa in 1955 (in Finnish), and by Nečesaný in 1956 (in Czech). In a report on his extensive research on compression wood, Ollinmaa (1959) also discussed the general properties of this wood. Dadswell (1958) has contributed a survey of the structure and properties of juvenile, mature, compression, and tension woods. The article by Špoljarič (1959) contains a brief section devoted to reaction wood. In a summary of reaction wood from the Philippines (Anonymous 1965), reaction wood is referred to as an abnormal wood. Its properties and utilization are briefly summarized.

In addition to his article of the structure of reaction wood, mentioned above, Casperson (1963a, b, 1965b) has contributed a general review of reaction wood in his dissertation, as well as two shorter summaries in published articles. Compression wood has also been reviewed by Böhlmann (1965) and Schultze-Dewitz (1969). Three brief reviews of compression wood have been prepared by Wooten (1967, 1968a, b). Peng's (1969) review (in Chinese) is based largely on Japanese research on compression wood and is part of a general treatise of the relationship between wood structure and wood quality. Compression wood is considered in Phillip's (1968) brochure on the identification of softwoods.

A brief outline of the properties of compression wood is found in an article by Kärkkäinen (1976) on weight measurements of pine and spruce logs. Compression and tension woods are the subjects of concise summaries in two Canadian reports on kiln-drying of lumber (Bramhall and Wellwood 1976; Cech and Pfaff 1978). The same topic was reviewed by Kennedy (1976) at a wood-engineering seminar in Canada. In his position paper on the topochemistry of wood, Côté (1977) discussed the ultrastructure of reaction wood and the distribution of cellulose, hemicelluloses, and lignin in normal and reaction woods. In a recent review of the structure of wood, Gray and Parham

(1982) described and illustrated the general anatomy of compression and tension woods.

General information on compression wood is also found in a large number of textbooks and monographs. Sometimes the subject is treated at some length, while in other instances compression wood is only mentioned in passing. Not surprisingly, the majority of these books fall within the purviews of wood anatomy, wood defects, and wood technology. The section on compression wood in Mork's (1946) book *Vedanatomi* (*Wood Anatomy*) is essentially an abbreviated version of the same author's 1928 article on this subject. Knuchel's (1954) book *Das Holz* outlines the properties of compression wood together with some illustrations. Jane (1956), in his monograph *The Structure of Wood*, discusses the anatomy of compression wood at considerable length with reference to numerous photomicrographs. Krzysik's (1957) *Nauka o Drewnie* (*Science of Wood*) contains a long section with illustrations and tables dealing with compression wood. The subject is also discussed in the second edition of this book (Krzysik 1975). Wagenführ's (1966) *Anatomie des Holzes* describes compression wood with illustrations previously published by Casperson (1965a). The subject is also covered in later editions of this book (Wagenführ 1980, 1984) and also in the recent *Mechanics of Wood and Wood Composites* by Bodig and Jayne (1982).

In his atlas of electron micrographs, *Wood Ultrastructure*, Côté (1967) devotes three micrographs and a brief description to compression wood. Tsoumis (1968), in his book *Wood as a Raw Material*, has included a fairly comprehensive review of compression wood in a chapter on abnormalities in wood. A delightful book *Inside Wood, Masterpiece of Nature* was published by Harlow in 1970. It contains a section on compression wood, including a discussion and illustration of the peculiar spiral compression wood. The latest edition of Jane's *The Structure of Wood*, revised by Wilson and White in 1970, has a large section devoted to reaction wood. The review of compression wood is very well written, and one can only regret that the numerous illustrations do not include any electron micrographs (Jane et al. 1970).

An atlas of scanning electron micrographs *Three-Dimensional Structure of Wood* was published in 1972 by Meylan and Butterfield. The structure of compression wood is briefly summarized together with four scanning electron micrographs, showing the ultrastructure of the helical cavities and checks. The second edition of this book, published in 1980 by Butterfield and Meylan, contains a concise description of compression wood with recent literature references and three scanning electron micrographs. The large *Holzatlas* by Wagenführ and Scheiber (1974) describes in detail not only compression wood but also the different ecological agencies that are indirectly responsible for its occurrence in trees. In the first volume of *Holzkunde*, Bosshard (1974a, 1982) offers a brief description of compression wood and a light micrograph. The occurrence, formation, and physiology and compression wood are allotted more space in Volume 2 (Bosshard 1974b). Compression wood as a wood defect is discussed in Volume 3 of the same book (Bosshard 1974c). In the second edition of their superbly illustrated *Wood. Structure and Identification*, Core et al. (1979) present an outline of the properties of compression wood, which is illustrated by three micrographs. Volume I of Isenberg's (1980) *Pulpwoods of the United States and Canada* includes a scanning electron micrograph of compression wood in *Pinus resinosa* but no description of this tissue. The *CRC Handbook of Materials Science Volume IV. Wood* edited by Summitt and Sliker (1980) devotes only a few lines to compression wood. It is not described in the book *Wood: Its Structure and Properties* edited by Wangaard (1981). The third edition of *Canadian Woods* contains a brief but informative description of the most important properties of compression wood as well as other wood defects, such as tension wood, cross grain, shakes and checks (Keith and Kellogg 1981). The admirable *The Practical Identification of Wood Pulp Fibers* by Parham and Gray (1982) contains a description of both compression and tension woods. The structure of the two tissues are briefly mentioned in *Identification of Modern and Tertiary Woods* by Barefoot and Hankins (1982). In their book *The Wood Properties of Radiata Pine*, Bamber and Burley (1983) offer an excellent summary of the occurrence and properties of compression wood, branch wood, and knot wood in this important species. The recent book *In Harmony with Wood* by Becksvoort (1983) contains an excellent and succinct description of reaction wood. Siau (1984), in his monograph *Transport Processes of Wood* describes the general properties of reaction wood.

As could be expected, almost all books concerned with defects in wood devote some space to compression and tension woods. The first edition of Knuchel's well-known *Holzfehler* from 1940 contains only a brief section on compression wood in contrast to his somewhat earlier article *Kampf den Holzfehlern* where compression wood is discussed and illustrated at some length (1938). In the second edition, published in 1947, the subject is treated in more detail, and there are several illustrations. The book also has sections devoted to several of the ecological agencies indirectly responsible for the formation of compression wood, such as wind, snow, hail, and mammals. One chapter deals with stem defects commonly associated with the presence of compression wood, namely crook, sweep, fork, and eccentric radial growth. The similar book by Paclt (1954) *Kazy Dreva* (*Wood Defects*), contains a large section on reaction wood. Both Knuchel and Paclt have included photographs of spiral compression wood. Compression wood is reviewed in Durst's (1955) *Taschenbuch der Fehler und Schäden des Holzes* and in de Carvalho's (1957) *Defeitos da Madeira*.

König's (1957) excellent *Fehler des Holzes* has already been mentioned. The basic aspects of eccentric radial growth, compression wood, and tension wood are discussed. The book contains several chapters dealing with stem defects, such as crooks, forking, and compression failures, König's book if inferior to Knuchel's in the quality of its illustrations, but it offers a more thorough treatment of the subject, especially of all those defects indirectly associated with the presence of compression wood. The book *Defects in Wood* by Ertfeld et al. (1964) is written in English, German, and Russian. Compression wood receives a very cursory treatment, but there are sections describing common wood defects. The book is lavishly illustrated.

Most of the many books dealing with forest utilization or wood technology give at least a brief account of compression wood. Older books, such as those by Wahlgren and Schotte (1928), Holtmann (1929), Garratt (1931), or Thiemann (1942, 1951) devote a page or two to compression wood, while Henderson

(1944) only mentions it in passing, and still others, for example Boulger (1908), Snow (1917), Betts (1919), Stone (1921), or Brough (1955), fail even to mention it. In his now classicial monograph *Das Holz als Rohstoff,* Trendelenburg (1939) reviewed the occurrence and properties of compression wood at some length. In the second edition, which was admirably revised by Mayer-Wegelin (Trendelenburg and Mayer-Wegelin 1955), compression wood receives more attention. In his book *Trä. Dess Byggnad och Felaktigheter (Wood. Its Structure and Defects)* Thunell (1945) refers to the occurrence, appearance, and structure of compression wood. Wangaard (1950) discusses it is more detail in *The Mechanical Properties of Wood.* Hale (1951) has contributed a summary in *Canadian Woods.*

Giordano discusses and illustrates compression wood in several places in his monograph on wood *Il Legno* from 1951. In his three-volume *Tecnologia del Legno,* an excellent work of enormous size and scope, the first part *La Materia Prima* from 1971 contains sections dealing in depth with the occurrence, anatomy, chemical composition, and physical properties of compression wood. There are also many illustrations and unpublished data.

In Kollmann's (1951) well-known *Technologie des Holzes und der Holzwerkstoffe* only one page is devoted to compression wood, the large size of this book notwithstanding. *Textbook of Wood Technology* by Brown et al. (1952) and *Holz als Werkstoff* by Bieler (1953) are more generous in this respect. In his books *Werkstoff Holz* Göhre (1954, 1961) describes various tree malformations and the properties of the associated compression wood. Compression wood is referred to only briefly in the two editions of *Wood Handbook* published by the Forest Products Laboratory at Madison in 1955 and 1974. This is also true of *Handbuch für den Holzschutz* by Langendorf (1961).

The fine monograph *Forstbenutzung* by Knigge and Schulz (1966) contains a section on reaction wood with many references. Silvester (1967) presents an unusually detailed review of compression wood in his book *Timber.* In the monograph by Kollmann and Côté (1968) *Principles of Wood Science and Technology. I. Solid Wood,* Côté has contributed a well-illustrated discussion of the structure of compression wood. A later section on its physical properties is very short. The unusually well-documented chapter on the chemistry of wood includes data on the chemical composition of compression wood. In one of the chapters by Kollmann there is mention of the shrinkage of compression wood, but little else.

The monumental monograph *Utilization of the Southern Pines* by Koch (1972) contains references to compression wood. The subject is briefly mentioned in Findley's (1975) *Timber: Properties and Uses* and also in *The Life Cycle of Wood* by Jullander and Stockman (1978). The fourth edition of *Textbook of Wood Technology* by Panshin and de Zeeuw (1980) contains the best and most comprehensive review of reaction wood found in any book concerned with wood and its utilization. The amount of information on the subject confined within only 21 pages is amazing. No aspect of compression wood has been neglected, and there are also numerous, excellent illustrations. Structure, chemistry, physical properties, and effects on wood utilization are all thoroughly treated. A table with data on the physical properties of compression wood is one of the most exhaustive ever published.

The sixth edition of the well-known *Timber. Its Structure, Properties and Utilization* by Desch and Dinwoodie (1980) contains somewhat more information on compression wood than does the fifth edition by Desch (1973). The wood-working characteristics of compression wood are well described in Hoadley's (1980) fine guide to wood technology *Understanding Wood.* The occurrence, mode of righting, and properties of compression wood are well covered in Kubler's (1980) similar book *Wood as a Building and Hobby Material.* Compression wood is mentioned only briefly in *Timber* by Dinwoodie (1980). The recent textbook *Forest Products and Wood Science* by Haygreen and Bowyer (1982) contains an excellent section on the formation and properties of reaction wood, although a micrograph illustrating the structure of compression wood would have been welcome.

Most books concerned with tree growth and tree physiology contain shorter or longer sections dealing with reaction wood although older books are often silent on this subject. As early as 1897, Büsgen in his well-known *Bau und Leben unserer Waldbäume* devoted half a page to the then recent results obtained by Cieslar (1896). The best-known edition of this book is that published by Büsgen and Münch in 1927 and translated into English by Thomson (Büsgen et al. 1929). Compression wood is fully described in this excellent book, and especially the physiological aspects of its formation. This monograph has long remained a classic, and a new treatise on this ubject did not appear until 1967 when Lyr et al. published their *Gehölzphysiologie.* Unfortunately, it is not a book that has managed to take the place of that by Büsgen and Münch. It includes sections on gravitropism and phototropism, but the role of reaction wood is only briefly mentioned. In his *Plant Morphogenesis,* Sinnott (1960) discusses the physiology of compression wood formation. Written by a scientist who himself had contributed much to this subject, this is a very thoughtful contribution, and one can only regret that it is so short. The second volume of Kozlowski's (1971) *Growth and Development of Trees* contains a summary of the most important facts of compression wood, including a number of good illustrations, albeit no electron micrograph.

In his book *The Growing Tree* Wilson (1970) frequently refers to the role of compression wood in tree growth, including its effect on branch orientation and the physiology of its formation. Related topics also discussed are apical control, epinasty, and gravitropism. Its clear and succinct presentation of complex subjects makes this book very stimulating reading. In the admirable *Trees. Structure and Function* by Zimmermann and Brown, the latter author (Brown 1971) has written an excellent review of reaction wood with emphasis on causative factors and physiology of formation. Brown has also contributed several sections closely related to the formation of compression wood, such as apical control of growth and form, growth correlations and form, gravitropism, and gravimorphism. The review of Münch's classical experiments on shoot growth is particularly noteworthy. Morey's (1973) *How Trees Grow* is a delightful little book that succeeds in imparting much information within its limited format. It contains a separate chapter devoted to reaction wood, its structure and distribution, and possible causes of its formation. *Physiology of Woody Plants* by Kramer and Kozlowski (1978) contains a brief review of reaction wood with several illustrations, including one of spiral compression wood.

Several monographs on general plant growth, plant anatomy, and plant cell walls discuss or at least mention compression wood. The subject is, for example, briefly referred to in Esau's (1965) *Plant Anatomy*. In his well-known monograph *Die Pflanzliche Zellwand*, Frey-Wyssling (1959) treats reaction wood at some length. The emphasis here is not on the anatomy or chemistry of compression wood but instead on its physical characteristics, especially its shrinkage and swelling, and on the mechanism of its action in the living tree. The more recent *The Plant Cell Wall* by the same author (Frey-Wyssling 1976) includes a summary of reaction wood.

In the well-known series *Encyclopedia of Plant Physiology*, Gessner (1961) has contributed a chapter on mechanical effects on plant growth with an interesting section on the structure and formation of reaction wood. The most important results obtained by Hartmann and by Sinnott are discussed and illustrated. In Fahn's (1967) *Plant Anatomy* and in the chapter on wood in the latest edition of Kirk-Othmer *Encyclopedia of Chemical Technology* (Wegner 1984), compression wood is only mentioned. Clowes and Juniper (1968) in their book *Plant Cells*, on the other hand, present a relatively detailed description of compression wood, including its ultrastructure. One of the very few books devoted to cambial growth in trees is *The Vascular Cambium* by Philipson et al. (1971). Its last chapter deals with the physiology of reaction wood formation. In his book on normal and compression xylem and phloem in *Picea abies*, Timell (1973) included a review of compression wood together with a considerable number of illustrations. The physiology of reaction wood formation is critically reviewed by Leopold and Kriedemann (1975) in their *Plant Growth and Development*.

Books dealing with the chemistry of wood often contain some information on compression wood. The last edition of Hägglund's *Chemistry of Wood* from 1951 treats compression wood at some length. Interestingly, the structure of the tracheids is illustrated with the drawings published by Hartig exactly 50 years earlier. They probably serve their purpose far better than the frequently poor light or electron micrographs found in similar chemistry books. Compression wood is only mentioned in *Wood Chemistry* by Wise and Jahn (Browning and Isenberg 1952) and in Sandermann's (1956) *Grundlagen der Chemie und chemischen Technologie des Holzes*, Stockman (1957) describes it very briefly in Treiber's *Die Chemie der Pflanzenzellwand*, and Isenberg (1963) devotes no more space to it in Browning's *The Chemistry of Wood*. In his book *Wood and Cellulose Science*, Stamm (1964) only mentions compression wood. Browning (1971) outlines the determination of compression wood by the TAPPI method in his *Methods in Wood Chemistry*. In the already classical monograph *Lignins. Occurrence, Formation, Structure, and Reactions* by Sarkanen and Ludwig there are several references to compression wood and especially to its lignin distribution (Sarkanen and Herget 1971).

In *Puukemia* (*Wood Chemistry*) edited by Jensen, reaction wood is reviewed by Ilvessalo-Pfäffli (1977) in her excellent chapter on wood anatomy. In his fine textbook *Wood Chemistry. Fundamentals and Applications,* Sjöström (1981) has included a concise, yet informative description of reaction wood together with electron micrographs. Reaction tissues are briefly described in the recent, comprehensive monograph *Wood. Chemistry, Ultrastructure, Reactions* by Fengel and Wegener (1984). *The Chemistry of Solid Wood,* edited by Rowell, contains an excellent chapter by Parham and Gray (1984) dealing with the formation and structure of wood. Despite the limited space available, the section on compression and tension woods offers an amazing amount of information together with many superb illustrations.

Compression wood is referred to, albeit usually far too briefly, in most books concerned with the manufacture of pulp and paper. Isenberg (1962) offers a short description of reaction wood in Libby's *Pulp and Paper Science and Technology.* A short but adequate review is found in Rydholm's (1965) monumental monograph *Pulping Processes.* Hale (1969) summarizes the properties of compression wood in the first volume of *Pulp and Paper Manufacture* edited by McDonald and Franklin. Browning (1970) in *Handbook of Pulp and Paper Technology* edited by Britt only mentions it. Wenzl (1970), by contrast, reviews compression wood in considerable detail in his *The Chemical Technology of Wood,* discussing its structure, chemistry, physical properties, and influence on wood utilization. There are also several photomicrographs, but it is unfortunate that none of them shows the structure of the most common type of compression wood. The pulp and paper books edited by Casey (1980) and by Rance (1980) contain descriptions of compression wood.

Tension wood has not been reviewed as extensively as compression wood, and there is no survey of this subject as inclusive and thorough as that by Westing (1965, 1968) on compression wood. The most exhaustive reviews are those by Hughes (1965) and by Wicker (1979). Three other relatively comprehensive summaries are those by Onaka (1949), Wardrop (1965), and Sachsse (1965). There are also a considerable number of less extensive reviews, for example by Dadswell and Wardrop (1955, 1956), Wardrop (1960, 1961, 1964), Rendle (1937, 1955), Anonymous (1956), Ollinmaa (1956), Vandevelde (1957), Dadswell et al. (1958), Knigge (1958), Correns (1961), Croon (1961), Perem and Clermont (1961), Perem (1963, 1964), Casperson (1965a), White (1965), Koch et al. (1968), Jutte (1969), Timell (1969), Scurfield (1973), and Detienne (1976). Several of these articles are also concerned with compression wood. Tension wood is described or at least mentioned in most of the books referred to previously.

References

Anonymous 1956 Reaction wood. (Tension wood and compression wood.) For Prod Res Lab GB Leafl 51, 16 pp
Anonymous 1965 Reaction wood. For Prod Res Inst Philipp Tech Note 71, 3 pp
Anonymous 1972 Reaction wood. (Tension wood and compression wood.) Dep Environ Build Res Establ GB Princes Risborough Lab Tech Note 57, 20 pp
Bamber RK, Burley J 1983 The wood properties of radiata pine. Commonw Agr Bur, Slough, 84 pp

Barefoot AC, Hankins FW 1982 Identification of modern and tertiary woods. Clarendon Press, Oxford, 189 pp
Becksvoort C 1983 In harmony with wood. Van Nostrand Reinhold, New York, 134 pp
Berg B 1931 Review of: Mork E 1928 Om tennar. (On compression wood.) J For 29:599–601
Betts HS 1919 Timber, its strength, seasoning, and grading. McGraw-Hill, New York, 234 pp
Bieler K 1953 Holz als Werkstoff, 2nd ed. Georg Westermann, Braunschweig, 158 pp
Boas IH 1947 The commercial timbers of Australia. Their properties and uses. Counc Sci Ind Res Aust, 344 pp
Bodig J, Jayne BA 1982 Mechanics of wood and wood composites. Van Nostrand Reinhold, New York, 712 pp
Böhlmann D 1965 Das Problem der Bildung von Reaktionsholz in Laub- und Nadelhölzern. Forst-Holzwirtsch 20(7):153–155
Bosshard HH 1974a Holzkunde. 1 Mikroskopie und Makroskopie des Holzes. Birkhäuser, Basel Stuttgart, 224 pp
Bosshard HH 1974b Holzkunde. 2 Zur Biologie, Physik und Chemie des Holzes. Birkhäuser, Basel Stuttgart, 321 pp
Bosshard HH 1974c Holzkunde. 3 Aspekte der Holzbearbeitung und Holzverwertung. Birkhäuser, Basel Stuttgart, 286 pp
Bosshard HH 1982 Holzkunde. 1 Mikroskopie und Makroskopie des Holzes, 2nd ed. Birkhäuser, Basel Stuttgart, 224 pp
Boulger GS 1908 Wood. Edward Arnold, London, 348 pp
Bramhall G, Wellwood RW 1976 Kiln drying of western Canadian lumber. Can For Serv Rep VP-X-159, 112 pp
Brough JCS 1955 Timbers for woodwork. Evans Brothers, London, 232 pp
Brown CL 1971 Secondary growth. 4 Reaction wood. In: Zimmermann MH, Brown CL Trees. Structure and Function. Springer, New York Heidelberg Berlin, 98–105
Brown HP, Panshin AJ, Forsaith CC 1952 Textbook of wood technology. II The physical, mechanical, and chemical properties of the commercial woods of the United States. McGraw-Hill, New York, 783 pp
Browning BL 1967 Methods of wood chemistry, Vol 1. Interscience, New York, 384 pp
Browning BL 1970 Wood Chemistry. In: Britt KW (ed) Handbook of pulp and paper technology, 2nd ed. Van Nostrand Reinhold, New York, 3–12
Browning BL, Isenberg IH 1952 Analytical data and their significance. In: Wise LE, Jahn EC (eds) Wood chemistry. Reinhold, New York, 1259–1277
Büsgen M 1897 Bau und Leben unserer Waldbäume. Gustav Fischer, Jena, 230 pp
Büsgen M, Münch E 1927 Bau und Leben unserer Waldbäume, 3rd ed. Gustav Fischer, Jena, 426 pp
Büsgen M, Münch E, Thomson T 1929 The structure and life of forest trees, 3rd ed. Chapman and Hall, London, 436 pp
Butterfield BG, Meylan BA 1980 Three-dimensional structure of wood, 2nd ed. Chapman and Hall, London, New York, 103 pp
Campredon J 1953 Le bois de réaction. Veine rouge. Cellules gélatineuses. Rev Bois Appl 8(2):3–7
Casey JP (ed) 1980 Pulp and Paper. Chemistry and chemical technology, Vol 1. Wiley, New York, 820 pp
Casperson G 1963a Chemische, anatomische und physiologische Eigenheiten des Reaktionsholzes. Chemiefasersymp 1962, Akademie Verlag, Berlin, 39–52
Casperson G 1963b Reaktionsholz. Seine Struktur und Bildung. Habilitationsschr, Humboldt-Univ, Berlin DDR, 116 pp
Casperson 1965a Zur Anatomie des Reaktionsholzes. Svensk Papperstidn 68:534–544
Casperson G 1965b Über die Entstehung des Reaktionsholzes bei Kiefern. In: Aktuelle Probleme der Kiefernwirtschaft. Tagungsber 75, Deutsch Akad Landwirtsch, Berlin DDR, 523–528
Cech MY, Pfaff F 1978 Wood moisture relations. In: Kiln operator's manual for eastern Canada. Can For Serv Rep OP-X-1992E, 15–24
Cieslar A 1896 Das Rothholz der Fichte. Cbl Ges Forstwes 22:149–165
Clarke SH 1940 The importance of the plant cell wall to the forester and the wood anatomist. For Abstr 2(1):1–4

References

Clowes FAL, Juniper BE 1968 Plant cells. Blackwell, Oxford Edinburgh, 546 pp
Core HA, Côté WA Jr, Day AC 1961 Characteristics of compression wood in some native conifers. For Prod J 11:356−362
Core HA, Côté WA Jr, Day AC 1979 Wood. Structure and identification, 2nd ed. Syracuse Univ Press, Syracuse, 182 pp
Correns E 1961 Über anormale Holzfasern. Pap Puu 43:47−62
Côté WA Jr 1967 Wood ultrastructure. Univ WA Press, Seattle London, 60 pp
Côté WA Jr 1977 Wood ultrastructure in relation to chemical composition. In: Loewus FA, Runeckles VC (eds) The structure, biosynthesis, and degradation of wood. Plenum New York, London, 1−44
Côte WA Jr, Day AC 1965 Anatomy and ultrastructure of reaction wood. In: Côté WA Jr (ed) Cellular ultrastructure of woody plants. Syracuse Univ Press, Syracuse, 391−418
Croon I 1961 Tryckved och dragved, morfologi och kemisk sammansättning. (Compression wood and tension wood, morphology and chemical composition.) Svensk Papperstidn 64:175−180
Dadswell HE 1958 Wood structure variations occurring during tree growth and their influence on properties. J Inst Wood Sci 1:11−33
Dadswell HE 1963 Tree growth−wood property inter-relationships. VIII Variations in structure and properties in wood grown under abnormal conditions. In: Maki TE (ed) Proc Spec Field Inst For Biol, Raleigh NC 1960, NC State Univ School For, 55−66
Dadswell HE, Wardrop AB 1949 What is reaction wood? Aust For 13:22−33
Dadswell HE, Wardrop AB 1955 The structure and properties of tension wood. Holzforschung 9:97−104
Dadswell HE, Wardrop AB 1956 The importance of tension wood in timber utilization. Appita 10:30−42
Dadswell HE, Wardrop AB, Watson AJ 1958 The morphology, chemistry and pulping characteristics of reaction wood. In: Bolam F (ed) Fundamentals of papermaking fibres. Tech Sect Brit Pap Board Makers' Assoc, London, 187−219
de Carvalho A 1957 Defeitos da madeira. I Estudos e divulgação técnica. Direcção Geral dos Seviços Florestais e Aquícolas, Lisbon, 143 pp
Desch HE 1973 Timber. Its structure and properties, 5th ed. St Martin's, New York, 424 pp
Desh HE, Dinwoodie JM 1980 Timber. Its structure, properties and utilization, 6th ed. Timber Press, Forest Grove 410 pp
Detienne P 1976 Recherche et nature de bois de tension dans quelques arbres tropicaux. Centr Tech For Trop Nogent-sur-Marne, 46 pp
Dinwoodie JM 1981 Timber. Its structure and behavior. Van Nostrand Reinhold, New York, 190 pp
Durst J 1955 Taschenbuch der Fehler und Schäden des Holzes. Fachbuchverlag, Leipzig, 150 pp
du Toit AJ 1964 Probable causes of compression wood formation. For S Afr 4:25−35
Ertfeld W, Mette HJ, Achterberg W 1964 Defects of wood. Leonard Hill, London, 76 pp
Esau K 1965 Plant anatomy. Wiley, New York, 767 pp
Fahn F 1967 Plant anatomy. Pergamon, Oxford London 534 pp
Fengel D, Wegener G 1984 Wood. Chemistry, ultrastructure, reactions. Walter de Gruyter, Berlin New York, 613 pp
Findley WPK 1975 Timber: properties and uses. Crosby Lockwood Staples, London, 224 pp
Forest Products Laboratory, Madison WI, US For Serv 1955 Wood Handbook. Agr Handb 72, 528 pp
Forest Products Laboratory, Madison WI, US For Serv 1974 Wood handbook. Wood as an engineering material. Agr Handb 72, 216 pp
Frey-Wyssling A 1959 Die pflanzliche Zellwand. Springer, Berlin Göttingen Heidelberg, 367 pp
Frey-Wyssling A 1976 The plant cell wall. Gebrüder Bornträger, Berlin Stuttgart, 294 pp
Garratt GA 1931 The mechanical properties of wood. Wiley, New York, 276 pp
Gessner F 1961 Die mechanischen Wirkungen auf das Pflanzenwachstum. In: Ruhland W (ed) Encyclopedia of plant physiology. Springer, Berlin Heidelberg New York, 16:634−667
Giordano G 1951 Il legno e le sue carratteristische. Editore Ulrico Hoepli, Milano, 759 pp

Giordano G 1971 Tecnologia del legno. 1 La materia prima. Unione Tipografico-Editrice Torinese, Torino, 1086 pp

Göhre K 1954 Werkstoff Holz. VEB-Verlag Technik, Berlin DDR, 366 pp

Göhre K 1961 Werkstoff Holz. Technologische Eigenschaften und Vergütung. VEB Fachbuchverlag, Leipzig, 454 pp

Gray RL, Parham RA 1982 A good look at wood's structure. Chemtech 12:232–241

Grossenbacher JG 1915 The periodicity and distribution of radial growth in trees and their relation to the development of "annual" rings. Trans WI Acad Sci Arts Lett 18:1–77

Hägglund E 1951 Chemistry of Wood. Academic Press, New York, 631 pp

Hale JD 1951 Variations in properties of wood caused by structural differences. In: McElhanney TA (ed) Canadian woods. Their properties and uses, 2nd ed. For Prod Lab Can, 57–78

Hale JD 1969 Structural and physical properties of pulpwood. In: Macdonald RG, Franklin JN (eds) Pulp and paper manufacture. I The pulping of wood, 2nd ed. McGraw-Hill, New York, 1–32

Hale JD, Perem E, Clermont LP 1961 Importance of compression wood in appraising wood quality. 13th IUFRO Congr Wien, Can Dep For Preprint O-186, 12 pp

Harlow WM 1970 Inside Wood. Masterpiece of nature. Am For Assoc, Washington DC, 120 pp

Hartig R 1901 Holzuntersuchungen. Altes und Neues. Julius Springer, Berlin, 99 pp

Haygreen JG, Bowyer JL 1982 Forest products and wood science. IA State Univ Press, Ames, 495 pp

Henderson FY 1944 Timber. Its properties, pests and preservation. Crosby Lockwood, London, 148 pp

Hoadley RB 1980 Understanding wood. A craftsman's guide to wood technology. The Taunton Press, Newtown, 256 pp

Holtmann DF 1929 Wood construction. Principles–practice–details. McGraw-Hill, New York, 711 pp

Hughes FE 1965 Tension wood. A review of literature. For Abstr 26:1–9, 179–186

Ilvessalo-Pfäffli MS 1977 Reaktiopuu. (Reaction wood.) In: Jensen W (ed) Puukemia. (Wood chemistry), 2nd ed. Akad Tekn Vetensk, Helsinki, 63–67

Isenberg IH 1962 Fibrous raw materials and wood structure. In: Libby CE (ed) Pulp and paper science and technology. Vol 1 Pulp. McGraw-Hill, New York, 20–53

Isenberg IH 1980 Pulpwoods of the United States and Canada. I Conifers. Inst Pap Chem, Appleton, 219 pp

Jane FW 1952 Abnormal wood. Wood 17:101–104

Jane FW 1956 The structure of wood. MacMillan, New York, 427 pp

Jane FW, Wilson K, White DJB 1970 The structure of wood. Adam and Charles Black, London, 478 pp

Jones N 1956–1957 Compression wood. Y Coedwigwr 3(1):12–17

Jullander I, Stockman L 1978 The life cycle of wood. National Swed Board Tech Dev STU 110-1978, 271 pp

Jutte SM 1969 A comparative study of normal and tension wood fibres in beech (Fagus silvatica L) and ash (Fraxinus excelsior L). Ph D Thesis Univ Leiden, 62 pp

Kärkkäinen 1976 Havutukkien painomittauksen edellytyksiä puurieteelliseltä kannalta. (Wood science. Prerequisites for the weight measurement of pine and spruce logs.) Commun Inst For Fenn 89.1, 59 pp

Keith CT, Kellogg RM 1981 The structure of wood. In: Mullins EJ, McKnight TS (eds) Canadian woods. Their properties and uses, 3rd ed. Univ Toronto Press, Toronto Buffalo London, 59–60

Kennedy RW 1976 Basic wood characteristics. In: Tayelor FA (ed) Wood engineering seminar 1975. Can For Serv Rep VP-X-162, 63 pp

Knigge W 1958 Das Phänomen der Reaktionsholzbildung und seine Bedeutung für die Holzverwendung. Forstarchiv 29:4–10

Knigge W, Schulz H 1966 Grundriß der Forstbenutzung. Paul Parey, Hamburg Berlin, 584 pp

Knuchel H 1938 Kampf den Holzfehlern. Schweiz Z Forstwes 89:241–255

Knuchel H 1940 Holzfehler. Büchler, Bern 144 pp

Knuchel H 1947 Holzfehler. Werner Classen, Zürich, 119 pp

Knuchel H 1954 Das Holz. Entstehung und Bau, physikalische und gewerbliche Eigenschaften, Verwendung. HR Sauerländer, Frankfurt, 472 pp
Koch CB, Li FF, Hamilton JR 1968 The nature of tension wood in black cherry. WV Univ Agr Exp Sta Bull 561, 14 pp
Koch P 1972 Utilization of the southern pines. US Dept Agr, Agr Handb 420, 1663 pp
Kollmann FFP 1951 Technologie des Holzes und der Holzwerkstoffe. Vol 1, 1050 pp
Kollmann FEP, Côté WA Jr 1968 Principles of wood science and technology. I Solid wood. Springer, Berlin Heidelberg New York, 592 pp
König E 1942 Fehler und Krankheiten des Holzes. Deutsch Holzwirtsch 59(153):1, 3
König E 1946 Exzenterwuchs — Entstehung und Beeinträchtigung der Holzqualität. Holz-Zentralbl 72(1):3—4
König E 1952 Die Holzfehler. Norddeutsch Holzwirtsch 6(50):3
König E 1956 Sortierung und Pflege des Holzes. Holz-Zentralblatt Verlag, Stuttgart, 244 pp
König E 1957 Fehler des Holzes. Holz-Zentralblatt Verlag, Stuttgart, 256 pp
Kozlowski TT 1971 Growth and development of trees. II Cambial growth, root growth, and reproductive growth. Academic Press, New York London 514 pp
Kramer PJ, Kozlowski TT 1978 Physiology of woody plants. Academic Press, New York, 811 pp
Krzysik F 1957 Nauka o drewnie (Science of wood.) Państwowe Wydawnictwo, Rolnicze i Leśne, Warszawa, 899 pp
Krzysik F 1975 Nauka o drewnie. (Science of wood.) Państwowe Wydawnictwo, Rolnicze i Leśne, Warszawa, 654 pp
Kubler H 1980 Wood as building and hobby material. Wiley, New York, 256 pp
Langendorf G 1961 Handbuch für den Holzschutz. VEB-Fachbuchverlag, Leipzig, 330 pp
Leopold AC, Kriedemann PE 1975 Plant growth and development. McGraw-Hill, New York, 545 pp
Linnaeus (von Linné), C 1732 Iter Laponicum. (The Lapland journey.)
Low AJ 1964 Compression wood in conifers: a review of literature. For Abstr 25(3, 4), 14 pp
Lukić-Simonović N 1955 (On knowledge of the compression wood.) Šumarstvo 8:474—479
Lyr H, Polster H, Fiedler HJ 1967 Gehölzphysiologie. VEB Gustav Fischer, Jena, 444 pp
Meylan BA, Butterfield BG 1972 Three-dimensional structure of wood. A scanning electron microscopy study. Syracuse Univ Press, Syracuse, 80 pp
Mombächer R 1971 In memoriam Ewald König, Holz Roh-Werkst 29:77—80
Morey PR 1973 How trees grow. Edward Arnold, London, 59 pp
Mork E 1928 Om tennar. (On compression wood.) Tidsskr Skogbr 36 (suppl):1—41
Mork E 1946 Vedanatomi. (Wood anatomy.) Johan Grundt Tanum, Oslo, 65 pp
Nečesaný V 1956 Struktura reakčního dřeva. (Structure of reaction wood.) Prelia 28:61—65
Ollinmaa PJ 1955 Havupuiden lylypuun rakenteesta ja ominaisuuksista. (On the structure and properties of coniferous compression wood.) Pap Puu 37:544—549
Ollinmaa PJ 1956 Vetopuun rakenteesta ja ominaisuuksista. (On the structure and properties of tension wood.) Pap Puu 38:603—611
Ollinmaa PJ 1959 Reaktiopuututkimuksia. (Study on reaction wood.) Acta For Fenn 72.1, 54 pp
Onaka F 1949 (Studies on compression and tension wood.) Mokuzai Kenkyo, Wood Res Inst Kyoto Univ 1, 88 pp. Transl For Prod Lab Can 93 (1956), 99 pp
Paclt J 1954 Kazy dreva. (Wood defects.). Státne Nakladatel'stvo Technickej Literatúry, Bratislava, 92 pp
Panshin AJ, de Zeeuw CH 1980 Textbook of wood technology. Structure, identification, properties, and uses of commercial woods of the United States and Canada, 4th ed. McGraw-Hill, New York, 722 pp
Parham RA, Gray RL 1982 The practical identification of pulp wood fibers. TAPPI Press, Atlanta, 212 pp
Parham RA, Gray RL 1984 Formation and structure of wood. In: Rowell R (ed) The chemistry of solid wood. Am Chem Soc Adv Chem Ser 207:3—56
Peng WT 1969 (The relationship of wood structure and wood quality.) Q J Chin For 2:175—207
Perem E 1963 Reaction wood in hardwoods. For Prod Res Br Can Contr P-36, 4 pp
Perem E 1964 Tension wood in Canadian hardwoods. Dept For Can Publ 1057, 38 pp

Perem E, Clermont LP 1961 Importance of tension wood in appraising wood quality. Proc 13th IUFRO Congr Wien 2(2):41/7, 7 pp

Philipson WR, Ward JM, Butterfield BG 1971 The vascular cambium, its development and activity. Chapman and Hall, London, 182 pp

Phillips EWJ 1968 Identification of softwoods by their macroscopic structure. For Prod Res GB Bull 22, 56 pp

Pillow MY 1930 Compression wood as a cause of distortion of softwood lumber. J For 28:1173 – 1177

Pillow MY 1933 Structure, occurrence, and properties of compression wood. M S Thesis, Univ WI, Madison, 83 pp

Pillow MY, Luxford RF 1937 Structure, occurrence, and properties of compression wood. US Dep Agr Tech Bull 546, 32 pp

Rance HD (ed) 1980 Handbook of paper science. 1 The raw materials and processing of papermaking. Elsevier, Amsterdam, 298 pp

Rendle BJ 1937 Gelatinous wood fibres. Trop Woods 52:11 – 19

Rendle BJ 1955 Tension wood. A nataural defect of hardwoods. Wood 20:348 – 351

Rendle BJ 1956 Compression wood. A natural defect of softwoods. Wood 21:120 – 123

Robards AW 1969 The effect of gravity on the formation of wood. Sci Progr (Oxf) 57:513 – 532

Rydholm SA 1965 Pulping processes. Interscience, New York London Sydney, 1269 pp

Sachsse H 1965 Untersuchungen über Eigenschaften und Funktionsweise des Zugholzes der Laubbäume. Schriftenr Forstl Fak Univ Göttingen 35, 110 pp

Sandermann W 1956 Grundlagen der Chemie und chemischen Technologie des Holzes. Akademische Verlagsgesellschaft, Leipzig, 498 pp

Sanio C 1860 Einige Bemerkungen über den Bau des Holzes. Bot Ztg 18:193 – 198, 201 – 204, 209 – 217

Sarkanen KV, Hergert HL 1971 Classification and distribution. In: Sarkanen KV, Ludwig CH (eds) Lignins. Occurrence, formation, structure and reactions. Wiley-Interscience, New York, 43 – 94

Schultze-Dewitz G 1969 Die Variation der Strukturelemente im Nadelholz. Holztechnologie 10:185 – 189

Scurfield G 1973 Reaction wood: its structure and function. Science 179:647 – 655

Siau JS 1984 Transport processes in wood. Springer, Berlin, Heidelberg, New York, 245 pp

Silvester D 1967 Timber. Its mechanical properties and factors affecting its structural use. Pergamon, Oxford, 152 pp

Sinnott EW 1960 Plant morphogenesis. McGraw-Hill, New York, 550 pp

Sjöström E 1981 Wood chemistry. Fundamentals and applications. Academic Press, New York, 223 pp

Snow CH 1917 Wood and other organic structural materials. McGraw-Hill, New York, 478 pp

Špoljarič Z 1959 Struktura i kvaliteta drva. (Structure and quality of wood.) Drvna Ind 10:105 – 113

Spurr SH, Hyvärinen MJ 1954 Compression wood in conifers as a morphogenetic phenomenon. Bot Rev 20:551 – 560

Stamm AJ 1964 Wood and cellulose science. Ronald, New York, 549 pp

Stockman L 1957 Der Einfluß morphologischer Faktoren auf die Herstellung von Zellstoff und Papier. In: Treiber E (ed) Die Chemie der Pflanzenzellwand. Springer, Berlin Göttingen Heidelberg, 22 – 43

Stone H 1921 A textbook of wood. William Rider, London, 240 pp

Summitt R, Sliker A 1980 CRC handbook of materials science. CRS Press, Boca Raton, 459 pp

Thiemann HD 1942 Wood technology. Pitman, New York, Chicago, 328 pp

Thiemann HD 1951 Wood technology. Constitution, properties, and uses. Pitman, London, 396 pp

Thunell B 1945 Trä. Dess byggnad och felaktigheter. (Wood. Its structure and defects.) Byggnadsstandardiseringen, Bröderna Lagerström, Stockholm, 103 pp

Tikka PS 1935 Puiden vikanaisuuksista pohjois-Suomen metsissä tilastollis-metsäpatologinen tutkimus. (On tree defects in the forests of northern Finland.) Acta For Fenn 41, 371 pp

Timell TE 1969 The chemical composition of tension wood. Svensk Papperstidn 72:173 – 181

Timell TE 1973 Ultrastructure of the dormant and active cambial zones and the dormant phloem associated with formation of normal and compression woods in Picea abies (L) Karst. SUNY Coll Environ Sci For Syracuse, Tech Publ 96, 94 pp

Trendelenburg R 1932 Über die Eigenschaften des Rot- oder Druckholzes der Nadelhölzer. Allg Forst-Jagdztg 108:1–14

Trendelenburg R 1939 Das Holz als Rohstoff. J F Lehmanns, München Berlin, 435 pp

Trendelenburg R, Mayer-Wegelin H 1955 Das Holz als Rohstoff. Carl Hanser, München, 541 pp

Tsoumis G 1968 Wood as a raw material. Pergamon, Oxford, 276 pp

Vandevelde R 1957 Bijdrage tot de studie van reactiehout. (Contribution to the study of reaction wood.) Meded Houttechnol Gent, 22:751–768

von Pechmann H 1973 Beobachtungen über Druckholz und seine Auswirkung auf die mechanischen Eigenschaften von Nadelholz. Proc IUFRO-5 Meet S Afr, 2:1114–1122

Wagenführ R 1966 Anatomie des Holzes. VEB Fachbuchverlag, Leipzig, 377 pp

Wagenführ R 1980 Anatomie des Holzes, unter besonderer Berücksichtigung der Holztechnik. VEB Fachbuchverlag, Leipzig, 328 pp

Wagenführ R 1984 Anatomie des Holzes, unter besonderer Berücksichtigung der Holztechnik. VEB Fachbuchverlag, Leipzig, 320 pp

Wagenführ R, Schreiber C 1974 Holzatlas. VEB Fachbuchverlag, Leipzig, 690 pp

Wahlgren A, Schotte G 1928 Sveriges skogar och huru vi utnyttjar dem. (The forests of Sweden and how we utilize them.) Hökerbergs, Stockholm, 1682 pp

Wangaard FF 1950 The mechanical properties of wood. Wiley, New York Chapman and Hall, London, 377 pp

Wangaard FF (ed) 1981 Wood: its structure and properties. EMMSE project, PA State Univ, University Park, 465 pp

Wardrop AB 1960 Scientific review: the structure and formation of reaction wood in angiosperms. IAWA Bull 1960 (1):2–8

Wardrop AB 1961 The structure and formation of reaction wood in angiosperms. In: Recent advances in botany, 9th Int Bot Congr, Montreal PQ, 1959, II:1325–1330

Wardrop AB 1964 The reaction anatomy of arborescent angiosperms. In: Zimmermann MH (ed) The formation of wood in forest trees. Academic Press, New York London, 405–456

Wardrop AB 1965 The formation and function of reaction wood. In: Côté WA Jr (ed) Cellular ultrastructure of woody plants. Syracuse Univ Press, Syracuse, 371–390

Wegner TH 1984 Wood. In: Kirk-Othmer Encyclopedia of Chemical Technology, 3rd ed. Wiley, New York, 24:579–611

Wenzl HFJ 1970 The chemical technology of wood. Academic Press, New York London, 692 pp

Westing AG 1965 Formation and function of compression wood in gymnosperms. Bot Rev 31:381–480

Westing AH 1968 Formation and function of compression wood in gymnosperms II. Bot Rev 34:51–78

White DJB 1965 The anatomy of reaction tissues in plants. In: Carthy JD, Duddington CL (eds) Viewpoints in biology IV. Butterworth, London, 54–82

Wicker M 1979 Le bois de tension: acquisitions récentes. Ann Biol 18:221–254

Wilson BF 1970 The growing tree. Univ MA Press, Amherst, 152 pp

Wilson BF 1981 The development of growth strains and stresses in reaction wood. In: Barnett JR (ed) Xylem cell development. Castle House, Tunbridge Wells, 307 pp

Wilson BF, Archer RR 1977 Reaction wood: induction and mechanical action. Ann Rev plant Physiol 28:23–43

Wilson BF, Archer RR 1979 Tree design: some biological solutions to mechanical problems. BioScience 29:293–298

Wooten TE 1967 Compression wood – a common defect in southern pine. For Farmer 28(3):14–15

Wooten TE 1968a The identification and properties of compression wood. MS For Prod Utiliz Lab Inform Ser 7, 7 pp

Wooten TE 1968b Structure and properties of wood. I Gross structural features of wood. MS For Prod Utiliz Lab, 15 pp

Zimmermann MH, Brown CL 1971 Trees. Structure and function. Springer, New York Heidelberg Berlin, 336 pp

Chapter 2 Historical Background

CONTENTS

2.1 Introduction 29
2.2 Occurrence and Appearance of Compression Wood 29
2.3 Structure and Other Properties of Compression Wood 35
2.4 General Conclusions 41
References 42

2.1 Introduction

Because of its universal occurrence in practically all coniferous forest trees, compression wood must have been known ever since wood began to be utilized by man and certainly as long as loggers and carpenters have existed, that is, for at least 4000 years. Actual mention of compression wood in the literature is of much more recent date. The Greek Theophrastos (370–285 B.C.) and the Roman Plinius the elder (23–79 A.D.) do not refer to it in their extensive descriptions of trees and forestry in general (Makkonen 1967, 1969).

Compression wood is most commonly associated with eccentric radial growth, and this phenomenon attracted attention in the scientific literature much earlier than did the tissue itself. Duhamel du Monceau (1753) in his book *La Physique des Arbres* discusses various reasons for eccentric growth in stems, but he does not mention anything about the wood. Additional observations were later made by Knight (1801), de Candolle (1833), and Treviranus (1835). In 1854 Schimper introduced the terms *epinasty* and *hyponasty* to signify increased growth on the upper or lower side of a plant organ, respectively. These and other early attempts to explain eccentric radial growth in trees are discussed in Chap. 11.2. The best known of these contributions are those made by the German botanists Nördlinger (1860, 1871, 1882), von Mohl (1862, 1869), Kny (1873, 1877, 1882), Detlefsen (1882), and Wiesner (1892a, b). Most of these early investigations have been reviewed by Grossenbacher (1915) and also, more briefly, by Ursprung (1906) and by Tischendorf (1943).

2.2 Occurrence and Appearance of Compression Wood

The fact that conifers develop a hard, dark type of wood on the lower side of branches and a leaning or crooked stem is so conspicuous that it must have been observed long ago. During his travel in Lapland in 1732, Carolus Linnaeus (Carl von Linné) noticed that in Jokkmok many of the pine (*Pinus syl-*

Fig. 2.1. Karl Gustav Sanio (1832–1891). Dozent, Universität Königsberg (1858–1866)

Fig. 2.2. Adolf Cieslar (1858–1934). Professor, Hochschule für Bodenkultur, Wien (1905–1927)

vestris) trees were crooked. He writes (in translation): "Crooked pines grew here in many places. Their lower side always becomes hard, like box wood (*Buxus*). The Laplanders make skis and sleigh runners from this wood which they call 'kör'." Undoubtedly, this represents one of the earliest recorded descriptions of compression wood and its utilization.

In his well-known book *Die technischen Eigenschaften der Hölzer* from 1860, Nördlinger discussed in detail subjects such as eccentric radial growth, branch wood, brash wood, and wood defects in general. Nowhere is here, however, anything that could possibly be contrued as a reference to compression wood.

In the same year, 1860, Sanio (Fig. 2.1) reported some observations on the highly lignified latewood present on the lower side of horizontal branches of *Picea abies*. This wood was red in color except for the few first-formed cells of earlywood. Two years later, Schacht (1862) described bands of dark wood with thick-walled cells on the under side of branches of *Araucaria angustifolia*. In his famous publication from 1873, *Anatomie der gemeinen Kiefer* (*Pinus silvestris* L), whose central subject is the organization and function of the vascular cambium, Sanio also included a detailed description of the occurrence, structure, and lignification of compression wood (Timell 1980). Sanio mentions that he has observed the reddish-brown, hard and brash compression wood, which he refers to as "differenziertes Holz," not only on the lower side of pine and spruce branches but also at the base of pine stems in the form of bands which sometimes extended into the earlywood portion of the growth rings. The compression wood was often associated with eccentric radial growth. Toward the top of

2.2 Occurrence and Appearance of Compression Wood

the tree, compression wood became more sparse. Sanio mentions that it is frequently found in the stem immediately below a branch base, and that this compression wood is a direct continuation of that in the branch. He also states that compression wood is never formed in roots as long as they remain buried in the ground but that it develops on exposure. Long a moot question, this observation was not to be attested until almost 100 years later (Chap. 10.7.3).

Nördlinger (1875) found that in an 80-year-old *Pinus sylvestris* tree the density of the wood decreased with increasing height above the ground. The stem of this tree was eccentric and on the side with the largest radius the maximum density was 0.673 compared to 0.555 on the opposite side. Nördlinger attributed the higher density value to the presence of "three very wide growth rings, consisting almost entirely of red, dense summerwood." Obviously, Nördlinger is referring here to compression wood, for he adds that such wood is regularly found on the under side of horizontal and inclined branches and on the lower side of leaning conifer stems. Typically, such wood appears "on the eccentric side" of stems when trees are released by thinning (Chap. 16.7.4.1).

According to Donner (1875) most *Pinus sylvestris* trees in northern Germany exhibit eccentric growth with a tendency for wider growth rings to occur on the side of the stem facing the north. The wood on this side is said to be harder, more brittle, and colored dark red. The same wood is found on the under side of *Picea abies* branches and on the convex side of stems forming a bow. There can be little doubt that Donner is referring to compression wood. His brief article is concluded by a few remarks by Hartig. Hartig later (1896, 1901) refers to certain facts in Donner's paper in almost identical words, and it would not be surprising if Hartig's later, great interest in compression wood received its first impetus from Donner's observations. In his important article on compression wood from 1888, Kononchuk mentions that his investigations on this tissue were triggered by Donner's paper.

Three years later, Nördlinger (1878) reported more detailed observations on wood formed on the upper and lower sides of a *Larix decidua* tree which had been weighted down by snow. The wood on the lower side, where the growth rings were wider, was red and had a density of 0.710 compared to only 0.565 for the upper (opposite) wood. He made similar observations with trees of *Abies alba*, *Picea abies*, and *Pinus sylvestris*, which likewise had been subjected to pressure from snow. Wood on the lower side of pine branches was also denser than that on the upper, and the average density of the entire branch wood was higher than that of the stem wood (Chap. 7.2.3).

In his book on the influence of external factors on the radial growth in trees, Kny (1882) observed the occurrence of wood containing tracheids with a thick cell wall and a narrow lumen, obviously compression wood, on the lower side of branches from ten conifer species, namely *Abies alba*, *A. nordmanniana*, *Juniperus communis*, *J. occidentalis*, *Larix decidua*, *Picea abies*, *Taxodium distichum*, *Taxus baccata*, *Thuja occidentalis*, and *Tsuga canadensis*. While investigating the rays in stem, root, and branch woods of *Picea abies*, Fischer (1885), to his surprise, found that in branches most of the wood was dark and hard. Contrary to the situation in the stem, even the widest growth rings in the branch consisted largely of "latewood." Obviously, Fischer is referring to compression wood.

A year later, Mer (1887) reported the occurrence of a hard type of wood, orange to red in color, in *Abies alba* and *Picea abies*. It was always present on the under side of the eccentric branches and on the lower side of crooks and bends in the stem. Large amounts of this "bois rouge" were produced, again on the lower side, when a vigorous branch, by bending upward, managed to replace a broken leader in the tree. Mer (1887), unlike Fischer (1885) was aware of the fact that his "bois rouge" was not ordinary latewood, since it was formed during almost the entire growing season. In subsequent publications Mer (1888a, b, 1889) reiterated his opinion that there is very little earlywood in growth rings containing compression wood but that the dominating "latewood" differs from the latewood found in normal wood. His many observations on the environmental factors that are associated with formation of compression wood are discussed in Chap. 15.

In the same year, Kononchuk (1888) published an extensive report on compression wood, which was to be largely neglected by later investigators, until Westing (1965) finally placed it in its proper perspective. Kononchuk observed compression wood only in eccentrically grown branches or inclined stems and always on the lower side. In branches it was more common in *Picea abies* than in *Pinus sylvestris*. Compression wood occurred on the lower (convex) side of the curvature in bent trees of these species (Fig. 10.40). Stems with spiral grain also contained compression wood but it was never present in roots. The bark on the compression wood side was thinner in *Pinus* and harder and stronger in *Picea*.

Early in 1896 Wiesner reported the results of some experiments carried out at his suggestion by Cieslar at Mariabrunn in Austria. Four 8-year-old *Picea abies* trees were tied down so that the lower part of the stems remained vertical while the upper was horizontal. It was found that increased radial growth was initiated not only on the lower side of the inclined main stem but also on the under side of the associated branches, in both locations with concomitant formation of compression wood. Wiesner points out that in coniferous branches a radial growth promotion always takes place on the under side, a phenomenon for which he suggested the term *hypotrophy* rather than "hyponasty" introduced by Schimper (1854).

The same year, 1896, saw the publication of two important and now classical contributions, both entitled *Das Rothholz der Fichte* (*Norway Spruce Compression Wood*), namely one by Cieslar (Fig. 2.2) with a complete description of his experiments and observations, and the other by Hartig. Actually, Hartig's article appeared in print about half a year earlier than Cieslar's, but since Cieslar's results had already been discussed by Wiesner (1896), it will be considered first here.

After a detailed review of Hartig's results, Cieslar (1896) describes his own experiment. Four 9-year-old *Picea abies* trees, referred to above, which were about 1.5 m high, were bent into a horizontal position 60–70 cm above the ground. The trees were bent in May 1893, and the upper part of the stem was kept in this position until the fall of 1895. At that time, sections were cut from the horizontal and vertical portions of the stem, and detailed data were obtained for the extent of growth in the different quarters of each increment, as well as for the location and nature of the compression wood formed. Cieslar's extensive data cannot be included here, and his photographic reproduction of eight cross sections do not lend themselves to reproduction. The lowest part of the stems exhibited hardly any eccentricity, but the remainder of the vertical portion had grown more in the direction in which the upper part had been bent, and this eccentric radial growth increased with increasing height. The bend and the horizontal portion were highly hypotrophic.

Compact zones of compression wood had been produced throughout the lower part of the horizontal stem and in the bend itself. Regions of compression wood were also present in the entire vertical part of the stem. In addition, both eccentric growth and formation of compression wood had occurred not only within the 1894 and 1895 growth rings but also within the latewood portion of the 1893 increment which had been completed more than half a year before the upper part of the tree had been bent. It is unlikely that this eccentric radial growth and formation of compression wood could have been caused by the later bending, as suggested by Cieslar, who seems to regard this puzzling phenomenon as an analogy to the formation of heartwood and vaguely mentions a possible change in the form and composition of the tracheids in the growth ring of 1893. The most probable explanation is that this compression wood had been formed during the 1893 season, when the tree was still growing vertically, presumably as a result of wind action (Chap. 15.2.1.3). It should perhaps be added here that both the nature of Cieslar's experiment and his own description of it make it clear that no examination of the stem wood of this tree was carried out until the entire experiment had been completed. It remains a fact, however, that Jaccard (1919) later claimed to have observed a similar phenomenon in a *Pinus sylvestris* tree that was continuously being bent down and then brought back to a vertical position. At the end of the experiment, the amount of compression wood formed was larger than the amount of wood produced, a seeming paradox. Jaccard's explanation, which is not unreasonable, is discussed in Chap. 11.3. It could perhaps also be applied to Cieslar's observation.

Cieslar (1896) found that in the horizontal part and within the curvature only a thin zone of seemingly normal earlywood had been produced, the remainder of the 1894 and 1895 growth rings consisting of compression wood. In the vertical portion, compression wood was restricted to the tissue produced during the latter part of the season. This wood was lighter in color than that present in the horizontal stem portion.

In discussing his results, Cieslar states that compression wood should be regarded as a tissue with the function of mechanical support. He attributes increased radial growth on the lower side to an abundance of nutrients here, nutrients that are supposed to sink to this side under the influence of gravity. He also argues that such an explanation does not suffice to explain the formation of compression wood in the same location, and concludes with the suggestion that many of the factors involved in the formation of latewood may also play a role in the formation of compression wood.

Cieslar's (1896) contribution, which included a detailed examination of the compression wood, discussed in Sect. 2.3, was of a pioneering nature. In his simultaneous investigation of compression wood in the same species Hartig (1896) for this time restricted himself to observations of naturally growing forest trees. Cieslar is the first to have artificially caused the formation of compression wood in a tree for scientific purposes, and in some respects his examination of this wood was more thorough than Hartig's.

Hartig's investigations on compression and opposite woods in conifers are described in three publications, namely from 1896, 1899, and 1901. He obviously became interested in compression wood during his last few years, for the phenomenon is seldom, if ever, mentioned in his numerous publications prior to 1896. Very likely, the articles by Donner (1875), Sanio (1873), and Mer (1887) served to draw his attention to compression wood. Once interested, he took up these new studies with his usual enthusiasm and thoroughness. Hartig's (1896) first paper contains only a brief reference to Sanio's (1873) earlier observations, but he devotes considerable space to Mer (1887), whose opinions he criticizes sharply. Undoubtedly he is justified in his complaint that Mer had failed to give a thorough account of the anatomy, technical properties, and importance of compression wood. Hartig's own opinion at this time was that compression wood develops where there exists a strong pressure in the longitudinal direction of a stem or branch. Wind exerts such a pressure and is, according to

Fig. 2.3. Location of compression wood (*CW*) in a *Picea abies* stem that had been turned into a loop. (Hartig 1896, 1898, 1901)

Hartig, a frequent cause of compression wood formation. In one case a tree exposed to a prevailing westerly wind had developed increased radial growth and compression wood on the eastern, leeward side of the stem despite the fact that most of the branches were located on the opposite, western side. A stem of *Picea abies*, which formed a loop as shown in Fig. 2.3, was cut open and inspected. The distribution of compression wood, indicated by the seven discs, according to Hartig was exactly what would be expected if the compression wood were caused by an axial pressure and had the function of a strengthening tissue.

Hartig (1896) also studied the occurrence of compression wood on the lower side of branches of *Picea abies* and *Pinus sylvestris*. He was one of the first to notice that the wood located on the upper side of a branch (opposite wood) has its own, characteristic structure and properties (Chap. 21). Compression wood was best developed near the stem, where the pressure from the weight of the branch was largest.

2.3 Structure and Other Properties of Compression Wood

The first to observe and describe the structure of compression wood was the brilliant German botanist Karl Gustav Sanio (Ascherson 1893, Mägdefrau 1975), a fact that seems to have remained unnoticed until very recently (Timell 1980). The information is found in a paper by Sanio published in 1860 under the title *Einige Bemerkungen über den Bau des Holzes*. It deals largely with the structure of the pits in wood, a subject much debated at the time by German plant anatomists, the "tertiary" wall, the intercellular region, and lignification. In a section largely devoted to the formation and distribution of lignin, Sanio reports some observations on the highly lignified latewood present on the lower side of horizontal branches of *Picea abies*. This wood was red in color, a fact attributed by Sanio to its high lignin content. The tracheids had a rounded outline, a thin, readily discernable primary wall, and a very thick secondary wall. Intercellular spaces were present between the cells. They were either empty or contained a "yellowish mass" (probably cytoplasm). Sanio's original drawing of this wood is shown in Fig. 2.4.

Sanio also made the interesting observation that one of the layers of the secondary wall seemed to be as highly lignified as the primary wall, adding that obviously no distinction can be made between primary and secondary walls on the basis of differences in their chemical composition. Although it cannot be fully ascertained, it is very likely that he is referring here to the outer portion of the S_2 layer, which is now known to have the same high lignin content as the primary wall.

Sanio (1873) returned to the subject of compression wood 13 years later in the second of his two, now classical articles on the nature of the wood in *Pinus sylvestris*. He reports that a typical growth ring of compression wood contained about five first-formed cells with an only moderately thick wall, together forming a narrow band of light-colored earlywood. Occasionally, these tracheids were thick-walled and dark in color. All other cells in the ring had the same size and the same thick wall except that the last-formed tracheids were radially flat-

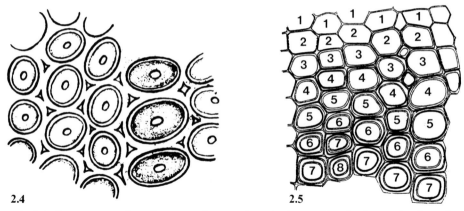

Fig. 2.4. Transverse section of compression wood on the lower side of a *Picea abies* branch, showing intercellular spaces and tracheids with a rounded outline and thick walls. This drawing represents the first attempt to illustrate the structure of compression wood. (Sanio 1860)

Fig. 2.5. Differentiating compression wood on the lower side of a *Pinus sylvestris* branch. Transverse section. Drawing. (Sanio 1873)

tened. Each tracheid possessed a primary and a secondary wall. Inside the former there was a layer that was notably thicker at the cell corners, thereby conferring a rounded outline on the next layer. What Sanio is referring to here must have been the S_1 layer. The following layer was of the same, moderate width, but, unlike the outer one, it was similar to the primary wall in both optical properties and chemical behavior. This layer coated the tracheids after the two outer layers had been removed by treatment with nitric acid. This seemingly isotropic layer, now referred to as the $S_2(L)$ layer, was not to be observed again until 70 years later, when it was rediscovered by Bailey and Berkley (1942) (Chap. 4.4.5.3). Sanio's conclusion that this layer was as highly lignified as the primary wall was not confirmed until a century later (Chap. 6.4.2).

The third layer of the secondary wall was quite thick. This is obviously what is now known as the inner portion of S_2. This part of the wall always contained spiral striations. When these striations crossed a pit, the pit aperture had the form of a long slit, extending beyond the pit border. Sanio does not comment any further on these spiral striations, now more often referred to as helical cavities. It is notable, however, that he realized that they do not extend into the $S_2(L)$ layer, a fact overlooked by many later investigators (Chap. 4.4.5.4). The S_2 layer was terminated on the lumen side by a tertiary inner lining, similar to that in normal tracheids. Since no S_3 layer is present in compression wood tracheids, it is possible that what Sanio observed here might have been the warty layer which was not to be discovered until 80 years later (Chap. 4.4.5.6).

A cross section of differentiating compression wood tracheids as observed and drawn by Sanio (1873) is shown in Fig. 2.5. The angular outline of the young, enlarging tracheids is clearly indicated in the cells marked 1. Deposition of the S_1 layer has commenced in cells 2. This layer gradually thickens at the

2.3 Structure and Other Properties of Compression Wood

corners in cells 3 so that the inner contour becomes rounded. The $S_2(L)$ layer has been formed in cells 4, and in cells 6 the inner portion of S_2 has been deposited. The presence of helical cavities is not indicated in this drawing, probably because they frequently cannot be seen in transversion sections under the light microscope. The somewhat angular outline of the tracheids and the associated absence of intercellular spaces are not typical compression wood features, but they are by no means unknown (Chap. 4.4.2).

Sanio (1873) also studied the deposition of lignin in compression wood tracheids, treating developing tissues with suitable lignin stains (Chap. 9.6.1.4.2). He noted that lignification lagged behind cell wall thickening. The primary wall, for example, was fully lignified when the secondary wall was still being formed. Lignification began in the primary wall at the cell corners, spread in radial and tangential directions, and gradually proceeded from the outside toward the inside of the secondary wall. There was little or no cellulose in the intercellular region. Sanio expressed the opinion that the cell wall probably was formed both by apposition and by intussusception, a remarkably modern view.

Sanio's observations on the occurrence, formation, and properties of compression wood have almost all been confirmed by later investigators. What makes this especially noteworthy is the fact that most of these investigations were carried out during the last 25 years of his life when he was living on his father's estate at Lyck in East Prussia and had very little scientific equipment at his disposal (Timell 1980).

Sanio's pioneering studies on compression wood have been nearly totally and entirely undeservedly neglected by later investigators. Hartig (1896) refers only briefly to Sanio and his "differentiated wood", while Cieslar (1896) and others do not cite him at all. In his comprehensive review of compression wood, Weting (1965) mentions Sanio's (1873) observation of dark bands on the lower side of pine branches, but he does not include Sanio's more important results on the ultrastructure of the tracheids, nor does he refer to his earlier article from 1860. Sanio clearly deserves recognition for having been the first to describe the occurrence of compression wood in the stem, branches, and roots of forest trees and also for being the first to study the ultrastructure, chemical composition, and lignification of the tracheids in this tissue.

Two years after Sanio's first description of compression wood, Schacht (1862) reported similar observations on the compression wood present on the lower side of branches of *Araucaria angustifolia*. He found the tracheids in this brownish wood to be "moderately long," namely 1.14–1.50 mm, and to possess a "rather thick wall." The tracheids had one row of small pits with oval or round apertures in their radial walls. The uniseriate rays were one to six cells high and were composed of thin-walled cells. The drawings in Fig. 2.6 show that the tracheids had a rounded outline in cross section, as pointed out by Schacht, but also that there were a few intercellular spaces, a fact that he does not mention.

Mer (1887) does not seem to have been interested in the fine structure of compression wood. He states that the tracheids were rounded in outline and thick-walled and that they had a narrow lumen. They were highly lignified and

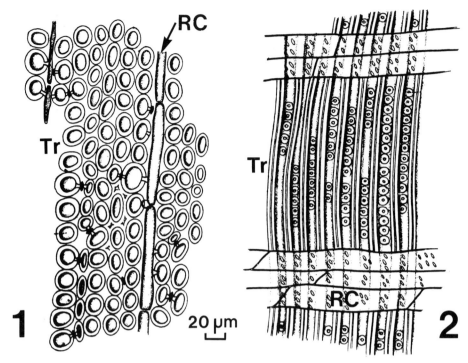

Fig. 2.6. Transverse (**1**) and radial (**2**) sections of compression wood on the lower side of an *Araucaria angustifolia* branch, showing tracheids (*Tr*) and ray cells (*RC*). Drawing. (Schacht 1862)

contained much resin. The ray cells were also thick-walled, a strange observation, and filled with resin. Later investigations have shown that only compression wood in dead branches and knots has a high resin content. Mer found that compression wood contained less water than does normal wood, but that it loses it more rapidly on drying (Chap. 7.3.1.2). If dried quickly at a high temperature, it became lighter in color than normal wood. In general, the color of compression wood was lighter the lower the water content (Chap. 3.3).

More than two decades were to pass after Sanio's second study of compression wood before serious attempts were made to characterize this tissue, namely in 1896, when both Cieslar and Hartig reported their observations on compression wood in *Picea abies*. Cieslar (1896) found that compression wood tracheids were round or oval in cross section and never angular. The cell wall was thick, varying between 4 and 8 µm, which is twice as thick as the wall of normal earlywood cells. The lumen was always narrow and within the range of 4–16 µm. The radial diameter of the tracheids varied between 12 and 25 µm, and the tangential one between 12 and 23 µm. Compared to normal wood tracheids, those in compression wood were 20–30% shorter, varying in length from 0.45 to 1.30 mm.

Cieslar speaks of "striations which actually are spiral thickenings" on the inner surface of the secondary wall, readily detected in tangential sections. He

2.3 Structure and Other Properties of Compression Wood

also mentions that these thickenings tended to protrude into the lumen. In comparing the spiral thickenings of compression wood with those present in normal latewood tracheids of certain species, Cieslar makes no distinction between these two cases and does not seem to have noticed that the "thickenings" in compression wood are winding around the long axis of the tracheids at a much steeper angle than do the helical thickenings in normal wood. He observed that the round tracheids touched one another over only small distances and that between them there were three- or four-sided intercellular spaces. The radial walls had fewer pits than normal tracheids. Within certain growth rings consisting of compression wood, Cieslar observed concentric rings with different, darkbrown colors. Like Sanio (1860), he noted that the first-formed rows of earlywood tracheids were lighter in color, and had a thinner wall and a larger lumen than the compression wood tracheids formed later, suggesting that these cells served to conduct water (Chap. 4.5). The last-formed zone of the latewood, on the other hand, always consisted of very dark compression wood. Cieslar offered the tentative suggestion that perhaps these variations in tracheid anatomy were associated with a varying water content of the cambium in the course of the growing season. Compared to normal wood, compression wood contained one third as many ray cells, which, according to Cieslar, was a consequence of the greater supply of nutrients available to the compression wood cambium.

Unlike Mer (1887), Cieslar came to the conclusion that compression wood contained less water than does normal wood, an opinion also expressed by Hartig (1896) (Chap. 7.3.1.2). The red color was reduced on drying, albeit not as much as had been claimed by Mer. It reverted on renewed wetting and remained permanent if the wood was treated with Vaseline. Cieslar (1896) found compression wood from the horizontal part of his spruce stem to have a specific gravity of 0.83 (dry volume) compared to only 0.45 for corresponding normal wood. The specific gravity of the compression wood located in the vertical part of the stem was lower. The wood was exceptionally hard and also brittle, and Cieslar attributed the small volumetric shrinkage of compression wood to this circumstance as well as to the high lignin content.

Cieslar was the first to attempt a determination of the lignin content of compression wood (Chap. 5.2.2.1). Unfortunately, the method used involved an estimation of the methyl content of the wood. Normal and compression woods were found to have the same methyl contents, and Cieslar concluded that the two tissues contained the same amount of lignin. Since the lignin in compression wood contains fewer methoxyl groups that that in normal softwood (Chap. 5.9.2), the similar methyl content of the two tissues instead indicates a higher lignin content for compression wood.

Hartig's (1896) description of the anatomy of compression wood is limited to the tracheids and was obviously based on long and careful observations. His results are best presented as he does it himself, namely with reference to two painstakenly and amazingly detailed microscopic drawings, shown here in Figs. 2.7 and 2.8. Especially the transverse section is remarkable, for there is little here that can be added from contemporary electron micrographs, yet Hartig depended on light microscopy to construct these drawings. The helical cavities, for example, are only barely seen in cross section under the light micro-

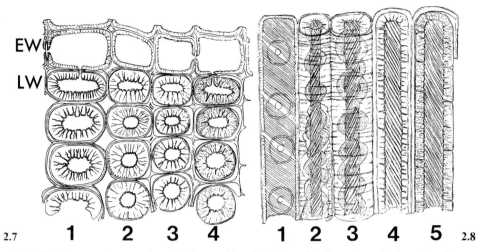

Fig. 2.7. Compression wood on the lower side of a *Picea abies* branch at the boundary between earlywood (*EW*) and latwood (*LW*), showing four different types of tracheids (*1–4*) in the latewood zone. Transverse section. Drawing. (Hartig 1896, 1901)

Fig. 2.8. Longitudinal section of the same compression wood as that shown in Fig. 2.7. Tracheids *1*, *4*, and *5* have been sectioned through their center, while tracheids *2* and *3* are unsectioned and are assumed to contain air. (Hartig 1896, 1901)

scope, yet they are reproduced here not only perfectly accurately but also with all known variations clearly depicted.

The transverse section in Fig. 2.7 contains four rows of fully developed compression wood tracheids, four in each row, and in addition a single row of first-formed tracheids. The latewood tracheids are rounded in outline to the left (row 1) and are separated by large intercellular spaces, but to the right (row 4) they are square, and the intercellular spaces are smaller. In this case, the S_1 layer is thickened at the corners, so that the inner surface of this layer, as well as the entire S_2 layer, are round in transverse section, as had also been observed by Sanio (1873). Compared to normal tracheids, those in the compression wood have a thick wall. In the first-formed earlywood, the tracheids are angular in outline and have thin walls, but small intercellular spaces are nevertheless present.

Hartig mentions that the primary cell wall develops a round outline on differentiation, an observation that was not to be confirmed until much later. In mature tracheids Hartig makes no distinction between the thin primary wall and the subsequent, relatively thick S_1 layer. Hartig was the first to give a complete description of the helical cavities and ribs in the inner portion of the S_2 layer of compression wood, and he devotes special attention to them in this paper. Fig. 2.7 shows how in some cells, illustrated in row 2, the cavities are very narrow and branch inward. In row 3, ribs are clearly visible, and in rows 1 and 4 they protrude into the lumen. It is evident from the drawing that the cavities do not extend as far as the S_2 layer, but unlike Sanio (1873) Hartig does not mention the existence of an isotropic region in the outer part of S_2. Shallow cavities and short ribs are also present in the S_2 layer of the first-formed, earlywood tracheids.

The longitudinal sections in Fig. 2.8 show in tracheids 1, 4, and 5 how the cavities and ribs spiral around the long axis of the cell with a pitch in the range of $30-50°$. Sections 2 and 3 according to Hartig, illustrate the presence of transverse striations in the primary wall. No such striations have since been observed in the primary wall of compression wood tracheids, and it is probable that what Hartig saw here was the S_1 layer with its almost horizontally oriented microfibrils. Section 1 also illustrates how the helical cavities widen when they pass around a pit, creating a trough or slit that extends beyond the pit border, all observations corroborated by later investigators (Chap. 4.10.1).

According to Hartig, compression wood tracheids are $20-30\%$ shorter than those in normal wood. In agreement with Cieslar (1896), he found that compression wood contains less water than does normal wood. He noted the small volumetric shrinkage of compression wood, a fact that he attributed to the presence of the intercellular spaces, the loose attachment of the cell wall layers to one another, and the occurrence of the spiral striations in the secondary wall.

When one recalls that Hartig was not a wood anatomist by profession and probably devoted only brief periods of his unusually active life to microscopic observations, his accurate description of the ultrastructure of the tracheids in compression wood becomes even more remarkable. In the article now discussed, Hartig (1896) also describes the fine structure of the tracheids in opposite wood (Chap. 21). He was the first to do so, and his observations have all been confirmed by later investigators.

2.4 General Conclusions

The year 1896, which saw the publication of Wiesner's note and Cieslar's and Hartig's two papers, is a suitable point in time for concluding this discussion of the origin of the compression wood concept. This brief history has previously been alluded to in several reviews, and it has also been outlined by Schmucker and Linnemann (1951) in their interesting *History of Wood Anatomy*. In 1899 and 1901, Hartig added new data to his earlier observations, this time including also results of experiments with trees. In the same year, Lämmermayr (1901) published an extensive study of the occurrence of compression wood in different tree parts and in different species. Three years later, Sonntag (1904) reported on the mechanical properties of compression wood. In subsequent years up to the present, compression wood has continued to attract the interest not only of wood scientists but also of tree physiologist and foresters. These and other early contributions are discussed in Chap. 11.1.

Although compression wood was first described in 1860 and relatively well characterized by 1896, general acceptance of its existence was slow. It is not mentioned in Schwarz's forest botany monograph from 1892, nor in Gayer's well-known textbook from 1894. One exception is Büsgen (1897), who included Cieslar's (1896) results in his textbook. In the edition from 1927, co-authored by Büsgen and Münch, compression wood is discussed in detail. Even today, one can find books dealing with wood technology, wood defects, or the effect of wind on forest trees that fail to mention compression wood. Between 1860 and 1900 there was seldom any discussion of reaction wood in publications concerned with eccentric growth in trees. Schwappach (1897, 1898), for example, treats compression wood very briefly. Schwarz (1899), in his monograph on *Pinus sylvestris*, devotes a long chapter to eccentric radial growth in this species with a reference to Cieslar's (1896) investigations, but without mentioning compression wood.

In 1896 Robert Hartig complained that many of the previous observations on the anatomy of wood were ignored by the anatomists of his day. Seventy years later, Frey-Wyssling (1966) pointed out that much of the earlier work on compression wood seemed to have been forgotten by later investigators. Many of the results obtained by Sanio and Hartig were reported again several decades later without any reference to their investigations (Timell 1980). The experiments on the formation of compression wood carried out in Zürich by Engler (1918) and by Jaccard (1919) were repeated 20–30 years later, often with identical results. The first report of the peculiar spiral compression wood, still not satisfactorily explained, has always been ascribed to Mork (1928). Actually, it was first described and illustrated 10 years earlier by Alice Spencer (1918). The discovery that exogenous indoleacetic acid causes formation of compression wood has generally been attributed to Wershing and Bailey (1942) despite the previous, extensive report by Onaka (1940). More examples could be cited, all of them demonstrating the truth of George Santayana's famous dictum that those who cannot remember the past are condemned to repeat it.

References

Ascherson P 1893 Karl Sanio. Verh Bot. Ver Brandenburg 34:xli–il
Bailey IW, Berkley EE 1942 The significance of x-rays in studying the orientation of cellulose in the secondary walls of tracheids. Am J Bot 29:231–241
Büsgen M 1897 Bau und Leben unserer Waldbäume. Gustav Fischer, Jena, 230 pp
Büsgen M, Münch E 1927 Bau und Leben unserer Waldbäume, 3rd ed. Gustav Fischer, Jena, 426 pp
Cieslar A 1896 Das Rothholz der Fichte. Cbl Ges Forstw 22:149–165
de Candolle AP 1833 Pflanzenphysiologie. Transl by J Röper, Stuttgart Tübingen
Detlefsen E 1882 Versuche einer mechanischen Erklärung des excentrischen Dickenwachsthums verholzter Ästen und Wurzeln. Arb Bot Inst Würzburg 2:670–688
Donner 1875 Die harte und weiche Seite der Kiefer. Z Forst Jagdwes 7:242–246
Duhamel du Monceau M 1753 La physiques des arbres. BL Guerin et LF Delatour, Paris, 49–51
Engler A 1918 Tropismen und exzentrisches Dickenwachstum der Bäume. Stiftung von Schyder von Wartensee, Zürich, 106 pp
Fischer H 1885 Ein Beitrag zur vergleichenden Anatomie des Markstrahlgewebes und der jährlichen Zuwachszonen im Holzkörper von Stamm, Wurzel und Ästen bei Pinus abies L. Flora 68:263–294, 302–309, 313–324
Frey-Wyssling A 1966 Editorial. IAWA Bull 1966(1):1–2
Gayer K 1894 Die Forstbenutzung. Paul Parey, Berlin, 676 pp
Grossenbacher JG 1915 The periodicity and distribution of radial growth in trees and their relation to the development of "annual" rings. Trans WI Acad Sci Art Let 18(1):1–77
Hartig R 1896 Das Rothholz der Fichte. Forstl-Naturwiss Z 5:96–109, 157–169

Hartig R 1899 Über die Ursachen excentrischen Wuchses der Waldbäume. Cbl Ges Forstwes 25:291−307

Hartig R 1901 Holzuntersuchungen. Altes und Neues. Julius Springer, Berlin, 99 pp

Jaccard P 1919 Nouvelles recherches sur l'accroissement en épaisseur des arbres. Fondation Schnyder von Wartensee, Zürich, 200 pp

Knight TA 1801 Account of some experiments on the ascent of the sap in the trees. Phil Trans R Soc London 1801:333−353

Kny L 1873 Über die Bedeutung der Florideen in morphologischer und histologischer Beziehung und den Einfluss der Schwerkraft auf die Koniferenblätter. Bot Ztg 31:433−435

Kny L 1877 Dickenwachsthum des Holzkörpers an beblätterten Sprossen und Wurzeln und seine Abhängigkeit von äusseren Einflüssen, insbesondere von Schwerkraft und Druck. Sitzungsber Ges Naturforsch Freunde zu Berlin 1877:23−50

Kny L 1882 Über das Dickenwachsthum des Holzkörpers in seiner Abhängigkeit von äusseren Einflüssen. Berlin, 136 pp

Kononchuk PI 1888 (On the loal or one-sided "hard-layerness" of trees.) Yearbook St Petersburg For Inst 2 (Unoff sect):41−56

Lämmermayr L 1901 Beiträge zur Kenntniss der Heterotrophie von Holz und Rinde. Sitzungsber Kaiserl Akad Wiss Mat-Naturwiss Cl Wien Pt 1 110:29−62

Linnaeus (von Linné) C 1732 Iter Lapponicum. (The Lapland journey.)

Mägdefrau K 1975 Sanio, Karl Gustav. In: Gillispie CC (ed) Dictionary of scientific biography, XII:99−100

Makkonen O 1967 Ancient forestry. An historical study. I Facts and information on trees. Acta For Fenn 82(3):1−84

Makkonen O 1969 Ancient forestry. An historical study. II The procurement and trade of forest products. Acta For Fenn 95:1−46

Mer E 1887 De la formation du bois rouge dans le sapin et l'épicea. CR Acad Sci 104:376−378

Mer E 1888a Des causes qui produisent l'eccentricité de la moelle dans les sapins. CR Acad Sci 106:313−316

Mer E 1888b Recherches sur les causes d'excentricité de la moelle dans les sapins. Rev Eaux For 27:461−471, 523−530, 562−572

Mer E 1889 Recherches sur les causes d'excentricité de la moelle dans les sapins. Rev Eaux For 28:19−27, 67−71, 119−130, 151−163, 201−217

Mork E 1928 Om tennar. (On compression wood.) Tidsskr Skogbr 36 (suppl):1−41

Nördlinger H 1860 Die technischen Eigenschaften der Hölzer für Forst- und Baubeamte, Technologen und Gewerbetreibende. JG Cotta'schen Buchhandlung, Stuttgart, 550 pp

Nördlinger H 1871 Der Holzring als Grundlage des Baumkörpers. Stuttgart, 47 pp

Nördlinger H 1875 Einfluss des Lichtstandes auf die Beschaffenheit des Föhrenholzes. Cbl Ges Forstwes 1:233−235

Nördlinger H 1878 Liegt an schiefen Bäumen das bessere Holz auf der dem Himmel zugekehrten oder auf der unteren Seite? Cbl Ges Forstwes 4:246−247, 494−495

Nördlinger H 1882 Ovale Form des Schaftquerschnitts der Bäume. Cbl Ges Forstwes 8:204−206

Onaka F 1940 (On the influence of auxin on radial growth, particularly regarding compression wood formation in trees.) J Jap For Soc 22:573−580

Sanio C 1860 Einige Bemerkungen über den Bau des Holzes. Bot Ztg 18:193−198, 201−204, 209−217

Sanio K 1873 Anatomie der gemeinen Kiefer (Pinus silvestris L). Jahrb Wiss Bot 9:50−126

Schacht H 1862 Über den Stamm und die Wurzel der Araucaria brasiliensis. Bot Ztg 20:409−414, 417−423

Schimper C 1854 Amtlicher Bericht über die 31. Versammlung deutscher Naturforscher und Ärzte zu Göttingen im September 1854. Göttingen 1860, 87

Schmucker R, Linnemann G 1951 Geschichte der Anatomie des Holzes. In: Freund (ed) Handbuch der Mikroskopie in der Technik V Mikroskopie des Holzes und des Papiers, Teil 1, Umschau, Frankfurt am Main, 1−78

Schwappach A 1897 Untersuchungen über Raumgewicht und Druckfestigkeit des Holzes wichtiger Waldbäume. I Die Kiefer. Julius Springer, Berlin, 130 pp

Schwappach A 1898 Untersuchungen über Raumgewicht und Druckfestigkeit des Holzes wichtiger Waldbäume II Fichte, Weisstanne, Weymouthkiefer und Rotbuche. Julius Springer, Berlin, 136 pp

Schwarz F 1892 Forstliche Botanik. Paul Parey, Berlin, 513 pp

Schwarz F 1899 Physiologische Untersuchungen über Dickenwachstum und Holzqualität von Pinus silvestris. Paul Parey, Berlin, 371 pp

Sonntag P 1904 Über die mechanischen Eigenschaften des Roth- und Weissholzes der Fichte und anderer Nadelhölzer. Jahrb Wiss Bot 39:71−105

Spencer A 1918 The spiral spruce. Am For 24:342

Timell TE 1980 Karl Gustav Sanio and the first scientific description of compression wood. IAWA Bull (ns) 1:147−153

Tischendorf W 1953 Über Gesetzmässigkeit und Ursache der Exzentrizität von Baumquerflächen. Cbl Ges Forstwes 69:33−54

Treviranus LC 1835 Physiologie der Gewächse. Bonn, 240 pp

Ursprung A 1906 Die Erklärungsversuche des exzentrischen Dickenwachstums. Biol Cbl 26:257−272

von Mohl H 1862 Einige anatomische und physiologische Bemerkungen über das Holz der Baumwurzeln. Bot Ztg 20:225−230, 233−239, 268−278, 281−287, 289−295, 313−319, 321−327

von Mohl H 1869 Ein Beitrag zur Lehre von Dickenwachstum des Stammes der dicotylen Bäume. Bot Ztg 27:1−15

Wershing HF, Bailey IW 1942 Seedlings as experimental material in the study of "redwood" in conifers. J For 40:411−414

Westing AH 1965 Formation and function of compression wood in gymnosperms. Bot Rev 31:381−480

Wiesner J 1892a Untersuchungen über den Einfluss der Lage auf die Gestalt der Pflanzenorgane. Sitzungsber Kaiserl Akad Wiss Mat-Naturwiss Cl, Pt 1, 101:657−705

Wiesner J 1892b Über das ungleichseitige Dickenwachsthum des Holzkörpers in Folge der Lage. Ber Deutsch Bot Ges 10:605−610

Wiesner J 1896 Experimenteller Nachweis paratonischer Trophieen beim Dickenwachstum des Holzes der Fichte. Ber Deutsch Bot Ges 14:180−185

Chapter 3 Taxonomy, Designation, General Characteristics, and Determination of Compression Wood

CONTENTS

3.1	Taxonomy	45
3.2	Designation	46
3.3	Color	51
3.4	Detection	58
3.5	Determination	62
3.6	Classification	65
3.7	General Conclusions	69
References		69

3.1 Taxonomy

Compression wood occurs almost exclusively in the gymnosperms but has also been reported to be present in a few arboreal dicotyledons. It has never been observed in any monocotyledon. The majority of the dicotyledons either produce tension wood or do not form reaction wood at all. Compression wood has been detected in all six families of the Coniferales, in the Taxales, and in the monotypical *Ginkgo biloba* (Ginkgoales). It has so far been observed in 45 of the 51 extant genera and, in addition, in some extinct species and genera (Chap. 8.2). A list of the gymnosperm genera known to form compression wood has been prepared by Westing (1968). Some of this information has later been confirmed by others, most recently by Yoshizawa et al. (1982). A summary is presented in Table 3.1. In 12 of these genera, the occurrence of compression wood was first reported by Westing (1968). Undoubtedly, compression wood is also formed in the few genera where it has not yet been observed.

Concerning the other gymnosperm orders, no information is available about the Cycadales. The cycads produce comparatively little xylem and probably lack the ability to form compression wood. The Gnetales, a late offshoot of the gymnosperms of disputed evolutionary and taxonomic significance, are unique in several respects. Their xylem contains vessels, has a polysaccharide composition typical of woody angiosperms (Melvin and Stewart 1969) and has a guaiacyl-syringyl, hardwood-type lignin (Gibbs 1958, Kawamura and Higuchi 1963, 1964, 1965a, b). Instead of compression wood, the Gnetales form tension wood (Höster 1970). Several families among the Ranales are considered to be primitive and several of them lack vessels (Chap. 8.4). Some of these primitive angiosperms have been reported to form compression wood, for example *Drimys aromatica* (Winteraceae) (Dadswell and Wardrop 1949), while others, such as *Drimys winteri* (Kucera and Philipson 1977a) and *Pseudowintera colorata* (Kucera and Philipson 1978, Meylan 1981) develop neither compres-

Table 3.1. Gymnosperm genera in which compression wood has been observed. (Westing 1968, Yoshizawa et al. 1982)

CONIFERALES	*Thujopsis*	TAXODIACEAE
	Widdringtonia	*Athrotaxis*
ARAUCARIACEAE	PINACEAE	*Cryptomeria*
Agathis	*Abies*	*Cunninghamia*
Araucaria	*Cedrus*	*Glyptostrobus*
	Keteleeria	*Metasequoia*
CEPHALOTAXACEAE	*Larix*	*Sciadopitys*
Cephalotaxus	*Picea*	*Sequoia*
	Pinus	*Sequoiadendron*
CUPRESSACEAE	*Pseudolarix*	*Taiwania*
Actinostrobus	*Pseudotsuga*	*Taxodium*
Callitris	*Tsuga*	
Chamaecyparis		GINKGOALES
Cupressus	PODOCARPACEAE	GINKGOACEAE
Diselma	*Acmopyle*	*Ginkgo*
Fitzroya	*Dacrydium*	
Fokienia	*Microcachrys*	TAXALES
Juniperus	*Pherosphaera*	TAXACEAE
Libocedrus	*Phyllocladus*	*Taxus*
Tetraclinis	*Podocarpus*	*Torreya*
Thuja	*Saxegothaea*	

sion wood nor tension wood. In 15 primitive dicotyledons, Kucera and Philipson (1977b) could observe tension wood (gelatinous) fibers in only three species. Compression wood has been claimed to occur in *Cotinus coggygria* (Onaka 1949), *Calluna vulgaris,* and *Hamamelis virginiana* (Nečesaný 1955a), although these observations would seem to need confirmation.

Buxus is a genus whose xylem contains both tracheids and vessels. Three *Buxus* species have been reported to develop compression wood, namely *B. japonica, B. microphylla* var. *sinica* (Onaka 1949), and *B. sempervirens* (Lämmermayr 1901, Nečesaný 1955a, Höster 1966, Höster and Liese 1966, Nečesaný and Oberländerova 1967, Timell 1981, 1982). Whether the thick-walled tracheids in *Buxus sempervirens* (Fig. 8.17), which lack intercellular spaces and helical cavities, should be classified as compression-wood tracheids is still open to doubt. The wood has the dark color characteristic of severe compression wood in the gymnosperms, as can be seen in Fig. 3.1. The occurrence of compression wood in primitive gymnosperms and angiosperms is further discussed in Chap. 8.4. Trees and shrubs with a xylem composed largely of tracheids tend to develop compression wood, whereas those with a xylem consisting mostly of fibers tend to produce tension wood (Höster and Liese 1966). In agreement with this, gelatinous (tension wood) fibers occur in the phloem of both angiosperms (Scurfield and Wardrop 1962, White 1965) and gymnosperms (Liese and Höster 1965).

3.2 Designation

The existence and gross properties of compression wood must have been known for thousands of years, and the many vernacular names by which it has been

3.2 Designation

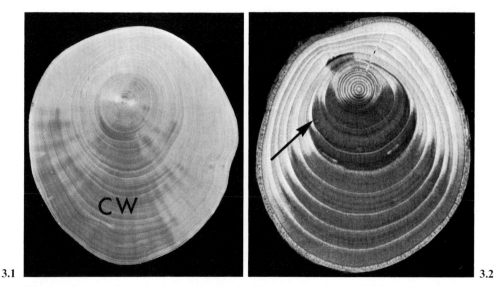

Fig. 3.1. Cross section of an eccentric branch of *Buxus sempervirens*, showing the presence of dark compression wood (*CW*) on the lower side. (Höster and Liese 1966)

Fig. 3.2. Cross section of an upward-bending branch of *Picea abies* with clearly delineated growth rings. Note that the first three rings with compression wood (*arrow*) are darker than the other

known to woodsmen and carpenters probably go far back in time. The French forester Émile Mer (1887) suggested the term *bois rouge* because of the more or less reddish color of all compression woods. A few years later, Robert Hartig (1896) translated *bois rouge* into the German *Rotholz* (spelled *Rothholz* at the time). This designation soon won widespread acceptance, also in English-speaking countries. The fact that *Rothholz* was originally used in German also for *Guilandia (Caesalpina) echinata* (Brazil wood) does not seem to have caused any confusion.

The corresponding English terms *red wood* used by Ewart and Mason-Jones (1906), and by White (1907), *red-wood* (Grossenbacher 1915, Büsgen et al. 1929), *redwood* (Burns 1920) or *"redwood"* (Wershing and Bailey 1942) are no longer in common use. Greguss (1955) defines compression wood as "Redwood: the latewood tracheids of some coniferous woods with pigment deposited in them resulting in a red colour" (Chap. 8.2). Several investigators, such as Amman (1970), Vyse (1971), Bryant (1974), Fedde (1974), Schooley and Oldford (1974), Page (1975), and Mullic (1977), use the terms *redwood* and *"redwood"* to designate the abnormal wood, very similar to compression wood, that is formed in trees infested by the balsam woolly aphid (*Adelges piceae*) (Chap. 20). It is clearly an unsatisfactory term, since *redwood* is the common name of *Sequoia sempervirens* and its wood and is also used for Himalayan spruce (*Picea smithiana*) (Glover 1920, Khan 1920). In French *bois rouge* or *veine rouge* are occasionally seen. The corresponding term in Italian is *legno rosso* (Messeri and Scaramuzzi 1960). The same word in Norwegian, however, *rødved*, means a partly decomposed wood.

Around the turn of the century it was commonly believed that the abnormal, dark wood formed on the lower side of branches and leaning stems of conifers was caused by compressive forces acting on this side. It was natural, therefore, for German investigators to refer to this wood as *Druckholz*, which, translated into English, became *compression wood*. The origin of the word *Druckholz* has been attributed to both Hartig (Janka 1909) and to Metzger (Robards 1969), but actually neither of these two scientists seems ever to have used it, preferring instead the established *Rotholz*. Sonntag (1904) states that *Druckholz* is an alternative name for *Rotholz*, but uses only the latter, just as he prefers *Weissholz* over the synonymous *Zugholz* for the wood formed on the upper side of conifer branches. Janka mentions *Druckholz* in 1909 and Jaccard used it together with *Rotholz* in 1912, but the term did not come into common use until 20 years laters. In 1917 Jaccard published an article on "bois de tension et bois de compression" in the branches of angiosperms. This "compression wood" was located on the lower side of branches and is now usually referred to as opposite wood. Jaccard believed that compressive stress influenced the properties of the wood formed on the lower side of branches in both gymnosperms and angiosperms and therefore referred to both tissues as *bois de compression*. It is obviously not a nomenclature to be recommended, and the same applies to the use of *tension wood* for the tissue present on the upper side of conifer branches.

Trendelenburg (1931, 1932) used *Rotholz* and *Druckholz* interchangeably, and as late as 1930 Rothe preferred *Rotholz*. After 1935, however, *Druckholz* became the accepted term in the German-language literature, although *Rotholz* or *rotholz* has remained in use not only in German, but, surprisingly, also in English until the present day, as can be seen from publications by Lee and Smith (1916), Johnsen and Hovey (1918), Thiemann (1942, 1951), Berkley and Woodyard (1948), Göhre (1954), Langendorf (1961), and Thimann (1964, 1972). In addition, *rotholz* is sometimes still used in English to denote the wood produced in trees infested by the balsam woolly aphid (Doerksen and Mitchell 1965, Mitchell 1967, Puritch 1971).

Probably because *redwood* was patently unsatisfactory, the term *compression wood* was quickly adopted in English-speaking countries. It was used in 1919 by Heck in the form *"compression" wood*, a designation also employed by Koehler (1924) in his monograph on wood. By 1929, *compression wood* was evidently gaining acceptance, for Dadswell and Hawley in that year first refers to this type of wood as *"Rotholz"*, but subsequently as *"compression" wood*, and finally as *compression wood*, a term used throughout the remainder of the article. One year earlier, Berg (1931), in reviewing Mork's (1928) treatise on *"tennar"*, translated this Norwegian word into *compression wood*. After 1930, at least all American authors consistently refer to *compression wood*, beginning with the early papers by Kienholz (1930) and by Pillow (1931).

The not quite normal wood formed in the narrow growth rings laid down on the upper side of conifer branches and leaning stems, opposite to the com-

3.2 Designation

pression wood, was at first called *Weissholz* because of its white color. Later, when it was realized that this tissue is under tension, it was referred to as *Zugholz* (Hartig 1901), a term now reserved for the reaction wood produced in angiosperms. The name now used is *opposite wood*.

During the period 1930–1940, Hartmann (1932, 1942, 1943), after much experimentation came to the conclusion that both compression wood and tension wood serve the function of maintaining a stem or branch in its equilibrium position (Chap. 11.4). According to Hartmann, trees form these regulatory tissues as a *reaction* against a change in a genetically predetermined orientation, and in 1932 he proposed the term *Reaktionsholz* for both. Both this term and the corresponding English *reaction wood* have won universal acceptance.

It is at present believed by most investigators that neither compressive nor tensile stresses cause formation of compression or tension woods (Chap. 11.5.5). These tissues are, however, in a state of compression and tension, respectively, in the living tree. Pointing to the fact that compression wood exerts a "dynamic pressure" in the tree, Münch (1938) favored retention of the term *Druckholz*. Jaccard (1938) in the same year also came to the conclusion that compression wood exerts an "active pressure" and agreed with Münch that *Druckholz* is a justified designation. Nečesaný (1955b), on the other hand, has pointed out that formation of both compression and tension woods is caused by changes in the concentration of growth hormones in the cambial zone. For this reason, he considers the two terms to be erroneous and argues that they should be replaced by the collective designation *reaction wood*. Practical aspects apart, this suggestion would have been more attractive were it not for the fact that the ultimate cause of both compression and tension woods is still very much a moot question. They might not have the same origin. Jane et al. (1970) emphasize that no causality is involved in the terms compression wood and tension wood. They instead argue that the two designations should be understood to indicate the position of the two tissues. This is not entirely correct, for compression wood can occur in a region which is under tensile stress, and tension wood can develop in an area subjected to compressive stress, especially in branches. *Compression wood* and *tension wood* remain meaningful terms because they clearly indicate that this type of wood is capable of exerting a compressive or tensile stress along the grain in the living tree (Knigge and Schulz 1966).

Staff (1974) has pointed out that many scientists seem to believe that compression and tension woods are caused by compressive and tensile stresses, respectively. For this reason, he has suggested that compression wood be renamed *L-reaction wood* and tension wood *G-reaction wood*, L signifying the higher than normal lignin content of compression wood and G the presence of a G-layer (gelatinous) layer in tension wood fibers. The terms proposed by Staff are rather similar, however, and could easily lead to mistakes. It should also be remembered that mild and moderate compression woods can have the same lignin content as normal wood, and the lignin contents of pronounced compression wood and normal woods in *Ginkgo biloba* are not very different (Chap. 8.3.1). Furthermore, the term *L-reaction wood* would not be appropriate when applied to the parenchyma cells in compression wood, as these cells have the same lignin content as ray cells in normal wood. In many arboreal angiosperms, for

Table 3.2. Glossary of terms used for compression wood in different languages

Language	Compression wood	Reaction wood	Other names
Czech	Tlakové dřevo	Reakční dřevo	Čevené dřevo
Dutch	Drukhout	Reactiehout	Roodhout
English	Compression wood	Reaction wood	Red-wood, glassy wood, hard streak, timber bind
Finnish	Lyly, lylypuu	Reaktiopuu	Janhus
French	Bois de compression	Bois de reaction	Bois rouge, veine rouge, bois raide
German	Druckholz	Reaktionsholz	Rotholz, Buchs, Buchsholz, Nagelhart, Nagelfest, Wetterholz
Italian	Legno di compressione	Legno di reazione	Legno rosso, canastro
Norwegian	Tennar, Tennarved	Reaksjonsved	
Polish	Drewno naciskowe, Drewno kompresyjne	Drewno reakcyjne	Twardzica
Portuguese	Lenho de compressão	Lenho de reação	
Russian	креневая древесина	реакцыонная древесина	крень
Serbo-Croatian	Drvno kompresijsko	Drvo reakcijsko	Drvno crljen
Slovak	Tlakové drevo	Reakčné drevo	Červené drevo, kremeň
Spanish	Leño de compresión	Leño de reacción	Leño viso, leño rojo, leño dura
Swedish	Tryckved	Reaktionsved	Tjurved, vresved

example *Lirodendron tulipifera* (Barefoot 1963) and *Tilia americana* (Onaka 1949, Perem 1964, Cheng 1970), tension wood fibers lack a gelatinous layer. For all these reasons, the new names suggested by Staff (1974) are unsatisfactory. As long as it is understood that compression wood is a tissue that exerts a dynamic compressive stress in a living tree, the long-accepted designation is perfectly satisfactory.

Compression wood and *tension wood* are at present the accepted terms in most languages. Collectively, they are referred to as *reaction wood* since they both are the result of a reaction of the tree in response to external factors. The designation and definition of compression, tension, and reaction woods, as well as other names in seven different languages are found in the latest edition of the glossary issued by the International Association of Wood Anatomists (1964). A larger list is presented in Table 3.2

The term *Richtgewebe* first coined by Trendelenburg (1939) (Trendelenburg and Mayer-Wegelin 1955), and later used by Knigge (1958), Knigge and Schulz (1966), Sachsse (1961, 1965, 1974, 1978, 1980a, b), and Sachsse and Mohrdiek (1980) is not entirely felicitous, for reaction wood can easily develop all around a trunk or a branch, and also because laterals sometimes perform non-righting movements.

Other names for compression wood have not stood the test of time. Sanio (1860, 1873) used the term *differenziertes Holz*, Gothan (1905) *verkerntes Holz*, Metzger (1908) *Unterseites-Holz*, and Burger (1932) *Druckrotholz*. Engler (1918, 1924) who, like Hartig, believed that one type of compression wood was induced by gravity, while another was caused by compressive stress (Chap. 11.3), called the first type of tissue *geotrophes Holz* and the second *Druckholz*. With the demise of the concept of two different types of compression wood, the former term ceased to exist, although it was used by Jaccard as late as 1938. *Pressure wood*, coined by Thomson in his English translation of the well-known monograph by Büsgen and Münch (1927) (Büsgen et al. 1929), has found little acceptance, although it was later used by Misra (1943) and Banks (1973) and most recently also by Abetz and Künstle (1982) and Atmer and Thörnqvist (1982). Fielding (1940) has referred to compression wood as *stimulus wood*. Perhaps the oddest of all scientific terms ever applied to compression wood originated with Berkley (1934), who suggested that compression wood tracheids should be referred to as *torquimural* in contrast to *concentriumural* (normal) tracheids. Nobody seems to have felt inclined to adopt these terms.

In Swedish, compression wood has long been known as *tjurved* or, less commonly, *vresved* (Nylinder 1958), both meaning refractive wood. The current scientific term is *tryckved* (compression wood). The Norwegian word *tennar* is unique to this language and has often been quoted in the literature, sometimes mispelled as "tenar" and often erroneously stated to be Swedish (Pillow 1933, Pillow and Luxford 1937, Dadswell and Wardrop 1949, White 1965, Jane et al. 1970, Ford-Robertson 1971). According to Wegelius (1939) compression wood is referred to in Finnish either by *lyly* or by *janhus*. Onaka (1949) has given a detailed description of the use of the vernacular *Ate* in Japanese.

To the woodsman, compression wood is a nuisance because it is present under compression in the living tree and tends to bind the saw on cutting due to an expansion of both the lower and upper sides of a leaning or crooked stem (Chap. 18.3). From this fact stems the name *timber bind*, often applied by woodsmen in the United States to compression wood (Kienholz 1930, Pillow and Luxford 1937), and *"back bind"* (King 1954). Other English expressions reflect the great hardness of compression wood, such as *glassy wood* (Verrall 1928, Kienholz 1930), and *hard streak* (Rendle 1956). Donner (1875), Hartig (1896) and Janka (1909) refer to a "hard side" and a "soft side" in trees. Several German expressions also reflect the hardness of compression wood, such as *Rothart* (Donner 1875, Witzgall 1928), *eichiges Holz* (Hartig 1896), and also *Nagelhart* (Hartig 1901) or *Nagelfest* (Münch 1938), both indicating that compression wood is difficult or impossible to nail.

3.3 Color

Mild compression wood is less deply colored than the moderate grade which in its turn is paler than the severe form. This is especially apparent in pines. The three inner growth rings indicated by an arrow in Fig. 3.2 (*Picea abies*) contain a more severe and darker compression wood than do the outer increments.

Fig. 3.3. Cross sections of six conifer stems. **1** *Tsuga canadensis.* Beginning of spiral compression wood in a stem with basal sweep (Fig. 10.74). **2** *Picea* sp. Spiral compression wood (Fig. 10.73). **3** *Pinus sylvestris.* Stem leaning 60°. **4** *Abies balsamea.* Horizontal portion of a stem with basal sweep (Fig. 10.32). **5** *Picea abies.* Horizontal base of an upward-bending branch that had developed into a stem. **6** *Larix laricina.* Horizontal stem buried in a bog (Fig. 5.3)

3.3 Color

In general appearance, pronounced compression wood is almost invariably distinctly different from normal wood, especially when freshly cut. In the sapwood it is always darker than normal earlywood and it is denser and harder, albeit not to the same extent as fully developed normal latewood. The color is brown or reddish-brown. Six examples are shown in Fig. 3.3 where the color of the compression wood varies between different shades of brown in five of the specimens but is reddish-brown in *Pinus sylvestris* (Fig. 3.3.3), a trait typical of all pines. In this disc, the compression wood present within the first half of most growth rings is lighter in color than that formed during the later part of the season. According to Trendelenburg (1932), compression wood is darker in *Abies* and *Picea* than in *Pinus* and *Pseudotsuga* (König 1957), an observation that agrees with my own. Westing (1965) states that a blue filter should be used when compression wood is photographed with panchromatic light. According to my own experience, this also serves the purpose of eliminating or at least reducing the yellow tinge assumed by normal softwood when photographed with incandescent light. I have also found that if one is using a black and white film, best results are obtained with a moderately sensitive film, if the film is slightly overexposed, and if a hard paper is used for printing.

Mer (1887) mentions that if a piece of wet compression wood is rapidly dried (he apparently did this in front of an open fire), the dark color fades more and more, until in the end the compression wood is as pale as normal wood. Hartig (1896) states that drying does not reduce the dark color of compression wood. Five years later, however, he must have changed his mind, for he now (Hartig 1901) claims that the red color disappears almost completely when the wood is dried. Cieslar (1896) found that on drying at room temperature compression wood remained dark brown. Only when dried at 100 °C did the wood become paler, assuming a shade of grayish-yellow. Lämmermayr (1901) states that the dark color of compression wood immediately appears when old wood specimens are wetted. My own experience is that prolonged drying at 105 °C in a drying oven renders compression wood much lighter in color, but that it remains brown to red. Wetting restores the original color.

Hartig (1901) believed that the reason for the pale color of dry compression wood was to be found in the fact that air penetrates into the helical cavities that line the lumen of the tracheids, pushing apart the ribs. The spiral lamellae swell in water, the air is displaced, and the dark color of the green wood is restored. The same effect can be obtained with Vaseline, a fact which seems to have been first observed by Cieslar (1896). As pointed out by Hartig (1901), dry compression wood, when coated with Vaseline, does not fully regain the dark color of the green wood, evidently because the hydrophobic Vaseline is only partly imbibed. If, however, the wood is first thoroughly wetted with water and then coated with Vaseline, maximum contrast between normal and compression woods is instantly attained, and the dark color becomes permanent.

My own experience is that if compression wood is thoroughly polished, the wood assumes the appearance it had in its green state. The only exception I have encountered so far is pine wood, where wetting is necessary to bring out the full contrast between normal and compression woods, as shown in Fig. 3.4 for *Pinus resinosa*. Coating with a clear varnish gives the same result. The rea-

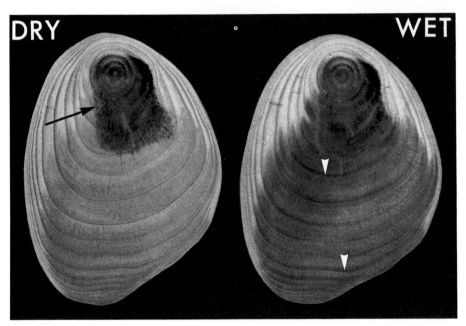

Fig. 3.4. Cross section of a stem of *Pinus resinosa*, photographed when dry and wet. Note the presence of darker bands of compression wood in the outer portion of each growth ring (*arrowheads*). The dark region on top (*arrow*) is caused by fungal decay

son why polishing darkens the color of compression wood is probably that all cavities in the wood become filled with fine sawdust, and a smooth surface scatters light less efficiently than a rough surface with many reflecting edges. Wood specimens that are repeatedly wetted and dried tend to develop cracks and fissures. Another reason why polishing is preferable to wetting is that the more highly polished the surface of the block the clearer the wood structure becomes (Fig. 3.5). Wetting, on the other hand, always blurs structural details, as can be seen in Fig. 3.4.

In longitudinal sections, as is shown in Fig. 3.6, compression wood has an appearance that has aptly been described as "dull and lifeless" (Pillow and Luxford 1937), a consequence of a lack of contrast between earlywood and latewood. Irrespective of whether the wood is wet or dry, the color is less intense on longitudinal than on transverse surfaces. An example comparing transverse and tangential surfaces of *Abies balsamea* is seen in Fig. 3.7. The reason for this difference is undoubtedly to be sought in the lesser scattering ability of transverse wood surfaces. Wilcox (1975a) found that in *Pinus taeda* brightness was much lower on transverse than on longitudinal surfaces.

Many different explanations have been offered for the difference in color between normal and compression woods. Its high lignin content is a characteristic feature of compression wood (Chap. 5.2.2.1). Sanio (1860, 1873) states that the red color of compression wood is a result of its high degree of lignification.

3.3 Color

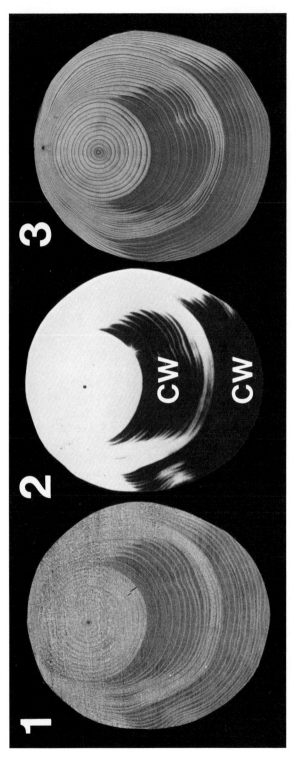

Fig. 3.5. Cross sections of a leaning stem of *Picea rubens* containing compression wood (*CW*). Unpolished section viewed in reflected light (**1**), the same section in transmitted light (**2**), and a similar, polished section viewed in reflected light (**3**). Compression wood is sharply contrasted with normal wood in transmitted light, but there is a lack of details which are best brought out in the polished section. (cf. Fig. 10.34)

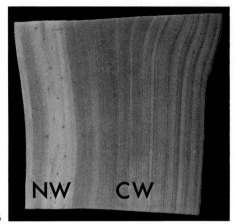 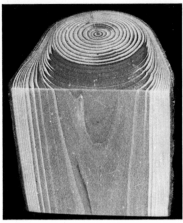

Fig. 3.6. Radial section of a stem of *Abies balsamea*, illustrating a lack of contrast between earlywood and latewood in the compression wood (*CW*) region. *NW:* Normal wood

Fig. 3.7. Transverse and tangential sections of a horizontal stem of *Abies balsamea*. The compression wood appears considerably darker in the cross section

Klason (1923) later made the same suggestion. Mork (1928) was of the same opinion, and Wegelius (1939) believed that the dark color of compression wood present in knots was caused by lignin. Similar statements have been made by several later investigators, for example Casperson (1962, 1963, 1965a, b), Thimann (1964), Westing (1965), Silvester (1967), and Molski (1969). Mer (1888 – 1889) attributed the dark color of compression wood to its high content of resin and tannin, and Onaka (1949) seems to have held a similar view. Actually, however, compression wood contains about the same amounts of these constituents as does normal wood except for knots, which have a very high content of resin (Chap. 5.2.8). Hartig (1901) attributed the color to tannins present in the cell wall. Trendelenburg (1940) pointed out that the color of compression wood probably had nothing to do with its degree of lignification. Instead, he suggested that the dark color was an entirely physical phenomenon. In the same year, Hale and Prince (1940) claimed that the color was due to the thick cell wall of the tracheids in compression wood, a statement later repeated by Hale (1951).

Although the color of wood has never been the subject of much research, it has attracted more attention in later years than previously, as witnessed both by reviews (Sullivan 1967, Lorås and Wilhelmsen 1973) and by original contributions (Gray 1961, Gupta and Mutton 1967, Jones and Heitner 1973, Wilcox 1975a, b, Douek et al. 1976, Douek and Goring 1976a, b). The color of wood is determined essentially by two factors, namely the light-scattering characteristics of the wood surface and the absorptive properties of the chemical constituents in wood, that is, cellulose, hemicelluloses, lignin, and extractives. Scattering

3.3 Color

coefficients and specific absorption are calculated with the aid of the Kubelka-Munk equations.

A wood surface scatters light in accordance with its physical and anatomical nature and less so the smoother the surface. Latewood is darker than earlywood because its thick-walled tracheids scatter light less than the thin earlywood cells. A transverse wood surface scatters light less than a longitudinal one. Light absorption is an intrinsic property of the wood, determined by its chemistry and independent of shape or structure. The absorption is positively correlated with the chromophoric groups present in wood. Many of the extraneous substances absorb strongly, creating the dark color typical of most heartwoods. The polysaccharides in wood have little or no absorptive capacity. Lignin, in contrast, contains many chromophoric groups, responsible for the yellow tinge of fresh sapwood, and especially conifer-aldehyde groups and phenyl-substituted quinones (Pew and Connors 1971, Hon and Glasser 1979). Absorption coefficients of wood have repeatedly been found to be directly related to the lignin content of wood (Lorås and Wilhelmsen 1973, Wilcox 1975a, b, Douek and Goring 1976a, b). Both Lorås and Wilhelmsen (1973) and Wilcox (1975b) have attributed the low brightness of compression wood to the high lignin content of this tissue.

Actually, lignin alone cannot be responsible for the dark color of compression wood even if it is one of the factors involved. The low light-scattering ability of compression wood must also be taken into account, a property caused by the uniformly thick wall of fully developed compression wood tracheids. This becomes evident when the first-formed tracheids in compression wood are considered (Sect. 4.5). This thin band is often as light in color as normal sapwood (Figs. 3.2 and 3.3). These tracheids have a thin wall but their lignin content is as high as that of the thick-walled cells formed later. Their high scattering ability must be responsible for the light color of this zone.

When normal softwood is ground to a powder, the difference in color between earlywood and latewood disappears, presumably because any differences in light-scattering ability have now been eliminated, while any possible differences in lignin content are too small to affect the absorption. Powdered compression wood, by contrast, is always darker than powdered normal wood. Obviously, the much higher lignin content of the compression wood is the decisive factor here.

The helical cavities and ribs in compression wood very likely have no effect on its color. This is evident from the fact that compression wood in genera lacking helical cavities, such as *Agathis, Araucaria, Cephalotaxus, Taxus,* and *Torreya,* and also in *Ginkgo biloba* (Fig. 8.7) is as dark as in genera with well-developed cavities. Exposed roots have a dark compression wood, their lack of cavities notwithstanding (Fig. 10.166). It has also been observed that sometimes first-formed compression wood containing no helical cavities can be as deeply colored as the more typical compression wood formed later (Yoshizawa et al. 1982) (Chap. 4.5). Actually, the helical cavities should lighten the color of compression wood since they ought to enhance the scattering of light. The reason why compression wood becomes much darker when wetted or when coated with Vaseline or varnish is probably that both cavities and lumina are filled in, resulting in a reduced scattering intensity. The darker color produced when compression wood is polished is undoubtedly caused by two factors. First, a smoother surface scatters less light than a rough one and, second, the fine sawdust produced fills all pores on the surface of the wood.

As a general rule, the intensity of the color of stem compression wood increases with increasing deviation of the stem from the vertical, that is, with increasing severity. Many exceptions to this trend have, however, been reported in the literature and most recently by Yumoto and Ishida (1982). These investigators found, for example, that the color of their compression wood was the same in trees displaced 90° and 45° from the vertical.

3.4 Detection

Unlike tension wood, which sometimes can be detected only with difficulty, especially on freshly cut sections, compression wood is usually readily recognized and particularly so in green wood. The presence of compression wood can, moreover, be suspected when the cross section of a stem is elliptic and has wide growth rings on one side. This, however, is not an infallible method, for compression wood can occur in perfectly concentric stems, and eccentric stems can be entirely devoid of compression wood (Chap. 10.3.6). Warped or bent lumber often contains compression wood, but such distortions can also be caused by the presence of juvenile wood or spiral grain. Mild and moderate compression woods are more difficult to detect than the pronounced type.

A report of the Forest Biology Subcommittee No. 2 of the Technical Association of the Pulp and Paper Industry (van Buijtenen 1968) deals with new methods for measuring wood and fiber properties in small samples. Included in these properties is the content of reaction wood and especially compression wood. In standing trees compression wood can be detected with the aid of an increment borer. Polge and Keller (1970) developed a new type of instrument which allows measurement of the torque required to drive an increment borer into a stem. This "torsiometer" can be used for evaluating wood-specific gravity and could obviously also serve the purpose of locating compression wood in standing trees.

Severe compression wood often occurs continuously over a large area of a stem cross section, as shown in Fig. 3.2. It can, however, also be present in concentric, crescent-shaped zones, normally concentrated in the latewood part of the growth rings. Mild and moderate forms frequently occur in this manner.

Best contrast between normal and compression woods is observed in species with a white sapwood, such as *Picea* spp. (Figs. 3.3.2 and 3.3.5) or *Abies balsamea* (Fig. 3.3.4). This also applies to black and white photographs, such as those in Fig. 3.2 (*Picea abies*) and 3.7 (*Abies balsamea*). In species with a deeply colored heartwood, for example *Juniperus virginiana*, compression wood areas are usually obscured (Rendle 1956). This is illustrated for *Larix laricina* in Fig. 3.3.6 and for *Taxus baccata* in Fig. 3.8, where the bands of compression wood can be only barely detected in the dark heartwood zone. Other investigators have also found it difficult to detect compression wood within heartwood

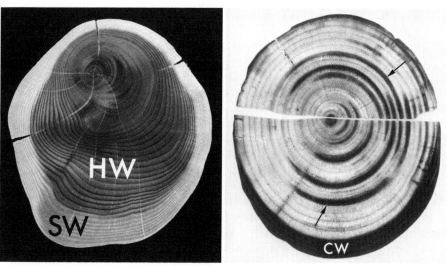

Fig. 3.8. Cross section of a stem of *Taxus baccata* with a sharp basal crook. Compression wood cannot be distinguished within the heavily stained heartwood (*HW*), only in the sapwood (*SW*)

Fig. 3.9. A thin cross section of a stem of *Picea sitchensis* photographed in transmitted light. Both major (*CW*) and minor (*arrows*) zones of compression wood are readily discernible. (Young et al. 1970). (cf. Fig. 10.45)

(Klem 1951, Cown 1974). Fungal and other stains have the same effect, as shown in Fig. 3.4.

Species where compression wood has been reported to be recognizable only with difficulty or not at all are *Juniperus* spp. (Lämmermayr 1901), and *Pseudotsuga menziesii* (Rendle 1956, von Pechmann 1969, 1973). Phillips (1937, 1940) was unable to detect compression wood in *Juniperus procera* on the basis of only color, even in the sapwood, and he had to resort to microscopic examination of the wood. In *Juniperus communis* the heartwood has almost the same color as compression wood, and an additional problem is the fact that in this genus rounded tracheids and intercellular spaces tend to occur also in normal wood (Chap. 4.4.2). Difficulties in detecting compression wood have also been reported for *Pinus strobus* (Anonymous 1966), *P. contorta* (Hallock 1969), and *Larix decidua*. According to Scott (1952) compression wood can be masked in *Pinus caribaea* by wide and pronounced zones of latewood. Ewart and Mason-Jones (1906) found that compression wood of *Pinus cembra* had to be wetted in order to bring out the dark color. Wooten (1968) has drawn attention to the fact that detection of compression wood becomes more difficult as its severity decreases, the mild form being least distinguishable.

Compression wood is sometimes difficult to discern in tropical or subtropical species lacking distinct growth rings, such as *Agathis australis, Araucaria klinkii,* and *Cupressus macrocarpa,* where it appears as a large area of dark wood (Dadswell and Wardrop 1949, Dadswell et al. 1958). *Araucaria angustifolia* (Parana pine), which is an important lumber species, often contains compression wood (Pillow 1951, Brown 1965) (Chap. 10.2.12). Detection is especially uncertain in this species and not only because of the lack of growth rings, but also because of the uniformly dark heartwood and the occurrence of red streaks (Rendle 1956, Anonymous 1956). According to Kingston (1953), much of the compression wood in *Araucaria bidwillii* and *A. cunninghamii* can only be ascertained microscopically. Molski (1969) found that compression wood had to be identified under the microscope when present in small amounts within a region of normal wood in *Pinus sylvestris.*

An accurate and convenient method for detecting compression wood was developed by Pillow (1941, Anonymous 1941, 1953, Anonymous 1943) at the Forest Products Laboratory in Madison, Wisconsin. The procedure is based on the observation that compression wood is opaque to transmitted light, whereas earlywood and latewood of normal wood are equally translucent. According to Pillow (1941) the helical cavities in the inner portion of the secondary wall in compression wood tracheids scatter the transmitted light. The more pronounced the compression wood, that is, the more numerous the helical cavities, the greater is the opacity. Because it often lacks cavities, the first-formed earlywood is more translucent than the fully developed compression wood produced later, as shown for *Picea rubens* in Fig. 3.5. Wood sections should be $1/8$ to $3/16$ inch (2.5–3.5 mm) thick, while veneer and plywood must be thinner. A convenient source of light is a wooden light box, fitted with an electric lamb bulb. Its construction has been described in detail. Simpler devices have been suggested by Ericson (1959) and by Core et al. (1961). Similar techniques are outlined in the Standard Methods of the Technical Association of the Pulp and Paper Industry, dealing with compression wood in pulpwood (1955, 1959, 1972) (Yankowski 1960).

Use of transmitted light is a reliable method for detecting compression wood, especially mild forms that are difficult to discern by any other means except a microscope, but it does not allow recognition of fine details, as is evident from Fig. 3.5. The method is especially useful for ascertaining the presence of small amounts of compression wood within larger areas of normal wood. An example of such a case is shown in Fig. 3.9. The technique has been widely applied ever since it was first developed, for example by Zobel et al. (1960), Zobel and Haught (1962), Anonymous (1966), Shelbourne et al. (1969), Chu (1972), Gaby (1972), Burdon (1975), Wilcox (1975b), Barger and Folliott (1976), Park et al. (1979), and Schulz and Bellmann (1982). Photographs have been published by Core et al. (1961), Hartler (1965), Wooten (1968), Young et al. (1970), and Boone and Chudnoff (1972). According to Low (1964), use of a light green filter increases the contrast between normal wood and compression wood.

Certain limitations exist. Sanded surfaces cannot be used, as the fine wood powder produced on polishing fills all cavities and renders also normal wood opaque. Compression wood in species with a high resin (pitch) content, such as certain pines, and especially so-called lightwood, is translucent. Wood that has become stained, for example on fungal decay, is opaque to transmitted light.

3.4 Detection

Moisture has less influence except that green wood is less translucent than partly dried one (Pillow 1941). According to Boas (1947) and Dadswell and Wardrop (1949), the method is unsatisfactory with several Australian conifer species, such as *Araucaria bidwillii*, *A. cunninghamii*, and *Athrotaxis selaginoides*, which are all highly colored. In such cases, compression wood must be identified microscopically.

The opaque appearance of compression wood in transmitted light has generally been attributed to the presence of the helical cavities in the inner portion of the S_2 layer of the tracheids. While this would appear to be the most likely explanation at present, it still remains to be explained why compression wood of species lacking helical cavities, such as *Ginkgo biloba* and *Taxus baccata*, nevertheless are opaque in transmitted light.

Staining can also be used for detecting compression wood. Most stains now used for wood stain the lignin rather than the cellulose and are therefore suitable for compression wood. Mergen (1958) used safranine green to differentiate compression wood from normal wood in terminal shoots of *Tsuga canadensis*. Malachite green in conjunction with methylene blue makes compression wood appear dark (Knigge 1958, Haasemann 1963). Bucher (1968) suggested the use of malachite green followed by staining with iodine for detecting lignin in wood fibers, including compression wood tracheids. According to von Pechmann (1972), compression wood is so readily detectable that no special stains are normally required. Among the stains available, acridine orange has the advantage of improving the optical resolution. According to Brodzki (1972), aniline blue gives a yellow fluorescence with compression wood tracheids, and resorcinol blue stains them blue. Normal tracheids respond to neither of these reagents, which can accordingly be used for distinguishing between normal and compression woods (Chap. 6.3.3). In a comprehensive survey of staining methods applicable to reaction wood and especially to tension wood, von Aufsess (1973) evaluated results obtained with phloroglucinol-hydrochloric acid, safranin, and Rhodamin B, which all gave a bright red color with compression wood, and malachite green, which stained it dark green. Westing (1959, 1965) recommends reflected light of a wavelength of 480 nm for viewing compression wood.

When compression wood is subjected to pulping, any compression wood tracheids in the pulp usually contain much more lignin than do normal ones. According to Jayme and Harders-Steinhäuser (1954), compression wood fibers can be detected in sulfite pulps by staining with p,p-azodimethylaniline, which reacts with the residual lignin, producing an intensive red color. These investigators also made use of malachite green.

In doubtful cases, examination has to be performed with a light microscope or a good stereomicroscope with a power of magnification up to 200. In cross sections, the rounded outline and thick cell wall of the compression wood tracheids and the intercellular spaces are readily recognizable. In longitudinal sections, the helical cavities are seen as flat, spiral striations, winding around the cell wall. Where they cross a pit, they are easy to detect because they always move apart in order to accommodate the pit opening between them (Cockrell 1974) (Chap. 4.10.1). The cavities are less readily discernible in cross section in ordinary light but can be readily seen in ultraviolet light (Fig. 6.15).

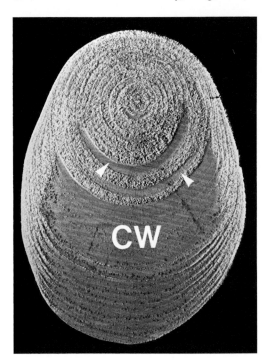

Fig. 3.10. Cross section of a horizontal stem of *Pinus strobus*. The compression wood (*CW*) areas are located below the surface of the regions of normal wood in this unpolished, air-dry disc. The core contains two separate bands of compression wood (*arrowheads*)

Sometimes, much larger checks are present in the cell wall of compression wood tracheids. As shown in Chap. 4.4.5.4.8, these checks are usually formed on drying and are accordingly artifacts. They are, nevertheless, of diagnostic value, for they do not develop in normal wood where the S_3 layer probably prevents this type of fission. Phillips (1937, 1940) used the occurrence of checks and cavities for detecting compression wood in *Juniperus procera*. Elliott (1966) made use of the extent of development of helical cavities in *Picea sitchensis* in formulating a compression wood index. Another means of detecting compression wood has been pointed out by Trendelenburg and Meyer-Wegelin (1955). A stem disc is cut in the green state and is then allowed to dry. Because compression wood shrinks much more than normal wood in the axial direction (Chap. 7.3.3.2), areas consisting of compression wood will sink considerably below the surface of the normal wood. An example of this is shown in Fig. 3.10.

3.5 Determination

Density (specific gravity) is one of the most important of all wood properties (Chap. 7.2). Since density is positively correlated with the proportion of latewood, many methods have been developed for determining the latter. The thick walls typical of latewood tracheids are also found in compression wood which as a consequence has a density that is approximately twice that of average normal wood (Chap. 7.2.3). Most methods for determining the density and proportion of latewood in normal wood can also be used for estimating the amount of compression wood present.

Techniques are now available which make possible rapid and accurate determinations of ring widths, intraring density, and proportions of earlywood and latewood (Kozlowski 1971, Parker and Kennedy 1973, Hughes and de Albuqerque Sardina 1975, Fritts 1976). Instruments for radiation densitometry based on the use of β-rays were first developed by Phillips and his co-workers (Cameron et al. 1959, Phillips 1960, Phillips et al. 1962) and later further elaborated by Keylwerth and Kleuters (1962) and Harris (1969). The method has been applied by several investigators, such as Sandermann et al. (1960), Harris and Birt (1972), and Harris (1973), for investigating wood density in pines. A second and now more widely used technique using x-rays was introduced by Polge (1963, 1965, 1966, 1970a, 1971a, Polge and Illy 1968, Polge and Keller 1969, Polge and Lutz 1969, Polge and Nicholls 1972) and later further developed by Parker (1972). It has found widespread use in later years (Smith and Worrall 1970, Smith 1977) by investigators such as Rudman (1968), Keller (1971), Echols (1973), Parker and Jozsa (1973), Parker et al. (1973, 1976), Heger et al. (1974), Lenz et al. (1976), and Cown and Parker (1978). The technique has proven especially useful in dendrochronological (Fritts 1976, Creber 1977) growth-ring analyses (Parker 1970a, b, Polge 1970b, 1971b, Jones and Parker 1970, Parker and Meleskie 1970, Parker and Henoch 1971, Thomas and Wooten 1973). The two methods, which both require extensive instrumentation, have been compared with each other by Harris and Polge (1967), Phillips (1968), and Parker and Kennedy (1973). The last-mentioned investigators believe that the x-ray technique is superior with respect to both accuracy and speed particularly if it incorporates a computer. According to Diaz-Vaz et al. (1975) it is now generally agreed that the x-ray method is more widely applicable than the β-ray procedure.

A scanning microphotometer using visible light was developed by Green (1964, 1965, 1966), based on measurement of light transmitted through thin transverse sections of wood. The instrument measures ring width and percentage of latewood and has the advantages of both convenience and low cost. An example of the results obtained is shown in Fig. 3.11, obtained by Green and Worrall (1964) with a specimen of *Picea mariana* wood containing bands of compression wood. The technique was further developed by Elliott and Brook (1967), who applied it to various softwoods, including compression wood in *Picea sitchensis*.

Methods are also available for estimating the total amount, either relative or absolute, of compression wood present in a stem or branch. In TAPPI standard methods T20-m-55 and T20-m-59 (Technical Association of the Pulp and Paper Industry 1955, 1959, 1972) a disc, ⅛–¼ inch (3–6 mm) thick, is sawn from the specimen to be examined, and areas of compression wood are outlined with a pencil. A specially constructed ground glass plate is placed on the disc, and total and compression wood areas are measured with a planimeter. Alternatively, a translucent sheet of paper is used instead of a plate. In this case, areas can be measured with a planimeter or, if the paper is uniform in thick-

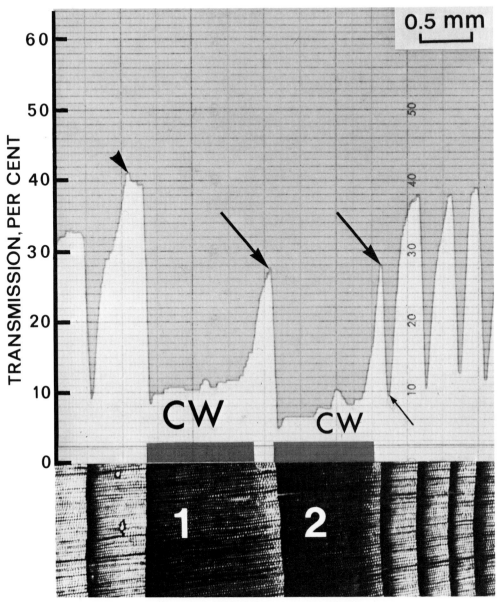

Fig. 3.11. Chart of a scan of nine growth rings in *Picea mariana* containing two rings consisting of compression wood (*CW, 1* and *2*). The narrow, first-formed wood in these increments (*large arrows*) has a lower transmission than normal earlywood (*arrowhead*). The compression wood has a transmission that is the same as (*1*) or lower than (*2*) that of the normal latewood (*small arrow*). (After Green and Worrall 1964)

ness, by weighing the excised sections. Low (1964) and Shelbourne (1966) examined thin discs in transmitted light, delineated areas of compression wood, and measured them with the aid of a dot grid. This technique was also used by Ladell et al. (1968), Einsphar et al. (1969), Young et al. (1970), and Boone and Chudnoff (1972). Zobel and Haught (1962) and Hans and Williamson (1973) traced areas of normal and compression woods onto graph paper. Holmes (1944) used a planimeter for area measurements. Bletchley and Tailor (1964) cut out the different areas and weighed them. Seth and Jain (1977, 1978) used a stereomicroscope fitted with a micrometer for estimating the amount of compression wood in stem discs. Archer and Wilson (1970) prepared drawings of stem cross sections with the aid of a stereomicroscope and a drawing tube, outlining the pith, first annual ring, extent of compression wood, and the outer boundaries of wood and bark.

In connection with his investigations of the occurrence of compression wood in *Pinus radiata* clones on different sites, Burdon (1975) developed a useful method for assessing the incidence of compression wood in stems. Discs were examined in transmitted light as described by Pillow (1941) in a moist condition. The opaque and reddish compression wood was classified into six different grades of severity, from 0 to 5, namely normal wood (0), latewood patchily opaque(1), latewood generally opaque (2), latewood opaque and earlywood partly opaque (3), latewood and earlywood generally opaque (4), and, finally, latewood and earlywood highly opaque (5). Boundaries and grades of compression wood were marked. A glass plate was placed over the disc, and disc circumference and boundaries and grades of the compression wood were traced on a gauged, translucent paper. Each zone was excised, and its area was measured by weighing.

A *compression wood rating* was defined as:

$$\frac{A \times G}{BA}$$

where A is the cross-sectioned area of the zone of compression wood, G ist the compression wood grade of that zone, and BA is the area of the disc. Percentages of total compression wood and of severe compression wood (grades 3−5) were also calculated.

An indirect method for determining the amount of compression wood in a shoot was developed by Little (1967). The procedure is based on the fact that compression wood shrinks much more longitudinally than does normal wood. The entire shoot is dried, and the change in curvature that takes place is measured. The extent of change is proportional to the amount of compression wood present.

3.6 Classification

Compression wood can occur in all gradations, ranging from only slightly modified normal wood to the fully developed type found on the lower side of branches and horizontally growing stems. Pillow and his co-workers (Pillow

1933, Pillow and Luxford 1937, Pillow and Bray 1935, Pillow et al. 1936, 1941) distinguished between *mild* and *pronounced* compression wood. Moore and Yorston (1945) used the expressions *slight, moderate,* and *strong* compression woods. Perem (1958, 1960) preferred *slight, intermediate,* and *pronounced.* The standard method of the Technical Association of the Pulp and Paper Industry (TAPPI) (1959) and its suggested standard method (1972) both distinguish between *border-line, intermediate,* and *severe* compression woods. Barefoot et al. (1964) graded various types of compression woods from *mild* to *severe.* Low (1964) distinguished between *slight, moderate,* and *pronounced* compression woods, a classification very similar to that used later by Shelbourne (1966) and Shelbourne and Ritchie (1968), namely *slight, moderate,* and *severe.* Little (1967) speaks of "modified compression wood tracheids", indicating weakly developed compression wood. Unlike most other investigators, Nichols (1982) distinguishes only between mild and severe compression woods. In this book, the terms *mild, moderate,* and *severe* or *pronounced* compression wood are used throughout. The grading system developed by Burdon (1975) is described in Sect. 3.5.

In connection with a study of the formation and properties of compression wood in *Pinus radiata,* Harris (1977) devised a grading system where the wood was classified visually on the basis of its darker color into four grades, namely normal wood (0), small crescents or patches of dark compression wood, not occupying more than half of the growth ring as measured radially (1), dark wood extending for at least 45° of an arc but not occupying the entire radial width of the increment (2), and, finally, dark wood extending across the entire radial width of a growth ring (3).

Yumoto et al. (1983) have recently made an ambitious attempt to develop a system for classifying compression wood according to its severity by microscopic means, far more detailed than the classicial subdivision into the mild, moderate and pronounced grades. As mentioned in Chap. 4.4.2, these investigators regard the presence of intercellular spaces and the absence of an S_3 layer as unreliable characteristics for diagnostic purposes, and probably rightly so. They therefore exclude these from primary consideration. Two features are deemed to be of primary importance, namely the occurrence of helical cavities and the high lignin content of the secondary wall in general and the $S_2(L)$ layer in particular. Both are considered to be independent of each other and of other characteristics. A third property of importance is the ratio of tracheid wall thickness to diameter when mesured in the middle of the growth ring. Based on these criteria, Yumoto et al. (1983) have suggested that compression wood be divided into six classes of decreasing severity, namely four major classes of which the second and third consist of two subclasses each. The system is outlined in Table 3.3

Compression wood tracheids belonging to the first class (I) have distinct helical cavities. In grade I the wall is very thick and the lignin content of the S_1 and the S_2 layer is so high that the contrast in ultraviolet light between these layers and the even more highly lignified $S_2(L)$ layer is only moderate. The lignin is distributed uniformly around the entire circumference of the latter layer. The boundary between S_1 and $S_2(L)$ is almost circular. Because of the thickness

3.6 Classification

Table 3.3. Grading system for compression wood tracheids in the middle of a growth ring, arranged in order of importance from the left toward the right. Helical cavities, ultraviolet absorption, and cell wall tickness are primary properties, while the outline of the boundary between S_1 and $S_2(L)$ are secondary traits. The presence or absence of intercellular spaces and an S_3 layer are variable features and therefore of least importance. (After Yumoto et al. 1983)

Grade	Helical cavities	UV absorption	Wall thickness	Boundary between S_1 and $S_2(L)$	Bordered pits	Variable features
I	Distinct	Strong. Low contrast between S_1 or inner S_2 and $S_2(L)$. Evenly distributed around circumference	Very thick	Nearly circular	Pit dome level or slightly raised	Intercellular spaces present
I'	Distinct	Strong but less in S_1 and inner S_2. $S_2(L)$ distinct	Thick	Round but not circular	Pit dome raised	Intercellular spaces present
II	Poorly developed	Relatively strong	Relatively thick	Round but variable depending on the presence or absence of intercellular spaces	No helical cavities around pit openings	Intercellular spaces present or absent
III	Absent	Strong and not confined to cell corners	Thicker than normal	Fairly round		Intercellular spaces usually absent. S_3 layer mostly absent
III'	Absent	Strong and not confined to cell corners	Slightly thicker, similar, or somewhat thinner than normal	Slightly round		Intercellular spaces generally absent. S_3 layer absent or present
IV	Absent	Slightly stronger than normal and evenly distributed over the entire S_2 layer	Similar to normal wood	Very faintly round		S_3 layer usually present

of the wall, the pit border does not protrude above the latter on the lumen side. Intercellular spaces are usually present. Although not mentioned by the investigators, one has to assume that these cells lack an S_3 layer.

Tracheids belonging to grade I' have a thinner wall and as a result a protruding pit border. The difference in ultraviolet absorbance between $S_2(L)$ and the less lignified inner S_2 is more pronounced than in grade I, the boundary between S_1 and $S_2(L)$ is less circular and can be angular when there are no intercellular spaces present.

The second class or grade II is characterized by poorly developed helical cavities which should be observed on both transverse and longitudinal surfaces or sections. The border between S_1 and $S_2(L)$ is variable in outline, depending on whether intercellular spaces are present or absent. The lignin content is high, and the cell walls are still thick.

Tracheids in the third class lack helical cavities. Its two subclasses cannot be readily distinguished. In grade III the wall is thicker than in normal tracheids, and the lignin concentration in $S_2(L)$ is still high. The outline is round, there are generally no intercellular spaces but usually no S_3 layer. Wall thickness cannot be used for characterizing the tracheids in grade III', since it is too variable, and an S_3 layer can be both absent and present. The lignin in the $S_2(L)$ layer is restricted to the cell corners. There are seldom any intercellular spaces. The grade IV tracheids are almost normal. They lack helical cavities, have walls of the same thickness as in normal wood, and usually possess an S_3 layer. The only features distinguishing them from normal tracheids are a slightly higher than normal ultraviolet absorption across the entire S_2 layer and a rounded outline.

As emphasized by Yumoto et al. (1983), these criteria for grading the severity of compression wood can only be applied to the tracheids present in the center of the growth ring and within the middle portion of each tracheid. The first-formed and last-formed cells do not fit into this system and the same applies to the characteristics shown of the wall near the tracheid tips. Among the features used, presence or absence of helical cavities has priority. Unfortunately such cavities are not present in genera such as *Ginkgo*, *Cephalotaxus*, *Taxus*, and *Torreya*, and they are usually absent from the middle portion of the growth ring in *Agathis* and *Araucaria* (Chaps. 4.4.5.4.4 and 8.3).

Fukazawa (1973) and Park et al. (1980) used the depth of the helical cavities as a measure of the severity of compression wood. According to Yumoto et al. (1983), this feature is not suitable for this purpose. One characteristic of compression wood which was not included among those considered by Yumoto et al. was the microfibril angle in S_2. Not only is this angle much larger than in normal wood, but it is also readily determined in the former wood from the slope of the helical cavities and ribs. The reason for this seeming omission is that Yumoto et al. (1983) examined only juvenile wood, a tissue where the microfibril angle is as large in normal wood as in compression wood (Chap. 6.2.1).

3.7 General Conclusions

Compression wood occurs in *Ginkgo biloba* and very likely in all species and genera of the Coniferales and Taxales. It is also formed by a few primitive arboreal angiosperms with tracheids as a major tissue element. The reaction woods present in gymnosperms and angiosperms are best referred to as *compression wood* and *tension wood*, respectively, and the tissue present on the opposite side as *opposite wood*. Compression wood and tension wood are names that have not only been used for a long time, but they also indicate the function of these tissues in the living tree. Old and established terms in wood science should not be abolished except for the most compelling reason. Compression wood occurs in grades ranging from only slightly modified normal wood to a fully developed reaction tissue. Three grades have been most frequently recognized, namely *mild, moderate,* and *pronounced* or *severe* compression wood. More exact grading systems have also been suggested.

Pronounced compression wood is always darker than normal earlywood and is brown or reddish brown in color. It also appears denser and harder. Mild and moderate forms are less colored and more difficult to detect. Wetting enhances the color of all compression woods. Compression wood is dark because it absorbs light more and scatters light less than normal earlywood. The former characteristic is a consequence of its high lignin content and the latter is caused by the thick wall of its tracheids. Because heartwood is often colored by extractives, compression wood is more readily discerned in the sapwood region.

The most reliable and convenient method for establishing the presence of compression wood is to examine a thin cross section of wood in transmitted light, when compression wood appears opaque, whereas normal earlywood and latewood are translucent. Equally reliable but more laborious is to identify compression wood microscopically. The contrast between normal wood and compression wood can be enhanced by staining. Compression wood can be measured by the β-ray and x-ray techniques now used for measuring intraring density and the width of earlywood and latewood. It has been determined with a scanning microphotometer using transmitted normal light. Quantitative methods are available for determining the total amount of compression wood in a specimen, using either direct observation or examination in a stereomicroscope of cross sections.

References

Abetz P, Künstle E 1982 Zur Druckholzbildung der Fichte. Allg Forst-Jagdztg 153:117–127
Amman GD 1970 Distribution of redwood caused by the balsam woolly aphid in Fraser fir of North Carolina. US For Serv Res Note SE-135, 4 pp
Anonymous 1941, 1953 A simple device for detecting compression wood. US For Serv FPL Rep 1390 (Revised 1953), 7 pp
Anonymous 1943 Compression wood: importance and detection in aircraft vener. US For Serv FPL Rep 1586, 14 pp
Anonymous 1956 Rection wood (Tension wood and compression wood). For Prod Lab GB Leafl 51, 16 pp
Anonymous 1966 Properties of the timber of home-grown Pinus strobus. Q J For 60:127–135

Archer RR, Wilson BF 1970 Mechanics of compression wood response. I Preliminary analyses. Plant Physiol 64:550–556

Atmer B, Thörnqvist T 1982 Fiberegenskaper i gran (Picea abies Karst) och tall (Pinus silvestris L). (The properties of tracheids in spruce (Picea abies Karst) and pine (Pinus silvestris L). Swed Univ Agr Sci Dep For Prod Rep 134, 78 pp

Banks CC 1973 The strength of trees. J Inst Wood Sci (32) 6 (2):44–50

Barefoot AC 1963 Abnormal wood in yellow poplar. For Prod J 13:16–22

Barefoot AC, Hitchings RG, Ellwood EL 1964 Wood characteristics and kraft paper properties of four selected loblolly pine. I Effect of fiber morphology under identical cooking conditions. Tappi 47:343–356

Barger RL, Ffolliott RL 1976 Factors affecting occurrence of compression wood in individual ponderosa pine trees. Wood Sci 8:201–208

Berg B 1931 Review of: Mork E 1928 Om tennar (On compression wood.) J For 29:599–601

Berkley EE 1934 Certain physical and structural properties of three species of southern yellow pine correlated with the ocmpression strength of their wood. Ann MO Bot Gard 21:241–338

Berkley EE, Woodyard OC 1948 Certain variations in the structure of wood fibers. Text Res J 18:519–525

Bletchley JD, Taylor JM 1964 Investigations on the susceptibility of home-grown Sitka spruce (Picea sitchesis) to attack by the common furniture beetle (Anobium punctatum Deg). J Inst Wood Sci 12:29–43

Boas IH 1947 The commercial timbers of Australia, their properties and uses. Counc Sci Ind Res Aust, 344 pp

Boone RS, Chudnoff M 1972 Compression wood formation and other characteristics of plantation-grown Pinus caribaea. US For Serv Res Pap ITF-13, 16 pp

Brodzki P 1972 Callose in compression wood tracheids. Acta Soc Bot Polon 41:321–327

Brown WH 1965 Overcoming the drawbacks of Parana pine. Woodwork Ind 22(11):43–44

Bryant DG 1974 A review of the taxonomy, biology and importance of the adelgid pests of true firs. Can For Serv Rep N-X-111, 50 pp

Bucher H 1968 Der mikroskopische Nachweis von Lignin in verholzten Fasern durch Jodmalachitgrün. Papier 22:390–396

Burdon RD 1975 Compression wood in Pinus radiata clones on four different sites. NZ J For Sci 5:152–164

Burger H 1932 Sturmschäden. Mitt Schweiz Anst Forstl Versuchswes 17:341–376

Burns GP 1920 Eccentric growth and the formation of redwood in the main stem of conifers. VT Univ Agr Exp Sta Bull 219, 16 pp

Büsgen M, Münch E 1927 Bau und Leben unserer Laubbäume, 3rd ed. Gustav Fischer, Jena, 426 pp

Büsgen M, Münch E, Thomson T 1929 The structure and life of forest trees, Chapman and Hall, London, 436 pp

Cameron JF, Barry PF, Phillips EWJ 1959 The determination of wood density using beta rays. Holzforschung 13:78–84

Casperson G 1962 Über die Bildung der Zellwand beim Reaktionsholz. Holztechnologie 3:217–222

Casperson G 1963 Reaktionsholz, seine Struktur und Bildung. Habilitationsschr, Humboldt-Univ, Berlin DDR, 116 pp

Casperson G 1965a Über die Entstehung des Reaktionsholzes bei Kiefern. In: Aktuelle Probleme der Kiefernwirtschaft. Int Symp Eberswalde 1964, Deutsch Akad Landwirtsch Berlin DDR, 523–528

Casperson G 1965b Zur Anatomie des Reaktionsholzes. Svensk Pappersidn 68:534–544

Cheng YC 1970 Shrinkage and related properties in tension wood of Tilia americana L. M S Thesis, SUNY Coll For, Syracuse, 105 pp

Chu LC 1972 Comparison of normal wood and first-year compression wood in longleaf pine trees. M S Thesis MS State Univ, 72 pp

Cieslar A 1896 Das Rothholz der Fichte. Cbl Ges Forstwes 22:149–165

Cockrell RA 1974 A comparison of latewood pits, fibril orientation, and shrinkage of normal and compression wood of giant sequoia. Wood Sci Technol 8:197–206

Core HA, Côté WA Jr, Day AC 1961 Characteristics of compression wood in some native conifers. For Prod J 11:356–362

Cown DJ 1974 Comparison of the effect of two thinning regimens on some wood properties of radiata pine. N Z J For Sci 4:540–551

Cown DJ, Parker ML 1978 Comparison of annual ring density profiles in hardwoods and softwoods by X-ray densitometry. Can J For Res 8:442–449

Creber GT 1977 Tree rings: a natural data-storage system. Biol Rev 52:349–383

Dadswell HE, Hawley LF 1929 Chemical composition of wood in relation to physical characteristics. Ind Eng Chem 21:973–975

Dadswell HE, Wardrop AB 1949 What is reaction wood? Aust For 13:22–33

Dadswell HE, Wardrop AB, Watson AJ 1958 The morphology, chemistry and pulping characteristics of reaction wood. In: Bolam F (ed) Fundamentals of papermaking fibres. Tech Sect Brit Pap Board Makers' Assoc, London, 187–219

Diaz-Váz JE, Echols R, Knigge W 1975 Vergleichende Untersuchungen der Schwankungen von Tracheidendimensionen und röntgenoptisch ermittelter Rohdichte innerhalb des Jahrrings. Forstwiss Cbl 94:161–175

Doerksen AH, Mitchell RG 1965 Effects of the balsam woolly aphid upon wood anatomy of some western true firs. For Sci 11:181–188

Donner 1875 Die harte und weiche Seite der Kiefer. Z Forst-Jagdwes 7:242–246

Douek M, Goring DAI 1976a Microscopical studies on the peroxide bleaching of Douglas fir wood. Wood Sci Technol 10:29–38

Douek M, Goring DAI 1976b The distribution of colour in Douglas-fir wood. CPPA Tech Sect Trans 2(3):83–87

Douek M, Heitner C, Lamandé L, Goring DAI 1976 The measurement of visible absorption of morphological elements in wood. CPPA Tech Sect Trans 2(3):78–82

Echols R 1973 Uniformity of wood density assessed from x-rays of increment cores. Wood Sci Technol 7:34–44

Einspahr DW, van Buijtenen JP, Peckham JR 1969 Pulping characteristics of ten-year loblolly pine selected for extreme wood specific gravity. Silvae Genet 18:57–61

Elliott GK 1966 Tracheid length and specific gravity distribution in Sitka spruce. Ph D Thesis, Univ Wales

Elliott GK, Brook SEG 1967 Microphotometric technique for growth-ring analysis. J Inst Wood Sci 18:24–43

Ericson B 1959 A mercury immersion method for determining the wood density of increment core sections. Medd Stat Skogsforskningsinst Rep 1, 31 pp

Engler A 1918 Tropismen and Exzentrisches Dickenwachstum der Bäume. Stiftung von Schyder von Wartensee, Zürich, 106 pp

Engler A 1924 Heliotropismus und Geotropismus der Bäume und deren waldbauliche Bedeutung. Mit Schweiz Centralanst Forstl Versuchswes 13:225–283

Ewart AJ, Mason-Jones AJ 1906 The formation of red wood in conifers. Ann Bot 20:201–204

Fedde GF 1974 A bark fungus for identifying Fraser firs irreversibly damaged by the balsam woolly aphid, Adelges piceae (Homoptera: Phylloxeridae). J GA Entomol Soc 9(1):64–68

Ford-Robertson FC (ed) 1971 Terminlogy of forest science, technology, practice and uses. Soc Am For, Washington DC, 349 pp

Fritts HC 1976 Tree rings and climate. Academic Press, New York London San Francisco, 567 pp

Fukazawa K 1973 Process of righting and xylem development in tilted seedlings of Abies sachalinensis. Res Bull Coll Exp For Hokkaido Univ 30:103–124

Gaby LI 1972 Warping in southern pine studs. US For Serv Res Pap SE-96, 8 pp

Gibbs RD 1958 The Mäule reaction, lignins, and the relationships between woody plants. In: Thimann KV (ed) The physiology of forest trees, Ronald Press, New York, 269–312

Glover HM 1920 Spruce redwood. Indian For 45:243–245

Göhre K 1954 Werkstoff Holz. VEB Verlag Technik, Berlin DDR, 366 pp

Gothan W 1905 Zur Anatomie lebender und fossiler Gymnospermen-Hölzer. Abhandl Königl Preuss Geol Landesanst 44:1–108

Gray VR 1961 The colour of wood and its changes. J Inst Wood Sci 8:35–57

Green HV 1964 Supplementary details and construction of the state and drive assembly of the scanning microphotometer. PPRIC Res Note 41

Green HV 1965 The study of wood characteristics by means of a photometric technique. Proc IUFRO Sect 41 Meet Melbourne, Vol 2, 17 pp

Green HV 1966 Wood characteristics. IX Some wood characteristics of three eastern Canadian coniferous species as determined by means of a scanning microphotometer. PPRIC Woodl Pap 12, 58 pp

Green HV, Worrall J 1964 Wood quality studies. I A scanning microphotometer for automatically measuring and recording certain wood characteristics. Tappi 47:419–428

Greguss P 1955 The identification of living gymnosperms on the basis of xylotomy. Akadémiai Kiadó, Budapest, 263 pp

Grossenbacher JG 1915 The periodicity and distribution of radial growth in trees and their relation to the development of "annual" rings. Trans WI Acad Sci Arts Lett 18:1–77

Gupta VN, Mutton DB 1967 The colour of wood and groundwood. Pulp Pap Mag Can 68 (Conv Issue):T107–T112

Haasemann W 1963 Bestimmung des Spätholzanteils bei Fichten- und Kiefernholz mit Hilfe des Auflichtmikroskopes. Holztechnologie 4:277–280

Hale JD 1951 The structure of Wood. In: McElhanney TA (ed) Canadian Woods. For Prod Lab Ottawa Can, 57–104

Hale JD, Prince JB 1940 Density and rate of growth in the spruce and balsam fir of eastern Canada. Dom For Serv Can Bull 94, 43 pp

Hallock H 1965 Sawing to reduce warp of lodgepole pine studs. US For Serv Res Pap FPL-102, 32 pp

Hans AS, Williamson JG 1973 Compression wood studies in provenances of Pinus kesiya Royle ex Gordon in Zambia. Commonw For Rev 52:27–30

Harris JM 1969 The use of beta rays in determining wood properties 1–5. NZ J Sci 12:395–451

Harris JM 1973 The use of beta rays to examine wood density of tropical pines grown in Malaya. In: Burley J, Nikles DG (eds) Selection and breeding to improve some tropical conifers. Commonw For Inst Oxford, Dep For Queensland 2:86–94

Harris JM 1977 Shrinkage and density of radiata pine compression wood in relation to its anatomy and mode of formation. NZ R For Sci 7:91–106

Harris JM, Birt DV 1972 Use of beta rays for early assessment of wood density development in provenance trials. Silvae Genet 21:21–25

Harris JM, Polge H 1967 A comparison of X-ray and beta-ray techniques for measuring wood density. J Inst Wood Sci 4:34–42

Hartig R 1896 Das Rothholz der Fichte. Forstl-Naturwiss Z 5:96–109, 157–169

Hartig R 1901 Holzuntersuchungen. Altes und Neues. Julius Springer, Berlin, 99 pp

Hartler N 1965 Der Einfluß von Faserschädigungen auf die Qualität der Sulfitzellstoffe. Papier 16:181–186

Hartmann F 1932 Untersuchungen über Ursachen und Gesetzmäßigkeit exzentrischen Dickenwachstums bei Nadel- und Laubbäumen. Forstwiss Cbl 54:497–517, 547–566, 581–590, 622–634

Hartmann F 1942 Das statische Wuchsgesetz bei Nadel- und Laubbäumen. Neue Erkenntnisse über Ursache, Gesetzmäßigkeit und Sinn des Reaktionsholzes. Springer, Wien, 111 pp

Hartmann F 1943 Die Frage der Gleichgewichtsreaktion von Stamm und Wurzel heimischer Waldbäume. Biol Generalis 17:367–418

Heck GE 1919 "Compression" wood and failure of factory roof-beam. Eng News-Rec 83:508–509

Heger L, Parker ML, Kennedy RW 1974 X-ray densitometry: a technique and an example of application. Wood Sci 7:140–148

Holmes WH 1944 The amount and distribution of compression wood in leaning white pine trees. M F Thesis Yale Univ New Haven, 26 pp

Hon DNS, Glasser W 1979 On possible chromophoric structures present in wood and pulps – survey of present state of knowledge. Polym-Plast Technol Eng 12(2):159–179

Höster HR 1966 Über das Vorkommen von Reaktionsgewebe in Wurzeln und Ästen der Dikotyledonen. Ber Deutsch Bot Ges 79:211–212

Höster HR 1970 Das Vorkommen von Reaktionsholz bei Tropenhölzern. Mitt BFA Reinbek, Holzwirtsch Symp, 225–231

Höster HR, Liese W 1966 Über das Vorkommen von Reaktionsgewebe in Wurzeln und Ästen der Dikotyledonen. Holzforschung 20:80–90

Hughes JF, de Albuquerque Sardinha RM 1975 The application of optical densitometry in the study of wood structure and properties. J Microsc 104:91–103

International Assocation of Wood Anatomistis, Comm Nomencl 1964 Multilingual glossary of terms used in wood anatomy. Konkordia, Winterthur, 186 pp. Mitt Schweiz Anst Forstl Versuchswes 40, 186 pp

Jaccard P 1912 Über abnorme Rotholzbildung. Ber Deutsch Bot Ges 30:670–678

Jaccard P 1917 Bois de tension et bois de compression dans les branches dorsiventrales des feuilles. Rev Gén Bot 29:225–244

Jaccard P 1938 Exzentrisches Dickenwachstum und anatomisch-histologische Differenzierung des Holzes. Ber Schweiz Bot Ges 48:491–537

Jane FW, Wilson K, White DJB 1970 The structure of wood. Adam and Charles Black, London, 478 pp

Janke G 1909 Untersuchungen über die Elastizität und Festigkeit der österreichischen Bauhölzer. III Fichte aus den Karpaten, aus dem Böhmerwalde, Ternovanerwalde und den Zentralalpen. Technische Qualität des Fichtenholzes im allgemeinen. Mitt Forstl Versuchsanst Österr 35, 127 pp

Jayme G, Harders-Steinhäuser M 1954 Optische Erfassung der Ungleichmäßigkeiten beim Sulfitaufschluß. Papier 8:509–520

Johnsen B, Hovey BW 1918 The determination of cellulose in wood. J Soc Chem Ind 37:132T–137T

Jones FW, Parker ML 1970 GSC Tree-ring scanning densitometer and data acquisition system. Tree Ring Bull 30:23–31

Jones HC, Heitner C 1973 Optical measurement of absorption and scattering properties of wood using the Kubelka-Munk equations. Pulp Pap Mag Can 74(5):T182–T186

Kawamura I, Higuchi T 1963 Studies on the properties of lignins of the plants in various taxonomical positions. I. On the UV absorption spectra of lignins. Mokuzai Gakkaishi 9:182–188

Kawamura I, Higuchi T 1964 Studies on the properties of lignins of the plants in various taxonomical positions. II. On the IR absorption spectra of lignins. Mokuzai Gakkaishi 10:200–206

Kawamura I, Higuchi T 1965a Studies on the properties of lignins of the plants in various taxonomical positions. III. On the color reactions, methoxyl contents and aldehydes which are yielded by nitrobenzene oxidation. Mokuzai Gakkaishi 11:19–22

Kawamura I, Higuchi T 1965b Comparative studies of milled wood lignins from different taxonomical origins by infrared spectroscopy. In: Chimie et biochimie de la lignine, de la cellulose and des hémicelluloses. Actes Symp Int Grenoble 1964, 439–456

Keller R 1971 Ermittlung der mechanischen Eigenschaften und der Dichte von Holz mit Hilfe von Röntgenstrahlen. Holztechnologie 12:225–232

Keylwert R, Kleuters W 1962 Beitrag zur isotopentechnischen Jahrringanalyse. Holz Roh-Werkst 20:173–181

Khan AH 1920 Redwood of Himalayan spruce (Picea morida Link) (Now: Picea smithiana Boiss). Indian For 45:496–498

Kienholz R 1930 The wood structure of "pistol-butted" mountain hemlock. Am J Bot 17:739–764

King WW 1954 Cause and remedy for warped southern pine 2×4's. South Lumberm 189(2361):31–34

Kingston RST 1953 The mechanical and physical properties of hoop pine and bunya pine. Aust J Appl Sci 4:197–234

Klason P 1923 Om granvedens halt av lignin. (On the lignin content of spruce wood.) Svensk Papperstidn 26:319–321

Klem GG 1951 Granvirke som råstoff for massaindustrien. (Spruce wood as a raw material for the pulp and paper industry.) Svensk Skogsvårsför Tidskr 49:329–341

Knigge W 1958 Das Phänomen der Reaktionsholzbildung und seine Bedeutung für die Holzverwendung. Forstarchiv 29:4–10

Knigge W, Schulz H 1966 Grundriß der Forstbenutzung. Paul Parey, Hamburg Berlin, 584 pp

Koehler A 1924 The properties and uses of wood. McGraw-Hill, New York, 354 pp

König E 1957 Fehler des Holzes. Holz-Zentralblatt Verlag, Stuttgart, 256 pp

Kozlowski TT 1971 Growth and development of trees. II Cambial growth, root growth, and reproductive growth. Academic Press, New York London, 514 pp

Kucera LJ, Philipson WR 1977a Growth eccentricity and reaction anatomy in branchwood of Drimys winteri and five native New Zealand trees. NZ J Bot 15:517–524

Kucera LJ, Philipson WR 1977b Occurrence of reaction wood in some primitive dicotyledonous species. NZ J Bot 15:649–654

Kucera LJ, Philipson WR 1978 Growth eccentricity and reaction anatomy in branchwood of Pseudowintera colorata. Am J Bot 65:601–607

Ladell JL, Carmichael AJ, Thomas GHS 1968 Current work in Ontario on compression wood in black spruce in relation to pulp yield and quality. In: Proc 8th Lake States For Tree Improv Conf, 1967, N Centr For Exp Sta, 52–60

Lämmermayr L 1901 Beiträge zur Kenntnis der Heterotrophie von Holz und Rinde. Sitzungsber Kaiserl Akad Wiss Mat-Naturwiss Cl Wien Pt 1 110:29–62

Langendorf G 1961 Handbuch für den Holzschutz. VEB Fachbuch-Verlag, Leipzig, 330 pp

Lee HN, Smith EM 1916 Douglas fir fiber, with special reference to length. Q J For 14:671–695

Lenz O, Schär S, Schweingruber FH 1976 Methodische Probleme bei der radiographisch-densitometrischen Bestimmung der Dichte und der Jahrringbreiten von Holz. Holzforschung 30:114–123

Liese W, Höster HR 1966 Gelatinöse Bastfasern im Phloem einiger Gymnospermen. Planta 61:245–258

Little CHA 1967 Some aspects of apical dominance in Pinus strobus L. Ph D Thesis Yale Univ, New Haven, 234 pp

Lorås V, Wilhelmsen G 1973 Lyshet i granved (Brightness in spruce wood.) Norsk Skogind 27:346–349

Low A 1964 A study of compression wood in Scots pine (Pinus silvestris L). Forestry 37:179–201

Melvin JF, Stewart CM 1969 The chemical composition of the wood of Gnetum gnemon. Holzforschung 23:51–56

Mer É 1887 De la formation du bois rouge dans le sapin et l'épicea. CR Acad Sci Paris 104:376–378

Mer É 1888–1889 Recherches sur les causes d'excentricité de la moelle dans les sapins. Rev Eaux For 27:461–471, 523–530, 562–572. 28:67–71, 119–130, 151–163, 201–217

Mergen F 1958 Distribution of reaction wood in eastern hemlock as a function of its terminal growth. For Sci 4:98–109

Messeri A, Scaramuzzi G 1960 Glossario internationale dei termini usati in anatomia del legno (Versione italiana). Publ Centro Speriment Agr For, Roma, 4:165–209

Metzger K 1908 Über das Konstruktionsprinzip des sekundären Holzkörpers. Naturwiss Z Forst-Landwirtsch 6:249–273

Meylan BA 1981 Reaction wood in Pseudowintera colorata – a vessel-less dicotyledon. Wood Sci Technol 15:81–92

Misra P 1943 Correlation between eccentricity and spiral grain in the wood of Pinus longifolia. Forestry 17:67–80

Mitchell RG 1967 Abnormal ray tissue in three true firs infested by the balsam woolly aphid. For Sci 13:327–332

Molski B 1969 The significance of compression wood in restoration of the leader in Pinus silvestris L damaged by moose (Alces alces). Acta Soc Bot Polon 38:309–338

Moore TR, Yorston FH 1945 Wood properties in relation to sulphite pulping. Pulp Pap Mag Can 46(3):161–164

Mork E 1928 Om tennar. (On compression wood.) Tidsskr Skogbr 36 (suppl):1–41

Mullick DB 1977 The non-specific nature of defence in bark and wood during wounding, insect and pathogen attack. In: Loewus F, Runeckles VC (eds) The structure, biosynthesis, and degradation of wood. Plenum, New York London, 395–441

Münch E 1938 Entstehungsursachen und Wirkung des Druck- und Zugholzes der Bäume. Forstl Wochenschr Silva 1938:337–341, 345–350

Nečesaný V 1955a Výskyt reačkního dřeva s hlediska taxonomického. (Occurrence of reaction wood from the taxonomic point of view.) Sborník vysoké školy zemědělské a lesnické fakulty, Brno, Sect C 3:131–149

Nečesaný V 1955b Vztah mezi reakčním drevem listnatých a jehličnatých dřevin. (The relationship between the reaction wood in broadleaved and needleleaved woods.) Biologia 10:642–647

Nečesaný V, Oberländerová A 1967 The analysis of causes of different formation of reaction wood in gymnosperms and angiosperms. Drev Vysk 12(2):61–71

Nichols JWP 1982 Wind action, leaning trees and compression wood in Pinus radiata D Don. Aust For Res 12:75–91

Nylinder P 1958 Synpunkter på granvirkets kvalitet med hänsyn till sulfitmassan. (Aspects of the quality of spruce wood with reference to the sulfite process.) Svensk Papperstidn 61(18B):712–717

Onaka F 1949 (Studies on compression and tension wood.) Mokuzai Kenkyo, Wood Res Inst Kyoto Univ 1, 88 pp. Transl For Prod Lab Can 94 (1956), 99 pp

Page G 1975 The impact of balsam woolly aphid damage on balsam fir stands in Newfoundland. Can J For Res 5:195–209

Park S, Saiki H, Harada H 1979 Structure of branch wood in Akamatsu (Pinus densiflora Sieb et Zucc). I Distribution of compression wood, structure of the annual ring and tracheid dimensions. Mokuzai Gakkaishi 25:311–317

Park S, Saiki H, Harada H 1980 Structure of branch wood in Akamatsu (Pinus densiflora Sieb et Zucc). II Wall structure of branch wood tracheids. Mem Coll Agr Kyoto Univ 115:33–44

Parker ML 1970a Preparation of x-ray negatives of tree-ring specimens for dendrochronological analysis. Tree Ring Bull 30:11–22

Parker ML 1970b Dendrochronological techniques used by the Geological Survey of Canada. In: Smith JHG, Worrall J (eds) Tree-ring analysis with special reference to northwest America. Univ BC Vancouver Fac For Bull 7:55–66

Parker ML 1972 Techniques in x-ray densitometry of tree-ring samples. 45th Ann Meet NW Sci Assoc Sect, West WA State Coll, Bellingham WA, 11 pp

Parker ML, Henoch WES 1971 The use of Engelmann spruce latewood density for dendrochronological purposes. Can J For Sci 1:90–98

Parker ML, Hunt K, Warren WG, Kennedy RW 1976 Effect of thinning and fertilization on intra-ring characteristics and kraft pulp yield of Dougls-fir. Appl Polym Symp 28:1075–1086

Parker ML, Jozsa LA 1973 X-ray scanning machine for tree-ring width and density analysis. Wood Fiber 5:237–248

Parker ML, Kennedy RW 1973 The status of radiation densitometry for measurement of wood specific gravity. Proc IUFRO-5 Meet S Afr, 17 pp

Parker ML, Meleskie KR 1970 Preparation of X-ray negatives of tree ring specimens for dendrochronological analysis. Tree Ring Bull 30:11–22

Parker ML, Schoorlemmer J, Carver LJ 1973 A computerized scanning densitometer for automatic recording of tree-ring width and density data from x-ray negatives. Wood Fiber 5:192–197

Perem E 1958 The effect of compression wood on the mechanical properties of white spruce and red pine. For Prod J 8:235–240

Perem E 1960 The effect of compression wood on the mechanical properties of white spruce and redwood. For Prod Lab Ottawa Can Tech Note 13, 22 pp

Perem E 1964 Tension wood in Canadian hardwoods. Dep For Can Publ 1057, 38 pp

Phillips EWJ 1937 The occurrence of compression wood in African pencil cedar. Emp For Rev 16:54–57

Phillips EWJ 1940 A comparison of forest- and plantation-grown African pencil cedar (Juniperus procera Hochst) with special reference to the occurrence of compression wood. Emp For Rev 19:282–288

Phillips EWJ 1960 The beta ray method of determining the density of wood and the proportion of summerwood. J Inst Wood Sci 5:16–28

Phillips EWJ 1968 A further contribution to the comparison of x-ray and beta-ray techniques for measuring wood density. J Inst Wood Sci (20)4(2):64–66

Phillips EWJ, Adams EH, Hearmon RFS 1962 The measurement of density variation within the growth rings in thin sections of wood using beta particles. J Inst Wood Sci 10:11−28

Pillow MY 1931 Compression wood records hurricane. J For 29:575−578

Pillow MY 1933 Structure, occurrence, and properties of compression wood. M S Thesis, Univ WI, Madison, 83 pp

Pillow MY 1951 A new method of detecting compression wood. J For 39:385−387

Pillow MY 1951 Some characteristics of Brazilian Parana pine affecting its use as millwork. Proc For Prod Res Soc 5:297−302

Pillow MY, Bray MW 1935 Properties and sulphate pulping characteristics of compression wood. Pap Trade J 101(26):31−34

Pillow MY, Chidester GH, Bray MW 1941 Effect of wood structure on properties of sulphate and sulphite pulps from loblolly pine. South Pulp J 4(7):6−12

Pillow MY, Luxford RF 1937 Structure, occurrence, and properties of compression wood. US Dep Agr Tech Bull 546, 32 pp

Pillow MY, Schafer ER, Pew JC 1936 Occurrence of compression wood in black spruce and its effect on properties of ground wood pulp. Pap Trade J 102(16):36−38

Polge H 1963 L'analyse densimétrique de clichés radiographiques. Ann Éc Nat Eaux For Sta Rech Exp 20(4):533−581

Polge H 1965 Study of wood density variation by densitometric analysis of X-ray negatives of samples taken with a Pressler auger. Proc IUFRO-5 Meet Melbourne Vol 2, 49 pp

Polge H 1966 Établissement des courbes de variation de la densité du bois par exploration densitometrique de radiographie d'échantillons prélevés a tarrière des arbres vivants. Applications dans les domaines technologiques et physiologiques. Ann Sci For 23:1−206

Polge H 1970a Biological aspects of wood quality research in France. For Prod J 20(1):8−9

Polge H 1970b The use of x-ray densitometric methods in dendrochronology. Tree Ring Bull 30:1−10

Polge H 1971a Héritabilité de la densité du bois de sapin pectiné. Ann Sci For 28:185−194

Polge 1971b Le "message" des arbres. La Rech 2:331−338

Polge H, Illy C 1968 Héritabilité de la densité du bois et corrélations avec croissance étudiés à l'aide de tests non destructives sur plants de pins maritimes de quatre ans. Silvae Genet 17:173−181

Polge H, Keller R 1969 La xylochronologie, perfectionnement logique de la dendrochronologie. Ann Sci For 26:225−256

Polge H, Keller R 1970 Première appréciation de la qualité du bois en forêt par utilisation d'un torsiomètre. Ann Sci For 27:197−223

Polge H, Lutz P 1969 Über die Möglichkeit der Dichtemessung von Spanplatten senkrecht zur Plattenebene mit Hilfe von Röntgenstrahlen. Holztechnologie 10:75−78

Polge H, Nicholls JWP 1972 Quantitative radiography and the densitometric analysis of wood. Wood Sc 6:51−59

Puritch GS 1971 Water permeability of the wood of grand fir (Abies grandis (Doug) Lindl) in relation to infestations by the balsam woolly aphid, Adelges piceae (Ratz). J Exp Bot 22:936−945

Rendle BJ 1956 Compression wood; a natural defect of softwoods. Wood 21:120−123

Robards AW 1969 The effect of gravity on the formation of wood. Sci Prog (Oxf) 57:513−532

Rothe G 1930 Druckfestigkeit und Druckelastizität des Rot- und Weissholzes der Fichte. Thar Forstl Jahrb 81:204−231

Rudman P 1968 Growth ring analysis. J Inst Wood Sci (20)4(2):58−63

Sachsse H 1961 Anteil und Verteilungsart von Richtgewebe im Holz der Rotbuche. Holz Roh-Werkst 19:253−259

Sachsse H 1965 Untersuchungen über Eigenschaften und Funktionsweise des Zugholzes der Laubbäume. Schriftenr Forstl Fak Univ Göttingen 35, 110 pp

Sachsse H 1974 Vorkommen und räumliche Verteilung von Richtgewebe im Wurzelholz der Populus×euramericana cv robusta. Holz Roh-Werkst 32:263−269

Sachsse R 1978 Vergleichende Untersuchung einiger Holzeigenschaften verschiedener Klonen der japanischen Lärche (Larix leptolepis Gord). Holz Roh-Werkst 36:61−67

Sachsse 1980a Über einige Holzeigenschaften der Carya ovata K Koch aus einem westdeutschen Versuchsbau. Holz Roh-Werkst 38:45−50

Sachsse H 1980b Der technologische Gebrauchswert des Holzes wichtiger Balsam- und Schwarzpappelhybriden. Forstarchiv 51:206–209

Sachsse H, Mohrdiek O 1980 Vergleichende Untersuchung technologisch wichtiger Holzeigenschaften der Schwarzpappelhybriden „Tannenhoeft", „I 45/51" und „Harff". Holz Roh-Werkst 38:285–296

Sandermann W, Schweers W, Gaudert P 1960 Messung der Holzdichte und Bestimmung der Holz-Jahrringbreite mit Hilfe von β-Strahlen. Holzarchiv 31:126–128

Sanio C 1860 Einige Bemerkungen über den Bau des Holzes. Bot Ztg 18:193–198, 201–204, 209, 215

Sanio K 1873 Anatomie der gemeinen Kiefer (Pinus silvestris L). Jahrb Wiss Bot 9:50–126

Schooley HO, Oldford L 1974 Balsam woolly aphid damage to the crowns of balsam fir trees. Can For Serv Rep N-X-121, 26 pp

Schulz H, Bellmann B 1982 Untersuchungen an einem durch Druckholz stark verkrümmten Fichtenbrett. Inst Holzforsch, Maximilian-Ludwig Univ München, 80 pp

Scott MH 1952 The quality of the wood of young trees of Pinus caribaea grown in South Africa. J S Afr For Assoc 22:38–47

Scurfield G, Wardrop AB 1962 Nature of reaction wood. VI Reaction anatomy of seedlings of woody perennials. Aust J Bot 10:93–105

Seth MK, Jain KK 1977 Relationship between percentage of compression wood and tracheid length in blue pine (Pinus wallichiana A B Jackson). Holzforschung 31:80–83

Seth MK, Jain KK 1978 Percentage of compression wood and specific gravity in blue pine (Pinus wallichiana A B Jackson). Wood Sci Technol 12:17–24

Shelbourne CJA 1966 Studies on the inheritance and relationships of bole straightness and compression wood in southern pines. Ph D Thesis, NC State Univ, Raleigh, 274 pp

Shelbourne CJA, Ritchie KS 1968 Relationships between degree of compression wood development and specific tracheid characteristics in loblolly pine (Pinus taeda L). Holzforschung 22:185–190

Shelbourne CJA, Zobel BJ, Stonecypher RW 1969 The inheritance of compression wood and its genetic and phenotypic correlations with six other traits in five-year-old loblolly pine. Silvae Genet 18:43–47

Silvester FD 1967 Timber, its mechanical properties and factors affecting its structural use. Pergamon, Oxford, 152 pp

Smith JHG 1977 Tree-ring analyses can be improved by measurement of component widths and densities. For Chron 53:91–95

Smith JHG, Worrall J (eds) 1970 Tree-ring analysis with special reference to northwest America. Univ BC Vancouver Fac For Bull 7, 125 pp

Sonntag P 1904 Über die mechanischen Eigenschaften des Roth- und Weissholzes der Fichte und anderer Nadelhölzer. Jahrb Wiss Bot 39:71–105

Staff IA 1974 The occurrence of reaction fibres in Xanthorrhoea australis R Br. Protoplasma 82:61–75

Sullivan JD 1967 Color characterization of wood: spectroscopy and wood color. For Prod J 17(7):43–48

Technical Association of the Pulp and Paper Industry, New York 1955, 1959 Compression wood in pulpwood. Standard method T 20-m-59. Tappi 38(1):174A–176A. 42(2):144A–145A

Technical Association of the Pulp and Paper Industry, Atlanta 1972 Compression wood identification in pulpwood. Proposed revision of T 20-m-59 as standard. Tappi 55:1119–1121

Thiemann HD 1942 Wood technology. Pitman, New York, Chicago, 328 pp

Thiemann HD 1951 Wood technology. Constitution, properties and uses, 3rd ed. Pitman, London, 396 pp

Thimann KV 1964 Discussion contribution. In: Zimmermann MH (ed) The formation of wood in forest trees. Academic Press, New York, London, 452–454

Thimann KV 1972 The natural plant hormones. In: Steward FC (ed) Plant physiology VIB Physiology of development: the hormones. Academic Press, New York London, 365 pp

Thomas WR, Wooten RE 1973 X-ray analysis of wood increment cores. Clemson Univ, Clemson, For Res Ser 26, 16 pp

Timell TE 1981 Recent progress in the chemistry, ultrastructure, and formation of compression wood. The Ekman-Days 1981 (Stockholm), SPCI Rep 38, Vol 1:99–147

Timell TE 1982 Recent progress in the chemistry and topochemistry of compression wood. Wood Sci Technol 16:83–122

Trendelenburg R 1931 Festigkeitsuntersuchungen an Douglasienholz. Mitt Forstwirtsch Forstwiss 2:132–208

Trendelenburg R 1932 Über die Eigenschaften des Rot- oder Druckholzes der Fichte. Allg Forst-Jagdztg 108:1–14

Trendelenburg R 1939 Das Holz als Rohstoff. Carl Hanser, München 435 pp

Trendelenburg R 1940 Über Faserstauchungen in Holz und ihre Überwallung durch den Baum. Holz Roh-Werkst 3:209–221

Trendelenburg R, Mayer-Wegelin H 1955 Das Holz als Rohstoff. Carl Hanser, München, 541 pp

van Buijtenen JP (chairman) 1968 TAPPI For Biol Subcomm 2. New methods of measuring wood and fiber properties in small samples. Tappi 51(1):75A–80A

Verrall AF 1928 A comparative study of the structure and physical properties of compression wood and normal wood. M S Thesis, Univ MN St Paul, 37 pp

von Aufsess H 1973 Mikroskopische Darstellung des Verholzungsgrades durch Färbemethoden. Holz Roh-Werkst 31:24–33

von Pechmann H 1969 Der Einfluss von Erbgut und Umwelt auf die Bildung von Reaktionsholz. Beih Z Schweiz Forstver (Festschr Hans Leibundgut) 46:159–169

von Pechmann H 1972 Das mikroskopische Bild einiger Holzfehler. Holz Roh-Werkst 30:62–66

von Pechmann H 1973 Beobachtungen über Druckholz und seine Auswirkungen auf die mechanischen Eigenschaften von Nadelholz. Proc IUFRO-5 Meet S Afr 2:1114–1122

Vyse AH 1971 Balsam woolly aphid. A potential threat to the BC forests. Can For Serv Rep BC-X-61, 54 pp

Wegelius T 1939 The presence and properties of knots in Finnish spruce. Acta For Fenn 48(1):1–191

Wershing HF, Bailey IW 1942 Seedlings as experimental material in the study of "redwood" in conifers. J For 40:411–414

Westing AH 1959 Studies on the physiology of compression wood formation in Pinus L. Ph D Thesis Yale Univ, New Haven, 177 pp

Westing AH 1965 Formation and function of compression wood in gymnosperms. Bot Rev 31:381–480

Westing AH 1968 Formation and function of compression wood in gymnosperms II. Bot Rev 34:51–78

White DJB 1965 The anatomy of reaction tissues in plants. In: Carthy JD, Duddington CL (eds) Viewpoints in biology IV, Butterworth, London, 54–82

White J 1907 The formation of red wood in conifers. Proc R Soc Victoria NS 20:107–124

Wilcox MD 1975a Measuring the brightness, ligh absorption coefficient, and light scattering coefficient of wood. Svensk Papperstidn 78:22–26

Wilcox MD 1975b Wood brightness variation in clones of loblolly pine. Silvae Genet 24:54–59

Witzgall 1928 Der Einfluß des Windes und sonstiger Druckstörungen auf die Qualität des Holzes. Deutsch Holzwirtsch 10:617–619

Wooten TE 1968 The identification and properties of compression wood. MS For Prod Util Lab Inf Ser 7, 7 pp

Yankowski AA 1960 VIII. Testing and control. Section 1 General considerations. In: Johnson EH (ed) Mechanical pulping manual. TAPPI Monogr Ser 21, 76–104

Yoshizawa N, Itoh T, Shimaji K 1982 Variation in features of compression wood among gymnosperms. Bull Utsunomiya Univ For 18:45–64

Young WD, Laidlaw RE, Packman DF 1970 Pulping of British-grown softwoods. VI. The pulping properties of Sitka spruce compression wood. Holzforsch 24:86–98

Yumoto M, Ishida S 1982 Studies on the formation and structure of the compression wood cells induced by artificial inclination in young trees of Picea glauca. III. Light microscopic observation on the compression wood cells formed under five different angular displacements. J Fac Agr Hokkaido Univ 60:337–351

Yumoto M, Ishida S, Fukazawa K 1983 Studies on the formation and structure of the compression wood cells induced by artificial inclination in young trees of Picea glauca. IV. Gradation of the severity of compression wood tracheids. Res Bull Coll Exp For Hokkaido Univ 40:409–454

Zobel BJ, Haught AD Jr 1962 Effect of bole straightness on compression wood of loblolly pine. NC State Coll School For Tech Rep 15, 13 pp

Zobel BJ, Thorbjornsen E, Henson F 1960 Geographic, site and individual tree variation in wood properties of loblolly pine. Silvae Genet 9:149–158

Chapter 4   The Structure of Compression Wood

CONTENTS

4.1	Introduction	82
4.2	Earlywood and Latewood	83
4.3	Tracheids. Gross Anatomy	98
4.3.1	Shape	98
4.3.2	Tracheid Length	104
4.3.2.1	Normal Wood	104
4.3.2.2	Compression Wood	107
4.3.2.3	Branch Wood	117
4.3.2.4	Factors Determining Tracheid Length in Normal and Compression Woods	122
4.3.3	Tracheid Diameter	126
4.3.4	Tracheid Wall Thickness	132
4.3.5	Abnormal Compression-Wood Tracheids	138
4.4	Tracheids. Ultrastructure	139
4.4.1	Introduction	139
4.4.2	The Middle Lamella	140
4.4.3	The Primary Wall	148
4.4.4	The S_1 Layer	150
4.4.5	The S_2 Layer	157
4.4.5.1	General Structure	157
4.4.5.2	Orientation and Nature of the Microfibrils	157
4.4.5.3	The $S_2(L)$ Layer	167
4.4.5.4	The Inner Portion of the S_2 Layer	171
4.4.5.4.1	Introduction	171
4.4.5.4.2	The Microscopic Evidence	173
4.4.5.4.3	The Origin of the Helical Cavities	179
4.4.5.4.4	Occurrence of the Helical Cavities	181
4.4.5.4.5	Size of the Helical Cavities and Ribs	185
4.4.5.4.6	Orientation of the Helical Cavities	186
4.4.5.4.7	The Helical Ribs	188
4.4.5.4.8	The Helical Checks	190
4.4.5.4.9	Conclusions	195
4.4.6	The S_3 Layer	195
4.4.7	Helical Thickenings	198
4.4.8	The Warty Layer	206
4.4.9	Trabeculae	207
4.5	The First-Formed Tracheids	210
4.6	The Last-Formed Tracheids	221
4.7	Rays and Ray Cells	222
4.7.1	Introduction	222
4.7.2	Frequency and Size of Rays and Ray Cells	223
4.7.3	Ultrastructure of the Ray Cells	231
4.7.4	Conclusions	234
4.8	Axial Parenchyma Cells	235
4.9	Resin Canals	236

4.10	The Pits in Compression Wood	245
4.10.1	The Bordered Pits Between Longitudinal Tracheids	245
4.10.2	The Half-Bordered Pits Between Longitudinal Tracheids and Ray Parenchyma Cells	253
4.10.3	Conclusions	254
4.11	Knot Wood	255
4.12	Root Wood	257
4.13	Bark	259
4.14	Anomalous Wood Tracheids	263
4.15	General Conclusions	266
References		267

4.1 Introduction

The structure of compression wood is radically different from that of normal wood, and it is not surprising that it has attracted the interest of many investigators for more than a century. It is largely the tracheids that display an abnormal anatomy, whereas the ray cells, resin canals, and epithelial cells are similar to or less different from those in normal wood. Since over 95% of the xylem in the Ginkgoales, Coniferales, and Taxales is composed of longitudinal tracheids, most of the xylem is modified in compression wood.

In the living tree, compression wood serves the function of effecting movements of orientation of stem and branches (Chap. 17.5) and its structure is well designed for this purpose. In the utilization of trees for lumber, the same structure renders compression wood inferior to normal wood in almost all respects (Chap. 18.4). When compression wood is converted into chemical pulp, its high lignin content results in low yields (Chaps. 19.4 and 19.5). The poor strength properties of paper made from such pulp are largely a consequence of the fine structure of the compression wood tracheids.

Many investigators have contributed to our present concept of the structure of compression wood. One of the first and foremost among them was Robert Hartig (1896, 1901). His observations, which he summarized in 1901, shortly before his death, have stood the test of time despite the much improved techniques that have since become available. Onaka (1949), who also prepared a summary of his results a short time prior to his death, studied in great detail the structure of compression wood from a large number of species, observing not only the longitudinal tracheids but all cell types present in compression wood. Later investigators have had the important advantage of having access to both transmission and scanning electron microscopes.

Compression wood, as mentioned in Chap. 3.1, is probably formed by all members of the Coniferales, Ginkgoales, and Taxales. Differences do occur with respect to tracheid structure between these orders, and there are also anatomical differences between families and orders, perhaps even species. All these differences, however, are only minor and are often restricted to the helical cavities. *Ginkgo biloba*, which lacks these cavities, has a primitive type of compression wood, described in Chap. 8.3.1. The Taxales, instead of developing cavities in their compression wood, retain the helical thickenings characteristic of their normal wood. On the whole, however, the structure of the compression wood tracheids is remarkably constant throughout the Coniferales and Taxales, and most of the features found in these orders are also found in the Ginkgoales.

More significant than these variations are the anatomical differences that occur between mild, moderate, and severe compression woods. Actually, only severe, that is, fully developed, compression wood tracheids exhibit all of the structural features usually considered typical of compression wood, and this applies especially to the rounded outline of the tracheids, the occurrence of intercellular spaces and helical cavities, and the absence of an S_3 layer. In the following, unless specifically stated otherwise, reference is always to pronounced compression wood.

In a growth ring containing only severe compression wood, the great majority of the tracheids are structurally the same. The only exceptions are those cells that are formed at the beginning of the growing season, usually 5–10 rows, and the few rows that are produced at the very end of the season. Moderate and mild compression woods almost invariably are formed only during the later part of the season and accordingly occur within the outer portion of each annual increment. The first-formed tracheids tend to have an angular outline and often lack helical cavities but are otherwise similar to those formed later. The few last-formed tracheids are flattened in the radial direction. In moderate and mild compression woods the tracheids are less different from those in normal wood. Typically, they are more angular, have thinner walls, and have shallow or no helical cavities. Intercellular spaces are often absent, and an S_3 layer can be present. Even in mild compression wood, however, the amount and distribution of lignin is sometimes the same as in the severe grade (Kennedy and Farrar 1965, Fukazawa 1974, Fujita et al. 1979, Park et al. 1980, Yumoto et al. 1982, 1983). The appearance on stem cross sections of mild, moderate, and severe compression woods is shown in Figs. 4.1 and 4.2.

4.2 Earlywood and Latewood

Trees of the temperate zones usually grow quite rapidly radially at the beginning of the growing season. The wood formed at this time is referred to as springwood or, better, earlywood. In the conifers, this wood is characterized by the presence of wide tracheids with thin walls. Later, when growth becomes slower, the wood formed consists of smaller tracheids with much thicker walls. This wood is referred to as summerwood or latewood. At the very end of the season, when growth is extremely slow, the cells become flattened in the radial direction. Latewood tracheids are generally longer than those in earlywood and contain fewer pits. The two zones also differ in chemical composition (Larson 1966), partly because the middle lamella comprises a larger part of the total tissue in the earlywood (Meier 1962) (Chap. 6.2.3). As can be expected, the physical properties of earlywood and latewood are quite different, for example in shrinkage and strength characteristics. Each type of wood contributes in its own way to the properties of paper (Chap. 19.2). The mode of transition from earlywood to latewood varies with environmental conditions and from one species to another. The subject of earlywood and latewood and the different chemical

84 4 The Structure of Compression Wood

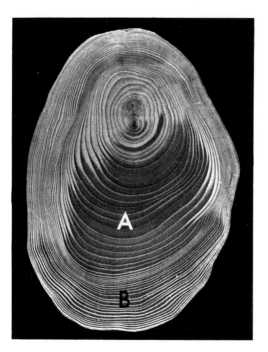

Fig. 4.1. Cross section of a stem of *Tsuga canadensis* with severe (*A*) and moderate (*B*) compression woods

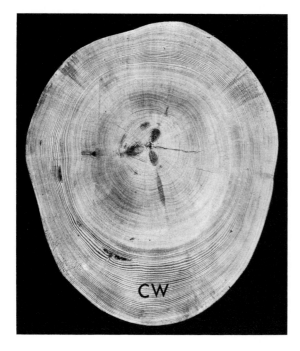

Fig. 4.2. Cross section of a stem of *Pinus ponderosa* which leaned 2–3° from the vertical. After the tree had been released, mild compression wood (*CW*) developed on the lower side of the stem. (Courtesy of US Forest Service) (Paul 1963)

4.2 Earlywood and Latewood

composition of these two portions of an annual increment are discussed in Chap. 6.2.3.

Statements in the literature concerning the occurrence of compression wood within the earlywood and latewood zones of a growth ring are generally uninformative and sometimes contradictory. Westing (1965) claims that compression wood forms in earlywood and latewood with equal facility, and Berkley (1934) notes that it can be formed in the earlywood, in the latewood, or in both. In agreement with this, Phillips (1940) found entire growth rings of *Juniperus procera* to consist of compression wood, and in *Araucaria* it is said to form a solid band over several increments (Dadswell and Wardrop 1949). Presence of compression wood throughout several growth rings has also been reported for *Pinus radiata* (Watson and Dadswell 1957), *P. resinosa* (Core 1962), and *P. strobus* (White 1965). Other cases are mentioned by Core et al. (1961). In contrast to these reports, other investigators have claimed that compression wood tends to occur in the latewood zone. In an early contribution, Mer (1888 – 1889) states that it forms only in the latewood region, and Cieslar (1896) a few years later found the same for *Picea abies*, as did Lämmermayr (1901). Casperson (1968) observed compression wood only in latewood of *Larix decidua*. Both Casperson and Hoyme (1965) and Hartler (1968) have claimed that compression wood seldom, if ever, is formed at the very end of the growing season (Sect. 4.6).

It has often been observed, and for the first time by Sanio (1873), that a clear distinction between earlywood and latewood, and thus between growth rings, often cannot be made when compression wood is present (Priestly and Tong 1925–1929, Kienholz 1930, Pillow and Luxford 1937, Wardrop and Dadswell 1952, Shelbourne and Ritchie 1968), especially in longitudinal sections (Dadswell and Wardrop 1949). Casperson (1962a, 1963a, 1965b) states that it is impossible to distinguish between earlywood and latewood in compression wood, and Molski (1969) believes that the entire concept needs further study. These difficulties notwithstanding, some investigators have tried to determine the proportions of earlywood and latewood in growth rings containing compression wood. When the transition from earlywood to latewood is gradual in normal wood, a distinction between the two zones is always uncertain, and several definitions of earlywood and latewood have been made. One of the best known is that suggested by Mork (1928b), according to which all tracheids with a double cell wall equal to or larger than the width of the lumen are assigned to the latewood. According to this definition, which has been widely accepted, practically all compression wood must be deemed to be latewood (Wiksten 1944–1945). It will be recalled (Chap. 3.5) that modern instrumentation developed for measuring wood density and proportions at earlywood and latewood registers compression wood as latewood.

Mork (1928a) investigated the location and properties of normal wood, opposite wood, and compression wood on the cross section of an eccentrically grown stem of *Picea abies*, shown in Fig. 4.3. The results obtained with opposite wood are discussed in Chap. 21.3. Table 4.1 and Fig. 4.4 summarize the data, which refer to ring width, percentage of latewood, and tracheid length. It will be noted that the zone of compression wood contained more latewood than the

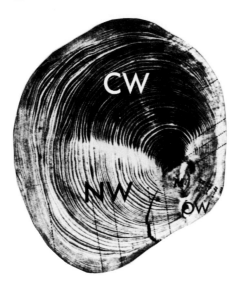

Fig. 4.3. Cross section of a stem of *Picea abies* containing normal (*NW*), opposite (*OW*), and compression (*CW*) woods. *M* indicates the location of the pith. (Mork 1928a)

Table 4.1. Width of growth rings, percentage of latewood, and average length of the tracheids in the zones of opposite, normal (side), and compression woods shown in Fig. 4.3. (Mork 1928a)

Age of ring, years	Distance of ring from pith, cm	Ring width, mm	Latewood, per cent	Tracheid length mm
Opposite wood				
5	0.5	0.550	34.0	1.20
30	1.8	0.867	5.2	1.75
42	2.0	0.567	33.2	1.82
55	2.5	0.184	32.6	1.89
63	2.7	0.483	15.5	1.90
81	3.5	1.020	18.4	2.02
113	4.2	0.161	32.9	2.29
Normal (side) wood				
30	2.1	1.167	15.1	1.95
42	3.7	3.050	6.1	2.04
55	6.8	1.967	12.4	1.96
63	8.3	4.633	14.4	1.39
81	13.1	2.333	40.2	1.85
113	15.5	0.500	34.6	2.16
Compression wood				
10	1.1	1.067	17.6	1.60
25	1.9	0.983	28.3	1.98
30	2.2	2.517	29.8	1.08
42	6.2	4.733	14.2	1.40
55	10.2	1.750	53.6	1.60
63	12.7	3.700	30.4	1.53
81	15.9	2.750	54.6	1.60
113	17.7	0.433	34.6	1.89

4.2 Earlywood and Latewood

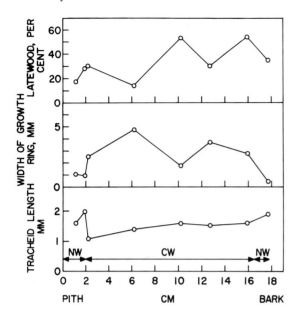

Fig. 4.4. Changes in tracheid length, ring width, and percentage of latewood with distance from the pith in a *Picea abies* stem containing normal and compression woods. (Drawn from Mork 1928a)

normal wood region, albeit not conspicuously so. The proportion varied considerably, ranging from as little as 14% to 55%. No correlation seemed to exist between ring width and percentage of latewood. Mork (1928a) concluded that the proportion of latewood is always larger in compression wood than in normal wood. It is independent of the width of the annual increment in compression wood and is instead directly proportional to the intensity of compression wood formation. Kienholz (1930), who examined "pistol-butted" *Tsuga mertensiana* trees growing on a steep slope, found the growth rings on the lower, compression-wood side of the stem to contain 65% latewood, with as much as 85% near the pith. Wood on the upper, opposite side consisted of 43% and side wood of 35% latewood.

Pillow and Luxford (1937) determined ring width and proportion of earlywood and latewood in normal wood and in mild and pronounced compression woods from several species. Their data, presented with other information in Table 4.2, indicate a positive correlation between ring width and proportion of latewood for normal wood but not for compression wood. Differences between species are obvious. In four cases, normal wood contained proportionately much less latewood than did compression wood, but in *Pinus taeda* there was slightly more latewood in the normal wood and also more latewood in mild than in pronounced compression wood.

In a leaning stem of *Abies alba*, Constantinescu (1956) found 20–50% latewood in normal wood and 50–90% in compression wood. Petrić (1962) studied a large *Abies alba* tree which had formed a sweep at the base. Pronounced compression wood had developed on the under side. Samples were removed from the entire perifery for examination. The proportions of latewood in the 23 samples are shown graphically in Fig. 4.5. The compression wood zone contained

Table 4.2. Anatomical properties of tracheids in normal wood and in mild and pronounced compression woods of five conifer species. (Courtesy of US Forest Service) (Pillow and Luxford 1937)

Species and type of wood	Intercellular spaces	Spiral checks	Microfibril angle, degrees		Ring width, mm	Latewood, per cent
			Earlywood	Latewood		
Abies concolor						
Normal wood	None	None	23.9	8.3	0.9	19
Mild compression wood	Few	Many	36.2	20.9	3.0	37
Pinus ponderosa						
Normal wood	None	None	19.6	3.9	0.8	18
Pronounced compression wood	Few–many	Many	30.1	24.7	2.0	34
Pinus taeda						
Normal wood	None	None	22.8	4.8	1.5	46
Mild compression wood	Few	Many	30.9	23.1	2.8	49
Pronounced compression wood	Many	Many	35.1	29.3	4.3	43
Pseudotsuga menziesii						
Normal wood	None	None	20.4	6.1	1.3	28
Mild compression wood	Few–many	Few–many	26.1	19.9	1.7	39
Pronounced compression wood	Many	Many	34.4	22.6	3.8	53
Sequoia sempervirens						
Normal wood	None	None	23.9	8.3	0.9	19
Pronounced compression wood	Many	Many	38.3	29.4	3.2	37

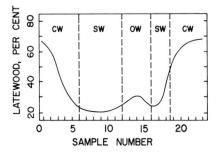

Fig. 4.5. Proportion of latewood in samples located along the periphery of a horizontal portion of a bent *Abies alba* stem containing compression (*CW*), side (*SW*), and opposite (*OW*) woods. (Redrawn from Petrić 1962)

about 70% latewood, a value that decreased to 20% on transition to the sidewood region, after which it increased to 30% in the opposite wood. According to Eskilsson (1972) the lower part of *Picea abies* branches contain 50–90% latewood, compared to only 20–30% in the upper portion. In their study of young, rapidly grown *Larix leptolepis* trees, Isebrands and Hunt (1975) found compression wood to be restricted to the latewood portion of the annual rings.

4.2 Earlywood and Latewood

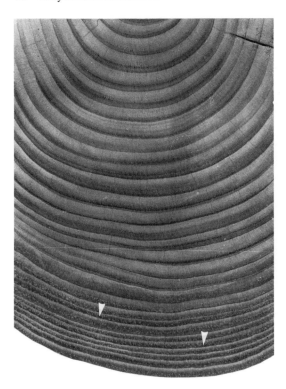

Fig. 4.6. Cross section of a stem of *Pinus banksiana*, showing the presence of mild compression wood only within the latter half of each growth ring (*arrowheads*)

Park et al. (1979), in their investigation of branch wood in *Pinus densiflora*, noted that Mork's definition of latewood was of limited usefulness with compression wood. Compared to opposite wood (Chap. 21.3) and side wood, the compression wood zone had a higher proportion of latewood, which increased from the pith toward the cambium and varied between 50% and 70% of the total ring area. The transition from earlywood to latewood was abrupt in 40% of the growth rings examined and gradual in the remainder. In the opposite wood and the side wood, the narrow rings had the highest proportion of latewood.

It is probable that some of the statements in the literature mentioned above were based on insufficient observations. The occurrence of compression wood within a given growth ring is obviously extremely variable. Mork's (1928a) opinion that compression wood formation is independent of ring width and instead determined by the intensity of the gravitational stimulus is undoubtedly correct. If a correlation is sometimes observed with ring width, this is a consequence of the fact the rapid growth and formation of compression wood, albeit under different hormonal controls, nevertheless often occur together.

Contrary to tension wood, which is seldom found in the latewood region of an annual increment, compression wood has a definite tendency to be absent from or only partly developed in earlywood, as is shown for *Pinus banksiana* in Fig. 4.6. Mild compression wood usually forms only in the latter half of each annual increment (Böning 1925), as shown here. Even in fully developed com-

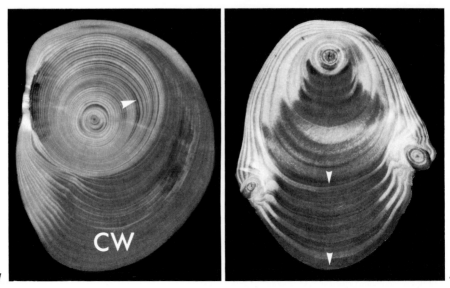

Fig. 4.7. Cross section of a horizontal stem of a slow-growing *Picea rubens* tree. In the eccentric portion, compression wood (*CW*) had formed over the entire growth rings. The more concentric core contains arcs of moderate compression wood (*arrowhead*)

Fig. 4.8. Cross section of a bent stem of *Pinus sylvestris*. Darker compression wood occurs in the latter half of most increments (*arrowheads*)

pression wood, the first few rows of cells in the earlywood generally do not have the appearance of typical compression wood tracheids, as is described in Sect. 4.5. Often, this region has a fairly constant width, and it can be recognized through its light color. In such cases, and they are very common, the growth rings can be readily detected thanks to these light-colored, narrow bands, as can be seen, for example, in Figs. 3.3.4 and 4.1.

Occasionally, compression wood forms over the entire annual increment, as is shown for a slowly growing *Picea rubens* in Fig. 4.7. In such cases, the rings are more difficult to distinguish, since the wood has the same dark appearance throughout the increment. In still other cases, the atypical, first-formed earlywood might be absent, but the last-formed latewood is noticeably darker than the remainder of the ring, thus making it possible to discern the increments fairly readily. Examples of this are seen in Fig. 3.3.3 and 4.8 for two *Pinus* spp., a genus where this is very common. Krieg's (1907) statement that compression should be found only in the latewood region is not correct. The fact that it is occasionally present across an entire growth ring has been pointed out by Core (1962). It is always easier to distinguish the annual increments in cross sections than in longitudinal, radial, or tangential sections of compression wood, as has been mentioned by several investigators and illustrated by Core et al. (1961).

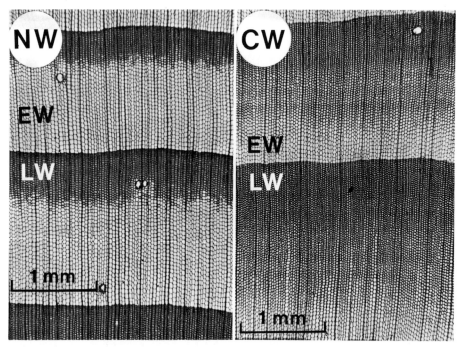

Fig. 4.9. Normal (*NW*) and compression (*CW*) woods of *Larix laricina* with an abrupt transition from earlywood (*EW*) to latewood (*LW*) in the normal wood but a very gradual transition in the compression wood. Transverse sections. LM. (Courtesy of A.C. Day)

Transverse sections of two to four growth rings in normal and compression woods are shown in Fig. 4.9 for *Larix laricina*. In both species the transition from earlywood to latewood is abrupt in normal wood and gradual in compression wood. An example where the transition is gradual in both types of wood is given in Fig. 4.10 for *Picea sitchensis*. In *Pinus strobus* (Fig. 4.11) the latewood zone is much more narrow in normal wood than in compression wood.

Jaccard (1912) has reported some interesting observations on a 9-year-old branch of *Pinus mugo* var. *rostrata*. Only normal wood had been produced in the first three growth rings on the lower side of the branch. At the beginning of the fourth growing season, however, typical compression wood tracheids had formed, and the entire increment consisted of pronounced compression wood. The fifth ring contained only normal tracheids, but the sixth began with 20 rows of compression wood tracheids, followed by four rows of normal earlywood and was terminated by 10 rows of normal latewood cells. The seventh ring consisted only of normal wood. In the eighth, the first-formed earlywood again consisted of compression wood, which, however, soon gave way to normal wood. The ninth and final increment contained only normal wood. It is likely that the growth pattern of this tree was unusual, for it was obviously a

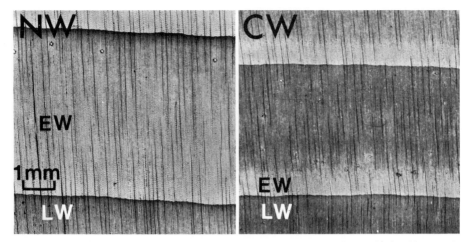

Fig. 4.10. Normal (*NW*) and compression (*CW*) woods of *Picea sitchensis*, both with a gradual transition from earlywood (*EW*) to latewood (*LW*). LM. (Young et al. 1970)

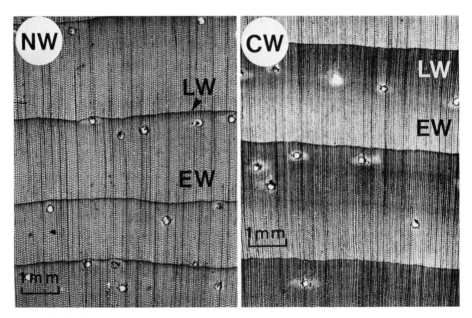

Fig. 4.11. Normal (*NW*) and compression (*CW*) woods of *Pinus strobus* with a very narrow zone of latewood in the normal wood (*LW, arrowhead*) and a much wider latewood area in the compression wood. Transverse sections. LM. (Core et al. 1961)

Fig. 4.12. Normal (*NW*) and compression (*CW*) woods of *Pinus albicaulis*. In the upper growth ring to the *right*, compression wood had developed at the very beginning of the growing season even though only normal tracheids had been formed in the previous year. *Arrows* indicate normal latewood. Transverse sections. LM. (Courtesy of A. C. Day)

dwarf form, only 2–3 m high, growing in a bog at an elevation of 1200 m, and its branches exhibited various bends and crooks.

A similar case, shown in Fig. 4.12, was observed many years later in *Pinus albicaulis* by Core et al. (1961). Here, fully developed compression wood tracheids had been formed immediately at the inception of the growing season despite the fact that normal wood had been produced throughout the previous year. It is likely that the part of the tree from which this photomicrograph was obtained had become displaced at some time after the end of the previous growing season and before the beginning of the next. Another case, which I have observed myself in *Juniperus virginiana*, is shown in Fig. 4.13. Krieg (1907) might have encountered the same situation in *Taxodium distichum*. Another pattern is illustrated in Fig. 4.14. In this *Pinus resinosa* tree, compression wood had formed across most of a growth ring except at the end of the season, when tracheids of the normal, latewood type had been produced. Formation of compression wood resumed the following spring.

In investigations also discussed in Chaps. 10.5.5 and 15.7.5, Molski (1969, 1971) studied the formation of earlywood, latewood, and compression wood in the growth rings of *Pinus sylvestris*. Two of the trees used in this careful and thorough investigation, the only one of its kind so far reported, had been repeatedly browsed by moose and had subsequently replaced their lost leader with a branch. Internode sections of the final, somewhat crooked stem were examined for distribution of compression wood, and the composition of each growth ring was established under the microscope. Two damaged specimens had produced less wood than an undamaged tree used as control. The more se-

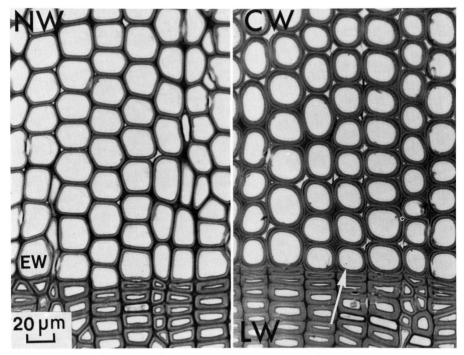

Fig. 4.13. Normal (*NW*) and compression (*CW*) woods of *Juniperus virginiana*. Intercellular spaces are present not only in the compression wood but also in the earlywood (*EW*) zone of the normal wood. The lower growth ring to the right consists of normal latewood (*LW*) tracheids. In the next increment, rounded compression wood tracheids occur next to the last-formed latewood (*arrow*). Transverse sections. LM. (Courtesy of A.C. Day)

vere the loss of leader and branches, the less were overall growth and amount of compression wood formed. As is generally the case, growth rings containing compression wood were wider than those without it, but there were also many equally wide or wider rings consisting only of normal wood, and Molski concluded that radial growth promotion and formation of compression wood are not causally related. The subject is further discussed in Chap. 10.3.2.

The pine tree used as control contained 71% earlywood and 29% latewood, while the most severely damaged tree contained 46% normal earlywood, 38% normal latewood, and 16% compression wood. The distribution of the three types of tissue at the different internodes of the first tree is shown in Fig. 4.15. The reduction in the proportion of earlywood was a consequence of the fact that compression wood had replaced this tissue 38 times, whereas the latewood had been replaced only 6 times. Compression wood occupied on the average only 45–50% of the growth rings. Increments consisting at any point of only compression wood were very rare and were found in only one of the trees. Inspection of the two tissue-distribution diagrams reveals that compression wood occurred in all possible combinations with earlywood and latewood except that

4.2 Earlywood and Latewood

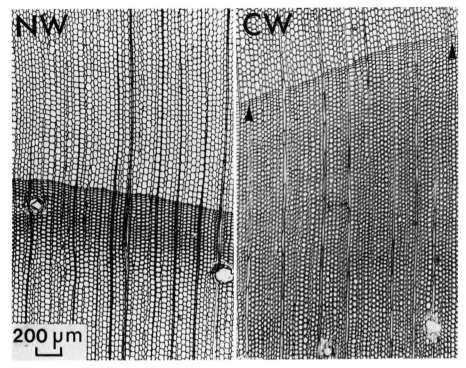

Fig. 4.14. Normal (*NW*) and compression (*CW*) woods of *Pinus resinosa*. Compression wood had ceased to form at the end of the growing season (*arrowheads*), an unusual situation. Transverse sections. LM. (Courtesy of A.C. Day)

it was never present in the middle of a latewood zone. The following sequences can be gleaned from the two diagrams:

Earlywood → Compression wood
Compression wood → Latewood
Compression wood → Earlywood → Latewood
Earlywood → Compression wood → Latewood
Earlywood → Latewood → Compression wood
Earlywood → Compression wood → Earlywood → Latewood

It is clear from Molski's results that compression wood can be formed at any time of the season, although it seldom develops inside a zone of normal latewood. Another important conclusion is that formation of compression wood is associated with a reduction in the proportion of earlywood and an increase in that of latewood. It is probably significant, as pointed out by Molski (1971), that the two tissues which contribute to the mechanical strength of the stem, namely compression wood and latewood, are both produced at the expense of the weak earlywood. Within the zone of juvenile wood, a region where com-

96 4 The Structure of Compression Wood

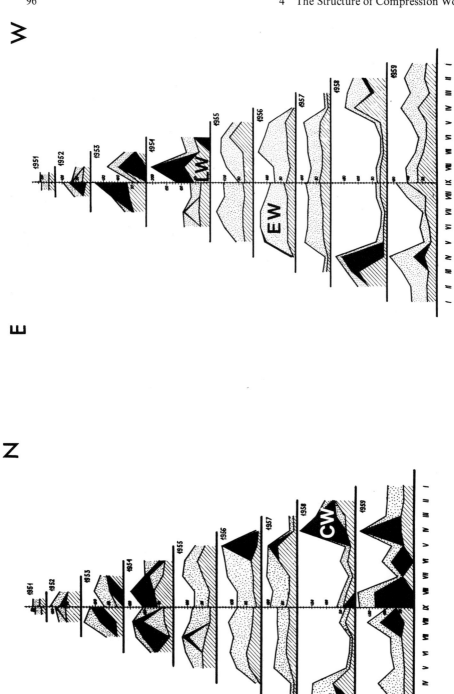

Fig. 4.15. Distribution of normal earlywood (*EW*) and latewood (*LW*) and of compression wood (*CW*) in the stem of a *Pinus sylvestris* tree that had been repeatedly browsed by moose (*Alces alces*). *Roman numerals* indicate heights, with *I* the highest and *IX* the lowest position examined. The four quarters are: South (*S*), North (*N*), East (*E*), and West (*W*). (Modified from Molski 1971)

4.2 Earlywood and Latewood

pression wood is common (Chaps. 6.2.1 and 10.3.6.2), the formation of compression wood profoundly changes the usual balance between earlywood and latewood. It would be of considerable interest if similar, comparative analyses of growth rings could be applied to a vertical and a leaning, but otherwise undamaged, conifer. In such an inclined stem, overall growth would be only moderately affected by gravity, and a more meaningful comparison would thus be possible.

It is often stated that the transition from earlywood to latewood is gradual and diffuse in compression wood. While this is generally the case for the mild and moderate grades, it is frequently not true of the pronounced type. Core et al. (1961) (Côté and Day 1965, Panshin and de Zeeuw 1980) claim that in conifers with an abrupt transition between normal earlywood and latewood, this transition is more gradual in compression wood. The investigators offered no explanation for this correlation. According to Harris (1976) the transition from earlywood to latewood is abrupt in normal and mild compression woods of *Pinus echinata* and gradual in severe compression wood.

Cieslar (1896) has drawn attention to what he considers to be a striking similarity between normal latewood and compression wood, pointing out that on visual inspection compression wood appears as an extension of the narrow zone of normal latewood. The reduced cell diameter, thick cell wall, and smaller lumen are, actually, features that distinguish both normal latewood and compression wood from normal earlywood. The similarity is, however, restricted to these properties. As emphasized by Trendelenburg (1932), compression wood actually resembles normal earlywood in the majority of its structural, chemical, and physical traits.

At this point it must also be emphasized that, when applied to growth rings consisting largely of compression wood, the earlywood-latewood terminology is not very meaningful or useful, as has been pointed out on several occasions, for example by Brazier (1969) and Timell (1972b). Earlywood and latewood, in the form in which these tissues occur in normal wood, are not produced in compression wood. What is usually taken for earlywood in compression wood is either normal wood or wood that is basically compression wood, although it often lacks some of the morphological characteristics of the latter. What is generally considered to be latewood is actually pronounced compression wood. Under optimal conditions, compression wood will form across an entire growth ring. The tracheids are then all of the same size except for the last few rows of cells, which tend to be flattened in the radial direction, as they also are in normal wood. More commonly, the first five to ten rows, produced at the beginning of vernal activity, lack certain of the traits of severe compression wood tracheids. This tissue should, nevertheless, still be regarded as a modified compression wood. The remainder of the increment consists of ordinary compression wood. The contention that compression wood should never be formed at the end of the growing season is erroneous. Under the influence of a weak gravitational stimulus, only mild or moderate compression woods are produced. No case seems to have been reported where, at constant gravistimulus, compression wood developed only at the beginning of the season. External agencies can, of course, easily create such a situation.

Statements to the effect that the transition between earlywood and latewood should be more gradual in compression wood than in normal wood or that this transition should occur sooner in compression wood are not very helpful. In his investigation on the chemical composition of the early growth rings in *Pinus resinosa*, Larson (1966) distinguished between earlywood and latewood only with normal wood but not with compression wood.

Summarizing, the earlywood-latewood concept, while of great theoretical and practical importance when applied to normal wood, is of less significance in dealing with compression wood. Its uncritical use has resulted in many contradictory statements in the literature. An example of this has been provided by Trendelenburg (1932). Drawing attention to the facts that compression wood has a high proportion of latewood and also a high lignin content, he concludes that normal latewood probably contains an unusually large amount of lignin. Actually, latewood has a lower lignin content than earlywood (Chap. 6.2.3), and the analogy between latewood and compression wood is false. The terms earlywood and latewood should be applied to compression wood only sparingly or not at all, since they do not have their usual meaning in this context. Compression wood is produced under conditions of prolonged and very high cambial activity, possibly as a result of a high concentration of auxin (Chap. 13.4). High concentrations of auxin are associated with the formation of earlywood in normal wood, whereas lack of auxin and presence of growth inhibitors in conjunction with an abundant supply of photosynthate result in formation of latewood (Chap. 13.3).

4.3 Tracheids. Gross Anatomy

4.3.1 Shape

Tracheids in normal softwood generally have an angular transverse section, usually square or rectangular. One of the most conspicuous anatomical features of compression wood is the fact that they are nearly circular or, occasionally, oval in outline. Sometimes, as seen in Fig. 4.16, the outline is somewhat irregular. Wergin and Casperson (1961) claim that compression wood tracheids in *Taxus* do not assume a circular form as readily as in *Picea*. The outline of the compression wood tracheids in *Ginkgo biloba* is less regular than in the Coniferales (Timell 1978b). Yumoto et al. (1983) have pointed out that the circular form of the tracheid outline is most conspicuous when there are intercellular spaces present. They have also drawn attention to the well-known fact that the boundary between the S_1 and S_2 layers and the innermost, lumen border of the S_2 can also be rounded when the surface outline of the cell is angular. This is possible because in such tracheids there is a significant increase in the thickness of the S_1 layer at the cell corners, as shown in Fig. 4.45. Yumoto et al. (1983) believe that the form of the boundary between S_1 and S_2 should be used for evaluating the severity of a compression wood (Chap. 3.6).

As was mentioned in Sect. 4.2, the tracheids formed at the very beginning and the very end of the growing season usually differ from the remainder of the

4.3.1 Shape

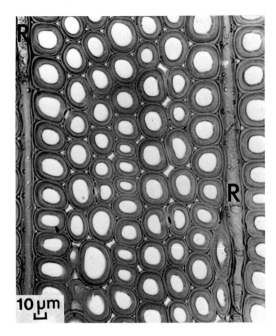

Fig. 4.16. Compression wood of *Picea abies* in the middle of a growth ring with rounded tracheids and two rays (*R*). Transverse section. LM. (Timell 1973d)

compression wood tracheids in a growth ring (Sect. 4.5). The first-formed cells tend to have an angular outline, as illustrated in Fig. 4.9. The two or three last-formed tracheids are flattened in the radial direction (Sect. 4.6). The rounded outline of the majority of the cells and the oblong shape of the last ones are shown in Figs. 4.17 and 4.18. Harris (1976) found that in *Pinus echinata* the earlywood tracheids were never rounded in mild or moderate compression woods and only occasionally in severe compression wood. They were circular in outline in all three types of tissues in the latewood portion. As the severity of compression wood gradually decreases, the tracheids slowly assume a more and more angular outline (Fujita et al. 1979, Yumoto and Ishida 1982, Yumoto et al. 1982, 1983).

In some conifer genera, the tracheids in normal wood can also have a rounded outline, with associated intercellular spaces, namely in *Agathis, Araucaria, Cupressus* (Butterfield and Meylan 1980), and *Juniperus* (McGinnes and Phelps 1972, Core et al. 1979). An example of this is shown in Fig. 4.19. The only visible feature distinguishing this normal wood of *Juniperus virginiana* from the first-formed compression wood with its angular tracheids, which lack helical cavities, is the presence of an S_3 layer in the former but not in the latter. A cross section of normal *Agathis australis* wood reported by Patel (1968) contains tracheids with an angular outline but a rounded lumen.

Compression wood probably effects some of its righting function (Chap. 17.5) by sliding or intrusive growth of the tracheids, which creates considerable growth stresses. As a result, compression wood tracheids are frequently distorted, especially at their tips. Verrall (1928) seems to have been the

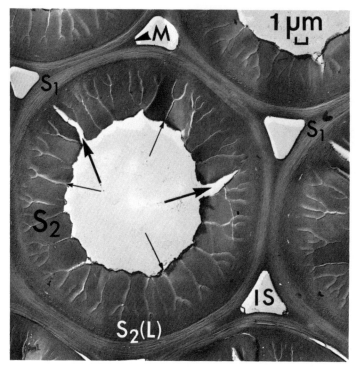

Fig. 4.17. Compression-wood tracheids in the middle of a growth ring in *Larix laricina* with a circular outline and intercellular spaces (*IS*). *Small arrows* indicate helical cavities and *large arrows* drying checks. *M* Middle lamella. Transversely oriented microfibrils are discernible in the S_1 layer. Transverse section. TEM. (Côté et al. 1966a, 1968b)

first to observe this. He reported distortions varying from a slight bending of the tracheid tip to instances where almost one third of the tracheid was horizontal as a result of gliding growth. Often, the tracheids appeared as if they had been subjected to longitudinal compression and had been bent over at the ends. Münch (1938), 10 years later, observed the same phenomenon and attributed it to the fact that in compression wood the tracheids grow intrusively, so that they become compressed and distorted for lack of sufficient space. Münch (1940) later reported further observations and showed that in species such as *Abies alba* and *Pinus peuce* the entire tracheids in compression wood can assume a wavy shape. He also presented some illustrations of severely crowded, distorted, and deformed tracheid tips in compression wood of *Pinus peuce*. Some of these drawings, which have been reproduced by Gessner (1961) and Wardrop (1965), are shown in Fig. 4.20. Trendelenburg (1940) has reported similar deformations in tracheids present in the callus tissue produced by conifers to cover compression failures (Chap. 15.2.1.5). Bailey and Berkley (1942) observed deformations in compression wood tracheids, especially in their upper and lower thirds. Onaka (1949) found that irregular bendings frequently occurred

4.3.1 Shape

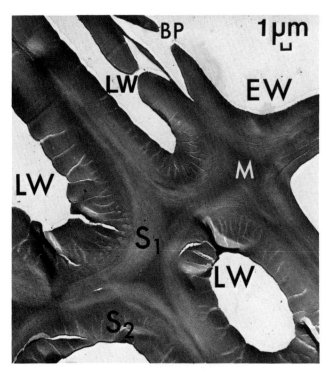

Fig. 4.18. Compression-wood tracheids at the earlywood-latewood boundary in *Larix laricina*. The last-formed tracheids (*LW*) have an oblong outline in the tangential direction and contain helical cavities but lack intercellular spaces. The first-formed tracheids (*EW*) have an angular outline and lack cavities. *BP* bordered pit; *M* middle lamella. Transverse section. TEM. (Côté et al. 1966a, 1968b)

where compression wood tracheids met rays. Figure 4.21 shows an example of truncated tracheid tips in compression wood of *Abies balsamea*.

Wardrop and Dadswell (1952, 1953), in a comprehensive study of the phenomenon, examined macerated compression wood tracheids rather than wood sections. Bifurcated, abnormal, and distorted tips were common in compression wood tracheids. Similar deformations were, however, also present in tracheids from a wide growth ring of normal wood, and while they were positively correlated with the severity of compression wood formation, it was evident that they were not restricted to this tissue. Obviously, such distortions are to be expected where tracheid tips have grown rapidly against obstructions, such as other tracheids or ray cells. Bannan and Whalley (1950) in a study of the elongation of the fusiform initials in *Chamaecyparis*, suggested that this growth was intrusive, elongating tips penetrating between adjacent initials and forcing them apart. They also observed that growing tips occasionally became stalled at obstructing rays. Similar views were later expressed by von Pechmann (1973).

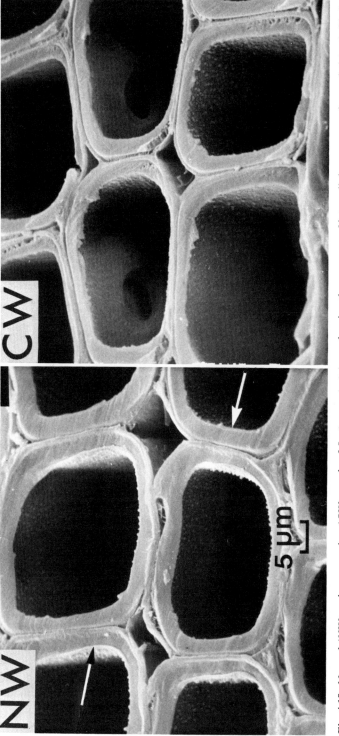

Fig. 4.19. Normal (*NW*) and compression (*CW*) woods of *Juniperus virginiana*, showing the presence of intercellular spaces and rounded tracheids in both types of tissue (cf. Fig. 4.13). *Arrows* indicate the thin S_3 layer in the normal wood. Transverse surface. SEM

4.3.1 Shape

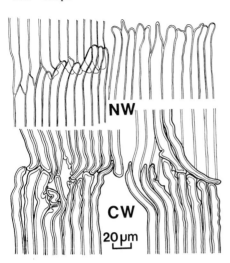

Fig. 4.20. Tracheid tips in normal (*NW*) and compression (*CW*) woods of *Pinus peuce*, the latter much distorted. Drawing. (Münch 1940)

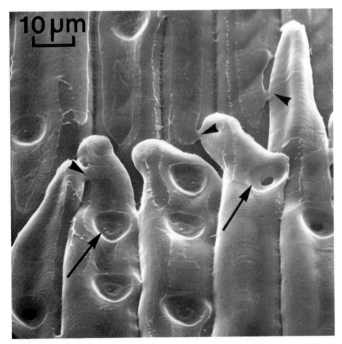

Fig. 4.21. Truncated tracheid tips in compression wood of *Abies balsamea*. *Arrows* indicate bordered pits and *arrowheads* parental primary walls. SEM. (Courtesy of A. C. Day)

4.3.2 Tracheid Length

4.3.2.1 Normal Wood

Length is an important property of tracheids and fibers in the conversion of wood to pulp and paper products. The actual size of these cells, the factors that influence it, and how the length varies with position in the tree are factors that have been extensively studied. It is now more than 110 years since Sanio (1872) reported his pioneering investigations on the length of the tracheids in *Pinus sylvestris,* the results of which he summarized in his five laws (Bailey and Shephard 1915). Sanio found that in both stems and branches the tracheids increased in length from the pith toward the bark until a definite size was reached which remained constant in subsequent annual increments. This final length increased with height above the ground, reached a maximum, and then again decreased. Tracheid length in branches was positively correlated with tracheid length in the stem portion where the branch originated.

Tracheid and fiber length in wood has been reviewed by Bisset (1949), Bisset et al. (1951), Spurr and Hyvärinen (1954), and Dinwoodie (1961). Bannan (1968c) considered the problem of sampling in studies of tracheid length in conifers. Seth and Agrawal (1984) have recently discussed the effect of age and distance from the pith on tracheid length. Briefer surveys have been presented by Brown (1970), Kozlowski (1971), Seth (1979), and by Panshin and de Zeeuw (1980). A critical review of the variations in size of the fusiform initials in the cambium and the variations in tracheid and fiber length has been contributed by Philipson et al. (1971).

The variation of tracheid length in conifers is strongly influenced by the occurrence of juvenile wood, a tissue referred to in Chap. 6.2.1. This wood, sometimes called core wood, is formed by a young vascular cambium and usually comprises the first 5–20 growth rings at the center of a stem. The tree top consists entirely of juvenile wood. As the cambium grows older, it begins to produce mature (adult) wood. Juvenile wood is often, albeit not always, associated with rapid growth and wide increments. It differs in many respects from mature wood. Among its major characteristics are a higher proportion of earlywood and accordingly a lower density, shorter tracheids (0.5–1.5 mm), thinner cell walls, and a larger microfibril angle (30°–50°). Juvenile wood contains less cellulose and galactoglucomannan and more galactan and lignin than mature wood. Its strength properties are poor, and its longitudinal shrinkage is high. Juvenile wood is clearly similar to compression wood in many respects (Watanabe et al. 1963, Zobel 1975). It is also characteristic of juvenile wood that it tends to have a high content of compression wood (Chaps. 6.2.1 and 19.2). The tracheids in juvenile wood can be as short as 0.5–1.0 mm (Sudo 1968a, 1970b). Their length increases rapidly from the pith toward the outside. The changes in tracheid length that occur within the zone of juvenile wood should be considered separately from the subsequent and less drastic changes that take place within the mature wood.

A few investigators (Gerry 1915, 1916, Fry and Chalk 1957) have claimed that the tracheids in the latewood are shorter than those in the earlywood, while

4.3.2.1 Normal Wood

others (Jackson 1959, Jackson and Morse 1965, Voorhies and Jameson 1969, Barger and Ffolliott 1976, Taylor and Moore 1981) have found no difference in this respect between the two tissues. All other investigators agree that the tracheids in the latewood are longer than those in the earlywood. Sanio (1872) found that in *Pinus sylvestris* tracheid length reached a maximum after a certain number of years and then remained constant, an observation recently confirmed by Atmer and Thörnqvist (1982). Some species require a very long time before this maximum is attained. The best known example is *Sequoia sempervirens* where this length is reached only after 300–350 years (Bailey and Tupper 1919, Bailey and Faull 1934, Resch and Arganbright 1968). Tracheid length in this species has a tendency to decrease after 400 years, when overmature wood begins to develop (Shephard and Bailey 1914, Bailey and Faull 1934). A reduction in tracheid length toward the bark after a maximum had been reached has also been reported for younger trees, such as *Abies pindrow* (Ahmad 1970), *Picea abies*, *Pinus sylvestris* (Helander 1933), and *Pseudotsuga menziesii* (Lee and Smith 1916).

A continuous, albeit slow, increase in tracheid length from pith to bark has been observed on several occasions, indicating that, contrary to Sanio's laws, no maximum is ever attained. Species for which this has been observed include several pines, namely *Pinus elliottii* (Jackson and Greene 1958), *P. radiata* (Uprichard 1971, Cown 1975), *P. strobus* (Foulger 1966), *P. sylvestris* (Petrić 1974), and *P. taeda* (Kramer 1957) but also *Picea abies* (Nečesaný 1961), *P. glauca* (Wang and Micko 1984), *Pseudotsuga menziesii* (Gerry 1916, Hejnowicz 1971), and *Thuja plicata* (Wellwood and Jurazs 1968). Erickson and Harrison (1974) found that in *Pseudotsuga menziesii* tracheid length increased rapidly with age for the first 15 years, following which the rate of increase decreased considerably. They estimated that after 30 years the increase would be very slight. The situation seems to be the same in *Metasequoia glyptostroboides* (Brazier 1963, Hejnowicz 1973).

Despite the fact that there are many irregularities, a general pattern can definitely be discerned in the variation of tracheid length with distance from the pith in conifers. Within the first few rings of juvenile wood, the length of the tracheids increases rapidly. As mature wood begins to form, the increase becomes more modest until a maximum has been reached. The age at which this happens increases with increasing longevity of the tree. In some species the tracheids retain their maximum length, while in others they gradually become shorter. In trees more than 400 years old, overmature wood tends to form, and tracheid length decreases with advancing years.

The variability of tracheid length with height in a conifer stem is more regular than that with distance from the pith. Sanio (1972) observed that in *Pinus sylvestris*, within a given growth ring, tracheid length first increased with increasing height above the ground and then, having reached a maximum, decreased toward the top. This has been verified by most later investigators and for a large number of conifer species. Usually, maximum tracheid length occurs at a level of 30–40% of the total height. A few exceptions to the pattern observed by Sanio have been reported. Voorhies and Jameson (1969) found no change in tracheid length in *Pinus ponderosa* above a height of 1.5 m. In both

Pinus radiata (Nicholls and Dadswell 1962) and *Thuja plicata* (Wellwood and Jurazs 1968) tracheid length has been observed to decrease continuously toward the top. According to Ohsako et al. (1973) tracheid length and diameter, as well as wall thickness, all decreased with increasing height in a stem of *Pinus massoniana* which was leaning and contained compression wood. Petrić (1974) found the length of both earlywood and latewood tracheids in *Pinus sylvestris* to increase from ground level to a height of 4 m and then to remain almost constant. Knigge and Wenzel (1982) have recently reported that in a 63-year-old *Sequoiadendron giganteum* tree there was no change in tracheid length between a height of 0.2 and 19.2 m, while the tracheids increased in length from the pith toward the cambium.

According to Sudo (1968a), long internodes in *Pinus densiflora* are associated with long tracheids in this region. Denne (1971) found that in seedlings of *Picea sitchensis* and *Pinus sylvestris* a doubling of shoot length caused an increase in tracheid length of 10%. He pointed out that tracheid length is determined not only by the rate of anticlinal divisions in the cambium but also by the rate of elongation of these initials. Root tracheids are shorter than those in the stem (Fegel 1941, Bailey and Faull 1934). Conifer branches always contain much compression wood and are discussed separately in Sect. 4.3.2.3.

Formation of compression wood is generally associated with a high radial rate of growth, and the variation of tracheid length with radial growth rate, usually expressed by ring width, is therefore of special interest in this context. A few investigators could not observe any correlation between tracheid length and ring width, for example Shephard and Bailey (1914) and Bailey and Shephard (1915) in *Pinus palustris* and *P. strobus*, Echols (1955) in *P. palustris* and Kramer (1957) in *P. taeda*. In the large majority of the investigations, however, tracheid length has been found to be inversely proportional to ring width, that is, the faster the growth, the shorter are the tracheids. Species for which this relationship has been established include *Picea abies* (Mork 1928b, Helander 1933, Nylinder and Hägglund 1954, Schultze-Dewitz 1959), *Picea engelmannii, P. glauca, P. mariana* (Bannan 1963a), *P. sitchensis* (Chalk 1930, Elliott 1960, Bannan 1963a, Dinwoodie 1963), *Pinus radiata* (Bisset et al. 1951, Wardrop 1951, Nicholls and Dadswell 1962), *P. pinaster* (Bisset et al. 1951), *P. strobus* (Bannan 1962), *P. sylvestris* (Helander 1933, Schultze-Dewitz 1965), and *Thuja occidentalis* (Bannan 1954).

On the basis of extensive investigations on the variations in tracheid length of *Pinus densiflora*, Sudo (1969b, 1970a) concluded that ring width is less decisive in determining tracheid length than is ring number from the pith. He also pointed out that maximum tracheid length is associated with an optimum ring width, very narrow and very wide rings having shorter tracheids. Bannan and Bindra (1970) arrived at a similar conclusion in a study of *Picea glauca*. In this species, maximum tracheid length was observed in rings 1–2 mm wide, the same as had been found earlier for *Picea* (Bannan 1963a). In *Cupressus* (Bannan 1963b) and *Pseudotsuga menziesii* (Bannan 1964a) maximum length occurred at a ring width of 1 mm and in *Pinus contorta* at 1–1.5 mm (Bannan 1964b). Bannan (1967a, b) has pointed out that the inverse relationship between tracheid length and ring width is valid only for increments larger than 1 mm. For rings 0.5 mm or less, the correlation is positive.

Whereas a few investigators have found no difference in tracheid length between fast and slow-growing trees, most have observed a positive correlation between tracheid length and the rate of height growth (Bannan 1965a). Usually, comparisons have been made between dominant and suppressed forest trees, but seedlings have also been used. Some of the species studied include *Picea abies* (Helander 1933, Schultze-Dewitz 1959, Stairs et al. 1966), *P. rubens* (MacMillan 1925), *P. sitchensis* (Elliott 1960, Dinwoodie 1963), *Pinus radiata* (Hartley 1960), *P. sylvestris* (Helander 1933), and *Pseudotsuga menziesii* (Lee and Smith 1916). Wellwood and his co-workers (Sastry and Wellwood 1971 a, b, 1974, 1975, Sastry et al. 1972, Wellwood et al. 1974) found that there exists a strongly positive and curvilinear correlation between tracheid length and weight of delignified coniferous wood. Evidently, the weight of the polysaccharides in each species depends only on the tracheid length and not on the species involved. Baas et al. (1984) recently studied the effect of dwarf growth on the structure of the wood. In several *Pinus* species the length of the axial stem tracheids was only 30–50% of that found for tracheids in trees with a normal growth rate in the axial and radial directions.

In summary, there appears to be more agreement concerning the variability of tracheid length in conifers than there is with respect to the similar variability of the specific gravity (Chap. 7.2.2). Perhaps this is only what could be expected, since cell length is a unique and well-defined anatomical property, whereas specific gravity is a complex parameter, dependent on several chemical and anatomical factors. Investigations on tracheid length and specific gravity in conifers have one thing in common: accidental inclusion of compression wood must be avoided if reliable results are to be obtained (Chap. 7.2.1). The short tracheids in compression wood and the high density of this tissue make this imperative.

4.3.2.2 Compression Wood

Sanio (1872, 1873), when examining the wood on the lower side of branches in *Pinus sylvestris*, was the first to notice the shortness of the tracheids in compression wood. Hartig (1896, 1901) measured the length of the tracheids in compression wood of *Picea abies* and found them to be 70–80% shorter than in opposite wood. Cieslar (1896) studied the formation of compression wood in four 9-year-old *Picea abies* trees, the stems of which must have consisted largely of juvenile wood. In two trees, compression wood tracheids were 0.74 and 0.97 mm long compared to 0.88 and 1.26 mm for opposite wood. Shephard and Bailey (1914) claimed that eccentric growth per se had no effect on tracheid length. Mature wood of *Pinus strobus* had an average tracheid length of 2.13 mm in the compression wood zone and 2.26 mm in the opposite wood, results later confirmed by Gerry (1915). Lee and Smith (1916) found that compression wood tracheids were shorter than those in opposite wood in *Pseudotsuga menziesii*.

Table 4.3. Length of tracheids in opposite, normal, and compression woods of various conifer species. All values in μm. (Verrall 1928)

Species	Opposite wood	Normal wood	Compression wood
Larix laricina	2.87	2.82	2.15
Picea mariana	2.21	1.91	1.78
Pinus banksiana	2.06	2.16	1.68
Pinus palustris	4.35	4.32	2.78
Thuja occidentalis	2.49	2.56	1.81
Thuja plicata	3.12	3.00	2.55
Average for all species	2.85	2.80	2.13

In 1928 Mork (1928a) and Verrall (1928) reported results of more comprehensive investigations on tracheid length in compression wood. Mork measured the length of the tracheids in side wood, opposite wood, and compression wood from an eccentrically grown *Picea abies* stem, shown in Fig. 4.3. Some of his extensive data are summarized in Table 4.1 and Fig. 4.4. Mork concluded that in normal spruce wood the length of the tracheids increased from the pith outward, the tracheids being about twice as long near the bark than around the pith. In compression wood, by contrast, tracheid length seemed to depend less on the distance from the pith or the width of the growth ring. As a rule, the tracheids were only 1.5 mm long. Mork emphasized that even at equal distance from the pith and equal ring width, compression wood tracheids nevertheless are shorter than in normal wood, which had not been unequivocally established prior to his investigation. A fact not commented upon by Mork is evident from Fig. 4.4, namely that the length of the compression wood tracheids increased slowly from pith to bark. Verrall (1928) measured tracheid length in normal, opposite, and compression woods of six conifer species, obtaining the results summarized in Table 4.3. The tracheids in compression wood were considerably shorter than those in either opposite wood or normal wood. The difference is very close to that found by Hartig (1896, 1901) for *Picea abies*.

Kienholz (1930) measured the lengths of 5600 tracheids in a "pistol-butted" *Tsuga mertensiana* tree. Average tracheid length was 1.81 mm on the lower, compression-wood side, compared to 1.99 mm in the opposite wood and 2.09 mm in a control tree. The tracheids increased in length in all three zones from the pith toward the bark, first rapidly and later more slowly and erratically, as shown in Fig. 4.22. Tracheid length also increased toward the top of the tree. Onaka (1949) compared the length of the tracheids in normal and compression woods of four conifer species. His data are listed in Table 4.4. The tracheids in the compression wood had a length that was only 80–90% of that in the normal wood, a finding in agreement with those of Hartig (1896, 1901), Mork (1928a), and Verrall (1928). Matsumoto (1950b) measured tracheid lengths in eccentrically grown discs of *Chamaecyparis obtusa* and *Thujopsis dolabrata*. Contrary to all earlier investigators he found that, for pronounced compression wood, the tracheids in the compression wood region were longer

4.3.2.2 Compression Wood

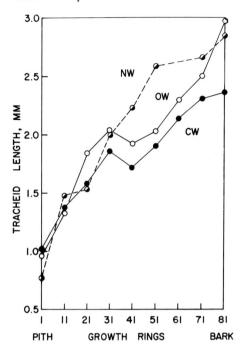

Fig. 4.22. Variation of tracheid length with distance from the pith in normal (*NW*), opposite (*OW*), and compression (*CW*) woods in *Tsuga mertensiana*. (Redrawn from Kienholz 1930)

Table 4.4. Length of tracheids in normal and compression woods of four conifer species. All values in µm. (Onaka 1949)

Species	Compression wood	Normal wood	Ratio
Abies veitchii	2.848	3.029	0.94
Cryptomeria japonica	1.304	1.726	0.76
Picea jezoensis	1.865	2.333	0.80
Pinus densiflora	2.430	3.144	0.77

than those in the opposite zone. The reverse was true in the case of moderate compression wood. These results would seem open to considerable doubt, notwithstanding the fact that they were based on a large number of measurements.

Dadswell and Wardrop (1949) pointed out that, in measuring tracheid length in compression wood, it is extremely difficult, if not impossible, to secure exactly comparable normal wood. They also claimed that in none of the earlier investigations had the basis of comparison been given, a criticism that certainly does not apply to the comprehensive data reported by Mork (1928a). Wardrop and Dadswell (1950) measured tracheid length in several specimens of *Pinus radiata*, comparing normal and compression woods. In one case, the average length of the tracheids in normal wood was 3.05 mm, in opposite wood 3.58 mm and 2.77 mm in compression wood. Care was taken in all measure-

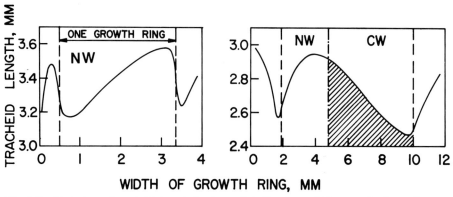

Fig. 4.23. Variation of tracheid length in the radial direction within growth rings of *Pinus radiata*, consisting of either normal wood (*NW, left*) or both normal wood and compression wood (*CW, right*). (Redrawn from Bisset and Dadswell 1950)

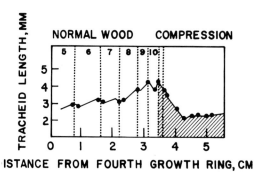

Fig. 4.24. Variation of tracheid length in the radial direction within growth rings of *Pinus pinaster* consisting of either normal or compression wood. (Redrawn from Bisset and Dadswell 1950)

ments to secure comparable data for normal and compression woods, and the investigators concluded that the tracheids in compression wood are appreciably shorter than those in corresponding normal wood (cf. Fig. 4.61).

Bisset and Dadswell (1950) examined the variation in tracheid length within a single growth ring at a constant height. Three of the species examined contained compression wood, namely *Picea sitchensis*, *Pinus pinaster*, and *P. radiata*. The effect of compression wood on the radial variation in tracheid length in a stem is shown in Fig. 4.23 (*Pinus radiata*) and Fig. 4.24 (*P. pinaster*). A distinct band of severe compression wood was present in both specimens. The latewood tracheids were always longer than those in the earlywood within the zone of normal wood, and average tracheid length increased toward the bark. When the band of compression wood was reached, the length of the tracheids decreased suddenly, in *Pinus pinaster* from 4.3 to 2.2 mm. Similar results were obtained with *Picea sitchensis*. Wardrop (1951), in comparing certain strength properties of normal and compression woods of *Pinus radiata*, found that the tracheids were consistently shorter in the latter tissue. An additional example, reported by Wardrop and Dadswell (1952), is shown in Fig. 4.25. In

4.3.2.2 Compression Wood

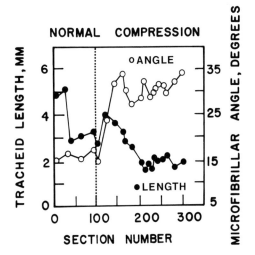

Fig. 4.25. Tracheid length and microfibril angle in a series of tangential sections of normal and compression woods of *Pinus pinaster*. (Redrawn from Wardrop and Dadswell 1952)

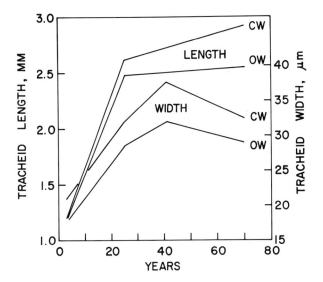

Fig. 4.26. Changes in tracheid length and width with distance from the pith in opposite (*OW*) and compression (*CW*) woods of *Larix decidua*. (Redrawn from Mariani 1955)

this particular specimen of *Pinus pinaster*, formation of compression wood resulted in a notable reduction in tracheid length and a simultaneous increase in microfibril angle.

Mariani (1955) examined opposite and compression woods in a very eccentric disc from *Larix decidua*. Like Matsumoto (1950b), but in contrast to all other investigators, she found the tracheids in the zone with compression wood to be longer than the opposite tracheids, as can be seen in Fig. 4.26, which also shows that both types of cells increased in length with distance from the pith, first quite rapidly and later, after 25 years, more slowly. According to a report from the Queensland Department of Forestry (Anonymous 1955), formation of

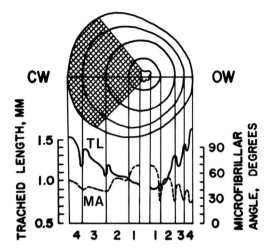

Fig. 4.27. Changes in tracheid length (*TL*) and microfibril angle (*MA*) across a disc from the stem of a 5-year-old *Pinus banksiana* tree containing compression (*CW*), normal and opposite (*OW*) woods. (Redrawn from Nečesaný 1955b)

compression wood in *Pinus elliottii* is associated with a reduction in tracheid length, but the effect is less marked than the increase in microfibrillar angle. A later report (Anonymous 1959) mentions the same effects for *Araucaria cunninghamii*, adding that the rate of increase in tracheid length from the pith and outward is less in compression wood than in normal wood.

In a study of the ultrastructure of the tracheids in normal and compression woods, Nečesaný (1955b) followed the changes in tracheid length and microfibril angle across a disc of *Pinus banksiana* which was eccentric and contained well-developed compression wood. A graphical summary of the results is shown in Fig. 4.27. At any given position in each growth ring, the tracheid length in the compression wood zone was less than in the zone of opposite wood. Within each ring, the length of the tracheids increased continuously from the beginning to the end of the growing season, but more rapidly in opposite wood than in compression wood. Average tracheid length increased in both tissues with increasing distance from the pith. The young tree stem obviously consisted of juvenile wood, and all tracheids were therefore short. Similar results were later obtained with *Picea abies* (Nečesaný 1961).

Ollinmaa (1955, 1959) measured 300 tracheids each from normal and compression woods of *Picea abies* and *Pinus sylvestris*. Average lengths for normal wood tracheids were 2.25 and 1.57 mm, respectively and for those in compression wood 1.82 and 1.26 mm. In spruce compression wood, latewood tracheids were only slightly longer than those in the earlywood. The tracheid length distributions are shown in Figs. 4.28 and 4.29. The distributions for both normal and compression wood tracheids were considerably wider in the spruce than in the pine. Nicholls and Dadswell (1962) found that slight amounts of compression wood did not affect the specific gravity of *Pinus radiata* wood, but that tracheid length was reduced by a few per cent. They considered this a fact worth noting, because formation of compression wood in *Pinus radiata* in Australia is the rule rather than the exception. Petrić (1962) determined the dimen-

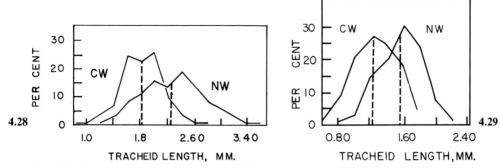

Fig. 4.28. Frequency distribution of tracheid length in normal (*NW*) and compression (*CW*) woods of *Picea abies*. (Redrawn from Ollinmaa 1959)

Fig. 4.29. Frequency distribution of tracheid length in normal (*NW*) and compression (*CW*) woods of *Pinus sylvestris*. (Redrawn from Ollinmaa 1959)

sions of the tracheids in a 70-year-old *Abies alba* tree which had first grown horizontally and then bent upward sharply. A disc from the bend revealed an exceedingly eccentric radial growth. The length of earlywood and latewood tracheids was determined from 23 samples taken around the periphery of the disc, giving the results shown graphically in Fig. 4.30. The compression wood tracheids were considerably shorter than those in normal wood, both in the earlywood and latewood. The tracheids in the opposite wood were, rather unexpectedly (Chap. 21.3.1), slightly shorter than those in the normal side wood. The average length in the compression wood was 2.66 mm for earlywood and 2.86 mm for latewood tracheids, compared to 3.65 and 4.04 mm for normal wood. In opposite wood, earlywood tracheids were 10% shorter than those in the latewood.

Riech (1966) and Riech and Ching (1970) subjected *Pseudotsuga menziesii* trees to bending stress. One of the changes associated with the stress was a reduction in average tracheid length. Riech failed to observe any relationship between severity of compression wood and tracheid lengths. Further aspects of this investigation are discussed in Chap. 11.4. Shelbourne and Ritchie (1968) measured tracheid lengths in eight *Pinus taeda* trees which contained normal wood and slight, moderate, and severe compression woods. Results obtained with three growth rings at a 2-m level are shown in Fig. 4.31. In the earlywood, the tracheids in all grades of compression wood were shorter than those in normal wood, but in latewood this applied only to the tracheids in mild compression wood. In both cases, the difference in tracheid length between normal wood and moderate to severe compression wood was only slight. This surprising result was considered to be an accidental consequence of sampling, the trees with most severe compression wood being long-tracheid trees. It is clear that great care has to be exercised when comparing tracheids obtained from different trees.

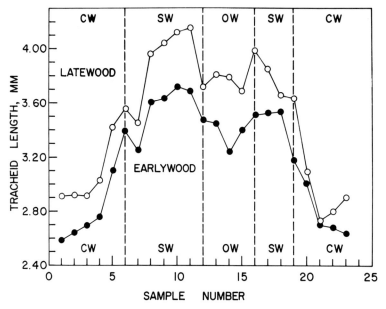

Fig. 4.30. Variation in the length of earlywood and latewood tracheids along the periphery of a horizontal stem of *Abies alba* containing compression (*CW*), side (*SW*), and opposite (*OW*) woods. (Redrawn from Petrić 1962)

In some investigations, where normal and compression woods were subjected to chemical pulping, average tracheid length was measured in the resulting pulps. Some of these data are discussed in Chaps. 19.4 and 19.5, including results obtained with *Pinus taeda* by Schafer et al. (1937, 1958) (Table 19.3) and with *Picea sitchensis* by Dinwoodie (1965) (Table 19.15) and Young et al. (1970) (Table 19.7).

Chu (1972) determined the tracheid length and microfibril angle in *Pinus palustris* at three heights above the ground. As can be seen in Table 4.5, the compression wood tracheids were shorter than those in normal wood. In both cases tracheid length decreased with increasing height above the ground, but the initial reduction was much more pronounced in normal wood than in compression wood. Sastry and Wellwood (1974) extended their earlier investigations (Sect. 4.3.2.1) on the relationship between tracheid length and weight to include delignified compression wood tracheids from *Pseudotsuga menziesii*. A correlation between tracheid length and weight of the type previously established for normal wood was found to apply also to compression wood. The relationships were very similar with respect to holocellulose but different for alpha-cellulose. The investigators concluded that, for the same tracheid length, compression wood tracheids contain more alpha-cellulose than those in normal juvenile wood but less than those in normal mature wood.

In their investigations on the variation of specific gravity (Chap. 7.2.3) and tracheid length in *Pinus wallichiana*, Seth and Jain (1976) found that in one growth ring consisting of 83% compression wood from a leaning stem, tracheid

4.3.2.2 Compression Wood

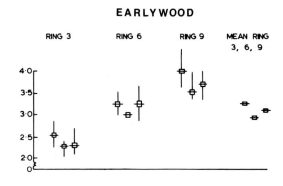
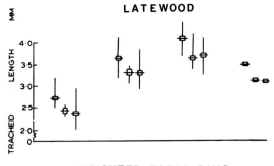
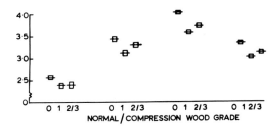

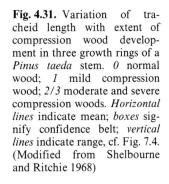

Fig. 4.31. Variation of tracheid length with extent of compression wood development in three growth rings of a *Pinus taeda* stem. *0* normal wood; *1* mild compression wood; *2/3* moderate and severe compression woods. *Horizontal lines* indicate mean; *boxes* signify confidence belt; *vertical lines* indicate range, cf. Fig. 7.4. (Modified from Shelbourne and Ritchie 1968)

Table 4.5. Tracheid length and microfibril angle in normal and compression woods of *Pinus palustris* at three different heights. (Chu 1972)

Vertical position	Tracheid length, mm		Microfibril angle, degrees	
	Normal wood	Compression wood	Normal wood	Compression wood
Base	3.80	2.60	13.3	30.0
Middle	5.00	2.97	12.8	27.5
Top	5.02	3.11	12.1	26.8

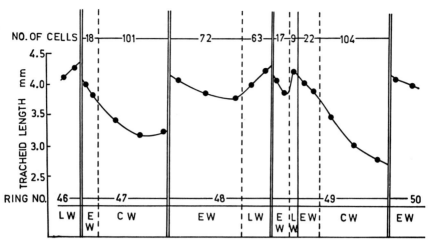

Fig. 4.32. Variation in tracheid length within three consecutive growth rings in a *Pinus wallichiana* tree containing compression wood (*CW*) and normal earlywood (*EW*) and latewood (*LW*). (Seth and Jain 1977)

length was only 2.73 mm compared to 3.65 mm for normal wood. They attributed this reduction in length to the higher rate of anticlinal divisions in compression wood. There was a linear, negative correlation between average tracheid length and proportion of compression wood:

$$Y = 4.1816 - 0.0180 \, X$$

where Y is average tracheid length in mm and X is percentage of compression wood (Seth and Jain 1977). With only compression wood tracheids present, the decrease in tracheid length would amount to 43%. The variation in tracheid length within three consecutive growth rings is shown in Fig. 4.32. In ring 47, which contained 76% compression wood, tracheid length decreased from 3.99 mm in the first-formed earlywood to 3.16 in the compression wood. The reduction in tracheid length was even greater in ring 49. The compression wood in this increment had no influence on the tracheid length in the normal earlywood of the next ring. In trees leaning up to 60°, the growth rings on the lower side were consistently wider than on the upper, while ring width on the flanks was independent of inclination (Jain and Seth 1980). Lean and ring width on the lower side were positively correlated in stems with a diameter at breast height of less than 50 cm. Average tracheid length within a given increment tended to be larger on the upper than on the lower side which is not surprising. The first-formed tracheids in each ring, however, had the same length around the circumference of the stem. Both tracheid length and specific gravity of the first-formed earlywood zone were unaffected by the extent of stem inclination and the presence of compression wood (Jain and Seth 1979, 1980). In close agreement with the findings reported by Jain and Seth, Nicholls

(1982) recently found normal tracheids in *Pinus radiata* (Chap. 15.2.1.3.6) to have an average length of 3.41 mm, while compression wood tracheids were only 1.82 mm long, a reduction of 47%.

In an investigation on the occurrence of compression wood in *Pinus ponderosa*, Barger and Ffolliott (1976) observed no difference in length between earlywood and latewood tracheids in either normal wood or compression wood, results that disagree with those obtained by Seth and Jain (1976, 1977) with *Pinus wallichiana*. Mean tracheid length was 2.37 mm for compression wood and 2.75 mm for normal wood. Harris (1977) found that in *Pinus radiata* the tracheids were slightly shorter in compression wood than in opposite wood except in the juvenile growth rings closest to the pith, where they were as short in the latter wood as in the former, an observation later confirmed by Park et al. (1979) with branch wood of *Pinus densiflora*. Within the earlywood zone of the compression wood, tracheid length increased gradually from 1.9 mm in the fourth increment to 2.7 mm in the 13th ring and from 2.1 mm to 2.8 mm in the latewood.

All investigators with the exception of Matsumoto (1950b) and Mariani (1955) have found that the tracheids in compression wood are shorter than those that would have been formed in normal wood of exactly the same age and position in the tree. The difference is directly related to the severity of compression wood development and is largest at maximum gravitational stimulus. Generally, compression wood tracheids have a length that is 60–80% of that of normal wood tracheids. In those cases where the entire growth ring consists of compression wood, tracheid length increases from the inception toward the cessation of seasonal growth. The tracheids in juvenile compression wood are very short. As in normal wood, there is a continuous increase in tracheid length from pith to bark. Little is known concerning the variation in the axial direction. Unlike the situation in normal wood, the width of an increment has little effect on average tracheid length in compression wood. On the whole, the location of the tracheids has less influence on their length in compression wood than in normal wood. The decisive factor is instead the intensity of compression wood development, which in its turn depends on the intensity of the stimulus.

4.3.2.3 Branch Wood

Many investigators have examined the tracheids present in branches, sampling either the entire branch or only the upper and lower portions. Most of the lower, abaxial part of a conifer branch consists of compression wood, while some of the upper, adaxial region is composed of opposite wood. The remainder of the branch wood is normal. Neither the total branch wood nor the lower and upper zones constitute homogeneous tissues, and for this reason branch wood is best considered separately from stem wood. Knot wood is a chemically modified branch wood (Chap. 10.6).

In his classical investigation on tracheid length in *Pinus sylvestris*, Sanio (1872) found that branches contained shorter tracheids than the adjacent stem, and that tracheid length first increased toward the branch tip, reached a maximum, and then decreased. Sanio mentions that the branches were considerably more developed (eccentric) toward the lower side and that he removed all his samples from this side. Obviously, it must have been largely compression wood tracheids that he was measuring, and it is hardly surprising that he found them to be shorter than in the stem. The increase in length from the pith toward the outside that he observed in all branches was later confirmed by Fujisaki (1975) and by Atmer and Thörnqvist (1982).

The earlier publications of Hartig (1896) and Cieslar (1896) notwithstanding, Baranetzky (1901) seems to have been unaware of the existence of compression wood or at least of its ubiquitous occurrence in conifer branches. In his lengthy report on the factors that determine the orientation of laterals in trees and bushes, he does not mention compression wood despite the fact that he measured tracheid length on the upper and lower sides of branches. In *Larix decidua, Pinus strobus*, and *P. sylvestris*, the tracheids on the upper side were always longer than those on the lower. The difference tended to decrease from the branch base toward the tip, where the upper and lower tracheids were of the same length. Compression wood, as we now know, reaches its maximum development at the branch base and is least developed at the tip. One *Picea abies* tree followed this pattern, but in another the longer tracheids were found on the lower side of the branch, a phenomenon that remains unexplained. Baranetzky (1901) also made the interesting observation that when the morphologically lower side of a branch was changed so that it became the upper with respect to the gravity vector, longer tracheids were formed. He contended that the tension produced on the upper side by the weight of the branch caused the tracheids on this side to increase in length.

The fact that the tracheids on the lower side of conifer branches are shorter than those on the upper has since been confirmed by many investigators. Because of the high proportion of compression wood in most conifer branches, the net effect is that branch wood has on the average shorter tracheids than stem wood, a situation observed by Bailey and Faull (1934) for *Sequoia sempervirens*, by Fegel (1938, 1941) for several species from northeastern United States, and by Hata (1949, 1950) for *Pinus densiflora*.

In his comprehensive study of *Pseudotsuga menziesii* and its wood, Göhre (1958) examined the tracheids present on the upper and lower sides of branches. Variations in tracheid length with distance from the stem are shown in Fig. 4.33. Tracheid length first increased in both opposite and compression woods toward the tip, reached a maximum after 2 m, and then declined for the remaining 1.5 m. The difference between the two sides was largest at the maximum and least at the branch base, the latter an unexpected result. Sudo (1968b, c) found a similar variation in tracheid length in 1-year-old branches of *Pinus densiflora*. Laterals in this species contain shorter tracheids than the stemwood (Watanabe et al. 1962).

Sanio (1872), Baranetzky (1901), Göhre (1958), and Atmer and Thörnqvist (1982) found the tracheids on the under side of a conifer branch to be shorter than those on the upper side. Unlike these investigators, Schultze-Dewitz et al. (1971), after extensive measurements, arrived at an average tracheid length of 2.31 mm for the upper and 2.34 mm for the lower side of branches in *Pinus syl-*

4.3.2.3 Branch Wood

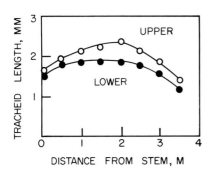

Fig. 4.33. Variation in tracheid length with distance from the stem on the upper and lower sides of branches of *Pseudotsuga menziesii*. (Redrawn from Göhre 1958)

vestris. This surprising result is difficult to explain, but the appearance of cross sections of two branches suggests that no pronounced compression wood was present. Contrary to Sanio's second law, the average length of the tracheids on the upper and lower sides did not increase with increasing height of the branch above the ground, nor did it change consistently within each branch with the distance from the stem.

In later years, many investigators who were interested in branches as a source of pulpwood have compared tracheid length in stem wood and branch wood. For *Pinus palustris*, Gleaton and Saydah (1956) found a tracheid length of 2.76 mm for stem-wood and 1.72 mm for branch wood tracheids. Average fiber length was 2.05 mm in pulps from stem wood of *Picea rubens* but only 1.14 mm in branch wood pulp (Young and Chase 1965) (Table 19.18). It has been observed on several occasions that tracheid length increases with increasing branch diameter, probably because small branches contain proportionally more juvenile wood, which outweighs the fact that older branches tend to have a higher content of compression wood. Eskilsson (1972) has reviewed earlier investigations and reported results obtained with *Picea abies*. Stem wood of this species had an average tracheid length of 3.2 – 3.5 mm, but in branch wood the length of the tracheids was only 1.3 – 1.6 mm. Compression wood constituted more than 50% of the lower half of these branches, which contained fibers that were slightly shorter than those on the upper side. Tracheid length decreased toward the branch tip and was positively correlated with branch diameter.

Hosia et al. (1971) studied the utilization of conifer branches for manufacture of hardboard (Chap. 18.5) and pulp (Chap. 19.9.2). Average tracheid lengths in stem and branch woods were 2.50 and 1.52 mm for *Picea abies* and 2.86 and 1.54 mm for *Pinus sylvestris*. In their studies on complete tree utilization, Keays and Hatton (1971) found a tracheid length of 3.23 mm in the mature bole of *Tsuga heterophylla* compared to only 1.48 mm in branch wood. These and other similar data are summarized in Table 4.6. Without exception, the tracheids in the branches have been found to be considerably shorter than in the bole.

In an investigation on *Pinus densiflora* branch wood, Park et al. (1979) found the tracheids in the stem to be 30 – 40% longer than those in the branch except near the pith. The tracheids within the compression-wood region on the lower side of the branches were consistently shorter than those in the opposite and side wood zones. Their length increased from 1.5 mm in ring 1 to 2.3 mm in ring 14.

In a study of specific gravity and tracheid length in *Pinus taeda* branches, Taylor (1979) found the average length of the tracheids to vary from 2.3 mm to 2.7 mm at different sampling points. The tracheids on the lower side of the branches, which contained much compression wood, were only 0.1–0.2 mm shorter than those on the upper side. Their length increased from pith to bark until a maximum of 2.7 mm had been reached about 20 growth rings from the pith and then decreased. The tracheids in the upper part of the branch began to decrease in length after the first 17–18 increments. Taylor suggested that this reduced tracheid length in the outer portion of older branches was the most probable reason for the fact that branch tracheids are shorter than stem tracheids in the southern pines. It would seem obvious, however, that the major reason for this must be that branch tracheids at any position in the branch are intrinsically shorter than those in the stem, except in the first few growth rings near the pith, and that this applies not only to the compression wood tracheids but also to the opposite and side wood tracheids.

Wegelius (1939) has reviewed the anatomy of knot wood tracheids. Since knots are branches encased in the stem, most properties of branches are also characteristic of knots. Tracheid length distributions for *Picea abies* knot wood reported by Wegelius (1939) indicate an average tracheid length of only 1.0–1.1 mm.

Variations in tracheid length with position in the tree occur in normal and compression woods within both stems and branches. Correlations between stem compression wood and branch wood are therefore always uncertain, especially as branch wood consists of three different tissues, namely normal (side) wood, opposite wood, and compression wood. A comparison between tracheid lengths in branch wood with corresponding data for stem compression wood nevertheless reveals differences so large that they must be real. Inspection of published data indicates that in *Picea*, for example, the tracheids in stem compression wood have an average length of 1.7–1.8 mm, whereas in branch wood the tracheids often reach only 1.0–1.1 mm in length. Tracheids in normal branch wood are usually shorter than in normal stem wood (Table 4.6), and tracheid length of branch compression wood is evidently less than that of stem compression wood. Information on this subject is, however, meager, and more measurements are needed.

Schultze-Dewitz and Götze (1973) determined the tracheid length of stem wood surrounding the knots in the nodal regions of *Picea abies*, *Pinus sylvestris*, and *Pseudotsuga menziesii*. Some of their numerous data are found in Fig. 4.34. As could be expected, average tracheid length of knot wood was lesser than that of stem wood. The tracheids in the wood closest to the knot were considerably shorter than those located further away, and this applied to both living and dead knots. Tracheids below an encased knot were shorter than those above. The investigators attributed this to pressure exerted on the stem wood by the knot above. Whether pressure can induce the cambium to produce shorter tracheids is, however, doubtful. One hundred years earlier Sanio (1873) had

4.3.2.3 Branch Wood

Table 4.6. Tracheid length of stem wood and branch wood of various conifer species

Species	Tracheid length, mm		Investigators
	Stem wood	Branch wood	
Picea abies	2.50	1.52	Hosia et al. (1971)
Picea abies	3.51	1.38 – 1.62	Eskilsson (1972)
Picea rubens	2.1	1.1	Young and Chase (1965)
Pinus palustris	2.8	1.7	Gleaton and Saydah (1956)
Pinus sylvestris	2.86	1.54	Hosia et al. (1971)
Tsuga heterophylla	3.0	1.7	Worster and Vinje (1968)
Tsuga heterophylla	3.23	1.48	Keays and Hatton (1971)

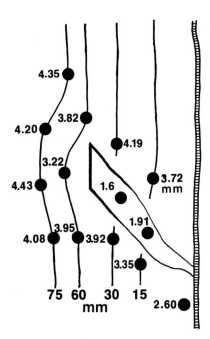

Fig. 4.34. Length of the tracheids surrounding a knot in a *Pinus sylvestris* stem. (Redrawn from Schultze-Dewitz and Götze 1973)

suggested that stem wood on the under side of knots should be considered as a continuation of the compression wood in the knot. Later investigations, discussed in Chap. 10.3.8.2, have shown that branches generally are associated with compression wood at their point of insertion in a stem. Very probably, this is the reason for the shorter tracheids in all wood surrounding a knot, especially since the reduction in length is observed not only below but also above a knot (Fig. 4.34). After a knot has become encased, the tracheids in the covering wood gradually assume the length found in normal stem wood.

4.3.2.4 Factors Determining Tracheid Length in Normal and Compression Woods

What determines the length of tracheids and fibers in trees has long been a moot question. A good review is found in the book by Philipson et al. (1971). The subject has been briefly summarized by Dinwoodie (1961) and by Bannan (1957c, 1967b).

Radial growth of trees is effected by cell divisions in the vascular cambium. Periclinal divisions add to the xylem and phloem in the radial direction. The cambium itself grows by anticlinal divisions in the tangential direction, a growth that makes it possible for the ring of cambial initials to increase its circumference as the radial expansion continues. Practically all of the anticlinal divisions of the fusiform initials are pseudotransverse. Bailey (1920, 1923) was the first to show that the length of the tracheids in conifers is closely related to the length of the fusiform initials formed by anticlinal division, subsequent growth of the initials being only 5–20%. Priestly (1930a, b, c), in a long and thoughtful treatise, developed the idea that very few anticlinal divisions should be required to maintain a coherent cambial ring, that the rate of anticlinal division in the cambium was low, and that what he called "symplastic growth" occurred, involving a slow, mutual adjustment of cell positions. This theory was widely accepted at the time and replaced earlier concepts of gliding growth.

It was not until 20 years later that Bannan and Whalley (1950) and Whalley (1950) were able to show that Priestly's concept of growth was based on a fundamental misunderstanding of the process of anticlinal division. What Priestly had failed to notice was that the apparent low rate of such divisions is a consequence of the fact that not only formation of new fusiform initials but also the disappearance of such initials is very rapid. Actually, the rate of loss is so high that the net increase in number of new fusiform initials is quite modest despite the high frequency of anticlinal divisions. In a typical case involving *Chamaecyparis thyoides,* Bannan (1950a) found that 1100 anticlinal divisions resulted in only 162 new, functional fusiform initials. In a long series of publications, which together constitute a unique contribution, Bannan and his co-workers elucidated these problems (Bannan and Whalley 1950, Bannan 1950a, 1951a, b, 1953, 1954, 1955, 1956, 1957a, b, c, 1960a, b, 1962, 1963a, b, 1964a, b, 1965a, b, c, 1966, 1967a, b, 1968a, b, c, Bannan and Bayly 1956, Bannan and Bindra 1970). Similar observations have also been reported by others, such as Hejnowicz (1961, 1963a, b, 1964, 1967), Srivastava (1963a) and Hejnowicz and Branski (1966).

The frequency of anticlinal divisions is higher in young trees than in old trees (Bannan 1950a), and it is accelerated by pressure (Bannan 1957b). The longest among the fusiform initials have a greater chance of survival (Bannan 1951b, 1957a, Bannan and Bayly 1956), whereas the shortest are usually lost. One of the reasons for this situation might be that longer cells have a larger area of contact with the rays and are thus better supplied with photosynthate and water (Bannan 1965b, Bannan and Bayly 1956). Bannan has also shown that fusiform initials at first elongate very rapidly and that this is effected by tip growth, each initial exhibiting intrusive growth (Sinnott and Bloch 1939), in-

4.3.2.4 Factors Determining Tracheid Length in Normal and Compression Woods

dependently of the others (Bannan 1956). As has already been mentioned, tracheid length frequently reaches a maximum in growth rings that are 1–2 mm wide, a fact demonstrated by Bannan (1962, 1963a, b, 1964a, b) for many conifer species. In rings wider than this size, the tracheids tend to be shorter the wider the annual increment, a consequence of an increased frequency of anticlinal divisions which does not allow enough time for elongation of the newly formed fusiform initials (Chalk and Ortiz 1961).

Tracheids are longer in latewood than in earlywood because elongation of the new initials is greater at the end of the growing season. In the first few growth rings closest to the pith, anticlinal divisions occur at a high rate, and tracheids are short. As the rate declines, the tracheids tend to become longer with increasing distance from the pith. Similarly, as also shown by Bannan (1966a, b), the frequency of anticlinal divisions decreases at first with increasing stem height, and cell length shows a corresponding increase. At this point it must be emphasized that factors other than those now described also determine the length of tracheids. Variation in cell length between different species is not uncommon. *Sequoia sempervirens*, for example, has unusually long tracheids, longer than the related *Sequiadendron giganteum* (Bannan 1966a, b). Ecotypic variations have been demonstrated (Dinwoodie 1961, 1963). It is also possible to control tracheid length genetically (Chap. 14.3.2.4).

Several investigators have attempted to explain why the tracheids in compression wood are consistently shorter than in normal wood. Hartig (1901) had noticed that in eccentrically grown stems, the tracheids were longer in the narrow than in the wide growth rings. His reasoning was that the faster the stem increased in size, the faster the cambial initials must divide and the shorter was the time available for them to stretch before a new division took place. Formation of compression wood is, according to Hartig, always associated with increased radial growth, and the tracheids in compression wood are therefore always short. Onaka (1949) objected to this hypothesis, pointing out that, following division, the initials increase in length and that accordingly the final tracheid length could not have much to do with the size of the mother cell. This criticism does not appear to be well founded, however, since it is known that the length of the mature tracheids is closely related to the length of the fusiform initials and does not exceed that of the latter by more than 5–20% (Bailey 1920, 1923). Obviously, however, Hartig's approach represents an oversimplification, since he considered only the periclinal divisions.

Priestly and Tong (1925–1929) discussed some physiological aspects of reaction wood formation. Pointing out that lignification and wall thickening take place more rapidly on the lower, compression wood side of a branch, they felt that the short tracheids on this side were a result of these facts. They also advanced the rather teleologic suggestion that the tracheids are shorter in compression wood in order better to resist compression. Mork (1928a) repeated Hartig's explanation of the short tracheids in compression wood but also considered the role of the anticlinal divisions and the subsequent gliding growth of the initials. Jaccard and Frey (1928), in a discussion of the microfibril angle in normal and reaction woods, argued that in compression wood periclinal divisions are more numerous than in normal wood, and that the initials as a conse-

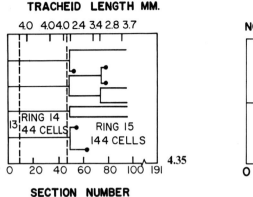
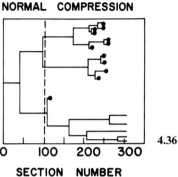

Fig. 4.35. Sequence of anticlinal division in four radial files of tracheids of *Pinus pinaster*, followed through three growth rings. A *black dot* represent elimination of a file of tracheids. (Redrawn from Wardrop and Dadswell 1952)

Fig. 4.36. Sequence of anticlinal divisions in a single radial file of tracheids extending through a zone of developing compression wood in *Pinus pinaster*. (Redrawn from Wardrop and Dadswell 1952)

quence have less time to elongate and therefore develop into short tracheids. This, of course, is the same explanation as that advanced earlier by Hartig (1901).

In the first of two important papers, Wardrop and Dadswell (1950) briefly discussed the problem of tracheid length in compression wood. Drawing attention to the earlier contributions of Priestly (1930a, b, c) they concluded that rapid, anticlinal divisions in conjunction with an increased number of transverse divisions were responsible for both the eccentric growth and the shorter tracheids characteristic of compression wood. The relation between rate of growth and frequency of anticlinal division was investigated by Wardrop and Dadswell (1952) in carefully planned experiments. One wood specimen was taken from a concentric *Pinus pinaster* stem containing 31 growth rings. The tree had been suppressed, and, when fertilized with superphosphate, had begun to grow rapidly, producing wide annual increments, beginning with the 15th ring. No compression wood was produced. In examining the sequence of anticlinal divisions, use was made by a technique originally developed by Klinken (1914) and later used by Whalley (1950) and by Bannan. This involved preparing thin, tangential sections throughout the entire width of three growth rings, after which each section was examined microscopically. The results are shown graphically in Fig. 4.35. No anticlinal divisions had occurred in the narrow ring (No. 14). In the subsequent wide growth ring (No. 15) there were six anticlinal divisions with five daughter cells surviving in the first 50 sections. Tracheid lengths, as can also be seen from Fig. 4.35, were notably lower in the wide ring (No. 15) than in the narrow one (No. 14). Obviously, any increase in radial growth involves a corresponding increase in the number of anticlinal divisions and a concomitant reduction in tracheid length.

4.3.2.4 Factors Determining Tracheid Length in Normal and Compression Woods

In another experiment, Wardrop and Dadswell (1952) selected a specimen of the same species which contained both normal wood and pronounced compression wood, following one radial file through 30 sections. Figure 4.25 shows how the tracheid length decreased on going from normal to compression wood zones. The sequence of anticlinal divisions are shown in Fig. 4.36. Before formation of compression wood had been initiated, only two anticlinal divisions took place in the 100 sections involved. After compression wood had begun to form, three occurred in the following 100 sections and 11 in the subsequent 306. Only five of the 17 daughter cells survived to reach maturity. The investigators concluded that the reduction in tracheid length in compression wood was a consequence of the larger number of anticlinal divisions in this tissue. The fact that the same sequence of events took place without formation of compression wood proves that an increase in the number of anticlinal divisions in the cambial zone and a concomitant reduction in length of the mature tracheid will occur whenever there is an increase in the rate of radial growth in a tree. This increase will result in shorter tracheids, irrespective of whether the tree exhibits eccentric growth and regardless of whether normal wood or compression wood is being produced.

In their 1950 communication, Wardrop and Dadswell (1950) expressed the opinion that the structure of compression wood tracheids should be regarded as a modification of that of the tracheids in normal wood, and that this structure is determined by the reduced tracheid length. Their later (Wardrop and Dadswell 1952) discovery that the same reduction in cell length can occur without concomitant formation of compression wood, rendered this view untenable. They now expressed the view instead that development of compression wood should be regarded as functional in character (Wardrop and Dadswell 1952, Wardrop 1954). Obviously, the change in growth rate cannot explain all of the anatomical properties of the tracheids in compression wood, not to mention their chemical composition.

Mariani (1955), who found the tracheids of *Larix decidua* compression wood to be longer than in corresponding normal wood, concluded that the factors that determine the rate of radial growth have no influence on the dimensions of the tracheids which are predetermined for each species and result from the activity of the cambium. Needless to say, this represents a view exactly opposite to that expressed by Wardrop and Dadswell (1950). It can, at best, be only partly correct, as both radial growth and tracheid length are determined by the activity of the cambium. In a discussion of the origin of growth stresses in trees, Bamber (1980) has suggested that the shorter tracheids of compression wood can be attributed to a lower turgor pressure at the time of cell development.

Dinwoodie (1961) has pointed out that it is not yet clear which factors are responsible for the shortness of the tracheids in compression wood. From the combined evidence now available it seems inescapable, however, that much of the reduction is a consequence of the increased number of anticlinal divisions. Whether this is the only factor involved remains to be proven. It would obviously be of considerable interest to establish whether or not normal and compression wood tracheids are of the same length when these tissues are formed at identical radial growth rates. The net gain of daughter initials might be the same in the two woods under these conditions, but it is quite possible that rates of appearance and disappearance of new initials could be higher in compression wood, resulting in shorter tracheids in this tissue. Lack of information of the type provided by Wardrop and Bannan makes further speculation futile at the present time.

The only other structural property of the tracheids in compression wood which is, at least indirectly, determined by the rate of growth is the microfibril angle. In the S_1 and S_2 layers this angle is negatively correlated with the length of the tracheid. This relationship is dealt with in Sect. 4.4.5.2.

4.3.3 Tracheid Diameter

Data reported by Johansson (1940), Smith and Miller (1964), Smith (1965), and Fengel (1969) indicate that in the earlywood of five conifer species the tangential diameter of the tracheids varied from 25 to 40 µm and the radial diameter from 30 to 57 µm. Corresponding values for latewood were 24–40 µm and 13–26 µm. The smaller cross sectional area of the latewood tracheids is almost entirely a consequence of the reduced radial diameter of these cells which tend to become flattened in the radial direction at the end of the growing season. Fergus et al. (1969) measured the changes in tracheid diameter across an entire growth ring of normal wood from *Picea mariana*. The radial diameter was 34 µm in the earlywood, decreased gradually in the transition zone, and finally reached a value of only 10 µm in the latewood. Similar measurements were made by Wood and Goring (1971) with wood from *Pseudotsuga menziesii*. Denne (1973) found that the radial diameter of the tracheids in *Picea abies* was positively correlated with shoot growth, ring width, and rate of xylem increment. This is in contrast to an earlier study by Skene (1972), who observed no such correlation in *Tsuga canadensis*. Cown (1975) found that the tangential tracheid diameter increased from pith to bark at a constant stem level in *Pinus radiata*.

Compared to the length, the other dimensions of the tracheids in compression wood have received little attention. Cieslar (1896) found that compression wood tracheids in *Picea abies* had a radial diameter of 20 µm and a tangenital diameter of 18 µm. Corresponding figures for opposite wood were 22 µm and 18 µm. Great variations were observed in compression wood, namely from 12 µm to 24 µm in both directions. Verrall (1928) measured the radial diameter of the tracheids in compression wood of five species. His results, summarized in Table 4.7 indicate that on the average, normal wood tracheids had a somewhat larger diameter. One exceptionally huge tracheid in *Picea mariana* compression wood measured 59×164 µm in cross section! Among the most extensive data on tracheid diameter in normal and compression woods as yet reported are those of Kienholz (1930), who examined *Tsuga mertensiana* trees growing on a steep hillside some of which had become "pistol-butted". Measurements were made with specimens located downhill, uphill, and sidehill at low levels of the stem, approximately corresponding to zones of compression, opposite and side

4.3.3 Tracheid Diameter

Table 4.7. Radial diameter of tracheids in normal and compression wood of five conifer species. All values in µm. (Verrall 1928)

Species	Normal wood	Compression wood
Larix laricina	29.7	24.7
Picea mariana	28.9	28.6
Pinus palustris	34.9	25.7
Thuja occidentalis	29.6	26.2
Thuja plicata	25.7	24.3

Table 4.8. Radial and tangential diameters of tracheids in compression, opposite, and side woods of a pistol-butted *Tsuga mertensiana* tree at a 50-cm level. All values in µm. (Kienholz 1930)

Ring number	Radial diameter						Tangential diameter		
	Earlywood			Latewood			Latewood		
	Compression wood	Opposite wood	Side wood	Compression wood	Opposite wood	Side wood	Compression wood	Opposite wood	Side wood
1–10	21.3	19.1	22.6	14.7	14.7	15.2	19.3	18.7	20.5
11–20	24.3	21.5	28.4	15.9	14.5	17.2	20.9	19.5	24.2
21–30	26.2	25.7	30.9	16.5	15.4	16.7	22.0	21.2	24.2
31–40	22.6	26.9	30.5	15.8	15.5	17.2	23.8	22.6	26.2
41–50	26.3	28.0	34.3	17.2	16.1	17.7	23.1	23.0	27.2
51–60	24.0	28.6	35.1	15.6	16.5	18.6	23.7	23.7	28.9
61–70	27.0	32.6	34.5	15.8	17.9	19.1	25.9	25.9	27.9
71–80	32.5	34.7		18.2	19.0		25.7	26.8	
Average	25.5	27.1	30.9	16.2	16.0	17.4	23.1	22.7	25.6

woods. Straight trees were used as checks. Results obtained at a 50-cm level are shown in Table 4.8. As had been found also by Verrall (1928), the tracheids in normal wood had a slightly larger diameter than those in compression wood. Opposite and compression woods were very similar in this respect.

Onaka (1949) has reported many data, listed in Table 4.9, on the tracheid diameter in the radial and tangential directions of normal and compression woods from 10 conifer species. In comparison with normal wood, the compression wood tracheids had a slightly smaller diameter in the radial direction in the earlywood zone, but they were larger in the latewood, a well-known fact. In the majority of the species, the tangential diameter was somewhat smaller in compression wood, and this applied to both earlywood and latewood, but the opposite was also observed, and it is doubtful if the differences noted were statistically significant. The average radial diameter in earlywood was least in

Table 4.9. Diameter and wall thickness of tracheids and pit diameter of normal and compression woods in earlywood and latewood of various gymnosperm species. All values in μm. (Onaka 1949)

Species and zone	Tracheid diameter				Tracheid wall thickness		Pit diameter	
	Radial		Tangential					
	Normal wood	Compression wood	Normal wood	Compression wood	Normal wood	Compression wood	Normal wood	Compression wood
Abies firma								
Earlywood	32	30	24	26	2.2	5.0	14	8
Latewood	12	15	24	30	6.0	5.5		
Abies mayriana								
Earlywood	32	30	24	26	2.2	4.0	15	12
Latewood	12	15	20	22	5.5	4.0		
Chamaecyparis obtusa								
Earlywood	32	22	26	22	2.5	5.0	13	5
Latewood	10	10	22	20	3.0	3.0		
Cryptomeria japonica								
Earlywood	30	24	20	19	1.2	4.2	12	4
Latewood	9	11	20	18	3.6	3.5		
Ginkgo biloba								
Earlywood	34	30	30	28	3.0	4.0	10	8
Latewood	13	15	30	24	3.5	3.0		
Larix gmelinii								
Earlywood	36	30	24	26	1.8	4.0	16	12
Latewood	15	17	24	22	6.0	5.5		
Picea jezoensis								
Earlywood	30	22	22	20	1.2	3.4	13	8
Latewood	12	16	20	18	4.5	4.5		
Pinus densiflora								
Earlywood	36	32	30	26	2.6	4.5	17	8
Latewood	18	20	30	26	6.0	5.5		
Pinus parviflora								
Earlywood	35	30	26	30	2.0	4.8	16	10
Latewood	18	18	30	28	5.5	4.8		
Podocarpus macrophyllus								
Earlywood	30	20	20	18	1.8	3.0	12	8
Latewood	12	15	20	18	3.6	3.4		

4.3.3 Tracheid Diameter

Table 4.10. Changes in tracheid diameter and wall thickness associated with formation of compression wood. All values in μm. (Onaka 1949)

Species	Zone	Tracheid diameter		Wall thickness	
		Earlywood	Latewood	Earlywood	Latewood
Cryptomeria japonica	Before lean	30.0	9.0	1.5	3.9
	After lean				
	Upper side, 1 year	15.6	6.0	1.2	2.7
	Upper side, 4 years	28.8	9.6	1.2	3.6
	Lower side, 1 year	21.6	9.6	4.2	2.7
	Lower side, 4 years	24.0	10.8	4.2	3.0
Pinus densiflora	Before lean	36.0	21.0	2.7	6.0
	After lean				
	Upper side, 1 year	30.0	15.6	1.5	3.3
	Upper side, 12 years	42.0	18.0	2.5	6.0
	Lower side, 1 year	38.5	19.2	5.4	6.0
	Lower side, 12 years	32.4	19.7	5.0	5.4

compression and largest in opposite wood, which is not surprising when it is recalled (Sect. 4.2) that true earlywood hardly exists in compression wood (Sect. 4.2 and Chap. 6.2.3). Both radial and tangential diameters in the latewood were about the same in normal and compression woods and slightly larger in opposite wood. The diameter increased notably from pith to bark, but especially so in the earlywood. It also increased with increasing height above the ground, but here the data are less meaningful, as the proportion of compression wood decreased rapidly in the same direction.

Onaka (1949) has also reported some interesting data on the radial tracheid diameter in a *Cryptomeria japonica* and a *Pinus densiflora* tree, before and after they had become inclined. One year after the displacement, as can be seen in Table 4.10, both opposite wood on the upper and compression wood on the lower side contained tracheids with reduced diameter. After 4 years, however, the tracheids in the opposite wood had regained or exceeded their original size, whereas those in the compression wood either were smaller or remained unchanged despite in the increase that must have occurred with increasing age. According to Berkley (1934), compression wood tracheids in southern pines often have greater radial diameters than adjacent normal latewood cells. In *Abies alba* Petrić (1962) found the average tracheid diameter in compression wood to be 26.6 μm compared to 33.0 μm in normal wood. Nečesaný and Oberländerová (1967) determined the radial diameter of the tracheids in compression and normal woods from seven conifer species, obtaining the ratios listed in Table 4.11. Except for *Abies alba* the tracheids were 5–20% wider in the compression wood.

Dinwoodie (1965) measured the fiber diameter in pulps prepared from normal and compression woods of four conifer species. The results, shown in Table

Table 4.11. Ratios between radial tracheid diameter, lumen diameter, and wall thickness in compression and normal woods of seven conifer species. (Nečesaný and Oberländerova 1967)

Species	Tracheid diameter	Lumen diameter	Wall thickness
Abies alba	0.96	0.73	1.87
Juniperus communis	1.21	0.71	2.06
Larix decidua	1.13	0.81	3.04
Picea abies	1.09	0.76	2.52
Pinus montana	1.05	0.71	2.40
Pinus nigra	1.07	0.73	2.38
Pinus sylvestris	1.11	0.68	2.29
Average	1.09	0.73	2.39

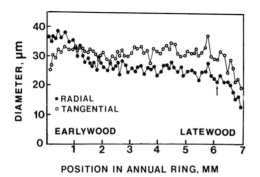

Fig. 4.37. Changes in radial and tangential diameter of compression wood tracheids across the seventh growth ring in a *Pinus thunbergii* stem. (Redrawn from Yoshizawa et al. 1981)

19.15, do not indicate any definite trend. In a similar study with *Picea sitchensis*, however, pulp fibers from compression wood had a consistently smaller diameter than in corresponding normal wood pulps (Young et al. 1970) (Table 19.7). Kibblewhite (1973) has reported values for the tracheid diameter in normal and compression woods of *Pinus radiata*. Normal earlywood tracheids had an average tangential diameter of 35.1 µm and a radial diameter of 46.9 µm. Corresponding values for normal latewood were 30.7 µm and 32.3 µm, respectively. Compression wood tracheids in the earlywood region had a tangential diameter of 33.0 µm and a radial diameter of 44.5 µm. Those in the latewood were smaller, with tangenital and radial diameters of 28.0 µm and 27.7 µm.

Yoshizawa et al. (1981) studied the variations in tracheid diameter and wall thickness in *Pinus thunbergii* compression wood. The changes in diameter across an early growth ring is shown in Fig. 4.37. The diameter in the tangential direction was relatively constant except in the first-formed earlywood, where it increased rapidly and in the last-formed latewood, where it decreased more gradually. The radial diameter, on the other hand, decreased continuously from earlywood to latewood. Except for the first-formed earlywood, the tangential diameter was consistently larger than the radial one. These results are all in

4.3.3 Tracheid Diameter

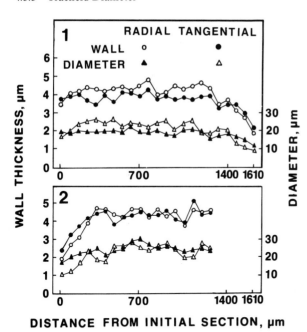

Fig. 4.38. Changes in the radial and tangential diameter and wall thickness along the length of two compression wood tracheids (**1** and **2**) in a *Pinus thunbergii* stem. (Redrawn from Yoshizawa et al. 1981)

agreement with those obtained by Kibblewhite (1973). Yoshizawa et al. (1981) found that the transverse dimensions of the compression wood tracheids also varied along the length of each tracheid. Two examples of such variations are shown in Fig. 4.38. In one tracheid type (1) the tangential diameter was larger than the radial one over most of the tracheid length, while in another type (2) there was no consistent trend. Many of the cells had an elliptic cross section with the longer axis in the tangential direction.

There are only few reports on the diameter of the tracheids in branch wood. Hägglund and Larsson (1937) found that tracheids in knot wood had a smaller diameter than normal tracheids. According to Fegel (1941), tracheid diameter is less in branch wood than in stem wood. Göhre (1958) found the diameter to vary in an inconsistent pattern in tracheids from the upper and lower sides of branches in *Pseudotsuga menziesii* and the same was later observed by Schultze-Dewitz et al. (1971) with branch wood of *Pinus sylvestris*. In a more extensive investigation, Park et al. (1979) measured the radial and tangential diameters of the tracheids in the earlywood and latewood regions of compression, opposite, and side woods in *Pinus densiflora* branches. The results are summarized in Table 4.12. In comparison with normal stem wood, the diameters were much smaller in the branch wood except in the first few rings of juvenile wood near the pith. There was no significant difference in tracheid diameter between compression, opposite, and side woods in the branches. The compression wood tracheids had a diameter that was slightly larger in the tangential than in the radial direction. They were consistently wider in the earlywood than in the latewood, especially radially. Tracheid diameter increased in all woods from pith to bark, and most rapidly so near the pith.

Table 4.12. Tracheid diameter and wall thickness in branch and stem woods of *Pinus densiflora* (Park et al. 1979)

Morphological region	Number of growth rings from the pith	Tangential diameter, μm		Radial diameter, μm		Wall thickness, μm	
		Early-wood	Late-wood	Early-wood	Late-wood	Early-wood	Late-wood
Branch							
Compression wood	1	15	16	16	8	2.0	1.9
	3, 5	22	20	23	17	3.2	3.7
	7, 10	22	21	25	17	2.9	3.9
	12, 14	24	22	26	19	2.9	4.2
Opposite wood	1	15	15	15	9	1.8	1.8
	3, 5	21	19	26	17	2.3	3.6
	7, 10	23	21	27	17	2.4	4.4
	12, 14	23	25	28	15	2.4	3.7
Side wood	3, 5	21	20	25	19	2.6	4.0
	7, 10	23	24	26	18	2.3	3.9
	12, 14	24	24	30	17	2.8	4.0
Stem							
Normal wood	1	16	20	18	14	2.0	2.3
	3	23	23	29	21	2.7	4.6
	7	28	30	37	23	3.0	5.1
	14	30	30	39	25	2.8	6.6

Although available data are few, it would appear that tracheid diameter in compression wood is slightly less than in normal wood. This conclusion applies, however, only when comparison is made with earlywood. Compared with normal latewood, the tracheids in compression wood are either equally wide or slightly wider. Actually, any comparison between normal and compression woods in this respect is of doubtful value for two reasons. First, tracheid diameter is very different in normal earlywood and latewood, and the earlywood-latewood concept is of limited usefulness when applied to compression wood. Second, compression wood tracheids have a circular or oval outline, whereas that of the tracheids in normal wood is square or rectangular. Direct comparisons under these conditions must at best be difficult and occasionally impossible.

4.3.4 Tracheid Wall Thickness

What was said above concerning tracheid diameter also applies to the thickness of the tracheid wall in compression wood. In comparing compression wood with normal earlywood and latewood, we are dealing with a tissue that does not possess any earlywood or latewood in the accepted sense of these terms. Many measurements of tracheid wall thickness have been reported. Fergus et al.

4.3.4 Tracheid Wall Thickness

(1969) determined the thickness of the tracheid wall across a growth ring in normal wood of *Picea mariana*. Tangential wall thickness increased continuously from $2-2.5\,\mu m$ in the earlywood to $6-7\,\mu m$ in the latewood. The thickness of the radial wall was $2\,\mu m$ in the first-formed earlywood, increased slowly to $3.5\,\mu m$ at the middle of the ring, and then increased more rapidly to reach a value of $8.5\,\mu m$ in the latewood. In *Pseudotsuga menziesii*, a species where the transition from earlywood to latewood is abrupt, tracheid wall thickness was $2.5\,\mu m$ in earlywood and then increased very rapidly to $8\,\mu m$ in the tangential and $9.5\,\mu m$ in the radial direction. Petrić (1974) found the thickness of the wall in earlywood tracheids of *Pinus sylvestris* to vary from 2 to $4\,\mu m$ with an average of $2.94\,\mu m$. The thickness of the latewood tracheid wall was in the range of $3-13\,\mu m$ with an average of $7.78\,\mu m$. It first increased with increasing distance from the pith and then gradually decreased.

Fengel and Stoll (1973) concluded that variations in thickness of the cell wall within a growth ring are largely a result of variations in the size of the layer S. Okumora et al. (1974) found that the S_2 layer in the tangential wall of latewood tracheids in *Pinus densiflora* was relatively uniform in the longitudinal direction of the cell. In the radial wall, on the other hand, S_2 became conspicuously thinner toward the tracheid tip. The S_2 layer was much thicker in the radial than in the tangential wall at mid section, whereas the opposite was true near the tip. In earlywood tracheids, S_2 remained constant in thickness in the axial direction and was always thicker in the tangential than in the radial wall.

Hartig (1896, 1901) did not report any values for the tracheid wall thickness in compression wood, nor did he discuss the matter. Cieslar (1896), however, mentions that the cell wall in *Picea abies* compression wood is $4-8\,\mu m$ thick compared to $2\,\mu m$ for earlywood and $4\,\mu m$ for latewood in opposite wood. The diameter of the lumen in the compression wood tracheids varied within the wide range of $4-16\,\mu m$. For opposite wood, the same range was $16-24\,\mu m$, figures that could hardly have included any latewood. White (1907), who studied the formation of compression wood in several conifer species, found compression wood tracheids to have thinner walls and larger lumina than corresponding normal tracheids.

Onaka's (1949) comprehensive data, summarized in Table 4.9, show clearly that it is only in comparison with normal earlywood that the tracheids in compression wood have thick cell walls. Compared to normal latewood, their walls are either of equal thickness or, more commonly, somewhat thinner. Onaka concluded from his data for compression wood that the thickness of the tracheid wall was essentially the same in earlywood and latewood. In his experiments with *Cryptomeria japonica* and *Pinus densiflora*, referred to above, Onaka (1949) measured the changes in cell wall thickness caused by inclination, obtaining the results listed in Table 4.10. Wall thickness changed less than cell diameter after displacement of the trees. In earlywood, the walls of the compression wood tracheids were $2-3.5$ times thicker than in opposite wood. In latewood, by contrast, the tracheids in the opposite wood had thicker walls than those in compression wood. An interesting feature was that during the first year after inclination, cell wall thickness in opposite wood was considerably reduced, probably due to gravimorphism (Chap. 12.4). In the course of the following

Table 4.13. Tangential and radial wall thickness of tracheids in earlywood and latewood zones of normal and compression woods of three conifer species. All values in μm. (Ollinmaa 1959)

Species	Normal wood							Compression wood				
	Earlywood			Latewood				Earlywood		Latewood		
	Tangential	Radial	Average	Tangential	Radial	Average		Average		Tangential	Radial	Average
Juniperus communis	4.01	4.11	4.07	4.00	5.04	4.53		6.17		4.57	6.23	5.41
Picea abies	2.90	3.52	3.25	4.69	6.23	5.45		6.85				
Pinus sylvestris	3.36	3.44	3.40	5.06	6.21	5.68		6.69		5.79	6.17	6.00

4.3.4 Tracheid Wall Thickness

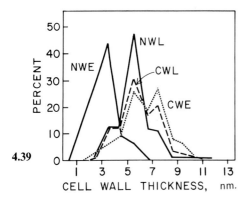
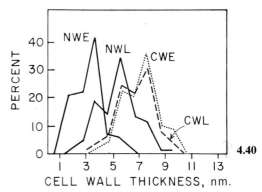

Fig. 4.39. Frequency distribution of tracheid wall thickness in normal earlywood (*NWE*), normal latewood (*NWL*), earlywood of compression wood (*CWE*), and latewood of compression wood (*CWL*) in *Picea abies*. (Redrawn from Ollinmaa 1959)

Fig. 4.40. Frequency distribution of tracheid wall thickness in *Pinus sylvestris*. For designations, see Fig. 4.39. (Redrawn from Ollinmaa 1959)

years, however, wall thickness increased to reach or exceed that existing before inclination. In compression wood, by contrast, wall thickness remained unchanged in one case, increased slightly in another, and decreased in another two. Onaka pointed out that it is actually only in the earlywood zone of compression wood that the cell wall is thicker than in normal tracheids. In latewood there is often no difference or, occasionally, the tracheid wall is thinner in compression wood, as can be seen from the data in Table 4.9.

The most extensive comparison of tracheid wall thickness in normal and compression woods so far is that reported by Ollinmaa (1955, 1959). His results with *Juniperus communis*, *Picea abies*, and *Pinus sylvestris* are shown in Table 4.13. Distributions of cell wall thickness for the last two species are seen in Figs. 4.39 and 4.40. Radial walls were always thicker than tangential walls. In compression wood of *Picea abies*, earlywood and latewood tracheids had approximately the same wall thickness. In the other two species the wall was thicker in the earlywood than in the latewood regions, demonstrating how misleading the earlywood-latewood concept can be when applied to compression wood. Compared to the tracheids in the latewood region of normal wood, those in the same region in compression wood had noticeably thicker walls. Most other investigators have found that in the latewood zone compression wood tracheids have a slightly thinner wall than corresponding normal tracheids. Ollinmaa's data must, nevertheless, be attributed considerable significance, for each numerical value was the result of as many as 100–1400 individual measurements.

Petrić (1962) found that earlywood tracheids in compression wood of *Abies alba* had an average wall thickness of 2.94 µm compared to only 1.86 µm for normal wood. Corresponding figures for latewood were 6.62 µm and 7.23 µm, respectively, indicating a slightly thicker wall in normal wood. The data reported by Nečesaný and Oberländerová (1967) show that the tracheid wall in the seven species examined was on the average 2.4 times thicker in compression wood than in normal wood. In three of the four species subjected to kraft pulping by Dinwoodie (1965) the resulting pulp fibers were thicker when derived from compression wood than from normal wood (Table 19.15). Compared to normal wood, all holocellulose and pulp fibers obtained by Young et al. (1970) from *Picea sitchensis* compression wood had considerably thicker cell walls (Table 19.7).

Göhre (1958) measured tracheid wall thickness on the upper and lower sides of branches from *Pseudotsuga menziesii*. No difference was noted between these branches at different heights above the ground. The tracheids on the lower, compression wood side were consistently 10–20% thicker than those in the opposite wood above. In both zones, the cell wall thickness decreased continuously toward the branch tip. Wegelius (1939) states that tracheids in knotwood of *Picea abies* have thick walls. Schultze-Dewitz et al. (1971) found that the tracheids on the lower side of the branches in *Pinus sylvestris* had a thicker wall than those on the upper, and that this applied to both earlywood and latewood. In *Pinus radiata*, Kibblewhite (1973) found compression wood tracheids in the earlywood region to have a wall thickness of 3.6 µm and in the latewood 4.2 µm. Corresponding values for normal earlywood and latewood tracheids were 2.9 µm and 4.0 µm, indicating a somewhat thicker wall in the compression-wood tracheids.

The variation in wall thickness of compression-wood tracheids in *Pinus thunbergii* reported by Yoshizawa et al. (1981) is shown in Fig. 4.41. Following an initial, brief decrease, both radial and tangential walls increased in thickness, remained constant over the central portion of the growth ring, and finally became thinner at the end of the season. Within each tracheid, wall thickness was relatively constant, with the walls in the radial direction exceeding those in the tangential in thickness. The peripheral variations in wall thickness in a disc, from compression wood to opposite wood, is illustrated in Fig. 4.42. Wall thick-

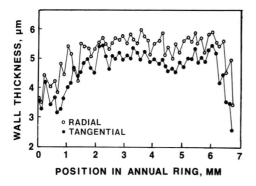

Fig. 4.41. Changes in radial and tangential wall thickness of compression-wood tracheids across the seventh growth ring in a *Pinus thunbergii* stem. (Redrawn from Yoshizawa et al. 1981)

4.3.4 Tracheid Wall Thickness

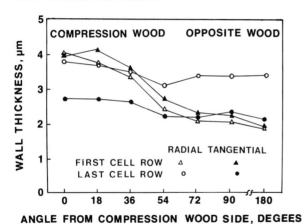

Fig. 4.42. Variation in radial and tangential wall thickness of the tracheids in the first and last rows in the outermost growth ring along the periphery of a *Pinus thunbergii* stem. (Redrawn from Yoshizawa et al. 1981)

ness decreased considerably in this direction in the first-formed earlywood but less so within the last-formed latewood. In the former zone the radial walls were thicker than the tangential ones but in the latter the difference was slight.

Compression wood tracheids within the latewood zone in *Pinus densiflora* branches were found by Park et al. (1979) to have a thicker wall than those in opposite or side woods, but the wall was considerably thinner than that of normal latewood in the stem (Table 4.12). It was thinner in the earlywood than in the latewood portion of the compression wood. Wall thickness increased throughout from pith to bark and most rapidly in the first few increments.

Evidence for the size of the lumen in compression-wood tracheids is conflicting. According to Berkley (1934), Pillow and Luxford (1937), and Pillow et al. (1936, 1959), compression-wood tracheids have a larger lumen than those in the earlywood region. Kienholz (1930), Ollinmaa (1959), and Nečesaný and Oberländerová (1967), by contrast, found the lumen to be smaller in compression wood. Any comparison is difficult because of the great variations in lumen size within a growth ring, especially in normal wood. It would nevertheless appear that the average lumen in pronounced compression-wood tracheids is slightly larger than in normal latewood tracheids.

Except for the narrow zones of first-formed earlywood and last-formed latewood, the tracheids in severe compression wood are of fairly uniform size and have a uniform wall thickness. They always have a much thicker wall and more narrow lumen than the tracheids in normal earlywood. The difference between normal latewood and compression wood in this respect is, however, very slight. It would appear that the wall generally is somewhat thicker and the lumen smaller in normal latewood than in compression wood. Normal latewood is also harder and heavier than compression wood. According to Yumoto et al. (1983), tracheid walls are thinner in mild compression wood than in corresponding normal wood.

4.3.5 Abnormal Compression-Wood Tracheids

Compression wood tracheids with a structure entirely different from that so far discussed have been reported on two occasions. On pulping of *Picea abies* wood, Bucher (1968) observed that some material, which might have consisted of knot wood, was difficult to delignify. Microscopic examination of sections that had been stained with iodine and malachite green showed the presence of round, thick-walled tracheids. A closer inspection revealed a concentric lamellation of the secondary wall, as seen in Fig. 4.43. Bucher (personal communication, 1970) was convinced that the cells were a kind of compression-wood tracheids containing concentric regions with a high lignin content. It should, however, be noted that intercellular spaces were missing, that many of the tracheids were angular rather than round in outline, and that they were in a highly swollen state and might have become deformed.

Compression wood tracheids of a similar, abnormal type were observed by Scott and Goring (1970) in a study of the concentration of lignin in the S_3 layer of *Picea abies*. In the lower portion of a branch they noticed, alongside and interspersed with compression wood tracheids of the usual kind, entirely different tracheids, an example of which is seen in Fig. 4.44. The round, thick-walled cells and the intercellular spaces indicate that this is a tissue related to ordinary compression wood, but here the similarity ends. These tracheids lack entirely the distinctly lamellar S_1 layer and the helical cavities chracteristic of ordinary compression wood tracheids. The extent of lignification, which is readily observable since ultraviolet light has been used, is very erratic, some cells apparently containing much less lignin than others. As in the similar tracheids from the same species observed by Bucher (1968), the cell wall is lamellar and contains concentric bands with a high concentration of lignin. Scott and Goring

Fig. 4.43. Abnormal tracheids observed on pulping of *Picea abies* wood. Note the concentric lamellation of the thick secondary wall. Transverse section. LM. (Bucher 1968)

4.4.1 Introduction

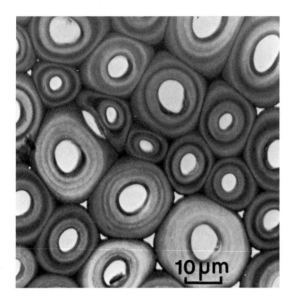

Fig. 4.44. Abnormal compression-wood tracheids in a branch of *Picea abies* with a lamellar and variably lignified secondary wall lacking helical cavities. Transverse section ULM. (Scott and Goring 1970)

drew attention to the occurrence of a highly lignified layer next to the lumen. This, however, might be an artifact, for this dark layer cannot be observed in all cells and in several cases extends only partly around the lumen border.

The peculiar tracheids observed by Bucher (1968) and by Scott and Goring (1970) are probably of the same type. It is very likely that both originated from branches. They might represent a kind of abnormal compression wood tracheid, but how and why they assumed their aberrant structure is unknown. A closer examination of these tracheids in the electron microscope should be rewarding. Other abnormal tracheids, in some respects resembling those now described, are discussed in Sect. 4.14.

4.4 Tracheids. Ultrastructure

4.4.1 Introduction

Cellular plant ultrastructure has made rapid progress during the past 30 years thanks to the development and perfection of the transmission and scanning electron microscopes and associated techniques. Woody tissues were long difficult to handle, but modern embedding and sectioning methods have overcome most of the earlier difficulties, and ultrathin sections of high quality can now be readily prepared. Ultramicrotomy, shadowing, staining, and replica techniques for woody tissues have been much improved. Use of the scanning electron microscope has made possible a direct, three-dimensional view of wood previously not attainable.

According to presently accepted terminology (Kerr and Bailey 1934), tracheids consist of a primary (P) and a secondary (S) cell wall. The secondary wall is composed of three layers, namely the outer (S_1), the middle (S_2), and the inner (S_3) layers, distinguishable from one another by the different orientation of their cellulose microfibrils. Following the S_3 layer is a warty layer, which terminates the secondary wall toward the lumen. Helical thickenings are sometimes part of the S_3 layer. The tracheids are separated from one another by the middle lamella (M). Both this intercellular region and the primary wall are encrusted by lignin so that it is often impossible to discern the primary wall. In such cases it is customary to distinguish between the true middle lamella between the cell corners and the (compound) middle lamella along the wall which also includes the primary wall (M + P).

The ultrastructure of wood has been reviewed in many books. Some of the more recent ones are those authored by Frey-Wyssling (1959, 1976), Roelofsen (1959), Côté (1967), Mark (1967), Kollmann and Côté (1968), Braun (1970), Jane et al. (1970), Bosshard (1974a, b, 1982), Preston (1974), Schweingruber (1978), Core et al. (1979), Butterfield and Meylan (1980), Panshin and de Zeeuw (1980), and Fengel and Wegener (1983) as well as those edited by Treiber (1957), Zimmermann (1964), Côté (1965), Robards (1974), Wangaard (1981), Baas (1982), and Rowell (1984).

4.4.2 The Middle Lamella

The intercellular region in compression-wood differs somewhat in composition from that in normal wood. Originally, the intercellular substance consists of pectin and hemicelluloses, some of which are derived from the cell plate. At a later stage, the intercellular zone and the primary wall become heavily encrusted with lignin. Both the true middle lamella (M) and the compound middle lamella (M + P) are less lignified in compression wood than in normal wood (Wood and Goring 1971, Fukazawa 1974). Timell (1980a) has attributed this to the fact that parental primary walls seem to traverse intercellular regions more frequently in compression wood (Chaps. 6.4.2 and 9.6.1.4). As a result, the middle lamella at the cell corners has a lignin concentration of 75% compared to 80–100% in normal wood. Along the wall, the lignin concentration of the middle lamella is only 50% because of the inclusion of the primary wall (Wood and Goring 1971). The proportion of the total cell volume occupied by the entire middle lamella seems to be less in compression wood than in normal wood. For earlywood and latewood of *Pseudotsuga menziesii* Wood and Goring (1971) found M + P to have a fractional tissue volume of 14.2% and 6.5%, respectively. The corresponding figure for pronounced branch compression wood was 7.0%, a much lesser proportion when it is recalled that in normal wood of this species earlywood occupies 70% or more of the growth rings. Inspection of light and electron micrographs reveals that the central part of the middle lamella along the wall in compression wood can be very thin. Figure 4.45, obtained with *Pinus resinosa* in a region free of intercellular spaces, shows a middle lamella which is only 0.1 μm thick along the wall.

The most characteristic feature of the middle lamella in compression wood, and one which sets it apart from the same region in normal wood, is the fact

4.4.2 The Middle Lamella

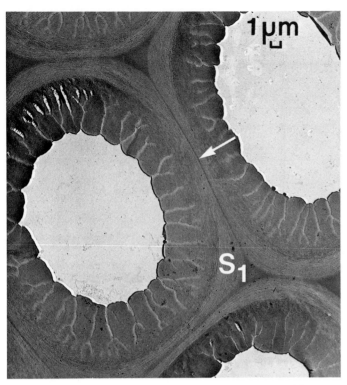

Fig. 4.45. Compression-wood tracheids in *Pinus resinosa* with an angular outline and an S_1 layer that is thicker at the cell corners than along the wall. The middle lamella separating the two center tracheids (*arrow*) is only 0.1 µm thick. Note the absence of intercellular spaces. Transverse section TEM

that it contains so-called *intercellular (interstitial) spaces*, empty regions where no polysaccharides or lignin have been deposited. A model of an entire compression-wood tracheid, developed by Scurfield and Silva (1969a) and seen in Fig. 4.46, shows the general appearance of the intercellular spaces which are always found at the cell corners where three or more tracheids meet. Hartig's (1896, 1901) drawing, reproduced in Fig. 2.7, also gives a good idea of the general appearance and location of these open spaces, including the fact that occasionally they occur also in the first-formed earlywood tracheids. Mio and Matsumoto (1982b) recently estimated that in *Pinus thunbergii* intercellular spaces occupied 2.5% of the compression-wood area. Large intercellular spaces between the tracheids at a ray crossing in compression-wood of this species are shown in Fig. 4.47. Together with the rounded outline of the tracheids and the spiral striations, the numerous intercellular spaces are a very useful diagnostic feature of compression-wood. Caution must, however, be exercised, for intercellular spaces, albeit never common, nevertheless do occur also in normal wood of several conifer genera (Chap. 8.2). Russow (1883) was the first to notice axial intercellular spaces between the tracheids in several conifer species,

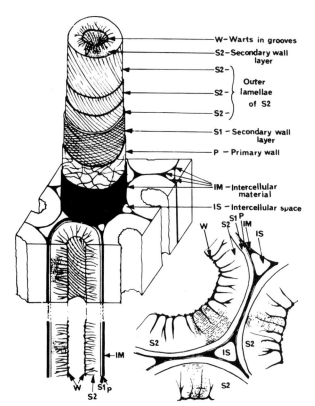

Fig. 4.46. Schematic model of a typical compression-wood tracheid. (Scurfield and Silva 1969a)

and later investigators have observed such spaces in normal wood of *Agathis* and *Araucaria* (Greguss 1955, Bolton et al. 1975), in latewood of *Pseudotsuga menziesii* (Göhre 1958), in *Juniperus* (White 1965, Jane et al. 1970, McGinnes and Phelps 1972, Core et al. 1979), and in *Cupressus* (Butterfield and Meylan 1980). Intercellular spaces are especially common and prominent in normal wood of *Juniperus virginiana,* and normal and compression woods of this species are strikingly similar in this respect, as can be seen from Fig. 4.19. Occasionally intercellular spaces are also present between the tracheids in normal wood of *Ginkgo biloba* (Eicke and Ehling 1965). According to Mio and Matsumoto (1979a), they occur in a number of conifer species. Bolton et al. (1975) found that the intercellular spaces in *Agathis* and *Araucaria* often were traversed by a fibrillar bridge which probably was the remnant of a perforated membrane. The investigators suggested that the spaces served in gaseous interchange at the time when the tissue was differentiating and respiring. According to my own observations, intercellular spaces are sometimes present also in opposite wood.

It should perhaps be noted at this point that radial intercellular spaces are also present in normal, coniferous wood. These spaces are located between the ray parenchyma cells and are believed to serve as gas canals (Russow 1883, Back 1969, Nyrén and Back 1960, Laming 1974, Mio and Matsumoto 1979b,

4.4.2 The Middle Lamella

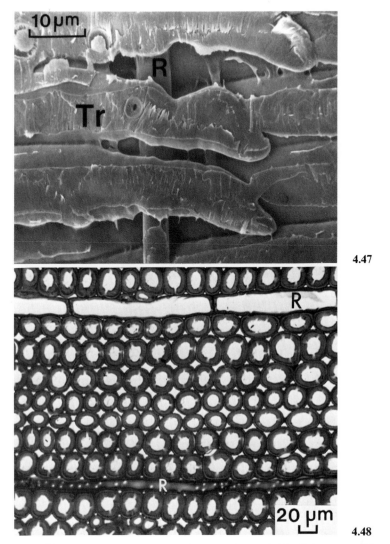

Fig. 4.47. Large intercellular spaces between the tracheids (*Tr*) at a ray (*R*) crossing in compression wood of *Pinus thunbergii*. (Mio and Matsumoto 1982b)

Fig. 4.48. Compression wood of *Abies balsamea* with a high frequency of intercellular spaces of various shapes and sizes. *R* ray cell. Transverse section. LM. (Courtesy of A.C. Day)

1982b). In *Picea abies* these radial spaces are generally triangular in cross section and are connected with the ray cells via blind pits (Laming 1974). Intercellular spaces can develop by collapse of unlignified ray parenchyma cells in the xylem of conifers (Bamber 1973). Longitudinal intercellular spaces are also present at the cell corners of axial parenchyma cells in many hardwood species (Mio and Matsumoto 1981).

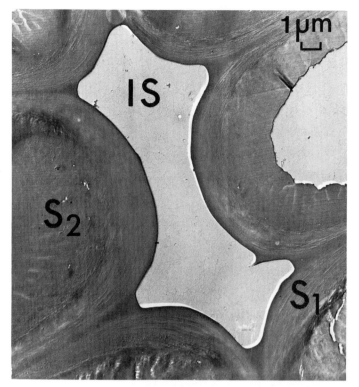

Fig. 4.49. Compression wood of *Picea abies* with an unusually large intercellular space (*IS*) bordering on six tracheids. Transverse section. TEM. (Timell 1973d)

An example of the distribution of intercellular spaces in compression wood is shown in Fig. 4.48. An intercellular space can border on as many as six compression wood tracheids, as seen in Fig. 4.49, but the most common numbers are three to four (Cieslar 1896) or two to five (Lämmermayr 1901). The spaces are more rare in first-formed and last-formed regions than in the remainder of a growth ring. According to Pillow and Luxford (1937), there are fewer intercellular spaces in mild than in pronounced compression wood (Table 4.2). Intercellular spaces also occur between tracheids and rays in compression wood, as can be seen from Figs. 4.48 and 4.124. Harris (1976) found intercellular spaces to be absent within the earlywood zone of mild and moderate compression woods and scarce in severe compression wood of *Pinus echinata*. Within the latewood region, intercellular spaces were sparse in mild, present in moderate, and abundant in severe compression woods. Harris (1977) has pointed out that intercellular spaces seem to occur independently of the other anatomical features of compression wood. In *Pinus radiata* they were sometimes present in mild compression wood, yet could be absent in pronounced compression wood. In such cases the lumen nevertheless remained rounded, resulting in thick walls at the tracheid corners. My own observations agree with this. In a specimen of

4.4.2 The Middle Lamella

Picea abies, intercellular spaces were conspicuously absent in severe compression wood, the rounded outline of the tracheids notwithstanding (Timell 1973). The last-formed tracheids in compression wood, which tend to be flattened in the radial direction, seldom border on intercellular cavities. In these tracheids, the S_2 layer has an oval outline, while that of the entire cell is more rectangular. The inevitable consequence of this is an S_1 layer that is much thicker at the cell corners than along the wall (Fig. 4.18). The same phenomenon is found close to the tip of the tracheids, where S_1 can be thicker than the adjacent S_2 layer. Absence of intercellular spaces in severe compression wood has recently also been reported by Mio and Matsumoto (1982b).

In their studies on the structure and formation of compression wood in tilted *Picea glauca* trees, Yumoto and Ishida (1982) and Yumoto et al. (1983) noted that the incidence of intercellular spaces seemed to be unrelated to the angle of inclination above a certain threshold value for the latter. In view of this and of the fact that intercellular spaces are known to occur also in normal wood, they suggested that the formation of these spaces and of compression wood might be unrelated events, and that gravity might have no influence on the former (Table 3.3) (Chaps. 3.6, 9.8, and 12.5.3).

It has often been stated that the intercellular spaces are a consequence of the rounded outline of the tracheids in compression wood. It is, of course, true that the region between the tracheids becomes much larger when the latter are round rather than angular. However, a causal relationship between tracheid form and failing lignification has never been established. In normal wood of *Pseudotsuga menziesii*, intercellular spaces are present notwithstanding the angular shape of the tracheids (Göhre 1958), and many cases are known where round compression-wood tracheids are surrounded by a fully lignified middle lamella. Intercellular spaces are as a rule both frequent and large in compression wood of *Larix*, *Picea*, and *Pinus*, but they seem to reach their highest development in *Pseudotsuga menziesii* (Fig. 4.50.1). An extreme case, observed by von Pechmann (1972), is seen in Fig. 4.50.2. This intercellular space, if it can be called that, is abnormally large, and one has the impression that four or five tracheids either had failed to develop in this region or had been destroyed.

One result of the inability of the middle lamella to become completely lignified is that the tracheids in compression wood are only loosely held together, especially along their radial walls. When compression wood is sectioned, radial rows of cells are therefore often set free, a fact commented upon by several investigators (Hartig 1896, Lämmermayr 1901, Münch 1938, Resch and Blaschke 1968).

Most frequently, an intercellular space extends across the entire middle lamella at the cell corners in compression wood. In no case that I have observed does it include the primary wall, which is always lignified. Another characteristic feature of these spaces is that they are sometimes traversed by narrow strips of lignin. These bars are not parental primary walls as claimed by Côté and Day (1965). A lignin band crossing an intercellular space in *Larix laricina* compression wood is seen in Fig. 4.51. Figure 4.52 shows that the connecting bars sometimes are quite numerous, that they are irregularly spaced, and that they can have the shape of a round rod. In some cases these rodlike structures

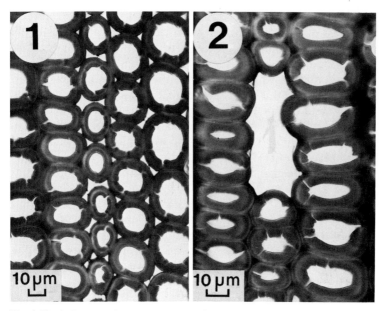

Fig. 4.50. 1 Compression wood of *Pseudotsuga menziesii* with numerous intercellular spaces (von Pechmann 1972). **2** An abnormally large intercellular space in compression wood of the same species. (Courtesy of A. von Pechmann). Transverse sections. LM

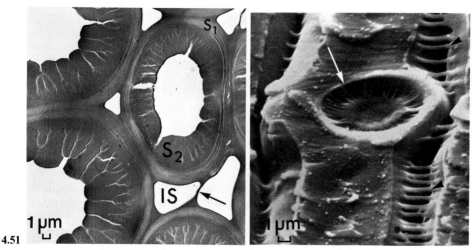

Fig. 4.51. Compression wood of *Larix laricina* with a band of lignin traversing a large intercellular space (*IS*) (*arrow*). Transverse section. TEM

Fig. 4.52. Compression wood of *Pinus mugo* with a bordered pit extending far into an intercellular space (*arrow*) and with numerous lignin bars traversing the intercellular spaces (*arrowheads*). Transverse surface. SEM. (Resch and Blaschke 1968)

4.4.2 The Middle Lamella

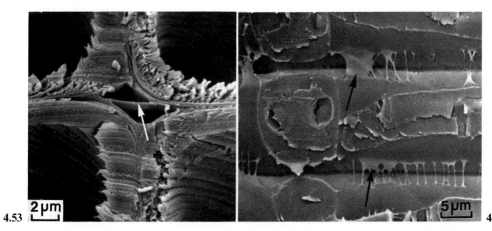

Fig. 4.53. Parent primary cell wall (*arrow*) traversing an intercellular space between compression-wood tracheids in *Pinus thunbergii*. Transverse surface. SEM. (Mio and Matsumoto 1982a)

Fig. 4.54. Torn parental primary walls (*arrows*) between compression-wood tracheids in *Pinus thunbergii*. Radial surface. SEM. (Mio and Matsumoto 1982a)

fail to traverse the intercellular space and remain short outgrowths on the outside of the tracheid wall. Intercellular spaces are not present in the dividing cambial cells of compression wood. They appear at a stage when the S_1 layer is being initiated (Chap. 9.6.1). According to Wardrop and Davies (1964), they are formed schizogenously at the time when the young cells separate along the middle lamella and assume a rounded outline. Until completion of the cell wall, the spaces are filled with what appears to be cytoplasmic ground substance (Wardrop and Davies 1964, Casperson and Zinsser 1965, Kutscha 1968). Cell organelles have also been claimed to occur in intercellular spaces (Foster and Marks 1968). Some investigators, such as Hartig (1899), Münch (1938), and Jacobs (1945) have expressed the view that the rounded outline of the tracheids and the intercellular spaces should be a consequence of an unusually high turgor pressure in developing compression wood. Bamber (1980), by contrast, has suggested that they are formed when the turgor pressure is reduced at this stage of xylem differentiation.

When cambial cells divide, each new daughter cell surrounds itself with a new primary wall, as is discussed in Chap. 9.4. As a result, parental primary walls traverse intercellular regions in the mature xylem. Such primary walls have been observed within the cambial zone of *Picea abies* (Timell 1980a). They have also been reported to cross intercellular spaces in *Pinus pinaster* (Wardrop 1952, Wardrop and Dadswell 1952, 1953) and in *Pinus sylvestris* (Casperson and Zinsser 1965) compression woods. More recently, Mio and Matsumoto (1982a) examined such parental primary walls in compression wood of *Pinus thunbergii*. They are seen traversing an intercellular space in Fig. 4.53. Fragments of torn wall membranes are bridging intercellular spaces in Fig. 4.54. They can also be seen in Fig. 4.21.

4.4.3 The Primary Wall

At the time of its formation, the primary wall in compression-wood tracheids, like that in normal wood, consists of a thin meshwork of cellulose microfibrils, embedded in a matrix of hemicelluloses and pectin. Using histochemical techniques, Scurfield (1967) showed that the primary wall in *Pinus radiata* is cellulosic in nature. As far as is now known, it has the same chemical composition in normal and compression woods (Côté et al. 1968b, Larson 1969). On lignification, the primary wall becomes heavily encrusted with lignin (Hartig 1901). Jurášek (1964), on the basis of staining with phloroglucinol, found the primary wall in compression wood tracheids of *Abies alba* to be only slightly lignified. This is contrary to all other evidence.

As described in Chap. 9.6.1, the fusiform initials in the resting cambium of compression wood are flattened in the radial direction and either oblong or rectangular in outline. When cell division begins, the newly formed cells soon become enclosed by a primary wall on whose inner surface the secondary wall is subsequently deposited in the form of concentric lamellae. The primary wall is observed to best advantage at this stage. The thickness of the primary wall, when measured from electron micrographs obtained before lignification has taken place, is on the average 0.15 µm in both normal and compression woods of *Abias balsamea* (Kutscha 1968). According to Onaka (1949), the primary wall is slightly thicker in compression wood.

Casperson and his co-workers originally claimed that the primary wall should be much thicker in compression wood than in normal wood (Casperson 1959a, b, 1962a, b, 1963a, b, c, d, 1964, 1965b, Correns 1961, Wergin and Casperson 1961, Wergin 1965, Casperson and Zinsser 1965). According to Casperson (1962a), the primary wall in compression-wood tracheids has a thickness varying from 0.5 µm along the wall to 1.5 µm at the cell corners. The range is stated to be 0.3–1.0 µm by Casperson and Zinsser (1965). Inspection of the electron micrographs published by these investigators reveals that the cell wall referred to is actually not the primary wall but the outer layer of the secondary wall (S_1). This opinion was later accepted by Casperson (1968) in a study of the ultrastructure of compression wood in *Larix decidua*. According to Casperson's later measurements, the primary wall in compression wood is approximately 0.1–0.2 µm thick.

At times, but certainly not often, the primary wall can be detected in transmission electron micrographs of untreated, fully lignified compression wood. One such case is shown in Fig. 4.139 (Sect. 4.10.1). In this transverse section of compression-wood in *Pinus sylvestris*, the clearly discernible primary walls are about 0.15 µm thick, which agrees well with Casperson's (1968) and Kutscha's (1968) measurements. It is obvious that the primary wall has the same thickness in compression wood as in normal wood.

The primary wall is only rarely accessible for study in mature wood because it is completely submerged in the lignin of the middle lamella. If thin cross sections of wood are treated with strong acid so that all polysaccharides are hydrolyzed and removed, the location of the former primary wall can sometimes be seen as an irregular streak of interrupted openings in the middle

4.4.3 The Primary Wall

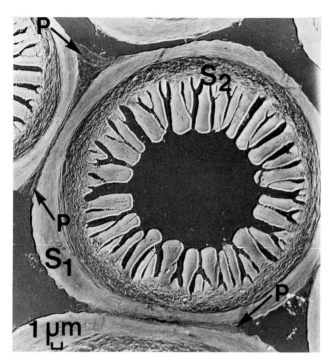

Fig. 4.55. Delignified compression-wood tracheids in *Larix laricina*. Microfibrils from the primary wall (*P*) can be seen immediately outside S_1 (*arrows*). (Cf. Figs. 6.16 and 6.25). Transverse section. TEM

lamella, close to the S_1 layer (Jayme and Fengel 1961a, b). Occasionally, as already mentioned, the primary wall can be traced in the intercellular region despite the overlying lignin. Another approach entails selective removal of the lignin, for example with acid chlorite. A transverse section of a compression-wood tracheid consisting only of cellulose and hemicelluloses is shown in Fig. 4.55. Here the loosely organized microfibrils of the primary wall can be seen immediately outside the more compacted S_1 layer. Primary wall material is still located within the narrow space between the walls of adjacent tracheids, and an original, parental primary wall can be seen connecting two cells.

One method for establishing the detailed structure of a cell wall layer in wood is to use some suitable replica technique in conjunction with transmission electron microscopy. Surprisingly few such studies have been concerned with compression wood. Early investigators who used this approach were Harada and Wardrop and their co-workers. Harada and Miyazaki (1952) and Harada et al. (1958) examined replicas of the primary wall in *Chamaecyparis obtusa* and *Pinus densiflora* compression woods from which the lignin had been eliminated by various means. The major microfibril orientation was found to be transverse, and the organization of the microfibrils was meshlike. Côté (1977) has reported an electron micrograph of a replica of the primary wall in *Larix laricina* compression wood from which both lignin and hemicelluloses had been removed. The remaining cellulose microfibrils show a random orientation. Obviously there is no difference in this respect between normal and compression woods.

4.4.4 The S_1 Layer

The outer layer of the secondary wall (S_1) in normal wood, which was first studied by Preston and Wardrop 1949, Hodge and Wardrop (1950), Emerton and Goldsmith (1956), Frei et al. (1957), Meier (1957), and Wardrop (1957, 1964), differs less from the S_1 layer in compression wood tracheids than does the S_2 layer. A well-known and frequently reproduced model, which was first published by Wardrop and Dadswell (1950), and which compares the cell wall organization of a normal and a compression wood tracheid, is seen in Fig. 4.56. According to this schematic model, the microfibrils in the S_1 layer are organized in a helix that is considerably flatter in compression wood than in normal wood, the angle being 80–90° according to Wardrop (1954) and 90° according to Dadswell (1963). Nečesaný (1955b), by contrast, in a very similar representation, indicates exactly the same microfibrillar orientation in the S_1 layer in the two tissues. That the average orientation of the microfibrils in S_1 is more or less transverse in both normal and compression woods is obvious from observations of the tracheids in polarized light, especially in transverse sections, such as those shown in Fig. 4.57. In this, as well as in all other of the several micrographs that have been reported since the first investigation of Bailey and Berkley (1942), the birefringence of the S_1 layer appears to be of equal intensity in the two tissues (Wardrop and Dadswell 1950).

Bailey and Berkley (1942) concluded that the orientation of the microfibrils in the outer layer of the secondary wall in compression-wood tracheids of *Taxodium distichum* was predominantly transverse. They also found, as did

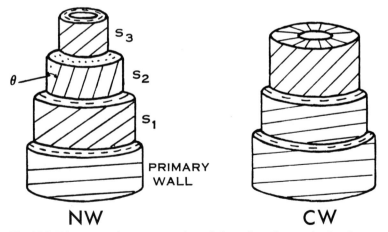

Fig. 4.56. Diagrammatic representation of the cell wall organization in normal (*NW*) and compression-wood (*CW*) tracheids. (Wardrop and Dadswell 1950)

4.4.4 The S₁ Layer

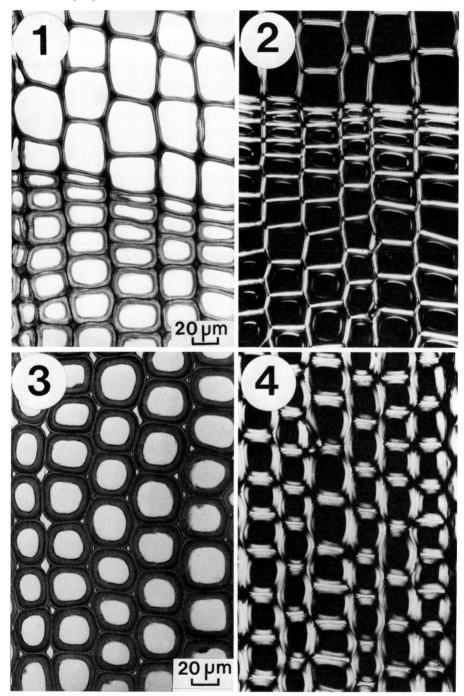

Fig. 4.57. Normal (**1** and **2**) and compression (**3** and **4**) woods of *Picea rubens* photographed in normal (**1** and **3**) and in polarized (**2** and **4**) light. The S_1 and S_3 layers are birefringent in the normal wood and the S_1 and S_2 layers in the compression wood. Transverse sections. (Courtesy of A.C. Day)

later Preston (1946), Wardrop and Preston (1947), and Wardrop and Dadswell (1950), that the S_1 layer did not contribute to the x-ray pattern of either normal or compression-wood tracheids. In contrast, Berkley and Woodyard (1948), Kantola and Seitsonen (1961), Kantola and Kähkönen, and Marton et al. (1972) have claimed that the x-ray pattern of compression-wood tracheids clearly indicates the presence of a cell wall layer, presumably S_1, where the microfibrils are oriented almost transversely to the fiber axis. Wardrop and Dadswell (1952), contrary to their results from 1950, later found that the S_1 layer does affect the x-ray pattern of the tracheids in compression wood and estimated the microfibril angle at $78-90°$.

It is now generally agreed (Harris 1977, Meylan 1981) that the x-ray technique most commonly adopted for measuring the microfibril angle in S_2 (Cave 1966, Meylan 1967, Sobue et al. 1971, Stewart and Foster 1976, Boyd 1977) can be used for determining this angle also in the S_1 layer, and especially so in compression wood. Procedures entailing the use of polarized light, such as those developed by Page (1969) or by Cousins (1972), measure the average microfibril angle in all cell wall layers. If the S_1 and S_3 layers are thick, the value obtained for S_2 will be in error (El Hosseiny and Page 1973, Harris 1977). Meylan (1981) found that in compression wood of *Pinus radiata* differences as large as $15°$ could occur between angles determined by this method and those indicated in the microscope.

After examination of compression wood of 13 conifer species by the x-ray method, Wardrop and Dadswell (1950) concluded that the microfibrillar helix in S_1 was very flat. Using a replica technique and electron microscopy, Harada and Miyazaki (1952) found that the microfibrils in the S_1 layer of compression wood tracheids were oriented at an angle of about $80°$ with respect to the fiber axis in *Chamaecyparis obtusa* and *Pinus densiflora*. Nečesaný (1955b) determined the microfibril angle in the various layers of the secondary wall in normal, opposite, and compression woods from several species. His data for *Pinus banksiana* and *Taxus baccata* are listed in Table 4.14. They indicate that the

Table 4.14. Microfibril angle in the secondary wall layers of normal and compression woods of two gymnosperm species. All values in degrees. (Nečesaný 1955b)

Species	Cell wall layer	Microfibril angle		
		Oposite wood	Normal wood	Compression wood
Pinus banksiana	S_1	70	73	68
	S_2			32
	S_3	87	81	
Taxus baccata	S_1	60–80	65–80	60–80
	S_2	13–25		40
	S_3	65–85	52–74	

4.4.4 The S_1 Layer

angle in the S_1 layer is the same in all three tissues, varying from 60° to 80°. Harada et al. (1958) presented further evidence for the large microfibril angle in the S_1 layer, using replicas of this layer in compression wood of *Pinus densiflora*, similar to those later prepared from compression wood of *Pinus radiata* by Dadswell et al. (1958) and Wardrop and Davies (1964). According to Casperson (1959b), who also used the replica technique, the angle is 85° in *Pinus pungens*. Nečesaný (1966) later reported angles varying from 40° to 60°, a very wide range. Marton et al. (1972) state that the microfibril angle in the S_1 layer of compression-wood tracheids in *Picea abies* is close to 90°. Tang (1973), on the basis of observations in polarized light, found that the angle was 78–80° in compression wood tracheids of *Pinus virginiana*. Stewart and Foster (1976) demonstrated how the microfibril angle in S_1 could be determined in compression wood from several species. This was possible thanks to the great thickness and large angle in this layer.

Notwithstanding the somewhat limited evidence, there seems to be no reason to believe that the orientation of the microfibrils in the S_1 layer of compression wood tracheids should be significantly different from that in normal tracheids. In both tissues the microfibrils are organized in a flat helix at an angle that varies from 70° to 90° with the longitudinal fiber axis. The angle decreases with increasing tracheid length in normal wood. The same negative correlation probably holds for compression wood. The fact that compression wood tracheids always are shorter than corresponding tracheids in normal wood apparently does not result in a larger microfibril angle in S_1 in compression wood. Possibly, if more careful and extensive measurements were made, involving tracheids with exactly the same location within the tree, the S_1 layer in compression wood tracheids might be found to have a larger microfibril angle than the same tracheids in normal wood. The experimental evidence available at present does not, however, suggest this. It is not known whether or not the S_1 has the fine structure suggested by Dunning (1969a, b), for this layer in normal wood, where it is believed to consist of several lamellae with widely different and sometimes crossed microfibril orientations.

Although Bailey and Kerr (1937) mention in passing that the outer layer of the secondary wall in compression wood tracheids is narrow, this obviously is meant only in relation to the S_2 layer. All investigators agree that S_1 is thicker in compression wood than in normal wood tracheids, and this is actually one of the most characterstic features of this tissue. Measurements of layer thickness in polarized light are, as pointed out by Bailey and Berkley (1942), highly unreliable. Better results are obtained with ordinary or ultraviolet light, but the difficulty here is to distinguish between the different layers. The best method is undoubtedly to carry out measurements on suitably enlarged transmission electron micrographs. The extensive data reported by Saiki (1970) demonstrate the usefulness of this approach. The S_1 layer is less highly lignified than the compound middle lamella on its outside and the outer portion of the S_2 layer on its inside (Chap. 6.4.2). Unlike these regions, S_1 also contains almost transversely oriented microfibrils which are readily discernible in transverse sections in the electron microscope. For these reasons, the S_1 layer is easy to recognize and delineate in compression wood tracheids.

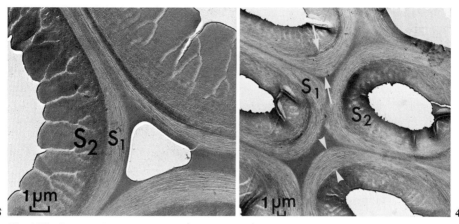

Fig. 4.58. Compression wood tracheids in *Pinus resinosa* at a cell corner region, showing S_1 layer of different thickness. Transverse section. TEM

Fig. 4.59. Last-formed compression wood tracheids in *Tsuga canadensis*. In one cell (*arrowheads*) S_1 equals S_2 in thickness (3.0 µm). In another cell (*arrows*) S_1 is 3.5 µm thick. Transverse section. TEM

The thickness of the S_1 layer varies a great deal in both normal latewood and compression wood, but the variation along the circumference of each tracheid is much greater in compression wood. According to Casperson, Correns, and Wergin, whose contributions are referred to in Sect. 4.4.3, the S_1 layer is very thin and frequently not even discernible in compression wood. The layer actually referred to is, however, the outer portion of S_2.

The varying size of the S_1 layer in compression wood is evident in Fig. 4.58, which also shows the typical increase in thickness at the cell corners. In this transverse section, the minimum width of S_1 is 0.30 µm and the maximum 2.00 µm. The occurrence of more massive walls at the cell corners is also found in normal wood and especially in the S_1 layer of ray cells (Fengel 1965). This increase in thickness probably entails both deposition of more microfibrils and a lesser compaction of the lamellae (Fengel 1966b).

The values in Table 4.15 were obtained from a number of electron micrographs of transverse sections of normal and compression woods in *Tsuga canadensis*. The S_1 layer in compression wood was two to four times thicker than in normal wood. In both tissues it was especially thick at the cell corners, as is shown for last-formed compression wood in Fig. 4.59. The lumen in these tracheids was larger in normal wood than in compression wood, a fact also observed by Kienholz (1930) and Onaka (1949). At one of the cell corners, the S_1 layer in Fig. 4.59 has exactly the same width as S_2, namely 3.0 µm, and it is 3.5 µm thick in another tracheid. Toward the tip of the cells the relative thickness of the two layers changed further, and along certain sections of the circumference the S_1 now became the dominating cell wall layer.

4.4.4 The S_1 Layer

Table 4.15. Thickness of the secondary wall layers along the wall and at the corners of earlywood and latewood tracheids in normal and compression woods of *Tsuga canadensis*. All values in μm

Wood and location	S_1	S_2	S_3
Normal wood			
Earlywood, middle	0.28	0.85	0.14
Earlywood, corner	0.50	0.70	0.21
Latewood, middle	0.40	3.4	0.40
Latewood, corner	1.5	3.4	0.80
Compression wood			
Earlywood, middle	1.2	4.0	
Earlywood, corner	2.0	5.1	
Latewood, middle	1.0	2.4	
Latewood, corner	2.6	2.7	

Among the three cell wall layers in compression tracheids, the S_1 is more distinctly lamellar in transverse sections, also in fully lignified tracheids. After removal of the lignin, the well-ordered lamellae can be seen more clearly, as shown in Fig. 6.16. In this holocellulose the lamellae are undoubtedly swollen and have become more tightly packed than they are in their native state. They are probably of the same size as in normal wood, that is, $0.10-0.15$ μm thick (Nečesaný 1955b, Casperson 1959b, Fengel 1966a). The entire S_1 layer is highly lignified (Wood and Goring 1971, Fukazawa 1974).

Evidence that the layer S_1 is less dense than S_2 has been presented by Wooten et al. (1967), who found that in *Pinus taeda* the S_1 layer occupied a larger proportion of the secondary wall than in normal wood, a fact that they could qualitatively correlate with the extent of longitudinal shrinkage (Chap. 7.3.3.2). When transverse sections were treated with iodine, microscopic observations revealed little change in normal wood. In compression wood tracheids, a dark band appeared in the outer portion of the secondary wall, as can be seen in Fig. 4.60. There can be no doubt that the investigators' conjecture that the dark band represents the S_1 layer is correct, and even the lamellae typical of this layer can be distinguished. Wooten et al. (1967) advanced two explanations for the fact that the S_1 layer was stained more deeply than S_2. According to one, the S_1 layer appears darker than S_2 because in the former the iodine was deposited in the form of long crystals, oriented in the same transverse direction as the microfibrils, whereas in the latter they would be oriented closer to the fiber axis and the direction of observation. If this were so, however, one would expect also the S_1 and S_3 layers in normal tracheids to stain darkly, which they did not. The second explanation, namely that the S_1 is more porous than the S_2 in compression wood tracheids, is therefore more plausible. That S_1 has a relatively porous structure is also in agreement with the fact that this layer is fairly accessible to both chemical reagents and microorganisms. The ease of removal of the galactan present in S_1 (Chap. 5.5.3) is an example of this, and another is the fact that on delignification of compression wood, the S_1

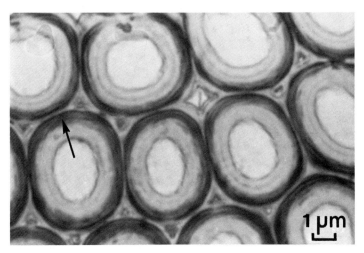

Fig. 4.60. Compression-wood tracheids in *Pinus taeda* stained with iodine and showing a high concentration of iodine crystals in the S_1 layer (*arrow*). Transverse section. LM. (Wooten et al. 1967)

seems to be affected first. The S_1 is also the first layer in compression wood to be destroyed by brown rot fungi (Chap. 5.11.2). From transverse sections of delignified compression wood tracheids, such as that shown in Fig. 6.16, one might have an impression of a closer packing of the cellulose microfibrils in S_1 than in the inner S_2. This might, however, be only apparent and be a consequence of the transverse orientation of the microfibrils in S_1. In S_2 the microfibrils are oriented in a steeper helix, and this layer therefore presents an image of seemingly lesser density (Côté 1967). The lamellar organization of S_1 is clearly discernible during the formation of this layer, before it has become lignified (Figs. 9.24, 9.25, 9.27, and 9.28). According to Kibblewhite (1972) there are no transitional lamellae between S_1 and S_2 in compression-wood tracheids of *Pinus radiata*, as is the case in normal wood (Harada et al. 1958, Imamura et al. 1972, Tang 1973).

In summary, the S_1 layer in compression-wood tracheids is considerably thicker than in normal wood, about four times as thick compared to earlywood and twice as thick in comparison with latewood. Its microfibrils are oriented in a flat helix with a pitch of $70-90°$ to the fiber axis, as in normal wood, and are organized in distinct lamellae. The S_1 is probably more porous than the S_2 layer. At the cell corners, the S_1 layer can be as thick as S_2, and at the cell tips it often exceeds S_2 in thickness. Obviously, the S_1 layer cannot be disregarded when the physical properties of compression wood are considered, for example the abnormally high longitudinal shrinkage of this wood. In this case, the S_1 layer probably does not act as a restraint on S_2 (Barber 1968) but, on the contrary, contributes itself actively to the shrinkage (Chap. 7.3.3.2). Because it constitutes a much larger proportion of the secondary wall, the S_1 layer undoubtedly plays a larger role in compression wood than it does in normal wood.

4.4.5 The S$_2$ Layer

4.4.5.1 General Structure

The S$_2$ layer, which in compression-wood tracheids is the inner layer of the secondary wall, has a structure which is unique to this tissue and has attracted attention almost from the time when compression wood was first described scientifically. More has been written about this layer than about all other morphological entities of compression wood combined. Since the helical cavities are the most characteristic, not to say peculiar, feature of the S$_2$ layer, it is not surprising that they have been the subject of the largest number of investigations, speculations, and comments.

The fact that the S$_2$ layer of compression wood tracheids is much thicker than the corresponding layer in normal, earlywood tracheids is evident from the values in Table 4.15. Compared to normal latewood, however, the S$_2$ layer is often of the same thickness or even thinner. This is especially so in the last-formed compression wood tracheids where, instead, the S$_1$ layer is particularly thick, so that the thickness of the entire secondary wall remains about the same.

The ultrastructure of the S$_2$ layer in compression wood tracheids is entirely different from that in normal tracheids and has attracted much attention over the years, in addition to causing a good deal of controversy. The major features are the orientation of the cellulose microfibrils, the seemingly isotropic, highly lignified layer in the outer S$_2$, the spiral striations or helical cavities, as they are now usually called, the origin and nature of the large, helical checks, the spiral or helical thickenings found in a few species, the warty layer, and the sculpturing of the pits. In the earlywood tracheids, formed at the very beginning of the growing season, the organization of the S$_2$ layer differs in some respects from that of the majority of the compression wood tracheids and to a lesser extent this also applies to the last-formed latewood tracheids.

4.4.5.2 Orientation and Nature of the Microfibrils

The microfibrillar angle, in older literature referred to as the micellar angle, is the angle at which the helix of cellulose microfibrils is oriented with respect to the longitudinal cell axis in a wall layer. Usually, unless otherwise is stated, the layer referred to is the S$_2$. This angle is an important characteristic, for it has a great influence on the physical characteristics of wood (Mark and Gillis 1973), and it has also been the subject of much research. Meylan and Probine (1969) have stated that the microfibril angle is the most important submicroscopic property of wood. In normal softwood, the microfibril angle in the S$_2$ layer of the axial tracheids varies between 20° and 30° in earlywood and between 5° and 10° in latewood. The helix is flatter in the radial than in the tangential wall, especially in earlywood, a fact attributed by Boyd (1974) to interactions during lignification.

Preston (1934, 1947, 1948) has shown that the average length of a tracheid (L) is related to its microfibril angle (α) by the simple relationship

$$L = a + b \cot \alpha$$

where a and b are constants, different for different cell wall layers. Originally developed for the S_2 layer, this equation has been found to be valid also for S_1 (Preston and Wardrop 1949, Wardrop 1957) and for S_3 and the helical thickenings attached to this layer (Wardrop and Dadswell 1951). According to this equation, longer tracheids have a steeper microfibril helix than shorter ones, and as the tracheids in a conifer increase in length from the pith toward the outside, the microfibril angle in S_2 becomes smaller. Preston (1934) found that the angle was greater on the radial than on the tangential cell wall, an observation later extended by Wardrop and Dadswell (1951) to the spiral thickenings in *Pseudotsuga menziesii*. Preston's equation requires that the number of turns of the microfibril helix per cell be constant. Wardrop and Dadswell (1951) observed, however, that in both *Pseudotsuga menziesii* and *Taxus baccata* the number of turns of the helical thickenings increased with increasing tracheid length. Wardrop and Dadswell (1952) a year later reported a similar situation with respect to the microfibrils in the S_2 layer of compression wood tracheids. They pointed out that the microfibril angle usually increases toward the tip of the tracheid. The result is that, despite the fact that the number of helical turns per cell does not remain constant, the average microfibrillar angle is less in longer tracheids. Preston's equation accordingly does not hold for individual cells and can only be regarded as an empirical relationship, valid as an average for a large number of fibers.

Early investigators did not determine the orientation of the microfibrils in compression wood tracheids since the existence of these microfibrils was unknown at the time. They did, however, observe under the light microscope the spiral striations in the inner part of S_2 and noted their orientation. Hartig's (1896, 1901) accurate and detailed drawing in Fig. 2.8 indicates a microfibril angle in the range of 30–45° for *Picea abies*. Krieg (1907) found angles from 37° to 48° for *Pinus sylvestris*, apparently the first measurement of what we now know is the orientation of the microfibrils in the S_2 layer of compression-wood tracheids.

In 1909 Sonntag reported extensive observations on spiral striations in a large number of different plants. In wood tracheids from the under side of *Picea abies* branches, the angle of the spiral striations, which could be readily observed where they crossed a pit, was 40.5°. In *Pseudotsuga menziesii* the angles varied from 48° to 71°. The largest angles, namely 62–71°, occurred in tracheids which contained both spiral striations and spiral thickenings. Sonntag observed that the orientation of the spiral striations always was the same as that of the micelles (microfibrils) in the layer where they occurred and that they reflected the inner structure of this layer (S_2). After a discussion of earlier observations on spiral striations by Correns and Dippel in Germany, Sonntag objected to the views of Gothan (1905), who had argued that the helical cavities in compression wood tracheids should be secondary phenomena, associated with the formation of heartwood (Chap. 9.6.1.3). No further observations seem to have been made for about 20 years, and it was not until 1928 that Verrall reported angles of 14.5–32.8° for compression wood tracheids of various species.

4.4.5.2 Orientation and Nature of the Microfibrils

In the same year, Jaccard and Frey (1928) reported values for the microfibril angle in compression wood tracheids and could demonstrate that the spiral striations were oriented in the same direction as the micelles. For compression wood tracheids of *Pinus nigra* and *Pseudotsuga menziesii* they obtained, using polarizing optics, a microfibril angle of 42.1°. They also concluded, anticipating by several years the later results of Preston (1934) and Wardrop and Dadswell (1950, 1952), that the flatter helix of these tracheids was caused by the increased number of anticlinal divisions in the cambium, which resulted in shorter fusiform initials.

Pillow and Luxford (1937) determined the microfibril angle in earlywood and latewood of normal and compression woods from five coniferous species, obtaining the results listed in Table 4.2. Using polarizing optics, they confirmed that the orientation of the microfibrils and the spiral striations were identical. Microfibril angles were consistently larger in pronounced than in mild compression wood, and in both they exceeded those in corresponding normal wood, and especially in normal latewood. In all tissues the helical pitch was flatter in earlywood than in latewood. The largest angle was 38.3°, observed in pronounced compression wood of *Sequoia sempervirens*. Münch (1938) found a microfibril angle of 32–38° in *Picea abies* compression wood.

In 1942, Bailey and Berkley reported results of a pioneering investigation where they used polarized light for determining the orientation of cellulose microfibrils in tracheids. In photomicrographs similar to those shown in Fig. 4.56, the S_1 and S_3 layers were both birefringent in normal wood but S_2, because of the steep helix of its microfibrils, exhibited complete extinction. The S_1 layer in compression wood tracheids of *Taxodium distichum* was strongly birefringent but the S_2 only moderately so, as could be expected from its microfibril angle of 30–50°. No S_3 layer could be observed. Transverse sections of compression wood tracheids photographed in polarized light have since been published by several investigators, such as Wardrop and Dadswell (1950, 1952), Wergin and Casperson (1961), Kantola and Kähkönen (1963), Wardrop and Davies (1964), Côté and Day (1965), Côté et al. (1967), Scurfield (1967), Casperson (1968), and Harris (1977). Using x-ray diffraction, Bailey and Berkley (1942) found the S_2 layer in compression wood tracheids of *Taxodium distichum* to have a helical pitch of 44–48°.

X-ray diffraction (Cave 1966, Meylan 1967) has been used by several later investigators for determining the microfibril angle in compression wood tracheids. Berkley and Woodyard (1948) found this angle to be 47.5° on the tangential and 56° on the radial walls in *Pinus ponderosa*. According to Vichrov (1949), it is 38° in earlywood but only 27° in latewood of compression wood tracheids in *Larix decidua*, a result in agreement with those obtained by Pillow and Luxford (1937). Lindgren (1958) also used x-ray techniques for studying the orientation of the microfibrils in compression wood tracheids of *Picea abies* which had a flatter helix than normal tracheids. For *Pinus sylvestris* Kantola and Seitsonen (1961) found an angle of 35°. In a more comprehensive investigation, Kantola and Kähkönen (1963) determined the orientation of the microfibrils in the S_2 layer of compression wood tracheids in three conifer species, obtaining for *Juniperus communis* angles of 34–40°, for *Picea abies*

Table 4.16. Microfibril angle in the S_2 layer of normal and compression wood tracheids in five gymnosperm species. All values in degrees. (Onaka 1949)

Species	Microfibril angle	
	Normal wood	Compression wood
Chamaecyparis obtusa	20.5	44.5
Cryptomeria japonica	36.8	44.8
Pinus thunbergii	28.9	33.0 – 46.8
Taxus cuspidata	24.8	44.5
Torreya nucifera	25.5	45.2

22.5 – 35.5°, and for *Pinus sylvestris* 29 – 40° (Kantola 1964). A similar study of compression wood from branches of *Picea abies* and *Larix decidua* has been reported by Kocoń (1967a, b). Onaka (1949) determined the orientation of the microfibrils in normal and compression wood tracheids from five coniferous species, obtaining the results summarized in Table 4.16. The orientation remained constant throughout the S_2 layer with respect to both direction and angle of inclination, and Onaka concluded that the microfibril angle in the S_2 layer of compression wood tracheids is approximately 45°.

Wardrop and Dadswell (1950) studied the ultrastructure of compression wood from 13 conifer species, namely *Araucaria bidwillii*, *A. cunninghamii*, *Athrotaxis selaginoides*, *Callitris rhomboidea*, *Dacrydium franklini*, *Diselma archeri*, *Pinus pinaster*, *P. radiata*, *Pherosphaera hookeriana*, *Phyllocladus asplenifolius*, *Pseudotsuga menziesii*, *Taxodium distichum*, and *Taxus baccata*. Examination of transverse sections in polarized light showed S_2 to be less birefringent than S_1. In both layers the microfibrils were parallel to the cell surface. When the tracheids were crushed and examined under the light microscope, the microfibrils were found to be inclined about 45° to the longitudinal cell axis. This result was confirmed from the spread of the equatorial arcs in the x-ray diffraction pattern. When the x-ray and optical data were compared, it was evident that the spiral striations followed the orientation of the microfibrils.

Wardrop and Dadswell (1950) also determined the variation in tracheid length and microfibril angle in the radial direction of two specimens of *Pinus radiata* which both contained compression wood on one side. The angle of orientation was obtained from the 020 diffraction arcs in the x-ray diagrams. The results obtained with one of three trees are shown in Figs. 4.61 and 4.62 in which it can be seen that tracheid length decreased and microfibril angle increased as soon as formation of compression wood had been initiated. When formation of compression wood ceased, tracheid length again began to increase and the microfibril angle to decrease. In opposite wood the length of the tracheids increased and the helical pitch decreased from the pith toward the outside, both gradually assuming a constant value. When the cotangent of the microfibrillar angle was plotted against tracheid length and the data were statistically evaluated, it was found that Preston's equation was valid for both nor-

4.4.5.2 Orientation and Nature of the Microfibrils

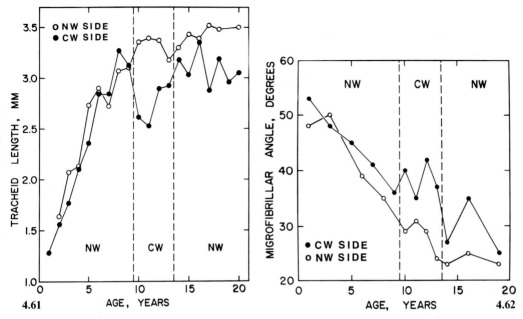

Fig. 4.61. Variation of tracheid length with age in normal (*NW*) and compression (*CW*) woods of *Pinus radiata*. (Redrawn from Wardrop and Dadswell 1950)

Fig. 4.62. Variation of microfibril angle with age in normal (*NW*) and compression (*CW*) woods of *Pinus radiata*. (Redrawn from Wardrop and Dadswell 1950)

mal and compression woods, as shown in Fig. 4.63. Wardrop and Dadswell concluded that the orientation of the microfibrils in the S_2 layer was the same in normal and compression wood tracheids of the same length. Their second conclusion, that the lesser length of the tracheids also determined other structural features of the tracheids in compression wood, soon proved to be untenable (Wardrop and Dadswell 1952, Wardrop 1954).

Two years later, Wardrop and Dadswell (1952) reported further results of a similar investigation. Again, it was found, this time with *Pinus pinaster*, that formation of compression wood was associated with a decrease in tracheid length and an increase in the microfibril angle in S_2, as shown in Fig. 4.25. In an earlier study, Wardrop (1951) had measured the number of helical turns in the S_3 layer by observing the spiral thickenings. They now determined the number of turns of the microfibrillar helix per cell in the S_2 layer by examining compression wood tracheids from *Pinus pinaster* and *Cupressus macrocarpa*. The results, shown graphically in Figs. 4.64 and 4.65, make it clear that the number of helical turns per cell increased with increasing tracheid length and accordingly did not remain constant, as required by Preston's equation.

Nečesaný (1955b) determined the tracheid length and microfibril angle of the tracheids in compression and opposite woods within each growth ring of a 5-year-old *Pinus banksiana* tree. The results, shown in Fig. 4.27, indicate that in

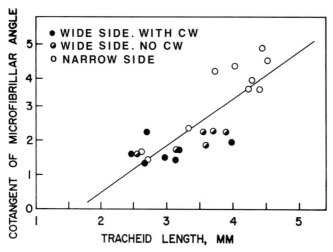

Fig. 4.63. Relationship between the cotangent of the microfibril angle and tracheid length in an eccentric stem of *Pinus radiata*. (Drawn from values reported by Wardrop and Dadswell 1950)

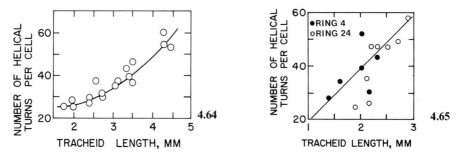

Fig. 4.64. Variation of the number of turns of the microfibrillar helix per cell with tracheid length in compression wood of *Pinus pinaster*. (Redrawn from Wardrop and Dadswell 1952)

Fig. 4.65. Variation of the number of turns of the microfibrillar helix per cell with tracheid length in compression wood of *Cupressus macrocarpa*. (Redrawn from Wardrop and Dadswell 1952)

opposite wood an increase in tracheid length was associated with a decrease in microfibrillar angle, but the relationship was less clear in compression wood, where the angle varied from 22° to 45°. The pitch of the helix was flatter at the beginning than at the end of the growing season in two increments but steeper in the next two. The difference was slight compared to that in opposite wood. The average angle in each growth ring decreased in both tissues from the pith toward the bark. It was 32° in *Pinus banksiana* and 40° in *Taxus baccata,* as can be seen from Table 4.14. In *Abies alba* it was 46° in earlywood and 35° in latewood. The models of typical tracheids of normal and compression woods presented by Nečesaný (1955b) in this paper are more accurate than those

4.4.5.2 Orientation and Nature of the Microfibrils

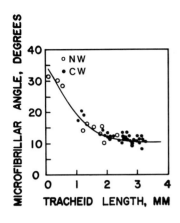

Fig. 4.66. Relationship between microfibril angle and tracheid length in normal wood (*NW*) and moderate compression wood (*CW*) of *Picea abies*. (Nečesaný 1961)

shown in Fig. 4.56. Nečesaný claimed that there exists a linear relationship between the size of the microfibrillar angle and the degree of lignification of compression wood. With constant tracheid length, this is what would be expected, since both microfibril angle and lignin content increase with increasing proportion and severity of the compression wood.

Nečesaný (1961) has reported similar studies of tracheid length and microfibril angle in moderate compression wood of *Picea abies*. When the angle was plotted against tracheid length, as shown in Fig. 4.66, the nonlinear relationship applied to both normal and compression wood tracheids. It will be noted that the use of the contangent rather than of the angle itself gave a linear relationship in the hands of Wardrop and Dadswell (1950) (Fig. 4.63). In a later contribution, Nečesaný (1966) reported that the microfibril angle in the S_2 layer of compression wood tracheids can vary from 35° to 75°, the latter an unusually high value.

Hiller (1964) studied wood from the hybrid *Pinus x attenuradiata*. She found that the microfibril angle decreased from the beginning of the earlywood until the end of the latewood zone in both normal and compression woods. The trend within an increment was not influenced by its width or proportion of latewood. Formation of compression wood in *Pinus elliottii* and *Araucaria cunninghamia* has a much greater effect in increasing the microfibril angle in S_2 than in reducing tracheid length (Anonymous 1955, 1959). The results obtained by Wardrop and Dadswell (1952) (Fig. 4.25) would seem to attest to this claim.

Ollinmaa (1959) determined microfibril angles in normal and compression woods of three conifer species, obtaining the values in Table 4.17. For *Abies alba* compression wood, Jurášek (1964) observed angles in the range of 30°–40°. Casperson (1959b, 1968) found 20°–50° for *Picea pungens* and 45° for *Larix decidua* compression woods. Dadswell (1963), in a review of the anatomy of compression wood, states that the microfibril angle in S_2 is 50°–60°. Suzuki (1968), who examined compression woods from *Abies firma*, *Cryptomeria japonica*, *Larix leptolepis* and *Pinus densiflora*, found that the microfibril angle increased from earlywood to latewood within each increment. The largest angles occurred in the region containing most compression wood. Marton et al.

Table 4.17. Microfibril angle in the S_2 layer of normal and compression wood tracheids of three conifer species. All values in degrees. (Ollinmaa 1959)

Species	Normal wood	Compression wood
Juniperus communis	23.7	36.5
Picea abies	25.9	33.7
Pinus sylvestris	26.8	40.4

(1972), on the basis of x-ray measurements, arrived at a microfibril angle of $38° - 40°$ in the S_2 layer of compression wood tracheids in *Picea abies*.

Cockrell (1974), in a study of the anatomical and physical properties of compression wood in *Sequoiadendron giganteum*, observed microfibrillar angles of only $16° - 32°$ in latewood tracheids, and he pointed out that the often quoted value of $45°$ probably represents a maximum. In a few, sporadically occurring tracheids, however, angles of $45° - 78°$ were observed, values as large as those reported for *Picea abies* by Sonntag (1909).

As discussed more fully in Chap. 17.4, Boyd and Foster (1974) attempted to correlate anatomical and physical properties with gravitationally imposed growth strains around the periphery of a bent *Pinus radiata* tree which contained compression wood on its lower side. The microfibril angle reached a maximum of $27°$ in the region with most compression wood, compared to a minimum of $12°$ in the opposite wood, as shown in Fig. 17.23. This maximum coincided with similar maxima in compression strength and density. There was a strong, negative correlation between microfibril angle and the externally induced strains around the stem periphery. According to the investigators, growth stresses generated within the stem were determined by the microfibril angles.

Harris (1977) measured the average microfibril angle of all tracheid wall layers in opposite and compression woods of *Pinus radiata*. The mean angle was $48°$ in both tissues. In severe compression wood, the angle was slightly smaller than in opposite wood. It is likely that the exceptionally large, average microfibril angles in the opposite wood were a consequence of the thick S_3 layer with its transversely oriented microfibrils in this wood. Park et al. (1980) made similar measurements of microfibril angles in opposite and compression woods in a branch of *Pinus densiflora*, using the same method of measuring the angles as had Harris (1977) (Cousins 1972). The occurrence of side (lateral), opposite, and compression woods and the variation in microfibril angle around the periphery are shown in Fig. 4.67. Maximum angles, $30° - 40°$ occurred in the compression wood and minimum values, $10° - 15°$, in the opposite wood region. In both tissues the angles were somewhat larger in earlywood than in latewood. The largest microfibril angles were observed in the center of each growth ring where they were associated with fully developed helical cavities. The angles were, however, still $30° - 35°$ in the first-formed earlywood, where helical cavities were either absent or only weakly developed.

4.4.5.2 Orientation and Nature of the Microfibrils

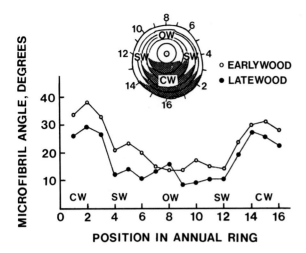

Fig. 4.67. Variation of microfibril angle around the periphery of a *Pinus densiflora* branch containing compression (*CW*), side (*SW*), and opposite (*OW*) woods. (Redrawn from Park et al. 1980)

As mentioned in Chap. 8.4, Meylan (1981) found the tracheids to have a larger microfibril angle on the lower (37°) than on the upper (26°) side of a branch in the primitive angiosperm *Pseudowintera colorata*. These tracheids had no other features typical of those in compression wood of the gymnosperms. Wilson (1981), on examination of compression wood from the lower side of a 60-year-old branch of *Pinus strobus* found the average microfibril angle to be 31° in the earlywood and 25° in the latewood.

It is clear that in mature wood the microfibrillar angle in the S_2 layer is considerably larger in compression wood than in normal wood tracheids. In the former it is usually 20°–30° in earlywood and 10° or less in latewood. The statement has often been made that the angle is 45° in compression wood, but it is probably more correct to say that it varies within a range of 30°–50°. According to Meylan (1974) (Harris and Meylan 1965) the mean microfibril angle is generally less than 40° in compression wood of *Pinus radiata*. Observations in polarized light indicate that the microfibrillar helix in the first-formed and last-wood zones in approximately the same as in the middle of a growth ring. The angle decreases within each increment from earlywood to latewood (Nečesaný 1955b, Hiller 1964, Park et al. 1980). As in normal wood, the microfibril angle in S_2 is very large in juvenile compression wood. In the second increment from the pith of *Abies alba*, Nečesaný (1955b) found an angle of 55°–65° (Fig. 4.27). In the opposite wood of the same increment it was 65° in earlywood but only 30° in the latewood. The S_2 helix in the compression wood gradually became steeper from pith to bark, the change being especially pronounced within the zone of juvenile wood. The largest microfibril angles reported for compression wood tracheids are all within a range of 71°–78° (Sonntag 1909, Nečesaný 1955b, Cockrell 1974).

The helical cavities and ribs in compression wood tracheids are always oriented in exactly the same direction as the cellulose microfibrils (Sect. 4.4.5.4.6). Direct observation of the cavities and ribs therefore gives a good indication of

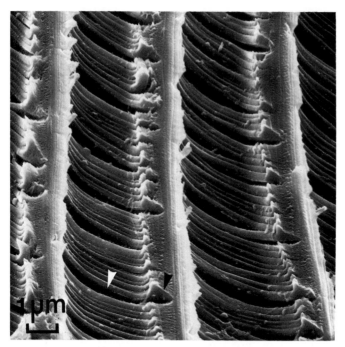

Fig. 4.68. Scanning electron micrograph of compression tracheids in *Pinus sylvestris*. Note the flat orientation of the helical cavities. *Arrowhead* indicates one of the numerous helical checks which all extend deeply into the S_2 (L) layer and are artifacts. Radial surface

the orientation of the microfibrils. Scanning electron micrographs are best suited for this purpose. Figure 4.68 shows an example of this, obtained with *Pinus sylvestris*. The helix of ribs and cavities and accordingly also the microfibrillar helix is unusually flat in this case. Yumoto et al. (1983) have recently reported an almost transverse orientation of the microfibrils in the S_2 layer of some tracheids in compression wood of *Picea glauca*.

Investigations by Wardrop and Dadswell (1950, 1952) and Nečesaný (1961) indicate that the large microfibril angle in the S_2 layer of the tracheids in compression wood is associated with the shortness of these tracheids, a feature that is a consequence of the increased frequency of anticlinal divisions in the compression wood cambium (Sect. 4.3.2). The close association between a small microfibril angle, short tracheids, and rapid growth in compression wood has recently been emphasized by Wilson (1981). The data in Table 4.5 reported by Chu (1972) show that the microfibril angle is considerably larger in compression wood than in normal wood of *Pinus palustris*, and that both decrease with increasing tracheid length. Preston's equation, which is applicable to both normal and compression woods, is, of course, only a formal relationship and offers no information of what causes rapid growth to be associated with a flat microfibrillar helix. It should not be interpreted as an indication of a causal relationship between microfibril angle and tracheid length (Wardrop 1954).

4.4.5.3 The $S_2(L)$ Layer

The large microfibril angle in the S_2 layer of compression wood is, as already mentioned, far from unique to this tissue. Pillow and Luxford (1937) mention that "an abnormal type of wood" is formed throughout the cross section of some vertical, rapidly growing conifers, particularly at the base of the stem. In this wood, the microfibrillar helix in the latewood is said to be flatter than normal, a value of 50° being quoted by Pillow et al. (1941). Bailey and Berkley (1942) have reported a case where in normal latewood of *Sequoia sempervirens* the microfibril angle in S_2 had the unusually large value of 30°–40°. In transverse section all three layers of these tracheids exhibited birefringence in polarized light, and the x-ray diffraction pattern was similar to that shown by compression wood tracheids. In the juvenile zone of normal wood, the tracheids are short and the microfibrillar helix in S_2 is flat (Chap. 6.2.1). An angle of 45° has been recorded near the pith in *Pinus elliottii* (Dadswell and Nicholls 1959) and in *P. radiata* (Watson and Dadswell 1964). In their studies of the factors influencing the longitudinal shrinkage of wood (Chap. 7.3.3), Harris and Meylan observed that in no case were the microfibril angles in the S_2 layer of compression wood tracheids outside the range observed for normal wood, namely up to 40°. Meylan (1968) found microfibril angles as large as 50° in normal wood of *Pinus jeffreyi*. The circumstance that both juvenile and compression woods have short tracheids and large microfibril angles in the S_2 layer led Wardrop and Dadswell (1950) to the conclusion that compression wood could be regarded as "physiologically younger" than normal wood. In as much as the frequency of the anticlinal divisions is high in both juvenile and compression woods this statement is correct, but it should not be extended to other properties of compression wood.

The pitch of the helix in the S_2 layer has a decisive influence on the shrinkage behavior of wood, as shown in Chap. 7.3.3, and this applies to both normal and compression woods. It also affects the strength properties of the tracheids and the paper-making characteristics of pulp fibers (Chap. 19.2).

Only limited microscopic evidence is available concerning the size and nature of the microfibrils in the S_2 layer of compression-wood tracheids. Practically all replicas examined in the electron microscope have been obtained from the inner, fissured wall of this layer. The microfibrils are deeply embedded in lignin in compression wood, and best results are achieved if the lignin is first partly removed, for example with acid chlorite. If all lignin is eliminated, the microfibrils tend to swell and become distorted (Fig. 4.55). The replica of the inner wall of a compression wood tracheid in *Larix laricina* shown in Fig. 4.69 indicates that each helical ridge consists of microfibrils which are 24–40 nm wide and all oriented in the same direction. According to Nečesaný (1955b), the microfibrils in the S_2 layer of compression wood tracheids have a diameter of 30–50 nm. Comparison with the microfibrils in the S_1 layer indicates that their width is the same in the S_1 and S_2 layers.

4.4.5.3 *The $S_2(L)$ Layer*

The S_2 layer in the compression wood tracheids in most of the Coniferales and Taxales is fissured by helical cavities, or as they were first called, spiral stri-

Fig. 4.69. Helical ribs in a compression-wood tracheid in *Larix laricina*, partly delignified with acid chlorite. Note the microfibrils in the ribs. Oblique (15°) section. Carbon replica, TEM. (Côté et al. 1968)

ations, so that it consists of ridges, which often protrude into the lumen. The cavities, albeit penetrating deeply, never reach the S_1 layer, but extend from the lumen to a certain distance from S_1 where they all terminate abruptly. This distance varies, as can be seen from Fig. 4.58, but measurements from a large number of electron micrographs indicate that it is usually 0.5 – 1.5 µm, the most common distance being the average of this approximate range, 1.0 µm. This outer portion of the S_2 layer differs in several respects from the inner part. It contains no helical cavities, and it also has a much higher lignin and lower cellulose content than the remainder of S_2. While this outer, solid region of S_2 is ontogenetically an integral part of this layer, it is nevertheless convenient to be able to refer to it with a shorter term than "the outer portion of the S_2 layer". Some years ago I therefore suggested that this part of the S_2 layer in compression-wood tracheids should be designated as the $S_2(L)$ *layer*, L being a reference to the exceptionally high lignin content of this zone (Côté et al. 1968a). Judging from later publications, this term has been found to be useful also by others (Robards 1969; Morey and Morey 1971a, b; Wood and Goring 1971, Morey 1973, Yoshizawa et al. 1982, 1984a, b, Yumoto et al. 1982, 1983). As pointed out by Yumoto et al. (1983), the $S_2(L)$ layer is the only compression-wood feature that is common to the Ginkgoales, Taxales, and Coniferales. It

4.4.5.3 The S₂(L) Layer

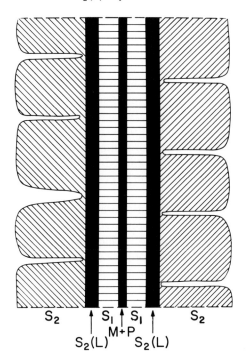

Fig. 4.70. Schematic model of a double cell wall in a compression-wood tracheid, indicating the location of the M + P, S₁, S₂(L), and inner S₂ layers as well as the orientation of the microfibrils in S₁ and S₂. Black symbolizes the high lignin contents of the isotropic M + P and S₂(L) regions

has never been observed in normal wood, and it is present not only in severe compression wood but also in the mild and moderate grades (Chap. 3.2). Yumoto et al. (1983) rightly consider the S₂(L) to be the anatomical feature most characteristic of compression wood.

The appearance and location of the S₂(L) layer is best brought out in transmission electron micrographs of transverse sections of wood (Figs. 4.17, 4.18, 4.45, and 4.51), holocellulose (Figs. 4.55, 6.15, 6.16, and 6.25), or lignin skeletons (Figs. 6.13 and 6.14). It can also be seen in light micrographs of stained sections (Fig. 6.12), but is more readily discernible in ultraviolet light thanks to its high lignin content (Figs. 6.17, 6.21, 6.22, and 9.53). A schematic model of a double cell wall in longitudinal section is shown in Fig. 4.70, illustrating the lignin-rich M + P and S₂(L) regions in black and with the approximate orientation of the microfibrils indicated by the transverse lines in S₁ and the diagonal lines in the inner part of the S₂ layer. The helical cavities terminate either at the border or the S₂(L) layer or at a certain, slightly variable distance therefrom.

The S₂(L) layer contains 50–60% lignin (Wood and Goring 1971; Fukazawa 1974) and has a fairly high concentration of galactan (Côté et al. 1968b) (Chap. 6.3.3). Because of its low cellulose content, immediately evident from electron micrographs of holocelluloses, the S₂(L) layer gives complete extinction in polarized light. Sanio observed the S₂(L) layer as early as 1860 and again in 1873 (Chap. 2.3), but the first to describe it in detail were Bailey and

Kerr (1937). They treated compression-wood tracheids of *Pinus roxburghii* with concentrated sulfuric acid and examined the remaining lignin skeleton under the microscope when they observed an "isotropic layer of non-cellulosic composition" between the narrow, outer and the broad, inner layers of the secondary wall. This result was confirmed in a later investigation by Bailey and Berkley (1942), who examined transverse sections of compression wood from *Taxodium distichum* under the polarizing microscope. Between the strongly birefringent S_1 and the moderately birefringent S_2 layers they could discern an intervening layer that showed only feeble birefringence or complete extinction. They considered the apparent isotropy of this zone to be caused, not by a longitudinal orientation of the cellulose microfibrils, but by the largely noncellulosic composition of this region.

Wardrop and Dadswell (1950) were unable to confirm Bailey's results with *Taxodium distichum*, as they could not observe any isotropic zone between S_1 and S_2. Inspection of their only published micrograph obtained in polarized light reveals a somewhat less than satisfactory resolution, notably inferior to that obtained by Bailey and Berkley (1942). Although they did obtain some evidence for the occurrence of an isotropic layer inside S_1 in *Pinus pinaster, P. radiata,* and *Pseudotsuga menziesii,* Wardrop and Dadswell (1950) concluded that the existence of an isotropic layer between S_1 and S_2 could not be regarded as characteristic of compression wood tracheids. Twelve years later, however, Wardrop and Davies (1964), using polarized and ultraviolet light microscopy, could demonstrate that a zone of low birefringence and with a high lignin content existed between S_1 and S_2, a result in complete agreement with Bailey's earlier observations. They also found that the weakly birefringent or isotropic layer had a low concentration of cellulose. The cellulose microfibrils were not oriented longitudinally but in the same direction as in the remainder of the S_2 layer and were organized in lamellae.

The seemingly isotropic zone observed by Bailey and Wardrop and their coworkers is neither located outside S_1 as claimed by Onaka (1949) nor is it identical with S_1, as believed originally by Casperson (1965b). It should perhaps also be mentioned that Lange, using ultraviolet light, had detected the lignin-rich $S_2(L)$ layer in 1954. The ultraviolet micrographs presented by Wergin (1965), although interpreted differently by the investigator, show the presence of a zone with a high concentration of lignin in the outer part of S_2.

The location and general appearance of the $S_2(L)$ layer were confirmed by Côté et al. (1966b) for *Abies balsamea* and *Picea rubens*. On the basis of what was obviously insufficient microscopic evidence these investigators suggested that the few cellulose microfibrils in the $S_2(L)$ layer might be organized transversely to the long axis of the tracheids. In a later and more comprehensive study of *Larix laricina*, Côté et al. (1968a) presented direct microscopic evidence for the presence in the $S_2(L)$ layer of loosely packed cellulose microfibrils, organized in lamellae. They could also demonstrate the high concentration of lignin in this zone, a conclusion later confirmed in an almost identical investigation by Parham and Côté (1971) of the lignin distribution in *Pinus taeda*. There was no indication that the microfibril angle in the $S_2(L)$ was different from that prevailing in the remainder of the S_2 layer.

That the $S_2(L)$ layer has a lamellar structure is evident from studies on the formation of the tracheids in compression wood of *Abies balsamea* (Côté et al. 1968c, Timell 1979) (Chap. 9.6.1.3). Before lignification begins, these lamellae can be readily observed in the developing tracheids, as can be seen in Figs. 9.25, 9.27 – 9.29, 9.37, 9.41, and 9.47. When the lignin is removed from mature tracheids by treatment with acid chlorite, the loose lamellae in the $S_2(L)$ layer are either completely destroyed (Wergin and Casperson 1961, Casperson 1963d) or are broken up into shorter fragments and become disorganized (Fig. 6.16). The lamellae can be seen in partly delignified tracheids, such as that in Fig. 6.15, and occasionally also in fully delignified cells, as shown in Fig. 4.55. Boyd (1973), in considering the orientation of the microfibrils in the $S_2(L)$ layer, has expressed the opinion that it would be unlikely if these microfibrils were oriented in the same direction as those in the remainder of S_2. Instead he assumes that the microfibrillar angle in $S_2(L)$ is 75°, without, however, citing any convincing experimental evidence.

There can be little doubt that the presence of two distinct zones in the S_2 layer of fully developed compression wood tracheids is of universal occurrence, no exception having been reported so far. In this connection it should be emphasized that the $S_2(L)$ layer is present also in tracheids lacking helical cavities in the inner part of S_2. Robard's (1969) statement that the S_2 often exists as a double layer in compression wood is an understatement. It always does.

Another example of an isotropic cell wall layer has been reported by Chafe (1974a, b, c) and by Chafe and Chauret (1974). They have found that in the xylem parenchyma cells of a number of hardwood species, for example *Populus tremuloides*, there are two separate, concentric secondary walls, separated by an optically isotropic layer which has a high lignin and pectin and a low cellulose content. This "isotropic layer" is always present in axial parenchyma cells and occurs occasionally also in ray parenchyma cells.

4.4.5.4 The Inner Portion of the S_2 Layer

4.4.5.4.1 Introduction. No anatomical feature of compression-wood tracheids has caused more controversy than the inner part of the S_2 layer with its characteristic helical cavities and ribs, and even today several questions remain unsettled. As is described in Chap. 9.6.1.3, there is also current disagreement concerning the mode of formation of the cavities. To a certain extent, the often widely different opinions that have been expressed regarding the origin and properties of the cavities can be attributed to a lack of agreement on nomenclature in this area. The glossary of the International Association of Wood Anatomists contains very little related to the anatomy of either compression or tension wood.

Before proceeding further, a word must be said here on definitions. What is being referred to in this book as *helical cavities* are deep fissures in the inner zone of the S_2 layer, spiraling around the tracheid in the same direction as the cellulose microfibrils, that is, 30° – 50° to the longitudinal fiber axis. These cavities are narrow, 0.1 – 0.2 μm wide and not always observable under the light microscope. They extend from the lumen to the $S_2(L)$ layer, becoming narrower and branched as they approach the latter. An example is found in Fig. 4.17. When compression wood is subjected to stress, the tracheid wall will tend to crack where it is weakest, which usually means along the helical cavities. These cracks or checks, which can be seen in Figs. 4.17, 4.18, 4.51, and 4.58, are often 1 µm or more in width and can penetrate as far as to the S_1 layer. In this book they are referred to as *helical checks*. The helical cavities might be partly schizogenous in origin, and the helical checks are entirely so. Both are present in the living tree but the cavities are formed while the cell wall is being laid down, whereas the checks are produced in the mature, fully lignified wall. In addition, the checks can be produced when wood is being processed, and especially on drying. The grooves seen in longitudinal sections from the lumen side in the region of a bordered pit are not to be regarded as checks but as areas where no cavities and ribs are present (Sect. 4.10.1).

Before the advent of the electron microscope, the helical cavities were referred to as *spiral striations* or *spiral checks*, for they can usually be seen in the light microscope as striations or checks in the cell wall when viewed in longitudinal section. They have also been termed *oblique striations, spiral cracks, spiral grooves, helical openings, helical checks, checks, microchecks,* and *helical splits*. The material between the cavities has been referred to as *helical ridges, ribs,* or *ribbons*. Robards (1969) calls them "ribbon-like helical thickenings" which is not to be recommended, for the helical cavities have nothing to do with the spiral or helical thickenings (Sect. 4.4.7). Some investigators, for example White (1965), make a clear distinction between the "fine spiral striations" and the "spiral checks", or "splits", but others treat them as equivalent. The statements by Côté and Day (1965), Côté (1967), and von Pechmann (1973) that the helical cavities traverse the entire S_2 layer and terminate at the border between S_1 and S_2 are erroneous. What these investigators are referring to are helical checks.

Theodor Hartig reported early that the inner wall of wood tracheids can possess a fine spiral striation or a coarse spiral folding, but it cannot be decided whether he was observing compression wood or normal wood with helical thickenings. Sanio mentions in 1873 that compression wood tracheids contain spiral striations in their innermost cell wall layer. Kraus (Lämmermayr 1901) undoubtedly saw them, for he observed spiral striations in branch wood of *Pinus cembra* and *P. sylvestris*. In 1896, Cieslar referred to them in passing as spiral thickenings of the tracheid wall in compression wood, but it was Robert Hartig (1896) who, in the same year, gave the first detailed description of the "peculiar" inner secondary wall of compression wood tracheids. After careful observations in the light microscope of thin sections of *Picea abies* compression wood, Hartig prepared composite drawings of typical tracheids as seen in transverse and longitudinal sections. He reproduced them in two later publications (Hartig 1899, 1901), but without the original, detailed description. These drawings, which leave little to be desired in fine detail and accuracy, are shown in Figs. 2.7 and 2.8.

Hartig (1896) distinguished between several different helical cavities. In the thin-walled, first-formed tracheids, the cavities were narrow and did not pen-

4.4.5.4 The Inner Portion of the S$_2$ Layer

etrate deeply into the cell wall. They did not cross the pit orifice, as can be seen in Fig. 2.8.1. This case is discussed in Sect. 4.5, and it should only be mentioned here that sometimes the cavities are missing altogether in this zone. In row 2 in Fig. 2.7, the latewood tracheids contained very fine cavities which remained as lines even at a magnification of 1400 and were arranged in a spiral around the lumen, as shown in Fig. 2.8. Hartig noted that the lines were branched and concluded that the number of cavities must have been larger when the developing wall was still thin. In the remaining cases (Fig. 2.7) the ribs (Hartig calls them "lamellae") were separated from one another and protruded into the lumen. Irregularly oriented, finer ribs appeared to be present inside some of the larger ribs. When it is remembered that Hartig had only light microscopes at his disposal, it must be deemed as remarkable that he nevertheless succeeded in establishing a structure for these tracheids that agrees so well with what is shown by the electron microscope. It could not have been done without both great skill and good judgement.

In an article on the anatomy of extant and fossil woods, Gothan (1905) devoted a long chapter to the occurrence of "spiral striations" in gymnosperm woods. Although he was apparently unaware of the earlier contributions of Hartig (1896, 1901) and Cieslar (1896), it is clear that Gothan was observing compression-wood tracheids, for he examined branch wood, and his description and one drawing, albeit erroneous, indicates that he was dealing with helical cavities. His basic idea was that spiral striations are only found in heartwood and never occur in sapwood. In order to prove this hypothesis, he used much ingenuity, even in cases where his own observations indicated something else. Gothan also seems to have believed that compression wood, because of its red color, must be heartwood! He was of the opinion that the striations are formed on drying of heartwood and also that they can be produced by hitting pieces of wood with a small hammer (Chap. 9.6.1.3). Krieg (1907), 2 years later, very convincingly demonstrated the untenable nature of Gothan's views, correctly emphasizing that presence of striations in compression wood and the transition of sapwood into heartwood are entirely unrelated, and pointing out that spiral striations occur in living trees. He also made the observation that checks or cracks only appear in tracheids containing striations, and that the latter cannot be artifically produced, all views which are now known to be correct.

Since this time, the helical cavities in compression wood tracheids have been studied by many investigators, sometimes only in passing, sometimes more thoroughly. Among the latter are Bailey and Kerr (1937), Pillow and Luxford (1937), Münch (1938), Jaccard (1940), Matsumoto (1950a), Wardrop and Dadswell (1950), Ollinmaa (1959), Casperson (1959a, b, 1962a, 1963a, c, d, 1964, 1965a, b, 1968), Wergin and Casperson (1961), Patel (1963), Wardrop and Davies (1964), Casperson and Zinsser (1965), Côté and Day (1965), Côté et al. (1967, 1968a, b, c), Scurfield and Silva (1969a, b), Parham and Côté (1971), Jutte and Levy (1972), Cockrell (1973, 1974), Fujita et al. (1973), and Fukazawa (1973, 1974).

4.4.5.4.2 The Microscopic Evidence. When compression-wood tracheids are viewed in cross section under the microscope in ordinary light, the helical cavi-

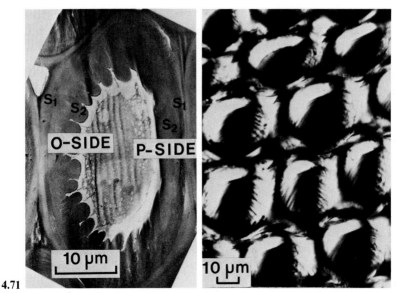

Fig. 4.71. Oblique (45°) section of compression wood in *Cryptomeria japonica*. The helical ribs and cavities and the tracheids are sectioned in an orthogonal direction on one side (*O*) and in a parallel direction on the opposite side (*P*). TEM. (Fujita et al. 1973)

Fig. 4.72. Oblique (45°) section of compression wood of *Picea pungens*, photographed in polarized light, showing the helical cavities and ribs in the tracheids. (Casperson 1959b)

ties frequently cannot be distinguished, as can be seen in Figs. 4.13 (*Juniperus virginiana*), 4.16 (*Picea abies*), 4.48 (*Abies balsamea*), 4.50 (*Pseudotsuga menziesii*), 4.57 (*Picea rubens*), and 4.60 (*Pinus taeda*). They are not visible in the transverse sections of *Sequoiadendron giganteum* published by Cockrell (1974). In other instances, the helical cavities can be only barely traced, for example in Figs. 6.9 and 6.12 (*Larix laricina*). They cannot be observed in transverse sections in polarized light, as shown in Fig. 4.57 (*Picea rubens*). It is obvious, however, that under optimum conditions, the cavities can be observed in many genera, such as *Abies, Cryptomeria, Larix, Picea, Pinus,* and *Tsuga*. The best example of this is the composite drawing prepared by Hartig (1896) (Fig. 2.7), while another is that reported by Mork (1928a) (Fig. 4.109).

One way of making the helical cavities better visible in the light microscope is to cut thin sections at an angle of about 45° to the longitudinal axis of the tracheids. Because the cavities are oriented approximately in this direction, they will, when observed in such an oblique section, have the appearance shown in the transmission electron micrograph in Fig. 4.71. Under these conditions the helical cavities and ribs will appear in true transverse section along part of the circumference of the tracheid. The technique was first applied to compression wood by Krieg (1907), and it has since been used by Münch (1940), Wardrop and Dadswell (1950), Wardrop (1954), and Fujita et al. (1973,

4.4.5.4 The Inner Portion of the S_2 Layer

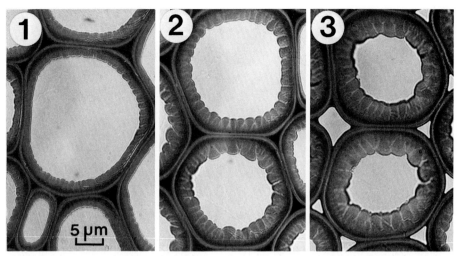

Fig. 4.73. Ultraviolet photomicrographs of compression wood in *Larix leptolepis*. The helical cavities in the tracheids are discernible in both earlywood (**1** and **2**) and latewood (**3**) cells. Note the absence of intercellular spaces in the earlywood. (Imagawa, personal communication, 1974)

1978). Casperson (1959b) has demonstrated that by this technique the helical cavities can be seen not only in ordinary but also in polarized light, as shown in Fig. 4.72.

The size of the helical cavities is only slightly above or at the threshold of the power of resolution of the ordinary light microscope, although they can often be detected with proper staining (von Aufsess 1973). In ultraviolet light with its shorter wavelength, the cavities can be readily discerned. In the first ultraviolet photomicrographs reported, they cannot be seen at all (Lange 1954) or only with difficulty (Dadswell et al. 1958, Wardrop and Davies 1964). However, by the improved techniques developed by Goring and his coworkers the cavities can be observed under the ultraviolet microscope in astonishing detail, as seen for branch wood of *Pseudotsuga menziesii* in Fig. 6.17 (Wood and Goring 1971). They are also visible in the ultraviolet photomicrographs prepared by Imagawa (personal communication, 1974) and not only in the latewood but also in the earlywood of *Larix leptolepis*, as shown in Fig. 4.73. The cavities are, however, only barely detectable in the similar micrographs published by Wergin (1965) (*Picea abies*) and by Fukazawa (1974) (*Abies sachalinensis*) (Figs. 6.21 and 6.22).

In longitudinal sections, the helical cavities can be observed surprisingly well under the light microscope, even at moderate magnifications. An example reported by Casperson (1965b) with *Pinus sylvestris* is shown in Fig. 4.74. Occasionally, it appears that the striations or cavities are oriented in two different directions within the same tracheid. This, as first pointed out by Pillow and Luxford (1937), is caused by the fact that the front and back of the tracheid can be seen in the microscope at the same time. The cavities are not readily dis-

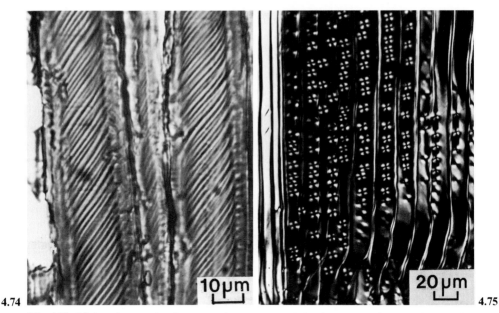

Fig. 4.74. Light micrograph of compression-wood tracheids in *Pinus sylvestris* with readily distinguishable helical cavities. Tangential section. (Casperson 1959a, 1965b)

Fig. 4.75. Compression wood of *Larix laricina*, photographed in polarized light. A Maltese cross appears where the helical cavities widen around a pit opening. (Courtesy of A.C. Day)

tinguished when viewed in polarized light in longitudinal section, as is illustrated in Fig. 4.75. Where they widen around a bordered pit, they give rise to a Maltese cross.

The great depth of field of the scanning electron microscope makes this instrument ideal for observing structural features in wood (Echlin 1968, Resch and Blaschke 1968, Findley and Levy 1969, Scurfield and Silva 1969a, b, Collett 1970, Treiber 1971, Butterfield and Meylan 1980). Scanning electron micrographs reveal the organization of the helical cavities and ribs in compression wood with great clarity. Such micrographs have been published by several investigators in recent years, such as Ishida et al. (1968), Scurfield and Silva (1969a, b), Meylan and Butterfield (1972), von Aufsess (1973), Fukazawa (1973), Core et al. (1979), Butterfield and Meylan (1980), Yoshizawa et al. (1981, 1982, 1984a, b), and Yumoto et al. (1982, 1983). Jutte and Levy (1973) discussed the application of electron microscopy to the study of the ultrastructure of compression wood, pointing out the usefulness of this technique. The helical cavities and ribs are best viewed on the longitudinal face when examined in the scanning electron microscope. Two examples are shown in Figs. 4.68 and 4.76. They can, however, also be readily observed with cross sections, when they are seen not only on the transverse face of the tracheid wall but also on the longitudinal surface of the lumen, as shown in Fig. 4.77.

4.4.5.4 The Inner Portion of the S$_2$ Layer

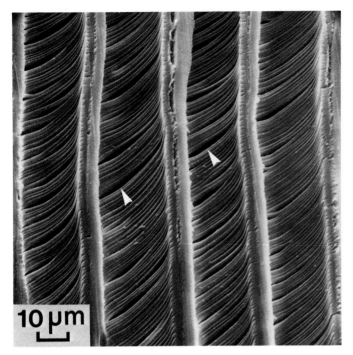

Fig. 4.76. Scanning electron micrograph of compression wood tracheids in *Abies balsamea* with uniformly fine, helical ribs. *Arrowheads* indicate helical checks. Radical surface. SEM

Several investigators have prepared replicas of the lumen wall of the S$_2$ layer in compression-wood tracheids and examined them in the transmission electron microscope. The technique was first applied to compression wood in 1952 by Harada and Miyazaki and has since been used by Harada et al. (1958), Casperson (1959a, b), Wardrop and Davies (1964), Liese (1965), Côté et al. (1968a), and Robards (1969). A carbon replica of an oblique section of a partly delignified compression wood tracheid in *Larix laricina* is seen in Fig. 4.69. For this particular purpose, the technique has in later years been largely superseded by direct observation in the scanning electron microscope.

Examination in the transmission electron microscope of longitudinal sections of compression wood for studying the helical cavities has only rarely been used. Côté and Day (1965) have reported one such micrograph. Transverse sections are more useful for this purpose, and such sections have been frequently studied by transmission electron microscopy. Figures 4.17 and 4.18 show clearly the fine, branched helical cavities together with a few larger, helical checks in *Larix laricina*. Another example, obtained with the same species, is seen in Fig. 4.78. Here some of the cavities have expanded, so that their width no longer is the original one. That this has happened is indicated by the strands of microfibrils that are seen crossing some of the fissures. Another artifact that must be recognized is that, through swelling or other distortion of the wall, the

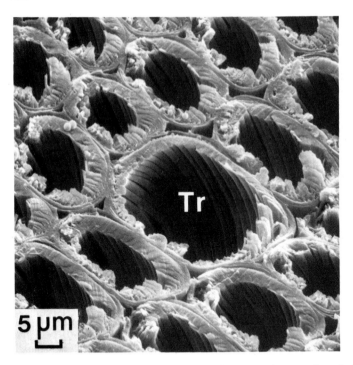

Fig. 4.77. Scanning electron micrograph of compression wood tracheids in *Cryptomeria japonica*. The helical cavities and ribs can be readily seen on the inner surface of the tracheids. They can also be discerned on the transverse surface of the cell walls. Note the large size of the tracheid in the center (*Tr*). Transverse surface. SEM

helical cavities easily can become closed and sometimes entirely obliterated. Examples of this are shown in Fig. 4.45.

Even greater changes occur when either the polysaccharides or the lignin are selectively removed from the cell wall and the resulting lignin skeletons or holocelluloses are sectioned, embedded, and examined in the electron microscope. Generally, when lignin skeletons are viewed, it is found that the cavities have expanded considerably, as shown in Figs. 6.13 and 6.14. In other cases the cavities have become partly obliterated, as can be seen in Fig. 6.13. Examples of this have been reported by Parham and Côté (1971). An artificial widening of the cavities is also observed in holocelluloses, accompanied by a swelling in the region close to the lumen, resulting in elimination of the cavities there. Both artifacts are exemplified in Figs. 4.55 and 6.16. In the holocellulose from *Pinus taeda* examined by Parham and Côté (1971), the portion of the cavities closest to the lumen had become artifically widened, while the remaining portion of them had partly disappeared.

Wood specimens to be sectioned by ultramicrotomy for subsequent examination in the transmission electron microscope are usually embedded in methacrylates. The great disadvantage of the methacrylates is that they shrink 15–20% on polymerization, severely disrupting the specimens.

4.4.5.4 The Inner Portion of the S$_2$ Layer

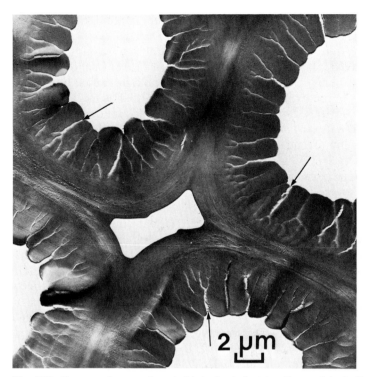

Fig. 4.78. Compression-wood tracheids in *Larix laricina*. Some of the helical cavities have become enlarged (*arrows*), probably on polymerization of the methacrylate used as an embedding medium. Transverse section. TEM

Explosion artifacts, caused by local swelling of the methacrylate during its polymerization, are also a serious problem (Côté and Day 1962, Jayme and Bergh 1968, Côté et al. 1969). For these reasons, one must be careful when interpreting transmission electron micrographs of morphological entities such as the helical cavities.

4.4.5.4.3 The Origin of the Helical Cavities. The formation of the helical cavities is discussed in Chap. 9.6.1.3. This section will be largely limited to a discussion of the few attempts that have been made to explain them as artifacts.

Among the several misconceptions entertained by Gothan (1905) was his view that the chemical reactions involved in the formation of heartwood caused a weakening of the bordered pit membrane so that, on subsequent drying, spiral striations were produced in the cell wall. A similar view has been attributed by Càsperson (1959a, b, 1963a) to Bailey and Kerr (1937), but this is incorrect, for what these authors are actually referring to are not the helical cavities but the checks that can develop on drying of mild or moderate compression wood. Matsumoto (1950a) failed to observe any "spiral cracks" in green compression wood of *Cryptomeria japonica* and *Thujopsis dolobrata*. From his illustrations it is clear that he is referring to helical cavities. Such cavities appeared only after the wood had been treated with chemicals, such as cuprammonium or 72% sulfuric acid. Matsumoto also noted the appearance of helical cavities after thin sections of compression wood had been cut with a dull microtome knife. Casperson (1959a, b, 1963a) and Casperson and Zinsser (1965) mention that according to Frey-Wyssling (1953) the helical cavities could be caused by internal stresses and strains in the living tree and might be regarded as initial compression failures. It is obvious, however, that, like Kerr and Bailey (1937), Frey-Wyssling (1953) is referring to the helical checks. In his well-known monograph, Mayer-Wegelin (Trendelenburg and Meyer-Wegelin 1955) also seems to have misunderstood Frey-Wyssling. The statement by White (1965) that the "fine spiral striations" are thought to be incipient compression failures probably has the same source.

It is now firmly established that the helical cavities are present in green wood, that is, in the living tree. Already Hartig (1896) doubted that the fine striations could be caused by drying, which, as he had observed, tended to produce much larger fissures. Krieg (1907) was the first to state emphatically that the cavities represent a preformed morphological feature, and the correctness of this view has since been confirmed by many other investigators, such as Pillow and Luxford (1937), Onaka (1949), Casperson (1959a, b, 1962a, b, 1963a, b, c, d, 1965a, b), Wergin and Casperson (1961), Core et al. (1961), Wardrop and Davies (1964), Casperson and Zinsser (1965), Kutscha (1968), Côté et al. (1968c), Jutte and Levy (1972), Timell (1972b, 1973), Cockrell (1973, 1974), Fujita et al. (1973), and Yumoto et al. (1982, 1983). The helical cavities are initiated at a stage in cell differentiation when only a part of the cell wall has been formed and prior to lignification of this region. According to Wardrop and Davies (1964), the cavities are a result of a tangential contraction of the inner part of the S_2 layer in a direction transverse to that of the microfibrils. Casperson and Zinsser (1965), on the other hand, concluded that the cavities arise as the helical ribs are directly formed by the cytoplasm. An intermediate position was taken by Côté et al. (1968c), who believed that at first the cavities arise on tangential contraction of the cell wall, but that the later, innermost part of them is formed by deposition of cell wall material onto the top of the already existing ridges. Fujita et al. (1973), in their studies on the formation of the helical cavities, came to the conclusion that these fissures are not of schizogenous origin but are caused by an irregular deposition of cell wall material from the cytoplasm. These as well as later contributions are discussed in Chap. 9.6.1.3. A good indication that the helical cavities cannot possibly be artifacts produced after the death of the cell and the disappearance of the cytoplasm is the fact that a warty layer frequently is seen to cover not only the top of the helical ribs but also the inside of the cavities for a short distance, as shown in Fig. 4.79. During differentiation of compression-wood tracheids, the cytoplasm can also be seen deep inside the cavities (Figs. 9.35 and 9.42).

Spiral striations, similar to those present in compression-wood tracheids have occasionally also been observed in tracheids of normal conifer wood. Jaccard (1940) has reported the sporadic occurrence of such striations in normal wood of *Pinus contorta*, *P. echinata*, and *P. ponderosa*. According to Onaka (1949), spiral striations can form in wood when it is dried suddenly. Greguss

4.4.5.4 The Inner Portion of the S$_2$ Layer

Fig. 4.79. Inner surface of a compression-wood tracheid in *Pinus densiflora*. Both ridges and cavities are covered with a warty layer. Longitudinal section. Carbon replica, TEM. (Harada et al. 1958)

(1967) observed spiral striations in several fossil, coniferous woods, which could not all have been compression wood (Chap. 8.2).

4.4.5.4.4 Occurrence of the Helical Cavities. Although the few published reports are partly conflicting, it would appear that the occurrence and size of the helical cavities are among the few morphological features of compression wood that are not the same in all gymnosperm families, genera, or species. Care must always be exercised in this connection, as the cavities tend to be either absent or only poorly developed in mild and moderate compression woods and frequently are missing in the first-formed earlywood even in pronounced compression wood. As already mentioned, the orientation of the cavities is sometimes such that they do not appear on all sides of a transversely sectioned compression-wood tracheid.

Lämmermayr (1901) was the first to examine the structural features of compression wood from a large number of species, whereas Hartig (1896, 1899, 1901) and Cieslar (1896) had restricted their observations to *Picea abies*. Using light microscopy, Lämmermayr found the spiral striations to be "most completely and beautifully" developed in the genus *Pinus*, but they were also very readily distinguished in *Larix* and *Podocarpus*. The majority of the 36 species examined contained cavities, the only exceptions being *Ginkgo biloba*, *Taxus baccata*, and members of the two genera *Agathis* and *Araucaria*. Lämmermayr (1901) did note, however, that in *Agathis*, *Araucaria*, and *Taxus*, the pits had a slitlike opening which extended beyond the pit border. Krieg (1907) mentions that the spiral striations can be detected only with great difficulty in *Taxodium distichum*. It is likely that he examined only mild or moderate compression wood.

Onaka (1949), also using light microscopy, examined compression wood from 24 species and found all of them to contain helical cavities, including *Ginkgo biloba* and *Taxus cuspidata.* Westing (1965) believes that he might have been observing spiral drying checks, which seems very likely. Wardrop and Dadswell (1950) examined 13 species and found helical cavities in all of them, including *Araucaria bidwillii, A. cunninghamii,* and *Taxus baccata.* The cavities were very coarse in the various species of the genus *Pinus* but extremely fine in *Araucaria cunninghamii, Diselma archeri,* and *Pherosphaera hookeriana.* Patel (1963), by contrast, saw no spiral striations, either fine or coarse, in *Taxus baccata* and *Torreya californica.*

Wergin and Casperson (1961) could not observe any spiral striations in longitudinal sections of *Taxus baccata* compression wood, even at the highest magnification of the light microscope. In the electron microscope, however, helical cavities could be distinguished. The sections used were considerably deformed, but it could nevertheless be estimated that the cavities must be 0.1 μm or less in width, that is, below the resolution power of the light microscope. The investigators claimed that the cavities in this species are only barely visible under the light microscope, which should explain the negative results reported by Lämmermayr (1901), Krieg (1907) and Patel (1963) and the positive identification by Onaka (1949) and by Wardrop and Dadswell (1950). Höster (1971) could not observe any helical cavities in compression wood tracheids of *Agathis* and *Podocarpus,* a fact that he attributed to the presence of only partly developed compression wood of the mild type. Takaoka and Ishida (1974) observed helical thickenings but no cavities in compression wood of *Cephalotaxus harringtonia, Taxus cuspidata,* and *Torreya nucifera.*

My own observations of transverse or longitudinal sections of compression wood in a large number of conifer species by transmission or scanning electron microscopy has given as a result that all tracheids contained helical cavities except four, namely *Ginkgo biloba, Agathis robusta, Araucaria bidwillii,* and *Taxus baccata.* The compression wood present in *Ginkgo biloba,* a tree of very ancient lineage, is discussed in connection with the evolution of compression wood in Chap. 8.3.1. Suffice it to say here that compression-wood tracheids in Ginkgo lack helical cavities and ribs (Figs. 8.10–8.12 and 8.14) (Timell 1978b, 1981, 1983). Early investigators, who reported the occurrence of spiral striations in this species, were probably observing drying checks. *Agathis* and *Araucaria* are regarded as primitive among the conifers. Even pronounced compression wood of the two species examined had no helical cavities (Fig. 8.15), although drying checks were not uncommon, as can be seen in Fig. 4.80. I also found (Timell 1978a) that helical thickenings were attached to the inner surface of the tracheids in both normal and compression woods of *Taxus baccata,* namely to the S_3 layer in normal and to S_2 in compression-wood cells. There were no helical cavities in the compression wood (Figs. 4.94–4.98). Compression wood formed in exposed roots also lacked helical cavities (Fig. 4.145). My results confirm those obtained by Lämmermayr (1901) and also agree with Patel's (1963) and Höster's (1971) observations with *Taxus* and *Agathis* compression woods. It is very easy to mistake drying checks for helical cavities, particularly

4.4.5.4 The Inner Portion of the S$_2$ Layer

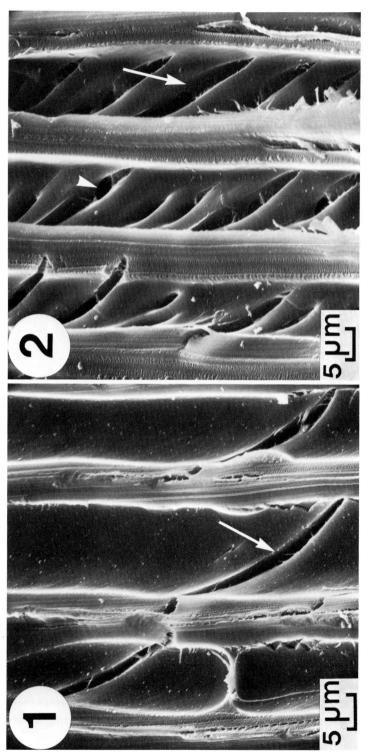

Fig. 4.80. Compression wood tracheids in *Agathis robusta* (**1**) and *Araucaria bidwillii* (**2**). *Arrows* indicate drying checks. Those in *Araucaria* had originated at pit apertures in the wall (*arrowhead*) (cf. Fig. 8.15). Radial surfaces. SEM

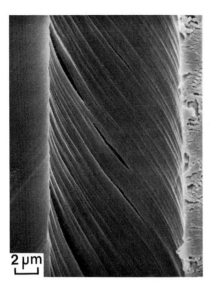

Fig. 4.81. Compression-wood tracheid in *Araucaria angustifolia*, formed at the end of the growing season and containing fine, shallow helical cavities. Radial surface. SEM. (Yoshizawa et al. 1982)

under the light microscope, and this fact might explain most of the contrary observations in the past.

Recently, Yoshizawa et al. (1982) have reported the absence of helical cavities in *Ginkgo biloba, Cephalotaxus harringtonia, Taxus cuspidata*, and *Torreya nucifera*. Contrary to Timell (1978b), they could observe helical cavities in an *Araucaria* species, namely *A. angustifolia*, but only in the last few tracheids formed at the end of the growing season (Fig. 4.81).

Shelbourne and Ritchie (1968) have reported an estimation of the frequency of the "checks or striations" in compression wood tracheids. Mild, moderate, and severe compression woods, as well as normal wood of *Pinus taeda*, were examined. In earlywood there was, not unexpectedly, little difference in this respect between normal wood (3.3%) and mild (7.3%), and severe (11.6%) compression woods. For latewood, on the other hand, corresponding figures were 2.0%, 14.0%, and 59.1%. Harris (1976) similarly found that in *Pinus echinata* there were no "cell wall checks" in mild and moderate compression woods and only a few in severe compression wood within the earlywood zone. In the latewood, such checks were present in mild and abundant in moderate and severe compression woods. Obviously, cavities and checks are more common in severe than in mild or moderate compression woods, and they also occur more frequently within the latewood than within the earlywood portion of the growth rings. Unfortunately, it is not clear whether a distinction was made in these two investigations between helical cavities and artificial drying checks.

In their investigations on the effect of the extent of inclination on the structure of the compression wood formed in tilted *Picea glauca* trees, discussed in detail in Chap. 12.5.3, Yumoto et al. (1983) found that helical cavities and ribs were either absent or incompletely developed in trees inclined 20° or less from the vertical. Compression-wood tracheids of the mild type, with an angular out-

4.4.5.4 The Inner Portion of the S₂ Layer

line and thin walls lacked helical cavities. According to Yumoto et al. (1983) helical cavities should be observed not only on the longitudinal surface of the lumen in the tracheids but also on transverse sections.

Very often the first-formed earlywood tracheids in a growth ring either lack helical cavities entirely, or the cavities are only poorly developed. Occasionally, the same is found for the last-formed, radially flattened tracheids in the latewood, as shown in Fig. 4.18. Sometimes the cavities are also missing at the tip of the tracheids, but in other cases they can be both numerous and well developed here. Wardrop and Dadswell (1952) and Wardrop (1954) have claimed that the helical cavities are either difficult to distinguish or entirely absent at the tip of the tracheids. There appear to be no helical cavities in the abnormal compression wood tracheids observed by Bucher (1968) and by Scott and Goring (1970) (Figs. 4.43 and 4.44). On the whole, the occurrence of the helical cavities is somewhat erratic, and sometimes they can be missing in fully developed compression wood for no obvious reason. Yumoto et al. (1983), by contrast, consider the helical cavities to be a stable feature of compression-wood tracheids, while they regard the presence of intercellular spaces and the absence of an S_3 layer as unstable characteristics.

4.4.5.4.5 Size of the Helical Cavities and Ribs. The width of the helical cavities in compression wood is variable and probably for this reason has attracted little attention. It has been claimed by some investigators that the cavities should be especially narrow in *Ginkgo, Agathis, Araucaria, Taxus,* and *Torreya,* but, as already mentioned, it is doubtful if these genera possess any cavities. According to Casperson (1959b, 1962a) the helical cavities have an average width in *Picea pungens* of 0.1 µm. In a transverse section of a compression wood tracheid in *Pinus resinosa,* which had never been dried from water but solvent-exchanged from its green state and subsequently embedded into methacrylate, the helical cavities had a fairly uniform width of 0.12 µm along their entire length. Their width was more variable after air-drying (Timell, unpublished results).

The depth of the helical cavities has been the subject of many contradictory statements in the literature. Sanio (1873), who was the first to describe them (Timell 1980b), noted that they did not extend into what is now known as the $S_2(L)$ layer. Hartig (1896) believed that they did not reach what is now referred to as the S_1 layer, and Krieg (1907) states that they do not reach the middle lamella, which probably means the same. According to Münch (1938), the cavities do not extend as far as the primary wall, but according to Correns (1961) they come close to it. Casperson (1962a) concluded that the spiral striations traverse the entire secondary wall. If his later nomenclature (Casperson 1968) is used, this means that they should terminate at the S_1 layer. Côté and Day (1965) and Côté (1967) also state that the helical cavities terminate at the S_1 layer, but the published electron micrograph clearly shows that the only cavity that reaches S_1 does so because it had been split open at the bottom, probably on drying or on polymerization of the methacrylate used for embedding. Boyd (1973) also seems to believe that the cavities can extend to the border between the S_1 and S_2 layers. Other investigators have found that the cavities terminate at the inner boundary of the $S_2(L)$ layer, and that accordingly the outer one third of the S_2 layer (S_2 (L)) is devoid of any fissures (Wardrop and Davies 1964, Côté et al. 1966 b, 1968 a). Scurfield and Silva (1969 a, b) have suggested that the cavities occasionally might reach S_1, but the evidence presented for this view cannot be deemed conclusive.

Inspection of the electron micrographs in this book of transverse and longitudinal sections of compression-wood tracheids and of lignin skeletons and holocelluloses of such tracheids shows that the helical cavities only extend to the S_2 (L) layer, and that they accordingly all come to an end at the same distance of about 1 µm from the inner surface of the S_1 layer. I have never yet come across a case where the cavities continued into the S_2 (L) layer. Reports to the contrary must have been due to the presence of artifacts, such as drying checks, insufficient observations, or misinterpretation of the ultrastructure of the tracheids in compression wood.

As mentioned in Chap. 3.2, Fukazawa (1973) has suggested that the depth of the helical cavities should be used as a measure of the severity of different compression woods. Park et al. (1980) later used this technique in their study of branch compression wood (Chap. 10). More recently, however, Yumoto et al. (1983) found this method to be unreliable.

Casperson (1959 b, 1962 a, b, 1965 b) and Casperson and Zinsser (1965) have on several occasions pointed out that the helical cavities are empty and not filled with lignin as, according to Casperson, has been claimed by several investigators, including Bailey and Kerr (1937), Dadswell and Ellis (1940), Dadswell and Wardrop (1949), Wardrop and Dadswell (1950), Wardrop (1954), and Nečesaný (1955 b). Careful perusal of the publications cited fails, however to unearth any such claim, and the entire controversy must be regarded as resulting from an unfortunate misunderstanding. What these investigators, and particularly Wardrop, have shown is that the radial pattern created by the radially oriented helical cavities persists in the lignin meshwork that remains after the polysaccharides have been removed from compression wood tracheids. Needless to say, the lignin is located within the ribs and not in the cavities which, as stressed by Wardrop and Dadswell (1950) and later by Boyd (1978), are empty. Scurfield and Silva (1969 a, b) have claimed that the helical cavities should be filled with a substance different in density from the material in the ribs. No microscopic or other evidence was, however, cited for this conclusion. In a similar vein, Brodzki (1972) has reported that callose should be located within the helical cavities. The evidence for this, which is unconvincing (Boyd 1978, Waterkeyn et al. 1982), is discussed in Chap. 6.3.3.

4.4.5.4.6 Orientation of the Helical Cavities. All investigators agree that the helical cavities in compression-wood tracheids follow exactly the orientation of the cellulose microfibrils in the S_2 layer (Sect. 4.4.5.2), and, if it is accepted that the cavities are at least partly schizogenous in origin, this is, of course, what one would expect. Their orientation was first determined by Krieg (1907), who found an average angle of 24° for *Chamaecyparis lawsoniana*, 30° for *Juniperus virginiana*, 41° for *Pinus sylvestris*, 34° for *Taxodium distichum*, and 40° for *Thuja orientalis*. Krieg was the first to show that the angle of orientation of the striations and of the slit-like openings across the pits was the same as that of the

4.4.5.4 The Inner Portion of the S₂ Layer

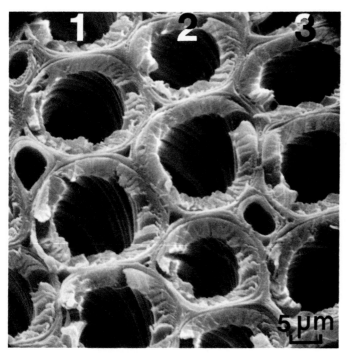

Fig. 4.82. Compression-wood tracheids in *Pinus resinosa*. The cavities and ribs are organized in a steeper helix in cell files *1* and *2* than in *3*, where it is unusually flat. Transverse surface. SEM

micelles (microfibrils), a fact which he discovered with the aid of the polarizing microscope. Jaccard and Frey (1928), who used the same technique, could confirm this 20 years later. Subsequent investigators, such as Wardrop and Dadswell (1950), have made the same observation. Onaka (1949) showed that the helix runs from left to right, that is, it is of the Z-type. In comparisons between the helical cavities and the microfibrils, the orientation of the latter has been established with the aid of the polarizing microscope, x-ray diffraction, and direct observation in the transmission or scanning electron microscopes.

Figure 4.82 illustrates a remarkably large difference in the orientation of the helical cavities between adjacent tracheids in compression wood of *Pinus resinosa*, with a steep helical pitch in rows 1 and 2 but a flat pitch in row 3. There is undoubtedly a similar difference in microfibril angle between the two cell rows. Tracheids in compression wood of *Pinus sylvestris*, where the helical cavities and ribs have an almost transverse orientation, are seen in Fig. 4.68. Yumoto et al. (1983) have reported a similar, very flat helix in *Picea glauca*.

In normal wood, helical thickenings are not formed in the area occupied by a bordered pit. The helical ridges in compression wood also tend to bend around a pit (Scurfield and Silva 1969a, b, Yumoto et al. 1983). Instead, one single cavity, crossing the pit diagonally in the middle, is notably widened, often conferring a slit-like appearance to the pit aperture (Sect. 4.10.1). These slits

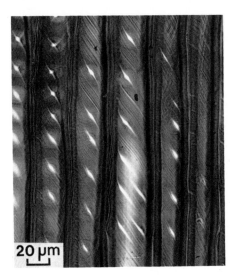

Fig. 4.83. Compression-wood tracheids in *Picea rubens* with (seemingly) slit-like pit openings. Longitudinal section. LM. (Courtesy of A.C. Day)

are among the characteristic features of compression wood and have often been used for identification purposes. They are unique to compression wood tracheids and can be readily observed under the light microscope in longitudinal sections, as shown in Fig. 4.83. The typical Maltese cross appearing in polarized light (Fig. 4.75) has already been referred to. It is caused by the fact that, like the striations, the slits on the front and on the back of the tracheid wall occasionally can be seen at the same time. It should perhaps be pointed out here that the pit aperture actually is circular in compression wood. The slit-like opening seen in the light microscope is caused by the fact that the round aperture is located at the bottom of a groove, as is shown in Sect. 4.10.1.

4.4.5.4.7 The Helical Ribs. Because of the presence of the helical cavities, the cell wall material in the inner two thirds of the S_2 layer in compression-wood tracheids is organized in the form of helical ribs or ridges. Their general appearance is clearly brought out in the cell wall model shown in Fig. 4.84, which is a slight modification of an earlier model proposed by Casperson (1962a) (Casperson and Zinsser 1965). Another modification of the same original model has been published by Dadswell and Wardrop (1960). According to Nečesaný (1955b) the ribs are massive bundles of microfibrils, 1.2–2.0 µm thick. Casperson (1959b, 1962a, b) also finds that the ribs have a width of 1–2 µm. Actually, the helical ribs are of variable width, but whether there are differences in this respect between species, genera, or families is not known at present. The ribs seen in Fig. 4.76 (*Abies balsamea*) are less than 0.05 µm wide, whereas those in Fig. 4.85 (*Pseudotsuga menziesii*) are 0.15–0.20 µm wide. The ribs can also vary considerably in width within a single tracheid, as shown in Fig. 4.68 for *Pinus sylvestris*. According to Takaoka (1975) the ribs increase in width with increasing development of the helical cavities.

4.4.5.4 The Inner Portion of the S$_2$ Layer

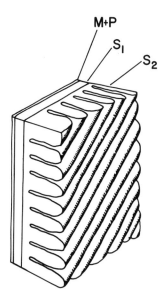

Fig. 4.84. Model showing a portion of the cell wall in a compression-wood tracheid. The orientation of the microfibrils and the helical cavities in S$_2$ is assumed to be 45°. (Modified from Casperson 1962a and Casperson and Zinsser 1965)

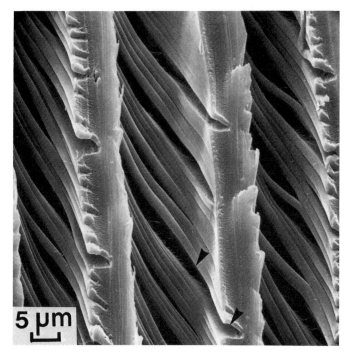

Fig. 4.85. Compression-wood tracheids in *Pseudotsuga menziesii* with coarse helical ribs (cf. Fig. 4.99). *Arrowheads* indicate a drying check. Radial surface. TEM. (Timell 1978a)

The top of the ribs often protrudes into the lumen in a characteristic manner already observed by Hartig (1896, 1899, 1901), as can be seen from Fig. 2.7. Other examples are found in Figs. 4.71 and 4.78. Sometimes the protruding ribs are also evident in both lignin (Fig. 6.14) and holocellulose (Figs. 4.55 and 6.16) skeletons. The phenomenon, which has been discussed by Wergin and Casperson (1961), appears to be especially common in *Larix*, *Picea*, and *Pinus*, while it seems to be more rare in *Abies* and *Tsuga*.

Scurfield and Silva (1969a, b), in a scanning electron microscopy study of compression wood in *Cupressus arizonica*, *Pinus canariensis*, and *P. radiata*, have claimed that the helical ribs should be multistranded. Both transmission and scanning electron micrographs published in later years clearly indicate, however, that the ribs are homogeneous and composed of regularly arranged microfibrils. It is true, as pointed out by Scurfield and Silva, that strands of microfibrils probably cannot be distinguished on the surface of walls still carrying the warty layer, such as the lumen surface shown in Fig. 4.79. However, in those cases where the warty layer and part of the lignin had been removed prior to examination, the strands, if existing, certainly should be observable. This, however, is not the case, at least not in the compression wood of *Larix laricina* shown in Fig. 4.69. The matter needs further investigation and preferably with more than the three species so far studied.

Examination of developing compression wood tracheids (Chap. 9.6.1.3) and also of delignified, mature tracheids indicates that the inner portion of the S_2 layer, like all other cell wall layers, is lamellar in nature. In developing tracheids, the lamellae are parallel to the tangential surface of the cell (Figs. 9.26, 9.28, 9.31, and 9.38), a fact first pointed out by Wardrop and Davies (1964). In fully differentiated tracheids, the situation seems to be more complex. The lamellar orientation is usually difficult to distinguish in transverse sections of lignified sections. In delignified specimens, the lamellae are, of course, readily observed but now they are often distorted as a result of swelling. In Fig. 4.58 the tangential orientation of the lamellae can be traced in the inner part of S_2 of these fully lignified tracheids. Figure 6.16 shows the orientation of the lamellae in a delignified compression wood tracheid. In these micrographs, the arrangement of the lamellae within the ribs appears to be somewhat concentric, and more so the closer they are to the top of the ridges. Wardrop and Davies (1964) have discussed this problem and come to the conclusion, based on their observations of developing tracheids, that all lamellae in the S_2 layer are parallel to the inner surface of the cell. Judging from their model, shown in Fig. 4.46, Scurfield and Silva (1969a) seem to share this view. In Chap. 9.6.1.3, by contrast, it is suggested that the lamellae in the inner S_2 are at first parallel with the lumen surface (Fig. 9.26.2) but later assume the more concentric pattern illustrated in Fig. 9.26.3. An example of this is seen in Fig. 9.39.

4.4.5.4.8 The Helical Checks. The helical checks, unlike the helical cavities, are not present in developing compression-wood tracheids (Chap. 9.6.1). They have, however, occasionally been reported to occur in green wood and might accordingly occur in living trees. Factors causing them to appear are stresses

4.4.5.4 The Inner Portion of the S_2 Layer

and strains of the same nature as those that cause microscopic compression failures (Chap. 15.2.1.5) or movements caused by drying, for example when sapwood is converted into heartwood. It must also be remembered that compression wood is in a state of active expansion in the living tree and that distortions of the cells are to be expected, not only the buckling of the tracheid tips, already referred to. Checks, cracks, or splits are, of course, also produced on drying of green compression wood when the large longitudinal and still considerable radial and tangential shrinkage cause the helical ribs to separate from one another (Verrall 1928). Treatment with chemicals that destroy either the lignin or the polysaccharides in the cell wall has the same effect. According to Cockrell (1973), splits extending from the pit aperture and spiral checks in the S_2 layer of compression wood tracheids of *Sequoiadendron giganteum* are not present in green wood, as can be seen in Fig. 4.133. Furthermore, they do not develop on drying, according to Cockrell.

Krieg (1907) was the first to discuss the formation of spiral or helical checks in compression wood. He noted that they were only formed in tracheids already containing spiral striations. The fact that the helical ribs, when strained tangentially, would tend to separate from one another along the helical cavities is only what could be expected, and Krieg (1907) mentions that he had actually observed the initiation of such separations. He also noted that the formation of a crack often began from the inner portion of the cell wall and not from the lumen. An example of this is shown in Fig. 4.45. The strands of microfibrils seen crossing the checks in this micrograph clearly indicate the schizogenous origin of the latter. Krieg also made the interesting observation that when compression wood tracheids are set free, the helical ribs separated to form larger and more irregularly distributed fissures than was the case when the tracheids were held together by the middle lamella in the wood. He interpreted this as a result of the greater restraint imposed within the wood upon the wall. The considerable distortions generally observable in delignified compression-wood tracheids indicate the correctness of this interpretation.

The apparent ease with which helical checks are produced in compression wood tracheids is due not only to the presence of the helical cavities but also to the absence of an S_3 layer. This layer, with its almost transversely oriented microfibrils, protects the tracheid wall in normal wood from radial fracture. In the absence of this layer, helical checks can develop in the S_2 layer, even if no helical cavities are present. Figure 4.80 shows this for *Agathis robusta* and *Araucaria bidwillii* and Figs. 8.11 and 8.12 for *Ginkgo biloba*. Checks are also present in the S_2 layer of the root wood tracheids shown in Fig. 4.145, their lack of any helical cavities notwithstanding. This is the reason why the helical checks, albeit artifacts, nevertheless are diagnostically useful in identification of compression wood. They are obviously only formed in the absence of an S_3 layer. Helical checks also develop in compression wood tracheids of *Taxus baccata*, where they are oriented in the same direction as the helical thickenings, as shown in Fig. 4.86. The helical check seen in Fig. 4.87 contains no warts on its inner surface, whereas warts are uniformly distributed over the adjacent cavities and ribs, demonstrating that this fissure had developed after the cell had ceased to differentiate. This is also obvious from Fig. 4.101.

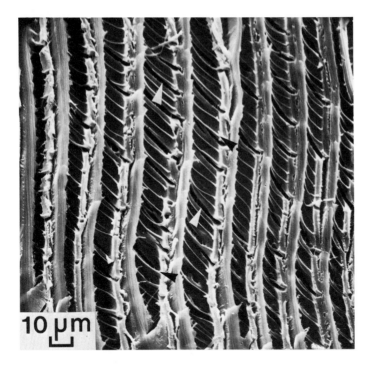

Fig. 4.86. Compression wood tracheids in *Taxus baccata* with helical thickenings (*black arrowheads*) and helical checks (*white arrowheads*) oriented in the same direction. Radial surface. TEM. (Timell 1978a)

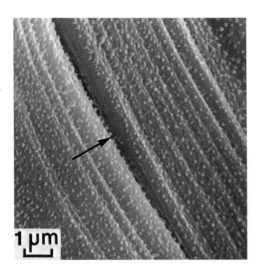

Fig. 4.87. Inner surface of a compression wood tracheid in *Abies sachalinensis*. Note that warts occur both on the helical ribs and inside the cavities but not inside the helical check (*arrow*). Longitudinal surface. SEM. (Ishida et al. 1968), (cf. Fig. 4.101)

4.4.5.4 The Inner Portion of the S_2 Layer

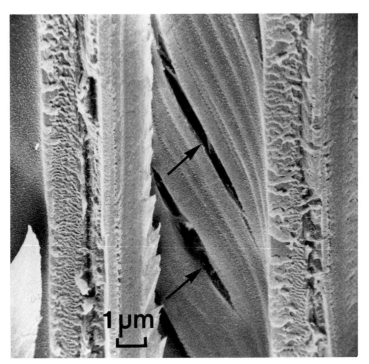

Fig. 4.88. Compression wood of *Sequiadendron giganteum* that had been boiled in water, sectioned, and dried, which caused helical checks (*arrows*) to develop. Longitudinal surface. SEM. (Cockrell 1973)

In many cases, the helical checks extend only as far inward from the lumen as the helical cavities, that is, to the inner surface of the $S_2(L)$ layer. Examples of this are seen in Figs. 4.17, 4.18, 4.51, 4.58, and 4.78. Frequently, however, the tangential contraction of the helical ribs is so great that the fracture continues beyond the cavities and enters the $S_2(L)$ layer. Rupture of the S_1 layer with its transversely oriented microfibrils is rare. Examples of this type of fracture are seen in Figs. 4.50, 4.51, 4.68, 4.80, 4.85, 4.86, and 5.26. On drying from liquid ammonia (Chap. 5.3.3), the helical cavities tend to widen into checks which often reach the S_1 layer, as shown for *Pinus taeda* by Parham (1970) and by Parham et al. (1972). Cockrell (1973) found that helical checks developed when compression wood of *Sequoiadendron giganteum* was boiled in water, cut when wet, and then dried. Figure 4.88 shows three large checks in addition to fine helical cavities. Occasionally, the helical checks can extend across the entire S_1 layer, reaching the middle lamella. An example of such a deep fissure is seen in Fig. 4.85.

The helical checks vary in width and shape. They are always wider and sometimes also deeper than the helical cavities, but they are seldom more than 1.5 µm wide. Because of their much larger size, the checks can be observed in transverse sections under the light microscope, even when the helical cavities

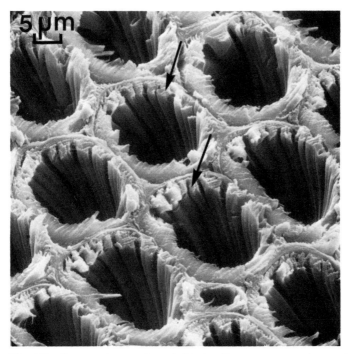

Fig. 4.89. Compression wood tracheids in *Cupressus sempervirens* with an unusually large number of helical checks, conferring a rope-like appearance on the steeply oriented helical ribs (*arrows*). Radial surface. SEM

are not visible. At times, the entire secondary wall appears to be broken up by helical fissures. An example of such excessive checking in *Cupressus sempervirens* is shown in Fig. 4.89.

Removal of lignin or polysaccharides from compression wood or treatment with strong swelling agents rarely, if ever, leaves the helical cavities unchanged. Usually, the ribs are pulled apart, producing large checks, or, with swelling agents, the cavities are obliterated when the ribs are pressed together. When the polysaccharides are eliminated by treatment with hydrofluoric acid, the lignin skeletons in the ribs contract laterally, producing large checks between them, as shown in Figs. 6.13 and 6.14. When the lignin is removed, for example in pulping (Chap. 19), the inner portion of the S_2 has a tendency to separate from the S_1 because of the low concentration of polysaccharides in the intervening $S_2(L)$ layer. The polysaccharide material in the ribs contracts on drying, obliterating the lamellae and creating large checks, as shown in Fig. 4.55. Sometimes the ribs expand radially into the lumen, as seen in Fig. 6.16. Hartig (1901) observed a similar effect on treating compression wood with concentrated sulfuric acid, a strong swelling agent for wood polysaccharides. Apparently, any chemical reagent used to eliminate lignin from compression wood tracheids causes the helical ribs to separate from one another. Acetic acid and hydrogen peroxide, used by Scurfield and Silva (1969a, b), causes such a separation. Miniutti (1967) exposed *Sequoia sempervirens* compression wood to ultraviolet radiation

4.4.6 The S₃ Layer

for 27 weeks. The radiation completely destroyed the ray cells and transformed helical cavities in the tracheids into checks. Fungal degradation (Bailey 1939) and weathering (Hiller and Brown 1967) of normal wood also produce checks in the wall of the tracheids.

4.4.5.4.9 Conclusions. Winding in a helix around the lumen of the tracheids, the helical cavities and ribs are the most conspicuous and peculiar among the ultrastructural features of compression wood. Of all properties of this tissue, they seem to require the most intense stimulus to develop. When compression wood formation ceases, the high lignin concentration and the round outline of the tracheids persist long after the helical cavities have disappeared. Similarly, when normal wood is gradually transformed into compression wood, the helical cavities are the last characteristics to appear (Chap. 9.8). They are well developed only in severe compression wood and are occasionally missing even in such wood. They are often absent in mild and moderate compression woods and in first-formed earlywood. It is possible that formation of cavities appeared later in evolution than the other ultrastructural characteristics of compression wood, since they are not formed in *Ginkgo biloba* or in some primitive conifer genera, as is discussed more fully in Chap. 8.

Helical checks and fissures develop after tracheid differentiation has been completed and are artifacts, most commonly produced on drying. Their formation is most probably made possible by the absence of a restraining S_3 layer. Whether they occur also in green, never dried compression wood is still a moot question.

All cavities and most checks terminate at the inner surface of the $S_2(L)$ layer and thus occupy $2/3-3/4$ of the total width of the S_2 layer. The cavities are approximately 0.1 µm wide, and the checks 0.5–1.5 µm. The helical ribs are 1–2 µm wide. Sometimes their top protrudes characteristically into the lumen. The significance of the cavities and ribs for the function of compression wood in the living tree is uncertain. Perhaps they contribute to the high resilience so characteristic of compression wood which makes it possible for displaced stems and branches to return rapidly to their original position. Perhaps they serve a function in making it possible for compression wood to expand in the living tree and thus cause the necessary movements of orientation. Whatever role the helical cavities and ribs may play in the living tree, there can be no doubt that they must have some effect on the physical properties of compression wood.

4.4.6 The S_3 Layer[1]

In 1896 Robert Hartig reported that compression-wood tracheids in *Picea abies* had a cell wall that was terminated toward the lumen by a thin "tertiary wall".

---

[1] The notion that S_3 should not be a part of the secondary wall in tracheids and fibers and therefore should be referred to as "the tertiary wall" is a concept that is patently erroneous, and not only from an ultrastructural and chemical but also from an ontogenetic point of view. The nomenclature introduced by I. W. Bailey 50 years ago is as valid today as it was then (Kerr and Bailey 1934).

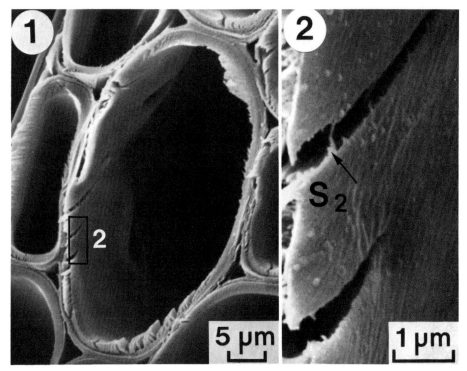

Fig. 4.90. Severe compression wood of *Pinus radiata* (**1**) with an enlarged portion (**2**), showing the presence of a thin S_3 layer. *Arrow* indicates where S_3 is crossing a drying check. Transverse surface. SEM. (Harris 1977)

In his last publication from 1901, however, he states that compression wood lacks a tertiary wall, a conclusion which he had reached after observing in the microscope compression-wood tracheids that had been treated with concentrated sulfuric acid. In this respect, these tracheids were quite different from those in opposite wood which contained an unusually thick S_3 layer (Chap. 21.3.2). Many investigators have since confirmed Hartig's later opinion that a "tertiary" wall is not formed in fully developed compression-wood tracheids. Some of those who have done so are Sonntag (1904) and Münch (1938), who both might be citing Hartig, Wardrop and Dadswell (1950, 1957), Harada and Miyazaki (1952), Wardrop (1954), Ollinmaa (1955, 1959), Nečesaný (1955b, 1957, 1966), Meier (1957), who states that it is "probably" absent, Harada et al. (1958), Casperson (1959a, b, 1962a, b, 1963a, d, 1965a, b, 1968), Wergin and Casperson (1961), Patel (1963), Jurášek (1964), Wardrop and Davies (1964), Casperson and Zinsser (1965), Wergin (1965), and Scurfield and Silva (1969a, b). Frey-Wyssling's (1976) statement that the tracheids in compression wood have a highly lignified S_3 layer must be due to a misunderstanding and be meant to apply to opposite wood.

Wardrop and Dadswell (1950) mention that a poorly developed S_3 layer sometimes occurs in compression-wood tracheids, illustrating this with a cross

4.4.6 The S₃ Layer

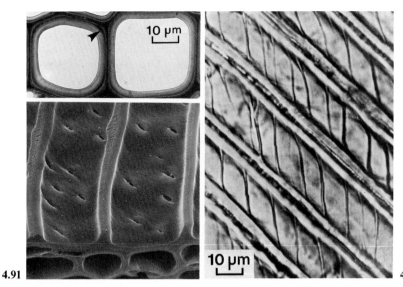

Fig. 4.91. Moderate compression wood tracheids in *Picea glauca* with a rounded outline, thick wall, and a highly lignified S$_2$ (L) layer (*arrowhead*) but also an innermost S$_3$ layer. Transverse section. ULM. Radial surface. SEM. (Yumoto et al. 1983)

Fig. 4.92. Compression wood tracheids in *Taxus baccata*. The helical thickenings have the same orientation (about 45°) as the microfibrils in S$_2$. Longitudinal section. LM. (Wergin and Casperson 1961)

section from *Pinus pinaster* which had been photographed in polarized light. The S$_1$ layer in this micrograph, however, exhibits complete extinction and if this is really a compression-wood tracheid, it must be a most unusual one. Nečesaný (1957) states that compression wood that is not fully developed might contain a thin S$_3$ layer. When Fan (1970) administered indoleacetic acid to seedlings of *Pinus banksiana*, the resulting compression wood seemed to contain an S$_3$ layer. Harris (1977) observed faint traces of an S$_3$ layer not only in mild and moderate but also in severe compression woods in *Pinus radiata*. The S$_3$ consisted of only a few strands of microfibrils oriented almost transversely to the tracheid axis, sometimes traversing drying checks, as shown in Fig. 4.90.

When normal wood is transformed into compression wood, the S$_3$ layer is the first ultrastructural feature to be eliminated (Fujita et al. 1979, Yumoto et al. 1982a, b, Yoshizawa et al. 1984a, b), as illustrated in Fig. 9.51. In the reverse process, the helical cavities and ridges are the first to disappear, following which an S$_3$ layer appears (Chap. 9.8). Mild and moderate compression woods have occasionally been reported to contain an S$_3$. An example, observed by Yumoto et al. (1983) in *Picea glauca*, is shown in Fig. 4.91. All normal wood tracheids observed by Yumoto et al. (1983) in *Picea glauca* had an S$_3$ layer. These investigators suggested that in those instances where an S$_3$ layer had been reported to be absent in normal wood tracheids (Wardrop 1964, Côté and Day 1965), what had been observed was probably very mild compression wood. Yumoto et al. found the occurrence of S_3 in moderate and mild compression woods to be exceedingly variable and even suggested that its absence in compression wood tracheids might be controlled by a factor other than gravity.

4.4.7 Helical Thickenings

Cieslar (1896) seems to have regarded the spiral striations in compression-wood tracheids as helical thickenings, which they definitely are not. Lämmermayr (1901) claimed to have observed helical thickenings in compression-wood tracheids of *Cupressus*. Gothan (1905) was probably the first to discuss the relationship between helical cavities in compression wood and the occurrence of helical thickenings. Although he had no clear concept of the nature of compression wood, he did point out that striations and thickenings must not be confused with each other, a warning repeated by several later authors, for example Kukachka (1960), Giordano (1971), Patel (1971), Bosshard (1974a), and Core et al. (1979).

Krieg (1907) pointed out that striations and thickenings sometimes occur together in *Larix, Pseudotsuga, Picea abies,* and *Taxus baccata,* but that they are never found together in one and the same tracheid: the two structures exclude each other. Krieg observed helical thickenings in opposite wood on the upper side of a branch from *Pseudotsuga menziesii*. In the compression wood on the lower side, on the other hand, he found that only the last-formed latewood tracheids of each growth ring possessed helical thickenings. The remainder contained spiral striations. Younger wood contained relatively more than older. *Larix decidua* and *Picea abies* gave similar results. In wood on the lower side of a branch from *Taxus baccata*, the helical thickenings were very narrow and oriented at an angle of only 21° to the fiber axis compared to 80° on the upper side. Normal wood of this species has almost transverse helical thickenings, and it is therefore very likely that Krieg here had observed for the first time the presence of thickenings in a compression-wood tracheid.

Two years later, Sonntag (1909) reported observations on spiral striations in a large number of different plants, including several arboreal species. He found that helical thickenings usually were absent in conifer tracheids with spiral striations but found *Pseudotsuga menziesii* to be an exception to this rule. In compression wood from the lower side of a branch of this species, he observed on the lumen side of the tracheids both spiral striations and spiral thickenings, both oriented at angles of 62°–71° with respect to the fiber axis. He concluded that the two features occasionally can occur together.

The observations of Krieg (1907) and Sonntag (1909) were destined to go unnoticed for a long time. Instead, most later investigators either repeated Krieg's statement that spiral striations and helical cavities are mutually exclusive or came to this conclusion on the basis of their own experience (Lee and Smith 1916, Münch 1938, Nečesaný 1955b, Boutelje 1965, 1966, Sudo 1968d). Verrall (1928) found that helical thickenings occurred in both normal and com-

4.4.7 Helical Thickenings

pression woods of *Taxus canadensis*. Pillow and Luxford (1937) noted that helical thickenings were present in compression-wood tracheids, for example in *Pseudotsuga menziesii*, although they were less frequent and not as well developed as in normal wood. They found that the thickenings were confined to the first-formed earlywood cells in pronounced compression wood and that their orientation had no relation to that of the microfibrils. Onaka (1949) observed helical thickenings in compression wood tracheids of *Cephalotaxus, Taxus,* and *Torreya*. Their pitch was 45°, and the direction is stated to be left.

Wergin and Casperson (1961), who believed that the helical cavities in compression-wood tracheids of *Taxus baccata* were too narrow to be discernible in the light microscope, observed that these tracheids contained helical thickenings. Since there is no S_3 in compression-wood tracheids, the thickenings would have to be attached to the S_2 layer. Wergin and Casperson (1961) could demonstrate that this was indeed the case, as shown in Fig. 4.92. Here the thickenings are spiraling around the lumen in the same direction as the microfibrils and with a helical pitch of $40° - 50°$. The thickenings have a width of $0.5-0.8$ μm and occur at a distance of $6-10$ μm from one another. In the electron microscope the width was estimated at $0.4-0.6$ μm and the distance at $3.5-7.5$ μm. Wergin and Casperson (1961) and Casperson (1962a, 1963a) concluded from their observations that the helical thickenings were only extended helical ribs. Every 6th to 12th rib was supposed to protrude into the lumen, while the others terminated at the same distance from the $S_2(L)$ layer so that the lumen surface between the thickenings appeared to be relatively smooth.

Patel (1963) examined helical thickenings in wood from the upper and lower sides of branches of *Taxus baccata* and *Torreya californica*. He assumed the upper side to consist of normal wood, although it was probably opposite wood and identified the tissue on the under side as compression wood. Both tissues contained helical thickenings throughout each growth ring, but the compression-wood thickenings were less abundant and distinct than those in normal wood. Their orientation was the same as that of the elongated pit openings, 45°. Patel suggested that this was possibly also the orientation of the microfibrils in the S_2 layer. Interestingly, he could not observe any "fine spiral striations" in the compression-wood tracheids, an observation in direct contrast to the results obtained by Wergin and Casperson (1961) with *Taxus baccata*.

Jutte and Levy (1973) have pointed out that according to the evidence presented by Wergin and Casperson (1961), the thickenings in compression-wood tracheids are oriented in an S helix (right to left, when viewed from the outside). These thickenings are attached to S_2, as already mentioned, and are generally believed to follow the orientation of the microfibrils in the wall layer to which they are attached. In normal wood, the microfibrils in S_2 are most commonly arranged in a Z helix (left to right, when viewed from the outside). There is accordingly a change from a Z to an S helix in the S_2 layer of the tracheids when normal wood is transformed into compression wood.

These arguments are not entirely correct. The light micrograph in Fig. 4.92 obtained by Wergin and Casperson (1961) is ambiguous, and the line drawing of the wall organization of a compression wood tracheid in *Taxus baccata* is erroneous. In normal wood, the thickenings are as a rule oriented in an S helix,

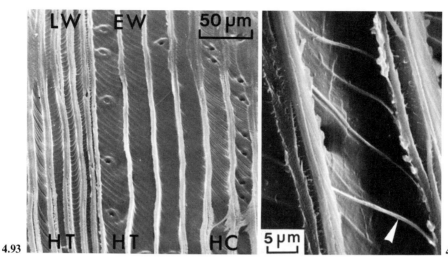

Fig. 4.93. Compression wood tracheids in *Pseudotsuga japonica* at the latewood (*LW*) – earlywood (*EW*) boundary. Helical thickenings (*HT*) attached to S_3 are present in the last-formed latewood tracheids and also in the first-formed earlywood where they occur on S_2. The rest of the tracheids contain helical cavities (*HC*). Longitudinal surface. SEM. (Takaoka and Ishida 1974)

Fig. 4.94. Compression wood tracheids in *Taxus baccata* with helical thickenings (*arrowhead*) of variable orientation. Note the absence of any helical cavities. Longitudinal surface. SEM

reflecting the orientation of the microfibrils in the innermost layer of the S_3 of which they are an integral part (Côté 1967). Possibly because of the complex organization of this layer (Dunning 1969a), a Z helix is by no means rare. Judging from published material, it can occur in species such as *Taxus baccata* (Timell 1978a) and *Torreya taxifolia* (Parham and Kaustinen 1973). It has also been observed that in *Taxus brevifolia* there can be a sudden change from an S to a Z helix from one tracheid to the next. In *Taxus cuspidata*, Yoshizawa et al. (1982) recently found thickenings oriented in a relatively steep S helix or a flat Z helix within one and the same tracheid. While the direction of the helical thickenings evidently is somewhat variable in normal wood, the thickenings in compression wood are always arranged in a Z helix. This is what one would expect, for these thickenings are attached to S_2, and, except for Jutte and Levy (1973), all investigators agree that the microfibrils in the S_2 layer in compression-wood tracheids are organized in a Z helix, as they usually are also in normal wood. I myself have never found any exception to this rule.

In an article published in Japanese in 1974 but made available to me in English only 8 years later (Yumoto, personal communication, 1982), Takaoka and Ishida (1974) reported on the occurrence of compression wood in conifer species with helical thickenings. In *Pseudotsuga japonica*, helical thickenings occurred in the first-formed earlywood and the last-formed latewood tracheids in compression wood, as shown in Fig. 4.93. These thickenings were deposited on the S_2 and had an orientation of 45° in the first-formed earlywood. In the last-

4.4.7 Helical Thickenings

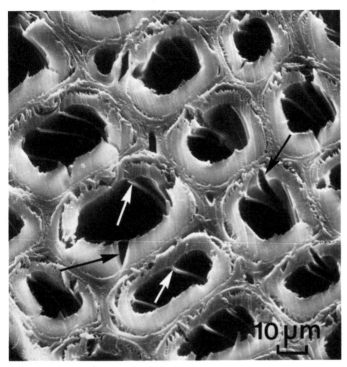

Fig. 4.95. Compression wood tracheids in *Taxus baccata* with rope-like helical thickenings (*white arrows*) but no cavities. *Black arrows* indicate deep helical checks. Transverse surface. SEM. (Timell 1978a)

formed latewood tracheids, the thickenings were attached to the S_3 layer present in these cells. The intervening portion of the growth ring had tracheids with well-developed helical cavities and ribs. Compression woods of *Cephalotaxus harringtonia*, *Taxus cuspidata*, and *Torreya nucifera* lacked helical cavities and ribs in their tracheids and instead had helical thickenings attached to the layer S_2. These thickenings were often branched, and, while usually oriented in the same direction as the microfibrils in S_2, were frequently organized in a somewhat flatter helix.

Timell (1978a) subjected the ultrastructure of compression wood tracheids in *Taxus baccata* to a renewed investigation, using scanning and transmission electron microscopy. The compression wood was not obtained from a branch as in the previous studies but from a forest tree with a basal stem crook. The pitch of the helical thickenings was relatively steep but also variable, as shown in Fig. 4.94. There were no helical cavities in the tracheids, but wide and deep helical checks were common, as can be seen in Fig. 4.95. The thickenings were attached to and part of the S_3 layer in normal wood, while in compression wood they were attached to S_2 (Fig. 4.96). The lignin skeletons in Fig. 4.97 show that the thickenings are an integral part of the S_2 layer in compression wood tracheids and also that these cells have all the characteristics typical of com-

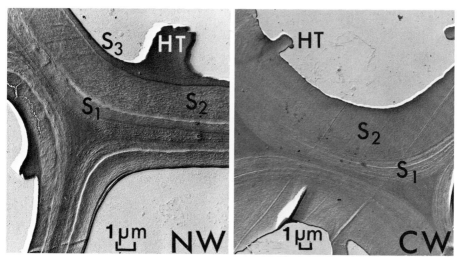

Fig. 4.96. Normal and compression wood tracheids in *Taxus baccata*. The helical thickenings (*HT*) are part of the S_3 layer in the normal wood but are attached to the S_2 layer in the compression wood. Transverse sections. TEM

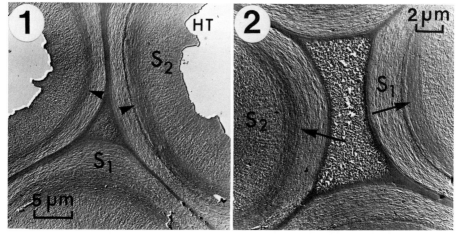

Fig. 4.97. Lignin skeletons of compression wood tracheids in *Taxus baccata*. The lignin pattern is the same in the helical thickenings (*HT*) as in the inner part of S_2 (**1**). Note the presence of the highly lignified S_2 (L) layer (*arrowheads and arrows*), immediately inside the wide S_1 layer (**1** and **2**). There are no helical cavities. Transverse sections. TEM.

pression wood tracheids, namely a thick secondary wall, a thick S_1 layer, and a highly lignified S_2(L) layer. The only exception is the absence of any helical cavities in the inner part of S_2. These results are in agreement with the earlier observations reported by Krieg (1907) and by Patel (1963), but they disagree with the conclusions drawn by Wergin and Casperson (1961). A close inspection of the micrographs published by these investigators fails to reveal any

4.4.7 Helical Thickenings

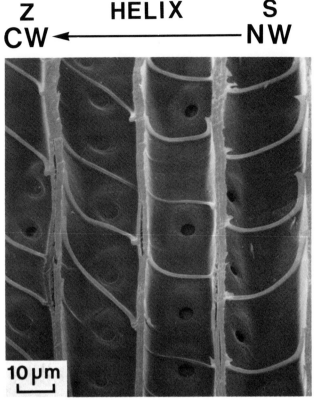

Fig. 4.98. Transformation of normal wood tracheids (*NW*) into compression wood tracheids (*CW*) in *Taxus cuspidata* (in the direction of the *arrow*) with an associated change in the organization of the helical thickenings from an *S*-helix to a *Z*-helix. Radial surface. SEM. (Yoshizawa et al. 1982)

helical cavities in their compression-wood tracheids in the light micrograph and in three of the electron micrographs. In the two electron micrographs where cavities are purported to occur, it is likely that the investigators were actually observing helical checks. Timell (1978a), unlike Patel (1963), found that the helical thickenings were as frequent and well developed in compression wood as in normal wood.

Yoshizawa et al. (1982) have recently reported interesting observations relating to the occurrence of helical cavities and thickenings in compression wood tracheids. They could confirm Onaka's original finding that compression wood tracheids in *Cephalotaxus*, *Taxus*, and *Torreya* contain helical thickenings but no cavities or ribs. Unlike Onaka, however, they came to the conclusion that the thickenings were always arranged in a Z helix. When normal wood was transformed into compression wood, the thickenings changed gradually in direction from an S to a Z helix, as illustrated in Fig. 4.98 for *Taxus cuspidata*. In these three species, helical thickenings and cavities never occurred together within one tracheid.

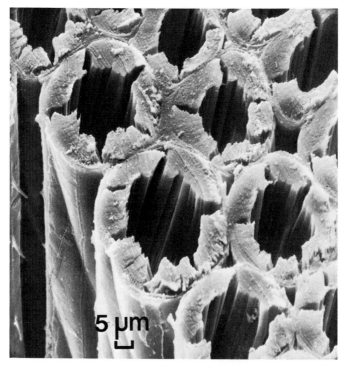

Fig. 4.99. Tracheids in pronounced compression wood of *Pseudotsuga menziesii* with coarse helical ribs and no thickenings (cf. Fig. 4.85). Transverse section. SEM

In their investigations on the transition between normal and compression woods in *Taxus cuspidata* and *Torreya nucifera*, discussed in Chaps. 9.8 and 12.5.2, Yoshizawa et al. (1984, 1985) found that the change between the S helix in normal wood and the Z helix in compression wood was gradual. A transitional tracheid, shown in Fig. 9.56, contained no S_3 layer. Although the thickenings in this cell were attached to the S_2 layer, they were organized in an S helix as in normal wood, crossing the S_2 microfibrils instead of being parallel with them. This is a striking example of the fact that helical thickenings are not always oriented in the same direction as the microfibrils in the cell wall layer to which they are attached. It is also obvious from Fig. 9.54 that the flat helical pitch of the thickenings in normal wood of *Taxus cuspidata* is more variable than the steeper pitch in compression wood, a fact that agrees with my own observations with *Taxus baccata*.

Tracheids in normal wood of *Pseudotsuga menziesii* have well-developed, often almost transverse helical thickenings. Those in fully developed compression wood, by contrast, have coarse helical cavities and ribs, as illustrated in Figs. 4.85 and 4.99 (Timell 1981). Yoshizawa et al. (1982) could demonstrate how helical cavities and ribs are changed into helical thickenings when compression wood is transformed into normal wood in *Pseudotsuga japonica*. As can be seen in Fig. 4.100, compression wood tracheid (1) contains helical cavities

4.4.7 Helical Thickenings

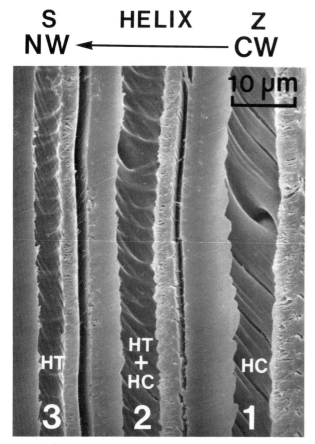

Fig. 4.100. Transformation of compression wood tracheids (*CW*) into normal-wood tracheids (*NW*) in *Pseudotsuga japonica* (in the direction of the *arrow* with an associated change of a Z-helix (*CW*) to an S-helix (*NW*). Cell *1* contains helical cavities (*HC*), in transition cell *2* helical thickenings (*HT*) have been deposited on top of cavities, and cell *3* contains helical thickenings only. Radial surface. SEM. (Yoshizawa et al. 1982)

organized in a Z-helix. In the intermediate cell (2), helical thickenings have been deposited on top of cavities. Normal tracheid (3), finally, has only helical thickenings, arranged in an S-helix. The nature of this gradual transition is such that thickenings and cavities can occur together within one and the same tracheid. The thickenings appeared to be covered by a thin layer whose microfibrils crossed them, but the nature of this layer could not be established. This is the only instance so far reported of a simultaneous occurrence of thickenings and cavities in a compression wood tracheid.

It is evident that helical cavities and thickenings never occur together within the same tracheid in *Cephalotaxus*, *Taxus*, and *Torreya*. In these genera, helical thickenings are instead attached to the S_2, usually, albeit not invariably, oriented in the same direction as the microfibrils in this layer. Compression wood of the phylogenetically younger *Pseudotsuga* develops helical thickenings only in the first-formed earlywood and in the last-formed latewood tracheids. Coarse helical cavities and ribs are formed in the remaining, major portion of each annual increment. Helical thickenings and cavities can occur within the same tracheid of mild compression wood of *Pseudotsuga* but not in the severe grade. In *Taxus baccata* the S_1 and $S_2(L)$ layers and the distribution of lignin are the same as in compression wood tracheids in the Coniferales. The possibility that the presence of helical thickenings may be a primitive trait and the occurrence of cavities and ribs an advanced feature of compression wood is discussed in Chap. 8.3.2.

4.4.8 The Warty Layer[2]

Despite the lack of an S_3 layer, severe compression wood tracheids contain not only helical thickenings but also a warty layer which lines the inner surface of S_2. The presence of such a layer in compression wood tracheids was first observed in 1952 by Harada and Miyazaki in *Chamaecyparis obtusa* and *Pinus densiflora*. Additional information was reported 6 years later by Harada et al. (1958). Warts were very few on the inner tracheid surface of *Picea jezoensis* but they were numerous in *Pinus densiflora*, where they were uniformly distributed over the cavities and ribs. In the *Sequoiadendron giganteum* compression wood, shown in Fig. 4.101 (Cockrell 1973), warts are present within the pit opening and in the pit groove but not within the checks. In the replica of *Callitris calcarata*, shown in Fig. 4.102 (Liese 1965), the warts are as numerous in the cavities as on the top of the ribs. Wardrop and Davies (1964) found that the warts in compression wood tracheids of *Pinus radiata* were smaller than those in *Actinostrobus pyramidalis*. Another difference was that in the pine, warts occurred inside the helical cavities, whereas in the latter species they were arranged along the top of the helical ribs.

Scurfield and Silva (1969 a, b), in a study of the ultrastructure of normal and compression woods of *Cupressus arizonica, Pinus canariensis,* and *P. radiata*, found that fewer warts occurred in compression wood than in normal wood of these species. In the sequence, normal wood, moderate compression wood, and pronounced compression wood, warts tended to become more sparse and confined to the helical cavities, as shown in Fig. 4.103. According to Knigge (1976) warts are more common inside the cavities than on top of the ribs. Figure 4.104 shows the presence of warts on the pit membrane in compression wood tracheids. The scanning electron micrographs in Figs. 4.105 and 4.106 illustrate the occurrence of numerous warts on the helical ribs, within the cavities, within the pit grove, and within the pit opening in *Abies sachalinensis* compression wood (Ishida et al. 1968, Ishida and Fujikawa 1970). Even though there is conflicting evidence in the literature, it would appear that warts are as frequent in compression as in normal wood tracheids. It is also clear that they can occur both on the top of the ribs and for a short distance inside the cavities.

---

[2] Ohtani et al. (1984) have recently suggested that the term *warty layer* be replaced by the term *vestured layer*.

4.4.9 Trabeculae

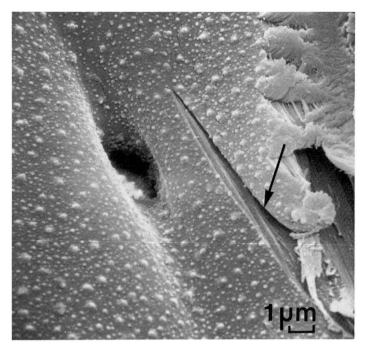

Fig. 4.101. Inner surface in a compression wood tracheid in *Sequoiadendron giganteum*. Warts are present in the pit groove and opening but not within the drying check (*arrow*). Longitudinal surface. SEM (Cockrell 1973), (cf. Fig. 4.87)

4.4.9 Trabeculae

Sanio's (1863) trabeculae are rod-like structures which traverse the lumen of tracheids and fibers from one tangential wall to the next. They often occur in a radial row as though they had been formed by a single cambial initial (Panshin and de Zeeuw 1980) and they may be of fungal origin (Hale 1923). Keith (1971), Keith et al. (1978), and Parameswaran (1979) have shown that trabeculae are confluent with the cell wall from which they originate. Their core, which sometimes is hollow, is otherwise identical with the middle lamella-primary wall, and they have the same wall layers as those in the tracheid or fiber wall. They can be terminated by a warty layer. In a species such as *Pseudotsuga menziesii*, the trabeculae can have helical thickenings attached to their surface. The cellulose microfibrils are almost invariably oriented parallel with the long axis of the trabeculae. Yumoto (1984), who has recently contributed a very thorough review of trabeculae and related structures, distinguishes between five basic types and two varieties. The former include trabeculae, plates, confluences, adhesions, and tangential confluences.

Observations on trabeculae in compression wood of *Picea glauca* were recently reported by Yumoto and Ohtani (1981) and Yumoto (1984) who found them to be very similar to those previously studied in normal wood. Like the

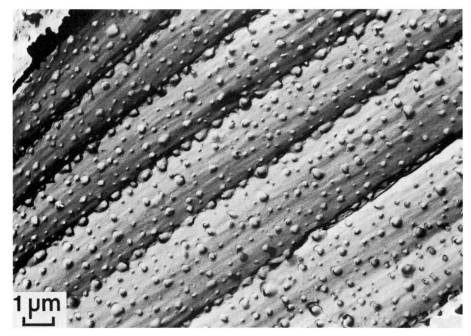

Fig. 4.102. Inner surface of a compression wood tracheid in *Callitris calcarata*. Warts are uniformly distributed over the helical cavities and ribs. Longitudinal surface. Carbon replica. TEM. (Liese 1965)

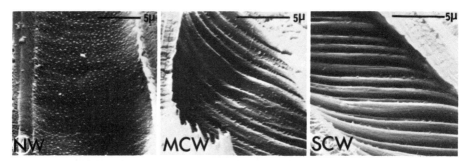

Fig. 4.103. Normal wood (*NW*), moderate compression wood (*MCW*), and severe compression wood (*SCW*) in *Pinus radiata*, showing a decreasing incidence of warts. Longitudinal section. SEM. (Scurfield and Silva 1969a, b)

latter, they occurred singly or in a radial row, and they were present in mild, moderate, and severe compression woods. Some of them were thin, while others were thick. Especially in pronounced compression wood, their surface was often fissured by helical cavities and ribs, as shown in Fig. 4.107.1, but in other instances there were no cavities (Figs. 4.107.2 and 4.107.3). Intercellular spaces were frequent where the trabeculae were attached to the middle lamella (Fig. 4.107.4). Some trabeculae were hollow, while others had a lignified core (Figs. 4.107.3 and 4.107.4).

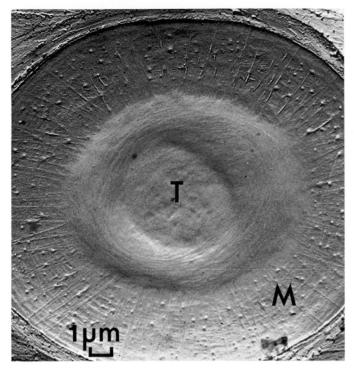

Fig. 4.104. Bordered pit in a compression wood tracheid in *Pinus taeda*, showing the torus (*T*) and the margo (*M*) which is covered by a warty layer. Longitudinal section. Carbon replica. TEM. (Parham 1970)

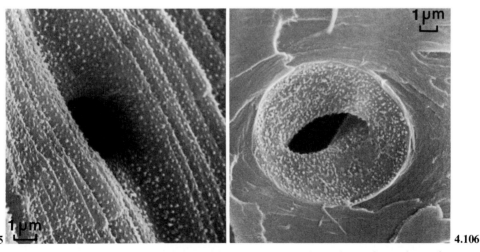

Fig. 4.105. Compression wood tracheid in *Abies sachalinensis* with an inner pit opening at the bottom of a groove. Ribs, cavities, and groove are all covered with a warty layer. Longitudinal surface. SEM. (Ishida et al. 1968, Ishida and Fujikawa 1970)

Fig. 4.106. Compression wood tracheid in *Abies sachalinensis* with an outer pit opening covered by a warty layer as seen from the middle lamella. The pit membrane was removed during the preparation of the specimen. Longitudinal surface. SEM. (Ishida and Fujikawa 1970)

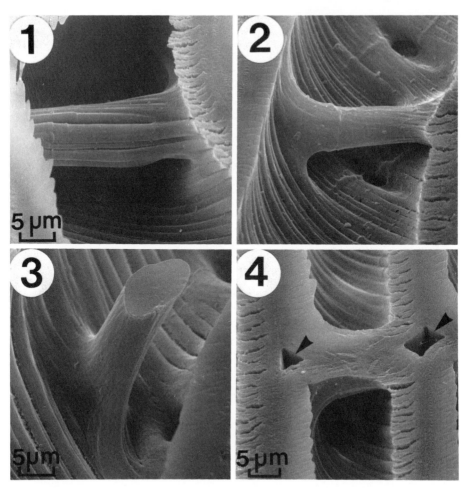

Fig. 4.107. Trabeculae in compression wood of *Picea glauca* with (**1**) or without (**2** and **3**) helical cavities. The trabeculae in **3** and **4** are solid. *Arrowheads* in **4** indicate intercellular spaces. Radial surface. SEM. (Yumoto and Ohtani 1981)

4.5 The First-Formed Tracheids

Even in pronounced, fully developed compression wood, the first few rows of cells formed at the very beginning of the growing season often do not have the appearance of those formed later. In such wood, 90% of a growth ring might consist of tracheids with all the characteristics generally associated with compression wood. The first-formed tracheids comprise perhaps 5–10% and the last-formed less than 2% of the increment.

When growth rings are readily discernible in compression wood, it is almost always because the first-formed earlywood lacks the dark color or the remainder of the ring, being as light as normal earlywood. Figure 4.108 shows an example of this in *Abies balsamea*, a species that tends to contain uniform and

4.5 The First-Formed Tracheids

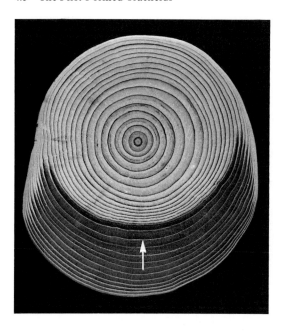

Fig. 4.108. Cross section of a horizontal *Abies balsamea* stem. The light-colored bands separating the growth rings in the compression wood (*arrow*) consist of first-formed earlywood. Note that these bands become wider as the entire increment narrows. (Côté et al. 1967)

conspicuous zones of compression wood. In this fir tree the rate of growth had begun to slow down, so that the last three increments are relatively narrow, including those on the lower side. In these rings, the light zone of first-formed wood is wider and occupies about one third of the ring. This also happens when formation of compression wood gradually ceases, as it had been doing on the two flanks of this cross section where the gravistimulus was less than in the center.

In a disc taken from the basal end of the stem of a 105-year-old *Pinus sylvestris* tree, Sanio (1873) observed that the ninth growth ring consisted of compression wood except for the first five rows of tracheids, which had thinner walls than the rest (Fig. 2.5). One side in another disc was brown in color but was subdivided into bands by thin, white layers of earlywood. Sanio also noted that occasionally these first-formed tracheids had the brown color of compression wood, their much thinner walls notwithstanding. In such cases, the entire increment was uniformly dark.

Cieslar (1896) found that the wood formed in the early spring differed from the rest of the compression wood in *Picea abies*. It was lighter in color and had thinner tracheid walls and larger lumina. He also made the observation, since then repeated by several others, that this zone undoubtedly serves the function of conducting water. Hartig (1896) in the same year, found that the thin-walled earlywood tracheids in the same species, shown in Fig. 2.7, contained helical cavities which, however, were quite shallow. Spiral checks developed over the bordered pits but never anywhere else in the cell wall. Later, Hartig (1901) refers to the occurrence of a fine, light band of wood in the first-formed earlywood. Krieg (1907), contrary to Hartig, could not detect any striations in the

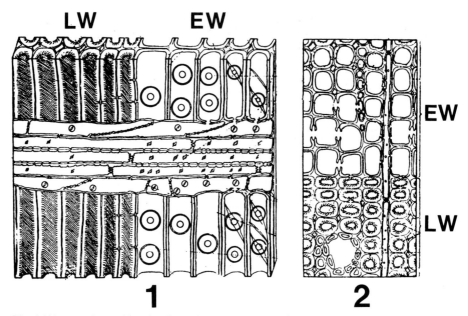

Fig. 4.109. Drawings of longitudinal (**1**) and transverse (**2**) sections of compression wood at the earlywood (*EW*) – latewood (*LW*) boundary in *Picea abies*. Note the angular outline, thin walls, and lack of helical cavities in the first four tracheids of the earlywood. (Mork 1928 a)

earlywood. Jaccard's (1912) studies on *Pinus montana* are referred to in Chap. 11.3. Jaccard and Frey (1928) observed tracheids with a rounded outline in earlywood of *Pinus nigra*, but in a later contribution Jaccard (1938) emphasized that compression wood tracheids do not develop in earlywood.

Verrall (1928) found that the tracheids formed in the spring had walls that were thinner than in the compression wood produced later and that they were thicker than in normal wood. Mork (1928 a), in the same year, pointed out that in the wood formed first in the spring, compression wood tracheids had a wall similar to that of normal tracheids. In his accompanying illustration, shown in Fig. 4.109, it can be seen that the first four rows of cells in the earlywood are square or rectangular in outline, have thin walls, and lack helical cavities. He also added, however, that diagonal fissures occurred across the bordered pits in these tracheids. Michels (1941) found that in *Abies alba* branches the entire growth ring consisted of compression wood except for a narrow zone formed early in the spring. Casperson (1963 d) has reported several cases where first-formed earlywood cells were angular in compression wood of *Picea abies*, while at the same time possessing helical cavities, as seen in Fig. 4.110. Green and Worrall (1964) confirmed Verrall's (1928) earlier finding, noting that the first-formed tracheids in *Picea mariana* had thicker walls than corresponding tracheids in normal wood. Molski (1969) made the same observation with *Pinus sylvestris*. Schultze-Dewitz et al. (1971) found that on the lower side of branches in *Pinus sylvestris* compression wood tracheids were not present in the first-formed earlywood. Some investigators have stated that the first-formed ear-

4.5 The First-Formed Tracheids

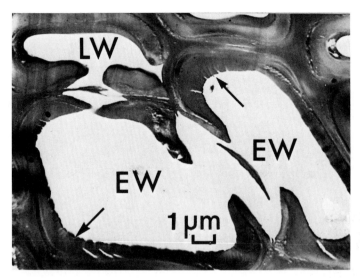

Fig. 4.110. Compression wood of *Picea abies* at the earlywood (*EW*) – latewood (*LW*) boundary. The two first-formed tracheids have an angular outline and thin walls but contain helical cavities (*arrows*). Transverse section. TEM. (Casperson 1963d)

lywood in a growth ring, otherwise composed of severe compression wood, should consist of entirely normal wood (Höster 1971, Seth and Jain 1977, 1978, Seth 1979, 1981, Jain and Seth 1979, 1980). Such a situation occurs, however, only rarely.

Côté et al. (1967) studied the histological, ultrastructural, and chemical (Chap. 6.3.3) properties of the first-formed earlywood in seven coniferous species namely *Abies balsamea, Larix laricina, Picea abies, P. rubens, Pinus strobus, P. sylvestris,* and *Tsuga canadensis*. The sections of normal and compression woods of *Pseudotsuga menziesii* var. *glauca* (Core et al. 1961) shown in Fig. 4.111 illustrate the differences between earlywood and latewood in these two tissues, namely the rounded outline, smaller size, and larger lumen of the latewood tracheids in compression wood. All tracheids in the earlywood zone of this tissue were angular in outline. In comparison with the normal tracheids, they had a more irregular outline and were less regularly arranged.

The two transverse sections of *Larix laricina* in Fig. 4.112 show that in the latewood tracheids the lumen is larger in compression wood than in normal wood, as already mentioned. Only the first two tracheid rows in the compression wood are angular. They are smaller and have a thicker wall than normal earlywood tracheids, as first observed by Verrall (1928). Casperson (1968) has reported similar results with *Larix decidua*. In Fig. 4.113, the first-formed, angular tracheids gradually assume a rounded outline in both *Picea abies* and *Tsuga canadensis*. The S_1 and S_2 layers are birefringent in polarized light. There is no indication of any S_3 layer. An electron micrograph of the border between two growth rings in compression wood of *Abies balsamea* is shown in Fig. 4.114. The first-formed tracheids have an angular outline and contain no helical cavi-

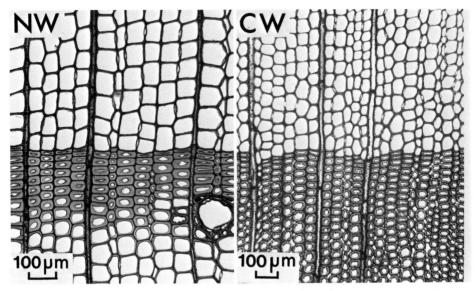

Fig. 4.111. Normal (*NW*) and compression (*CW*) woods in *Pseudotsuga menziesii* var. *glauca*. The first-formed tracheids in the normal wood have a square outline, are uniform in size, and are regularly arranged, whereas those in the compression wood have an irregular, angular outline, vary in size, and are less ordered. Transverse section. LM. (Core et al. 1961)

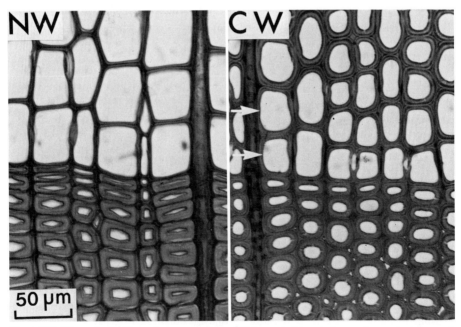

Fig. 4.112. Normal (*NW*) and compression (*CW*) tracheids at the earlywood–latewood boundary in *Larix laricina*. Only the first two rows in the earlywood tracheids in the compression wood have an angular outline (*arrows*). Note that in the latewood zone the normal tracheids have a smaller lumen than do those in the compression wood. Transverse section. LM. (Courtesy of A.C. Day)

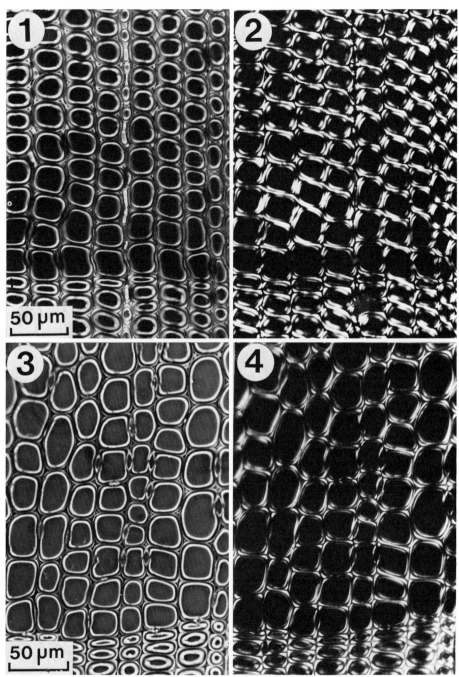

Fig. 4.113. Compression wood tracheids in *Picea abies* (**1** and **2**) and *Tsuga canadensis* (**3** and **4**), photographed in ordinary (**1** and **3**) and polarized (**2** and **4**) light. The transition from angular to rounded tracheids is more gradual in *Picea abies*. Note that both S_1 and S_2 are birefringent in polarized light also in the first-formed earlywood. Transverse sections. LM. (**1** and **2** Côté et al. 1967; **3** and **4** Courtesy of A.C. Day)

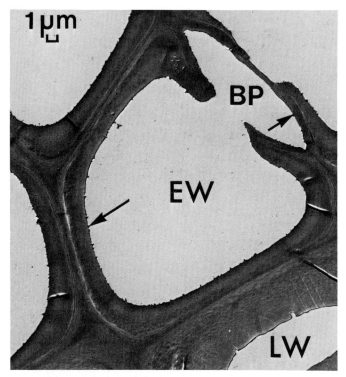

Fig. 4.114. Latewood (*LW*) and first-formed earlywood (*EW*) tracheids in *Abies balsamea*. The first-formed tracheids have an angular outline and a thin wall and lack helical cavities and an S_3 layer. No intercellular spaces are present. Note the occurrence of warts, also inside the chamber of the bordered pit (*BP*) (*arrows*). Transverse section. TEM. (Côté et al. 1967)

ties. No S_3 layer is present, and a warty layer is attached to S_2 as well as to the initial pit border in the pit chamber. The last-formed tracheids in *Larix laricina*, shown in Fig. 4.18, contain helical cavities, but none are present in the first- formed cell. In the *Tsuga canadensis* tracheids seen in Fig. 4.115, both the last-formed and the first-formed tracheids are devoid of any cavities.

Unlike the other genera examined, *Pinus* was found to have well-developed helical cavities and protruding ribs in all earlywood tracheids. Figures 4.116 and 4.117 illustrate for *Pinus resinosa* and *P. sylvestris* how all cells are fissured by numerous cavities except where latewood tracheids have been sectioned close to the tip. The first-formed earlywood tracheids in these species differ from those in pronounced compression wood only in their angular outline and the absence of intercellular spaces. *Pinus strobus* and *Pinus resinosa* were similar to *P. sylvestris* in all these respects.

According to Côté et al. (1967) the first-formed earlywood tissue has the high lignin and galactose contents typical of severe compression wood (Chaps. 6.3.3 and 6.4.2). The first-formed compression wood tracheids also have the same distribution of lignin as the cells produced later. The lignin

4.5 The First-Formed Tracheids

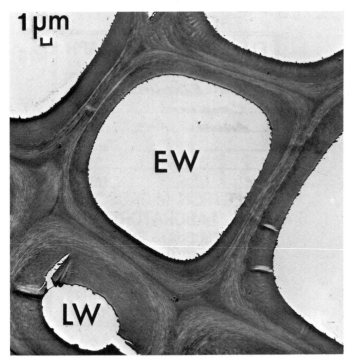

Fig. 4.115. Last-formed (*LW*) and first-formed (*EW*) tracheids in compression wood of *Tsuga canadensis*, all lacking helical cavities. Transverse section. TEM. (Côté et al. 1967)

skeleton from the earlywood-latewood boundary in *Larix laricina* shown in Fig. 4.118 illustrates this. The three earlywood tracheids have a moderately lignified S_1 layer, a highly lignified $S_2(L)$ layer, and a somewhat less lignified inner zone of S_2, all features characteristic of severe compression wood tracheids. Later investigations by Wood and Goring (1971) (Fig. 6.17) and by Fukazawa (1974) (Figs. 6.21 and 6.22) have also made it clear that in compression woods of *Pseudotsuga menziesii* and *Abies sachalinensis* the distribution of lignin in the first-formed tracheids is that typical of the more fully developed cells produced later in the season.

In the differentiation of compression wood tracheids, the cambial cells have an angular outline until cell expansion has ceased, when they become round, a transition that terminates during formation of the S_1 layer. Examples of this gradual development of the typical compression-wood tracheids are found in Chap. 9 (Figs. 9.11–9.13, 9.20 and 9.22). In the first-formed earlywood, the original, angular shape of the cambial cells never changes. Differentiating first-formed tracheids in compression wood of *Picea abies* are seen in Fig. 9.10. These cells have a rectangular outline in transverse section.

One of the characteristic features of compression wood is that all tracheids within a growth ring usually are affected (Westing 1965), albeit not always to the same extent (Harris 1977). This is in marked contrast to tension wood,

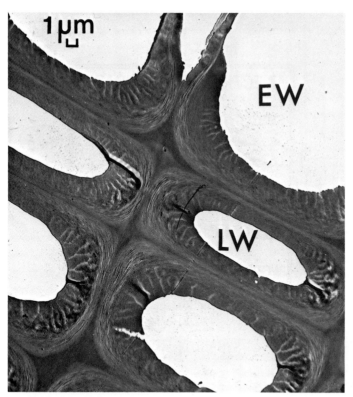

Fig. 4.116. Compression wood tracheids at the earlywood (*EW*)–latewood (*LW*) boundary in *Pinus resinosa*. The last-formed cells, except for being oblong in outline, are typical compression-wood cells. The first-formed tracheids contain helical cavities. Transverse section. TEM. (Timell 1972b)

where gelatinous fibers are frequently absent from the outer portion of each ring and the inner part often contains both tension and normal wood fibers. Cases where the entire increment consisted of pronounced compression wood are not unknown. The first example of this was described by Sanio (1873) for *Pinus sylvestris*. Three other examples of the formation of severe compression wood at the beginning of the growing season were cited earlier, namely in *Pinus montana* (Jaccard 1912), *P. albicaulis* (Fig. 4.12) and *Juniperus virginiana* (Fig. 4.13) (Core et al. 1961).

Yoshizawa et al. (1981, 1982) recently subjected the first-formed tracheids in compression wood to a renewed investigation. The specimens probably consisted of pronounced compression wood for they had intercellular spaces and rounded tracheids with a highly lignified $S_2(L)$ layer and, where formed at all, helical cavities. No S_3 layer could be observed. Of the 38 gymnosperm species examined, only 16 had lighter colored wood zones at the beginning of their growth rings. In some of these species, such as *Cedrus* and *Glyptostrobus*, the first-formed earlywood seemed to be normal wood. A second group, which in-

4.5 The First-Formed Tracheids

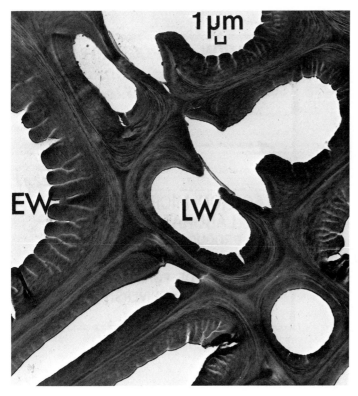

Fig. 4.117. Compression wood tracheids in *Pinus sylvestris* at the earlywood (*EW*)–latewood (*LW*) boundary. The first-formed cell (*EW*) contains well-developed helical cavities with protruding ribs. Transverse section. TEM

cluded *Abies, Larix, Picea,* and *Pseudotsuga,* had a relatively thick tracheid wall, an $S_2(L)$ layer, and no S_3, but the cells had an angular outline and lacked helical cavities. The tracheids in a third group had a slightly angular outline but otherwise displayed all the features of severe compression wood, including helical cavities. *Pinus* was the most typical member of this group, while the cavities were less well developed in *Tsuga*. The first-formed earlywood zone in *Pinus thunbergii* lacked intercellular spaces (Yoshizawa et al. 1981). A majority of the species examined had no light-colored earlywood zones, namely those belonging to the genera *Araucaria, Cephalotaxus, Chamaecyparis, Cupressus, Juniperus, Libocedrus, Thuja, Thujopsis, Keteeleria, Podocarpus, Cryptomeria, Cunninghamia, Sciadopitys, Sequoia, Sequoiadendron, Taiwania, Taxodium, Taxus, Torreya,* and *Ginkgo*. In most of these genera fully developed helical cavities were present throughout the growth ring. The first-formed tracheids in *Ginkgo, Cephalotaxus, Taxus,* and *Torreya,* which do not develop helical cavities, lacked this feature in their first-formed earlywood, as could be expected, but cavities were also absent in *Araucaria, Keteeleria, Sciadopitys,* and *Thuja orientalis*.

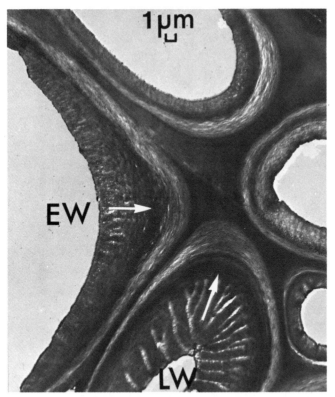

Fig. 4.118. Lignin skeletons of compression wood tracheids at the earlywood (*EW*) – latewood (*LW*) boundary in *Larix laricina*. Both types of cells have the lignin distribution typical of compression wood tracheids, including a highly lignified $S_2(L)$ layer (*arrows*). Transverse section. TEM. (Côté et al. 1967)

In *Picea glauca* trees inclined at an angle of 90°, Yumoto and Ishida (1982) found that the first-formed earlywood tracheids had a rounded outline and a thick, highly lignified wall. Intercellular spaces did not appear, however, until 10 cell rows from the growth ring boundary. The tracheid walls were thinner in trees inclined at 45°. As the angle of inclination decreased, the compression wood features of the first-formed tracheids became less conspicuous until, at an angle to the vertical of 5°, normal wood was formed at the beginning of the season.

It is evident from the observations reported by Côté et al. (1967) and by Yoshizawa et al. (1982) that light-colored zones of first-formed compression wood are readily formed in the Pinaceae. The more recent results (Yoshizawa et al. 1982) indicate that such earlywood bands are more rare in the other families of the Coniferales, Taxales, and Ginkgoales. The phenomenon is, however, affected not only by internal but also by external factors, such as the intensity of the gravitropic stimulus. It is well known that under conditions of less than maximum stimulus, compression wood tends to develop only within the outer portion of a growth ring (Chap. 10.3.6.3).

The first-formed tracheids in *Abies, Larix, Picea,* and *Tsuga* all have the angular outline of normal conifer tracheids and also lack intercellular spaces. Often, helical cavities are entirely missing or they are, at best, few and poorly developed. In other respects, however, these cells are similar to ordinary compression wood tracheids. They have the thick, lamellar S_1 layer typical of the latter, possess a highly lignified zone in the outer portion of S_2, and lack an S_3 layer. The orientation of the microfibrils in S_2 is that characteristic of fully developed compression wood tracheids, and so are chemical composition and distribution of lignin. In the genus *Pinus* only the absence of intercellular spaces and the angular outline of the tracheids distinguish the first-formed cells from those produced later. It is not known why the first-formed earlywood tracheids wood differ in their structure from those produced later. The few hypotheses that have been proposed, such as that suggested by Höster (1974), are discussed in Chap. 9.6.1.7.

4.6 The Last-Formed Tracheids

It was mentioned previously that the one or two last-formed tracheids in a compression wood increment usually are flattened in the radial direction, as they also are in normal latewood. Sanio (1873) noted that in one growth ring of *Picea abies* compression wood, the last-formed latewood tracheids lacked spiral striations. Sonntag (1904) mentions that even the last, flattened tracheids in compression wood of the same species have pits that are traversed by a slit-like opening. Krieg (1907), however, states that in *Pseudotsuga menziesii* the last-formed tracheids lack spiral striations and instead possess spiral thickenings, an observation confirmed almost 70 years later by Takaoka and Ishida (1974). Boutelje (1965, 1966) observed that the last four to seven tracheids on the lower side of knots in *Picea abies* contained spiral thickenings. Sometimes these cells were angular, and no intercellular spaces were present. According to Boutelje, all three features point to the presence of normal tracheids. Similar results were obtained with knot wood of *Pinus sylvestris.* Hartler (1968) states that the last-formed tracheids in compression wood are normal latewood tracheids and that they are transparent to transmitted light and not opaque. He also points out that a widening of the latewood zone does not necessarily mean that compression wood should be present. Schultze-Dewitz et al. (1971) claim that in branch compression wood of *Pinus sylvestris,* the last-formed latewood consists of normal wood.

Timell (1972b) examined transverse sections of the last-formed tracheids in compression wood of *Larix laricina, Picea abies, Pinus sylvestris, Sequoia sempervirens,* and *Tsuga canadensis* in the transmission electron microscope. Helical cavities were observed in the last two rows of radially flattened cells except for *Sequoia sempervirens,* where none could be detected. All other ultrastructural features with the exception of the oval rather than round outline of the cells and the absence of intercellular spaces were those of typical compression wood tracheids. Yoshizawa et al. (1981) frequently observed last-formed compression wood tracheids without helical cavities in *Pinus thunbergii,* adding that the severity of compression wood could not be judged from inspection of such tracheids. This is, of course, correct. Last-formed latewood in compression wood from *Picea glauca* trees inclined at an angle of 90° were found by Yumoto and Ishida (1982) to lack intercellular spaces in the last three to five rows of tracheids.

The factors responsible for the formation of the last-formed tracheids in normal wood are undoubtedly also operative in compression wood. In both tissues the last few rows of cells are much flattened in the radial direction so that they have an oblong outline, possibly because of the lack of auxin and the presence of growth inhibitors. The last-formed tracheids in compression wood assume an oval form with a thick wall and a narrow lumen. Examples of this are seen in the excellent drawings prepared by Hartig (1901) (Fig. 2.7) and by Mork (1928a) (Fig. 4.109) from *Picea abies* and in the micrographs obtained with *Larix laricina* (Fig. 4.112), *Picea abies* (Figs. 4.113 and 9.10), *Pseudotsuga menziesii* (Fig. 4.111), and *Tsuga canadensis* (Figs. 4.59 and 4.113). Except for their oval rather than round outline, these compression-wood tracheids are often in every other respect identical with those deposited earlier. In other instances there are minor ultrastructural differences between the two types of cells. As already mentioned, the S_1 layer tends to be very thick at the cell corners in the last-formed tracheids, so that it sometimes exceeds S_2 in thickness (Figs. 4.18, 4.59, 4.115–4.118). Intercellular spaces are far more rare than normal in last-formed latewood, probably because of an abundant supply of photosynthate.

Last-formed tracheids often lack helical cavities close to their tip, as can be seen from Figs. 4.115 and 4.117, or these cavities are only weakly developed (Fig. 4.59). Cavities are also absent or only barely discernible at the mid section of the tracheids in *Abies balsamea* shown in Fig. 4.114. Elsewhere, however, there are fully developed cavities in this region of the last-formed tracheids in species such as *Larix laricina* (Figs. 4.18 and 4.118), *Picea abies* (Figs. 9.3–9.5), *Pinus resinosa* (Fig. 4.116), and *P. sylvestris* (Fig. 4.117). As can be seen in Fig. 4.118, the last-formed cells have the distribution of lignin characteristic of all compression-wood tracheids, including a well-developed $S_2(L)$ layer.

In summary, the last-formed tracheids are typical compression wood cells except for their oblong outline, frequent absence of intercellular spaces, and occasional lack of helical cavities at their tip. Statements that these cells should be normal softwood tracheids are incorrect. Even a casual inspection of compression wood in transmitted light reveals that it is the first-formed tracheids in each growth ring that are transparent, not the last-formed cells.

4.7 Rays and Ray Cells

4.7.1 Introduction

The longitudinal tracheids in compression wood differ fundamentally in structure from those in normal wood, and not only histologically but also, and perhaps even more so, in their ultrastructure. Compared to the considerable attention that has been devoted to the tracheids, the rays and the ray cells in compression wood have been rather neglected. Some investigators have observed or measured the number of rays per unit area in compression and normal woods, and a few have also estimated the size of the individual ray cells. Very little information is available concerning the ultrastructure of either the ray parenchyma or the ray tracheids in compression wood.

In a study of the ultrastructure of the ray parenchyma cells in normal wood of *Chamaecyparis obtusa, Cryptomeria japonica, Pinus densiflora,* and *Picea jezoensis,* Harada et al. (1958) came to the conclusion that the secondary wall in these cells consisted of three layers with a different orientation of the microfibrils. Later, Harada and Wardrop (1960) claimed that an S_4 layer was also present. Chafe (1974b) studied the ultrastructure of the cell wall of the xylem parenchyma cells in *Cryptomeria japonica.* He found that the organization of both ray and axial parenchyma walls in this species lacked the layering typical of the tracheids, and that no S_4 layer was present. The optically homogeneous, secondary wall consisted of many crossed lamellae, inclined 35° and 45° to the cell axis. Wall thickenings in the form of transverse bars in the parenchyma and longitudinal ribs in the axial ray cells had a similar structure. Both types of cell had a high lignin content with the lignin distributed in the form of lamellae.

A thorough study of the ultrastructure of the ray parenchyma cells in 33 gymnosperm species was carried out by Fujikawa and Ishida (1975), who found that the wall structure of these cells varied considerably but could be assigned to five major types. Ray cells belonging to the *Sciadopitys* and *Cryptomeria* types consisted of a middle lamella, a primary wall, and what the investigators termed a "protective layer" which lined the lumen. The cell wall was lignified in *Cryptomeria* but not in *Sciadopitys.* Confirming the earlier results of Bailey and Faull (1934), Fujikawa and Ishida (1975) concluded that a secondary wall is present only in the ray parenchyma cells of the Pinaceae. Within the diploxylon pines the cells matured and lignified slowly until they consisted of a middle lamella, primary wall, protection layer, and S_1. The rays in the haploxylon pines matured more rapidly within the sapwood and were composed of middle lamella, primary wall, S_1, and protection layer. Ray cells belonging to the *Abies* type, finally, consisted of middle lamella, primary wall, S_1, S_2, and protection layer. These cells were more highly lignified than those of the haploxylon type.

In a later study of the maturation of ray parenchyma cells in five *Pinus* species, Yamamoto (1982) found that these cells could be classified into three types, namely those that thickened and lignified in the sapwood, those that thickened and lignified within the intermediate wood, and those that lignified without thickening in this region. A similar classification could be applied to the ray parenchyma cells in other conifer genera. Lignification took place long after the primary wall had been completed.

4.7.2 Frequency and Size of Rays and Ray Cells

Although it is stated in the early German literature that rays tend to be more frequent on the lower side than on the upper side of conifer branches, Hartig

Table 4.18. Frequency and height of rays in five conifer species. (Verrall 1928)

Species	Number per mm²		Height in μm		Height in number of cells	
	Compression wood	Normal wood	Compression wood	Normal wood	Compression wood	Normal wood
Larix laricina	42.1	36.6	164	162	10.3	9.0
Picea mariana	41.6	39.2	155	128	8.7	8.1
Pinus palustris	36.1	29.4	124	117	7.4	6.5
Thuja occidentalis	39.8	38.2	127	94	6.3	5.0
Thuja plicata	43.0	42.6	78	79	4.2	4.1
Average	40.5	37.2	130	116	7.4	6.5

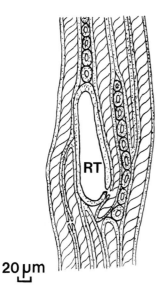

Fig. 4.119. An abnormally large ray tracheid (*RT*) in *Picea* sp. compression wood. Tangential section. Projection drawing. (Verrall 1928)

(1896, 1901) never mentions them. Cieslar (1896) found that in *Picea abies* opposite wood contained only 75% as many rays as did compression wood, a fact that he attributed to the greater availability of photosynthate on the compression wood side. In *Abies alba, Picea abies,* and *Sequoia sempervirens,* Jaccard (1915) observed that the rays were more numerous in branch wood than in stem wood. In redwood branches he also found that the frequency of rays was higher on the lower side than on the upper.

Verrall (1928) determined the number of rays per unit tangential area in five conifer species and also measured their height in terms of actual distance and number of cells. His results, which are summarized in Table 4.18, show that in all five species examined, rays were slightly, but consistently, more frequent in compression wood than in normal wood. The rays in the compression wood were considerably higher, both in distance and in number of cells. One ray tracheid, shown in Fig. 4.119, was exceptionally large and measured 59×134 μm

4.7.2 Frequency and Size of Rays and Ray Cells

in cross section. The possibility, mentioned by Verrall himself, that this ray actually may be an artifact, caused by gliding growth, cannot be excluded. Bannan (1937), confirming Jaccard's (1915) earlier results, found that rays were more numerous on the lower than on the upper side in branches of *Tsuga canadensis*. Fegel (1938, 1941), who studied eight species of conifers native to northeastern United States, obtained an average value of 7 rays per mm in branch wood compared to 5.5 for stem wood. The ray volume, however, was 5.1 – 5.2% in both cases, implying that the rays in the branch were smaller than those in the stem.

Onaka (1949), contrary to all earlier investigators, found the frequency of the rays to be lower in compression wood than in normal wood of ten coniferous species, as can be seen in Table 4.19. In some species the difference was slight, whereas in others it was substantial. The different results obtained with the two *Larix* specimens are puzzling. In seven species the average height of the rays was larger in compression wood, and in eight species there were fewer ray cells per unit area in this tissue. Onaka (1949) concluded that compression wood contains fewer ray cells than does normal wood. Ollinmaa (1959) confirmed this for *Picea abies*, where he observed 40 rays per mm^2 in normal wood, compared to only 34 in compression wood. The ray volume, however, was 5.7% in both tissues. Kučera and Nečesaný (1970) also observed fewer rays in compression wood on the lower side of branches from *Abies alba* than on the upper side and noted that each ray on the average contained fewer cells. Kennedy (1970), in contrast, found that in *Pseudotsuga menziesii*, compression wood had a much higher incidence of rays than normal wood. Some of the rays were biseriate, which is rare in normal wood of this species. A comparison between the radial sections presented by Sudo (1968d) reveals no difference in ray frequency or size for normal and compression woods of *Picea montigena*.

It should perhaps be pointed out that while some investigators consider biseriate rays to be absent or very rare in normal wood of all conifers except in *Juniperus* and *Sequoia*, others believe that such rays occur in all conifers. According to Greguss (1955) they are found in all gymnosperm families except the Ginkgoaceae. This view has been supported by Kucera and Bosshard (1975), who could demonstrate the presence of biseriate rays in *Abies alba*. Kucera and Bariska (1972) examined ray cells on the upper and lower side of branches in *Abies alba*. Statistical analysis indicated that the rays on the upper side contained 35% more cells than did those on the lower, compression wood side. Frequency distributions of ray heights exhibited only one maximum on the under side, were narrow, and were skewed to the left. On the upper side, these distributions were wider and contained several peaks. No exact explanation could be given for this interesting difference, but it was assumed that it was caused by differences in concentration of "substances" transported to and stored in the branch. It was also assumed that there existed a correlation between the ray distribution and the eccentricity of the branches, but no data were presented to indicate this.

Timell (1972a) examined rays and ray cells in normal and compression woods of seven coniferous species. The tangential sections in Fig. 4.120 show clearly that, in this *Pinus resinosa* specimen, compression wood contained more

Table 4.19. Number of rays per mm² of tangential surface, average height of rays in number of cells, and number of individual ray cells per mm² of tangential surface in normal and compression woods of various gymnosperm species. (Onaka 1949)

Species	Number of rays			Average height of rays			Number of ray cells		
	Compression wood	Normal wood	Ratio	Compression wood	Normal wood	Ratio	Compression wood	Normal wood	Ratio
Abies firma	37.9	41.6	0.91	8.64	5.76	1.50	326	240	1.36
Abies mayriana	41.3	64.6	0.64	3.86	3.94	0.98	159	254	0.63
Chamaecyparis obtusa	53.5	72.6	0.74	5.80	4.54	1.28	311	329	0.95
Ginkgo biloba	54.4	59.5	0.91	2.23	2.20	1.01	121	131	0.92
Larix gmelinii var. japonica	41.1	42.2	0.97	5.04	6.25	0.81	206	264	0.78
Larix gmelinii var. japonica	41.6	84.6	0.49	4.94	3.74	1.32	206	317	0.65
Picea jezoensis	49.6	61.4	0.81	5.52	5.14	1.07	274	316	0.87
Pinus densiflora	31.2	33.3	0.94	4.85	5.17	0.94	151	172	0.88
Pinus densiflora	35.8	46.5	0.77	5.50	4.41	1.25	197	205	0.96
Podocarpus macrophyllus	33.3	33.3	1.00	2.84	2.71	1.05	95	90	1.06

4.7.2 Frequency and Size of Rays and Ray Cells

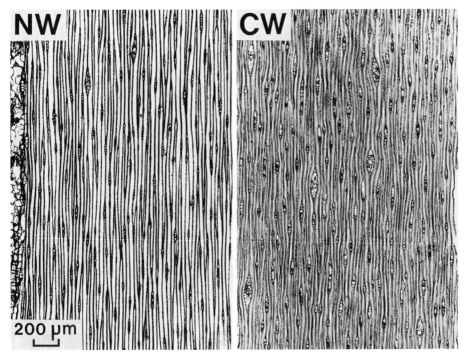

Fig. 4.120. Normal (*NW*) and compression (*CW*) woods of *Pinus resinosa*, showing the presence of more numerous and larger rays in the compression wood. Tangential sections. LM. (Timell 1972a)

rays than did normal wood, and that both uniseriate and fusiform rays were larger in the former tissue. The increase in size was mostly a result of the greater width of the rays, as can be better seen in Fig. 4.121 where the parenchyma cells are oval in transverse section in normal wood but almost circular in compression wood. In compression wood from *Abies balsamea*, *Larix laricina*, *Picea rubens*, *Pseudotsuga menziesii* (Fig. 4.122), and *Tsuga canadensis*, the rays in the compression wood did not differ from those in the normal wood with respect to either frequency or size. Biseriate rays, previously observed by Kennedy (1970) in *Pseudotsuga menziesii*, could not be detected in compression wood of this species. In no instance did compression wood contain fewer rays than corresponding normal wood.

Abnormal rays have frequently been observed in normal wood of conifers. In the wood that covers compression failures in conifers (Chap. 15.2.1.5.3), Trendelenburg (1940) found twice as many rays as in the adjacent normal and compression woods, which were similar in this respect. Gerry (1942) observed uniseriate rays in *Picea* which were higher than normal and contained a red, gummy material. Bannan examined abnormal rays in wood of *Chamaecyparis lawsoniana* (Bannan 1950b), *C. nootkatensis*, *C. thyoides* (Bannan 1950a), *Juniperus virginiana* (Bannan 1942), *Libocedrus decurrens* (Bannan 1944), and *Thuja occidentalis* (Bannan 1941). In *Chamaecyparis* the abnormal rays oc-

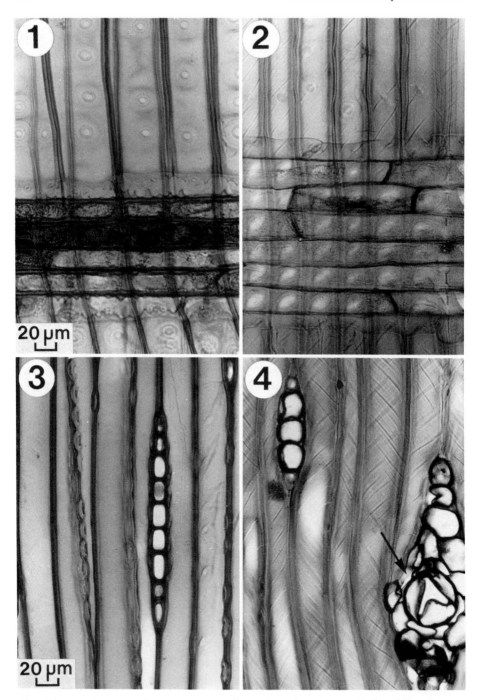

Fig. 4.121. Radial (**1** and **2**) and tangential (**3** and **4**) sections of normal (**1** and **3**) and compression (**2** and **4**) woods of *Pinus resinosa*. Note the wider rays and the larger and more rounded ray cells in the compression wood. *Arrow* indicates a fusiform ray. LM. (Timell 1972a)

4.7.2 Frequency and Size of Rays and Ray Cells

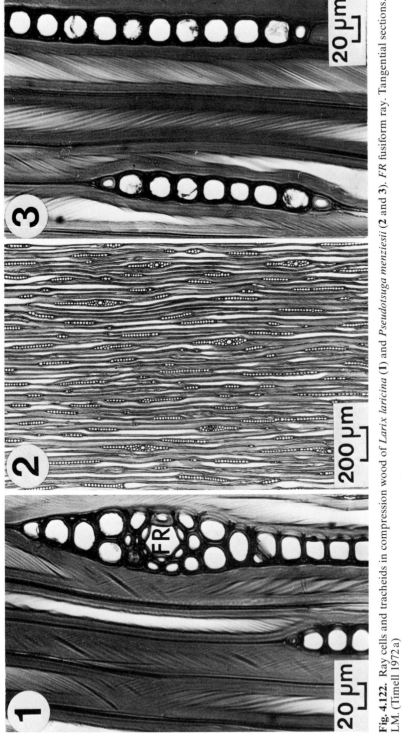

Fig. 4.122. Ray cells and tracheids in compression wood of *Larix laricina* (**1**) and *Pseudotsuga menziesii* (**2** and **3**). *FR* fusiform ray. Tangential sections. LM. (Timell 1972a)

curred most frequently in the outer wood on the lower side of branches, that is, in a region that probably consisted of compression wood, but they were rare on the upper side. Some abnormal rays were also present in stem wood of young trees. None occurred in wood from older stems or roots. The rays were larger than normal, and the ray cells were more irregular in shape. Another difference was that the ray tracheids were present in the central row of the rays. Often, the abnormal rays gradually became multiseriate and very wide. Their distribution was erratic and could not be related to any injury to the trees.

Enlarged rays containing copious amounts of resinous matter have been reported by Rendle (1952) to occur in *Picea sitchensis* wood. The number of rays in "hazel" wood of *Picea abies* is 40–50% larger than in undisturbed zones (Ziegler and Merz 1961). Fengel (1966a) and Süss and Müller-Stoll (1970) found that in diffuse-porous hardwoods not only the rays but also the ray cells increase in size at the growth ring boundary, while the number of cell rows remains unchanged.

Halbwachs and Kisser (1967) examined wood produced in a dwarf form of *Picea abies* which had developed as a result of exposure to industrial fumes containing hydrogen fluoride. A cross section of the stem revealed the presence of irregular growth rings and large amounts of severe compression wood. The xylem (presumably not compression wood) contained 110 rays per mm^2 compared to only 37 in wood from a normal spruce tree. Fusiform rays were also more common. The rays were lower and the ray cells were larger in the dwarf tree than in a normal tree. Halbwachs and Ritter (1969) have also reported the occurrence of a similar wood in so-called fasciated shoots of *Picea abies*, that is, shoots that have become flattened so that they are no longer cylindrical or polygonal as is usually the case (Kienholz 1932, White 1948). Such wood had more rays and ray cells than normal wood, and instead of 2–6 cells per average normal ray, these rays contained up to 32 cells. There was also an increase in the incidence of biseriate rays.

In wood infected by dwarf mistletoe (*Arceuthobium*), a dicotyledoneous parasite of conifers (Chap. 15.10), the rays undergo profound changes (Srivastava and Esau 1961). Instead of uniseriate rays, this wood seems to have multiseriate rays which are abnormally wide and large. The frequency of the rays is also increased. Enormously long ray cells are formed in xylem infected by the fungal wound parasite *Diplodia pinea* (Foster and Marks 1968) (Chap. 15.12.1.4). Pine wood infected by the fungus *Cronartium fusiforme* develops high rays containing more and larger cells than normal (Jackson and Parker 1958, Jewell et al. 1962) (Chap. 15.12.5). Abnormally large ray cells are also formed in the tumors that occasionally occur in *Picea sitchensis* (Rickey et al. 1974) (Chap. 15.13).

Bamber and Lyon (1971) examined abnormal rays in stained heartwood of *Pseudotsuga menziesii*. This stain was darkly reddish in color and extended radially through the wood. The rays in the stained region were multiseriate and much larger than in normal wood of this species, where the rays are generally uniseriate. The increase in size of the abnormal rays was a result of a larger number of cells per ray. A dark material was present, as had also been observed by Gerry (1942) and Rendle (1952), and was the reason for the macroscopic stain. The presence of fungal hyphae in both tracheids and ray cells indicated that the tree had suffered some injury, but the exact nature of the latter could not be ascertained.

Fir trees infested by the balsam woolly aphid (*Adelges piceae*) produce an abnormal wood which in many respects resembles compression wood (Chap. 20.8.2). As first observed by Balch (1952), such wood often contains twice as many rays as normal wood. Biseriate rays are also common in both xylem (Mitchell 1967, Smith 1967) and phloem (Saigo 1976) of infested fir trees. Doerksen and Mitchell (1965) and Mitchell (1967) have shown that the ray cells are notably higher and wider than those in normal fir wood, as can be seen in Figs. 20.30 – 20.32.

Gregory and Romberger (1975) have demonstrated that there exists a direct relationship between rate of radial growth and frequency of ray cells. In 50-year-old *Picea abies* trees they found a positive, linear correlation between number of tracheid rows in the growth rings and number of ray cells per unit tangential area. As the ray frequency remained constant, it was the number of cells per ray, and especially ray height, that increased with increasing growth rate. These findings were confirmed with a 73-year-old *Abies balsamea* tree, recently released by nearby clearcutting. The resulting higher rate of radial growth was associated with a 70% higher incidence of ray cells. In these investigations, both size and number of rays were found to increase when the rate of growth accelerated following wounding. As pointed out by Gregory and Romberger, this is apparently typical of wounded tissues. When only increased radial growth is involved, the only change is an increase in the number of ray initials per unit area of the cambium (White and Robards 1966). Imamura (1978) found that the rays in stem flutings in *Cryptomeria japonica* were more frequent and irregular in outline than in normal stem wood. Biseriate rays were also more common, and the ray cells were larger. Application of pressure to the stem side produced the same effects.

It would appear from the examples now mentioned that many physiological disturbances, resulting in abnormal growth patterns in trees, are associated with departures from the normal mode of formation of rays in wood. Characteristically, the number of both rays and ray cells increases considerably, as does also the frequency of biseriate, multiseriate, and fusiform rays. The rays are often higher and wider than under normal conditions, and the individual ray cells are usually larger. These changes seem always to be the same, irrespective of the nature of the disturbances that cause them. Except for compression wood, the tracheids are unaffected or only slightly affected. Rapid radial growth, often associated with formation of compression wood, tends to result in an increase in the incidence of rays in conifers.

4.7.3 Ultrastructure of the Ray Cells

The ultrastructure of the ray parenchyma and ray tracheids in compression wood were briefly examined by Timell (1972a), who also alluded to this subject in a later publication (Timell 1973). The wall in the compression wood ray cells

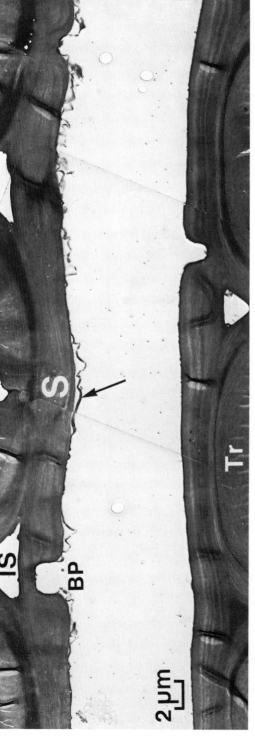

Fig. 4.123. A ray parenchyma cell in compression wood of *Picea abies*. The secondary wall (*S*) is lamellar and has simple pits. The terminal lamella has become partly detached from the cell wall (*arrow*). There are intercellular spaces (*IS*) between the tracheids and the ray cell and a blind pit (*BP*) in the ray cell wall. Transverse section. TEM. (Timell 1973d)

4.7.3 Ultrastructure of the Ray Cells

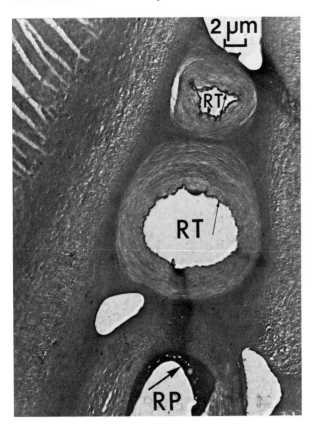

Fig. 4.124. Lignin skeleton of compression wood in *Larix laricina*, showing a ray parenchyma cell (*RP*) with an electron-dense terminal lamella (*large arrow*) and two ray tracheids (*RT*) with a warty layer (*small arrows*). Tangential section. TEM

was distinctly lamellar, a feature also demonstrated by Harada and Wardrop (1960) and by Harada (1965a) for ray parenchyma cells in normal wood of *Cryptomeria japonica*. As was noted also by Harada and Wardrop (1960), the individual layers in the secondary wall of the ray cells could be distinguished only with difficulty. In his investigations on the xylem parenchyma cells in *Chamaecyparis nootkatensis* and *Cryptomeria japonica*, Chafe (1974a, b) also found the cell wall to have a lamellar structure. Several ultrastructural features of ray parenchyma cells in compression wood can be seen in Fig. 4.123, including the lamellation of the secondary wall and the presence of intercellular spaces and blind pits between axial tracheids and ray cells, all features also found in normal wood.

The lignin skeleton from compression wood of *Larix laricina* seen in Fig. 4.124 shows two ray tracheids with a warty layer and a ray parenchyma cell containing a thick, electron-dense zone next to the lumen. This zone, originally termed "S_4" by Harada (1962), "tertiary wall" by Fengel (1965), and "protective layer" by Fujikawa and Ishida (1975), is probably not a part of the secondary wall and is perhaps best referred to as "terminal lamella", as suggested by Preusser et al. (1961). First observed in ray cells by Harada (1953, 1962), this

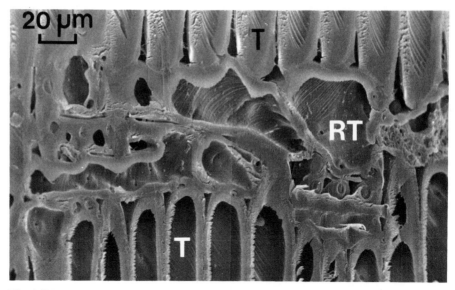

Fig. 4.125. Compression wood of *Picea glauca* with axial tracheids (*T*) and ray tracheids (*RT*). Like the longitudinal tracheids, the ray tracheids have a thick wall and helical cavities, here observed for the first time in such cells. Radial section. SEM. (Yumoto et al. 1982)

lamella might consist of cytoplasmic debris (Harada and Wardrop 1960, Fengel 1965, Harada 1965b). It separates easily from the secondary wall, as shown for normal wood of *Fagus sylvatica* by Fengel (1965) and for compression wood of *Larix laricina* and *Picea abies* by Timell (1972a, 1973) (Fig. 4.124). After removal of the lignin, it can be seen that the secondary wall of the ray cells consists of seemingly densely packed cellulose microfibrils, as seen in compression wood of *Larix laricina* in Fig. 6.25.

Ray tracheids with helical cavities were recently observed for the first time by Yumoto et al. (1982) in compression wood of *Picea glauca*. As can be seen in Fig. 4.125, these cells had a thick wall and contained helical cavities and ribs similar to those in the adjacent axial tracheids. Thin-walled ray tracheids lacking helical cavities were, however, also present.

4.7.4 Conclusions

Although reports in the literature are often contradictory, a majority of investigators have found that rays are more frequent in compression wood than in normal wood and also that both rays and ray cells are larger. In analogy with the results reported by Gregory and Romberger (1975), the higher incidence of rays in some compression woods might be associated with the fact that compression wood is formed at a higher rate than normal wood. Bannan (1937, 1954) found that ray volume and ray height increase with increasing width of the growth rings in certain conifers. White and Robards (1966), in a study of ring-porous hardwoods, noted the occurrence of more and wider rays in wide than in narrow growth rings. They pointed out that an increase in radial growth could result in stretching of the cambial cells and a higher rate of anticlinal divisions. A mature ray consequently would be expected to be composed of more cells. The same reasoning could, of course, also be applied to the ray cells in compression wood in analogy with the arguments used by Wardrop and Dadswell (1950) to explain the fact that compression wood tracheids are shorter than those in corresponding normal wood (Sect. 4.3.2.2).

Hoffmann and Timell (1972a, b) have shown that the ray cells have the same chemical composition in normal and compression woods (Chap. 6.3.2), and it is also evident that the distribution of lignin is the same (Chap. 6.4.3). The ultrastructure of the ray cells is probably also the same in the two tissues. As was already stated, the chemical and ultrastructural changes associated with the transition from normal to compression wood formation are largely limited to the fusiform initials in the cambium, whereas the ray initials probably remain unchanged.

4.8 Axial Parenchyma Cells

The longitudinal or axial parenchyma cells in compression wood have attracted little attention, possible because they are present only in some of the Coniferales and are absent in the Taxales. They occur most widely in the Taxodiaceae but are relatively frequent also in the Cupressaceae and Podocarpaceae.

Onaka (1949) examined longitudinal parenchyma cells in three coniferous species, obtaining the results summarized in Table 4.20. Axial parenchyma cells were evidently half as numerous per unit area in normal wood as in compression wood. However, when measured per unit length in the tangential direction in the first growth ring, they were more frequent in compression wood than in normal wood and notably so in *Cryptomeria japonica*. Onaka could observe no difference in shape or size of the axial parenchyma cells in normal and compression woods.

Table 4.20. Incidence of axial parenchyma cells in normal and compression woods of three conifer species. (Onaka 1949)

Species	Number per mm^2 of transverse surface			Number per cm in tangential direction of the first growth ring		
	Compression wood	Normal wood	Ratio	Compression wood	Normal wood	Ratio
Chamaecyparis obtusa	10	18	0.56	78	32	2.44
Cryptomeria japonica	25	39	0.64	112	23	4.87
Podocarpus macrophyllus	217	303	0.72	248	214	1.16

4.9 Resin Canals

Resin canals, often also referred to as resin ducts, are a normal feature of the genera *Larix, Picea, Pinus,* and *Pseudotsuga,* and can occur after wounding also in other conifer genera, such as *Abies, Cedrus, Pseudolarix,* and *Tsuga.* Traumatic resin canals can also form in the first-mentioned four genera (Nylinder 1951). Lately, there has been a trend to regard all traumatic resin canals as being associated with injury (Bannan 1936, Werker and Fahn 1969, Fahn and Zamski 1970). Matsuzaki (1972), in a study of wood and bark of *Abies sachalinensis,* noted that traumatic resin canals as well as hypertrophy of ray parenchyma cells could be caused by various environmental agencies, such as aphids, pathogenic fungi, cold damage, and mechanical injury. Kuroda and Shimaji (1983) recently found that pins inserted into the stem of *Tsuga sieboldii* induced formation of traumatic resin canals which could be used for marking xylem growth.

There are two kinds of resin canals, longitudinal or axial and transverse or radial canals, which are located inside fusiform rays. Resin canals are postcambial and schizogenous in origin. Horizontal resin canals are more frequent in juvenile wood than in mature wood (Mergen and Echols 1955). Wide growth rings tend to contain more vertical resin canals per unit area than narrow rings (Stephan 1967, Flotyńzki and Tafliński 1971). Resin canals are lined with an inner layer of epithelial cells and an outer layer consisting of axial parenchyma cells and strand tracheids, most recently studied by Sato and Ishida (1982). The isolation, chemical composition, and intercellular bonding of the epithelial cells have been investigated by Thompson and his co-workers (Thompson et al. 1966, Thompson and Hankey 1968, Kibblewhite et al. 1971). Werker and Fahn (1969) studied the resin canals in *Pinus halepensis.* The anastomosis of axial and transverse resin canals has been the subject of a thorough study by Hug (1979) and Bosshard and Hug (1980). Takahara et al. (1983) recently studied the ultrastructure of the epithelial and other cells lining the axial resin canals in *Picea abies.* Longitudinal, traumatic resin canals are usually arranged in a tangential row and tend to be confined to the earlywood portion of an increment. The transverse, traumatic canals occur only within fusiform rays. Both canals and rays are larger than normal. Bamber (1972), in a study of the resin canal tissue in *Pinus lambertiana* and *P. resinosa,* found that the epithelial cells developed a secondary wall only when sapwood was transformed into heartwood, and then only in the radial resin canals of the former species, a most unusual situation. Lignification was delayed until the beginning of heartwood formation. Not surprisingly, the sapwood resin canals as a result were susceptible to collapse.

Data in the literature concerning the occurrence of resin canals in compression wood are few and frequently contradictory. Often it is impossible to tell whether normal or traumatic canals were examined. Petersen (1914) found that resin canals were associated with bands of compression wood in branches of *Picea abies* which were in the process of replacing a lost leader. Since the canals occurred in a tangential row it is probable that they were of the traumatic type. In a horizontally growing stem of *Pinus cembra,* shown in Fig. 11.34, Jaccard

4.9 Resin Canals

Table 4.21. Incidence of vertical resin canals in normal and compression woods of three conifer species. (Onaka 1949)

Species	Number per mm² of transverse surface			Number per cm in tangential direction of the first growth ring		
	Compression wood	Normal wood	Ratio	Compression wood	Normal wood	Ratio
Picea morrisonicola	0.89	0.63	1.41	19.6	5.1	3.84
Pinus densiflora	0.56	2.14	0.26	9.5	6.4	1.48
Pinus parviflora	0.56	1.00	0.56	16.2	3.0	5.40

Table 4.22. Number of horizontal resin canals per cm² of tangential surface in normal and compression woods of five conifer species. (Onaka 1949)

Species	Compression wood	Normal wood	Ratio
Larix gmelinii var. japonica	90	69	1.30
	41	31	1.32
Picea jezoensis	124	168	0.74
Pinus densiflora	91	91	1.00
	76	94	0.81
	70	62	1.13
Pinus parviflora	53	69	0.77

(1919) observed a large number of resin canals on the lower, compression wood side. Verrall (1928), like Petersen (1914), noted a tangential row of resin canals in the outer part of the annual increments in *Picea abies*, extending across the entire zone of compression wood. All five conifer species studied by Fegel (1938, 1941), namely *Larix laricina*, *Picea mariana*, *P. rubens*, *Pinus resinosa*, and *P. strobus*, contained about twice as many resin canals in branch wood as in stem wood. Their diameter was 85 µm in the former and 126 µm in the latter.

A more systematic study of the frequency of the resin canals in compression wood was carried out by Onaka (1949). His data, summarized in Table 4.21, show that resin canals were fewer in compression wood than in opposite wood when measured per unit area but more numerous when measured per unit tangential length of a growth ring. It is of interest to note in this connection that several investigators, such as Münch (1919–1921), Bannan (1933a, b), Mergen and Echols (1955), and Reid and Watson (1966), have observed an inverse relationship between rate of growth and incidence of longitudinal resin canals. The frequency of transverse resin canals seems to be the same in normal and compression woods, as can be seen from Onaka's (1949) data in Table 4.22. Sato and Ishida (1983) found the longitudinal resin canals in *Larix leptolepis* to be less frequent in compression wood than in normal wood. They were usually present within the outer portion of the annual increments in both tissues. In the compression wood, small canals occurred in the narrow zone of the radially flattened, last-formed tracheids. Traumatic resin canals could not be detected.

When conifer stems have suffered compression failures, a subject discussed in Chap. 15.2.1.5, the zone of failure is usually covered by a layer which partly consists of compression wood and partly of callus tissue. Trendelenburg (1940) observed in such a layer in *Picea abies* tangential rows of what must have been traumatic resin canals. Mergen and Winer (1952) have reported the presence in such compression wood from *Pinus strobus* many and conspicuous transverse resin canals of the traumatic type. Mergen (1958) also noted the occurrence of traumatic, longitudinal canals associated with the compression wood that always forms in the drooping leader of *Tsuga canadensis* (Chap. 10.3.9.3). Ollinmaa (1959) found the number of longitudinal resin canals to be less in compression wood than in normal wood of *Picea abies* and *Pinus sylvestris*, but the canals were slightly wider in compression wood. Confirming Onaka's (1949) results, Ollinmaa noted that with increasing size of the growth rings the number of resin canals per unit area decreased, while it increased when measured per unit length in the tangential direction.

In their review of compression wood, Core et al. (1961) referred to the presence of rows of traumatic resin canals in compression woods from *Pinus albicaulis* and *Pseudotsuga menziesii* var. *glauca*. An example from the latter species is shown in Fig. 4.126. The investigators suggested the possibility that the cambium had become injured by deformation and that this had caused the formation of the traumatic resin canals. Scurfield (1967), in a histochemical study of compression wood in *Pinus radiata*, could detect only very few or no resin canals.

Fig. 4.126. Compression wood of *Pseudotsuga menziesii* var. *glauca*. The lower growth ring has a gradual and the upper a more abrupt transition from earlywood (*EW*) to latewood (*LW*). A row of traumatic resin canals (*arrow*) is present in the earlywood of the upper increment. Transverse section. LM. (Core et al. 1961)

4.9 Resin Canals

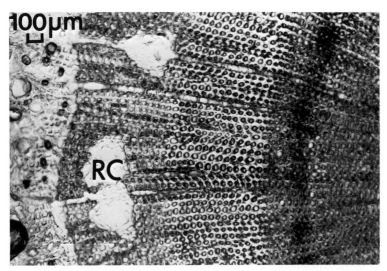

Fig. 4.127. Compression wood in *Pinus halepensis* with a high incidence of traumatic resin canals (*RC*). Transverse section. LM. (Fahn and Zamski 1970)

Fahn and Zamski (1970) studied the effect of pressure, wind, wounding, and growth substances on the formation of resin canals in *Pinus halepensis*. All factors were found to cause formation of resin canals, but only after a considerable induction time. In one experiment, plants were placed for 6 months at distances of 30 cm and 80 cm from a fan. As could be expected, considerable formation of compression wood took place. Compared to controls, the number of resin canals was three times as large in the plants closest to the fan and one and a half times as large in those further away. Figure 4.127 shows the appearance of the compression wood and the resin canals. The number of longitudinal resin canals was also increased by application of indoleacetic acid, naphthaleneacetic acid, and 2,4-dichlorophenoxyacetic acid. At the same time, there was an increase in the rate of radial growth and formation of compression wood. An example of the numerous resin ducts is shown in Fig. 4.128, where it can be seen that the canals were surrounded by a large number of parenchyma cells which apparently had remained unlignified. All three growth substances also caused formation of many large multiseriate rays and thin-walled tracheids. Naphthaleneacetic acid was most effective in causing production of resin canals in the summer, while 2,4-dichlorophenoxyacetic acid was best in the winter.

As already mentioned, wounding of conifer xylem can result in formation of abnormally large or traumatic resin canals. A cambium that has been exposed to gamma irratiation produces traumatic resin ducts (Ladza and Odintzov 1968). The balsam woolly aphid (*Adelges piceae*), when feeding on *Abies* trees, stimulates formation of pathological resin canals in the xylem (Balch 1952, Oechssler 1962) (Chap. 20.8.2.3). Cerezke (1972) observed a large number of traumatic, longitudinal resin canals in *Pinus contorta* var. *latifolia* above wounds caused by the weevil *Hylobius warreni*. Stem wood of *Pinus thunbergii* also develops traumatic resin canals when infected by *Bursaphelenchus lig-*

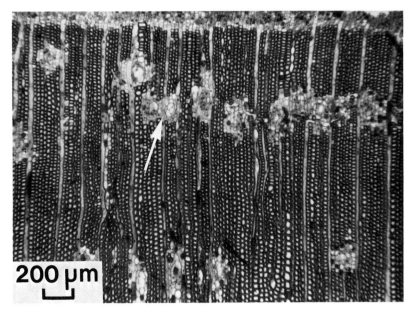

Fig. 4.128. Compression wood of *Pinus halepensis* with numerous resin canals surrounded by unlignified parenchyma cells (*arrow*). Transverse section. LM. (Fahn and Zamski 1970)

nicolus, a nematode (Sugawa 1978). Traumatic resin ducts are present in the wood of *Tsuga mertensiana* trees infected by dwarf mistletoe (*Arceuthobium*) (Srivastava and Esau 1961) (Chap. 15.10) and abnormal canals occur in wood of *Pinus radiata* infected by the fungus *Diplodia pinea* (Foster and Marks 1968) (Chap. 15.12.1.4). Wood infected by southern fusiform rust (*Cronartium fusiforme*) contains a larger than normal number of resin canals (Jackson and Parker 1958, Jewell et al. 1962) (Chap. 15.12.5). Numerous, traumatic resin ducts are present in the xylem of the tumors growing on *Picea glauca* and *P. sitchensis* (Tsoumis 1965, Rickey et al. 1974) (Chap. 15.13). Traumatic resin canals develop in the xylems of *Abies balsamea*, *Larix laricina* (Kiatgrajai et al. 1976a), and *Tsuga canadensis* (Kiatgrajai et al. 1976b) on treatment with paraquat, a herbicide that, when administered to pine trees, greatly increases their formation of oleoresin (Roberts 1973, Sioumis and Lau 1976).

Mechanical injury often causes formation of traumatic resin canals. Such ducts developed when pins were inserted into the cambial zone of *Abies firma* (Shimaji and Nagatsuka 1971). Sachsse (1969, 1971, 1983) found that when *Pinus sylvestris* and *Pseudotsuga menziesii* trees were mechanically pruned with a saw that climbed the stem, the pressure exerted on the cambial zone and the young xylem induced formation of traumatic resin canals in the wood.

Until quite recently, little was known about the epithelial and other cells that line the resin canals in compression wood. Verrall (1928) states that the epithelial cells have thicker walls in compression wood than in normal wood, and that sometimes there are no spaces between them. Timell (1972a) found that the epithelial cells lining the horizontal resin canals in both normal and compression woods were thin-walled in *Pinus resinosa* (Fig. 4.121) but thick-walled in *Larix laricina, Picea rubens,* and *Pseudotsuga menziesii* (Fig. 4.122). Yumoto et al. (1982) have recently reported that on transition of normal wood into compression wood in *Picea glauca,* the flat, thin-walled epithelial cells of the horizontal resin canals in the former tissue became wider and developed thick walls, as shown in Fig. 4.129. No investigation has yet been reported on the formation of the secondary wall and the lignification of the epithelial cells in compression wood, such as that carried out with normal wood of *Pinus* by Bamber (1972).

Results of a thorough examination of the longitudinal resin canals in compression wood of *Larix leptolepis* were recently reported by Sato and Ishida (1983). In normal wood of this species, the axial resin canals are lined with an inner and an outer layer of longitudinal cells, as illustrated in Fig. 4.130 (Sato and Ishida 1982, 1983). The inner layer, or epithelium, is composed of epithelial cells with either thin or thick walls (E.1 and E.2). The outer layer consists of three types of cells, namely two kinds of axial parenchyma cells (P.1 and P.2) and strand tracheids (S.T.). Only the inner layer is cylindrical. According to Sato and Ishida (1983), the axial parenchyma cells and the strand tracheids are both derived from similar fusiform cambial cells, a suggestion that is in accord with views previously expressed by Laming (1974) and by Panshin and de Zeeuw (1980).

All longitudinal cells lining the axial resin canals in normal wood of *Larix leptolepis* were also present in compression wood of this species (Sato and Ishida 1983). There were, however, several minor differences between the two tissues. Resin canals lacking the outer layer of cells were more common in compression wood than in normal wood, and where this layer was present, it consisted of a higher proportion of strand tracheids in the compression wood. Many of these strands contained helical cavities and ribs, while others had a smooth lumen surface, as illustrated in Fig. 4.131.1. All three of the outer cell types were rectangular in longitudinal section. In transverse section, the P.1 parenchyma cells formed a crescent, whereas the P.2 cells and the strand tracheids had a square outline (Fig. 4.131.2). In these respects, the three cell types were similar to those present in normal larch wood.

As in normal wood, the inner layer consisted of epithelial cells, some with a thin wall (0.5 µm) and others with a thick one (5 µm), but the proportion of thick-walled cells was much higher in the compression wood. These epithelial cells had a more oval outline in transverse section than those in normal wood (Fig. 4.131.2). As shown in Fig. 4.131.3, thin-walled cells were often adjacent to thick-walled ones. Unlike normal wood, compression wood frequently contained what the investigators refer to as "cylinders" which traversed the longitudinal resin canals in the radial direction. They consisted of one or several longitudinal epithelial cells with either a thick (Fig. 4.131.4) or a thin (Fig. 4.131.6) wall. Some of the longitudinal epithelial cells were quite large (Fig. 4.131.5). The fact that these cylinders never traversed a canal tangentially was interpreted as an indication that the longitudinal epithelial cells separate radially from one another in compression wood.

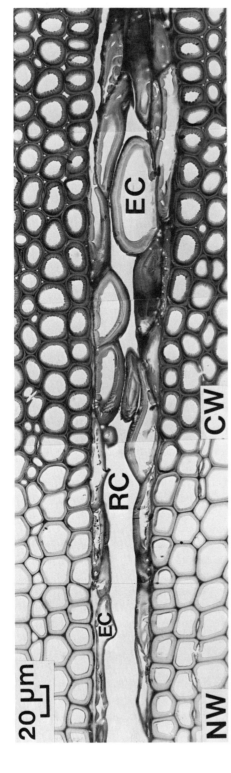

Fig. 4.129. Transition from normal wood (*NW*) to compression wood (*CW*) in *Picea glauca*. The flat, thin-walled epithelial cells (*EC*) lining the horizontal resin canal (*RC*) in the normal wood become larger, wider, and thick-walled in the compression wood. Transverse section. ULM. (Yumoto et al. 1982b)

4.9 Resin Canals

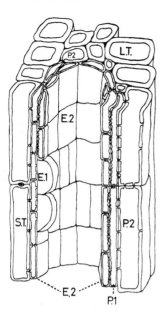

Fig. 4.130. Model showing the structure of a vertical resin canal complex in normal wood of *Larix leptolepis*. The inner layer consists of epithelial cells with either a thin (*E.1*) or a thick (*E.2*) wall. Parenchyma cells (*P.1* and *P.2*) and strand tracheids (*S.T.*) constitute the outer layer. *L.T.* longitudinal tracheid. Drawing. (Sato and Ishida 1982)

Sato and Ishida (1983) concluded from their observations that the scarcity of thin-walled epithelial cells is a characteristic structural feature of the longitudinal resin canals in compression wood. They also considered it significant that compression wood contained more epithelial cells with an irregular outline.

Summarizing, it would appear from the information currently available that resin canals of the normal type are less frequent in compression wood than in normal wood. Possibly, this is a consequence of the increased radial growth in compression wood. In *Larix leptolepis*, the longitudinal resin canals are lined with the same types of cells as in normal wood, albeit with several modifications. Axial parenchyma cells are less and strand tracheids more frequent in the outer layer in compression wood, a layer that is occasionally absent in this tissue. In the inner layer, a higher proportion of axial epithelial cells have a thick wall in compression wood. There can be no doubt that formation of compression wood is frequently associated with the appearance of longitudinal resin canals of the traumatic type, almost invariably arranged in a tangential row that extends over much of the compression wood zone. If they are caused by cambial disturbances, as thas has been suggested by Core et al. (1961), is unknown. The experiments carried out by Fahn and Zamski (1970) suggest that an increased concentration of auxin, known to be involved in the formation of compression wood (Chap. 13.4), might be the causative agent in this case.

244   4 The Structure of Compression Wood

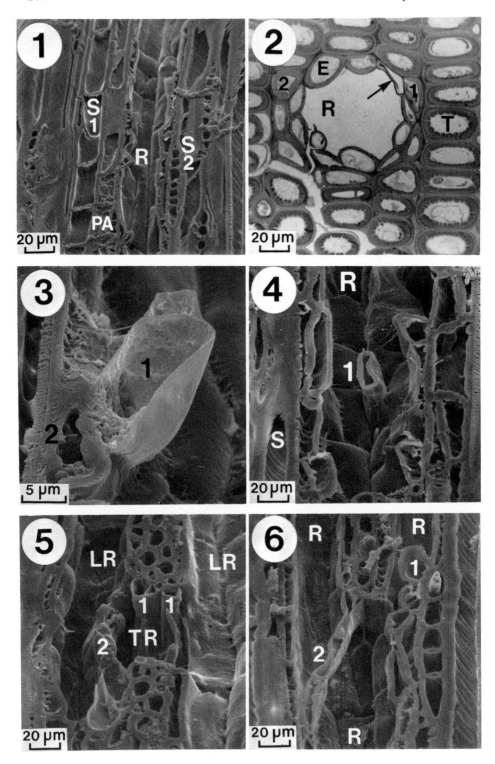

4.10 The Pits in Compression Wood

4.10.1 The Bordered Pits Between Longitudinal Tracheids

The pits connecting the axial tracheids in compression wood are among the most conspicuous features of this tissue and attracted the attention of many of the early investigators. When viewed in longitudinal section in the light microscope the pits appear to be crossed diagonally by a slit which can extend beyond the pit annulus, as can be seen in Fig. 4.83. Even a casual inspection reveals that the slits are parallel to the spiral striations permeating the inner tracheid wall, and the slope of the slits can therefore be used as a measure of the microfibril angle in the S_2 layer. The different appearance of the pits in normal and compression woods under the light microscope is illustrated in Fig. 4.132. Sometimes the pits in compression wood seem to be crossed by two symmetrically diagonal slits. This happens when the pit has been cut so that both the front and the rear slit in a tracheid wall have become visible. When viewed in longitudinal section in polarized light, the pit gives rise to a characteristic Maltese cross, as shown in Fig. 4.75.

The lens-shaped opening, first observed Sanio (1873), is clearly indicated in Hartig's (1896) first drawing of compression wood tracheids shown in Fig. 2.8. The spiral striations are bending around the pit aperture here, but the relationship between the pit border and the slit is not clear. Except for a single reference to the fact that the slit can extend far beyond the pit, Hartig made no further comment on the subject, nor does he mention the pits in his last publication. Cieslar (1896) in the same year states that the tracheids in compression wood of *Picea abies* contain far more pits on their radial walls than do normal wood tracheids. Lämmermayr (1901) drew attention to the fact that if longitudinal sections are cut sufficiently thin, only one slit is seen across the pits in compression wood tracheids. The slit reached beyond the pit border in *Araucaria*, *Dammara*, and *Taxus baccata*.

Sonntag (1904), in his study of compression and opposite woods, paid more attention to the pits and their structure. He found that in branch wood of *Picea abies*, the pits in the first-formed earlywood had the same appearance as pits in normal wood, a fact also evident from Mork's (1928a) later drawing

Fig. 4.131. Resin canals and associated cells in compression wood of *Larix leptolepis*. **1** Longitudinal resin canal (*R*) complex, showing strand tracheids (*S*) without (*1*) and with (*2*) helical cavities. *PA* parenchyma cell P.2. **2** A longitudinal resin canal with wide, thick-walled (*E*) and flat, thin-walled (*arrow*) axial epithelial cells. *1* parenchyma cell P.1 with a crescent outline. *2* parenchyma cell P.2 with a more angular outline. *T* longitudinal tracheid. **3** A longitudinal, thin-walled epithelial cell (*1*) next to a thick-walled (*2*) one. **4** Longitudinal resin canal (*R*) with a hollow "cylinder" (*1*). *S* strand tracheid with helical cavities. **5** Longitudinal (*LR*) and transverse (*TR*) resin canals with both small (*1*) and large (*2*), thin-walled epithelial cells. **6** Longitudinal resin canals (*R*) with a thick-walled epithelial cell (*1*) and a thin-walled epithelial cell ("cylinder") (*2*), traversing one of the resin canals. **2** Transverse section. UVLM. All other figures: tangential surface. SEM. (Modified from Sato and Ishida 1983)

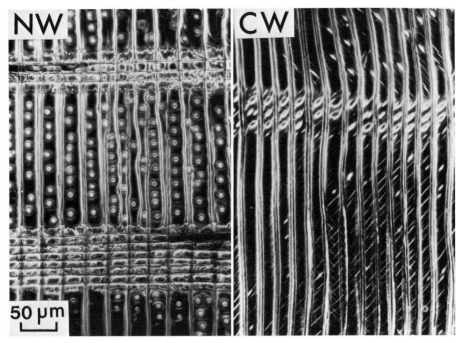

Fig. 4.132. Normal (*NW*) and compression (*CW*) woods of *Pinus resinosa*, showing the different appearance of the bordered pits in the two tissues. Radial sections. LM. (phase contrast) (Courtesy of A.C. Day)

(Fig. 4.109). In tracheids formed at a later stage, Sonntag observed that the aperture was drawn out in a slit which cut through the pit border, thus rendering the pit aperture larger than the border. Such pits occurred throughout the remainder of the growth ring, including the last-formed latewood. Pits were especially numerous in radial walls close to rays. The angle between the oval pit apertures and the longitudinal tracheid axis varied between 22° and 55° with an average value of 40.5°. Gothan (1905) noted that the spiral striations in *Taxodium distichum* were widest when they crossed a pit. He considered the openings to be schizogenous in origin and believed that they were formed when sapwood was transformed into heartwood.

A thorough investigation of the pits in compression wood was reported in 1907 by Krieg. Confirming the earlier observations of Hartig (1896) and Sonntag (1904), Krieg noted that the slit could extend beyond the pit annulus, and that such pits also occurred in the last few tracheids in each growth ring. Contrary to the earlier investigators, Krieg (1907) emphatically rejected the notion that the oval openings in the pits should be real slits. He believed that they were actually grooves which extended outward from the pit aperture in the same direction as the microfibrils, a fact evident when the tracheids were observed in polarized light. These grooves or furrows could continue beyond the pit border, while in other cases they terminated within this border. Each groove sloped regularly on both sides of the pit aperture toward the latter. According to Krieg

(1907), pit grooves are only found in tracheids which possess spiral striations in their cell walls. In *Pinus sylvestris*, by contrast, he claimed to have observed round or oval pit openings without any grooves in tracheids containing spiral striations. Compared to normal tracheids, pits were fewer in tracheids with striations. Similar grooves at the bordered pits of compression wood tracheids were observed 2 years later by Sonntag (1909).

Trendelenburg (1932) drew attention to the fact that compression wood tracheids contained fewer pits than normal ones and that this probably contributed to the larger compressive strength of compression wood. Phillips (1937, 1940) found the oblique pit apertures useful for detecting compression wood in *Juniperus procera*. Onaka (1949) measured the diameter of the pits in normal and compression woods of ten coniferous species. As shown in Table 4.9, the latewood pits were larger and the earlywood pits smaller in compression wood than in normal wood. The pits were stated to have a lens-like opening, extending along the spiral striations. Harada and Miyazaki (1952) were the first to examine the pits in compression wood in the electron microscope. Their micrographs show a structure very similar to that of normal bordered pits, including a torus supported by thick, radiating microfibrils and a warty layer covering the initial pit border on the pit chamber side.

Because of their complex, three-dimensional structure, bordered pits are ideal subjects for examination in the scanning electron microscope. Several investigators have also used this instrument for studying the pits in compression-wood tracheids. The first were Resch and Blaschke (1968), who found that the bordered pits in compression wood of *Pinus mugo* occasionally extended far into the intercellular spaces, as shown in Fig. 4.52, a situation also illustrated for *Abies balsamea* in Fig. 4.21. This observation was more recently confirmed by Ohtani and Ishida (1981) with compression wood of *Abies sachalinensis*. Ishida et al. (1968) and Ishida and Fujikawa (1970) studied the bordered pits in compression wood of *Abies sachalinensis*. Their electron micrographs clearly show how the pit aperture is located at the bottom and center of a groove, exactly as demonstrated 60 years earlier by Krieg (1907). A warty layer covers the entire groove and the helical ribs (Sect. 4.4.8). The outer surface of a bordered pit is shown in Fig. 4.106. The membrane had been removed here, and a part of the groove can be seen through the opening. Scurfield and Silva (1969b) made similar observations with compression wood of *Pinus radiata*, pointing out that helical cavities and ribs do not occur in the neighborhood of pits. Jutte and Levy (1972) examined the helical cavities and pits in compression wood of *Pinus ponderosa*. Fully developed compression-wood tracheids contained helical cavities and pit apertures sunk in grooves that extended through and beyond the pit border.

Cockrell (1973), using scanning electron microscopy, studied the effect of pretreatment on the tracheid pits in compression wood of *Sequoiadendron giganteum*. He came to the conclusion that green compression wood did not contain any helical checks ("slits") in the wall or across the pit apertures (Fig. 4.133). After the wood had been boiled in water and dried, checks had developed, continuing beyond the pit apertures, as shown in Fig. 4.134. Cockrell found the pit openings to be flared toward the lumen, with the inner aperture

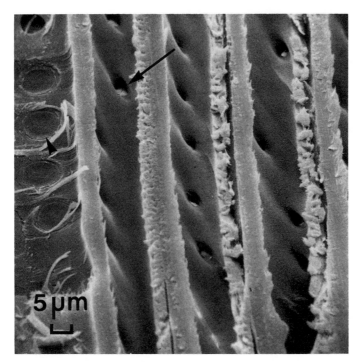

Fig. 4.133. Compression wood of *Sequoiadendron giganteum* (cut in the dry condition), with pit openings (*arrow*) and pit membranes (*arrowhead*). The inner surface of the S_2 layer is covered with fine warts. Note the absence of any slits at the pit openings. Radial surface. WEM. (Cockrell 1973)

aligned with the microfibrils in S_2. The outer aperture was oval and less steeply oriented than the inner one (Fig. 4.101). Similar observations were reported in a later publication (Cockrell 1974).

Parham (1970) examined surface replicas of the bordered pits in compression wood of *Pinus taeda*, using the transmission electron microscope. The pit shown in Fig. 4.104 is structurally indistinguishable from the intertracheid pits in normal pine wood. In their investigation of pit membranes in gymnosperms, Bauch et al. (1972) also examined pits in compression wood. These pits, on the whole, resembled those in corresponding normal wood except that the torus often was slightly thinner than in normal tracheids. Similar observations with *Picea* were later reported by Jutte et al. (1977).

Ohtani and Ishida (1981) and Yumoto et al. (1983) have recently reported observations on bordered pits in compression woods of *Abies sachalinensis* and *Picea glauca*, respectively. The inner pit aperture was often difficult to distinguish, as had also been found by Cockrell (1974), because the helical ribs covered part of the outline, as shown in Figs. 4.105 and 4.135, where a part of the circular pit opening is hidden. Figure 4.135 illustrates the different size and shape of pit grooves in compression-wood tracheids, including a wide groove

4.10.1 The Bordered Pits Between Longitudinal Tracheids 249

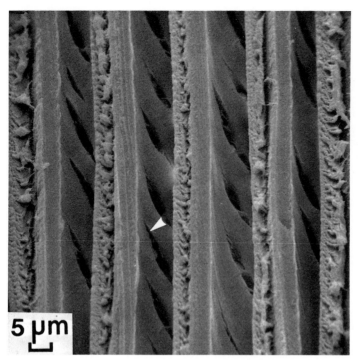

Fig. 4.134. Compression wood of *Sequoiadendron giganteum* that had been boiled, dried, and finally cut in a wet condition. Note the extensions of the inner pit apertures (slits) (*arrowhead*). Radial surface. SEM. (Cockrell 1973)

extending across the entire inner S_2 layer (1), a narrow, shorter groove (2), and a very short groove (3). The pit apertures seen in Fig. 4.136 are all located at the bottom of long and deep grooves, which, because of their great length, had not been split on drying. As can be seen in Fig. 4.137, the diameter of the bordered pits in normal wood of *Abies sachalinensis* was larger in the earlywood than in the latewood (Ohtani and Ishida 1981). In compression wood, the pit diameter in the first-formed earlywood (Sect. 4.5) was as large as in corresponding normal wood, but it decreased rapidly thereafter and remained constant throughout the rest of the growth ring. The pit border did not protrude above the wall surface of the tracheids in pronounced compression wood as it did in mild and moderate grades with their thinner walls. Yumoto et al. (1983) concluded that whatever differences may exist between the bordered pits in normal and compression woods, these differences are probably only a result of the different anatomy of the tracheid walls in these two tissues. This conclusion agrees entirely with my own observations.

Casperson (1968) has drawn attention to the fact that the cell wall organization of the pit border in compression wood must be simpler than in normal wood since the S_3 layer is missing. On the basis of detailed studies of the bor-

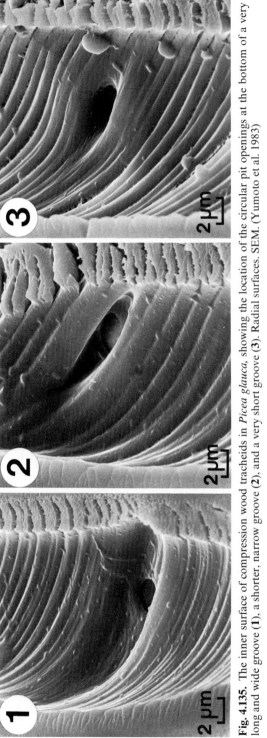

Fig. 4.135. The inner surface of compression wood tracheids in *Picea glauca*, showing the location of the circular pit openings at the bottom of a very long and wide groove (**1**), a shorter, narrow groove (**2**), and a very short groove (**3**). Radial surfaces. SEM. (Yumoto et al. 1983)

4.10.1 The Bordered Pits Between Longitudinal Tracheids

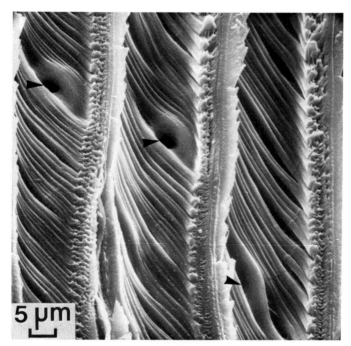

Fig. 4.136. Compression wood of *Chamaecyparis nootkatensis* with pit openings at the bottom of very long grooves (*arrowheads*). Note the absence of any splits in the grooves and the low incidence of warts. Radial surface. SEM

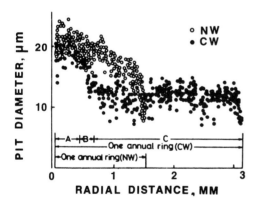

Fig. 4.137. Variation in the diameter of the bordered pits within one growth ring in *Abies sacchalinensis* containing normal (*NW*) and compression (*CW*) woods. *A* inner zone; *B* transition zone; *C* compression wood zone. (Ohtani and Ishida 1981)

dered pits connecting compression wood tracheids in *Larix decidua*, Casperson suggested the structure shown in Fig. 4.138. Except for the absence of an S_3 layer on the lumen side, this is the same organization as that proposed by Harada and Côté (1967) and Hirakawa and Ishida (1981) for intertracheid pits in normal wood. The S_2 layer terminates at the pit aperture, but the S_1 layer extends partly into the pit border. On the pit chamber side, the border is lined by the initial pit border. Although not indicated in the drawing, the pit membrane in *Larix* contains a torus.

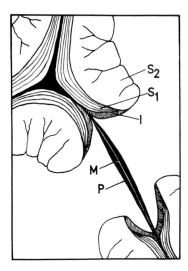

Fig. 4.138. A bordered pit connecting two longitudinal tracheids in compression wood of *Larix decidua*. *I* initial pit border. *M* middle lamella. *P* primary wall. The torus is not indicated in this drawing. Transverse section. (Modified from Casperson 1968)

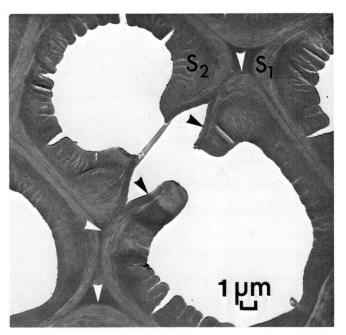

Fig. 4.139. Compression wood tracheids in *Pinus sylvestris* with an aspirated bordered pit. *Black arrowheads* indicate the initial pit border. This is one of the rare cases where the primary walls (*white arrowheads*) can be readily observed in untreated compression wood (Sect. 4.4.3). Transverse section. TEM

4.10.2 The Half-Bordered Pits

My own observations of intertracheid pits in compression woods of both *Ginkgo biloba* and the conifers confirm the correctness of the structure suggested by Casperson (1968). In the aspirated pit in *Pinus sylvestris* compression wood in Fig. 4.139 the clearly distinguishable S_1 layer extends half way into the pit border. It is followed on the pit chamber side by the initial pit border. In other cases, a warty layer could be seen, lining both S_2 and the initial pit border. The primary walls seen in this figure are discussed in Sect. 4.4.3.

According to Onaka (1949), the simple pits connecting the ray parenchyma cells in compression wood have lens-like, protruding inner openings. They appear to be the same in all coniferous species and cannot be used, as in normal wood, for diagnostic purposes. According to my own observations, the bordered pits connecting ray tracheids in compression wood have the same structure as in normal wood.

4.10.2 The Half-Bordered Pits Between Longitudinal Tracheids and Ray Parenchyma Cells

The only mention of the structure of half-bordered pits in compression wood is in a study of *Larix decidua* by Casperson (1968) (Fig. 4.140). Foster and Marks (1968), who studied the effect of the fungus *Diplodia pinea* on *Pinus radiata*, observed that the compression wood tracheids formed in the diseased trees contained an unusually large number of blind pits toward the ray cells. Actually, it

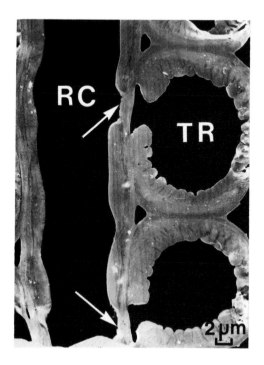

Fig. 4.140. Compression wood of *Larix decidua* with half-bordered pits (*arrows*) between tracheids (*TR*) and a ray parenchyma cell (*RC*). Transverse section. TEM. (Casperson 1968)

seems to be a typical feature of the pits between axial tracheids and ray parenchyma cells in all compression woods that many of them are blind. Jutte et al. (1977) also observed numerous blind pits in ray cells of *Picea* compression wood. Such a blind pit is seen in Fig. 4.123.

4.10.3 Conclusions

It is evident that not all structural characteristics of the pits are entirely clear at the present time. Nečesaný and Oberländerová (1967) have claimed that the number of pits in compression wood cannot be determined accurately. This is probably an exaggeration. Most investigators agree that the pits in compression wood are fewer than in normal wood. One exception to this was recently reported by Savidge (1983), who found the incidence of bordered pits to be abnormally high in tracheids induced by exogenous indoleacetic acid (Chap. 13.4). It would appear that bordered pits in tracheids tend to occur in the neighborhood of rays. The location of the bordered pits in the wall of the longitudinal tracheids is the most unique feature of the pits in compression wood. They occur here at the bottom of a groove or furrow which actually is nothing but an unusually wide helical cavity, oriented in the same direction as the ordinary cavities, that is $30° - 50°$ to the longitudinal fiber axis. This is an entirely different arrangement than that which is found in tracheids containing spiral thickenings on their inner wall. In this case, as can be seen in Figs. 4.93, 4.94, and 4.98, the thickenings deflect around the pit. Frequently, the groove extends considerably beyond the pit annulus, cutting through the pit border in a wide swath. When viewed in radial sections in the light microscope, the groove appears as a diagonal slit across the pit.

It has often been stated in the past that the pit aperture should be "slit-like" in compression wood tracheids. Light micrographs, such as that seen in Fig. 4.83, do give this impression. I believe, however, that the notion of an oblong or elongated pit opening is based on a misinterpretation of these light micrographs. What is observed here is not the pit aperture alone but also the pit groove and/or a wall check created on drying. All available transmission electron microscope replicas of bordered pits in compression wood, such as that shown in Fig. 4.104, indicate the presence of a circular pit aperture and torus. Inspection of scanning electron micrographs, for example those in Figs. 4.21, 4.52, 4.105, 4.107, 4.135, and 4.136 as well as the many micrographs published by Yumoto et al. (1982, 1983) also show a perfectly circular pit opening. Bailey (1958) has drawn attention to the fact that a slit-like pit opening would not permit complete closure by aspiration, since the torus is circular, while the aperture would be oblong, resulting in a defective valve action. Actually, both the torus and the aperture are circular in compression wood, and there is no reason to believe that pit aspiration should be less efficacious in preventing intertracheid transport than in normal wood. It should perhaps also be added here that compression wood is less and not more permeable than normal wood (Chap. 7.3.2). The reason for this does not reside in the structure of the bordered pits but instead in the facts that compression wood tracheids have fewer pits and a smaller lumen than those in normal wood.

The exact organization of the pit border in compression-wood tracheids is not known with certainty at the present time. Some facts are clear. The S_1 layer extends fairly deeply into the border, but it does not line the initial pit border on the lumen side. The S_2 layer lines the pit aperture but does not appear to continue into the pit chamber. Such a structure would be close to that suggested for the pits in normal wood tracheids by Murmanis and Sachs (1969a, b).

4.11 Knot Wood

A knot is a former branch, encased in the stem of a tree. It differs from a branch in its high resin content but is otherwise similar to a branch in its chemical composition (Chap. 10.6). Anatomically, knot wood and branch wood could be expected to be similar, yet certain differences do seem to exist. Whether they are real or not cannot be decided at the present time, largely because so few studies have been devoted to the ultrastructure of either branch wood or knot wood. Brief mention of the histology of knot wood has been made by Klem (1928) and by Nylinder and Hägglund (1954). Hägglund and Larson (1937), who studied the chemical composition and pulping behavior of *Picea abies* knot wood, found that no diagonal slit occurred across the pits in this wood. However, it is not clear whether they were observing the lower side of the knots, which contains compression wood, or the upper which consists of opposite wood.

In his comprehensive investigation of the occurrence and properties of knots in *Picea abies*, Wegelius (1939, 1946) studied the gross anatomy of opposite, normal, and compression woods. He emphasized the heterogeneous nature of knot wood and also noted that it contains much compression wood and latewood. It is clear from his description of the latewood, however, that most of the latter was actually compression wood. A transverse section of knot wood shows pronounced compression wood with intercellular spaces, rounded tracheids, and thick cell walls. Only the first-formed tracheids in each growth ring are angular in outline. The thickness of the tracheid wall in knot compression wood ranged from 3 µm to 10 µm with an average value of 5.5 µm. The distribution curve was skewed to the right. For corresponding opposite wood, the wall thickness varied uniformly from 5 µm to 55 µm, with an average value of 30 µm. Other measurements by Wegelius are difficult to evaluate, as he made no distinction between opposite, normal, and compression woods.

The most detailed investigation on the anatomy of knot wood so far reported is that of Boutelje (1965, 1966), who examined knots in *Picea abies* and *Pinus sylvestris*, including opposite wood on the upper side of the knots and compression wood on the lower. The location of the specimens are shown in Figs. 7.5 and 7.6. A transverse section from the lower side of a knot in *Picea abies* is shown in Fig. 4.141. The first two thirds of the central growth ring undoubtedly consisted of compression wood, whereas the last one third appears to be normal latewood. The reason for this change is unknown, but obviously the gravitational or other stimulus ceased to cause formation of compression wood. The last four to seven rows of tracheids contained helical thickenings, while the

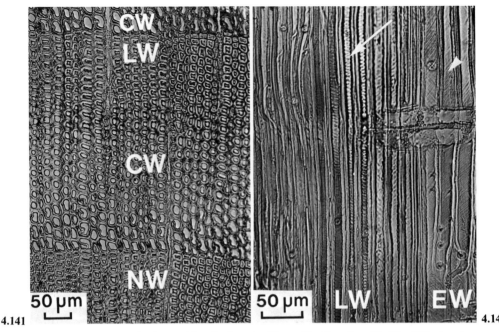

Fig. 4.141. Wood on the lower side of a knot in *Picea abies* (location 6 in Fig. 7.5), showing the presence of compression wood (*CW*) and normal latewood (*LW*) tracheids in the middle growth ring. Transverse section. LM. (Boutelje 1966)

Fig. 4.142. Wood on the lower side of a knot in *Picea abies* (location 7 in Fig. 7.5). Helical thickenings are present in the last-formed latewood (*LW*) tracheids (*arrow*), while the first-formed earlywood (*EW*) tracheids contain spiral striations (*arrowhead*). Radial section. LM. (Boutelje 1966)

following earlywood tracheids had helical cavities, as can be seen in Fig. 4.142. The cells with the helical thickenings have all the appearance of opposite wood tracheids. The simultaneous occurrence of such tracheids and compression wood tracheids on the lower side of a branch or knot is unusual. Boutelje never observed helical thickenings and helical cavities (spiral striations) within one and the same tracheid.

In knot wood of *Pinus sylvestris*, the occurrence of compression wood was irregular, and such wood was often found also on the upper side, a not uncommon situation in branches (Chap. 10.6). The middle growth ring in Fig. 4.143 consists of compression wood and a thin band of normal latewood. In the radial section in Fig. 4.144 all tracheids contain spiral striations except the last three to four in the latewood. Judging from the transverse section (Fig. 4.143), these tracheids would appear to be of the normal type.

Obviously, knots, having once been part of living branches, should have the same structure as branches. In leaning or bent stems, growth is usually rapid on the lower side. In branches, except for the first few years of their life, radial growth tends to be slow. Perhaps some of the less familiar properties of com-

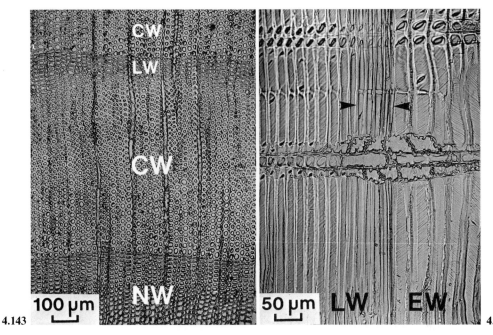

Fig. 4.143. Wood on the lower side of a knot in *Pinus sylvestris* (location *10* in Fig. 7.6). The middle growth ring consists of compression wood (*CW*) and normal latewood (*LW*). Transverse section. LM. (Boutelje 1966)

Fig. 4.144. Wood on the lower side of a knot in *Pinus sylvestris* (location *8* in Fig. 7.6), showing the presence of spiral striations in all tracheids except the last 3–4 in the latewood (*arrowheads*). Radial section. LM. (Boutelje 1966)

pression wood in branches and knots are due to this fact. The investigations by Park et al. (1979, 1980) on branch wood in *Cryptomeria japonica* did not reveal any unusual features of the compression wood. These tracheids were smaller than those in the stem wood, but otherwise had all the characteristics of typical compression-wood tracheids, such as a large microfibril angle in S_2 and well-developed helical cavities. There were also tracheids transitional between normal and compression wood cells, but such tracheids are also known to occur in stem wood (Fujita et al. 1979).

4.12 Root Wood

The structure of root wood has received relatively little attention, and no attempt will be made here to review what information is available. Eskilsson (1969) has reported that coarse and fine roots of *Picea abies* differ anatomically. Compared to the tracheids in slender roots, those in coarse roots are shorter, have a smaller cross section, and have thicker cell walls. The proportion

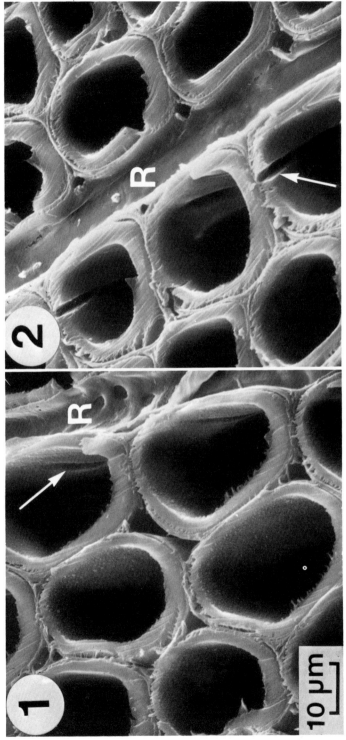

Fig. 4.145. Severe compression wood in an underground root of *Abies balsamea* (**1**) and *Tsuga canadensis* (**2**). Note the complete absence of any helical cavities. The inner surface of the tracheids is covered by a warty layer. *Arrows* indicate drying checks. *R* ray. Transverse surface. SEM

of latewood is high in the coarse roots but remarkably low in the fine ones, and as a result the density of the former root wood is higher than that of the latter.

As described in Chap. 10.7.3, compression wood occurs very irregularly and sporadically in roots. It is seldom, if ever, found in buried roots but develops readily when roots are exposed to light. Such compression wood has the same macroscopic appearance as that present in stems and branches of conifers. According to my own observations, compression wood on the under side of exposed roots of *Abies balsamea* and *Tsuga canadensis* had the same structure as that formed in stems, except for the fact that the tracheids did not contain any helical cavities, as can be seen in Fig. 4.145. Whether this is a general feature of all root compression woods cannot be decided until more specimens and species have been examined.

4.13 Bark

The term *reaction wood*, as pointed out by Höster and Liese (1966), should not be applied to bark. *Reaction tissue*, first used by White (1965), is to be preferred. Scurfield and Wardrop (1962) use the designation *reaction phloem*. There seems to be no reason why the terms *reaction bark*, *compression bark*, and *tension bark* could not also be used. Excellent reviews of bark structure are found in Chang's (1954) study of pulpwood species, in Srivastava's (1963b) monograph on the secondary phloem in the Pinaceae and in den Outer's (1967) report on the secondary phloem in gymnosperms. The phloem is well described in a paper by Howard (1971) on the southern pines. The structure of the sieve elements in the phloem has been ably reviewed by Parthasarathy (1975). The standard, classical work on the phloem is still Katherine Esau's monumental book *The Phloem*, published in 1969.

The outward appearance of bark on the lower side of branches or leaning stems of conifers has been mentioned by several investigators. Detlefsen (1878) found bark on the lower side of *Pinus sylvestris* to be more wrinkled than the upper bark. Kononchuk (1888) noted that lower bark from *Picea abies* was harder and contained smaller scales than the upper. When Onaka (1949) forced the young branch of a pine tree to grow horizontally, the bark formed on the under side became extremely thick. Bark on the lower side of *Pinus densiflora* was more scalelike, the upper more platelike. My own observations of several conifers indicate a considerable difference in appearance of the two types of bark. On leaning or horizontal stems of *Abies balsamea*, for example, the bark on the upper side is gray, rough, and similar to bark from vertical trees of the same age. The bark on the under side is frequently light brown and smooth. It is, of course, possible that the more rapid growth on the lower side of the stem, lack of illumination, or less exposure to rain and snow could account for these differences.

Onaka (1949) found radial growth in the bark to be larger on the lower, compression wood side of *Chamaecyparis obtusa* and *Cryptomeria japonica* than on the upper, although the difference was less than in the wood. Bark fibers were notably less developed on the lower side and were also shorter. Table 4.23

Table 4.23. Number of rows of bark fibers per 550 μm in normal and compression barks of *Chamaecyparis obtusa* and *Cryptomeria japonica*. (Onaka 1949)

Type of fiber	*Chamaecyparis obtusa*				*Cryptomeria japonica*			
	1		2		1		2	
	Compression bark	Normal bark	Compression bark	Normal bark	Compression bark	Normal bark	Compression bark	Normal bark
Fibers with continuous cells and thick walls	2	6	1	4	1	6	1	5
Fibers with discontinuous cells and thick walls	2	1	3	1	1	1	4	2
Fibers with continuous cells and thin walls	13	8	13	9	14	9	10	6
Total number of fibers	17	15	17	14	16	16	15	13

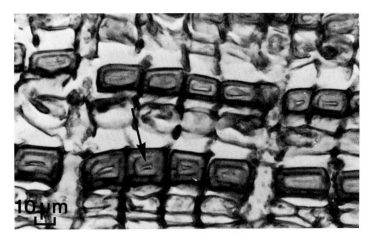

Fig. 4.146. Secondary phloem in a root of *Juniperus communis*, containing bast fibers with a thick, gelatinous inner layer (*arrow*). Transverse section. LM. (Liese and Höster 1966)

indicates that fibers with continuous cells and thick walls were fewer in compression bark than in normal bark of both species, whereas fibers with discontinuous cells or thin walls were more numerous. Outer bark on the lower side tended to split and fall off as a consequence of the larger radial growth of the compression wood, and compression bark as a result was thin. My own observations by light microscopy have failed to reveal any differences in general anatomy between bark from the upper and the lower, compression wood, side of *Picea rubens*. A chemical analysis of normal and compression barks indicated that the two tissues had the same chemical composition (Côté et al. 1966a) (Table 5.42).

Nečesaný (1955a, 1956, 1957) argued that it is the nature of the tissue that determines whether compression or tension wood is going to be formed. This concept was further developed by Höster and Liese (1966), who pointed out that compression wood is formed in tissues consisting primarily of tracheids and tension wood in tissues composed of libriform fibers. If this is true, secondary phloem, rich in bast fibers, should be expected to form tension, rather than compression, phloem. This has, indeed, been found to be the case in those gymnosperms which have a phloem with bast fibers, namely those belonging to the Cupressaceae, Podocarpaceae, Taxodiaceae, and Taxaceae.

Liese and Höster (1966) examined phloem in roots from *Juniperus communis* and also phloem in horizontal branches of this species as well as *Cryptomeria japonica*, *Metasequoia glyptostroboides*, *Podocarpus salignus*, and *Sequoiadendron giganteum*. All fibers in the roots of *Juniperus communis* contained a gelatinous inner cell wall layer. These fibers were uniformly distributed over the entire cross section of the phloem. The section in Fig. 4.146 contains two rows of thick-walled fibers. Staining demonstrated that the inner portion of the cell wall was unlignified and probably consisted of cellulose. The electron micrograph in Fig. 4.147 shows the structure of the fibers in more de-

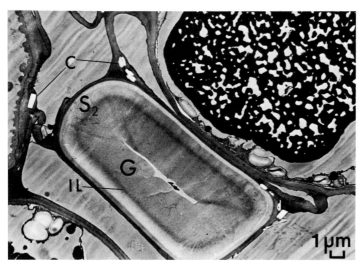

Fig. 4.147. Secondary phloem in a root of *Juniperus communis*, containing bast fibers with inorganic crystals (*C*), S_2, an intermediate layer (*IL*), and a thick, gelatinous inner layer (*G*). Transverse section. TEM. (Liese and Höster 1966)

tail. Evidently, the cell wall consists of primary wall, S_1, S_2, and a thick inner layer. Between the latter and the S_2 is a narrow layer (IL) which is more electron-dense than either of them. The lumen is very narrow. Liese and Höster (1966) considered the thick, unlignified inner layer to be equivalent to the gelatinous layer in tension wood fibers. The gelatinous fibers in the phloem were uniformly distributed around the periphery also in the branches, although in two of the species examined some of the bast fibers were of the normal, lignified type.

In many conifers, a continuous band of tannin-filled parenchyma cells is formed each growing season. Kutscha et al. (1975) observed that in compression phloem of *Abies balsamea* these bands of tannin cells were discontinuous (Fig. 9.1) instead of continuous as in normal wood. They also noted that old sieve cells tended to be more crushed in compression phloem than normal, probably because of the rapid radial growth of the compression wood. In contrast to these findings, Timell (1973) and Höster (1974) both found the tangential rows of tannin cells to be continuous in compression phloem of *Picea abies* (Chap. 9.5).

Thick, unlignified layers have also been observed in fibers present in tension phloem, for example in *Eucalyptus* (Dadswell and Wardrop 1955), *Quercus alba* and *Q. falcata* (Nanko and Côté 1980), and *Tilia cordata* (Böhlmann 1971). Recently, Nanko et al. (1982) reported the presence of such fibers in tension phloem of *Populus euramericana*. These cells were composed of lignified S_1 and S_2 layers, followed by a variable number of alternatively arranged unlignified and lignified inner layers. The cellulose microfibrils were parallel or almost parallel to the fiber axis in the inner layers.

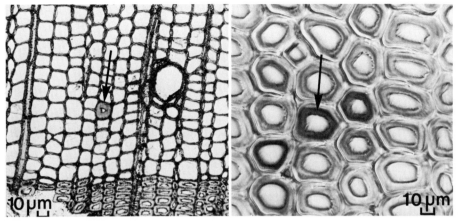

Fig. 4.148. Normal wood in the stem of *Pseudotsuga menziesii* that had been subjected to compression along the grain. Note the presence of a single, anomalous tracheid with a thick wall (*arrow*). Transverse section. LM. (Courtesy of Canadian Forestry Service). (Kennedy and Adamovich 1968)

Fig. 4.149. Normal wood of *Larix leptolepis* with a few gelatinous tracheids with a thick, inner layer (*arrow*). Transverse section. LM. (Höster 1970)

4.14 Anomalous Wood Tracheids

Anomalous tracheids in the wood of gymnosperms have been observed occasionally over the years. Lately, several such cases have been reported, all in members of the Coniferales. Balatinecz (1966) observed in *Pinus strobus* a lens-like cluster of abnormal tracheids around the resin canals in the latewood. These cells had a large lumen and thin walls and resembled earlywood tracheids. As a result, they were more permeable than normal latewood tracheids with their narrow lumina. There is at present no explanation for the occurrence of these tracheids.

Kennedy and Adamovich (1968) studied the effect of compression along the grain on the growth of four *Pseudotsuga menziesii* trees during three growing seasons. In one of the trees, they observed a single, anomalous tracheid. As can be seen in Fig. 4.148, this cell was of the same size as the surrounding, normal earlywood tracheids but had a wall considerably thicker than that of normal latewood cells. Closer examination revealed the presence of a thick, highly lignified layer inside S_3 where the cellulose microfibrils had an axial orientation. The investigators pointed out the similarity of this anomalous tracheid to those found in compression wood. It would, however, seem to be more closely related to the gelatinous fibers of tension wood, at least in its cell wall structure. It is true that the gelatinous layer in tension wood generally is unlignified, but a few cases of lignified G-layers have been reported (Scurfield and Wardrop 1962, 1963, Casperson 1967) and more might exist. Similar anomalous tracheids were later observed by Hejnowicz (1971) in Douglas-fir wood.

Höster (1970) observed anomalous tracheids in the xylem of *Larix leptolepis*, shown in Fig. 4.149. These cells contained a thick, inner layer which failed to respond to various staining reagents for lignin. The investigator suggested the designation *gelatinous tracheids* for these cells. Their wall structure varied somewhat. In some, only the thin S_3 layer was unlignified, while in others also the inner portion of S_2 was free of lignin. In still another case, the gelatinous layer contained fissures reminiscent of the helical cavities in compression wood tracheids. No gelatinous tracheids were formed after the larch tree had been bent over and formation of compression wood tracheids had begun on the lower side.

Kucera (1972) studied traumatic xylem and phloem present in *Taxus baccata* trees that had been wounded by woodpeckers (*Picidae*). Some of the xylem tracheids, shown in Fig. 4.150, had abnormal secondary walls, sometimes containing additional wall layers, while at other times there had developed a thick S_3 layer which was only weakly birefringent when viewed in cross section under polarized light.

Jacquiot and Trenard (1974) have reported the presence of gelatinous tracheids in the wood of conifers. In *Abies alba* a thick, gelatinous layer had been formed inside the S_3 layer, as shown in Fig. 4.151. It appeared to be unlignified but was terminated on the lumen side by a seemingly lignified membrane. Sometimes, the layer S_2 also lacked lignin. Radial striations could be observed in the gelatinous layer. The abnormal tracheids were located within the latewood zone of several adjacent growth rings. The sequence was usually: ordinary compression wood with helical cavities, disorganized parenchyma tissue, and a region containing gelatinous tracheids. It should be noted that neither rounded tracheids nor intercellular spaces were present in the third zone. Similar gelatinous tracheids also occurred together with compression wood in a

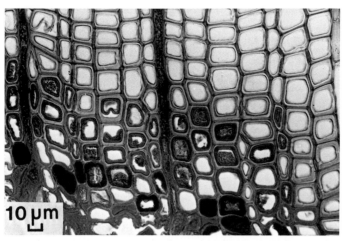

Fig. 4.150. Normal wood of *Taxus baccata* that had been wounded by woodpeckers. Various abnormal tracheids are present, some of them with an additional, thick inner wall layer. Transverse section. LM. (Kucera 1972)

4.14 Anomalous Wood Tracheids

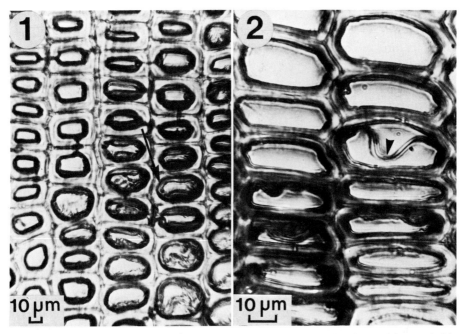

Fig. 4.151. 1 "Tension wood" in *Abies alba*, stained with Congo red and containing tracheids with an inner, unlignified, gelatinous layer (*arrow*). **2** Similar "tension wood" in *Picea abies* with an easily detached gelatinous layer in some of the tracheids (*arrowhead*). Transverse sections. LM. (Jacquiot and Trenard 1974)

sample of *Pinus sylvestris*. Occasionally, wood of *Picea abies* was also lignified less than normal and some tracheids contained a gelatinous layer, which, in contrast to those in the other two species, became readily detached from the remainder of the cell wall (Fig. 4.153b). Jacquiot and Trenard (1974) pointed out that their gelatinous tracheids always appeared close to anatomically disordered regions or adjacent to zones of compression wood. Gelatinous and compression wood tracheids occurred together in *Pinus sylvestris*, and the investigators suggested that both types of tissues probably were associated with an uneven distribution of auxin, perhaps induced by an attack by parasites.

The observations reported by Kennedy and Adamovich (1969), Höster (1970), Hejnowicz (1971), Kucera (1972), and Jacquiot and Trenard (1974) all make it clear that tracheids containing a gelatinous, inner wall layer can occur in the xylem of conifers. The statement by Höster and Liese (1966) that compression wood develops in tissues consisting largely of tracheids and tension wood in tissues composed of fibers should perhaps be modified to include these exceptions. Jacquiot and Trenard (1974) refer to their tissue as *tension wood*. This is likely to cause confusion, as it fails to take into account the anatomical differences between this tissue and the tension wood present in angiosperms. *Gelatinous tracheids*, suggested by Höster (1970), is a better term.

4.15 General Conclusions

Of all features of compression wood, its structure is probably that which is best known. The anatomy of the tracheids, by far the dominant cell type, has been studied in great detail by many investigators for more than a 100 years. Ray parenchyma, ray tracheids, axial parenchyma, epithelial cells, and resin canals have been examined less frequently.

The major histological characteristics that set compression wood tracheids apart from those in normal wood are their lesser length, rounded outline, distorted tips, and thick wall. The middle lamella is only occasionally fully lignified, and large intercellular spaces often surround each tracheid, which as a result is connected to its neighbors only by a thin strip of intercellular material. Fully developed compression wood tracheids have a thick wall, albeit not as thick as that of normal latewood tracheids, which also have a smaller lumen. A rapid radial growth with an increased rate of anticlinal cambial divisions is probably responsible for the reduced length of the tracheids.

Ultrastructurally, compression wood tracheids differ significantly from those in normal wood. Compared to the latter, the S_1 layer, which contains almost transversely oriented microfibrils, is much thicker, especially at the cell corners. The outer portion of the thick S_2 layer has a high concentration of lignin but is not devoid of cellulose and hemicelluloses. The remainder of S_2 is deeply fissured so that the wall consists of helical ribs. Cavities and ribs follow the orientation of the microfibrils in S_2 which is in the range of $30-50°$ to the fiber axis. The cavities do not penetrate into the $S_2(L)$ layer and accordingly never reach S_1. No cavities and ribs are formed in *Ginkgo biloba*, and they seem to develop only rarely in *Agathis* and *Araucaria*. There are also indications that the compression wood present in exposed roots lacks helical cavities. No S_3 layer is formed in pronounced compression wood, but a warty layer is usually present. Helical thickenings instead of helical cavities and ribs occur in *Taxus* and *Torreya* but not in *Pseudotsuga*. The intertracheid, bordered pits are located at the bottom of a groove, which usually extends beyond the pit border so that the pits appear to have a slit-like opening. The aperture is, in reality, circular, and the bordered pits are basically the same in normal and compression woods.

The first rows of tracheids formed at the beginning of the growing season often lack a rounded outline, intercellular spaces, and helical cavities, but are in all other respects similar to the more fully developed compression wood tracheids produced later. As in normal wood, the last few tracheids deposited at the end of the season are flattened in the radial direction. They also tend to lack intercellular spaces but are otherwise identical with the more circular cells formed earlier.

In contrast to tension wood, compression wood develops more readily at the middle or end than at the beginning of the growing season. Mild or moderate compression wood tracheids, formed under the influence of a weak stimulus, are therefore always found in the later portion of an annual increment, while the early part consists of normal wood. Mild and moderate compression woods sometimes differ in their ultrastructure from the severe type. Such tracheids might, for example, lack helical cavities, or they might contain an S_3 layer. Occasionally, the presence of a highly lignified $S_2(L)$ layer is the only feature that sets them apart from normal tracheids. There exists a continuous transition in structure from normal tracheids to those in pronounced compression wood and vice versa (Kennedy and Farrar 1965, Fujita et al. 1979, Yumoto et al. 1982, 1983).

Only limited information is available concerning the rays in compression wood. Occasionally, the rays are more numerous and larger than in normal wood, but in other instances no differences have been observed between normal and compression woods in these respects. The ultrastructure of the ray parenchyma cells and the ray tracheids is on the whole the same in the two tissues. Some exceptions to this do exist, however, and more might become known in the future. The ray parenchyma cells are sometimes more numerous and larger in compression wood, and the ray tracheids in this tissue can contain helical cavities and ribs. Compression wood tends to have fewer ordinary resin canals than normal wood, but resin ducts of the traumatic type are not uncommon. Among the different types of cells that line the longitudinal resin canals the strand tracheids are more frequent in compression wood and are able to develop helical cavities and ribs. The proportion of axial parenchyma cells is correspondingly lower. There are also more thick-walled epithelial cells than in corresponding normal wood.

Compression wood on the lower side of branches or knots may differ anatomically from stem compression wood. At present, the evidence on this point is contradictory. Phloem of branches and roots occasionally contains circumferentially distributed gelatinous fibers in those genera that possess bast fibers in their bark. On several occasions, so-called gelatinous tracheids have been observed in otherwise normal xylem, occasionally in association with adjacent compression wood.

The question of what causes the cambium to produce compression wood rather than normal wood has not yet been answered. Studies on the ultrastructural and cytological changes that occur when compression wood is formed (Chap. 9) have not provided much information on this subject. Currently, it would seem likely that the factors governing the differentiation of both normal and compression woods are of a biochemical nature.

References

Ahmad SS 1970 Variation in tracheid dimensions within a single stem of fir (Abies pindrow Spach). Pak J For 20:89–109
Anonymous 1955 Ann Rep Dep For Queensland 1954/1955: 45
Anonymous 1959 Ann Rep Dep For Queensland 1958/1959: 45
Atmer B, Thörnqvist T 1982 Fiberegenskaper i gran (Picea abies Karst) och tall (Pinus silvestris L.) (The properties of tracheids in spruce (Picea abies Karst) and pine (Pinus silvestris L.) Swed Univ Agr Sci Dep For Prod Rep 134, 78 pp
Baas P (ed) 1982 New perspectives in wood anatomy. Martinus Nijhoff, The Hague, 252 pp
Baas P, Chenglee L, Xinying Z, Keming C, Yuefen D 1984 Some effects of dwarf growth on wood structure. IAWA Bull (ns) 5:45–63
Back EL 1969 Intercellular spaces along the ray parenchyma. The gas canal system of living wood? Wood Sci 2:31–34

Bailey IW 1920 The cambium and its derivative tissues. II. Size variations of cambial initials in gymnosperms and angiosperms. Am J Bot 7:355–367

Bailey IW 1923 The cambium and its derivative tissues. IV. increase in girth of the cambium. Am J Bot 10:499–509

Bailey IW 1939 The microfibrillar and microcapillary structure of the cell wall. Bull Torrey Bot Club 66:201–213

Bailey IW 1958 The structure of tracheids in relation to the movement of liquids, suspensions, and undissolved gases. In: Thimann KV (ed) The physiology of forest trees. Ronald, New York, 71–82

Bailey IW, Berkley EE 1942 The significance of x-rays in studying the orientation of cellulose in the secondary wall of tracheids. Am J Bot 29:231–241

Bailey IW, Faull AF 1934 The cambium and its derivative tissues. IX. Structural variability in the redwood, Sequoia sempervirens, and its significance in the identification of fossil woods. J Arnold Arbor 15:233–254

Bailey IW, Kerr T 1937 The structural variability of the secondary wall as revealed by "lignin" residues. J Arnold Arbor 18:261–272

Bailey IW, Shephard HB 1915 Sanio's laws for the variation in size of coniferous tracheids. Bot Gaz 60:66–71

Bailey IW, Tupper WW 1919 Size variations in tracheary cells. Proc Am Acad Arts Sci 54:149–204

Balatinecz JJ 1966 An anomaly in tracheid anatomy in the latewood of white pine. For Prod J 16(8): 46

Balch RE 1952 Studies on the balsam woolly aphid Adelges piceae (Ratz) (Homoptera:Phylloxeridae) and its effect on balsam fir, Abies balsamea (L) Mill. Can Dep Agr Publ 867, 76 pp

Bamber RK 1972 Properties of the cell walls of the resin canal tissue of the sapwood and heartwood of Pinus lambertiana and P radiata. J Inst Wood Sci (31) 6 (1):32–35

Bamber RK 1973 The formation and permeability of interstitial spaces in the sapwood of some pine species. J Inst Wood Sci (32) 6 (2):36–38

Bamber RK 1980 The origin of growth stresses. IUFRO Conf Laguna, Philippines, 1978. For Comm NSW Repr 715 WT, 7 pp

Bamber RK, Lanyon JW 1971 Abnormal rays in stained heartwood of Pseudotsuga menziesii (Mirb) Franco. J Inst Wood Sci (29) 5 (5):28–30

Bannan MW 1933a Vigour of growth and traumatic resin cyst development in the hemlock Tsuga canadensis (L) Carr. Trans R Soc Can 27:197–202

Bannan MW 1933b Factors influencing the distribution of vertical resin canals in the wood of the larch Larix laricina (Du Roi) Koch. Trans R Soc Can 27:203–218

Bannan MW 1936 Vertical resin ducts in the secondary wood of the Abietineae. New Phytol 35:11–46

Bannan MW 1937 Observations on the distribution of xylem-ray tissue in conifers. Ann Bot 1:717–726

Bannan MW 1941 Vascular rays and adventitious root formation in Thuja occidentalis L. Am J Bot 28:457–463

Bannan MW 1942 Wood structure of the native Ontario species of Juniperus. Am J Bot 29:245–252

Bannan MW 1944 Wood structure of Libocedrus decurrens. Am J Bot 31:346–351

Bannan MW 1950a The frequency of anticlinal division in fusiform cambial cells of Chamaecyparis. Am J Bot 37:511–519

Bannan MW 1950b Abnormal xylem rays in Chamaecyparis. Am J Bot 37:232–237

Bannan MW 1951a The reduction of fusiform cambial cells in Chamaecyparis and Thuja. Can J Bot 29:57–67

Bannan MW 1951b The annual cycle of size changes in the fusiform cambial cells of Chamaecyparis and Thuja. Can J Bot 29:421–437

Bannan MW 1953 Further observations on the reduction of fusiform cambial cells in Thuja occidentalis L. Can J Bot 31:63–74

Bannan MW 1954 Ring width, tracheid size, and ray volume in stem wood of Thuja occidentalis L. Can J Bot 32:466–479

References

Bannan MW 1955 The vascular cambium and radial growth in Thuja occidentalis L. Can J Bot 33:113–138
Bannan MW 1956 Some aspects of the elongation of fusiform cambial cells in Thuja occidentalis L. Can J Bot 34:175–196
Bannan MW 1957a Girth increase in white cedar stems of irregular form. Can J Bot 35:425–434
Bannan MW 1957b The relative frequency of the different types of anticlinal divisions in conifer cambium. Can J Bot 35:875–884
Bannan MW 1957c The structure and growth of the cambium. Tappi 40:220–225
Bannan MW 1960a Cambial behavior with reference to cell length and ring width in Thuja occidentalis L. Can J Bot 38:177–183
Bannan MW 1960b Ontogenetic trends in conifer cambium with respect to frequency of anticlinal division and cell length. Can J Bot 38:795–802
Bannan MW 1962 Cambial behavior with reference to cell length and ring width in Pinus strobus L. Can J Bot 40:1057–1062
Bannan MW 1963a Cambial behavior with reference to cell length and ring width in Picea. Can J Bot 41:811–822
Bannan MW 1963b Tracheid size and rate of anticlinal division in the cambium of Cupressus. Can J Bot 41:1187–1197
Bannan MW 1964a Tracheid size and anticlinal division in the cambium of Pseudotsuga. Can J Bot 42:603–631
Bannan MW 1964b Tracheid size and anticlinal division in the cambium of lodgepole pine. Can J Bot 42:1105–1118
Bannan MW 1965a The rate of elongation of fusiform initials in the cambium of Pinaceae. Can J Bot 43:429–435
Bannan MW 1965b Ray contacts and rate of anticlinal divisions in fusiform cambial cells of some Pinaceae. Can J Bot 43:487–507
Bannan MW 1965c The length, tangential diameter and length/width ratio of conifer tracheids. Can J Bot 43:967–984
Bannan MW 1966 Cell length and rate of anticlinal division in the cambium of the sequoias. Can J Bot 44:209–218
Bannan MW 1967a Anticlinal divisions and cell length in conifer cambium. For Prod J 17(6):63–69
Bannan MW 1967b Sequential changes in rate of anticlinal division, cambial cell length, and ring width in the growth of conifer trees. Can J Bot 45:1359–1369
Bannan MW 1968a Anticlinal division and the organization of conifer cambium. Bot Gaz 129:107–113
Bannan MW 1968b Polarity in the survival and elongation of fusiform initials in conifer cambium. Can J Bot 46:1005–1008
Bannan MW 1968c The problem of sampling in studies of tracheid length in conifers. For Sci 14:140–147
Bannan MW, Bayly IL 1956 Cell size and survival in conifer cambium. Can J Bot 34:769–776
Bannan MW, Bindra M 1970 Variations in cell length and frequency of anticlinal division in the vascular cambium throughout a white spruce tree. Can J Bot 48:1363–1371
Bannan MW, Whalley BE 1950 The elongation of fusiform cambial cells in Chamaecyparis. Can J Res C-28:341–355
Baranetzky J 1901 Über die Ursachen welche die Richtung der Äste der Baum- und Straucharten bedingen. Flora 89:138–239
Barber NF 1968 A theoretical model of shrinking wood. Holzforschung 22:87–103
Barger RL, Ffolliott PF 1976 Factors affecting occurrence of compression wood in individual ponderosa pine trees. Wood Sci 8:201–208
Bauch J, Liese W, Schultze R 1972 The morphological variability of the bordered pit membranes in gymnosperms. Wood Sci Technol 7:165–184
Berkley EE 1934 Certain physical and structural properties of three species of southern yellow pine correlated with compression strength of their wood. Ann MO Bot Gard 21:241–338
Berkley EE, Woodyard OC 1948 Certain variations in the structure of wood fibers. Text Res J 18:519–525

Bisset IJW 1949 Bibliography of references on the variations of tracheid and fibre lengths and their distribution in angiosperms and gymnosperms. Summarized data on the variation of fibre and tracheid lengths and their distribution in angiosperms and gymnosperms. CSIRO Aust Div For Prod Bibl Ser 37, 10 pp

Bisset IJW, Dadswell HE 1950 The variation in cell length within one growth ring of certain angiosperms and gymnosperms. Aust For 14:17 − 29

Bisset IJW, Dadswell HE, Wardrop AB 1951 Factors influencing tracheid length in conifer stems. Aust For 15:17 − 30

Böhlmann D 1971 Zugbast bei Tilia cordata Mill. Holzforschung 25:1 − 4

Bolton AJ, Jardine P, Jones GL 1975 Interstitial spaces. A review and observations on some Araucariaceae. IAWA Bull 1975(1):3 − 12

Böning K 1925 Über den inneren Bau horizontaler und geneigter Sprosse und ihre Ursachen. Mitt Deutsch Dendrol Ges 35:86 − 102

Bosshard HH 1974a Holzkunde. 1. Mikroskopie und Makroskopie des Holzes. Birkhäuser, Basel, Stuttgart, 224 pp

Bosshard 1974b Holzkunde. 2. Zur Biologie, Physik und Chemie des Holzes. Birkhäuser, Basel, Stuttgart, 312 pp

Bosshard HH 1982 Holzkunde. 1. Mikroskopie und Makroskopie des Holzes, 2nd ed. Birkhäuser, Basel, Stuttgart, 224 pp

Bosshard HH, Hug UE 1980 The anastomoses of the resin canal system in Picea abies (L) Karst, Larix decidua Mill and Pinus sylvestris L. Holz Roh-Werkst 38:325 − 328

Boutelje JB 1965 The anatomical structure, moisture content, density, shrinkage, and resin content of the wood in and around knots of Swedish redwood (Pinus silvestris L) and Swedish whitewood (Picea abies Karst). Proc IUFRO Sect 41 Meet Melbourne, Vol 2, 17 pp

Boutelje JB 1966 On the anatomical structure, moisture content, density, shrinkage, and resin content of the wood in and around knots in Swedish pine (Pinus silvestris L) and Swedish spruce (Picea abies Karst). Svensk Papperstidn 69:1 − 10

Boyd JD 1973 Helical fissures in compression-wood cells: causative factors and mechanics of development. Wood Sci Technol 7:92 − 111

Boyd JD 1974 Relating lignification to microfibril angle differences between tangential and radial faces of all wall layers in wood cells. Drev Vysk 19:41 − 54

Boyd JD 1977 Interpretation of X-ray diffractograms of wood for assessment of microfibril angles in fibre cell walls. Wood Sci Technol 11:93 − 114

Boyd JD 1978 Significance of laricinan in compression-wood tracheids. Wood Sci Technol 12:25 − 35

Boyd JD, Foster RC 1974 Tracheid anatomy changes as responses to changing structural requirements of the tree. Wood Sci Technol 8:91 − 105

Braun HJ 1970 Funktionelle Histologie der sekundären Sprossachse. Gebrüder Bornträger, Berlin Stuttgart, 190 pp

Brazier JD 1963 The timber of young plantation grown Metasequoia. J For 57:151 − 153

Brazier JD 1969 Some considerations in appraising within-ring density. FAO/IUFRO Sec Consult For Tree Breeding, Washington DC, FO-FTB-69-4/2, 9 pp

Brodzki P 1972 Callose in compression-wood tracheids. Acta Soc Bot Polon 41:321 − 327

Brown CL 1970 Physiology of wood formation in conifers. Wood Sci 3:8 − 22

Bucher H 1968 Der mikroskopische Nachweis von Lignin in verholzten Fasern durch Jodmalachitgrün. Papier 22:390 − 396

Butterfield BG, Meylan BA 1980 Three-dimensional structure of wood. An ultrastructural approach. Chapman and Hall, London New York, 103 pp

Casperson G 1959a Elektronenmikroskopische Untersuchungen des Zellwandaufbaues beim Reaktionsholz der Coniferen. Ber Deutsch Bot Ges 72:230 − 235

Casperson G 1959b Mikroskopischer und submikroskopischer Zellwandaufbau beim Druckholz. Faserforsch Textiltech 10:536 − 541

Casperson G 1962a Über die Bildung der Zellwand beim Reaktionsholz. Holztechnologie 3:217 − 223

Casperson 1962b Der submikroskopische Bau der verholzten Zellwand. In: Holzzerstörungen durch Pilze. Symp Eberswalde, Akademie-Verlag, Berlin DDR, 63 − 69, 373, 376

Casperson G 1963a Chemische, anatomische und physiologische Eigenheiten des Reaktionsholzes. Chemiefasersymp 1962, Akademie-Verlag, Berlin DDR, 39–52

Casperson G 1963b Über die Bildung der Zellwand beim Reaktionsholz. 2. Zur Physiologie des Reaktionsholzes. Holztechnologie 4:33–37

Casperson G 1963c Die Bildung der Zellwand als morphologisch-physiologisches Problem. In: Perspektiven der Grundlagenforschung des Holzes, Medzinárodné Kolokvium, Bratislava, 49–54

Casperson G 1963d Reaktionsholz. Seine Struktur und Bildung. Habilitationschr Math-Naturwiss Fak Humboldt-Univ Berlin DDR, 116 pp

Casperson G 1964 Über den lamellaren Aufbau der verholzten Zellwand. 3rd Reg Eur Conf El Microsc, Praha, 169–170

Casperson G 1965a Über die Entstehung des Reaktionsholzes bei Kiefern. In: Aktuelle Probleme der Kiefernwirtschaft. Int Symp Eberswalde 1964, Deutsch Akad Landwiss Berlin DDR, 523–528

Casperson G 1965b Zur Anatomie des Reaktionsholzes. Svensk Papperstidn 68:534–544

Casperson G 1967 Über die Bildung von Zellwänden bei Laubhölzern. 4. Untersuchungen an Eiche (Quercus robur L). Holzforschung 21:1–6

Casperson G 1968 Anatomische Untersuchungen an Lärchenholz (Larix decidua Miller). Faserforsch Textiltech 19:467–476

Casperson G, Hoyme E 1965 Über endogene Faktoren der Reaktionsholzbildung. 2. Untersuchungen an Fichte (Picea abies Karst). Faserforsch Textiltech 16:352–358

Casperson G, Zinsser A 1965 Über die Bildung der Zellwand bei Reaktionsholz. 3. Zur Spaltenbildung im Druckholz von Pinus sylvestris L. Holz Roh-Werkst 23:49–55

Cave ID 1966 Theory of X-ray measurement of microfibril angle in wood. For Prod J 16(10):37–42

Cerezke HF 1972 Effects of weevil feeding on resin duct density and radial increment in lodgepole pine. Can J For Res 2:11–15

Chafe SC 1974a Cell wall thickenings in the ray parenchyma of yellow cypress. IAWA Bull 1974(2):3–10

Chafe SC 1974b Cell wall structure in the xylem parenchyma of Cryptomeria. Protoplasma 81:63–76

Chafe SC 1974c Cell wall formation and "protective layer" formation in the xylem parenchyma of trembling aspen. Protoplasma 80:335–354

Chafe SC, Chauret G 1974 Cell wall structure of the xylem parenchyma of trembling aspen. Protoplasma 80:129–147

Chalk L 1930 Tracheid length with special reference to Sitka spruce (Picea sitchensis Carr). Forestry 4:7–14

Chalk L, Ortiz M 1961 Variation in tracheid length within the ring in Pinus radiata D Don. Forestry 34:119–124

Chang YP 1954 Anatomy of common North American pulpwood barks. TAPPI Monogr Ser 14, 249 pp

Chu LC 1972 Comparison of normal wood and first-year compression wood in longleaf pine trees. M S Thesis MS State Univ, 72 pp

Cieslar A 1896 Das Rothholz der Fichte. Cbl Ges Forstwes 22:149–165

Cockrell RA 1973 The effect of specimen preparation on compression wood and normal latewood pits and wall configurations of giant sequoia. IAWA Bull 1973(4):19–23

Cockrell RA 1974 A comparison of latewood pits, fibril orientation and shrinkage of normal and compression wood of giant sequoia. Wood Sci Technol 8:197–206

Collett BM 1970 Scanning electron microscopy: a review and report on research in wood science. Wood Fiber 2:113–133

Constantinescu A 1956 Certăre preliminare asupra formatiunii de lemn de compresiune la bradul (Abies alba) din pădurea Chilerei-Valea Timisului. (Preliminary investigations on the formation of compression wood in Abies alba in Chilerei forest in the Timis valley.) Ind Lemn 5:455–457

Core HA 1962 Variables affecting the formation of compression wood in plantation red pine. Ph D Thesis, NY State Coll For Syracuse, 64 pp

Core HA, Côté WA, Jr, Day AC 1961 Characteristics of compression wood in some native conifers. For Prod J 11:356–362

Core HA, Côté WA, Jr, Day AC 1979 Wood. Structure and identification, 2nd ed. Syracuse Univ Press, Syracuse, 182 pp

Correns E 1961 Über anormale Holzfasern. Pap Puu 43:47–62

Côté WA, Jr (ed) 1965 Cellular ultrastructure of woody plants. Syracuse Univ Press, Syracuse, 603 pp

Côté WA, Jr 1967 Wood ultrastructure. An atlas of electron micrographs. Univ WA Press, Seattle London, 60 pp

Côté WA, Jr 1977 Wood ultrastructure in relation to chemical composition. In: Loewus FA, Runeckles VC (eds) The structure, biosynthesis, and degradation of wood. Plenum, New York, 1–44

Côté WA, Jr, Day AC 1962 The G-layer in gelatinous fibers. Electron microscopic studies. For Prod J 12:333–338

Côté WA, Jr, Day AC 1965 Anatomy and ultrastructure of reaction wood. In: Côté WA, Jr (ed) Cellular ultrastructure of woody plants. Syracuse Univ Press, Syracuse, 391–418

Côté WA, Jr, Day AC, Kutscha NP, Timell TE 1967 Studies on compression wood. V. Nature of the compression wood formed in the early springwood of conifers. Holzforschung 21:180–186

Côté WA, Jr, Day AC, Timell TE 1968a Studies on compression wood. VII. Distribution of lignin in normal and compression wood of tamarack (Larix laricina (Du Roi) K Koch). Wood Sci Technol 2:13–37

Côté WA, Jr, Day AC, Timell TE 1969 A contribution to the ultrastructure of tension wood tracheids. Wood Sci Technol 3:257–271

Côté WA, Jr, Kutscha NP, Simson BW, Timell TE 1968b Studies on compression wood. VI. Distribution of polysaccharides in the cell wall of tracheids from compression wood of balsam fir (Abies balsamea (L) Mill). Tappi 51:33–40

Côté WA, Jr, Kutscha NP, Timell TE 1968c Studies on compression wood. VIII. Formation of cavities in compression-wood tracheids of Abies balsamea L. Holzforschung 22:138–144

Côté WA, Jr, Simson BW, Timell TE 1966a Studies on compression wood. 2. The chemical composition of wood and bark from normal and compression regions of fifteen species of gymnosperms. Svensk Papperstidn 69:547–558

Côté WA, Jr, Timell TE, Zabel RA 1966b Studies on compression wood. I. Distribution of lignin in compression wood of red spruce (Picea rubens Sarg). Holz Roh-Werkst 24:432–438

Cousins WJ 1972 Measurement of mean microfibril angles of wood tracheids. Wood Sci Technol 6:58

Cown DJ 1975 Variation in tracheid dimensions in the stem of a 26-year-old radiata pine tree. Appita 28:237–245

Dadswell HE 1963 Tree growth–wood property inter-relationships. VIII. Variations in structure and properties in wood grown under abnormal conditions. In: Maki T (ed) Special field institute in forest biology, Raleigh NC 1960, 55–66

Dadswell HE, Ellis J 1940 Study of the cell wall. I. Methods of demonstrating lignin distribution in wood. J Counc Sci Ind Res Aust 13:44–54

Dadswell HE, Nicholls JWP 1959 Asessment of wood qualities for tree breeding. CSIRO Aust Div For Prod Technol Pap 4, 16 pp

Dadswell HE, Wardrop AB 1949 What is reaction wood? Aust For 13:22–33

Dadswell HE, Wardrop AB 1955 The structure and properties of tension wood. Holzforschung 9:97–104

Dadswell HE, Wardrop AB, Watson AJ 1958 The morphology, chemistry and pulping characteristics of reaction wood. In: Bolam F (ed) Fundamentals of papermaking fibres. Tech Sect Brit Pap Board Makers' Assoc, London, 187–206

Denne MP 1971 Tracheid length in relation to seedling height in conifers. Wood Sci Technol 5:135–146

Denne MP 1973 Tracheid dimensions in relation to shoot vigour in Picea. Forestry 46:117–124

den Outer RW 1967 Histological investigations of the secondary phloem of gymnosperms. Meded Landbouwhogeschool, Wageningen 67(7), 119 pp

Detlefsen E 1878—1882 Versuch einer mechanischen Erklärung des excentrischen Dickenwachstums verholzter Achsen und Wurzeln. Arb Bot Inst Würzburg 2:670—688
Dinwoodie JM 1961 Tracheid and fibre length in timber. A review of literature. Forestry 34:125—144
Dinwoodie JM 1963 Variation in tracheid length in Picea sitchensis Carr. Dep Ind Sci Res London, For Prod Res Spec Rep 16, 55 pp
Dinwoodie JM 1965 Tensile strength of individual compression wood fibres and its influence on the properties of paper. Nature 205:763—764
Doerksen AH, Mitchell RG 1965 Effects of the balsam woolly aphid upon wood anatomy of some western true firs. For Sci 11:181—188
Dunning CE 1969a The structure of longleaf-pine latewood. I. Cell wall morphology and the effect of alkaline extraction. Tappi 52:1326—1335
Dunning CE 1969b The structure of longleaf-pine latewood. II. Intertracheid membranes and pit membranes. Tappi 52:1335—1341
Echlin P 1968 The use of the scanning reflection electron microscope in the study of plant and microbial material. J R Microsc Soc 88:407—418
Echols RM 1955 Linear relation of fibrillar angle to tracheid length, and genetic control of tracheid length in slash pine. Trop Woods 102:11—22
Eicke R, Ehling E 1965 Die Ausbildung der jungen Tracheiden im Holz von Ginkgo biloba L. Ber Deutsch Bot Ges 78:326—337
El-Hosseiny F, Page DH 1973 The measurement of fibril angle of wood fibers using polarized light. Wood Fiber 5:208—214
Elliott GK 1960 Distribution of tracheid length in a single stem of Sitka spruce. J Inst Wood Sci 5:38—47
Emerton AW, Goldsmith V 1956 The structure of the outer secondary wall of pine tracheids from kraft pulps. Holzforschung 10:108—115
Erickson HD, Harrison AT 1974 Douglas-fir wood quality studies. I. Effects of age and stimulated growth on wood density and anatomy. Wood Sci Technol 8:207—226
Esau K 1969 The Phloem. Gebrüder Borntraeger, Berlin Stuttgart, 505 pp
Eskilsson S 1969 Fibre properties in the spruce root system. Cellul Chem Technol 3:409—416
Eskilsson S 1972 Whole tree pulping. 1. Fibre properties. Svensk Papperstidn 75:397—402
Fahn A, Zamski E 1970 The influence of pressure, wind, wounding and growth substances on the rate of resin duct formation in Pinus halepensis wood. Isr J Bot 19:429—446
Fan WYH 1970 Induction of tracheid development in decapitated jack pine seedlings (Pinus banksiana Lamb) with indole-3-acetic acid and gibberellic acid. Ph D Thesis Univ Toronto, 122 pp
Fegel AC 1938 A comparison of the mechanical and physical properties and the structural features of wood-, stem-, and branch-wood. Ph D Thesis, NY State Coll For Syracuse, 66 pp
Fegel AC 1941 Comparative anatomy and varying physical properties of trunk, branch, and root wood in certain northeastern trees. NY State Coll For Syracuse Bull 14, No 2-b, Tech Publ 55, 20 pp
Fengel D 1965 Elektronenmikroskopische Beiträge zum Feinbau des Buchenholzes (Fagus sylvatica L). 1. Untersuchungen an Markstrahl-Parenchymzellen. Holz Roh-Werkst 23:257—263
Fengel D 1966a Elektronenmikroskopische Beiträge zum Feinbau des Buchenholzes (Fagus sylvatica L). 2. Weitere Beobachtungen an Markstrahlzellen der Buche. Holz Roh-Werkst 24:177—185
Fengel 1966b Entwicklung und Ultrastruktur der Pinaceen-Hoftüpfel. Svensk Papperstidn 69:232—241
Fengel D 1969 The ultrastructure of cellulose from wood. 1. Wood as the basic raw material for the isolation of cellulose. Wood Sci Technol 3:203—217
Fengel D, Stoll M 1973 Über die Veränderungen des Zellquerschnitts, der Dicke der Zellwand und den Wandschichten von Fichtenholz-Tracheiden innerhalb eines Jahrringes. Holzforschung 27:1—7
Fengel D, Wegener G 1983 Wood. Chemistry, Ultrastructure, Reactions. Walter de Gruyter, Berlin New York, 613 pp

Fergus BJ, Procter AR, Scott JAN, Goring DAI 1969 The distribution of lignin in sprucewood as determined by ultraviolet microscopy. Wood Sci Technol 3:117−138

Findlay GWD, Levy JF 1969 Scanning electron microscopy as an aid to the study of wood anatomy and decay. J Inst Wood Sci (23) 4 (5):57−63

Flotiński J, Tafliński J 1971 Związek między szerokością przyrostów rocznych a liczebnością pionowych przewodów żywicznych u sosny pospolitej (Pinus sylvestris L). (Relationship between the width of annual rings and number of vertical resin canals in the Scots pine (Pinus sylvestris L.) Sylwan 65:29−42

Foster RC, Marks GC 1968 Fine structure of the host−parasite relationship of Diplodia pinea on Pinus radiata. Aust For 32:211−225

Foulger AN 1966 Longitudinal shrinkage pattern in eastern white pine stems. For Prod J 16(12):45−47

Frei E, Preston RD, Ripley GW 1957 The fine structure of the wall of conifer tracheids. VI. Electron microscope investigations of sections. J Exp Bot 8:139−146

Frey-Wyssling A 1953 Über den Feinbau der Stauchlinien in überbeanspruchtem Holz. Holz Roh-Werkst 11:283−288

Frey-Wyssling A 1959 Die pflanzliche Zellwand. Springer, Berlin Göttingen Heidelberg, 367 pp

Frey-Wyssling A 1976 The plant cell wall, 3rd ed. Gebrüder Bornträger, Berlin Stuttgart, 294 pp

Fry G, Chalk L 1958 Variation in density in the wood of Pinus patula grown in Kenya. Forestry 30:29−45

Fujikawa S, Ishida S 1975 Ultrastructure of ray parenchyma cell wall of softwood. Mokuzai Gakkaishi 21:445−456

Fujisaki K 1975 Studies on the branch wood of Sugi (Cryptomeria japonica D Don). I. On the variation of tracheid length in the branches of 46 years old Sugi. Bull Ehime Univ For 12:37−46

Fujita M, Saiki H, Harada H 1973 The secondary wall formation of compression-wood tracheids. On the helical ridges and cavities. Bull Kyoto Univ For 45:192−203

Fujita M, Saiki H, Harada H 1978 The secondary wall formation of compression-wood tracheids. II. Cell wall thickening and lignification. Mokuzai Gakkaishi 24:158−163

Fujita M, Saiki H, Sakamoto J, Araki N, Harada H 1979 Cell wall structure of transitional tracheids between normal and compression wood. Bull Kyoto Univ For 51:247−256

Fukazawa K 1973 Process of righting and xylem development in tilted seedlings of Abies sachalinensis. Res Bull Coll Exp For Hokkaido Univ 30:103−124

Fukazawa K 1974 The distribution of lignin in compression- and lateral-wood of Abies sachalinensis using ultraviolet microscopy. Res Bull Coll Exp For Hokkaido Univ 31:87−114

Gerry E 1915 Fiber measurement studies: length variations: where they occur and their relation to the strength and uses of wood. Science 41:179

Gerry E 1916 Fiber measurement studies: a comparison of tracheid dimensions in long-leaf pine and Douglas fir with data on the strength and length, mean diameter and thickness of wall of the tracheids. Science 43:360

Gerry E 1942 Radial streak (red) and giant resin ducts in spruce. US For Serv FPL Rep 1391, 2 pp

Gessner F 1961 Die mechanischen Wirkungen auf das Pflanzenwachstum. In: Ruhland E (ed) Encyclopedia of plant physiology, Springer, Berlin Heidelberg New York, 16:634−667

Giordano G 1971 Tecnologia del legno: La materia prima. I. Unione Tipografico-Editrice Torinese, Torino, 1068 pp

Gleaton EN, Saydah L 1956 Fiber dimensions and papermaking properties of the various portions of a tree. Tappi 39(2):157 A−158 A

Göhre K 1958 Die Douglasie und ihr Holz. Akademie-Verlag, Berlin DDR, 596 pp

Gothan W 1905 Zur Anatomie lebender und fossiler Gymnospermenhölzer. Abhandl Königl Preuss Geol Landesanst 44:1−108

Green HV, Worrall J 1964 Wood quality studies. I. A scanning microphotometer for automatically measuring and recording certain wood characteristics. Tappi 47:419−427

Gregory RA, Romberger JA 1975 Cambial activity and height of uniseriate vascular rays in conifers. Bot Gaz 136:246−253

Greguss P 1955 Identification of living gymnosperms on the basis of xylotomy. Akadémiai Kiadó, Budapest
Hägglund A, Larsson S 1937 Om grankvistens kemiska sammansättning och dess förhållande vid sulfitkokningsprocessen. (On the chemical composition of spurce knots and their behavior on sulfite pulping.) Svensk Papperstidn 40:356 – 360
Halbwachs G, Kisser J 1967 Durch Rauchimmissionen bedingter Zwergwuchs bei Fichte und Birke. Cbl Ges Forstwes 84:156 – 173
Halbwachs G, Richter H 1969 Xylotomische Untersuchungen an einem fasziierten Fichtensproß. Mikroskopie 25:127 – 135
Hale JD 1923 Trabeculae of Sanio – their origin and distribution. Science 57:155
Harada H 1953 The electron-microscopic observation of xylary ray cells of conifer woods. J Jap For Soc 35:194 – 199
Harada H 1962 Electron microscopy of ultrathin sections of beech wood (Fagus crenata Blume). Mokuzai Gakkaishi 8:252 – 258
Harada H 1965a Ultrastructure and organization of gymnosperm cell walls. In: Côté WA, Jr (ed) Cellular ultrastructure of woody plants. Syracuse Univ Press, Syracuse, 215 – 233
Harada H 1965b Ultrastructure of angiosperm vessels and ray parenchyma. In: Côté WA, Jr (ed) Cellular ultrastructure of woody plants. Syracuse Univ Press, Syracuse, 235 – 249
Harada H, Côté WA, Jr 1967 Cell wall organization in the pit border region of softwood tracheids. Holzforschung 21:81 – 85
Harada H, Miyazaki Y 1952 (Electron-microscopic observation of compression wood.) Bull Gov For Exp Sta Tokyo 54:101 – 108
Harada H, Miyazaki Y, Wakashima T 1958 (Electronmicroscopic investigation on the cell wall structure of wood.) Bull Gov For Exp Sta Tokyo 104, 115 pp
Harada H, Wardrop AB 1960 Cell wall structure of ray parenchyma cells of a softwood (Cryptomeria japonica). Mokuzai Gakkaishi 6:34 – 41
Harris JM 1977 Shrinkage and density of radiata pine compression wood in relation to its anatomy and mode of formation. NZ J For Sci 7:91 – 106
Harris JM, Meylan BA 1965 The influence of microfibril angle on longitudinal and tangential shrinkage in Pinus radiata. Holzforschung 19:144 – 153
Harris RA 1976 Characterization of compression-wood severity in Pinus echinata Mill. IAWA Bull 1976(4):47 – 50
Hartig R 1896 Das Rothholz der Fichte. Forstl-Naturwiss Z 5:96 – 109, 157 – 169
Hartig R 1899 Über die Ursachen excentrischen Wuchses der Nadelbäume. Cbl Ges Forstwes 25:291 – 307
Hartig R 1901 Holzuntersuchungen. Altes und Neues. Julius Springer, Berlin, 99 pp
Hartler N 1968 Något om tryckved. (On compression wood.) Svensk Papperstidn 71:54
Hartley WR 1960 Nutrients and tracheid length in seedlings of Pinus radiata D Don. Emp For Rev 39:474 – 482
Hata K 1949 Studies on the pulp of Akamatsu (Pinus densiflora Sieb et Zucc). I On the length, diameter, and length-diameter ratio of tracheids in Akamatsu wood. Kawaga-Ken Agr Coll Tech Bull 1:1 – 35
Hata K 1950 Studies on the pulp of "Akamatsu" (Pinus densiflora S et Z) wood. 1. On the length, diameter, and the length/diameter ratio of tracheids in "Akamatsu" (Pinus densiflora S et Z) wood. J Jap For Soc 32:1 – 7
Hejnowicz A 1971 Anatomical studies on the wood of Pseudotsuga menziesii (Mirb) Franco. Abor Kórnickie 16:169 – 197
Hejnowicz A 1973 Anatomical studies on the development of Metasequoia glyptostroboides He et Cheng wood. Acta Soc Bot Polon 42:473 – 491
Hejnowicz Z 1961 Anticlinal divisions, intrusive growth, and loss of fusiform initials in nonstoried cambium. Acta Soc Bot Polon 30:729 – 758
Hejnowicz Z 1963a Wzrost intruzywny, podzialy poprzeczne i skośne we wrzecianowatych komórkach inicjalnych zranionej miazgi modrzewia. (Intrusive growth, transverse and pseudotransverse divisions in fusiform initials and wounded cambium of Larix europaea.) Acta Soc Bot Polon 32:493 – 503
Hejnowicz Z 1963b Udzial wzrostu intruzywnego w procesie zrastania sie miazgi po poprzecznym nacieciu u modrzewia. (Significance of intrusive growth in establishment of cambium union after transverse incision in Larix.) Acta Soc Bot Polon 32:625 – 630

Hejnowicz Z 1964 Orientation of the partition in pseudotransverse division in cambia of some conifers. Can J Bot 42:1685–1691

Hejnowicz Z 1967 Interrelationship between mean length, rate of intrusive elongation, frequency of anticlinal divisions and survival of fusiform initials in cambium. Acta Soc Bot Polon 36:367–378

Hejnowicz Z, Branski S 1966 Quantitative analysis of cambium growth in Thuja. Acta Soc Bot Polon 35:395–340

Helander AB 1933 (Variations in tracheid length of pine and spruce.) Found For Prod Res Finl Publ 14, 75 pp

Hiller CH 1964 Pattern of variation of fibril angle within annual rings of Pinus attenuradiata. US For Serv Res Note FPL-34, 11 pp

Hiller CH, Brown RS 1967 Comparison of dimensions and fibril angles of loblolly pine tracheids formed in wet or dry growing seasons. Am J Bot 54:453–460

Hirakawa Y, Ishida S 1981 A scanning and transmission electron microscopic study of layered structure of wall in pit border region between earlywood tracheids in conifers. Res Bull Coll Exp For Hokkaido Univ 38:249–264

Hodge AJ, Wardrop AB 1950 Electron microscope investigation of the cell wall organization of conifer tracheids and conifer cambium. Aust J Sci Res B-3:265–269

Hoffmann GC, Timell TE 1972a Polysaccharides in ray cells of normal wood of red pine (Pinus resinosa). Tappi 55:733–736

Hoffmann GC, Timell TE 1972b Polysaccharides in ray cells of compression wood of red pine (Pinus resinosa). Tappi 55:871–873

Hosia M, Lindholm CA, Toivanen P, Nevalainen K 1971 Undersökningar rörande utnyttjandet av barrträdsgrenar som råmaterial för kemisk massa och hård fiberskiva I. (Investigations on utilizing softwood branches as pulp and hardboard raw material.) Pap Puu 53:49–66

Höster HR 1970 Gelatinöse Tracheiden im sekundären Xylem von Larix leptolepis (S and Z) Gord. Holzforschung 24:4–6

Höster HR 1971 Das Vorkommen von Reaktionsholz bei Tropenhölzern. Mitt Bundesforschungsanst Forst-Holzwirtsch Reinbek 82:225–231

Höster HR 1974 On the nature of the first-formed tracheids in compression wood. IAWA Bull 1974(1):3–9

Höster HR, Liese W 1966 Über das Vorkommen von Reaktionsgewebe in Wurzeln und Ästen der Dikotyledonen. Holzforschung 20:80–90

Howard ET 1971 Bark structure of the southern pines. Wood Sci 3:134–148

Hug UE 1979 Das Harzkanalsystem im juvenilen Stammholz von Larix decidua Miller. Beih Z Schweiz Forstver 61, 127 pp

Imamura Y 1978 (Abnormal ray tissues in excrescence featured wood of Sugi (Cryptomeria japonica D Don).) Mokuzai Gakkaishi 24:71–74

Imamura Y, Harada H, Saiki H 1972 Electron microscopic study on the formation and organization of the cell wall in coniferous tracheids. Crisscrossed and transition structures in the secondary wall. Bull Kyoto Univ For 44:182–190

Isebrands JG, Hunt CM 1975 Growth and wood properties of rapid-grown Japanese larch. Wood Fiber 7:119–128

Ishida S, Fujikawa S 1970 Study on the pit of wood cells using scanning electron microscopy. 2. Pit membrane of the tracheid bordered pit in a living tree trunk of Todo-Matsu, Abies sachalinensis Fr Schm. Res Bull Coll Exp For Hokkaido Univ 28:355–372

Ishida S, Ohtani J, Fujikawa S 1968 Application of the scanning electron microscope to the study of wood structure. Proc 1st Ann Meet Hokkaido Br Jap Wood Res Soc No 1

Jaccard P 1912 Über abnorme Rotholzbildung. Ber Deutsch Bot Ges 30:670–678

Jaccard P 1915 Über die Verteilung der Markstrahlen bei den Coniferen. Ber Deutsch Bot Ges 33:492–498

Jaccard P 1919 Novelles recherches sur l'accroissement en épaisseur des arbres. Fondation Schnyder von Wartensee, Zürich, 200 pp

Jaccard P 1938 Exzentrisches Dickenwachstum und anatomisch-histologische Differenzierung des Holzes. Ber Schweiz Bot Ges 48:491–537

Jaccard P 1940 Sur les épaississements spiralés et les striations des parois des fibres, des vaisseaux ou des trachéides du bois et leur signification. Ber Schweiz Bot Ges 50:285–292

Jaccard P, Frey A 1928 Einfluß von mechanischen Beanspruchungen auf die Micellarstruktur, Verholzung und Lebensdauer der Zug- und Druckholzelemente beim Dickenwachstum der Bäume. Jahrb. Wiss Bot 68:844–866

Jackson LWR 1959 Loblolly pine tracheid length in relation to position in tree. J For 57:366–367

Jackson LWR, Greene JT 1958 Tracheid length variation and inheritance in slash and loblolly pine. For Sci 4:316–318

Jackson LWR, Morse WE 1965 Tracheid length variation in single rings of loblolly, slash and shortleaf pine. J For 63:110–112

Jackson LWR, Parker JN 1958 Anatomy of fusiform rust galls on loblolly pine. Phytopathology 48:637–640

Jacobs MR 1945 The growth stresses of woody stems. Commonw For Bur Aust Bull 28, 64 pp

Jacquiot C, Trenard Y 1974 Note sur la présence de trachéides à parois gélatineuses dans des bois résineux. Holzforschung 28:73–76

Jain KK, Seth MK 1979 Intra-incremental circumferential variation in tracheid length at breast height as a basis for sampling in straight trees of blue pine. Silvae Genet 28:79–83

Jain KK, Seth MK 1980 Effect of bole inclination on ring width, tracheid length and specific gravity of wood at breast height in blue pine. Holzforschung 34:52–60

Jane FW, Wilson K, White DJB 1970 The structure of wood. Adam and Charles Black, London, 478 pp

Jayme G, Bergh NO 1968 Über die Vermeidung von Polymerisationsartefakten bei der Methakrylateinbettung für Dünnschnitt-Elektronenmikroskopie. Holz Roh-Werkst 26:427–429

Jayme G, Fengel D 1961a Beitrag zur Kenntnis des Feinbaus der Frühholztracheiden. Beobachtungen an Ultradünnschnitten von Fichtenholz. Holz Roh- Werkst 19:50–55

Jayme G, Fengel D 1961b Beitrag zur Kenntnis des Feinbaus der Fichtenholztracheiden. II. Beobachtungen an Ultradünnschnitten von delignifiziertem Holz und Ligingerüsten. Holzforschung 15:98–102

Jewell FF, True RP, Mallett SL 1962 Histology of Cronartium fusiforme in slash pine seedlings. Phytopathology 52:850–858

Johansson D 1940 Über Früh- und Spätholz in schwedischer Fichte und Kiefer und über ihren Einfluß auf die Eigenschaften von Sulfit- und Sulfatzellstoff. Holz Roh-Werkst 3:73–78

Jurášek L 1964 Změny V mikroskopické struktuře při rozkladu dřeva dřevokaznými houbami. (Changes in the microstructure at the destruction of wood by wood destroying fungi.) Drev Vysk 9:127–144

Jutte SM, Jongebloed WL, Sachs IB 1977 Influence of water environment on normal and compression wood of Picea species observed by scanning electron microscopy (SEM). Scan El Microsc 1977/II, Proc on other biological applications of the SEM/STEM, ITT Res Inst Chicago, 683–690

Jutte SM, Levy JF 1972 Compression wood in Pinus ponderosa Laws. A scanning electron microscopic study. IAWA Bull 1972(2):3–7

Jutte SM, Levy JF 1973 Helical thickenings in the tracheids of Taxus and Pseudotsuga as revealed by the scanning electron microscope. Acta Bot Neerl 23:100–105

Kantola M 1964 Röntgenografische Untersuchungen über die Orientierung der Kristallite in Holzfasern. Faserforsch Textiltech 15:587–590

Kantola M, Kähkönen H 1963 Small-angle X-ray investigation of the orientation of crystallites in Finnish coniferous and deciduous wood fibers. Ann Acad Sci Fenn A-VI Phys 137, 14 pp

Kantola M, Seitsonen S 1961 X-ray orientation investigations on Finnish conifers. Ann Acad Sci Fenn A-VI Phys 80, 15 pp

Keays JL, Hatton JV 1971 Complete-tree utilization studies. I. Yield and quality of kraft pulp from the components of Tsuga heterophylla. Tappi 54:99–104

Keith CT 1971 Observations on the anatomy and fine structure of the trabeculae of Sanio. IAWA Bull 1971(3):3–11

Keith CT, Godkin SE, Grozdits GA, Chauret C 1978 Further observations on the anatomy and fine structure of the trabeculae of Sanio. IAWA Bull 1978(2/3):47

Kennedy RW 1970 An outlook for basic wood anatomy research. Wood Fiber 2:182–187

Kennedy RW, Adamovich L 1968 An anomalous tracheid in Douglas-fir earlywood. Can Dep For Bi-Mon Res Notes 24(3):22
Kennedy RW, Farrar JL 1965 Tracheid development in tilted seedlings. In: Côté WA, Jr (ed) Cellular ultrastructure of woody plants. Syracuse Univ Press, Syracuse, 419–453
Kerr T, Bailey IW 1934 The cambium and its derivative tissues. X. Structure, optical properties and chemical composition of the so-called middle lamella. J Arnold Arbor 15:327–349
Kiatgrajai P, Conner AH, Rowe JW, Peters W, Roberts DR 1976a Attempts to induce lightwood in balsam fir and tamarack by treating with paraquat. Wood Sci 9:31–36
Kiatgrajai P, Conner AH, Rowe JW, Peters W, Roberts DR 1976b Attempts to induce lightwood in eastern hemlock by treating with paraquat. Wood Sci 8:170–173
Kibblewhite RP 1972 Effects of beaters and wood quality on the surface and internal structure of radiata pine kraft fibres. Pap Puu 54:709–714
Kibblewhite RP 1973 Effects of beating and wood quality on radiata pine kraft paper properties. NZ J For Sci 3:220–239
Kibblewhite RP, Thompson NS, Williams DG 1971 A study of the bonding forces between the epithelial cells surrounding the resin canals of slash pine (Pinus elliottii Engelm) holocellulose. Wood Sci Technol 5:101–120
Kienholz R 1930 The wood structure of "pistol-butted" mountain hemlock. Am J Bot 17:739–764
Kienholz R 1932 Fasciation in red pine. Bot Gaz 94:404–410
Klem GG 1928 Kvalitetsundersøkelser i granskog og på grantømmer. (Quality investigations in spruce forests and on spruce lumber.) Medd Norsk Skogforsøksves 3(13):399–452
Klinken J 1914 Über das gleitende Wachstum der Initialen im Kambium der Koniferen und den Markstrahlenverlauf in ihrer sekundären Rinde. Bibl Bot 19:1–37
Knigge W 1976 Some remarks on SEM-observations on wood defects. Paper presented at the Fifth Plenary Meeting of the International Academy of Wood Science, Copenhagen, Denmark
Knigge W, Wenzel B 1982 Über die Variabilität der Faserlänge innerhalb eines Stammes von Sequoiadendron giganteum (LindL) Buchholz. Forstarchiv 53:94–99
Kocoń J 1967a Rentgenograficzna analiza orientacji oszarów krystalicznych celulozy w drewnie sosny. (X-ray radiographic analysis of the orientation of crystalline zones of cellulose in pine wood.) Sylwan 111(9):25–34
Kocoń J 1967b Badanie promieniami X orientacji obszarów krystalicznych celulozy w drewnie gałęzi swierka i modrzewia. (X-ray examination of the orientation of crystalline areas of cellulose in the wood of spruce and larch branches.) Sylwan 111(11):49–51
Kollmann FFP, Côté, Jr 1968 Principles of wood science and technology. I. Solid wood. Springer, Berlin Heidelberg New York, 592 pp
Kononchuk PI 1888 (On the local or one-side "hard-layerness" of tree.) Yearb St. Petersburg For Inst 2 (Unoff Sect):41–56
Kozlowski TT 1971 Growth and development of trees. II. Cambial growth, root growth, and reproductive growth. Academic Press, New York London, 514 pp
Kramer PR 1957 Tracheid length variation in loblolly pine. TX For Serv Tech Rep 10, 22 pp
Krieg W 1907 Die Streifung der Tracheidenmembran im Koniferenholz. Bot Zbl Beih 21:245–262
Kucera L 1972 Wundsgewebe in der Eibe (Taxus baccata L). Diss 4804, ETH Zürich. Vierteljahrschr Naturforsch Ges Zürich 116:445–470
Kucera L, Bariska M 1972 Einfluß der Dorsiventralität des Astes auf die Markstrahlbildung bei der Tanne (Abies alba Mill). Vierteljahrschr Naturforsch Ges Zürich, 117:305–313
Kucera L, Bosshard HH 1975 The presence of biseriate rays in fir (Abies alba Mill). IAWA Bull 1975(4):51–56
Kučera L, Nečesaný V 1970 The effect of dorsiventrality on the amount of wood rays in the branch of fir (Abies alba Mill) and poplar (Populus monilifera Henry). I. Some wood ray characteristics. Drev Vysk 15:1–6
Kukachka BF 1960 Identification of coniferous woods. Tappi 43:887–896
Kuroda K, Shimaji K 1983 Traumatic resin canal formation as a marker of xylem growth. For Sci 29:653–659
Kutscha NP 1968 Cell wall development in normal and compression wood of balsam fir, Abies balsamea (L) Mill. Ph D Thesis State Univ Coll For Syracuse NY, 231 pp

Kutscha NP, Hyland F, Schwarzmann JM 1975 Certain seasonal changes in balsam fir cambium and its derivatives. Wood Sci Technol 9:175 – 188

Ladza EE, Odintsov PN 1968 (Anatomy of resin ducts and rays of pine formed by irradiated cambium.) Izv Akad Nauk Latv SSR 10:84 – 92

Laming PB 1974 On intercellular spaces in the xylem ray parenchyma of Picea abies. Acta Bot Neerl 23:217 – 22

Lämmermayr L 1901 Beiträge zur Kenntnis der Heterotrophie von Holz und Rinde. Sitzungsber Kaiserl Akad Wiss Math-Naturwiss Cl Wien Pt 1, 110:29 – 62

Lange PW 1954 The distribution of lignin in the cell wall of normal and reaction wood from spruce and a few hardwoods. Svensk Papperstidn 57:525 – 532

Larson PR 1966 Changes in chemical composition of wood cell walls associated with age in Pinus resinosa. For Prod J 16(4):37 – 45

Larson PR 1969 Incorporation of ^{14}C in the developing walls of Pinus resinosa tracheids: compression wood and opposite wood. Tappi 52:2170 – 2177

Lee HN, Smith EM 1916 Douglas fir fiber with special reference to length. For Q 14:671 – 695

Liese W 1965 The warty layer. In: Côté WA, Jr (ed) Cellular ultrastructure of woody plants. Syracuse Univ Press, Syracuse, 251 – 269

Liese W, Höster HR 1966 Gelatinöse Bastfasern im Phloem einiger Gymnospermen. Planta 69:338 – 346

Lindgren PH 1958 X-ray orientation investigations on some Swedish cellulose materials. Arkiv Kemi 12(38):437 – 452

MacMillan WB 1925 A study of comparative lengths of tracheids of red spruce grown under free and suppressed conditions. J For 23:34 – 42

Mariani P 1955 Accrescimento e caratteri istologici di un fusto eccentrico di Larix europaea D C. Ital For Mont 10:216 – 224

Mark RE 1967 Cell wall mechanics of tracheids. Yale Univ Press, New Haven London, 310 pp

Mark RE, Gillis PP 1973 The relationship between fiber modulus and S2. Tappi 56:164 – 167

Marton M, Rushton P, Sacco JS, Sumiya K 1972 Dimensions and ultrastructure of growing fibers. Tappi 55:1499 – 1504

Matsumoto T 1950a (Studies on the compression wood. 1. On the spiral cracks in the tracheid wall.) Mokuzai Gakkaishi 32:16 – 20

Matsumoto T 1950b (Studies on the compression wood. 2. On the tracheid length of compression wood.) Mokuzai Gakkaishi 32:21 – 27

Matsuzaki SI 1972 Structural changes in tree tissue affected by environmental factors, with special reference to those in the bark structure of Todo-fir (Abies sachalinensis Mst). J Jap For Soc 54:287 – 294

McGinnes EA, Jr, Phelps JE 1972 Intercellular spaces in eastern redcedar (Juniperus virginiana L). Wood Sci 4:225 – 229

Meier H 1957 Discussion of the cell wall organization of tracheids and fibres. Holzforschung 11:41 – 46

Meier H 1962 Chemical and morphological aspects of the fine structure of wood. Pure Appl Chem 5:37 – 52

Mer É 1888 – 1889 Recherches sur les causes d'excentricité de la moelle dans les sapins. Rev Eaux For 27:461 – 471, 523 – 530, 562 – 572. 28:67 – 71, 119 – 130, 151 – 163, 201 – 217

Mergen F 1958 Distribution of reaction wood in eastern hemlock as a function of its terminal growth. For Sci 4:98 – 109

Mergen F, Echols RM 1955 Number and size of radial resin ducts in slash pine. Science 121:306 – 307

Mergen F, Winer HI 1952 Compression failures in the boles of living conifers. J For 50:677 – 679

Meylan BA 1967 Measurement of microfibril angle by X-ray diffraction. For Prod J 17(5):51 – 58

Meylan BA 1968 Cause of high longitudinal shrinkage in wood. For Prod J 18(4):75 – 78

Meylan BA 1974 Compression wood force generation. NZ J For Sci 4:116

Meylan BA 1981 Reaction wood in Pseudowintera colorata – a vessel-less dicotyledon. Wood Sci Technol 15:81 – 92

Meylan BA, Butterfield BG 1972 Three-dimensional structure of wood. A scanning electron microscopy study. Syracuse Univ Press, Syracuse, 80 pp

Meylan BA, Probine MC 1969 Microfibril angle as a parameter in timber quality assessment. For Prod J 19(4):30−34

Michels P 1941 Feuchtigkeitsverteilung im Holz des Weißtannenstammes, Gewicht und Schwindmaß des Weißtannenholzes. Mitt Forstwirtsch Forstwiss 12:295−329

Miniutti VP 1967 Microscopic observations of ultraviolet irradiated and weathered softwood surfaces and clear coatings. US For Serv Res Pap FPL-74, 32 pp

Mio S, Matsumoto T 1979a (Morphological observation on longitudinal intercellular spaces in normal softwoods.) Bull Kyushu Univ For 51:1−12

Mio S, Matsumoto T 1979b (Morphological observation on intercellular spaces in wood ray.) Bull Kyushi Univ For 51:13−18

Mio S, Matsumoto T 1981 Intercellular spaces in the axial parenchyma of hardwoods. Mokuzai Gakkaishi 27:626−632

Mio S, Matsumoto T 1982a A note on parent cell walls in coniferous woods. IAWA Bull (ns) 3:55−58

Mio S, Matsumoto T 1982b On intercellular spaces in compression wood of Kuromatsu (Pinus thunbergii Parl). Bull Kyushu Univ For 52:107−114

Mitchell RG 1967 Abnormal ray tissue in three true firs infested by the balsam woolly aphid. US For Serv Res Note PNW-46, 17 pp

Molski BA 1969 The significance of compression wood in the restoration of the leader in Pinus silvestris L damaged by moose (Alces alces). I. Distribution and function of compression wood in the stems. Acta Soc Bot Polon 38:309−338

Molski BA 1971 The significance of compression wood in the restoration of the leader in Pinus silvestris L damaged by moose (Alces alces). II. Structure of growth rings in regenerating stems in relation to juvenile wood formation. Acta Soc Bot Polon 40:315−340

Morey PR 1973 How trees grow. Edward Arnold, London, 59 pp

Morey PR, Morey ED 1971a Anatomy of a lignitized wood from Senftenberg. Am J Bot 58:621−626

Morey PR, Morey ED 1971b The cell wall residue of fossil wood from Senftenberg. Am J Bot 58:627−633

Mork E 1928a Om tennar. (On compression wood.) Tidsskr Skogbr 36 (suppl):1−41

Mork E 1928b Die Qualität des Fichtenholzes unter besonderer Rücksichtnahme auf Schleif- und Papierholz. Papier-Fabr 26:741−747

Münch E 1919−1921 Naturwissenschaftliche Grundlagen der Kiefernharznutzung. Biol Reich Land-Forstwirtsch 10:1−140

Münch E 1938 Statik und Dynamik des schraubigen Baues der Zellwand, besonders des Druck- und Zugholzes. Flora 32:357−424

Münch E 1940 Weitere Untersuchungen über Druckholz und Zugholz. Flora 34:45−57

Murmanis L, Sachs IB 1969a Structure of pit border in Pinus strobus L. Wood Fiber 1:7−17

Murmanis L, Sachs IB 1969b Seasonal development of secondary xylem in Pinus strobus L. Wood Sci Technol 3:177−193

Nanko H, Côté WA, Jr 1982 Bark structure of hardwoods grown on southern pine sites. Syracuse Univ Press, Syracuse, 56 pp

Nanko H, Saiki H, Harada H 1982 Structural modification of secondary phloem fibers in the reaction phloem of Populus euramericana. Mokuzai Gakkaishi 28:202−207

Nečesaný V 1955a Vztah mezi reakčním dřevem listnatých a jehličnatých dřevin. (The relationship between the reaction wood in gymnosperms and angiosperms.) Biológia 10:642−647

Nečesaný V 1955b Submikroskopická morfologie buněčných blan reakčního dřeva jehličnatých. (Submicroscopic morphology of the cell walls in the reaction wood of conifers.) Biológia 10:647−659

Nečesaný V 1956 Struktura reakčního dřeva. (The structure of reaction wood.) Preslia 28:61−65

Nečesaný V 1957 The nature of the so-called tertiary lamella. Svensk Papperstidn 60:8−16

Nečesaný V 1961 Bewertung „normalen" Holzes vom Standpunkt der Struktur. Faserforsch Textiltech 12:169−178

Nečesaný V 1966 Variabilita orientace celulosových mikrofibril ve zdřevnatělé buněčné bláne. (Orientation variability of the cellulose microfibrils within a lignified cell wall.) Drev Vysk 11:1−26

Nečesaný V, Oberländerová A 1967 The analysis of causes of different formation of reaction wood in gymnosperms and angiosperms. Drev Vysk 12:61−71

Nicholls JWP 1982 Wind action, leaning trees and compression wood in Pinus radiata D Don. Aust For Res 12:75−91

Nicholls JWP, Dadswell HE 1962 Tracheid length in Pinus radiata D Don. CSIRO Aust Div For Prod Technol Pap 24, 19 pp

Nylinder P 1951 Om patologiska hartskanaler. (On pathological resin canals.) Medd Stat Skogsforskningsinst 40(7), 12 pp

Nylinder P, Hägglund E 1954 Ståndorts − och trädegenskapers inverkan på utbyte och kvalitet vid framställning av sulfitmassa av gran. (The influence of stand and tree properties on yield and quality of sulphite pulp from Swedish spruce (Picea excelsa).) Medd Stat Skogsforskningsinst 44(11), 184 pp

Nyrén V, Back EL 1960 Characteristics of parenchymatous cells and tracheidal ray cells in Picea abies Karst. The resin in parenchymatous cells and resin canals in conifers VI. Svensk Papperstidn 63:501−509

Oechssler G 1962 Studien über die Saugschäden mitteleuropäischer Tannenläuse im Gewebe einheimischer und ausländischer Tannen. Z Angew Entomol 50:408−454

Ohsako Y, Kato H, Nobuchi T 1973 Studies on properties of natural bending wood formed in tree growth. I. On Taiwan-Akamatsu (Pinus massoniana Lamb) planted in Kyoto University Experimental Forest in Kamigamo. Bull Kyoto Univ For 45:238−251

Ohtani J, Ishida S 1961 Study on the pit of wood cells using scanning electron microscopy. 6. Pits of compression wood tracheids of Abies sachalinensis. Proc Hokkaido Br Jap Wood Res Soc 13:1−4

Ohtani J, Meylan BA, Butterfield BG 1984 Vestures on warts − proposed terminology. IAWA Bull (ns) 5:3−8

Okumora S, Harada H, Saiki H 1974 The variation in the cell wall thickness along the length of a conifer tracheid. Bull Kyoto Univ For 46:162−169

Ollinmaa PJ 1955 Havupuiden lylypuun rakenteesta ja ominaisuuksita. (On the structure and properties of coniferous compression wood.) Pap Puu 37:544−549

Ollinmaa PJ 1959 Reaktiopuututkimuksia. (Study on reaction wood.) Acta For Fenn 72, 54 pp

Onaka F 1949 (Studies on compression and tension wood.) Mokuzai Kenkyo Wood Res Inst Kyoto Univ 1, 88 pp. Trans For Prod Lab Can 93 (1956), 99 pp

Page DH 1969 A method for determining the fibrillar angle in wood tracheids. J Microsc 90:137−143

Panshin AJ, de Zeeuw CH 1980 Textbook of wood technology, 4th ed. McGraw-Hill, New York, 722 pp

Parameswaran N 1979 A note on the fine structure of trabeculae in Agathis alba. IAWA Bull 1979(1):17−18

Parham RA 1970 Structural effects of ammonia treatment of the wood of Pinus taeda L. Ph D Thesis SUNY Coll For Syracuse, 248 pp

Parham RA, Côté WA, Jr 1971 Distribution of lignin in normal and compression wood of Pinus taeda L. Wood Sci Technol 5:49−62

Parham RA, Davidson RW, de Zeeuw CH 1972 Radial-tangential shrinkage of ammonia-treated loblolly pine wood. Wood Sci 4:129−136

Parham RA, Kaustinen H 1973 On the morphology of spiral thickenings. IAWA Bull 1973(2):8−18

Park S, Saiki H, Harada H 1979 Structure of branch wood in Akamatsu (Pinus densiflora Sieb et Zucc). I. Distribution of compression wood, structure of annual ring and tracheid dimensions. Mokuzai Gakkaishi 25:311−317

Park S, Saiki H, Harada H 1980 Structure of branch wood in Akamatsu (Pinus densiflora Sieb et Zucc). II. Wall structure of branch wood tracheids. Mem Coll Agr Kyoto Univ 115:33−44

Parthasarathy MV 1975 Sieve-element structure. In: Zimmermann MH, Milburn JA (eds) Encyclopedia of plant physiology Vol 1 (ns) Transport in plants. I. Phloem transport, 3−38

Patel RN 1963 Spiral thickening in normal and compression wood. Nature 198:1225−1226

Patel RN 1968 Wood anatomy of Cupressaceae and Araucariaceae indigenous to New Zealand. NZ J Bot 6:9−18

Patel RN 1971 Anatomy of stem and root wood of Pinus radiata D Don. NZ J For Sci 1:37−49

Petersen OG 1914 Forandring i vedbygning ved grenrejsning hos rødgran. (Changes in wood structure when branches bend upward in Norway spruce (Picea excelsa.)) Bot Tidsskr 33:354–361

Petrić B 1962 Varijacije u strukturi normalnog i kompresijskog drva jelovine. (Wood structure variation of normal and compression fir wood.) Drvna Ind 13(1/2):12–23

Petrić B 1974 Utjecaj starosti i širine goda na strukturu i volumnu težinu bijele borovine. (Influence of the age and width of the annual ring on the structure and density of Scots pine wood.) Ann Pro Exp For (Zagreb) 17:157–228

Philipson WR, Ward JM, Butterfield BG 1971 The vascular cambium. Its development and activity. Chapman and Hall, London, 182 pp

Phillips EWJ 1937 The occurrence of compression wood in African pencil cedar. Emp For J 16:54–57

Phillips EWJ 1940 A comparison of forest- and plantation-grown African pencil cedar (Juniperus procera Hochst) with special reference to the occurrence of compression wood. Emp For J 19:282–288

Pillow MY, Chidester GH, Bray MW 1941 Effect of wood structure on properties of sulphate and sulphite pulps from loblolly pine. South Pulp Pap J 4(7):6–12

Pillow MY, Luxford RF 1937 Structure, occurrence, and properties of compression wood. US For Serv Bull 546, 32 pp

Pillow MY, Schafer ER, Pew JC 1936 Occurrence of compression wood in black spruce and its effect on properties of ground wood pulp. Pap Trade J 102(16):36–38

Pillow MY, Schafer ER, Pew JC 1959 Occurrence of compression wood in black spruce and its effect on properties of groundwood pulp. US For Serv FPL Rep 1288

Preston RD 1934 The organization of the cell walls of the conifer tracheids. Phil Trans R Soc London 224:131–136

Preston RD 1946 The fine structure of the wall in the conifer tracheid. I. The X-ray diagram of conifer wood. Proc R Soc London B-133:337–348

Preston RD 1947 The fine structure of the cell wall of the conifer tracheid. II. Optical properties of dissected walls in Pinus insignis. Proc R Soc London B-134:202–218

Preston RD 1948 The fine structure of the wall of the conifer tracheid. III. Dimensional relationships in the central layer of the secondary wall. Biochim Biophys Acta 2:370–383

Preston RD 1974 The physical biology of plant cell walls. Chapman and Hall, London, 491 pp

Preston RD, Wardrop AB 1949 The fine structure of the wall of the conifer tracheid. IV. Dimensional relationships in the outer layer of the secondary wall. Biochim Biophys Acta 3:585–592

Preusser HJ, Dietrichs HH, Gottwald H 1961 Elektronenmikroskopische Untersuchungen an Ultradünnschnitten des Markstrahlparenchyms der Rotbuche – Fagus sylvatica L. Holzforschung 15:65–78

Priestley JH 1930a Studies in the physiology of cambial activity. I. Contrasted types of cambial activity. New Phytol 29:56–73

Priestley JH 1930b Studies in the physiology of cambial activity. II. The concept of sliding growth. New Phytol 29:96–140

Priestley JH 1930c Studies on the physiology of cambial activity. III. The seasonal activity of the cambium. New Phytol 29:316–354

Priestley JH, Tong D 1925–1929 The effect of gravity upon cambial activity in trees. Proc Leeds Phil Lit Soc Sect 1:199–208

Reid RW, Watson JA 1966 Sizes, distributions, and numbers of vertical resin ducts in lodgepole pine. Can J Bot 44:519–525

Rendle BJ 1952 "Strawberry mark" in selected Sitka spruce wood. Selected Gov Res Rep Vol 8, London

Resch A, Blaschke R 1968 Über die Anwendung des Raster-Elektronenmikroskopes in der Holzanantomie. Planta 78:85–88

Resch H, Arganbright DG 1968 Variation of specific gravity, extractive content, and tracheid length in redwood trees. For Sci 14:148–155

Rickey RG, Hamilton JK, Hergert HL 1974 Chemical and physical properties of tumor-affected Sitka spruce. Wood Fiber 6:200–210

Riech FP 1966 Influence of static bending stress on growth and wood characteristics of nine-year-old Douglas-fir from two geographic sources. M S Thesis OR State Univ Corvallis, 58 pp

Riech FP, Ching KK 1970 Influence of bending stress on wood formation of young Douglas-fir. Holzforschung 24:68–70

Robards AW 1969 The effect of gravity on the formation of wood. Sci Progr (Oxf) 57:513–532

Robards AW (ed) 1974 Dynamic aspects of plant ultrastructure. McGraw-Hill, Maidenhead, 546 pp

Roberts DR 1973 Inducing lightwood in pine trees. US For Serv Res Note SE-191, 4 pp

Roelofsen PA 1959 The plant cell wall. Gebrüder Bornträger, Berlin, 335 pp

Rowell R (ed) 1984 The chemistry of solid wood. Am Chem Soc Adv Chem Ser 207, 614 pp

Russow E 1883 Zur Kenntnis des Holzes, insbesondere des Coniferenholzes. Bot Cbl 13:134–144

Sachsse H 1969 Sind Baumschädigungen durch die Klettersäge möglich? Holz-Zentralbl 95:1460–1462

Sachsse H 1971 Anatomische und physiologische Auswirkungen maschineller Ästung auf lebende Nadelbäume. Holz Roh-Werkst 29:189–194

Sachsse H 1983 Untersuchungen über die Nebenwirkungen der Klettersäge „KS 31" auf Gesundheitszustand und Holzgüte von Douglasien. Forstarchiv 54:62–69, 107–114

Saigo RH 1976 Anatomical changes in the secondary phloem of grand fir (Abies grandis) induced by the balsam woolly aphid (Adelges piceae). Can J Bot 54:1903–1910

Saiki H 1970 Proportion of component layer in tracheid wall of early wood and late wood of some conifers. Mokuzai Gakkaishi 16:244–249

Sanio C 1860 Einige Bemerkungen über den Bau des Holzes. Bot Ztg 18:193–198, 201–204, 209–215

Sanio C 1863 Vergleichende Untersuchungen über die Elementarorgane des Holzkörpers. Bot Ztg 21:85–91, 93–98, 101–111, 113, 118, 121, 128

Sanio K 1872 Über die Größe der Holzzellen bei der gemeinen Kiefer (Pinus silvestris L). Jahrb Wiss Bot 8:401–420

Sanio K 1873 Anatomie der gemeinen Kiefer (Pinus silvestris L). Jahrb Wiss Bot 9:50–126

Sastry CBR, Kozak A, Wellwood RW 1872 New approach for evaluation of wood from fertilized trees. Can J For Res 2:417–426

Sastry CBR, Wellwood RW 1971a Individual tracheid weight–length relationships in Douglas-fir. Tappi 54:1686–1690

Sastry CBR, Wellwood RW 1961b A quartz ultra-microbalance for weighing single tracheids. Wood Sci 3:179–182

Sastry CBR, Wellwood RW 1974 Individual tracheid weight–length relationships in coniferous woods. Wood Sci Technol 8:266–274

Sastry CBR, Wellwood RW 1975 Some cellular characteristics of earlywood and latewood in conifers. Mokuzai Gakkaishi 21:207–211

Sato K, Ishida S 1982 Resin canals in the wood of Larix leptolepis Gord. II. Morphology of vertical resin canals. Res Bull Coll Exp For Hokkaido Univ 39:298–316

Sato K, Ishida S 1983 Resin canals in the wood of Larix leptolepis Gord. V. Formation of vertical resin canals. Res Bull Coll Exp For Hokkaido Univ 40:727–740

Savidge RA 1983 The role of plant hormones in higher plant cellular differentiation. II. Experiments with the vascular cambium, and sclereid and tracheid differentiation in the pine, Pinus contorta. Histochem J 15:447–466

Schafer ER, Pew JC, Curran CE 1937 Grinding of loblolly pine. Relation of wood properties and grinding conditions to pulp and paper quality. Pap Trade J 105(21):41–48

Schultze-Dewitz G 1959 Variation und Häufigkeit der Faserlänge der Fichte. Holz Roh-Werkst 17:319–326

Schultze-Dewitz G 1965 Variation und Häufigkeit der Faserlänge der Kiefer. Holz Roh-Werkst 23:81–86

Schultze-Dewitz G, Götze H 1973 Untersuchungen zur Faserlänge, Raumdichte und Druckfestigkeit inter- und circumnodialen Holzes der Baumarten Kiefer (Pinus sylvestris L), Fichte (Picea abies Karst) und Douglasie (Pseudotsuga menziesii Franco). Drev Vysk 18:33–44

Schultze-Dewitz G, Götze H, Günther B, Luthardt H 1971 Eigenschaften und Verwertung des Astholzes von Kiefer. Holztechnologie 12:214–221

Schweingruber FH 1978 Microscopic wood anatomy. Zürcher AG, Zug, 226 pp

Scott JAN, Goring DAI 1970 Lignin concentration in the S_3 layer of softwoods. Cellul Chem Technol 4:83–93

Scurfield G 1967 Histochemistry of reaction wood differentiation in Pinus radiata D Don. Aust J Bot 18:377–392

Scurfield G, Silva S 1969a Scanning electron microscopy applied to a study of the structure and properties of wood. Proc 2nd Ann El Microsc Symp IIT Res Inst, Chicago IL, 10 pp

Scurfield G, Silva S 1969b The structure of reaction wood as indicated by scanning electron microscopy. Aust J Bot 17:391–402

Scurfield G, Wardrop AB 1962 The nature of reaction wood. VI. The reaction anatomy of seedlings of woody perennials. Aust J Bot 2:93–105

Scurfield G, Wardrop AB 1963 The nature of reaction wood. VII. Lignification of reaction wood. Aust J Bot 11:107–116

Seth MK 1979 Studies on the variation and correlation among some wood characteristics in blue pine (Pinus wallichiana A B Jackson). Ph D Thesis Himachal Pradesh Univ, Simla, 168 pp

Seth MK 1981 Variation in tracheid length in blue pine (Pinus wallichiana A B Jackson). 2. Radial pattern of variation in tracheid length in the first-formed earlywood from pith to bark. Wood Sci Technol 15:275–286

Seth MK, Agrawal HO 1984 Variation in tracheid length in blue pine (Pinus wallichiana A B Jackson). Holzforschung 38:1–6

Seth MK, Jain KK 1976 Variation in tracheid length in blue pine (Pinus wallichiana A B Jackson). I. Intra-increment variation in tracheid length. Indian Acad Wood Sci 7(1):1–9

Seth MK, Jain KK 1977 Relationship between percentage of compression wood and tracheid length in blue pine (Pinus wallichiana A B Jackson). Holzforschung 31:80–83

Seth MK, Jain KK 1978 Percentage of compression wood and specific gravity in blue pine (Pinus wallichiana A B Jackson). Wood Sci Technol 12:17–24

Schafer ER, Pew JC, Curran CE 1958 Grinding of loblolly pine. Relation of wood properties and grinding conditions to pulp and paper quality. US For Serv FPL Rep 1163, 18 pp

Shelbourne CJA, Ritchie KS 1968 Relationships between degree of compression wood development and specific gravity and tracheid characteristics in loblolly pine (Pinus taeda L). Holzforschung 22:185–190

Shephard HN, Bailey IW 1914 Some observations on the variation in length of coniferous fibers. Proc Soc Am For 9:522–527

Shimaji K, Nagatsuka Y 1971 Pursuit of the time sequence of annual ring formation in Japanese fir (Abies firma Sieb et Zucc). Mokuzai Gakkaishi 17:122–128

Sinnott EW, Bloch R 1939 Changes in intercellular relationships during the growth and differentiation of living plant tissues. Am J Bot 26:625–634

Sioumis AA, Lau LS 1976 Paraquat-induced resinosis in Pinus radiata. Appita 29:272–275

Skene DS 1972 The kinetics of tracheid development in Tsuga canadensis Carr and its relation to vigour. Ann Bot 36:179–187

Smith DM 1965 Rapid measurement of tracheid cross-sectional dimensions of conifers: its application to specific gravity measurements. For Prod J 15:325–334

Smith DM, Miller RB 1964 Methods of measuring and estimating tracheid wall thickness of redwood (Sequoia sempervirens (D Don) Endl). Tappi 47:599–604

Smith FH 1967 Effects of balsam woolly aphid (Adelges piceae) infestation on cambial activity in Abies grandis. Am J Bot 54:1215–1223

Sobue N, Hirai N, Asano I 1971 Studies on structure of wood by X-ray. II. Estimation of the orientation of micelles in cell wall. Mokuzai Gakkaishi 17:44–50

Sonntag P 1904 Über die mechanischen Eigenschaften des Roth- und Weißholzes der Fichte und anderer Nadelhölzer. Jahrb Wiss Bot 39:71–105

Sonntag P 1909 Die duktilen Pflanzenfasern, der Bau ihrer mechanischen Zellen und die etwaigen Ursachen der Duktilität. Flora 99:203–259

Spurr SH, Hyvärinen MJ 1954 Compression wood in conifers as a morphogenetic phenomenon. Bot Rev 20:561–575

Srivastava LM 1963a Cambium and vascular derivatives of Ginkgo biloba. J Arnold Arbor 44:165–192

Srivastava LM 1963b Secondary phloem in the Pinaceae. Univ CA Publ Bot 36, 142 pp

Srivastava LM, Esau K 1961 Relation of dwarf mistletoe (Arceuthobium) to the xylem tissue of conifers. II. Effect of the parasite on the xylem anatomy of the host. Am J Bot 48:205–215

Stairs GR, Marton R, Brown AF, Rizzio M, Petric A 1966 Anatomical and pulping properties of fast- and slow-grown Norway spruce. Tappi 49:296–300

Stephan G 1967 Untersuchungen über die Anzahl der Harzkanäle in Kiefern (Pinus silvestris). Arch Forstwes 16:461–470

Steward CM, Foster RC 1976 X-ray diffraction studies related to forest products research. Appita 29:440–448

Sudo S 1968a Variation in tracheid length in Akamatsu (Pinus densiflora Sieb et Zucc). I. Variation in tracheid length within a young tree stem. Mokuzai Gakkaishi 14:1–5

Sudo S 1968b Variation in tracheid length in Akamatsu (Pinus densiflora Sieb et Zucc). II. Variation in tracheid length in one-year-old branches of a tree. Mokuzai Gakkaishi 14:6–10

Sudo S 1968c Variation in tracheid length in Akamatsu (Pinus densiflora Sieb et Zucc). III. Variation in tracheid length in one-year-old branches taken from ten trees from a fifty-year-old stand. Mokuzai Gakkaishi 14:70–74

Sudo S 1968d Anatomical studies on the wood of species of Picea with some consideration on their geographical distribution and taxonomy. Bull Gov For Exp Sta Tokyo 215:39–126

Sudo S 1969 Variation in tracheid length in Akamatsu (Pinus densiflora Sieb et Zucc) from a stand in Tohoku district. VI. Relation between tracheid length and growth. Mokuzai Gakkaishi 15:241–246

Sudo S 1970a Variation in tracheid length in Akamatsu (Pinus densiflora Sieb et Zucc). VII. Variation in tracheid length in Akamatsu from a stand in the Kansai district. Mokuzai Gakkaishi 16:162–167

Sudo S 1970b Variation in tracheid length in Akamatsu (Pinus densiflora Sieb et Zucc). VIII. Relation between tracheid length of juvenile and adult wood in Akamatsu from a stand in the Karsai district. Mokuzai Gakkaishi 16:209–212

Sugawa T 1978 (Occurrence of traumatic resin canals in the stem of Japanese black pine (Pinus thunbergii) seedlings suffering from pine wood nematode (Bursaphelencus lignicolus).) J Jap For Soc 60:460–463

Süss H, Müller-Stoll R 1970 Änderungen der Zellgrößen und der Anteil der Holzelemente in zerstreutporigen Hölzern innerhalb einer Zuwachsperiode. Holz Roh-Werkst 28:309–317

Suzuki M 1968 The relationship between elasticity and strength properties and cell structure of coniferous wood. Bull Gov For Exp Sta Tokyo 212:89–149

Takahara S, Nobuchi T, Harada H, Saiki H 1983 Wall structure of parenchyma cells surrounding axial resin canals in the wood of European spruce. Mokuzai Gakkaishi 29:355–360

Takaoka A 1975 Scanning electron microscopic observations on compression wood. M S Thesis Hokkaido University, Sapporo.

Takaoka A, Ishida S 1974 (Reaction wood in conifers with spiral thickenings.) Proc Hokkaido Br Jap Wood Res Soc 6:5–8

Tang RC 1973 The microfibrillar orientation in cell-wall layers of Virginia pine tracheids. Wood Sci 5:181–186

Taylor FW 1979 Variation of specific gravity and tracheid length in loblolly pine branches. J Inst Wood Sci (46)8(4):171–175

Taylor FW, Moore JS 1981 A comparison of earlywood and latewood tracheid lengths of loblolly pine. Wood Fiber 13:159–165

Thompson NS, Hankey JD 1968 The isolation of epithelia of conifers. Tappi 51:88–93

Thompson NS, Heller HH, Hankey JD, Smith O 1966 The isolation and the carbohydrate composition of the epithelial cells of longleaf pine (Pinus palustris Mill). Tappi 49:401–405

Timell TE 1972a Beobachtungen an Holzstrahlen im Druckholz. Holz Roh-Werkst 30:267–273

Timell TE 1972b Nature of the last-formed tracheids in compression wood. IAWA Bull 1972(4):10–19

Timell TE 1973 Ultrastructure of the dormant and active cambial zones and the dormant phloem associated with formation of normal and compression woods in Picea abies (L) Karst. SUNY Coll Environ Sci For Syracuse, Tech Publ 96, 94 pp

Timell TE 1978a Helical thickenings and helical cavities in normal and compression woods of Taxus baccata. Wood Sci Technol 12:1−15

Timell TE 1978b Ultrastructure of compression wood in Ginkgo biloba. Wood Sci Technol 12:89−103

Timell TE 1979 Formation of compression wood in balsam fir (Abies balsamea). II. Ultrastructure of the differentiating xylem. Holzforschung 33:181−191

Timell TE 1980a Organization and ultrastructure of the dormant cambial zone in compression wood of Picea glauca. Wood Sci Technol 14:161−179

Timell TE 1980b Karl Gustav Sanio and the first scientific description of compression wood. IAWA Bull (ns) 1:147−153

Timell TE 1981 Recent progress in the chemistry, ultrastructure, and formation of compression wood. The Ekman Days 1981 (Stockholm), SPCI Rep 38 Vol 1:99−147

Timell TE 1983 Origin and evolution of compression wood. Holzforschung 37:1−10

Treiber E (ed) 1957 Die Chemie der Pflanzenzellwand. Springer, Berlin Göttingen Heidelberg, 511 pp

Treiber E 1971 Svepelektronmikroskopi som hjälpmedel inom träforskningen. (Scanning electron microscopy as a tool in wood research.) Svensk Papperstidn 74:509−514

Trendelenburg A 1932 Über die Eigenschaften des Rot- oder Druckholzes der Nadelhölzer. Allg Forst-Jagdztg 108:1−14

Trendelenburg R 1940 Über Faserstauchungen in Holz und ihre Überwallung durch den Baum. Holz Roh-Werkst 3:209−221

Trendelenburg R, Mayer-Wegelin H 1955 Das Holz als Rohstoff, 2nd ed. Carl Hanser, München, 541 pp

Tsoumis G 1965 Structural deformities in an epidemic tumor on white spruce, Picea glauca. Can J Bot 43:176−181

Uprichard JM 1971 Cellulose and lignin content in Pinus radiata D Don. Within-tree variation in chemical composition, density, and tracheid length. Holzforschung 25:97−105

Verrall AF 1928 A comparative study of the structure and physical properties of compression wood and normal wood. M S Thesis, Univ MI, St Paul, 37 pp

Vichrov VE 1949 (Structure and physical and mechanical properties of Larix sibirica.) Trud Inst Lesn 4:174−194

von Aufsess H 1973 Mikroskopische Darstellung des Verholzungsgrades durch Färbemethoden. Holz Roh-Werkst 31:24−33

von Pechmann H 1972 Das mikroskopische Bild einiger Holzfehler. Holz Roh-Werkst 30:62−66

von Pechmann H 1973 Beobachtungen über Druckholz und seine Auswirkungen auf die mechanischen Eigenschaften von Nadelholz. IUFRO-5 Meet S Afr 2:1114−1122

Voorhies G, Jameson DA 1969 Fiber length in southwestern young-growth ponderosa pine. For Prod J 18(5):52−55

Wang EIC, Micko MM 1984 Wood quality of white spruce from north central Alberta. Can J For Res 14:181−185

Wangaard FF (ed) 1981 Wood. Its structure and properties. EMMSE Project, PA State Univ, College Place, 465, pp

Wardrop AB 1951 Cell wall organization and the properties of the xylem. I. Cell wall organization and the variation of breaking load in tension of the xylem in conifer stems. Aust J Sci Res B-4:391−414

Wardrop AB 1952 The formation of new cells in cell division. Nature 170:329

Wardrop AB 1954 The fine structure of the conifer tracheid. Holzforschung 8:12−29

Wardrop AB 1957 The organization and properties of the outer layer of the secondary wall in conifer tracheids. Holzforschung 11:102−110

Wardrop AB 1964 The structure and formation of the cell wall in xylem. In: Zimmermann MH (ed) The formation of wood in forest trees. Academic Press, New York London, 87−134

Wardrop AB 1965 The formation and function of reaction wood. In: Côté WA, Jr (ed) Cellular ultrastructure of woody plants. Syracuse Univ Press, Syracuse, 371−390

Wardrop AB, Dadswell HE 1950 The nature of reaction wood. II. The cell wall organization of compression wood tracheids. Aust J Sci Res B-3:1−13

Wardrop AB, Dadswell HE 1951 Helical thickenings and micellar orientation in the secondary wall of conifer tracheids. Nature 168:610

Wardrop AB, Dadswell HE 1952 The nature of reaction wood. III. Cell division and cell wall formation in conifer stems. Aust J Sci Res B-5:385−398

Wardrop AB, Dadswell HE 1953 The development of the conifer tracheid. Holzforschung 7:33−39

Wardrop AB, Dadswell HE 1957 Variation in the cell wall organization of tracheids and fibres. Holzforschung 11:33−41

Wardrop AB, Davies GW 1964 The nature of reaction wood. VIII. The structure and differentiation of compression wood. Aust J Bot 12:24−38

Wardrop AB, Preston RD 1947 Organization of the cell walls of tracheids and wood fibres. Nature 160:911−913

Watanabe H, Tsutsumi J, Kanagawa H 1962 (Properties of branch wood, especially on specific gravity, tracheid length, and appearance of compression wood.) Bull Kyushu Univ For 35:91−96

Watanabe H, Tsutsumi J, Kòjima K 1963 Studies on juvenile wood. I. Experiments on stems of Sugi trees (Cryptomeria japonica D Don). Mokuzai Gakkaishi 9:225−230

Waterkeyn L, Caymaex S, Decamps E 1982 La callose des trachéides du bois de compression chez Pinus silvestris et Larix decidua. Bull Soc R Bot Belg 115:149−155

Watson AJ, Dadswell HE 1957 Paper making properties of compression wood from Pinus radiata. Appita 11(3):56−70

Watson AJ, Dadswell HE 1964 Influence of fibre morphology on paper properties. 4. Micellar spiral angle. Appita 17:151−156

Wegelius T 1939 The presence and properties of knots in Finnish spruce. Acta For Fenn 48, 191 pp

Wegelius T 1946 Det finska granvirkets egenskaper och kvalitetsvariationer. (The properties and variations in quality of Finnish spruce wood.) Svensk Papperstidn 49:51−60

Wellwood RW, Jurazs PE 1968 Variation in sapwood thickness, specific gravity and tracheid length in western red-cedar. For Prod J 18(12):37−46

Wellwood RB, Sastry CBR, Micko MM, Paszner L 1974 On some possible specific gravity, α-cellulose tracheid weight/length and cellulose crystallinity relationships in a 500-year-old Douglas-fir tree. Holzforschung 28:91−94

Wergin W 1965 Über Entstehung und Aufbau von Reaktionsholzzellen. 4. Nachweis der Ligninverteilung in den Zellwänden des Druckholzes durch Untersuchungen im UV-Licht. Flora 156:322−331

Wergin W, Casperson G 1961 Über Entstehung und Aufbau von Reaktionsholzzellen. 2. Morphologie der Druckholzzellen von Taxus baccata L. Holzforschung 15:44−49

Werker E, Fahn A 1969 Resin ducts of Pinus halepensis. Their structure, development and pattern of arrangement. Bot J Linn Soc 62:379−411

Westing AH 1965 Formation and function of compression wood in gymnosperms. Bot Rev 31:381−480

Whalley BE 1950 Increase in girth of the cambium in Thuja occidentalis L. Can J Res C-28:331−340

White DJB 1965 The anatomy of reaction tissues in plants. In: Carthy JD, Duddington CL (eds) Viewpoints in biology IV, Butterworth London, 54−82

White DJB, Robards AW 1966 Some effects of radial growth rate upon the rays of certain ring-porous hardwoods. J Inst Wood Sci 17:45−52

White J 1907 The formation of red wood in conifers. Proc R Soc Victoria 30(2):107−124

White OE 1948 Fasciation. Bot Rev 14:319−358

Wiksten Å 1944−1945 Metodik vid mätning av årsringens vårved och höstved. (Methods for measuring the earlywood and latewood in an annual ring.) Medd Stat Skogsförsöksanst 34:451−490

Wilson BW 1981 The development of growth strains and stresses in reaction wood. In: Barnett JR (ed) Xylem cell development. Castle House, Tunbridge Wells, 275−290

Wood JR, Goring DAI 1971 The distribution of lignin in stem wood and branch wood of Douglas fir. Pulp Pap Mag Can 72(3):T 95−T 102

Wooten TE, Barefoot AC, Nicholas DD 1967 The longitudinal shrinkage of compression wood. Holzforschung 21:168–171

Worster HE, Vinje MG 1968 Kraft pulping of western hemlock tree tops and branches. Pulp Pap Mag Can 69(7):T308–T311

Yonezawa Y, Itoh A, Kikuchi F, Usami K, Takano I, Takamura N 1962 On the branches and top part of tree as the raw materials for pulp and fiberboard. Bull Gov For Exp Sta Meguro 146:119–131

Yonezawa Y, Murata T, Kayama T, Usami K, Takamura N, Takano I 1959 On the branches and top part of tree as the raw materials for pulp and fiberboard. Bull For Exp Sta Tokyo 113:145–152

Yoshizawa N, Itoh T, Shimaji K 1982 Variations in features of compression wood among gymnosperms. Bull Utsunomiya Univ For 18:45–64

Yoshizawa N, Koike S, Idei T 1984 Structural changes of tracheid wall accompanied by compression wood formation in Taxus cuspidata and Torreya nucifera. Bull Utsunomiya Univ For 20:59–76

Yoshizawa N, Koike S, Idei T 1985 Formation and structure of compression wood tracheids. induced by repeated inclination in Taxus cuspidata. Mokuzai Gakkaishi 31:325–333

Young HE, Chase AJ 1965 Fiber weight and pulping characteristics of the logging residue of seven tree species in Maine. ME Agr Exp Sta Tech Bull 17, 43 pp

Young WD, Laidlow RA, Packman DF 1970 Pulping of British-grown softwoods. VI. The pulping properties of Sitka spruce compression wood. Holzforschung 24:86–98

Yumoto M 1984 The trabeculae and its related structures. Res Bull Coll Exp For Hokkaido Univ 41:205–259

Yumoto M, Ishida S 1982 Studies on the formation and structure of the compression wood cells induced by artificial inclination in young trees of Picea glauca. III. Light microscopic observation on the compression-wood cells formed under five different angular displacements. J Fac Agr Hokkaido Univ 60:337–351

Yumoto M, Ishida S, Fukazawa K 1982 Studies on the formation and structure of the compression-wood cells induced by artificial inclination of young trees of Picea glauca. II. Transition from normal to compression wood revealed by a SEM-UVM combination method. J Fac Agr Hokkaido Univ 60:312–335

Yumoto M, Ishida S, Fukazawa K 1983 Studies on the formation and structure of the compression-wood cells induced by artificial inclination of young trees of Picea glauca. IV. Gradation of the severity of compression-wood tracheids. Res Bull Coll Exp For Hokkaido Univ 40:409–454

Yumoto M, Ohtani J 1981 Trabeculae in compression-wood tracheids. Proc Hokkaido Br Jap Wood Res Soc 13:9–12

Ziegler H, Merz W 1961 Der „Hasel"wuchs. Über Beziehungen zwischen unregelmäßigem Dickenwachstum und Markstrahlverteilung. Holz Roh-Werkst 19:1–8

Zimmermann MH (ed) 1964 The formation of wood in forest trees. Academic Press, New York London, 562 pp

Zobel BJ 1975 Using the juvenile wood concept in the southern pines. South Pulp Pap Manuf 38(9):14–16

Chapter 5 Chemical Properties of Compression Wood

CONTENTS

5.1	Introduction	290
5.2	General Chemical Composition	291
5.2.1	Introduction	291
5.2.2	Lignin	291
5.2.2.1	Compression Wood	291
5.2.2.2	Branch Wood and Knot Wood	298
5.2.3	Cellulose	303
5.2.3.1	Compression Wood	303
5.2.3.2	Branch Wood and Knot Wood	305
5.2.4	Galactose Residues	306
5.2.4.1	Compression Wood	306
5.2.4.2	Branch Wood and Knot Wood	306
5.2.5	Mannose Residues	306
5.2.5.1	Compression Wood	306
5.2.5.2	Branch Wood and Knot Wood	307
5.2.6	Pentose Residues (Pentosans)	307
5.2.6.1	Compression Wood	307
5.2.6.2	Branch Wood and Knot Wood	308
5.2.7	Acetyl Groups	308
5.2.8	Extractives	309
5.2.8.1	Compression Wood	309
5.2.8.2	Branch Wood and Knot Wood	312
5.2.8.3	Conclusions	313
5.2.9	Inorganic Constituents	313
5.2.9.1	Compression Wood	313
5.2.9.2	Branch Wood	314
5.2.10	Solubility Properties	314
5.2.11	Relative Sugar Composition	315
5.2.11.1	Compression Wood	315
5.2.11.2	Branch Wood	316
5.2.12	Absolute, Summative Composition	317
5.2.12.1	Compression Wood	317
5.2.12.2	Branch Wood	324
5.3	Cellulose	324
5.3.1	General Properties	324
5.3.2	Degree of Crystallinity	326
5.3.3	Effect of Liquid Ammonia	328
5.4	Laricinan	329
5.4.1	Introduction	329
5.4.2	Occurrence	333
5.4.3	Isolation	333
5.4.4	Structure	335
5.4.5	Molecular Properties	336
5.4.6	Crystal Structure	337
5.4.7	Conclusions	337
5.5	Galactan	338

5.5.1	Introduction	338
5.5.2	Occurrence	339
5.5.3	Isolation	339
5.5.4	Structure	344
5.5.5	Molecular and Other Properties	346
5.5.6	Conclusions	346
5.6	Galactoglucomannan	348
5.7	Xylan	350
5.8	Arabinogalactan	353
5.8.1	Introduction	353
5.8.2	Structure	354
5.8.3	Molecular Properties	356
5.8.4	Conclusions	357
5.9	Lignin	359
5.9.1	Introduction	359
5.9.2	Lignin in Compression Wood	360
5.9.3	Lignin-Carbohydrate Complexes from Compression Wood	379
5.9.4	Conclusions	380
5.10	Extractives	381
5.11	Microbial Degradation of Compression Wood	383
5.11.1	Introduction	383
5.11.2	Biodegradation of Compression Wood	384
5.11.3	Conclusions	388
5.12	Bark	390
5.13	General Conclusions	391
References		394

5.1 Introduction

The chemistry of compression wood has until recently received less attention than its structure, physical properties, and physiology. Many reviews of compression wood treat the subject in a cursory manner, while others omit it entirely. The fact that the tracheids in compression wood differ chemically from those of corresponding normal wood is, nevertheless, of importance, for it shows that the two types of cells are fundamentally different. The chemical composition of compression wood has a pronounced influence on the utilization of this type of wood in the manufacture of pulp and paper (Chap. 19). It also influences its physical properties (Chap. 7), contributing to the undesirable characteristics of compression wood in lumber (Chap. 18).

Wood is composed of cellulose, hemicelluloses, and lignin which are the skeletal, matrix, and encrusting substances, respectively, common to all higher land plants. Coniferous wood (softwood) generally contains 40–45% cellulose, 25–30% hemicelluloses, and 25–35% lignin and also small amounts of inorganic salts (ash) and varying quantities of so-called extractives. Compression wood differs from this normal wood not only in the proportion of the major constituents but also in their distribution in the tracheid wall (Chap. 6). The chemical differences between normal and compression woods seem to be restricted to the longitudinal tracheids, while the ray cells have the same chemical composition in the two woods.

An excellent monograph on wood chemistry was published in 1981 by Sjöström, the first book on this topic to appear in almost 20 years. The more recent and much larger volume by Fengel and Wegener (1983) treats the same subject in more detail. The standard work on cellulose is still the monograph by Ott et al. (1954), later expanded and updated by Bikales and Segal (1971). Timell (1964, 1965a, b, 1967) has reviewed the chemistry of the wood hemicelluloses. The most comprehensive treatment of the chemistry of lignin is found in the admirable monograph edited by Sarkanen and Ludwig (1971), with some of the best chapters authored by Sarkanen. A later contribution has been made by Adler (1977) in an excellent review paper. The biosynthesis of lignin has been treated by many authors, for example Freudenberg and Neish (1968), Gross (1977, 1980), and Higuchi (1980, 1981). No book on wood extractives has appeared since the pioneering work edited by Hillis (1962).

5.2 General Chemical Composition

5.2.1 Introduction

Until 30 years ago, knowledge of the chemistry of compression wood was limited to information adduced by application of standard methods of wood analysis. Some of these procedures, which have been reviewed by Browning (1967), are highly unreliable. The most characteristic chemical features of compression wood are its high contents of lignin and galactan and low contents of cellulose and galactoglucomannan. Chromatographic methods of sugar analysis were applied for the first time to compression wood in 1959 by Bouveng and Meier. In analyses carried out prior to this date, only the lignin values can be accepted without reservations. Fengel and Grosser (1975) have compiled published data for the chemical composition of various softwood and hardwood species. Results obtained in comparisons between normal and reaction woods are unfortunately not included in this survey. The chemistry of compression wood has been reviewed by Timell (1981, 1982). A brief summary is also found in a recent book by Siau (1984).

5.2.2 Lignin

5.2.2.1 *Compression Wood*

The most characteristic and important chemical property of compression wood is its high lignin content. This seems to have been mentioned first by Cieslar (1896, 1897), although the values quoted, namely 47.6% for normal and 48.2% for compression wood of *Picea abies*, are far from being correct and do not reflect the true difference in lignin contents between the two types of wood. Hartig (1901) mentions only that "the primary wall" in compression-wood tracheids is strongly lignified.

Sonntag (1904) appears to have been the first to discuss the degree of lignification of compression wood at some length. Using phloroglucinol-hydrochloric acid as a selective color reagent for lignin, Sonntag observed that in

Table 5.1. Chemical composition of normal and compression woods of *Picea sitchensis* and *Pseudotsuga meziesii*. All values in per cent of oven-dry wood. (Dadswell and Hawley 1929)

Constituent	*Picea sitchensis*		*Pseudotsuga menziesii*					
			Sapwood		Heartwood		Compression wood	
	Normal wood	Compression wood	Normal wood	Compression wood	Normal wood	Compression wood	Earlywood[a]	Latewood
Lignin	25.8	30.9	31.7	36.2	32.2	37.4	34.5	36.9
Cellulose	60.6	53.7	55.0	46.6	44.3	38.0	49.2	45.5
Pentosans	8.3	8.9	11.7	11.6	10.4	9.3		
Solubility in:								
Cold water	5.2	4.2	5.0	7.2	11.5	11.6	14.4	4.1
1% sodium hydroxide	13.7	13.5	12.0	18.2	25.5	24.5	24.7	15.4

[a] Calculated

compression-wood tracheids both the outer and the inner cell wall layers were strongly lignified, and the former more so that the latter. Sonntag removed the lignin with a solution containing potassium chlorate and hydrochloric acid, a reagent that obviously also destroyed some hemicelluloses. On the basis of these rather unsatisfactory experiments, he could, nevertheless, conclude that compression wood contained 20% more lignin than did normal wood, a result much closer to the truth than Cieslar's.

Johnsen and Hovey (1918), using 72% sulfuric acid to eliminate polysaccharides, found a lignin content of 24.6% for normal and 33.6% for compression wood of *Abies balsamea*[1]. Klason (1923) obtained a lignin content of 35.8% for compression wood from a stem of *Picea abies*. Normal wood of this species contained on the average 28.1% lignin (Klason 1931). Similar values, 28% and 34%, were obtained by Ulfsparre (1928) for normal and compression woods of this species. In the same year, Mork (1928) reported 28.2% and 39.7% lignin for the same tissues in *Picea abies*. This was probably the first correct determination of the lignin content of pronounced compression wood. In more detailed analyses, Dadswell and Hawley (1929) determined the chemical composition of various normal and compression woods of *Picea sitchensis* and *Pseudotsuga menziesii*. Some of their data are summarized in Table 5.1. It is noteworthy that the earlywood in the compression wood region of the redwood contained less lignin than did the latewood. This is contrary to the situation in normal wood, where the earlywood, because of its thinner cell walls, has a higher lignin content than the latewood (Wu and Wilson 1967). In compression wood, the first-formed earlywood is often not fully developed and resembles normal wood in several respects (Chap. 4.5).

---

1 The values attributed by Onaka (1949) to Schorger are those reported by Johnsen and Hovey (1918).

5.2.2.1 Compression Wood

Table 5.2. Chemical composition of normal and compression woods of *Picea abies*. All values in per cent of unextracted wood. (Hägglund and Ljunggren 1933)

Component	Normal wood	Compression wood
Lignin	28.0	38.0
Acetyl	1.4	0.8
Resin, ash, protein, and remainder	41.5	27.3
Acid-resistant polysaccharides of:		
Mannose	2.9	2.1
Xylose	2.2	2.4
Fructose	1.2	0.9
Total	6.3	5.4
Acid-labile polysaccharides of:		
Galactose	0.7	Nil
Mannose	7.7	1.9
Xylose	3.1	4.0
Fructose	0.7	Nil
Glucose and uronic acids	5.8	18.6
Total	18.0	24.5

When the pulping characteristics of compression wood began to attract interest in the early 1930's, the chemical composition of this wood was frequently compared to that of normal wood. Hägglund and Ljunggren (1933) subjected normal and compression woods of *Picea abies* to a sequence of chemical analyses which were based on the earlier, extensive studies of Hägglund and his coworkers. The results are summarized in Table 5.2[2]. Compression wood contained 36% more lignin than did normal wood. The increase in lignin content of the compression wood had not occurred at the expense of the pentosans. Compression wood from another spruce tree was found to contain 38.8% of lignin compared to only 26.2% for opposite wood. In the same year, Klem (1933) determined the lignin content of normal and compression woods of the same species. The lowest lignin content of the former was 26.8% and the highest of the latter 39.8%, a difference of almost 50% based on normal wood. In a later investigation, Stockman and Hägglund (1948) reported lignin contents of 27.1% and 37.5% for normal and compression woods from the same species.

In connection with their pulping studies, Pillow and Bray (1935) and, somewhat later, Curran (1936, 1937) determined the chemical composition of normal and compression woods of *Pinus taeda*, as shown in Tables 5.3 and 5.4. These analyses were carried out in the same laboratory, and some of the values reported obviously had the same origin. Hata (1951), in a series of studies on the pulping characteristics of *Pinus densiflora*, determined the chemical composition of normal and compression woods of this species, as shown in Table 5.5.

---

2 Some of these analytical data were later tabulated by Zherebov (1946), Correns (1961), and Kürschner (1964), in the latter two cases without any statement as to their origin.

Table 5.3. Chemical composition of normal and compression woods of *Pinus taeda*. All values in per cent of unextracted wood. (Pillow and Bray 1935)

Component	Normal wood	Compression wood
Extractives	2.7	2.5
Lignin	28.3	35.2
Cellulose	45.7	34.6
Pentosans	12.4	12.2
Solubility in:		
Hot water	1.8	2.0
1% Sodium hydroxide	9.9	12.6

Table 5.4. Chemical composition of normal and compression woods of *Pinus taeda*. All values in per cent of unextracted wood. (Curran 1936, 1937)

Constituent	Normal wood		Compression wood
	Earlywood	Latewood	
Extractives	4.2	2.5	2.7
Lignin	28.8	27.4	35.2
Cellulose	43.9	46.4	34.6
Pentosans	17.9	12.6	12.2
Solubility in 1% sodium hydroxide	14.0	12.9	12.6

Table 5.5. Chemical composition of normal and compression woods of *Pinus densiflora*. All values in per cent of unextracted wood. (Hata 1951)

Component	Normal wood	Compression wood
Extractives	3.23	3.14
Ash	0.23	0.28
Lignin	26.6	36.3
Cellulose	42.0	32.3
Pentosans	11.3	13.9
Solubility in:		
Cold water	1.27	1.29
Hot water	1.98	3.95
1% sodium hydroxide	11.0	12.8

5.2.2.1 Compression Wood

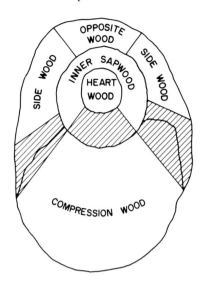

Fig. 5.1. Cross section of a *Pinus radiata* stem, showing the location of side wood, opposite wood, compression wood, sapwood, and heartwood. Wood located within the *hatched areas* was rejected. (Watson and Dadswell 1957, Bland 1958a)

Table 5.6. Chemical composition of wood from different zones of a stem of *Pinus radiata* (Fig. 5.1). (Watson and Dadswell 1957, Bland 1958a, Schwerin 1958)

Zone	Lignin, per cent	Methoxyl in lignin, per cent	Vanillin in per cent of lignin	Pentosan, per cent	Galactose residues, per cent
Opposite wood	25.8	14.1	20.2	8.8	2.0
Side wood	25.5	14.0	20.4	8.9	1.8
Normal wood	24.2	15.0	20.2	10.0	2.4
Compression wood	34.4	12.6	17.7	7.2	13.2

Watson and Dadswell (1957) investigated the papermaking properties of *Pinus radiata*, including opposite wood, side wood, normal wood, and compression wood, located as shown in Fig. 5.1. Analytical data for the four tissues are summarized in Table 5.6. Some of these values were published a year later by Bland (1958a) and by Schwerin (1958). Compression wood contained considerably more lignin than the other three woods, all of which had essentially the same chemical composition. In an investigation of a galactan from compression wood of *Picea abies*, Bouveng and Meier (1959), found this wood to contain 38.8% lignin compared to 28.0% for normal wood. In a study which remains unique in this respect, Bletchley and Taylor (1964) could not find any difference in cellulose and lignin contents between normal and compression woods of *Picea sitchensis*, nor did they find any difference with respect to other chemical components. Possibly, the compression wood analyzed was of a very mild type. Jurášek (1964a) found 38.5% lignin in compression wood of *Abies alba* and 30.3% in normal wood.

Table 5.7. Various properties of sodium lignosulfonates from opposite, normal, and compression woods of *Picea mariana*. (Yean and Goring 1965)

Characteristic	Opposite wood	Normal wood	Compression wood
Lignin, per cent of wood	26.8	27.0	38.6
Pulp yield, per cent	44	45	30
Sulfur, per cent of lignosulfonate	7.0	7.7	7.4
Methoxyl, per cent of lignosulfonate	13.7	14.5	10.7
Extinction coefficient, cm^{-1} g^{-1} at 280 nm	14.5	14.7	15.1
Nondialyzables, per cent	66	55	60
Weight-average molecular weight	19 000	19 000	27 000

Table 5.8. Various properties of wood from different zones of a *Picea mariana* stem. (Ladell et al. 1968)

Zone	Lignin, per cent	Pentosan, per cent	Specific gravity	Moisture content, per cent	Tracheid length, mm
Opposite wood	25.2	7.6	0.408[a]	8.8	3.47
Side wood	25.3	7.4		8.6	2.98
Compression wood	31.1	7.2	0.528	9.2	3.08
Control	27.0	7.3	0.396	8.6	3.40

[a] Includes side wood.

Table 5.9. Chemical composition of normal (side) and compression woods of *Picea sitchensis*. All values in per cent of unextracted wood or in relative per cent. (Young et al. 1970)

Component	Normal wood	Compression wood
Extractives	0.8	3.9
Lignin	27.9	38.9
Total carbohydrates	71.3	57.2
Sugar residues in relative per cent:		
Galactose	3.4	13.8
Glucose	71.8	65.5
Mannose	16.0	10.4
Arabinose	3.0	2.7
Xylose	5.8	7.6

5.2.2.1 Compression Wood

Yean and Goring (1965), in an investigation on lignosulfonates from various zones of *Picea mariana*, obtained the results summarized in Table 5.7. Both the published photograph of a stem cross section and the high lignin content of the compression wood indicate that the latter was of the severe type. Data reported for the same species by Ladell et al. (1968) (Table 5.8) suggest that in this case only moderate compression wood was involved. In an extensive study of the chemistry of the lignin present in compression wood of *Pseudotsuga menziesii*, Latif (1968) found compression wood from both the stem and a branch to contain 36.4% lignin compared to 27.0% for normal wood.

Hasegawa et al. (1970) studied the formation of wood under restraint in *Cryptomeria japonica*. Compression wood always developed on the lower side of bent stems and in one case it was found that the upper, normal part of the stem contained 24% lignin and compression wood formed on the lower side, 46% the highest lignin content ever reported for any compression wood and representing an increase of 92% from the value for the normal wood. Parham (1970) determined the lignin content of *Pinus taeda* compression wood by ultraviolet spectrophotometry. The value obtained, 39.1%, might be slightly too high, as the technique used afforded higher lignin contents than did the standard Klason method. Since coniferous wood contains almost no acid-soluble lignin, the reason for this discrepancy is unknown.

In a pulping study of normal and compression woods of *Picea sitchensis*, Young et al. (1970) obtained the data in Table 5.9 for the chemical composition of these woods. The lignin content of the compression wood (38.9%) was much higher than the value reported by Dadswell and Hawley (1929) (30.9%). In the same year, Philipp et al. (1970) studied the chemical composition and pulping characteristics of normal and compression woods of *Larix decidua*. The chemical data in Table 5.10 indicate a lower lignin content in the latewood than in the earlywood zones of the compression wood. The opposite, previously observed by Dadswell and Hawley (1929), would have been expected.

According to Wood and Goring (1971), an area of compression wood in a *Pseudotsuga menziesii* branch contained 42.0% Klason lignin. The normal sidewood had a lignin content of 31.6%. In his monograph on wood technology, Giordano (1971) has listed data for the cellulose, lignin, pentosan, and ex-

Table 5.10. Chemical composition of normal and compression woods of *Larix decidua*. All values in per cent of oven-dry wood. (Philipp et al. 1970)

Constituent	Normal wood		Compression wood	
	Earlywood	Latewood	Earlywood	Latewood
Lignin	28.1	26.9	37.4	34.8
Pentosans	7.8	5.8	5.9	6.8
Chlorite holocellulose	58.6	65.3	57.8	61.3
Hexosans	51.2	56.2	45.9	48.3

Table 5.11. Cellulose (Norman and Jenkins), lignin (Klason), pentosan, and extractives (alcohol-benzene) content of normal and compression woods from the same cross sections of four conifer species. All values in per cent. (Giordano 1971)

Species	Cellulose	Lignin	Pentosan	Extractives
Abies alba				
Normal wood	56.4	28.6	12.6	2.25
Compression wood	47.8	34.4	13.6	2.43
Larix decidua				
Normal wood	56.4	27.3	11.6	1.00
Compression wood	49.7	35.2	11.1	3.59
Picea abies				
Normal wood	62.2	27.9	10.6	5.63
Compression wood	43.8	34.9	11.7	6.66
Pinus nigra				
Normal wood	60.8	26.7	11.0	2.45
Compression wood	48.2	36.9	11.8	2.75

tractive content of normal and compression stem woods of four conifer species. As can be seen in Table 5.11, compression wood of all four species contained considerably more lignin than did corresponding normal wood. Morohoshi and Sakakibara (1971a) and Yasuda and Sakakibara (1975a) found compression woods of *Abies sachalinensis* and *Larix leptolepis* to contain 35.3 and 39.3% lignin, respectively, compared to 28.3 and 28.1% for normal woods. According to Chu (1972), compression wood of *Pinus palustris* has an average lignin content of 38.0% and normal wood 28.9%. Using ultraviolet microscopy, Fukazawa (1974) studied the distribution of lignin in normal and compression woods of *Abies sachalinensis* (Chap. 6.4.2). He estimated from his absorbance measurements that the former wood contained 26.2% lignin and the latter as much as 46.1%. Normal and compression woods of *Picea sitchensis* have been reported to have 27.3 and 35.4% lignin, respectively (Rickey et al. 1974) (Chap. 15.13).

It is clear that compression wood always contains more lignin than does normal wood, and in the case of pronounced compression wood, generally 40–50% more. Obviously, the lignin content increases in the order mild, moderate, and severe compression woods. Concerning the first-formed earlywood, it should be noted that Côté et al. (1967a) found nearly the same high lignin content (37–39%) in this tissue as in the more typical compression wood zone formed later (Table 6.2). Additional comparisons of the lignin contents of normal and compression woods are discussed in connection with the summative analysis of these tissues.

5.2.2.2 Branch Wood and Knot Wood

Most branches of conifers contain compression wood in their lower (abaxial) portion and opposite and normal woods in the upper (adaxial). Stem wood sur-

5.2.2.2 Branch Wood and Knot Wood

rounding a knot often contains compression wood (Chap. 10.6.4). The major chemical difference between branch and knot woods is the often extremely high resin content of the knot, whereas branch wood generally contains no more resins than stem wood. In comparing the chemical composition of stem, branch, and knot woods it is best to base all data on extractive-free wood.

Branch wood and knot wood have to be treated separately from compression wood per se. Some investigators in studying these tissues have simply reported the chemical composition of the entire branch or knot, while others, and this includes the majority, have analyzed the upper and lower parts separately. Even in the latter case, there is no assurance that homogeneous tissues were examined. As shown in Chap. 10.4, normal wood is frequently present in the lower part of branches. As a result of changes in the orientation of a branch, its upper part or flanks can contain compression wood. Such exceptions should, however, not be allowed to obscure the fact that in the great majority of instances, the upper portion of branches and knots consists largely of opposite and normal woods and the lower portion of compression wood. In Chap. 21 it is shown that opposite and compression woods, while differing in anatomy, are chemically indistinguishable. With respect to its chemical composition, the upper part of a branch can therefore be regarded as consisting of normal wood.

Cleve von Euler (1923) observed that branches of *Pinus sylvestris* contained 6% more lignin than the stem and that the lower side had 2–3% more lignin than the upper. She expressed the opinion that this must be due to the fact that the branches are located closer to the site of formation of the lignin precursors. Her views were criticized by Klason (1923), who pointed out that branches have a high lignin content because they contain compression wood on their lower side. In the upper part of a fairly large branch of *Picea abies*, Klason (1923) found a lignin content of 28.1% compared to 37.1% in the lower portion. In a later paper, Klason (1932) discussed the lignin present in branches and in crooked stems, suggesting that it be referred to as a "reserve lignin".

Hägglund and Larsson (1937) determined the chemical composition of knot wood of *Picea abies* which contained 15% resin. The analytical data were compared with those obtained previously for normal and compression woods of the same species by Hägglund and Ljunggren (1923). All values are summarized in Table 5.12. Based on resin-free substance, the knot wood had a lignin content of

Table 5.12. Chemical composition of normal, compression, and knot woods of *Picea abies*. All values in per cent of resin-free wood. (Hägglund and Larsson 1937)

Component	Normal wood	Compression wood	Knot wood
Lignin	28.6	38.8	33.0
Cellulose and glucan	48.3	46.9	31.2
Acetyl	1.4	0.8	1.1
Galactan	0.7	Nil	10.1
Mannan	10.8	4.0	9.2
Pentosans	5.4	6.6	6.6

300   5 Chemical Properties of Compression Wood

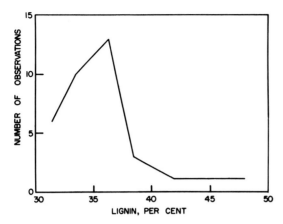

Fig. 5.2. Frequency distribution curve for the observed lignin content of knots in *Picea abies*. (Wegelius 1939)

Table 5.13. Chemical composition of normal sapwood, normal heartwood, and knot wood of *Pinus sylvestris*. All values in per cent of extractive free wood except those for extractives. (Jayme and Blischnok 1938)

Constituent	Normal wood		Knot wood
	Sapwood	Heartwood	
Extractives	3.47	5.88	29.5
Lignin	28.9	29.1	31.4
Galactan	0.30	0.50	1.63
Mannan	12.0	9.75	7.04
Pentosans	10.0	11.1	13.4

Table 5.14. Chemical composition of the upper and lower portions of a branch of *Pinus sylvestris* at different distances from the stem. All values in per cent. (Zherebov 1946)

Constituent	Zone	Distance from stem, m						Average
		0.08	1	2	3	4	5	
Cellulose	Upper	45.2	46.6	47.5	46.4	47.3	45.4	46.4
	Lower	40.0	41.1	42.3	40.7	40.9	39.7	40.8
Pentosans	Upper	16.1	16.4	16.1	16.9	15.8	17.1	16.4
	Lower	16.3	15.7	15.2	16.2	16.0	15.5	15.8
Lignin	Upper	31.9	31.4	31.3	30.8	30.1	30.1	30.9
	Lower	38.5	37.0	34.5	34.0	35.7	34.6	35.7
Ash	Upper	0.25	0.31	0.55	0.54	0.42	0.22	0.36
	Lower	0.24	0.28	0.55	0.46	0.30	0.29	0.35
Resin and fats	Upper	6.31	5.00	5.45	6.40	5.90	6.13	5.86
	Lower	6.32	4.90	5.34	6.50	5.72	8.57	6.23

5.2.2.2 Branch Wood and Knot Wood

33.0%, intermediate between those of normal and compression woods. Evidently, half of the branch consisted of compression wood.

A thorough investigation was undertaken by Jayme and Blischnok (1938) of the chemical composition of various wood tissues from *Pinus sylvestris*, namely normal sapwood and heartwood, knot wood, wood surrounding knots, and wood at the "root" of the knot. Some of the data are presented in Table 5.13. The knot wood had a surprisingly low lignin content.

In his extensive investigations on knots in *Picea abies*, Wegelius (1939, 1946) determined the lignin content of knot wood. The results were reported in the form of the frequency distribution curve shown in Fig. 5.2. Most of the values fell in the range of 33–38%. According to Brax (1936), the lowest lignin content was 33.2% and the highest 49.9%, a considerable spread. No compression wood containing 50% lignin is known, and it is inconceivable anyway that a knot could consist entirely of compression wood. The analysis for lignin was not carried out by the Klason method, and there is reason to suspect that polysaccharide material could have been included in the lignin.

Zherebov (1946) studied the chemical and mechanical properties of a 6-m-long branch of *Pinus sylvestris* that had grown horizontally. It was eccentric in cross section and contained compression wood on the under side. The results were summarized in numerous tables, one of which, dealing with five chemical constituents, is shown in Table 5.14. Throughout its length, the branch contained more lignin on the lower side than on the upper. In both zones, the lignin content decreased slowly and regularly from the trunk toward the branch tip. Zherebov's ideas concerning the physiological and mechanical roles of the cellulose, hemicelluloses, and lignin, which he assumed to enter into different complexes with one another, are now only of historical interest. In their studies on the properties of stem and branch woods of *Pinus densiflora*, referred to previously, Watanabe et al. (1962) found most branches to have a higher lignin content on their lower than on their upper side. Jarceva (1969) carried out a detailed chemical analysis of various parts of *Pinus cembra* var. *sibirica* trees. Stem wood contained 25–26% lignin and wood from branches and twigs usually 28–30%.

Hosia et al. (1971) reported some chemical data for stem and branch wood of *Picea abies* and *Pinus sylvestris*, listed in Table 5.15. Compression wood was present in the lower portion of the branches. For pine, the lignin content of the branch wood, even when based on extractive-free wood, is surprisingly low. Its carbohydrate composition, and especially its high galactose content, indicate that this wood must have contained as much or perhaps even more compression wood than the spruce.

Kowalski (1972) determined the physical, mechanical, and chemical properties of branch wood from *Pinus sylvestris*. These characteristics were found to vary considerably with the size of the branch and with the location of the sample analyzed. As could be expected, different results were obtained with wood from the upper and under side of a branch. Howard (1973) reported the chemical composition of stem and branches from *Pinus elliottii* var. *elliottii*. Her data, shown in Table 5.16, indicate a lower content of cellulose and higher contents of hemicelluloses and lignin for the branches. In connection with an in-

Table 5.15. Chemical composition of stem and branch woods of *Picea abies* and *Pinus sylvestris*. All values in per cent of unextracted wood or in relative per cent. (Hosia et al. 1971)

Component	Picea abies		Pinus sylvestris	
	Stem	Branch	Stem	Branch
Extractives	2.1	1.3	5.7	7.1
Ash	0.3	0.4	0.2	0.2
Lignin	25.9	32.1	26.7	28.5
Carbohydrates	71.7	66.2	67.4	62.4
Sugar residues in relative per cent:				
Galactose	2.2	8.5	3.4	9.8
Glucose	66.7	62.3	63.5	57.0
Mannose	16.8	13.0	16.8	12.2
Arabinose	3.3	3.6	4.4	6.0
Xylose	11.0	12.6	12.0	15.0

Table 5.16. Proportion and chemical composition of stem and branches from three *Pinus elliottii* var. *elliottii* trees. All values in per cent. (From data reported by Howard 1973)

Component	Stem	Branches
Relative amount of total	58.5	3.5
Extractives	8.7	14.3
Alpha-cellulose[a]	51.1	36.9
Hemicelluloses[a]	26.8	33.7
Lignin[a]	28.1	35.0

[a] Based on extractive-free wood.

Table 5.17. Chemical composition of conifer stem and branch woods. All values in per cent of unextracted wood. (Nacu and Constantinescu 1973)

Component	Stem wood	Branch wood
Extractives	2 – 3	4 – 6
Ash	0.4 – 0.5	0.7 – 0.8
Lignin	28 – 30	36 – 37
Pentosans	6.5	7.8
Material soluble in cold water	2	6 – 13
Material soluble in 1% sodium hydroxide	11 – 14	21 – 23

vestigation on the pulping properties of coniferous stem and branch woods, Nacu and Constantinescu (1973) determined the chemical composition of these tissues. Some of their numerous data are presented in Table 5.17. Evidently, the branch wood contained considerably more lignin than the stem wood.

Poller (1978) has reported the chemical composition of stem, branch, and root woods from a 25-year-old and a 100-year-old *Pinus sylvestris* trees. Wood from both small and large branches of the former tree had a notably higher density than the stem wood (Sect. 7.2.4). Strangely, the lignin content of the branch wood was practically the same as that of the stem wood (27.8%). Branch wood from the 100-year-old tree, on the other hand, was no denser than the stem wood, but in this case contained on the average 12% more lignin than the stem wood. These results cannot be readily rationalized. Although not mentioned by the investigator, it is likely that the branches must have contained at least some compression wood. It would therefore have been expected that all branch woods should have had both a higher density and a higher lignin content than the stem woods.

Since branch and knot woods usually contain compression wood it is not surprising that they have almost invariably been found to be more highly lignified than stem wood. The relative amount of compression wood in a branch varies considerably, and it is only to be expected that the proportion of lignin will vary accordingly. The same applies, albeit to a lesser extent, when the lower part of a branch is analyzed separately.

Branch wood, as is discussed in Chap. 19.9, has attracted considerable attention in later years as a potential raw material for pulp and paper. Their relatively high lignin content is only one of several drawbacks of branches as a source of pulpwood.

5.2.3 Cellulose

5.2.3.1 Compression Wood

Early estimates of the cellulose content of wood suffered from a lack of a clear definition of this polysaccharide and also from a want of methods for separating cellulose and hemicelluloses. According to the now accepted concept, cellulose consists exclusively of $(1 \rightarrow 4)$-linked β-D-glucopyranose residues, the number of which is about 10 000 in wood (Goring and Timell 1962). In the well-known but now outdated Cross and Bevan and Norman and Jenkins methods for estimating cellulose (Browning 1967), the plant tissue is treated with chlorine or hypochlorite, respectively, and subsequently extracted several times. The residual *Cross and Bevan cellulose* and *Norman and Jenkins cellulose* are both preparations that may contain as much as 20–30% hemicelluloses. In isolation of *alpha-cellulose*, delignified wood is extracted with aqueous sodium hydroxide. This preparation also contains varying amounts of hemicelluloses, albeit less than the Cross and Bevan cellulose. When either of these methods for isolating cellulose is applied to normal and compression woods, a strict comparison is probably not possible for, as will be shown later, normal and compression woods do not have the same hemicellulose composition, and the proportion of hemicelluloses that remain associated with a Cross and Bevan or an alpha-cellulose is not necessarily the same for the two types of wood. The higher lignin content of the compression wood usually makes it necessary to resort to a more drastic delignification, which results in removal of more hemicelluloses. In both delignification and extraction, the galactan in compression wood is removed more readily than are the xylan and galactoglucomannan. The net result is that Cross and Bevan and alpha-celluloses from compression wood can be expected to contain a higher proportion of cellulose than do similar preparations from normal wood (Sect. 5.5.3 and Chap. 6.3.3).

The early, extensive investigations of Hägglund and his co-workers (Hägglund 1951) showed that normal wood of conifers on the average contains $42 \pm 2\%$ cellulose. Johnsen and Hovey (1918) found compression wood from *Abies balsamea* to have only 39.1% Cross and Bevan cellulose compared to 50.4–52.9% for normal wood. Klason does not appear to have reported the cellulose content of compression wood. He realized that it must be low, however, for in a discussion of how cellulose ought to be defined he mentions (Klason 1924) that the presence of compression wood considerably reduces the yield of cellulose on pulping. Dadswell and Hawley (1929) applied the Cross and Bevan method for determining cellulose and reported for compression wood from *Picea sitchensis* and *Sequoia sempervirens* cellulose contents of 53.7% and 46.6%, respectively, compared to 60.6% and 55.0% for the normal woods (Table 5.1). In the earlywood zone of redwood, compression wood contained 49.2% cellulose but in the latewood area only 45.5%. Hägglund and Ljunggren (1933) used a more modern approach in their attempt to determine the cellulose content of normal and compression woods from *Picea abies*, obtaining values of 41.5% and 27.3%, respectively (Table 5.2).

The alpha-cellulose isolated by Pillow and Bray (1935) and by Curran (1936, 1937) from *Pinus taeda* evidently contained only minor quantities of hemicelluloses, as can be seen from Tables 5.3 and 5.4. For compression wood, a cellulose content of 34.6% was noted. Hata (1951) obtained 42.0% alpha-cellulose for normal wood and 32.3% for compression wood of *Pinus densiflora*, values probably close to the true cellulose content of the two woods (Table 5.5). For side wood and compression wood of *Pinus radiata*, Schwerin (1958) obtained 51.0% and 45.1% alpha-cellulose, respectively. When corrected for admixed galactan, these values become 49.2% and 38.4%. Both are still too high and for the obvious reason that if galactan was present, xylan and galactoglucomannan must have remained to an even greater extent, especially in the alpha-cellulose from the side wood. The values for the cellulose content of normal and compression woods of *Abies alba* reported by Jurášek (1964a), namely 49.3% and 41.5% are probably also spuriously high.

In a study of variations in wood properties of *Pinus taeda*, Zobel et al. (1960) arrived at the conclusion that there is a highly significant negative correlation between compression wood content and yield of cellulose. The lower hexosan content of compression wood from *Larix decidua* in comparison with normal wood, which was noted by Philipp et al. (1970), is, of course, also a result of the lower cellulose content of compression wood. Chu (1972) determined the holocellulose and alpha-cellulose contents of normal and compression woods of *Pinus palustris*. The values for normal wood were 71.6% and 47.8%, respectively, and 62.3% and 34.9% for compression wood. The alpha-celluloses probably still contained hemicelluloses, for both values are noticeably higher than would have been expected.

In spite of the inadequacy of most of the early methods used for determining the cellulose content of compression wood, they nevertheless suggest that this tissue contains considerably less cellulose than normal wood. Whether this is due to the presence of an increased amount of lignin or represents an actual reduction in the quantity of cellulose present per tracheid cannot be decided on the basis of the information now reviewed. The answer to this question had to await the advent of more refined methods of wood analysis.

5.2.3.2 Branch Wood and Knot Wood

Branches of *Picea abies*, according to Kinnman (1923), have a low cellulose content, especially on the lower side. Knot wood of *Picea abies* was found by Hägglund and Larsson (1937) to contain only 31.2% cellulose and "glucan" compared to 48.3% for normal wood (Table 5.12). Jayme and Blischnok (1938) found only slightly less cellulose in knot wood of *Pinus sylvestris* than in normal wood. Their values, however, cannot possibly represent pure cellulose. According to Wegelius (1939), the upper portion of a knot in *Picea abies* had a normal cellulose content. For the lower part he found cellulose values varying from 38% to 48% with two maxima at 29% and 40%.

The branch of *Pinus sylvestris* analyzed by Zherebov (1946) contained 46.4% cellulose in the upper and 40.8% in the lower part. Watanabe et al. (1962) have reported extensive analytical data for the content of cellulose, hemicelluloses, and lignin in stem wood and in the upper and lower portions of branch wood of *Pinus densiflora*. Stem wood was found to contain more cellulose than branch wood. The upper portion of the branches had a consistently higher cellulose content than the lower. Jarceva (1969) found that branches and especially twigs of *Pinus cembra* var. *sibirica* contained less cellulose than did the stem. Stem wood of *Pinus elliottii* var. *elliottii* has been reported to contain 51.1% alpha-cellulose compared to only 36.9% in branch wood (Howard 1973). In his analysis of wood from *Pinus sylvestris* trees, Poller (1978) noted cellulose contents ranging from 34.4% to 44.3% in wood from small and large branches compared to 48.5% for stem wood.

The relatively low cellulose content of branches is, of course, a direct result of their often quite high proportion of compression wood. Branch wood containing 50% by weight of compression wood could be expected to have a cellulose content of only 36% compared to 42% for normal stem wood. This situation renders the utilization of branches for pulping unattractive (Chap. 19.9.2).

5.2.4 Galactose Residues

5.2.4.1 Compression Wood

Besides its high lignin and low cellulose contents, compression wood is characterized chemically by its high proportion of galactose residues, now known to be derived from an acidic galactan. The abnormally high galactose content of compression wood escaped the attention of Hägglund and Ljunggren (1933), who reported 0.7% galactan in normal but none in compression wood of *Picea abies*. The value for normal wood is about three times lower than the now-accepted value of 2%, and it is obvious that in the case of compression wood, something went wrong in the attempt to oxidize galactose to insoluble mucic acid.

Four years later, Hägglund and Larsson (1937) found that knot wood of the same species contained as much as 10.1% galactan (Table 5.12) and they expressed their surprise at this difference between knot wood and branch wood. It was left to Stockman and Hägglund (1948), who used the same method for determination of galactose residues, to show for the first time that compression wood is unusually rich in galactan. Their value for the galactan content of pronounced compression wood of *Picea abies* was 9.6% compared to only 1.9% for normal wood.

Ten years later, Schwerin (1958), still using the mucic acid method, obtained a value of 13.2% for the galactan content of compression wood from *Pinus radiata* (Table 5.6). Her claim that much of this galactan should be closely associated with cellulose in the cell wall has not been confirmed by later investigators (Schreuder et al. 1966, Côté et al. 1968).

5.2.4.2 Branch Wood and Knot Wood

The investigation of Hägglund and Larsson (1937) was referred to above. Jayme and Blischnok (1938) found 1.63% galactan in *Pinus sylvestris* knot wood and only 0.3–0.5% in normal wood. Both sets of values are low. For upper and lower portions of branches from the same species Zherebov (1946) found a galactose content of 3%. According to Hans (1970), wood of young branches of *Thuja occidentalis* contains two to three times as many galactose residues as does stem wood. Branch wood of *Abies dahlemensis* gave similar results. The same investigator (Hans 1971) has also reported values for the relative sugar composition, including galactose, of young branches of *Taxodium distichum*. Much of the galactan was lost on delignification of the wood, a well-known phenomenon (Sect. 5.5.3).

5.2.5 Mannose Residues

5.2.5.1 Compression Wood

Hägglund and Ljunggren (1933) (Table 5.2) determined the mannose content of normal and compression woods of *Picea abies* by precipitation from the wood

5.2.6.1 Compression Wood

hydrolyzate of the insoluble mannose phenylhydrazone, obtaining values of 10.6% and 4.0%, respectively, based on the original wood. The same method was later applied by Stockman and Hägglund (1948) to the same species. Compression wood was now found to contain 4.1% mannan compared to 7.5% for normal wood. The considerably lower content of mannose residues in compression wood is clearly indicated in both cases. In a later study of Kibblewhite (1973), compression wood of *Pinus radiata* was found to contain one third less mannose than corresponding normal wood.

5.2.5.2 Branch Wood and Knot Wood

Hägglund and Larson (1937) found knot wood of *Picea abies* to contain 9.2% mannan compared to 10.8% for normal and 4.0% for compression wood (Table 5.12). In view of its high galactan content (10.1%), it is surprising that knot wood had more than twice as much mannan as compression wood and only slightly less than did normal wood. The knot wood of *Pinus sylvestris* examined by Jayme and Blischnok (1938) had a mannan content of 7.0% (Table 5.13), while normal sapwood contained 12.0%. Zherebov (1946) noted a 25% higher mannan content in the lower than in the upper portion of a branch of *Pinus sylvestris*. This observation is contrary to all other evidence and is open to considerable doubt. Jarceva (1969) found stem wood of *Pinus cembra* var. *sibirica* to contain approximately twice as many mannose residues as did branches and twigs.

5.2.6 Pentose Residues (Pentosans)

5.2.6.1 Compression Wood

The time-honored method for determining pentose residues in wood is to convert them with strong acid to furfural. Since both softwoods and hardwoods contain considerably more xylose than arabinose residues, the pentosan analysis is an approximate measure of the amount of xylan in these woods.

As can be seen from Table 5.1, Dadswell and Hawley (1929) found the same amounts of pentosan in normal and compression woods of sapwood from *Sequoia sempervirens*, slightly less in compression wood from the heartwood region, and somewhat more in compression wood of *Picea sitchensis*. Within the compression wood, latewood contained more pentosan than earlywood.

For two trees of *Picea abies*, Hägglund and Ljunggren (1933) noted a slightly higher pentosan content in compression wood than in opposite and normal woods, whereas the data reported by Stockman and Hägglund (1948) would seem to indicate equal amounts of pentosans in the two woods. Similar pentosan contents were also observed for normal and compression woods of *Pinus taeda* by Pillow and Bray (1935) and by Curran (1936, 1937), the values obtained all being within the narrow range of 12.2–12.6% (Tables 5.3 and 5.4). Normal earlywood, however, contained as much as 17.9% pentosan (Curran 1936, 1937), showing the variations that can occur within the same growth ring.

In compression wood of *Pinus densiflora,* Hata (1951) found 13.9% pentosans, compared to only 11.3% in normal wood (Table 5.5). According to Watson and Dadswell (1957), however, compression wood of *Pinus radiata* contains less pentosans than side wood, opposite wood, and normal wood (Table 5.6). Similar tissues from *Picea mariana* have been reported to have the same pentosan contents of 7.2–7.6% (Table 5.8) (Ladell et al. 1968). The pentosan data for *Larix decidua* in Table 5.10 obtained by Philipp et al. (1970) are interesting for they indicate considerable differences between earlywood and latewood. In normal wood, the earlywood contained more pentosan than the latewood, as is usually the case. In compression wood, the latewood had a somewhat higher pentosan content. Three of the conifer species analyzed by Giordano (1971) contained slightly more pentosans in normal wood than in compression wood, while the situation was the reverse in the fourth species (Table 5.11).

5.2.6.2 Branch Wood and Knot Wood

Both compression wood and knot wood in *Picea abies* were found by Hägglund and Larsson (1937) to contain 6.6% pentosans, the value for normal wood being 5.4% (Table 5.12). Jayme and Blischnok (1938) also observed a higher pentosan content in knot wood of *Pinus sylvestris* than in normal wood, as shown in Table 5.13. Wegelius (1939) determined the pentosan content of a large number of knots. Most of his data are within the wide range of 6–12%, the average value being about 8.5%.

As can be seen from Table 5.14, the lower portion of the branch of *Pinus sylvestris* examined by Zherebov (1946) contained slightly less pentosan than the upper. There was no discernible trend in the variation in pentosan content along the length of the branch. The numerous data reported by Jarceva (1969) for the chemical composition of different parts of *Pinus cembra* var. *sibirica* indicate a higher xylan content for wood from branches and twigs than from stem wood. Nacu and Constantinescu (1973) found conifer branch wood to contain somewhat more pentosans than did the stem wood (Table 5.17).

5.2.7 Acetyl Groups

Among early investigators, only Hägglund and his co-workers seem to have determined the acetyl content of compression wood. Such tissue of *Picea abies* contained 0.8% acetyl groups, normal wood 1.4% (Hägglund and Ljunggren 1933), and knot wood 1.1% (Hägglund and Larsson 1937). These values agree with those obtained by later investigators (Côté et al. 1966b, Timell 1957c, 1981, 1982).

5.2.8 Extractives

5.2.8.1 Compression Wood

The so-called "extractives" or "extraneous substances" in wood are an extremely heterogeneous group of substances, most of them of low molecular weight and extractable from wood with various organic solvents. In most cases, these compounds are extracellular, but sometimes they occur within the cell wall, for example in *Sequoia sempervirens* (Tarkow and Krueger 1961, Kuo and Arganbright 1980a, b). Their amount varies considerably, ranging from only traces to 30–40% by weight of the wood. Important among the extractives are terpenes, resin acids, starch, fats, waxes, tannins, and a large number of polyphenols, such as the lignans, and tropolones. The arabinogalactan present in larch heartwood could also be listed among the extractives, although it is more convenient to regard it as a hemicellulose.

Mork (1928) mentions that it had earlier been common to regard compression wood as being rich in resin (Thunell 1945). According to his own results, the opposite is true. Extraction of normal woods of *Picea abies* with ethyl ether yielded 1.44% of material, whereas compression wood gave only 1.06%. This agrees with the earlier results of Johnsen and Hovey (1918), who found 2.05% and 2.85% alcohol-solubles in normal wood of *Abies balsamea* but only 1.38% in compression wood. In the same year, Ulfsparre (1928) reported that compression wood of *Picea abies* contained 0.47% of substances soluble in ethyl ether and 0.36% soluble in benzene. Corresponding figures for normal wood were 0.82% and 0.81%, respectively. Dadswell and Hawley (1929), on the other hand, found 3.45% alcohol-solubles in normal wood and 3.75% in compression of *Sequoia sempervirens*. According to Hägglund and Ljunggren (1933), extraction of *Picea abies* compression wood with acetone afforded 2.0% of material, while normal wood yielded 2.1%. Sherrard and Kurth (1933) noted that compression wood of *Sequoia sempervirens* had an abnormally low content of extractives. It should perhaps be noted at this point that ethanol, ethyl ether, acetone, and benzene do not remove the same extractives or the same amounts of extractives from wood.

Extraction of normal earlywood and latewood and compression wood of *Pinus taeda* with ethanol-benzene gave 4.2%, 2.5%, and 2.7% of substances, respectively (Tables 5.3 and 5.4) (Pillow and Bray 1935, Curran 1936, 1937). Material soluble in ethyl ether amounted to 3.6%, 1.5%, and 1.3%, respectively. Using the same solvent mixture, Hata (1951) (Table 5.5) found 3.1–3.2% extractives in both normal and compression woods of *Pinus densiflora*. The values reported by Watson and Dadswell (1957) for ethanol-benzene solubles from *Pinus radiata* vary considerably, being in one case zero and in another 0.6% for compression wood, 0.5% for side wood, and 0.1% for normal wood.

In a study of the specific gravity and tracheid anatomy of compression wood from *Pinus taeda*, Shelbourne and Ritchie (1968) determined the amount of extractives (ethanol-benzene) in normal wood, and in moderate and moderate to severe compression woods. Values from 1.4% to 3.8% were obtained. No trends could be observed, except that earlywood contained more extractives than late-

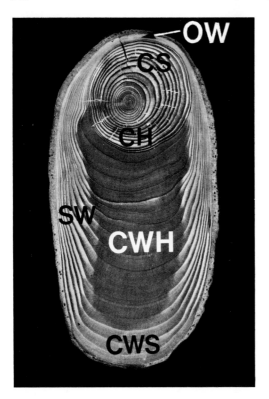

Fig. 5.3. Cross section of a horizontal stem of *Larix laricina* (Fig. 3.3.6), showing the location of normal sapwood (*CS*) and heartwood (*CH*) in the original core, side wood (*SW*), opposite wood (*OW*), and compression wood in the heartwood (*CWH*) and sapwood (*CWS*) zones

wood. The investigators concluded that there is no reason to assume that there should exist any correlation between formation of compression wood and extractives.

Côté et al. (1966 a), in a study of the distribution of arabinogalactan in larch wood, determined the amounts of ethanol-benzene extractives and arabinogalactan in the various zones of a stem of *Larix laricina*. The trunk was growing along the ground in a horizontal position and exhibited extreme eccentricity. Six zones were delineated, as shown in Fig. 5.3, five of which were analyzed, giving the results shown in Table 5.18. In the sapwood zones, compression wood had a lower content of extractives than normal wood and side wood, but there was no difference in the heartwood. Arabinogalactan, which occurs only in the heartwood of larches, was present in larger amounts in normal than in compression wood, an observation confirmed in a later investigation (Côté and Timell 1967). This is in agreement with an earlier observation by Chochieva et al. (1958). All four conifer species analyzed by Giordano (1971) contained notably less alcohol-benzene extractives in their normal wood than in their compression wood (Table 5.11).

Sakakibara and his co-workers determined the content of extractives in normal and compression woods of two conifer species. Wood of *Abies sachalinensis*

5.2.8.1 Compression Wood

Table 5.18. Amounts of extractives and arabinogalactan present in different zones of a stem of *Larix laricina*. All values in per cent of unextracted wood. (Côté et al. 1966a)

Zone	Extractives	Arabino-galactan
Normal core wood		
Sapwood	1.66	0.59
Heartwood	1.40	10.5
Sidewood, Sapwood	2.18	0.15
Compression wood		
Sapwood	1.18	0.32
Heartwood	1.43	7.03

Table 5.19. Extractives obtained on successive extraction of normal and compression woods of *Abies sachalinensis* (Morohoshi and Sakakibara 1971a) and *Larix leptolepis* (Yasuda and Sakakibara 1975a, Yasuda et al. 1975)

Extract	Abies sachalinensis		Larix leptolepis	
	Normal wood	Compression wood	Normal wood	Compression wood
Ethyl ether	2.99	4.01	1.19	0.92
Ethanol-benzene	0.98	0.98	2.54	3.82
95% Ethanol	0.52	0.59	0.29	1.10
Acetone-water	0.14	0.45		
Total	4.63	6.03	4.02	5.84

was successively extracted with ethyl ether, ethanol-benzene, 95% ethanol, and acetone-water, giving the values summarized in Table 5.19 (Morohoshi and Sakakibara 1971a). Compression wood contained more extractives than did normal wood, particularly of the ether-soluble type. A similar extraction of *Larix leptolepis* wood gave the results shown in the same table (Yasuda and Sakakibara 1975a, Yasuda et al. 1975). The extractives content of compression wood was almost 50% higher than that of normal wood, but in this case it was material soluble in ethanol-benzene and ethanol that was more abundant in compression wood. The nature of some of the extractives from larch wood is discussed in Sect. 5.10. Chu (1972) found compression wood of *Pinus palustris* to contain 2.70–3.05% of extractives compared to only 1.75–2.65% for normal wood. Rickey et al. (1974) reported a 1.3% content of petroleum ether solubles, identified as resin, in compression wood of *Picea sitchensis*. Corresponding normal wood contained only 0.3%.

Values for alcohol-benzene soluble extractives in different zones of opposite and compression woods in *Pinus densiflora* were found to vary considerably by

Table 5.20. Chemical composition of opposite and compression woods of *Pinus densiflora*. All values in per cent of unextracted wood or relative per cent. (Takashima et al. 1974)

Component	Opposite wood	Compression wood
Ash	0.3	0.4
Lignin	28.8	37.7
Cellulose	51.9	34.3
Galactan	0.2	2.4
Pentosans	11.6	13.6
Solubility in:		
Cold water	0.3	0.4
Hot water	1.2	1.4
1% Sodium hydroxide	3.3	2.4
Alcohol-benzene	16.0	14.2
Sugar residues in relative per cent:		
Galactose	2.4	9.6
Glucose	64.8	58.6
Mannose	21.9	14.5
Arabinose	2.9	2.9
Xylose	8.0	14.4

Takashima et al. (1974), and the authors concluded that there was no distinct difference in this respect between the two tissues. Wood within the first 15 growth rings from the pith contained much more extractives than the more mature wood, namely 14.6% in the zone of opposite wood but only 8.9% in the zone of compression wood. Mature xylem of both types of wood had only 3.5 – 3.7% extractives (Table 5.20).

5.2.8.2 Branch Wood and Knot Wood

While conifer branches usually contain only a few percent of resin, knots can have a resin content as high as 30 – 40% (Hägglund and Larsson 1937, Köster 1934, Jayme and Blischnok 1938, Harris 1961, 1963, Kininmoth 1961, Boutelje 1966, Lindgren and Norin 1969, Dietrich 1973, Anderegg and Rowe 1974) (Chap. 10.6.3). As can be seen from Table 5.14, Zherebov (1946) noted no difference in the content of resins and fats between the upper and lower portions of his *Pinus sylvestris* branch. Jarceva (1969) found branches and twigs of *Pinus cembra* var. *sibirica* to have a higher content of ether extractives than did the stem. According to Hosia et al. (1971), branch wood of *Pinus sylvestris* contained more extractives than the stem wood (Table 5.15). Branches of *Pinus elliottii* var. *elliottii* were found by Howard (1973) to contain twice as much extractives as the stem (Table 5.16). Taylor (1979), on the other hand, claims that in *Pinus taeda* branches have a lower extractives content than the stem, namely on the average 4%. The lower part of the branches, where compression wood was abundant, had the same extractives content as the upper. Nacu and Constantinescu (1973) observed twice as much extractives in branch wood than in stem wood of various conifer species. Eskilsson and Hartler (1973) found that branch wood of *Picea abies* also had an extractives content twice that of stem wood, whereas in *Pinus sylvestris* the two tissues did not differ in this respect.

In a study of the distribution of lignans in *Picea abies* trees, Ekman (1979, 1980) found that the stem heartwood below a height of 1.5 m contained up to 0.5% lignans but only 0.1% above this level. The lignan content of stem crooks with eccentric radial growth and pronounced compression wood was 0.3% above 1.5 m. Branches and roots had a high lignan content, namely 4−6% and 2−3%, respectively. According to Ekman, tissues formed under external stress, such as wood in stem crooks, branches, and roots, seem to contain more lignans than the wood of straight stems.

Knots of *Araucaria angustifolia* contain 30% of a resin that is recovered commercially in Brazil. About 90% of this resin consists of various lignans, including pinoresinol and its dimethyl ether (Dryselius and Lindberg 1956) as well as its monomethyl ether, hinokiresinol, isolariciresinol, and secoisolariciresinol (Anderegg and Rowe 1974).

5.2.8.3 Conclusions

Judging from the information available, it is evident that stem compression wood sometimes contains more and sometimes less extractives than corresponding normal wood. Branches have almost consistently been reported to contain more extractives than stem wood. The high resin content of knots is typical also of conifer species that do not possess resin canals (Chap. 10.6.3). Most of this resin is formed by senescent branches in processes that are probably similar to those causing transformation of sapwood into heartwood.

5.2.9 Inorganic Constituents

5.2.9.1 Compression Wood

The overall inorganic content of conifer wood is low, and even moderate variations cannot be attributed much significance. For normal and compression woods of *Picea abies*, Ulfsparre (1928) found ash contents of 0.35% and 0.33%, respectively, corresponding values obtained by Mork (1928) being 0.50% and 0.61%. Compression wood of *Picea sitchensis* has been reported to contain 0.60% ash and normal wood 0.20% (Dadswell and Hawley 1929).

Onaka (1949) determined the ash content of compression wood from *Chamaecyparis obtusa*, *Pinus densiflora*, and *Ginkgo biloba*, obtaining values of 0.45%, 0.34%, and 0.38%, respectively. Corresponding data for normal wood were 0.32%, 0.34%, and 0.38%. Compression wood of *Pinus densiflora* according to Hata (1951) contains 0.28% ash compared to 0.23% for normal wood (Table

5.5). Kuroyanagi (1953) noted no difference between normal and compression woods with respect to their content of inorganic substances.

5.2.9.2 Branch Wood

Branch wood generally has a higher ash content than stem wood, and according to Wegelius (1939), the more the higher its content of compression wood. Hägglund (1925) found branch wood of *Picea abies* to contain 0.37% ash, while stem wood contained 0.21%. The upper and lower portions of the *Pinus sylvestris* branch examined by Zherebov (1946) had similar ash contents (Table 5.14), and the same applies to the stem and branch woods studied by Hosia et al. (1971) (Table 5.15). Jarceva (1969), on the other hand, found that compared to stem wood, branches from *Pinus cembra* var. *sibirica* contained more ash, and twigs had twice as much ash as the branches. Similarly, Nacu and Constantinescu (1973) found that branch wood of conifers had a considerably higher ash content than the stem wood (Table 5.17). It is known that young wood generally has a higher ash content than more mature tissue. Although all published data do not agree with this conclusion, it would appear that compression wood contains more inorganic constituents than does normal wood.

5.2.10 Solubility Properties

Several investigators have determined the solubility of normal and compression woods in cold and hot water and 1% sodium hydroxide. Such data are summarized in Tables 5.1, 5.3–5.5, and 5.17. In the majority of these cases, compression wood was more soluble in both water and dilute alkali than was normal wood. The results of Hata (1951) with *Pinus densiflora* (Table 5.5) are typical in this respect. The higher solubility of compression wood in water and dilute alkali can at least partly be attributed to the presence in this wood of an accessible, water-soluble galactan (Sect. 5.5.3). Cold water alone removes very little of this hemicellulose (Schreuder et al. 1966), and the higher yield noted on extraction with hot water is probably a result of acid-catalyzed hydrolysis of the polysaccharides in wood. Alkali swells wood, thus rendering it more accessible, and is also a good solvent for many hemicelluloses. As a result, about ten times more material is dissolved by alkali than by cold water.

The data of Nacu and Constantinescu (1973) in Table 5.17 indicate an unusually high solubility of branch wood in cold water and in dilute, aqueous alkali. Compared to stem wood, branch wood contained three to six times more material soluble in cold water and twice as much material soluble in alkali. Clearly, this difference cannot be entirely attributed to the larger amounts of compression wood usually present in branches.

5.2.11.1 Compression Wood

Table 5.21. Relative percentage of neutral sugar residues in normal and compression woods of six conifer species. (Meier 1965)

Species	Galactose	Glucose	Mannose	Arabinose	Xylose
TAXODIACEAE					
Sequoiadendron giganteum					
Normal wood	4.1	64.2	8.0	4.3	19.5
Compression wood	19.7	55.2	8.8	3.3	13.0
Taxodium distichum					
Normal wood	6.0	65.8	13.0	2.2	13.0
Compression wood	22.9	58.0	9.8	1.4	7.8
CUPRESSACEAE					
Chamaecyparis lawsoniana					
Normal wood	7.3	64.5	11.5	4.2	12.6
Compression wood	26.2	51.9	8.5	2.9	10.6
PINACEAE					
Cedrus atlantica					
Normal wood	4.8	67.6	12.4	1.7	13.5
Compression wood	18.7	60.0	10.0	1.2	10.0
Picea abies					
Normal wood	2.1	70.7	17.2	1.3	8.7
Compression wood	19.6	55.6	11.5	2.6	10.7
Pinus sylvestris					
Normal wood	5.4	63.8	18.8	2.7	9.3
Compression wood	20.6	53.9	11.7	2.9	10.9

5.2.11 Relative Sugar Composition

5.2.11.1 Compression Wood

With the advent of paper chromatography for separation of sugars, around 1950, it became possible to determine the relative amounts of sugar residues in plant tissues with an ease and accuracy that had not been attainable before. Since then, separation and quantitative determination of sugars by gas-liquid chromatography has become an even more powerful analytical tool.

The first chromatographic determination of the relative sugar composition of compression wood was carried out by Bouveng and Meier (1959) in connection with a study of a galactan in compression wood of *Picea abies* (Sect. 5.5.3). Further analyses were later reported by Meier (1965), who determined the proportions of major sugar residues in normal and compression woods of six conifer species. The data, which are presented in Table 5.21, were obtained with trees from three different families of the Coniferales. Although the proportion of glucose residues is not equivalent to the content of cellulose, the larger part of these residues are derived from this polysaccharide. The results show, as pointed out by Meier (1965), that compression wood always contains less cellulose than normal wood. Among the species investigated, the difference was largest for *Picea abies* and smallest for *Taxodium distichum*. Obviously, this could be due to variations in the severity of the compression woods analyzed.

Compression wood of all six species had a much higher proportion of galactose residues than did normal wood, with the difference again largest for *Picea abies*. In *Chamaecyparis lawsoniana* more than a quarter of the total number of sugar residues consisted of galactose. Five of the species contained fewer mannose residues in compression wood than in normal wood, but the difference was pronounced only for *Picea abies* and *Pinus sylvestris*. Compression wood of *Sequoiadendron giganteum* had a higher mannose content than normal wood, the only instance where this has been reported. This situation is due to the abnormally low mannose content of the normal wood, for the amount present in the compression wood is comparable with that found in the other compression woods.

Normal wood of the giant sequoia is also exceptional in its exceedingly high proportion of xylose residues, approached among the conifers, which generally contain twice as much mannose and xylose, only by *Libocedrus decurrens* (Thompson and Kaustinen 1964, Thompson et al. 1966). Because of this situation, compression wood of this species contained much less xylose than did its normal wood, notwithstanding the fact that it had the highest xylose content of the six compression woods. It should perhaps be noted in this connection that the values reported by Meier (1965) are only relative and were not computed on the basis of wood weight, as had been done, for example, in the analysis for pentosan. When computed on an absolute basis, the number of xylose residues is larger for normal wood in all six species.

It is now known that the arabino-4-*O*-methylglucurono-xylan in compression wood contains only half the number of arabinose side chains typical of this polysaccharide when it occurs in normal wood (Hoffmann and Timell 1972b) (Sect. 5.7). For this reason, the arabinose content of compression wood is usually lower than that of normal wood. On an absolute basis, this was the case with five of the species examined by Meier (1965), but for *Picea abies*, compression wood was reported to contain 70% more arabinose residues than normal wood, an unexpected finding, difficult to explain. Relative carbohydrate data have also been reported by Young et al. (1970) for normal and compression woods of *Picea sitchensis*. As can be seen from Table 5.9, these values agree with those obtained by Meier (1965). In this case, the absolute amount of xylose residues would be somewhat higher in the compression wood.

5.2.11.2 Branch Wood

The relative carbohydrate composition of branch wood was first determined by Hosia et al. (1971), who examined stem wood and branch wood of *Picea abies* and *Pinus sylvestris*. Their data, shown in Table 5.15, indicate for the branches a sugar composition typical of compression wood. The difference in chemical composition between stem wood and branch wood is, of course, less pronounced than that between normal wood and pure compression wood. Judging from the galactose and lignin contents, about half of the branch wood must have consisted of compression wood.

5.2.12 Absolute, Summative Composition

5.2.12.1 *Compression Wood*

The first summative analysis of compression wood was reported in 1966 by Côté et al. (1966 b). A summative analysis of wood entails a quantitative determination of all cell wall constituents in wood: cellulose, hemicelluloses, lignin, and inorganic salts. Usually the results are based on extractive-free wood. Four constituents, namely lignin, ash, acetyl groups, and uronic acid residues are directly determined in weight per cent of the wood. The major polysaccharide-building units, namely galactose, glucose, mannose, arabinose, and xylose are obtained as free, reducing sugars after hydrolysis of the wood, and their proportions are estimated by chromatographic methods. Since wood largely consists of these nine constituents, and the weight percentages of the first four are known, the five sugar residues can also be expressed in weight per cent. With a knowledge of the composition of each polysaccharide, the actual composition of the wood can be computed in terms of individual, polymeric constituents. It should be noted that a few, minor wood sugars are not included in this method of analysis, namely the rhamnose and fucose residues which originate from pectin in the primary cell wall. No attempt is made in this method of analysis to determine the small amounts of pectic acid, arabinogalactan, and xyloglucan derived from the primary cell wall.

Côté et al. (1966 b) subjected normal and compression woods from *Ginkgo biloba* and 14 conifer species to a summative analysis, using methods previously adopted by Timell (1957 b). Additional species were later analyzed by Timell (1981, 1982). The specimens were collected from horizontal or strongly leaning or bent stems, but branch wood was used for *Pinus palustris, Pseudotsuga menziesii,* and *Taxodium distichum.* The total number of species studied included all conifers native to north-eastern United States and Sweden. All trees exhibited eccentric growth and had developed, continuous, severe compression wood on the wider, lower side. Normal wood was collected either from the side wood or from the outer rings of the concentric inner core which had formed while the tree was still in an upright, straight position. An example is shown in Fig. 5.4. The results, based on extractive-free wood are summarized in Table 5.22 except for the data obtained with *Ginkgo biloba,* which are listed in Tables 8.1 and 8.2 and discussed in Sect. 8.3.1.

Inspection of the Table 5.22 reveals that, in agreement with all previous investigations, the lignin content of compression wood was without exception much higher than that of corresponding normal wood. The highest lignin content, 40.9%, was exhibited by *Tsuga canadensis,* but actually many species had lignin contents within the narrow range of 39–40% and the others within 34–38%. Pronounced compression wood probably contains $40 \pm 1\%$ lignin.

Table 5.22. Summative chemical composition of normal and compression woods of 27 gymnosperm species. All values in per cent of extractive-free sapwood. The ash content of all specimens was 0.4±0.1%

Species	Lignin	Acetyl	Uronic anhydride	Residues of:				
				Galactose	Glucose	Mannose	Arabinose	Xylose
CONIFERALES								
CUPRESSACEAE								
Chamaecyparis nootkatensis								
Normal wood	30.7	1.3	4.8	1.8	42.3	11.0	1.1	6.6
Compression wood	37.7	0.8	4.4	10.6	31.4	5.6	0.9	8.2
Chamaecyparis pisifera								
Normal wood	30.1	1.1	4.2	1.7	45.7	7.8	1.4	7.6
Compression wood	40.2	0.6	3.8	12.9	29.6	4.5	0.8	7.2
Juniperus communis								
Normal wood	31.4	2.2	5.4	3.0	40.7	9.1	1.0	6.9
Compression wood	37.2	1.4	5.4	8.2	34.3	6.3	0.8	6.1
Juniperus virginiana								
Normal wood	30.7	1.5	5.0	2.1	41.0	9.2	1.3	8.8
Compression wood	39.3	1.2	4.9	7.7	31.9	5.5	1.0	8.1
Thuja occidentalis								
Normal wood	33.3	0.9	5.8	1.5	45.2	7.4	1.7	3.8
Compression wood	39.4	0.5	5.3	7.2	33.5	5.1	1.6	6.9
PINACEAE								
Abies balsamea								
Normal wood	30.0	1.3	3.5	1.0	45.8	10.8	2.3	4.9
Compression wood	38.5	1.1	4.0	9.1	35.1	5.1	1.0	5.7
Cedrus libani								
Normal wood	32.5	1.0	6.1	3.3	36.4	7.1	2.3	10.9
Compression wood	37.3	0.6	5.6	11.2	29.6	4.7	1.6	9.0
Larix decidua								
Normal wood	26.2	1.4	4.8	2.0	46.1	10.5	2.5	6.3
Compression wood	38.3	0.9	4.3	9.4	34.2	6.4	0.9	5.3
Larix laricina								
Normal wood	27.0	1.6	2.8	2.4	46.3	12.3	1.3	6.0
Compression wood	38.1	1.0	3.1	10.0	33.2	5.2	1.1	7.9
Picea abies								
Normal wood	29.1	1.2	5.3	2.3	43.3	9.5	1.4	7.4
Compression wood	40.8	0.8	5.1	7.7	32.7	4.8	0.9	6.8
Picea glauca								
Normal wood	28.6	1.2	4.4	1.9	44.4	12.0	1.1	7.0
Compression wood	39.4	0.7	4.3	10.9	33.3	4.3	0.5	6.2
Picea mariana								
Normal wood	30.2	1.3	5.1	2.0	44.2	9.4	1.5	6.0
Compression wood	39.5	0.7	5.2	8.7	32.8	4.9	1.2	6.7
Picea rubens								
Normal wood	28.1	1.4	4.7	2.2	44.2	11.5	1.4	6.2
Compression wood	37.0	0.9	4.1	11.5	34.5	5.2	0.8	5.6

Table 5.22 (continued)

Species	Lignin	Acetyl	Uronic anhydride	Residues of:				
				Galactose	Glucose	Mannose	Arabinose	Xylose
Pinus banksiana								
Normal wood	29.3	1.3	5.2	2.7	42.8	11.5	1.4	5.4
Compression wood	39.4	0.6	4.4	10.7	31.2	6.1	0.9	6.3
Pinus elliottii								
Normal wood	30.5	1.4	4.8	2.7	40.0	9.8	1.5	8.9
Compression wood	38.3	1.0	4.3	11.3	30.2	5.9	1.0	7.6
Pinus resinosa								
Normal wood	29.1	1.2	6.0	1.8	42.4	7.4	2.4	9.3
Compression wood	39.7	0.6	4.8	11.6	39.1	5.2	1.1	4.9
Pinus rigida								
Normal wood	28.1	1.2	4.0	1.4	47.2	9.8	1.3	6.6
Compression wood	37.6	0.7	3.9	11.5	34.5	5.5	1.0	5.0
Pinus strobus								
Normal wood	29.0	1.2	5.2	3.8	43.6	8.1	1.7	7.0
Compression wood	39.4	0.8	4.8	11.0	32.2	3.8	1.0	6.7
Pinus sylvestris								
Normal wood	26.1	1.6	5.0	1.9	44.5	12.4	1.5	6.4
Compression wood	37.9	0.8	4.9	11.6	33.5	5.0	1.1	4.9
Pinus taeda								
Normal wood	30.0	1.4	4.8	2.2	40.4	10.5	1.6	8.7
Compression wood	38.0	0.9	4.3	11.4	32.0	4.6	1.1	7.3
Pseudotsuga menziesii								
Normal wood	26.4	1.5	3.7	2.3	46.6	11.5	1.2	6.4
Compression wood	36.3	0.9	3.3	10.3	34.6	7.1	0.8	6.3
Tsuga canadensis								
Normal wood	33.7	1.4	4.7	1.8	43.1	10.6	1.0	3.3
Compression wood	40.9	0.7	4.7	12.9	30.4	4.3	0.7	5.0
TAXODIACEAE								
Cryptomeria japonica								
Normal wood	33.2	0.9	4.8	2.0	40.2	7.6	1.8	9.1
Compression wood	40.3	0.7	4.5	11.5	30.9	3.8	0.7	7.2
Metasequoia glyptostroboides								
Normal wood	30.9	1.0	4.3	1.3	41.3	9.7	1.3	9.8
Compression wood	40.7	0.7	4.8	8.8	32.5	5.3	0.5	6.3
Sequoia sempervirens								
Normal wood	34.0	1.0	4.1	2.5	40.3	7.8	1.4	8.5
Compression wood	40.5	0.6	3.9	10.7	30.7	5.3	0.6	7.3
Taxodium distichum								
Normal wood	33.6	1.4	4.1	1.5	41.0	8.5	1.3	8.2
Compression wood	39.6	1.0	4.5	11.6	30.3	5.6	0.8	6.2
TAXALES								
TAXACEAE								
Taxus baccata								
Normal wood	30.1	1.4	4.1	1.7	42.5	11.9	1.3	6.6
Compression wood	37.1	1.1	4.1	8.0	34.1	7.3	1.0	6.9

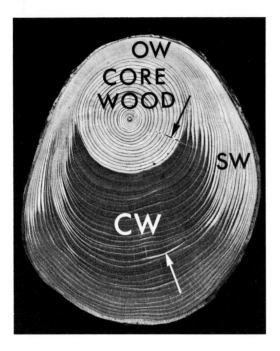

Fig. 5.4. Cross section of an inclined stem of *Picea rubens* at ground level. Note the crack in the normal core (*black arrow*) and a small band of normal wood in the compression wood (*CW*) region (*white arrow*). *OW* opposite wood. *SW* side wood

Normal wood of some conifers has more lignin than others. Among the present species, lignin contents extended from about 26% for *Larix decidua* and *Pseudotsuga menziesii* to 34% for *Tsuga canadensis* and *Taxodium distichum*. Since the lignin content of compression wood is approximately the same for all species, the difference in degree of lignification between normal and compression woods will be largest for those species that contain least lignin in their normal wood. In the present case, the maximum increase in lignin content, 46%, based on normal wood, was exhibited by *Larix decidua*.

As could be expected, the glucose content of compression wood was consistently lower than that of normal wood. This was partly caused by the lower content of galactoglucomannan in compression wood but was largely the result of a lesser amount of cellulose. In addition, it is now known (Hoffmann and Timell 1970a, 1972a) that some of the glucose residues in compression wood are derived from laricinan, a noncellulosic, $(1 \rightarrow 3)$-linked β-D-glucan, present to the extent of $3 \pm 1\%$ in compression wood (Sect. 5.4).

The data in Table 5.22 indicate that in the species examined, normal wood on the average contained 44–46% glucose residues, compared to only 32–34% for compression wood. Assuming an overall glucose to mannose ratio of 1:3 in galactoglucomannans, and correcting for the presence of 2% glucose residues in laricinan, the number of glucose units derived from cellulose alone can be computed. Some cellulose contents calculated in this way are listed in Table 5.23. The values for normal wood agree with results obtained by direct nitration or isolation of pure cellulose from wood (Timell 1957a). Those pertaining to compression wood are in accordance with some of the earlier estimates and also

5.2.12.1 Compression Wood

Table 5.23. Calculated cellulose content of normal and compression woods of 14 conifer species. All values in per cent of extractive-free wood

Species	Normal wood	Compression wood
Juniperus communis	37.7	30.2
Thuja occidentalis	42.8	29.8
Abies balsamea	42.2	31.2
Larix decidua	42.6	30.1
Larix laricina	42.2	29.5
Picea abies	40.2	29.1
Picea glauca	40.4	29.9
Picea mariana	41.1	29.2
Picea rubens	40.4	30.8
Pinus resinosa	40.0	28.2
Pinus rigida	43.9	30.7
Pinus strobus	40.9	28.9
Pinus sylvestris	40.4	29.8
Tsuga canadensis	39.6	27.0
Average	41.0	29.6

agree with yields noted on later isolation of compression wood celluloses. Normal wood on the average contained $40 \pm 2\%$ cellulose and compression wood only $30 \pm 1\%$, a reduction of 20–25%, much of it a result of the higher lignin content of compression wood. A recently reported (Panshin and de Zeeuw 1980) value of 34.9% for the average cellulose content of compression wood is too high, probably because it includes glucose residues derived from the $(1 \rightarrow 3)$-glucan and galactoglucomannan.

The galactose content of the compression wood specimens was very high throughout, sometimes eight but more commonly five to six times as high as in corresponding normal wood. Because of the higher lignin content of the compression wood, the increase would be larger if it were based on the total number of sugar residues, as was done by Meier (1965).

The proportion of mannose residues is only about half as high in compression wood as in normal wood. The difference is less if only the carbohydrates are considered, but it applies to all species to approximately the same extent. The acetyl content of compression wood is also half as high as that of normal wood, a fact first observed by Hägglund and Ljunggren (1933). This is what would be expected, since all acetyl groups in softwoods are attached to galactoglucomannans (Koshijima 1960, Meier 1961, Katz 1965, Lindberg et al. 1973, Kenne et al. 1975).

The uncertainty in the literature concerning the content of xylose residues in compression wood compared to normal wood appears again in Table 5.22. Some species contained more and others less xylan in their compression wood, while in other cases there was no difference. Compression wood of all pines

Table 5.24. Average summative chemical composition of normal and compression woods of 27 gymnosperm species (Table 5.22). All values in per cent of extractive-free wood

Constituent	Normal wood	Compression wood
Lignin	30.1	38.9
Acetyl	1.31	0.84
Uronic anhydride	4.69	4.47
Residues of:		
Galactose	2.10	10.3
Glucose	43.0	32.5
Mannose	9.80	5.27
Arabinose	1.63	0.94
Xylose	7.13	6.58

contained fewer xylose residues than did normal wood, and especially so *Pinus resinosa* and *P. sylvestris*. For *Tsuga canadensis* and *Thuja occidentalis* the opposite situation prevailed. The arabinose content was throughout lower for compression wood than for normal wood, reflecting the lower frequency of arabinose side chains in the compression wood xylan (Sect. 5.7).

Uronic anhydride includes largely 4-*O*-methylglucuronic acid, which occurs as a side chain in the xylan. Minor amounts of galacturonic and lesser quantities of glucuronic acid are also included. The former is derived from pectin and galactan, and the latter largely from the glucan. Compression wood on the whole contained slightly less uronic acid than did normal wood, especially in those species where there were fewer xylose residues. Average values based on all 27 species analyzed are shown in Table 5.24.

In a study of the distribution of arabinogalactan in *Larix occidentalis*, Côté et al. 1966a determined the extractives and arabinogalactan contents (Table 5.18), as well as the summative chemical composition of the different zones of an eccentric stem of *Larix laricina*, shown in Fig. 5.3. Core and compression woods were resolved into sapwood and heartwood, and the four zones, together with the side wood, which consisted of sapwood, were analyzed. The results are shown in Table 5.25. In this case, core wood and side wood had the same chemical composition. The difference between normal and compression sapwoods were the same as those in Table 5.22. The higher lignin contents of the two heartwood sections in comparison with the sapwoods was probably caused by the presence in the heartwood of tannins, which would be included in the lignin determination. The larger number of galactose residues in the heartwood was due to the occurrence here of an arabinogalactan (Sect. 5.8).

Summative data for the chemical composition of normal and compression woods have also been reported by Schreuder et al. (1966) for *Picea rubens*, by Côté et al. (1967b) for *Abies balsamea*, and by Hoffmann and Timell (1972d, 1972e) for *Pinus resinosa*. In the first case, the values obtained for normal wood of red spruce, which contained 28.1% lignin, were recalculated for an assumed

5.2.12.1 Compression Wood

Table 5.25. Summative chemical composition and contents of extractives and arabinogalactan of various zones in a stem of *Larix laricina*. All values in per cent of extractive-free wood except those for extractives. (Côté et al. 1966a)

Component	Normal core		Side wood	Compression wood	
	Sapwood	Heartwood	Sapwood	Sapwood	Heartwood
Lignin	26.0	28.6	25.9	37.2	38.8
Ash	0.3	0.5	0.3	0.3	0.6
Acetyl	1.7	1.4	1.4	0.9	0.7
Uronic anhydride	5.0	4.9	4.7	5.2	5.0
Residues of:					
Galactose	1.7	10.9	1.8	9.6	13.0
Glucose	47.0	35.0	46.5	32.6	29.7
Mannose	11.6	9.0	12.0	5.9	4.6
Arabinose	1.3	3.7	1.7	1.9	1.5
Xylose	5.4	6.0	5.7	6.4	6.1
Extractives:					
Ethanol-benzene	1.66	1.40	2.18	1.18	1.43
Arabinogalactan (water)	0.59	10.5	0.15	0.32	7.03

Table 5.26. Summative data for the chemical composition of normal and compression woods of *Picea rubens* and data calculated for the normal wood assuming it to have the lignin content of the compression wood. All values in per cent of extractive-free wood. (Schreuder et al. 1966)

Component	Normal wood Found	Compression wood Found	Normal wood Calculated
Lignin	28.1	37.0	(37.0)
Ash	0.33	0.36	0.25
Acetyl	1.43	0.87	1.08
Uronic anhydride	4.70	4.10	3.60
Residues of:			
Galactose	2.2	11.5	1.7
Glucose	44.2	34.5	33.6
Mannose	11.5	5.2	8.8
Arabinose	1.4	0.8	1.1
Xylose	6.2	5.6	4.7

lignin content of 37.0%, the value found for the compression wood. The data thus obtained, shown in Table 5.26, represent relative amounts of the different constituents *per tracheid*. This approach was adopted previously by Wardrop and Dadswell (1948), Jayme and Harders-Steinhäuser (1950, 1953), and Timell (1969) for comparing the chemical composition of normal and tension woods.

The data in Table 5.26 show that, based on equal lignin contents, normal and compression woods contain the same number of glucose residues, that is, cellulose, whereas the amounts of uronic acid residues are somewhat higher for compression wood which also contains much more galactose. The only components that suffer an absolute decline on transition from normal wood to compression wood are the acetyl groups and the mannose residues, that is, the *O*-acetyl-galactoglucomannan is reduced in amount.

Takashima et al. (1974) reported various analytical data for the chemical composition of opposite wood and compression wood from *Pinus densiflora*, some of which are summarized in Table 5.20[3]. Many of these values differ considerably from those found in Table 5.22. The cellulose content of the normal wood is abnormally high, while the galactan content of both normal and compression woods is many times lower than previously reported. The relative amounts of galactose, glucose, and mannose residues in the compression wood also differ from most earlier values for these sugar residues. The results do confirm that, in comparison with normal wood, compression wood contains more lignin and galactan and less cellulose and galactoglucomannan. In this case, its content of xylan (pentosan) was higher.

5.2.12.2 Branch Wood

Eskilsson and Hartler (1973) determined the chemical composition of various parts of *Picea abies* and *Pinus sylvestris* trees, namely stem, stump, roots, branches, needles, and bark, calculating the extractives and polymeric constituents of each tissue. Data for stem and branch woods are shown in Table 5.27. The results obtained indicate that in both species much of the branch wood must have consisted of compression wood. Evidence for this are the high lignin and galactan contents as well as the lesser proportions of cellulose and glucomannan in the branches. Since pronounced compression wood contains about 30% cellulose, the branches appear to have consisted entirely of such wood. The relatively low lignin content of the branch wood in the pine, on the other hand, does not agree with such a conclusion, and the paucity of galactan also militates against it.

5.3 Cellulose

5.3.1 General Properties

Little is known concerning the cellulose present in compression wood, since no pure, undegraded cellulose has so far been isolated from such wood. In two investigations (Bouveng and Meier 1959, Schreuder et al. 1966), the residue re-

---

3 Of the three zones analyzed in this study, only the middle one is considered here. The innermost zone, comprising the first 15 growth rings, undoubtedly consisted of juvenile wood, and the compression wood in the outermost zone was probably less pronounced than that in the intermediate one. It is unfortunate that the investigators did not include any data for the normal side wood.

5.3.1 General Properties

Table 5.27. Chemical composition of stem and branch woods of *Picea abies* and *Pinus sylvestris*. All values in per cent of unextracted wood. (Eskilsson and Hartler 1973)

Constituent	*Picea abies*		*Pinus sylvestris*	
	Stem wood	Branch wood	Stem wood	Branch wood
Undetermined	1	2	1	1
Extractives	0.8	1.7	3.0	3.3
Lignin	28.2	37.0	28.1	31.2
Total carbohydrates	70.0	59.4	67.9	64.5
Cellulose	42.7	29.1	40.8	32.1
Glucomannan	16.1	10.2	15.4	10.6
Galactan	1.2	7.6	1.1	5.1
Arabinan	0.5	0.5	0.8	1.5
Xylan	7.4	9.9	7.7	13.0
Uronic acid not in xylan	2.1	2.1	2.1	2.1

maining after removal of lignin, pectin, and hemicelluloses still contained residues of galactose, mannose, and xylose. When compression wood of *Abies balsamea* was subjected to the same treatment (Côté et al. 1967b) the residual material, obtained in a yield of 32% of the wood (compared to 41% for corresponding normal wood), was a pure, albeit indoubtedly much degraded cellulose. A similar extraction of a delignified compression wood from *Pinus resinosa* gave a 96% pure cellulose in a yield of 30.4%, corresponding to 29.1% pure cellulose. These yields agree well with the previously computed value of $30 \pm 1\%$ for the average cellulose content of compression wood.

The molecular weight of the cellulose present in compression wood is not known. For hardwoods it has been found that no difference exists in this respect between normal and tension wood celluloses (Timell 1969). In kraft pulps from compression woods, the fibers have a lower intrinsic viscosity than in similarly prepared pulps from normal woods (Dinwoodie 1965). In a later investigation, however, it was shown (Young et al. 1970) that this difference in intrinsic viscosity was due to an association of unknown nature between cellulose and hemicelluloses in normal wood, a phenomenon that is discussed more fully in Chap. 19.5. The original observation therefore sheds no light on the molecular size of compression wood cellulose.

It is doubtful if compression wood celluloses can be quantitatively isolated by direct nitration of the wood, the only method available for isolation of undegraded wood celluloses (Timell 1955). Earlier investigations (Timell 1957a, d, e, Goring and Timell 1960, 1962) have shown that with woods containing more than 30% lignin, the time required to achieve complete nitration of the cellulose is so long that depolymerization becomes unavoidable. Tanaka et al. (1981) found that the weight-average degree of polymerization of the cellulose was 3000 ± 500 in both opposite and compression woods of *Pinus densiflora*.

These are, however, values far below the degree of polymerization of 10 000 found by Goring and Timell (1962) to be typical of undegraded cellulose in normal wood of conifers.

5.3.2 Degree of Crystallinity

Cellulose, whether native or regenerated, is always partly crystalline and partly amorphous. For native celluloses, the degree of crystallinity varies with their botanical origin. Cotton and ramie celluloses are highly crystalline, and wood cellulose less so. The relative crystallinity can be determined by a number of methods, such as x-ray diffraction, density determination, exchange with deuterium or tritium oxides, and from studies of moisture regain. In the last three methods, the size of the crystallites has to be known in order to convert accessibility to crystallinity. Generally, wood cellulose has been found to be 60–70% crystalline.

In a review of x-ray investigations on the structure of wood, Caldwell and Lark-Horowitz (1937) mention that compression wood and earlywood can be shown by this technique to have a less pronounced fiber structure than latewood. The first detailed investigation of compression wood using x-rays was carried out by Bailey and Berkley (1942), who did not determine the crystallinity of the cellulose microfibrils in the different cell wall layers. The same applies to Wegelius (1946), who reported x-ray diffraction patterns of normal and compression woods of *Picea abies*, and to Berkley and Woodyard (1948), who recorded the diffraction pattern of compression wood in *Pinus ponderosa*. Lindgren (1958), confirming an earlier observation of Holzer and Lewis (1950), found that, within the same growth ring, latewood of *Picea abies* had a sharper x-ray fiber diagram than did earlywood, indicating a higher degree of crystallinity of the cellulose, since pectin, hemicelluloses, and lignin are all amorphous in their native state in wood. Indications were obtained that cellulose in compression wood regions was less crystalline than in normal wood and had a helical orientation.

Lee (1961), who also used an x-ray technique, likewise noted that latewood was more crystalline than earlywood. He found that the crystallinity of both holocellulose and pulp from compression wood of *Pseudotsuga menziesii* was significantly lower than that of corresponding materials from normal wood. Pulps from normal and compression woods had an average crystallinity index of 54.3 and 46.4, respectively. Lee postulated that compression wood cellulose contained a large area with a low lateral order on the surface of the crystallites. The same low crystallinity was exhibited by juvenile wood, which was shown to become more crystalline with age. Lee pointed out the several similarities which exist between this type of wood and compression wood, such as high lignin and low cellulose contents, short tracheids, a large microfibril angle, and a high longitudinal shrinkage. According to Lee compression wood is to be considered as more juvenile than normal wood of equal age, an observation made earlier by Wardrop and Dadswell (1950).

Kantola and co-workers (Kantola and Seitsonen 1961, Kantola and Kähkönen 1963, Kantola 1964) have reported x-ray studies on normal and compres-

5.3.2 Degree of Crystallinity

sion woods of *Juniperus communis, Picea abies,* and *Pinus sylvestris.* Their results, which concerned the orientation of the microfibrils, are referred to in Chap. 4.4.5.2.

Parham (1971 a), corroborating earlier investigators, found the crystallinity of mature wood of *Pinus taeda* to increase in the order compression wood, earlywood, latewood. Cellulose from these tissues had the same crystallite width of 27.64 Å, but the crystallinity was 28.7%, 37.6%, and 41.8%, respectively. These are values considerably lower than those reported by Lee (1961), probably because in the latter case the amorphous compound middle lamella was not included as the measurements were made with pulp and holocelluloses and not with wood.

Marton et al. (1972) compared the relative crystallinity of opposite and compression woods of *Picea abies.* The crystallites in the compression wood were thinner and much shorter and had a lower crystallinity than in opposite wood. This was attributed to the higher lignin content of compression wood, and it was suggested that the substructural elements of the cell wall are restrained by the rigid lignin network. Opposite and compression woods from *Pinus densiflora* have been reported to have a crystallinity index of 37.2 and 25.7, respectively (Takashima et al. 1974).

Tanaka et al. (1981) (Tanaka and Koshijima 1984) have recently reported data for the degree of crystallinity of cellulose in normal, opposite, and compression woods of *Pinus densiflora* trees exposed to wind. The degrees of crystallinity in the three types of tissue were 50%, 50–60%, and 45–50%, respectively. All crystallites were 3.2 nm wide but they varied in length, namely from 12 nm in compression wood to 17–33 nm in opposite wood. The investigators attributed the short crystallites in compression wood to the fact that in this tissue, which is under compressive stress, the cellulose chains are laterally ordered for only a short distance. The crystallites in compression wood were oriented at an angle of 30° to the stem axis, compared to an angle of 0° in opposite wood. These differences between opposite and compression woods were believed to be related to the fact that the former tissue is under tensile stress, whereas the latter is under compressive stress.

Cellulose is evidently less crystalline in compression wood than in normal and opposite woods. Interestingly, the cellulose microfibrils in the gelatinous layer of tension wood has an exceptionally high degree of crystallinity, comparable to that of cellulose in ramie fibers (Preston and Ranganathan 1948, Wardrop and Dadswell 1948, 1955, Dadswell and Wardrop 1955, Kantola 1964, Goto et al. 1975, 1978). This layer contains lignin only in exceptional cases. "Rubbery wood" is an abnormal, weak wood formed in apple trees attacked by a virus. It has been shown that such wood is incompletely lignified (Beakbane and Thomson 1955, Sondheimer and Simpson 1962, Nelmes and Preston 1968, Nelmes et al. 1973). Scurfield and Bland (1963) noted that the cellulose present in this wood had a high degree of crystallinity, possibly as high as that of tension wood cellulose. Crystallinity of cellulose seems to be inversely related to the degree of lignification of the cell wall. Since the crystalline width of cellulose does not change on removal of lignin and hemicelluloses from wood, it is generally believed that neither of these substances is able to penetrate the cellulose crystallites. It is also clear that wood tissues subjected to tensile stress, such as opposite wood, have a more highly ordered cellulose than those under compressive stress, such as compression wood.

5.3.3 Effect of Liquid Ammonia

The treatment of wood with either gaseous or liquid ammonia was developed by Schuerch and his co-workers into a useful process for molding or bending solid wood (Schuerch 1963, 1964, 1966, Koch 1972). Liquid ammonia acts as an excellent plasticizer on wood. It is strongly hydrogen-bonding and has the ability to penetrate both amorphous and crystalline regions of cellulose. It swells the hemicelluloses, forming acetamide from those that contain acetyl groups (Wang et al. 1964). Ammonia is an excellent plasticizer for lignin, much superior to water in this respect. It plasticizes the lignin in the middle lamella but also causes chemical and physical changes in the cell wall (Nečesaný and Lábsky 1973). Pieces of wood treated with ammonia can be bent and formed in a manner not possible on treatment with water. After evaporation of the ammonia, the wood has a higher density and is more hygroscopic and flexible than untreated wood (Bariska 1969, 1975, Bariska et al. 1969, 1970, Davidson and Baumgardt 1970, Pollisco et al. 1971, Schuerch and Davidson 1971, Bariska and Popper 1971, 1975, Mahdalik et al. 1971, Parham et al. 1972, Bariska and Strasser 1976, Berzinsh et al. 1976, Rocensh 1976, Bariska and Schuerch 1977). Anhydrous liquid ammonia swells wood strongly. Upon removal of the ammonia, the wood shrinks considerably and well beyond its original dimensions. This partial collapse has been shown to be caused by shrinkage of the S_2 layer (Coles 1973) and also by a reduction in the size of the cell lumina and the pores (Bariska 1975).

Parham (1970, 1971a, b) studied the effect of ammonia on the crystallinity and ultrastructure of normal and compression woods from *Pinus taeda*. Discs of wood were treated with ammonia vapor at room temperature and then dried in the air until free of ammonia. Normal wood was found to be crimped circumferentially, leading to buckling in the S_1 and S_3 layers, probably as a result of shrinkage of S_3. Compression wood was less deformed, but crimpings were observed in the thick S_1 layer as seen in Fig. 5.5.

Unexpectedly, an increase in crystallinity of the cellulose occurred on treatment with liquid ammonia, and more so for compression wood than for normal latewood and earlywood. That the cellulose itself should become more crystalline as a result of the treatment with liquid ammonia is unlikely. Parham (1971a) offered several possible explanations for the phenomenon, one involving a conversion of some of the hemicelluloses from an amorphous to a crystalline state. This explanation is entirely probable, for it is known that ammonia eliminates acetyl groups in wood, and isolated native wood hemicelluloses have a tendency to crystallize once their acetyl groups have been removed (Yundt 1951, Marchessault and Timell 1960, Lindberg and Meier 1957, Koshijima 1962). It is also known that on treatment of wood with aqueous alkali, crystallization of the amorphous xylan is readily induced (Marchessault et al. 1967). In

5.4.1 Introduction

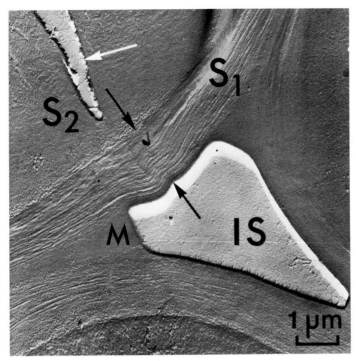

Fig. 5.5. Compression wood of *Pinus taeda* after treatment with ammonia vapor, showing buckling in the S_1 layer (*black arrows*) next to an intercellular space (*IS*). *White arrow* indicates a drying check. Transverse section. SEM. (Parham 1970)

his later contribution, Parham (1971b) expressed the opinion that the compaction of the cell wall, induced by the shrinkage in S_2, was probably responsible for the increase in crystallinity.

5.4 Laricinan

5.4.1 Introduction

The first hint that compression wood might contain a non-cellulosic glucan is found in a study of the chemical composition of normal and compression woods of *Picea abies* reported in 1933 by Hägglund and Ljunggren (Table 5.2). These investigators found "glucose" contents of 5.8% and 18.6% for normal and compression woods, respectively. Although uronic acids were included, we now know that both figures are much too high. Hägglund and Ljunggren were, nevertheless, correct in their conclusion that compression wood contains more "glucan" than normal wood.

An indication that compression wood may contain a glucan unrelated to cellulose or starch can be gleaned from the much later fractionation data re-

Table 5.28. Fractional extraction of wood and holocellulose from *Picea abies* compression wood. (Bouveng and Meier 1959)

Material	Yield, g	Sugar residues in relative per cent				
		Galactose	Glucose	Mannose	Arabinose	Xylose
EXTRACTIVE-FREE WOOD	391.0	19.6	55.6	11.5	2.6	10.6
Water extract	1.3	74.3	4.5	16.1	4.1	0.9
Dissolved in chlorite liquor						
Barium hydroxide precipitate	21.4	97.3	Trace	2.7	Trace	Trace
Soluble portion	5.5	72.1	2.2	1.9	3.7	20.1
HOLOCELLULOSE	208.0	7.5	65.8	11.9	2.4	12.5
Water extract						
Barium hydroxide precipitate	5.4	45.2	11.6	41.3	1.0	0.9
Soluble portion	2.5	33.1	8.4	2.8	5.8	49.9
14% Potassium hydroxide extract	40.7	19.0	5.2	12.4	6.2	57.2
24% Potassium hydroxide- 3% borate extract						
Barium hydroxide precipitate	6.6	6.7	23.3	68.3	0.8	0.9
Residue		1.5	91.3	6.1		1.0

Table 5.29. Fractional extraction of a chlorite holocellulose from *Picea rubens* compression wood. Yield in per cent of extractive-free wood. Sugar residues in relative per cent. (Schreuder et al. 1966)

Fraction	Yield	Galactose	Glucose	Mannose	Arabinose	Xylose
Chlorite liquor	3.5	83.1	6.0	6.2	1.7	3.0
Water	5.4	76.8	7.9	12.0	0.7	2.6
Ammonium oxalate	2.1	80.5	5.0	7.2	1.2	6.1
Sodium carbonate	5.9	60.6	Trace	Trace	3.0	36.4
Potassium hydroxide, 1%	4.5	30.4	34.4	0.8	2.7	31.7
Potassium hydroxide, 5%	5.4	13.1	14.4	1.8	2.7	68.0
Potassium hydroxide, 24%	6.0	15.5	31.4	19.3	1.6	32.2
Sodium hydroxide-borate	5.9	9.8	24.6	57.7	2.0	5.9
Residue	41.2	4.2	93.1	0.5		2.2

ported by Bouveng and Meier (1959), also for compression wood of *Picea abies*. Their data, shown in Table 5.28, include a fraction (water extract, soluble portion) consisting largeley of galactan and xylan but also containing 8.4% glucose and 2.8% mannose residues. Of the glucose, only 1.0−1.2% could have been combined with the mannose in a glucomannan, and the remainder must accordingly have had a different origin. The manner in which this particular fraction was obtained makes it unlikely that either cellulose or starch could have been present.

5.4.1 Introduction

Table 5.30. Fractional extraction of chlorite holocelluloses from normal and compression woods of *Abies balsamea*. Yield and uronic anhydride in per cent of extractive-free wood. Sugar residues in relative per cent. (Côté et al. 1967b)

Fraction	Yield	Galactose	Glucose	Mannose	Arabinose	Xylose	Uronic anhydride
Holocellulose							
Normal wood	72.9	1.3	68.2	19.2	2.3	9.0	10.6
Compression wood	65.2	9.1	64.0	13.4	1.4	12.1	8.6
Chlorite liquor							
Normal wood	5.8	19.4	13.5	33.1	14.0	20.0	16.6
Compression wood	7.4	74.1	11.2	14.7	Nil	Nil	16.5
Water							
Normal wood	5.0	4.1	14.8	40.8	6.2	34.6	14.7
Compression wood	5.7	58.0	12.4	22.4	Trace	7.2	11.3
Ammonium oxalate							
Normal wood	3.9	5.0	7.6	22.1	8.4	56.9	6.6
Compression wood	4.6	52.6	9.2	17.6	Trace	20.6	11.5
Sodium carbonate							
Normal wood	2.1	21.6	2.5	4.1	13.0	58.8	22.8
Compression wood	5.3	65.1	6.0	Trace	5.2	23.7	14.1
Potassium hydroxide, 1%							
Normal wood	0.4	9.4	4.1	4.2	9.2	73.1	16.4
Compression wood	2.6	25.5	25.7	Trace	4.7	44.1	16.7
Potassium hydroxide, 5%							
Normal wood	3.3	3.2	5.2	10.7	7.9	73.0	15.0
Compression wood	4.4	10.0	6.1	2.8	3.5	77.6	15.4
Potassium hydroxide, 24%							
Normal wood	4.6	1.5	13.6	37.5	3.3	44.1	11.1
Compression wood	5.0	17.3	11.2	30.5	Trace	41.0	12.8
Sodium hydroxide-borate							
Normal wood	8.0	1.2	33.0	63.5	0.7	1.6	4.5
Compression wood	5.4	6.5	25.4	27.9	Trace	Trace	Nil
Residue							
Normal wood	40.7	Nil	100.0	Trace	Nil	Nil	Nil
Compression wood	32.0	Trace	100.0	Nil	Nil	Nil	Nil

Schreuder et al. (1966) isolated several fractions from a compression wood holocellulose of *Picea rubens* (Table 5.29). Two of them contained more glucose than mannose residues and one, obtained by extraction with 1% aqueous potassium hydroxide, consisted of equal parts of galactose, glucose, and xylose units but almost no mannose. Complexing with barium hydroxide gave a material containing equal numbers of glucose and xylose residues, which gave two distinct peaks on electrophoresis, indicating that it was not homogeneous, but attempts to resolve it were unsuccessful.

Côté et al. (1967b) a year later reported similar fractionations of two holocelluloses from *Abies balsamea*, one from normal wood and the other from compression wood. Their data are summarized in Table 5.30. The largest difference between the two woods is the high proportion of galactose residues in many of the fractions from the compression wood and also the lower yield of residual cellulose. Another difference, however, is the higher proportion of glucose residues in several of the compression wood fractions. The extract obtained with 1% potassium hydroxide from normal wood contained 4.1% glucose and 4.2% mannose residues, whereas 25.7% glucose and only a trace of mannose residues were present in corresponding extract from compression wood. The investigators speculated that the glucose residues could possibly be derived from an as yet unknown glucan, perhaps the same glucan (callose) that occurs in bark of conifers.

Three years later, in 1970, Hoffmann and Timell (1972d, 1972e) established the presence of small amounts of a $(1 \rightarrow 3)$-linked β-D-glucan in ray cells of normal and compression woods from *Pinus resinosa*. Following this observation, a systematic search was made for such a glucan in various compression woods.

The first investigation concerned with the isolation of a $(1 \rightarrow 3)$-β-D-glucan was carried out with compression wood from *Larix laricina*. Hoffmann and Timell (1970a) suggested that the polysaccharide be named *laricinan* to distinguish it from callose, a polysaccharide located in the phloem sieve elements of gymnosperms and angiosperms, and from similar glucans, such as curdlan, laminaran, lentinan, leucosin, pachyman, and paramylon, present in algae, bacteria, and fungi.

In 1972 Brodzki reported the presence of callose in compression wood tracheids from a considerable number of gymnosperm species. The polysaccharide was located by observing its fluorescence at 366 nm in sections that had been stained with aniline blue. It was not removed by treatment with cuprammonium hydroxide. Since aniline blue gives the same fluorescence with all $(1 \rightarrow 3)$-linked glucans, it is likely that the material observed by Brodzki was actually laricinan. It was found to be present in *Abies concolor, Cedrus atlantica, Chamaecyparis pisifera, Cryptomeria japonica, Cupressus macrocarpa, Juniperus communis, Larix decidua, L. polonica, Picea abies, Pinus cembra, P. strobus, Pseudotsuga menziesii, Taxodium distichum, Thuja occidentalis,* and *Tsuga diversifolia*. None occurred in *Ginkgo biloba, Taxus baccata,* and *Torreya nucifera,* and its occurrence was doubtful in *Agathis alba* and *Araucaria bidwillii*. No callose was observed in the epithelial cells, and only small amounts appeared to be present in the last-formed and first-formed tracheids at the latewood–earlywood boundary. In *Pinus strobus*, callose did not occur in all compression wood tracheids. Brodzki's observations are also discussed in Chap. 6.3.3. Waterkeyn (1981) has shown that callose is deposited onto the innermost wall layer bordering the lumen in developing cotton fibers. It could be localized with the aid of a similar fluorescence method, using aniline blue. Most recently Waterkeyn et al. (1982) have reported that the 1,3-glucan is located together with cellulose within the inner, fissured part of the S_2 layer. Wloch and Hejnowicz (1983), by contrast, claim that it occurs between the helical ribs (Chap. 6.3.3). It is worth noting that among the species where no 1,3-glucan could be observed, *Ginkgo, Taxus,* and *Torreya* never form helical cavities and ribs in their compression wood tracheids and *Agathis* and *Araucaria* only very

5.4.2 Occurrence

Although it has so far been isolated from only a few species, it is likely that a $(1 \rightarrow 3)$-β-D-glucan occurs in most compression woods. Hoffmann and Timell (1970a, 1972a) obtained a pure laricinan from compression wood of *Larix laricina* in a yield of only 0.71%, but they estimated the total content at 1.7%. The same investigators also isolated laricinan from compression woods of *Picea rubens* and *Abies balsamea*, which probably contained 3−4% and 2.5%, respectively, of this polysaccharide. It was also obtained in a low yield from normal wood of *Abies balsamea*. It can probably be assumed that compression wood contains 3 ± 1% laricinan, whereas only trace amounts occur in normal softwood. The location of laricinan in the cell wall is uncertain. The fact that it can be completely removed from delignified wood by extraction with dilute alkali would suggest that it is not strongly associated with the cellulose microfibrils, as is the glucomannan.

5.4.3 Isolation

Laricinan, like all other compression wood polysaccharides, except the extracellular arabinogalactan (Sect. 5.8), cannot be extracted from the fully lignified wood. Hoffmann and Timell (1972a), on extraction of severe compression wood of *Larix laricina* with water, obtained an arabinogalactan in a yield of 3.9%. The wood was delignified with acid chlorite, giving a holocellulose which was subjected to fractional extraction, as shown in Table 5.31. Laricinan and a xylan (Sect. 5.7) were isolated from the 5% potassium hydroxide extract and a galactoglucomannan (Sect. 5.6) from the sodium hydroxide-borate extract.

A number of methods were tried to obtain a pure glucan from the 5% KOH extract, where it occurred together with a galactan, a galactoglucomannan, and a xylan. The procedures used included complexing with barium hydroxide (Meier 1958a), which removed the galactoglucomannan, and treatment with cetyltrimethylammonium hydroxide (Bouveng and Lindberg 1958), which failed to give any pure laricinan. In view of present knowledge, this is not surprising, for the latter reagent precipitates acidic polysaccharides, and not only the xylan but also the galactan (Sect. 5.5) and the laricinan are now known to be acidic. Complexing with iodine according to Gaillard et al. (1969) and Gaillard and Thompson (1971) was the only technique that afforded a pure product, as shown in Fig. 5.6. A much better and simpler method, however, was to extract the fraction with water which dissolved all polysaccharides present except the laricinan. Based on the estimated total amount of glucan in the wood, the yield was about 40%. This method of isolation is simple, and unlike complexing with iodine, lends itself to isolation of laricinan on a large scale. It can undoubtedly be applied to compression wood of any species.

Table 5.31. Fractional extraction of a chlorite holocellulose from compression wood of *Larix laricina*. Yield in per cent of extractive-free wood. Sugar residues in relative per cent. GalA: Galacturonic acid, GlcA: Glucuronic acid, 4-Me-GlcA: 4-O-Methylglucuronic acid. (Hoffmann and Timell 1972a)

Extract	Yield	Galactose	Glucose	Mannose	Arabinose	Xylose	Rhamnose	GalA	GlcA	4-Me-GlcA
Ammonium oxalate, 0.5%	4.5	60.0	9.7	14.0	3.2	13.1	Trace	+++	Trace	+
Sodium carbonate, 10%	4.4	52.6	0.7	3.3	6.4	32.2	4.8	+++	+	+++
Potassium hydroxide, 5%	8.7	18.3	20.1	2.4	5.6	53.6		+	Trace	+++
Potassium hydroxide, 24%	4.9	10.4	16.4	33.1	3.6	36.5		+	Trace	++
Sodium hydroxide, 25%, borate, 5%	3.8	4.9	26.7	66.8	Trace	1.6			Trace	
Residue	30.4	1.9	95.5	2.0		0.6				

5.4.4 Structure

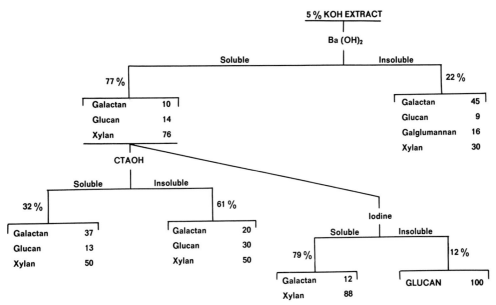

Fig. 5.6. Outline of an extraction sequence for isolation of a pure (1 → 3)-β-D-glucan from delignified compression wood of *Larix laricina*. CTAOH cetyltrimethylammonium hydroxide. (Hoffmann and Timell 1972a)

5.4.4 Structure

Laricinan from *Larix laricina* had a specific rotation of +17.6° in 1% sodium hydroxide and −16.1° in dimethyl sulfoxide. These facts as well as its infrared diagram indicate the occurrence of β-D-glucose residues (Hoffmann and Timell 1972a). On acid hydrolysis, laricinan yielded D-glucose, but also a small amount of glucuronic acid and an even lesser quantity of galacturonic acid. Enzymic hydrolysis with an endo-(1 → 3)-β-D-glucanase gave laminaribiose and laminaritriose. The consumption of periodate was very low. Partial, acid hydrolysis afforded a polymer-homologous series of laminaribiose oligosaccharides but also a small amount of cellobiose (Hoffmann and Timell 1972a). The combined evidence shows that the polysaccharide is largely composed of 1,3-linked β-D-glucopyranose residues.

This was confirmed by application of the classical methylation technique. The methylated glucan contained on the average 115 glucose residues and was found to consist of glucose units substituted at positions 2,3-, 2,4-, 2,6-, 4,6, 2,3,6, 2,4,6, and 2,3,4,6 in a ratio of 0.11:0.51:0.50:0.57:1.2:16.6:1.00. The preponderance of 2,4,6-tri-*O*-methyl glucose shows that most residues were (1 → 3)-linked, but the presence of a lesser amount of the 2,3,6-derivative makes it clear that 6−7% of the linkages were (1 → 4)-linked, which agrees with the formation of some cellobiose on partial hydrolysis. The average, methylated glucan macromolecule contained 4.6 branch points. Hoffmann and Timell were not able to establish the attachment of the galacturonic and glucu-

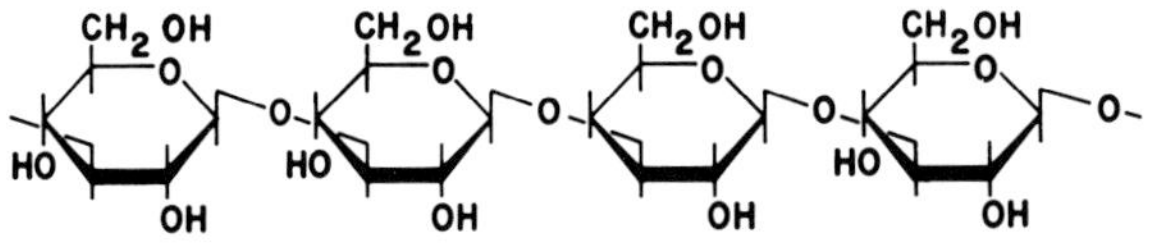

Fig. 5.7. Structure of an unbranched portion of laricinan

ronic acid residues to the main chain. Most likely, they were present as single unit side chains, perhaps linked to C-6 of the glucose residues. A portion of the main chain of the laricinan is shown in Fig. 5.7 and a more complete structure in the schematic Formula (**1**).

$$\rightarrow 3\text{-}\beta\text{-}\text{D-Glcp-1} \rightarrow 3\text{-}\beta\text{-}\text{D-Glcp-1} \rightarrow 3\text{-}\beta\text{-}\text{D-Glcp-1} \rightarrow 3\text{-}\beta\text{-}\text{D-Glcp-1} \rightarrow$$
$$6?$$
$$\uparrow$$
$$1$$
$$\text{D-GlcpA}$$

1

Young et al. (1972) studied the kinetics of the alkaline degradation (peeling) of laricinan and other $(1 \rightarrow 3)$-β-D-glucans. Theoretically, polysaccharides of this type should be rapidly and completely converted to glucometasaccharinic acid. Laricinan, however, was only partially degraded by alkali. The reason for this behavior might reside in the presence of a few 1,4-linkages, the mode of attachment of the acid side chains, or the occurrence of branch points, all of which could prevent β-alkoxy elimination at C-3.

5.4.5 Molecular Properties

The number-average degree of polymerization of laricinan, determined from osmotic pressure measurements with the nitrate derivative, was approximately 200 (Hoffmann and Timell 1972a). The chlorous acid used in delignification causes depolymerization (Glaudemans and Timell 1957), however, and the value for the original polymer is probably closer to 300, a size comparable to that of compression wood galactan (Schreuder et al. 1966), tension wood galactan (Kuo and Timell 1969), and larch arabinogalactan (Simson et al. 1968) but larger than that of wood xylans and galactoglucomannans.

Because of its acidic side chains, laricinan behaves as a typical polyelectrolyte in solution. In anhydrous dimethyl sulfoxide, the reduced viscosity of the polysaccharide increases exponentially on dilution instead of showing a linear decrease as do neutral polysaccharides (Fig. 5.8). The computed, extended chain viscosity (Fuoss and Strauss 1948) was 8.3 dl g^{-1}. In the presence of aqueous sodium chloride, the effect disappeared, and the viscosity could be estimated at 0.4–0.5 dl g^{-1}.

5.4.7 Conclusions

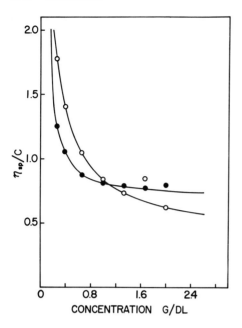

Fig. 5.8. Relationship between reduced viscosity and concentration in anhydrous dimethyl sulfoxide for two samples of laricinan. (Hoffmann and Timell 1972a)

5.4.6 Crystal Structure

Whether laricinan can be brought to crystallize remains to be established. Similar (1 → 3)-β-D-glucans, for example curdlan (Marchessault et al. 1977) and lentinan (Bluhm and Sarko 1977a, b) have been shown to crystallize in the form of a triple helix, a structure also assumed by (1 → 3)-β-D-xylans (Jelsma 1974, Jelsma and Kreger 1975).

5.4.7 Conclusions

Glucans of the laricinan type, with a main chain of (1 → 3)-linked β-D-glucose residues, are widely distributed in nature although they are never dominating polysaccharides in higher land plants. They occur in enormous quantities as laminaran in the *Laminaria* seaweeds and as leucosin or chrysoleucosin in other algae (Beattie et al. 1961, Archibald et al. 1963, Ford and Percival 1965). Paramylon is a (1 → 3)-β-D-glucan present as a storage polysaccharide in the unicellular alga *Euglena gracilis* (Clarke and Stone 1960, Marchessault and Deslandes 1979), in *Paranema trichophorum* (Archibald et al. 1963), and in *Astasia ocellata* (Manners et al. 1966). An inner layer of the cell wall of the bacterium *Neurospora crassa* contains a (1 → 3)-β-D-glucan as a matrix substance (Mahadevan and Tatum 1967). A similar polysaccharide, designated as curdlan, occurs in the bacterium *Alcaligenis faecalis* (Harada et al. 1968, Saito et al. 1968, Marchessault and Deslandes 1979, Deslandes et al. 1980). (1 → 3)-β-D-Glucans are common in the cell wall of fungi and have been isolated, for example, from *Aureobasidium (Pullularia) pullulans* (Brown and Lindberg

1967), *Phyophthora cinnamoni* (Zevenhuizen and Bartnicki-Garcia 1969), and *Piricularia oryzae* (Nakajima et al. 1972). The fungus *Poria cocos*, which grows on tree roots, contains pachyman, a $(1 \rightarrow 3)$-β-D-glucan (Warsi and Whelan 1957, Saito et al. 1968, Hoffmann et al. 1971). Lentinan, another $(1 \rightarrow 3)$-β-D-glucan, occurs in the fungus *Lentinus elodes*.

Herth et al. (1974) have reported the isolation of alkali-resistant fibrils from pollen tubes of *Lilium longiflorum*. The major component of these fibrils was a crystalline $(1 \rightarrow 3)$-β-D-glucan, very similar to pachyman. Fulcher et al. (1976) made some interesting observations with actively dividing plant cells which were stained with aniline blue and examined by fluorescence microscopy. An intense, yellow fluorescence appeared in the cell plate and also in new, transverse cell walls. The investigators suggested that a $(1 \rightarrow 3)$-β-D-glucan probably is associated with cell plate formation in cytokinesis. Several investigators have in later years observed the presence of a $(1 \rightarrow 3)$-β-D-glucan in developing cotton fibers (Delmer 1977, Meinert and Delmer 1977, Huwyler et al. 1979, Meier et al. 1981, Waterkeyn 1981, Jaquet et al. 1982).

As already mentioned, callose is a $(1 \rightarrow 3)$-linked-β-D-glucan that occurs in the sieve cells of the gymnosperms and in the sieve tubes of the angiosperms, where it is deposited within the pores of the sieve plates. Callose is present as dormancy callose in living sieve elements and as definitive callose in dead elements. It is also readily formed on wounding. Interestingly, callose, like laricinan, contains a few uronic acid groups (Aspinall and Kessler 1957, Kessler 1958, Fu et al. 1972). Unlike laricinan, callose, once it has been extracted and dried, usually cannot be redissolved in either water or aqueous alkali.

The presence of some $(1 \rightarrow 4)$-linked residues in largely $(1 \rightarrow 3)$-linked β-D-glucans is relatively common and has been reported for glucans from the alga *Peridinium westii* (Nevo and Sharon 1968), the fungus *Phytophthora cinnamoni* (Zewenhuizen and Bartnicki-Garcia 1969), and the mesocarp of *Mangifera indica* (Das and Rao 1964, 1965). Oat and barley $(1 \rightarrow 3)$-β-D-glucans and lichenan from Iceland moss (*Cetraria islandica*) contain a high proportion of $(1 \rightarrow 4)$-linked glucose residues.

5.5 Galactan

5.5.1 Introduction

A $(1 \rightarrow 4)$-β-D-galactan occu!s in considerable quantities in compression wood, whereas normal wood is either devoid of or contains only minute amounts of this polysaccharide. Because of this situation and probably also because of the relative ease with which the galactan can be isolated, this polymer has attracted more attention than the other hemicelluloses present in compression wood.

As has already been mentioned (Sect. 5.2.4.1), Hägglund and Ljunggren (1933) failed to detect any galactose residues in compression wood of *Picea abies*. Four years later, however, Hägglund and Larsson (1937) found extractive-free knot wood of the same species to contain 10% galactose units. In view of present knowledge there can be little doubt that these residues must

5.5.3 Isolation 339

have originated from a compression wood galactan. Unequivocal evidence that compression wood is rich in galactose residues was first presented by Stockman and Hägglund (1948), who concluded that compression wood of *Picea abies* might contain 10% of a galactan.

The first actual isolation of a galactan from compression wood was accomplished in 1959 by Bouveng and Meier, who also used wood from *Picea abies*. Meier (1965) has referred to unpublished studies on a galactan from compression wood of *Pinus sylvestris*. A galactan has also been isolated from compression wood of *Abies balsamea* (Côté et al. 1967b). The first structural investigation was carried out by Bouveng and Meier (1959). Later studies have been reported by Schreuder et al. (1966) on a compression wood galactan from *Picea rubens* and by Jiang and Timell (1972) on a similar polysaccharide from *Larix laricina*.

5.5.2 Occurrence

All compression woods so far examined have been found to contain considerable amounts of galactose residues. Since only a small proportion of these residues can be derived from galactoglucomannans, it is likely that compression wood of all Coniferales and Taxales and also *Ginkgo biloba* contain a galactan as one of their major hemicellulose constituents. The galactose content of compression wood can be used as a fairly accurate measure of the quantity of galactan present. The small amounts of galactose residues associated with galactoglucomannans are probably compensated for by the occurrence in the galactan of a few uronic acid residues. Judging from the values in Table 5.22, pronounced compression wood very likely contains $10 \pm 1\%$ galactan.

The uniformly high galactan content of all compression woods is in contrast to the highly variable galactan content of tension woods. Tension wood of some species does contain an acidic galactan, but in other species there is no more galactan than in normal wood, that is, extremely little or none (Meier 1965, Timell 1969) (Chap. 1.1).

5.5.3 Isolation

Bouveng and Meier (1959) found that direct extraction of *Picea abies* compression wood with water gave in a yield of 0.3% a fraction containing 74% galactose residues. When Schreuder et al. (1966) subjected compression wood of *Picea abies* to a similar extraction, a pure galactan was obtained in a yield of 0.1%, corresponding to less than 1% of the total galactan in the wood. No meaningful structural analysis is possible with a material obtained in such a low yield, and all investigators have removed the encrusting lignin from the wood before attempting any isolation of polysaccharides.

In their pioneering investigation on the galactan in compression wood of *Picea abies*, Bouveng and Meier (1959) first eliminated the lignin with chlorous acid. They noticed that in the course of the delignification, not only lignin but also significant amounts of polysaccharides dissolved in the hot chlorite liquor.

The polysaccharides were recovered and were found to consist largely of galactan, as seen from Table 5.28. The mixture was treated with barium hydroxide which forms insoluble complexes with galactans and galactoglucomannans but not with arabinogalactans and xylans (Meier 1958a, Lindberg 1962). The insoluble portion consisted of 97.3% galactose and 2.7% mannose residues, which were assumed to be derived from a contaminating glucomannan. This fraction, which also contained 12.9% uronic acid residues, was used in subsequent structural studies (Sect. 5.5.4). The portion remaining in solution on treatment with barium hydroxide consisted of three parts of galactan and one part of xylan. Evidently, 16% of the galactan had failed to give an insoluble complex, perhaps because it was more highly branched and/or had a much lower molecular weight.

Subsequent extractions were carried out by methods which had previously been used by Lindberg and Meier (1957) with normal spruce wood. The holocellulose was extracted in succession with hot water, 14% aqueous potassium hydroxide, and 24% potassium hydroxide containing 3% borate (Jones et al. 1956). The first and third fractions were further resolved by precipitation with barium hydroxide. As can be seen from Table 5.28, none of the fractions contained more than 45% galactose residues. The precipitate from the water extract was probably a mixture of equal parts of galactan and galactoglucomannan, while xylan predominated in the next two fractions. The last fraction consisted of a fairly pure galactoglucomannan of the alkali-soluble type. The residue was cellulose, still admixed with some hemicelluloses. It will be noticed that 25% of the galactan extracted from the holocellulose with water formed a soluble complex with barium hydroxide.

The galactan from the chlorite liquor contained about 9% lignin, which was removed by treatment with chlorine and 2-aminoethanol (Meier 1958b). The material could not be resolved with cetyltrimethyl-ammonium hydroxide, indicating that it did not contain a separate galactan and a pectic acid, and that accordingly at least some of the uronic acid residues were attached to the galactan. It could be shown that the galactan had not been produced by selective removal of acid-labile arabinofuranose groups from an arabinogalactan during the treatment with acid chlorite.

Schreuder et al. (1966) made an attempt to develop a better method for isolating compression wood galactans, using wood from *Picea rubens*. The original aim was to prevent loss of galactan in the preparation of the holocellulose. Variuos methods for selective removal of lignin from wood were accordingly tried, including the chlorine-2-aminoethanol method of Timell and Jahn (1951) and the procedure of Thomas (1945), the latter involving chlorination of moist wood meal in carbon tetrachloride rather than in water. Several chlorite procedures were also tried, but in no case was there any improvement in retention of hemicelluloses. The chlorination procedures, somewhat unexpectedly, caused removal of as much galactan as did the more drastic chlorite methods. When all water was excluded, no lignin could be removed. Evidently, dissolution of a sizable part of the galactan during delignification was unavoidable.

When wood is delignified, removal of the encrusting lignin gradually exposes more and more of the polysaccharides to the action of the oxidizing re-

5.5.3 Isolation

agents used and lignin-carbohydrate bonds are possibly also broken. The delignifying agents all have the ability to cleave glycosidic bonds, and as a result, the accessible, low-molecular weight hemicelluloses tend to dissolve in the various liquids used for removing the oxidized lignin. Accordingly, if excessive loss of hemicelluloses is to be avoided, some lignin must be left in the holocellulose (Browning and Bublitz 1953).

Timell and Jahn (1951), in a study of the delignification of normal wood of *Betula papyrifera*, found that the various chlorite procedures caused larger losses of hemicelluloses than did a modification of the original chlorine-3-aminoethanol method. Ahlgren and Goring (1971) studied the mechanism of chlorite delignification of *Picea mariana* wood. Not unexpectedly, they found that chlorous acid is selective in removing lignin only during the first 60% of the delignification. After this stage has been reached, hemicelluloses are also removed and especially galactoglucomannans, whereas xylan is hardly affected. Morphological factors involved in the removal of the lignin are discussed in a later contribution (Ahlgren et al. 1971).

When normal and compression woods of *Abies balsamea* were delignified with chlorous acid, 5.8% of normal and 7.4% of compression wood polysaccharides dissolved in the chlorite liquor, the latter fraction consisting of three parts of galactan and one part of galactoglucomannan (Côté et al. 1967b). The larger loss of polysaccharides from compression wood must be attributed to the presence in this wood of the water-soluble, accessible galactan.

Unexpectedly, Schreuder et al. (1966) found their spruce compression wood to be readily delignified by chlorous acid. It was, for example, fully delignified in half the time required to remove the lignin under identical conditions from normal wood of *Populus tremuloides*, which contained only 23% lignin compared to 37% for the compression wood. Presumably, the compression wood meal was highly permeable to the aqueous chlorite solution. It is possible that the numerous intercellular spaces offered a pathway for the chlorite liquor, in which case much of the compound middle lamella and the S_1 layer would be directly exposed to the reagent. As shown in Sects. 19.4 and 19.5, defibration is readily effected in chemical pulping of compression wood, a good indication that the lignin in the compound middle lamella must be fairly accessible to aqueous reagents. Fungi utilize not only the lumina but also the intercellular spaces in their penetration of compression wood (Sect. 5.11.2).

Schreuder et al. (1966) observed that 3–5% of the wood dissolved in the chlorite liquor. The material to go into solution first was an almost pure galactan, but later other polysaccharides were also removed from the wood. The residual holocellulose was subjected to a series of sequential extractions, shown in Table 5.29. The polysaccharide material dissolved during the delignification consisted to over 80% of a galactan admixed with some galactoglucomannan and lesser amounts of xylan. The water and ammonium oxalate extracts also contained about 80% galactan. All three fractions had a uronic anhydride content of 10–13%. The sodium carbonate extract consisted of two parts of galactan and one part of xylan. Polysaccharides other than galactan dominated in the remaining fractions all of which, however, still contained considerable amounts

Table 5.32. Occurrence of galactan in the various fractions in Table 5.29. (Scheuder et al. 1966)

Fraction	Galactan	
	Per cent of wood	Relative per cent
Chlorite liquor	2.78	19.3
Water extract	3.80	26.2
Ammonium oxalate extract	1.60	11.0
Sodium carbonate extract	3.20	22.9
1% Potassium hydroxide extract	1.23	8.6
5% Potassium hydroxide extract	0.60	4.2
24% Potassium hydroxide extract	0.52	3.6
Sodium hydroxide-borate extract	0.20	1.4
Residue	0.41	2.8

of this polysaccharide, as had also been observed by Bouveng and Meier (1959). The residual cellulose was still contaminated by some galactan.

The distribution of the galactan over the fractions in Table 5.32 was estimated by assuming that the galactose residues were derived from three different polysaccharides, namely from a galactan, from a water-soluble galactoglucomannan with a ratio between galactose, glucose, and mannose of 1:1:3, and from an alkali-soluble galactoglucomannan with a sugar ratio of 0.1:1:3. The second hemicellulose was probably present in the first seven extracts and the third in the sodium hydroxide-borate fraction.

Experience has shown that any polysaccharide which can be removed by extraction with water from a fully lignified softwood must be located outside the cell wall. If this applies to compression wood, only 0.1% of the galactan was extracellular. It has already been mentioned that on delignification with acid chlorite, hemicelluloses, and probably also pectin, are extracted from the wood to a greater extent with compression wood than with normal wood, and the material removed consists largely of galactan. The fact that this hemicellulose is water-soluble is probably of little significance in this respect, for the arabinoglucuronoxylan and some galactoglucomannan are also soluble in water, yet they are leached out to a much lesser extent. A more likely explanation is that much of the galactan in compression wood is located in the outer, more accessible regions of the cell wall. Inspection of the data in Table 5.32 shows that 57% of the galactan was removed on treatment of the holocellulose with water and acidic or neutral aqueous solutions, and that as much as 80% could be extracted without resort to aqueous alkali. Obviously, the galactan is much more accessible than either the galactoglucomannan or the arabinoglucuronoxylan. In a subsequent investigation, referred to in Sect. 6.3.2, Côté et al. (1968) were able to offer direct proof for the correctness of this assumption.

All polysaccharide fractions in Table 5.29 contained 2–5% Klason and 11–20% total lignin. Attempts to remove the lignin by treatment with chlorine and 2-aminoethanol (Meier 1958b, Croon et al. 1959) were unsuccessful.

5.5.3 Isolation

Schreuder et al. (1966) isolated galactans not only from the chlorite liquor, but also from the combined water-ammonium oxalate and from the sodium carbonate extracts. Treatment with barium hydroxide, as suggested by Bouveng and Meier (1959), did not give a pure galactan, but a mixture of galactan and galactoglucomannan, leaving the noncomplexing xylan in solution. When Fehling's solution was added to the mixture, part of the galactan and all of the galactoglucomannan formed an insoluble copper complex. An acidic galactan could be recovered from the solution in a fairly pure state. The yield of galactan obtained from the three extracts in this way was only fair. Further purification was effected by repeated complexing with barium hydroxide until the patterns obtained on boundary electrophoresis and sedimentation in the ultracentrifuge indicated that a homogeneous polysaccharide had been obtained.

Côté et al. (1967b) compared normal and compression woods of *Abies balsamea* with respect to general chemical composition and behaviour on delignification and fractional extraction. As has already been mentioned, the two woods were delignified with chlorous acid and then subjected to the same sequence of fractional extractions, resulting in eight extracts and a residue that consisted of a pure cellulose. Yield and composition of the fractions are summarized in Table 5.30. The values for uronic anhydride are too high because of the presence of residual lignin. Treatment with acid chlorite introduces carboxyl groups into lignin (Sarkanen et al. 1962, Dence et al. 1962), and these groups give rise to carbon dioxide on treatment with hot, concentrated acid, thus resulting in spuriously high values for uronic anhydride when the latter is determined by a decarboxylation method (Browning and Bublitz 1953, Miyazaki et al. 1971).

The major difference between normal and compression woods was, as could be expected, the much higher proportion of galactose residues in all extracts from the latter. As was the case with *Picea rubens* compression wood, the first four fractions contained most of the galactan. The fraction dissolved during delignification was a mixture of a galactan and a galactoglucomannan which could be resolved by formation of the insoluble copper complex of the galactoglucomannan. The sodium carbonate extract consisted of 85% galactan and 15% xylan. When an attempt was made to isolate the galactan by complexing with barium hydroxide, as much as 70% of this fraction remained in solution.

Jiang and Timell (1972) were able to obtain a pure galactan from compression wood of *Larix laricina* by the method developed by Bouveng and Meier (1959). During isolation of the holocellulose about 10% of the wood could be recovered from the chlorite liquor. This material was found to consist of 85% galactan, 5% xylan, and 10% galactoglucomannan. Treatment with barium hydroxide afforded an insoluble complex which consisted of 95% galactose, 5% galacturonic acid, and traces of glucuronic acid residues. The approximate yield, based on the original wood was 9.3%.

The galactan obtained by Bouveng and Meier (1959) represented 44% of the total amount present in the wood. The yield obtained from *Larix laricina* compression wood, 85%, is the highest so far reported for this polysaccharide. It is clear that the best procedure for isolating a compression wood galactan is to allow as much as possible of this polysaccharide to go into solution during delignification and then to purify the galactan via its insoluble barium hydroxide complex.

5.5.4 Structure

Compression wood galactan is freely soluble in water and has a specific rotation in this solvent of +57°. On hydrolysis it yields D-galactose, but also D-galacturonic acid as well as small amounts of D-glucuronic acid, a fact reported by several investigators (Bouveng and Meier 1959, Schreuder et al. 1966, Jiang and Timell 1972). The uronic acid content has been stated to be 10–13%, but this, as already mentioned, includes oxidized lignin. A galactan from *Larix laricina* was found by Jiang and Timell (1972) to contain only 5.3% galacturonic acid residues. The specific rotation and the fact that the infrared spectrum of the galactan has a strong band at 886 cm^{-1} (Bouveng and Meier 1959) both indicate that the polysaccharide is composed of β-D-galactose residues. The galactan is devoid of O-acetyl groups.

The first proof of structure was furnished by Bouveng and Meier (1959) who subjected their galactan from *Picea abies* compression wood to partial acid hydrolysis. After removal of acidic oligosaccharides, seven neutral oligomers were obtained, all belonging to the same polymer-homologous series of galactose oligosaccharides. The first member of the series, a galactobiose, was identical with a 4-O-β-D-galactopyranonyl-D-galactose, first isolated by Gillham and Timell (1958) and identified by Gillham et al. (1958). Galactans from compression wood of *Picea rubens* (Schreuder et al. 1966), *Abies balsamea* (Côté et al. 1967b), and *Larix laricina* (Jiang and Timell 1972) have been found to give the same oligosaccharides on partial acid hydrolysis. These results show that all galactans consist of (1 → 4)-linked β-D-galactopyranose residues.

Further structural details have been clarified by application of the methylation technique, which, however, in this case suffers from the fact that (1 →4)-linked β-D-galactans are virtually impossible to methylate to completion, probably as a result of the presence of an axial hydroxyl group at C-4. When a galactan from *Picea rubens* (Schreuder et al. 1966) was methylated and methanolyzed, there were obtained the methyl glycosides of mono-O-methylgalactose, di-O-methylgalactose, 2,3,6-tri-O-methylgalactose, and 2,3,4,6-tetra-O-methyl-D-galactose in a ratio of 0.8:5.4:28.2:1.0. The mono- and di-O-methylated sugars were of little structural significance because the galactan was not fully methylated. The large amount of 2,3,6-tri-O-methyl-D-galactose again proved that the polysaccharide was composed of (1 → 4)-linked β-D-galactopyranose residues.

The tetra-O-methylgalactose originated from nonreducing end groups in the polysaccharide. The number of glucose residues per average macromolecule, that is, the number-average degree of polymerization, was determined by osmometry and found to be 98. Since there was one tetra-O-methylgalactose unit, representing a nonreducing end group, per 35 galactose residues, the average methylgalactan contained 2.8 such groups. One of these was present in the main chain, while the remaining 1.8 must have terminated branches attached to the

5.5.4 Structure

main chain. The methylated galactan accordingly contained on the average 1.8 branch points. Some degradation is unavoidable during methylation of most polysaccharides. Prior to methylation, the polysaccharide contained about 110 galactose residues which, provided the degratation had been random, would correspond to 2.0 branches per average macromolecule. Schreuder et al. (1966) subjected four different galactan preparations to a structural analysis of the type now described, obtaining essentially the same results whith all four.

Jiang and Timell (1972) analyzed their galactan from *Larix laricina* compression wood by the methylation technique. The methylated galactan was found to contain on the average 104 residues, consisting of 2,3-di-, 2,3,6-tri- and 2,3,4,6-tetra-*O*-methyl-D-galactose in a ratio of 9.3:90.0:5.2. There were accordingly 4.2 branches per molecule, corresponding to 5.1 branches for the original polysaccharide with a degree of polymerization of 126.

While it has long been clear that the galactan in compression wood consists of a slightly branched chain of $(1 \rightarrow 4)$-linked β-D-galactopyranose residues, the amount, nature, and mode of attachment of the acidic groups long remained obscure. Bouveng and Meier (1959) noted that three acidic oligosaccharides were produced on partial acid hydrolysis of their galactan. All contained galactose, and, in addition, one contained a galacturonic acid and the other two a glucuronic acid residue. Schreuder et al. (1966) obtained several aldobiouronic acids on similar hydrolysis of their galactan but were unable to isolate pure chemical compounds. In the methylation analysis carried out by Jiang and Timell (1972), evidence was adduced for the occurrence of both 2,3-di-*O*-methylgalactose and a methylated galacturonosyl galactose, indicating that some of the galactose residues were substituted at C-6 with a galacturonic acid residue. An aldobiouronic acid was obtained on partial acid hydrolysis, which, after reduction, methylation and hydrolysis afforded 2,3,4,6-tetra-*O*-methyl-D-galactose and 2,3,4-tri-*O*-methyl-D-galactose. The aldobiouronic acid was accordingly 6-*O*-(β-D-galactopyranosyluronic acid)-D-galactose (**2**).

2

This compound has previously been isolated from a larch arabinogalactan (Aspinall and Nicolson 1960) and from a gum of *Salmalia malabarica* (Bose and Dutta 1963).

Although better evidence would be desirable, it is likely that all compression wood galactans contain residues of β-D-galacturonic acid residues, attached as single-unit side chains to C-6 of some of the galactose residues, one acid side chain being present per 20 galactose residues. The mode of attachment of the less numerous glucuronic acid residues remains unknown.

5.5.5 Molecular and Other Properties

Schreuder et al. (1966) found galactan trinitrate to be incompletely soluble in all solvents and therefore used the triacetate derivative for osmotic pressure measurements. The highest degree of polymerization obtained was 120, but a value of 280 was noted for the undegraded galactan extracted in low yield with water from the original compression wood. A value for the degree of polymerization of the entire, native polysaccharide is not available, but it is probably in the range of 200–300.

Intrinsic viscosities in cupriethylenediamine varied from 0.17 to 0.25 dl g^{-1}, values considerably lower than would be expected for linear polysaccharides of the same size in this solvent, such as the hardwood xylan and the degraded celluloses studied by Koshijima et al. (1965). Possibly, therefore, the branches in the galactan are fairly long, and certainly longer than in hardwood xylans. Bouveng and Meier (1959), on the other hand, observed that their galactan from *Picea abies*, which had a degree of polymerization of only 52, could be cast into a fairly strong film, a feature characteristic of a linear polysaccharide. A highly branched structure, such as that of larch arabinogalactan, is therefore out of the question. The molecular-weight distribution of the galactan has not been investigated. On ultracentrifugation, the galactan from *Picea rubens* (Schreuder et al. 1966) gave a maximum which broadened with time, indicating the presence of a polydisperse polymer. The conformation of the galactan is not known.

5.5.6 Conclusions

Although galactans have been isolated from compression wood of only five species, namely *Larix laricina*, *Picea abies*, *P. rubens*, and *Pinus sylvestris*, as well as in small amounts from ray cells in compression wood of *Pinus resinosa* (Hoffmann and Timell 1972e), it is highly likely that the same polysaccharide occurs in compression wood of all Coniferales and Taxales and probably also in *Ginkgo biloba* (Tables 5.21 and 5.22). Approximately 80% of this galactan is readily removed from the delignified wood, whereas a part of the remainder is so firmly held in the cell wall that it cannot be freed even with solvents capable of extracting the glucomannan completely. Evidently, this small portion of the galactan is closely associated with the cellulose microfibrils. Further evidence for the distribution of the galactan in the tracheid wall is presented in Chap. 6.3.3. The fact that some of the galactan very probably is chemically linked to lignin is discussed in Sect. 5.9.3.

The evidence so far adduced indicates that compression wood galactan consists of a main chain of 200–300 (1 → 4)-linked β-D-galactopyranose residues. There are probably 4–6 branches per average macromolecule. The length of these branches is unknown. One galactose residue out of 20 carries attached to C-6, a terminal, single residue of β-D-galacturonic acid. A few residues of glucuronic acid may also occur as side chains, but their number and mode of attachment are unknown. No acetyl groups are present. The galactan is

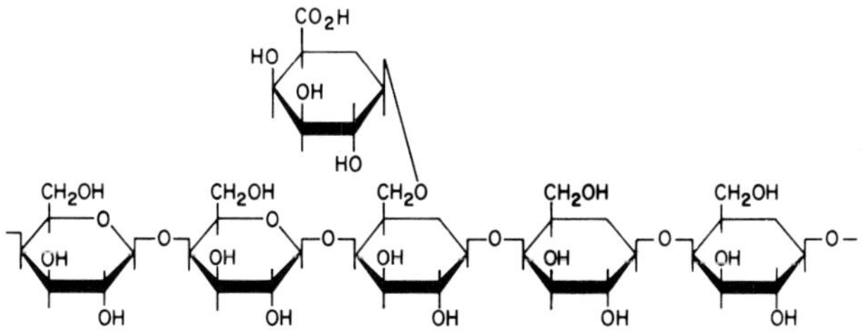

5.5.6 Conclusions

Fig. 5.9. Structure of the repeating unit of a compression wood galactan

amorphous. A simplified structural formula is shown below (**3**) and in Fig. 5.9. Mukoyoshi et al. (1981) have suggested that the galactan is slightly branched at C-2, C-3, and C-6, and that most of it is chemically linked to lignin in compression wood of *Pinus densiflora* (Sect. 5.9).

$$\rightarrow 4\text{-}\beta\text{-D-Galp-}1 \rightarrow 4\text{-}\beta\text{-D-Galp-}1 \rightarrow 4\text{-}\beta\text{-D-Galp-}1 \underset{6}{\overset{}{\rightarrow}} \left[4\text{-}\beta\text{-D-Galp-}1 \rightarrow \right]_{17}$$

$$\uparrow$$
$$1$$
$$\beta\text{-D-GalpA}$$

3

According to Beall (1969), compression wood galactan is degraded at an exceptionally high rate on combustion analysis in comparison with other wood hemicelluloses. Wood (1980) has reported that compression wood galactan does not interact with direct dyes, such as Congo red.

Galactans of the β-D-$(1 \rightarrow 4)$-linked type are fairly rare in nature. They have long been assumed to be members of the so-called "pectic triad", and the first $(1 \rightarrow 4)$-linked galactan to be examined was isolated from the pectin component of *Lupinus albus* seeds by Hirst et al. (1947) (Hay and Gray 1966). Later, it has also been found to occur in seeds of *Lupinus luteus* (Tomada and Murayama 1965) and of *Strychnos nux-vomica* (Andrews et al. 1954). Aspinall and his co-workers (Aspinall et al. 1967b) isolated from soybean cotyledons (Aspinall et al. 1967a, c) and hulls (Aspinall et al. 1967d) complex pectic polysaccharides containing fairly long sequences of $(1 \rightarrow 4)$-linked β-D-galactose residues, both in the backbone and in the side chains of the macromolecules. A $(1 \rightarrow 4)$-β-D-galactan has been recovered by Das and Das (1980) from garlic bulbs. This polysaccharide is more highly branched than the galactan in compression wood, since every fourth galactose residue carries a branch of unknown length attached to C-6.

Minor amounts of a $(1 \rightarrow 4)$-linked galactan may be present also in normal softwood. Roudier and Eberhard (1965) and Roudier and Galzin (1966) have reported evidence for the occurrence of a $(1 \rightarrow 4)$-linked galactan in wood of *Pinus pinaster*. Fractions rich in galactose residues have been isolated by Thompson and Kaustinen (1964, 1966) from various softwoods. The isolation of a galactan with unusual solubility properties from mature wood of *Pinus banksiana* has been reported by Thompson et al. (1968 a, b). A galactan has also been found to be present in inner bark of *Picea glauca* (Painter and Purves 1960).

Inspection of the data in Table 5.30 indicates that also in normal wood of *Abies balsamea* not all galactose originated from galactoglucomannans. The fractions dissolved in the chlorite liquor and the sodium carbonate extract contained approximately 20% galactose residues, which, especially in the latter fraction, could not have been derived only from galactoglucomannans. Some might have occurred in arabinogalactans, but at least in the sodium carbonate extract, where most, if not all, of the arabinose was probably associated with the xylose residues, the presence of a true galactan is strongly indicated. No pure galactan has as yet been isolated from normal wood of a conifer.

If a $(1 \rightarrow 4)$-β-D-galactan does occur in small quantities in normal wood of conifers, it follows that the occurrence of much larger amounts of this galactan in compression wood sets apart this wood quantitatively, rather than qualitatively, from a normal softwood. The acidic galactan, first isolated by Meier (1962) from tension wood of *Fagus sylvatica*, has been found to occur in both normal and tension woods of *Fagus grandifolia* (Kuo and Timell 1969).

The function of the galactan in compression wood is difficult to ascertain at the present time. It is probably more related to the hemicelluloses than to the pectic substances in wood. It differs from the former in its higher degree of polymerization, which is in the range of 200–300, whereas the hemicelluloses have degrees of polymerization of 100–200. Larch arabinogalactans have number-average degrees of polymerization varying from 300–500 (Simson et al. 1968) (Sect. 5.8).

5.6 Galactoglucomannan

Approximately 15–18% of normal conifer wood consists of various *O*-acetylgalactoglucomannans, ranging in composition from a galactose:glucose:mannose ratio of 0.1:1:3 to 1:1:3. Arabino-4-*O*-methyl-glucurono-xylan accounts for approximately 8–12% of the wood. Few exceptions to this composition are known at present, two being *Libocedrus decurrens* (Thompson and Kaustinen 1964) and *Sequoiadendron giganteum* (Meier 1965) both of which have an unusually high xylan content.

In compression wood, the proportion of galactoglucomannan is considerably less, generally only 8–9% of the wood (Table 5.22). The polysaccharide with a sugar ratio of 1:1:3 is soluble in water and is obtained by extraction of delignified wood with aqueous potassium hydroxide. The so-called glucomannan dissolves only in sodium hydroxide and has to be extracted with this alkali, preferably in the presence of borate. The compression wood data in Tables 5.28 and 5.29 show two exceptions to this rule in that some of the glucomannan could be extracted from the two holocelluloses with potassium hydroxide.

Anderson et al. (1941) were the first to report the isolation of a mannose-containing polysaccharide from compression wood. They found that compres-

5.6 Galactoglucomannan

sion wood of *Pinus strobus* gave larger amounts of hemicelluloses and more "mannan" free of uronic acid residues than did corresponding normal wood. Some mannan was still present after the final extraction of the wood. A pectinic acid, probably similar to that present in hardwoods, was also isolated. There is, of course, no indication in this early report that glucose and galactose residues should be a part of the so-called mannan.

Côté et al. (1967b) isolated pure galactoglucomannan and glucomannan from the last two fractions of *Abies balsamea* compression wood shown in Table 5.30. The polysaccharides contained residues of galactose, glucose, and mannose in ratios of $0.5:1:3$ and $0.3:1:3.3$. Both afforded β-D-$(1 \rightarrow 4)$-linked mannobiose, mannotriose, mannotetraose, mannosyl glucose, and glucosyl mannose on partial acid hydrolysis. It is clear from these results that the two hemicelluloses were similar to the galactoglucomannan and glucomannan present in normal conifer wood.

Hofmann and Timell (1972c) isolated a galactoglucomannan of the alkali-soluble type (glucomannan) from compression wood of *Larix laricina*. The polysaccharide was obtained from a sodium-hydroxide extract of the delignified wood (Table 5.31) in a yield of 3.8% based on the original wood. It consisted of residues of galactose, glucose, and mannose in a ratio of $0.2:1:2.5$. The specific rotation in alkali was $-30°$, and the degree of polymerization was about 70.

The average methylated galactoglucomannan contained 32 hexose residues. The molar composition was 2,3-di-*O*-methylglucose (2.43), di-*O*-methylmannose (0.36), 2,3,6-tri-*O*-methylglucose (7.06), 2,3,6-tri-*O*-methylmannose (20.8), 2,3,4,6-tetra-*O*-methylgalactose (0.52) and 2,3,4,6-tetra-*O*-methylated glucose and mannose (1.21). These results show that the galactoglucomannan consisted of at least 70 β-D-glucopyranose and β-D-mannopyranose residues, linked together to a chain containing about 0.4 branches. A few α-D-galactopyranose units were attached to C-6 of the glucose and mannose. This polysaccharide is very similar to galactoglucomannans present in normal wood of both *Larix laricina* (Kooiman and Adams 1961) and *Larix decidua* (Aspinall et al. 1962). Like several other conifer galactoglucomannans (Hoffmann and Timell 1970b), it is slightly branched. In normal conifer wood, the total amount of galactoglucomannans is relatively constant, but the ratio between the water-soluble and the alkali-soluble types varies from one species to another (Timell 1961c). Possibly, the same variation occurs in compression wood. According to Schreuder et al. (1966), *Picea abies* compression wood contains approximately 4% of the water-soluble and 7% of the alkali-soluble type of these polysaccharides.

It is not known whether the galactoglucomannans are partly acetylated in compression wood, as they are in normal wood. The fact that the ratio between the mannose residues in normal and compression woods of the species listed in Table 5.22 throughout is about the same as the ratio between the acetyl groups suggests that the latter are attached to the galactoglucomannans in both types of wood. Compression wood contains only half as much galactoglucomannan as does normal conifer wood. As far as can be judged at the present time, the structure and properties of this family of polysaccharides are the same in the

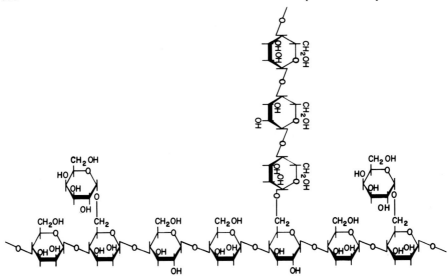

Fig. 5.10. Structure of the repeating unit of a compression wood galactoglucomannan

two tissues. The lesser amount of galactoglucomannan in comparison with normal wood corresponds closely to the larger amount of galactan in compression wood, the absolute percentage being about 10% in both cases. A formula of the galactoglucomannan is shown below (**4**) and in Fig. 5.10.

$$\rightarrow 4\text{-}\beta\text{-D-Glc}p\text{-}1 \rightarrow 4\text{-}\beta\text{-D-Man}p\text{-}1 \rightarrow 4\text{-}\beta\text{-D-Man}p\text{-}1 \rightarrow 4\text{-}\beta\text{-D-Man}p\text{-}1 \rightarrow$$
$$6 \qquad\qquad\qquad 2(3)$$
$$\uparrow \qquad\qquad\qquad |$$
$$1$$
$$\alpha\text{-D-Gal}p \qquad\qquad \text{Acetyl}$$

4

5.7 Xylan

It has already been pointed out that compression wood contains as much and occasionally somewhat more xylan than normal wood (Table 5.2). Anderson et al. (1941) were the first to obtain an acidic xylan from compression wood. Such wood of *Pinus strobus* gave a hemicellulose consisting of a monomethyluronic acid linked to five to six xylose residues, undoubtedly the polysaccharide now known as an arabino-4-*O*-methylglucuronoxylan. The investigators believed that this xylan was linked to a mannan. It is, of course, now obvious that they were dealing with a mixture of a xylan and a galactoglucomannan, two hemicelluloses that can only with difficulty be completely separated from each other (Meier 1958a). On several occasions, fractions consisting largely of xylan have been obtained on fractional extraction of delignified compression wood. The 14% potassium hydroxide extract recovered by Bouveng and Meier (1959)

5.7 Xylan

from *Picea abies* contained 57% xylose residues (Table 5.28). A 5% potassium hydroxide extract from *Picea rubens* had a 68% xylose content (Table 5.29) (Schreuder et al. 1966). The same extract from *Abies balsamea* contained 78% xylose residues (Table 5.30) (Côté et al. 1967b). If the arabinose and 4-*O*-methylglucuronic acid units are included, the xylan content of this fraction becomes 85%.

In xylans from normal conifer woods, the ratio between xylose and arabinose is 1:8 to 1:9. This is also the ratio between arabinose and xylose in the 1% and 5% potassium hydroxide extracts from normal wood of *Abies balsamea* (Table 5.30). In contrast, the 5% potassium hydroxide extract from compression wood of this species had an arabinose to xylose ratio of 1:22. Similarly, the same extract from *Picea rubens* compression wood (Table 5.29) had a ratio of 1:25. It would appear from these data that, compared to the xylan in normal wood, that in compression wood contains fewer arabinose side chains. In a study of the development of the cell wall in tracheids of *Pinus resinosa*, Larson (1969) found that the recovery ratio of ^{14}C in incorporation between arabinose and xylose was much less for compression wood than for normal wood.

Attempts by Côté et al. (1967b) to isolate a pure xylan from compression wood of *Abies balsamea* were not successful. Repeated treatment with barium hydroxide finally left a soluble fraction which contained only 50% xylose residues. Later, Hoffmann and Timell (1972b) succeeded in isolating a pure xylan from compression wood of *Larix laricina*, albeit in low yield. The fact that compression wood xylan cannot be obtained in a pure state by use of the barium hydroxide method alone was confirmed in this investigation. When applied to normal conifer woods, this technique almost invariably allows isolation of a pure xylan. Another case where it failed, has, however, been recorded by Fu and Timell (1972c), who were unable to obtain a pure xylan from phloem of *Pinus sylvestris* by this technique.

The 5% potassium hydroxide extract from *Larix laricina* compression wood contained 67% xylan. The water-insoluble portion of this extract was the (1 → 3)-β-D-glucan already discussed (Sect. 5.4). The water-soluble part consisted of a galactan, a galactoglucomannan, a xylan, and some glucan. Treatment with barium hydroxide eliminated the galactoglucomannan, and subsequent complexing with iodine removed the glucan and much of the galactan. The remaining mixture of galactan and xylan was acetylated, and the acetates were separated by fractional precipitation, the acetate of the pentosan being less soluble than that of the hexosan. After deacetylation, the pure xylan on hydrolysis yielded xylose, arabinose, 4-*O*-methylglucuronic acid, and 2-*O*-(4-*O*-methyl-α-D-glucopyranosyluronic acid)-D-xylose (**5**).

5

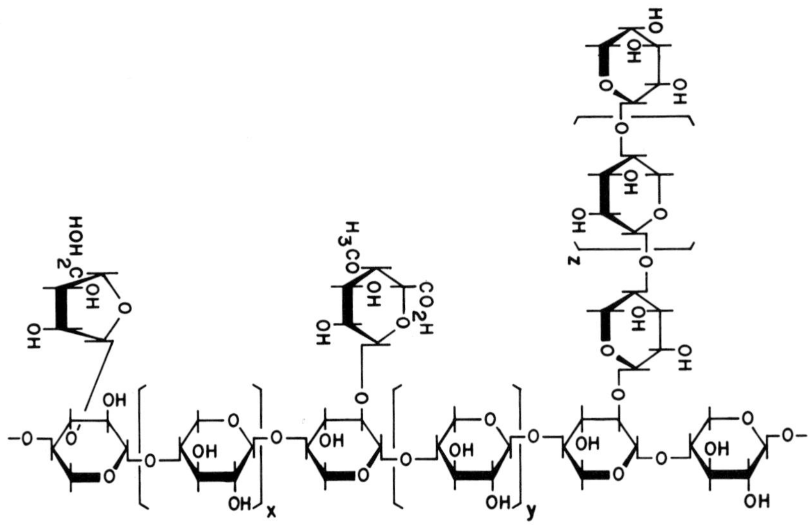

Fig. 5.11. Structure of the repeating unit of a compression wood arabino-4-*O*-methylglucuronoxylan

The methylated polysaccharide contained on the average 85 xylose residues and, after methanolysis, gave as methyl glycosides (moles): 2-*O*-methylxylose (7.1), 3-*O*-methylxylose (9.7), 2,3-di-*O*-methylxylose (49.0), 2,3,4-tri-*O*-methylxylose (5.2), 2,3,5-tri-*O*-methylarabinose (4.9), and 2-*O*-(2,3,4-tri-*O*-methylglucuronosyl)-3-*O*-methylxylose (14.2). The xylan accordingly consisted of (1 → 4)-linked β-D-xylopyranose residues which formed a framework containing about 4 branches per 85 xylose residues with the branch point probably at C-2. Residues of 4-*O*-methyl-α-D-glucuronic acid and α-L-arabinofuranose were attached to the main chain through C-2 and C-3, respectively. The hemicellulose contained only one arabinose side chain per 17 xylose residues, which agrees with the earlier evidence on this point. The structure of the xylan is shown below (**6**) and in Fig. 5.11.

$$\rightarrow 4\text{-}\beta\text{-D-Xylp-}1 \left[\rightarrow 4\text{-}\beta\text{-D-Xylp-}1 \right]_2 \rightarrow 4\text{-}\beta\text{-D-Xylp-}1 \rightarrow 4\text{-}\beta\text{-D-Xylp-}1 \rightarrow 4 \left[\text{-}\beta\text{-D-Xylp-}1 \right]_{11} \rightarrow$$

$$\begin{array}{ccc}
\uparrow & & \uparrow \\
1 & & 1 \\
4\text{-}O\text{-Me-}\alpha\text{-D-GlcpA} & & \alpha\text{-L-Araf}
\end{array}$$

6

Latif (1968) has suggested that in compression wood of *Pseudotsuga menziesii* one portion of the xylan should contain a large number of arabinose side chains, while another part should carry relatively few, but the evidence adduced for this is not convincing.

Although the structural proof could be more complete, it is evident that the xylan present in compression wood is very similar to that occurring in normal wood of conifers, the only difference being the lower frequency of arabinose side chains in the compression wood xylan. Hoffmann and Timell (1972b) did not determine the exact proportion of acid side chains in their xylan. The uronic acid content of holocellulose extracts is not a measure of the frequency of the side chains as these extracts always contain oxidized lignin.

5.8 Arabinogalactan

5.8.1 Introduction

Arabinogalactans are highly branched polysaccharides present together with xyloglucan, pectin, and cellulose in the primary cell wall of plants (Chap. 6.3.1). An arabinogalactan linked to protein has been isolated from the cambial zone of *Populus tremuloides* and *Tilia americana* (Simson and Timell 1978). A somewhat different, albeit closely related, arabinogalactan occurs in the heartwood of all members of the genus *Larix* in amounts that vary within the wide limits of 5% to 40% by weight of the wood. As is shown in Chap. 6.3.3, larch arabinogalactan is located in the lumen of the tracheids and ray cells in the heartwood. None occurs in the sapwood. Several reviews of the properties and utilization of larch arabinogalactan are available, such as those by Adams and Douglas (1963) and Timell (1965a).

Arabinogalactan is unique among wood hemicelluloses in being entirely extracellular and highly branched, and there are good reasons to consider it as a gum or mucilage rather than as a hemicellulose. One component of the gum exuded by *Araucaria bidwillii* is an arabinogalactan with almost the same structure as the larch arabinogalactan (Aspinall and Fairweather 1965, Aspinall and McKenna 1968, Aspinall et al. 1970). Larch arabinogalactan is also remarkable in the ease with which it is degraded within the living tree. As trees grow older, acetyl groups are gradually hydrolyzed, forming acetic acid, with the result that the acidity of the wood, and especially the heartwood, increases until a pH of 3 or lower is reached. The extracellular, amorphous, and water-soluble arabinogalactan is not only fully accessible but also acid-sensitive and, as the wood ages, the arabinogalactan is drastically reduced in molecular size and loses its acid-labile arabinose groups (Côté et al. 1967c, Jones et al. 1968). The chemical composition and molecular properties of larch arabinogalactan can therefore vary considerably, depending on the age of the wood from which the polysaccharide had been isolated. Much controversy in the literature on this point can probably be attributed to this lability toward acids within the living tree.

Arabinogalactan is also unique among wood polysaccharides in consisting of two components, customarily referred to as A and B (Mosimann and Svedberg 1942). Structurally, there seems to be little difference between the two components (Bouveng and Lindberg 1958, Bouveng 1959, 1961), although arabinogalactan A is probably more highly branched than B (Swenson et al. 1969). The distinguishing feature of the two polysaccharides is their widely different molecular weights, a fact first observed by Mosimann and Svedberg

(1942). According to Simson et al. (1968) the number-average molecular weight of arabinogalactan A is 70 000 whereas that of component B is only 11 000. Swenson et al. (1969) found components A and B to have molecular weights of 37 000 and 7500, respectively.

The proportions of arabinogalactans A and B vary considerably, although A always predominates. Some investigators have failed to observe any arabinogalactan B at all, while others have detected amounts up to 30% of the total. Simson et al. (1968) showed that the proportion of component B tends to increase with increasing age of the wood as a result of acid-catalyzed fragmentation of polysaccharide A.

Much effort has been devoted to establishing the structure of larch arabinogalactan. The polysaccharide is quite complex and is surpassed in this respect among wood polysaccharides only by the galactan present in tension wood. Many structural details therefore remain obscure. There is general agreement that the polysaccharide has a main chain of $(1 \rightarrow 3)$-linked β-D-galactopyranose residues, all of which carry a side chain at C-6. A large number of these side chains consist of $(1 \rightarrow 6)$-linked β-D-galactopyranose residues. While structural evidence suggests that most of these branches contain only two galactose residues, Young et al. (1972) have shown that some of them must be much longer. The extremely low viscosity of arabinogalactan in water also points to the presence of a large number of long branches.

The ratio between arabinose and galactose is usually 1:6. The majority of the arabinose residues occur in side chains of 3-O-β-L-arabinopyranosyl-L-arabinofuranose which are attached to the main chain, and the others are linked directly to this chain as L-arabinofuranose. The presence in the same polysaccharide of both arabinopyranose and the more common arabinofuranose is unusual. A few of the galactose residues are also directly attached to the main chain as single unit side chains. Finally, a few β-D-glucuronic acid residues are also present, all linked $(1 \rightarrow 6)$ to galactose. There is structural evidence also for other features, but most of the latter require additional investigation before being accepted. The absence of any $(1 \rightarrow 4)$-linked galactose is notable and shows that larch arabinogalactan and the galactan present in compression wood are entirely different polysaccharides.

5.8.2 Structure

It has already been mentioned that compression wood of *Larix laricina* seems to contain less arabinogalactan than does corresponding normal wood (Table 5.18) (Côté et al. 1966a, Côté and Timell 1967). Following an early investigation by Wise et al. (1933), the structure of the arabinogalactan in normal wood of *Larix laricina* was established by Adams (1960), Haq and Adams (1961), Urbas et al. (1963), and Lynch et al. (1968).

The constitution of the arabinogalactan present in compression wood of *Larix laricina* was studied by Fu and Timell (1972a). As already mentioned, the polysaccharide was obtained in a yield of 3.9% by extraction with water of the same larch wood from which the laricinan had been isolated (Table 5.31). It

5.8.2 Structure

contained arabinose and galactose residues in a ratio of 1:7.5 and had a specific rotation of +15°. Arabinogalactans from normal larch wood have been reported to have rotations within a range of 11°−18°. Small amounts of glucuronic acid were also formed on hydrolysis.

The structure of the arabinogalactan was determined by the techniques of partial acid hydrolysis, methylation, and Smith degradation. On partial hydrolysis the arabinogalactan yielded, besides arabinose and galactose, 3-*O*-*β*-L-arabinopyranosyl-L-arabinose, 3-*O*-*β*-D-galactopyranosyl-D-galactose, and 6-*O*-*β*-D-galactopyranosyl-D-galactose in an approximate ratio of 2:3:5. The arabinogalactan was methylated and subjected to methanolysis, and the methyl glycosides were separated by gas-liquid chromatography. The following methylated sugars were present (mole per cent): 2,5-di-*O*-methylarabinose (7.3), 2,3,4-tri-*O*-methylarabinose (6.1), 2,3,5-tri-*O*-methylarabinose (5.6), 2,4-di-*O*-methylgalactose (33.4), 2,3,4-tri-*O*-methylgalactose (16.7), 2,4,6-tri-*O*-methylgalactose (3.9), and 2,3,4,6-tetra-*O*-methylgalactose (27.0). These relative amounts of methylated sugars are very similar to those reported by Adams (1960) for an arabinogalactan from normal wood of *Larix laricina*.

The nature of the main chain in the arabinogalactan was established by Smith degradation (Aspinall et al. 1968), carried out in three successive stages. The procedure involves oxidation with periodic acid, reduction of the aldehyde groups produced to alcohol, and partial hydrolysis with preferential cleavage of acetal rather than glycosidic bonds. The degradation eliminates the arabinose groups and any side chains composed of (1 → 6)-linked galactose residues but leaves the main chain largely intact if the latter is composed of (1 → 3)-linked galactopyranose residues. When the compression wood arabinogalactan was degraded by this technique, the final, periodate-stable residue, gave 3-*O*-*β*-D-galactopyranosyl-D-galactose with only a small amount of a (1 → 6)-linked galactobiose on partial hydrolysis. The methylated residue consisted mostly of 2,4,6-di-*O*-methylgalactose together with some 2,4-di-*O*-methylgalactose and 2,3,4,6-tetra-*O*-methylgalactose. These results prove that the main chain of the polysaccharide consisted of a (1 → 3)-linked *β*-D-galactan. The same has been found for the arabinogalactan present in normal wood of *Larix laricina* (Lynch et al. 1968). The evidence obtained by Fu and Timell (1972a) makes it clear that the arabinogalactan present in compression wood of *Larix laricina* is structurally identical with that formed in corresponding normal wood. A simplified outline of the complete structure is shown below (7).

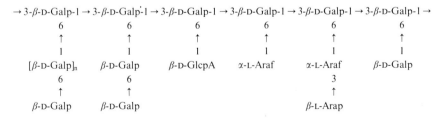

7

Again, it must be pointed out that the length of the (1 → 6)-linked galactose side chains is not known with certainty and that the location of the D-glucuronic acid residues is only tentative. The possibility that the galactan side chains are themselves branched cannot be excluded. The low viscosity of the arabinogalactan indicates that it is a spherical macromolecule.

5.8.3 Molecular Properties

In their study on the molecular properties of various larch arabinogalactans, Simson et al. (1968) examined an arabinogalactan from compression wood of *Larix laricina*. This polysaccharide was obtained in a yield of 7% and contained residues of arabinose and galactose in a ratio of 1:6 and also some glucuronic acid. Number-average molecular weights were determined by osmometry, using aqueous dimethylformamid as a solvent. Weight-average molecular weights were obtained by the sedimentation-equilibrium method. Some of the results are summarized in Table 5.33. Because the molecular weights of any larch arabinogalactan will depend on the age of the heartwood from which the polysaccharide was isolated, data obtained with arabinogalactans from *Larix occidentalis* and *L. decidua* have also been included.

When the variation in molecular weight due to acid hydrolysis is taken into account, it must be concluded that the arabinogalactan in compression wood probably has the same molecular properties as that in normal larch wood. The greater variation in the number-average than in the weight-average molecular weights in Table 5.33 is caused by the fact that the former are always more affected by random depolymerization than are the latter. The high ratio between weight- and number-average molecular weights would seem to indicate a high polydispersity. This ratio, however, results from the presence of two polysaccharides of widely different molecular sizes. When determined for each component separately, the $\bar{M}_w/\bar{M}_n$ ratio is only 1.1 for arabinogalactan A and

Table 5.33. Age of wood, number-average (M_n) and weight-average (M_w) molecular weight, and relative percentage of arabinogalactan B in larch wood. (Simson et al. 1968)

Species and wood	Age of wood, years	M_n	M_w	M_w/M_n	Arabinogalactan B, per cent
Larix laricina					
Normal wood	1 – 75	36 000	93 800	2.61	4.3
Compression wood	1 – 17	52 000	94 600	1.82	4.5
Larix occidentalis					
Normal wood	1 – 320	58 500	91 000	1.56	5.1
Larix decidua					
Normal wood	1 – 25	29 600	82 800	2.80	2.3

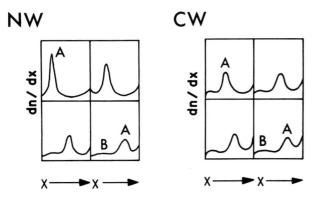

Fig. 5.12. Sedimentation diagrams of arabinogalactans in normal (*NW*) and compression (*CW*) woods of *Larix laricina*. (Simson et al. 1968)

1.3 for B (Simson et al. 1968), and even lower values have been reported by Swenson et al. (1969).

Arabinogalactans A and B can be separated from each other by paper electrophoresis, free boundary electrophoresis, sedimentation in the ultracentrifuge, and gel permeation chromatography. An example of sedimentation diagrams for arabinogalactans from normal and compression woods of *Larix laricina* is shown in Fig. 5.12. For preparative purposes, gel permeation is the method of choice. Simson et al. 1968 subjected arabinogalactans from normal and compression woods of *Larix laricina* to gel permeation on Sephadex, a cross-linked dextran. Both polysaccharides were found to consist of two components. The elution diagram of the arabinogalactan in the compression wood is shown in Fig. 5.13. Undoubtedly, the larger component was arabinogalactan A and the minor arabinogalactan B, although their separation was not complete in this case. The shoulder to the left of the elution diagrams is caused by phenolic impurities (Ettling and Adams 1968).

5.8.4 Conclusions

Conifers other than larches contain small amounts, generally less than 1–2% of the wood, of an arabinogalactan which is similar, albeit not entirely identical, to that occurring in larch wood. The major difference seems to be that, while in *Larix* some of the arabinose residues occur as nonterminal L-arabinofuranose or terminal L-arabinopyranose units, all arabinose in the other conifers is present as terminal L-arabinofuranose residues. An arabinogalactan from *Pinus sylvestris* has also been found to possess nonreducing end groups of D-xylose (Aspinall and Wood 1963). When softwoods other than larch wood are extracted with water, the arabinogalactan is usually obtained together with a galactoglucomannan.

Although no arabinogalactan of this type has ever been isolated from compression wood, there is every reason to believe that such a polysaccharide is present in this wood. The fraction obtained in a low yield by Bouveng and

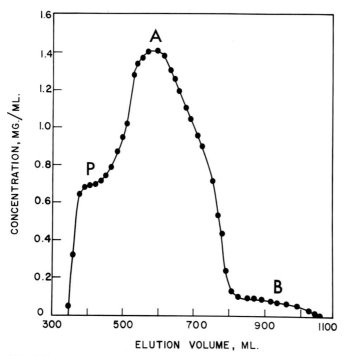

Fig. 5.13. Elution diagram obtained on gel permeation chromatography of an arabinogalactan from compression wood of *Larix laricina*. The diagram indicates the presence of arabinogalactan *A*, a smaller amount of arabinogalactan *B*, and an impurity (*P*), probably a polyphenolic tannin. (Simson et al. 1968)

Meier (1959) on extraction of compression wood of *Picea abies* with water probably consisted of an arabinogalactan, a galactan, and a galactoglucomannan. A similar extract from compression wood of *Picea rubens*, on the other hand, consisted only of galactan (Schreuder et al. 1966). There is a strong possibility that this type of arabinogalactan is derived from the primary cell walls in wood.

In members of the genus *Larix*, both normal and compression woods contain from 5 to 25% arabinogalactan, depending on the location in the stem of the wood (Côté et al. 1966a, Côté and Timell 1967, Hashizume and Takahashi 1974). Apparently, compression wood contains lesser amounts than does normal wood. It should perhaps be noted in this connection that arabinogalactan is a heartwood constituent, most probably formed in the ray cells at the sapwood-heartwood boundary some time after the wood in which it will ultimately be deposited has been laid down. In larches, sapwood is transformed into heartwood within the relatively short time of 5–10 years.

Unlike the tracheids, the ray cells in compression wood differ little from those in normal wood in their anatomy (Chap. 4.7) and not at all in the chemical composition of their cell walls (Chap. 6.3.2). It would accordingly appear that ray cells function in the same way in compression wood as in normal wood,

5.9 Lignin

5.9.1 Introduction

and it is therefore perhaps not surprising that the same arabinogalactan is produced in the two types of wood. Whether the same applies to the many extractives that are formed in the ray cells on transition from sapwood to heartwood is not known, but results obtained by Yasuda et al. (1975) (Sect. 5.10) with normal and compression woods of *Larix leptolepis* would indicate that this is indeed the case.

5.9.1 Introduction

Its high lignin content is the foremost chemical characteristic of compression wood. Despite this fact, the structure of this lignin has attracted attention only within the last 25 years. As is discussed in Sect. 6.4, the distribution of lignin in compression wood is quite different from that in normal wood, and this aspect has attracted the interest of several investigators over a fairly long period of time. The chemistry of the lignin present in compression wood, by contrast, was not seriously studied until 1958.

Lignin is a three-dimensional polymer consisting of phenylpropane (**8**) units linked together by carbon-carbon and carbon-oxygen bonds (Sarkanen and Ludwig 1971, Adler 1977). The three monomers that polymerize, largely by dehydrogenation (Freudenberg and Neish 1968), to form different types of lignin are p-coumaryl alcohol (**9**), coniferyl alcohol (**10**), and sinapyl alcohol (**11**). In almost all gymnosperm species the lignin is of the guaiacyl (**13**) type in normal wood, while all angiosperms have a guaiacyl (**13**)-syringyl (**14**) lignin of a composition that varies between fibers, vessels, and ray cells. The lignin in compression wood and possibly also that located in the middle lamella-primary wall region of normal softwood are guaiacyl (**13**)-p-hydroxyphenyl (**12**) lignins, formed on copolymerization of coniferyl and p-coumaryl alcohols. A schematic, two-dimensional model of the constitution of the lignin in *Picea abies* according to Freudenberg (1964, 1965, 1968) is shown in Fig. 5.14.

5.9.2 Lignin in Compression Wood

In their pioneering investigation on the chemistry of compression wood in *Picea abies*, Hägglund and Ljunggren (1933) included a brief examination of the lignin component of this wood and its behavior on sulfonation, an aspect discussed in Chap. 19.4.1. Compression wood lignin, isolated by the hydrochloric acid method, was found to contain only 13.5% methoxyl groups compared to 17.3% for lignin in normal spruce wood. Lignin from compression wood gave 2.9% formaldehyde on distillation with hydrochloric acid, whereas normal wood lignin yielded only 0.9%. The investigators suggested that some of the methoxyl groups in normal wood lignin had been replaced by methylene oxide groups. Half of the lignosulfonic acid obtained after sulfite digestion of the compression wood could not be precipitated with naphthylamine, compared to only 10% when normal wood was used. Hägglund and Ljunggren (1933) attributed this different behavior to differences in molecular size of the two lignosulfonates.

Hata (1951) isolated two hydrochloric acid lignins from normal and compression woods of *Pinus densiflora*. Their methoxyl contents were 14.3% and 12.7%, respectively, thus confirming the earlier results of Hägglund and Ljunggren (1933).

Bland (1958a, b, 1961), in a series of investigations, isolated lignins from normal and compression woods of *Pinus radiata* and compared their chemical properties. In the first investigation (Bland 1958a), wood was used from sections of an eccentrically grown stem, shown in Fig. 5.1. Lignin was extracted at 150 °C with methanol and subsequently separated by chromatography from low-molecular weight contaminations. The ultraviolet absorption spectra of normal and compression wood lignins were identical and exhibited a maximum at 279 – 280 nm, as is typical of conifer lignins.

Klason lignins were also isolated from the five zones and were analyzed for their methoxyl content. Oxidation with nitrobenzene and alkali yielded vanillin. Acetaldehyde was obtained by distillation with alkali, while distillation with hydrochloric acid gave formaldehyde. The results are summarized in Table 5.34. Data obtained with heartwood have been omitted, as it is now known that heartwood lignin tends to be contaminated with polyphenolics (Sarkanen et al. 1967b). Compression wood had a considerably lower methoxyl content than any of the other woods, a fact also noted by Wardrop and Bland (1961), and afforded less vanillin and more p-hydroxybenzaldehyde on oxidation. The yield of acetaldehyde was the same from all zones. Formaldehyde, however, was obtained in lower yield from compression wood than from normal wood, a result contrary to that reported by Hägglund and Ljunggren (1933). Wood from the different zones were subjected to sulfite cooking (Watson and Dadswell 1957) (Chap. 19.4.1), and the lignosulfonic acids formed were

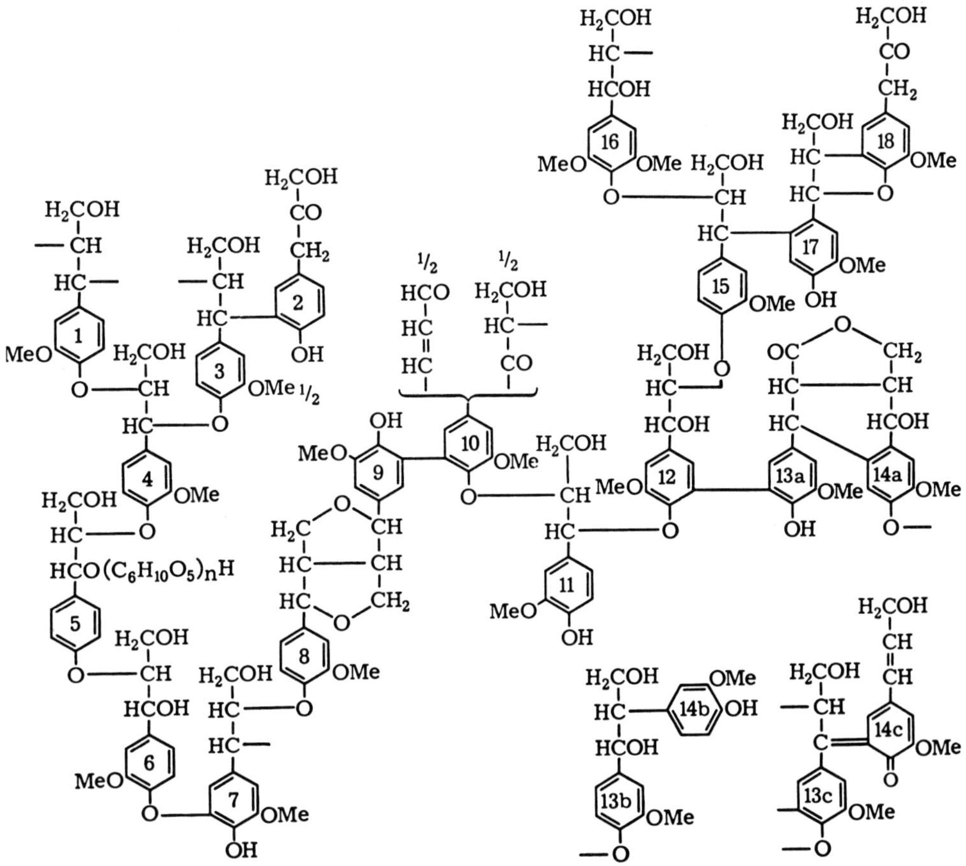

5.9.2 Lignin in Compression Wood

Fig. 5.14. Schematic, two-dimensional structure of the lignin present in *Picea abies*. (Freudenberg 1964, 1965, 1968)

subjected to chemical analysis. Nitrogen, sulfur, and methoxyl contents of the lignosulfonic acids from compression wood were all lower than those from normal woods. Bland concluded from his results that compression wood lignin has the same basic structure as lignin in normal wood except that it contains fewer guaiacylpropane and instead more p-hydroxyphenylpropane units.

In a subsequent investigation, Bland (1961) isolated milled wood lignins from normal and compression woods of *Pinus radiata* by milling in a vibratory ball mill for 48 h, followed by extraction with dioxane containing 10% water. The yields of lignin were low, and it is likely that the grinding time was too short. The lignin from compression wood could not be completely dissolved in a mixture of ethanol and ethylenedichloride, whereas lignin from normal wood was fully soluble. Some chemical data for these lignins are shown in Table 5.35. The Klason lignin content of the three lignin preparations was low (Latif 1968). As usual, the compression wood lignin contained fewer methoxyl groups. Xylan was present throughout.

Table 5.34. Properties and decomposition products of lignins isolated from different regions of *Pinus radiata* stem wood. (Bland 1958a)

Region	Klason lignin, per cent of wood	Methoxyl, per cent of lignin	Vanillin, per cent of lignin	Acetaldehyde, per cent of lignin	Formaldehyde, per cent of lignin
Inner sapwood	24.3	15.0	20.2	0.33	0.75
Side wood	25.5	14.0	20.5	0.37	0.69
Opposite wood	25.8	14.1	20.2	0.36	0.72
Compression wood	34.4	12.6	17.7	0.34	0.38

Table 5.35. Yield and chemical data for milled wood lignins isolated from normal and compression woods of *Pinus radiata*. (Bland 1961)

Wood and lignin	Yield, per cent of wood	Per cent of milled wood lignin		
		Klason lignin	Methoxyl	Xylan
Normal wood				
Reprecipitated lignin	3.2	73.0	14.2	2.6
Compression wood				
Reprecipitated lignin	3.5	82.0	12.4	0.7
Insoluble residue	4.6	77.4	11.3	1.2

Table 5.36. Yield of aromatic aldehydes from milled wood lignins isolated from normal and compression woods of *Pinus radiata*. All values in per cent of lignin. (Bland 1961)

Aldehyde	Normal wood	Compression wood
p-Hydroxybenzaldehyde	Nil	3.3
Vanillin	16.1	13.3
Total aldehydes	16.1	16.6

On oxidation with nitrobenzene and alkali, the lignins gave the aldehydes shown in Table 5.36. Like the Klason lignin, milled wood lignin from compression wood afforded less vanillin than did lignin from normal wood. Significantly, it yielded 3.3% p-hydroxybenzaldehyde, a compound not obtained with normal wood lignin. The lignin remaining in the compression wood, by contrast, gave no p-hydroxybenzaldehyde on oxidation and also afforded the same amount of vanillin as did the residual lignin in the normal wood, suggesting

5.9.2 Lignin in Compression Wood

that only the milled wood lignin was deficient in methoxyl groups. Milled wood lignin was obtained with ease from compression wood, and Bland (1961) concluded that this "extra" lignin must be readily accessible.

The yield of formaldehyde was the same from the two milled wood lignins which also contained the same proportions of carbonyl and p-hydroxybenzyl alcohol groups. Infrared spectra for the two lignins showed that they were quite similar except for the absence of a band at 5.8 nm (1725 cm^{-1}) in the lignin from compression wood. According to Sumiya et al. (1969) compression wood of *Cryptomeria japonica* differs from normal wood in the region 1060 – 1030 cm^{-1} in its infrared spectrum, a range where conifer lignins exhibit absorption (Sarkanen et al. 1967a, b).

Yean and Goring (1965) sulfonated opposite, normal, and compression woods of *Picea mariana* with acid sodium sulfite and characterized the sodium lignosulfonates obtained (Table 5.7). Compared to the other two woods, the compression wood lignosulfonate had a lower methoxyl content. Its weight-average molecular weight (27 000) was 30% higher than that of the other preparations (19 000), but the investigators did not attribute much importance to this difference. Interestingly, latewood of *Tsuga heterophylla* gave a sodium lignosulfonate with a molecular weight of 30 200 compared to 19 700 for earlywood.

Lebedev (1965a) studied the ozonization of compression wood from the lower portion of branches (probably from *Pinus sylvestris*). The lignin was ozonized extensively and almost half the wood dissolved, but it was evident that the wood polysaccharides were also affected. Lebedev (1965b) found that compression wood, presumably the lignin, had a higher content of double bonds than normal wood.

Extensive investigations of compression wood lignin in *Pseudotsuga menziesii* were carried out by Sarkanen and his co-workers Latif (1968) and Lee (1968). Since neither of these interesting studies is readily available and both contain much valuable information, they will be discussed in some detail here.

Latif (1968) obtained normal and compression woods from stem sapwood of Douglas-fir with lignin contents of 27.0% and 36.5%, respectively. Branch compression wood gave the same results as that from the trunk. Milled wood lignin was isolated by first grinding wood meal in a standard porcelain grinding jar for 6 weeks. The ground wood was extracted with dioxane containing 10% of water, and the extract was dissolved in ethanol-1,2-dichloroethane in which 74% of the lignin was soluble, a much larger proportion than that noted by Bland (1961). The remaining wood powder was extracted with 50% aqueous acetic acid to give a fraction consisting of both lignin and polysaccharides and commonly referred to as a "lignin-carbohydrate complex." Yields of lignin and complex, based both on the wood and the Klason lignin content for various times of grinding, are summarized in Table 5.37. The effect of grinding on the yield of milled wood lignin and the lignin-carbohydrate complex is shown graphically in Fig. 5.15.

The curves indicate that at first the yield of the complex increased more rapidly than that of the lignin, reaching a maximum of about 30% of the total (Klason) lignin content after 10 days. Subsequently, the yield of the complex

Table 5.37. Yields of milled wood lignin and lignin-carbohydrate complex on grinding and extraction of normal and compression woods of *Pseudotsuga menziesii*. All values in per cent except time. (Latif 1968)

Grinding time, days	Milled wood lignin				Lignin-carbohydrate complex			
	Normal wood		Compression wood		Normal wood		Compression wood	
	Based on wood	Based on Klason lignin	Based on wood	Based on Klason lignin	Based on wood	Based on Klason lignin	Based on wood	Based on Klason lignin
9	5.4	20.0	6.7	18.8	6.1	22.6	10.9	30.3
13	8.3	30.6	10.0	26.6	6.5	24.1	9.8	27.3
16	10.4	38.5	12.8	35.0	5.9	21.8	8.9	24.6
23	15.2	55.2	18.4	51.5	4.3	15.9	5.7	18.1
30	16.0	58.0	20.2	55.3	4.2	15.2	5.1	16.9

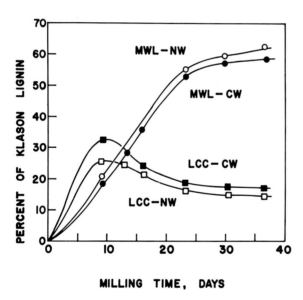

Fig. 5.15. Effect of milling time on the yield of milled wood lignin (*MWL*) and lignin-carobohydrate complex (*LCC*) from normal (*NW*) and compression (*CW*) woods of *Pseudotsuga menziesii*. (Redrawn from Latif 1968)

declined to a constant level of 15%. The yield of lignin increased at a constant rate during the first 25 days and then increased more slowly, until a value of about 60% had been reached after 40–50 days. The yield of milled wood lignin was throughout higher for compression wood than for normal wood. Based on the lignin contents of the woods, however, normal wood gave slightly more lignin than did the compression wood. Larger amounts of the lignin-carbohydrate complex were obtained throughout from compression wood than from normal wood. The yields of milled wood lignin reported by Latif (1968) were much higher than those obtained by Bland (1961). Attempts to isolate more lignin by treating the residual wood with cellulolytic enzymes were not successul. The chemical composition of the milled wood lignins is shown in Table 5.38. Compression wood lignin contained slightly less Klason lignin but more acid-soluble lignin than did the lignin from normal wood. It also had a higher polysaccharide content.

Björkman (1957) had found that the carbohydrate portion of the lignin-carbohydrate complex consisted of hemicelluloses that were present in the same proportions as in the original wood. The small amounts of hemicelluloses present in milled wood lignins, by contrast, contained more than twice as much galactose and arabinose residues as did the wood. It is known that, compared to the secondary wall, the middle lamella and primary wall have a much higher concentration of galacturonic acid, galactose, arabinose and rhamnose residues derived from pectin. This, as pointed out by Lai and Sarkanen (1971) indicates that some of Björkman's milled wood lignin may have originated from the compound middle lamella. However, as also emphasized by the same authors, the compound middle lamella of conifers contains only 25% of the total lignin in wood, and with yields of 50–60%, lignin from the secondary wall must also have been present. Presumably, the lignin released first on grinding comes from the intercellular region, while subsequently wall lignin is also released.

Whiting and Goring (1981) have demonstrated that when normal wood of *Picea mariana* is ball-milled and extracted with aqueous dioxane, four times more lignin is removed from the secondary wall than from the middle lamella region. Although no such data are available for any compression wood, it is likely that most of the milled wood lignin obtained from compression wood also comes from the secondary wall, and especially since only 5–10% of the total lignin in this tissue is located in the middle lamella (Chap. 6.4.2).

The carbohydrates in the normal, milled wood lignin isolated by Latif (1968), as can be seen from Table 5.38, contained a high proportion of galactose and rhamnose residues, suggesting the presence of pectin, but the arabinose content was only half of that reported by Björkman (1956, 1957). The carbohydrate composition of the compression wood lignin is interesting. Compression wood contains much more galactan, less galactoglucomannan, and about the same amount of xylan as normal wood (Table 5.43). It is also known that the galactan is located largely in the outer portion of the cell wall and that the middle lamella-primary wall has the same polysaccharide composition as in normal wood tracheids (Chap. 6.3) (Côté et al. 1968a). It is not surprising, then, to find that the carbohydrate part of the compression wood lignin contained 52% galactose but only 15% mannose residues. Some of the galactose and most of the arabinose and rhamnose residues were possibly derived from pectic material in the middle lamella-primary wall. The explanation offered by Latif (1968), namely that some of the compression wood xylan should contain a larger number of arabinose side chains, is not convincing, as it is now known that the xylan in compression wood has only half the number of such side chains compared to the xylan in normal softwood (Hoffmann and Timell 1972b). Meshitsuka et al. (1982) have suggested that the galacturonic acid residues present in a lignin-carbohydrate complex from birch wood most probably were

Table 5.38. Yield and chemical composition of milled wood lignins from normal and compression woods of *Pseudotsuga menziesii*. All values in per cent. (Latif 1968)

Zone	Yield		Klason lignin	Total lignin	Total carbo-hydrate	Sugar residues, relative per cent					
	Based on wood	Based on Klason lignin				Galactose	Glucose	Mannose	Arabinose	Xylose	Rhamnose
Normal wood	15.6	58.1	89.9	96.1	3.9	13.7	12.8	20.1	5.5	9.2	17.6
Compression wood	19.4	56.2	84.1	94.4	5.6	52.2	6.1	13.4	9.5	10.1	7.4

5.9.2 Lignin in Compression Wood

Table 5.39. Lignin content and yields of p-hydroxybenzaldehyde and vanillin on oxidation with nitrobenzene and alkali of original wood, milled wood lignin, and residual wood from normal and compression woods of *Pseudotsuga menziesii*. All values in per cent. (Latif 1968)

Material	Klason lignin	p-Hydroxy-benzaldehyde	Vanillin
Normal wood			
Wood meal	27.1		25.2
Milled wood lignin	89.9		36.8
Residual wood meal	19.1		16.8
Compression wood			
Wood meal	36.4	Trace	21.6
Milled wood lignin	84.1	13.3	23.2
Residual wood meal	19.8		15.3

derived from pectic material in the middle lamella-primary wall region. Minor (1982) believes that the galactan and arabinan present in a lignin-carbohydrate complex from *Pinus taeda* wood are also of pectic origin.

Recent investigations by Eriksson and Lindgren (1977) and by Eriksson et al. (1980) have shown that there are chemical links in *Picea abies* and *P. mariana* woods between lignin and both cellulose and hemicelluloses. Their results indicate that lignin is linked to C_2 and C_3 in the arabinose side chain of the xylan and to C_3 in the galactose side chain of the glucomannan. Benzyl ester bonds also exist between the lignin and the 4-*O*-methylglucuronic acid side chains in the xylan. Results obtained with complexes between carbohydrates and compression wood lignin are discussed at the end of this section.

Latif's (1968) milled-wood lignin from normal wood contained 13.7% methoxyl but that from compression wood only 10.9%, a result in agreement with those of previous investigators. Original wood, milled wood lignin, and wood remaining after the latter had been removed were oxidized with nitrobenzene and alkali, giving the results shown in Table 5.39. For unknown reasons, compression wood gave only a trace of p-hydroxybenzaldehyde, but the milled wood lignin afforded as much as 13%, considerably more than reported by Bland (1961). None was obtained from the residual wood, thus confirming Bland's conclusion that the p-coumaryl units occur only in that part of the lignin which is released on milling.

The compression wood preparations all gave less vanillin than did those from normal wood. The yields of vanillin were highest from the milled wood lignins, and especially so with normal wood. It is known that no vanillin is formed from condensed lignin units, that is, units chemically linked to another unit through C-5 in the guaiacylpropane residue. Uncondensed lignin units are obviously more frequent in that part of the lignin which can be removed on milling than in that which remains in the wood. Methoxyl groups recovered as vanillin amounted to 54.5% for normal wood but was only 43.8% for compression wood lignin. Ethanolysis of milled wood lignins and examination of the Hibbert's ketones formed provided evidence that the p-hydroxyphenyl groups in compression wood lignin probably originated from p-hydroxy-phenylglycerol-β-aryl ether units (**15**). According to Latif (1968) lignin in normal wood of

Pseudotsuga menziesii consists of 88% guaiacyl propane and 12% p-hydroxyphenylpropane units, corresponding figures for lignin in compression wood being 70% and 30%, respectively.

Infrared spectra of milled wood and thioglycolic acid lignins from normal and compression woods are shown in Figs. 5.16 and 5.17. For the milled wood lignins, the two spectra are very similar. The major differences are found in the region 1735 cm^{-1}, which is assigned to unconjugated carboxylic acid and carboxylate (acetate) ester groups, and at 1715 cm^{-1}, a band assigned to conjugated esters of aromatic acids such as vanillic, p-hydrobenzoic, and ferulic acid groups (Sarkanen et al. 1967b). Judging from the infrared spectra, it would appear that compression wood lignin contains fewer ester groups than normal wood lignin. Sarkanen et al. (1967b) have proposed that the ester groups may exert a plasticizing effect on the lignin by reducing the number of sites available for internal hydrogen bonding. They point out that the virtual absence of ester groups in compression wood lignin might render this lignin more rigid than that in normal wood.

Sarkanen (1971) has suggested that theoretically two types of lignin can be expected on dehydrogenative polymerization of coniferyl alcohol to lignin. The first case corresponds to the "Zulaufverfahren" used by Freudenberg (1964, 1965, 1968) when all monomer is added at once to the reaction mixture. The second is represented by his "Zutropfverfahren" in which monomer and catalyst are added dropwise. The lignin polymers produced have been called "bulk polymer" and "end-wise polymer" by Sarkanen (1971). In the formation of the bulk polymer, the concentration of coniferyl alcohol radicals is high and chances for their coupling with one another are accordingly also high. When the end-wise polymer is formed, coniferyl alcohol radicals are fewer and are as a result able to diffuse through the cell wall and react with lignin that has already been formed. The two types of lignin can be expected to have different proportions of the various types of linkages by which the phenylpropane units are bound together to form the lignin polymer.

Lee (1968) suggested that in a bulk polymer the proportion of linkages should decrease (Fig. 5.14) in the order phenylcoumaran (17–18) >

5.9.2 Lignin in Compression Wood

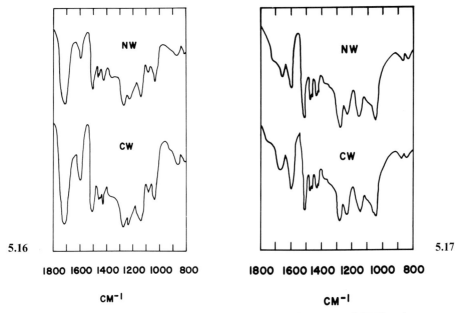

Fig. 5.16. Infrared absorption spectra of milled wood lignins from normal (*NW*) and compression (*CW*) woods of *Pseudotsuga menziesii*. (Redrawn from Latif 1968)

Fig. 5.17. Infrared absorption spectra of thioglycolic acid lignins from normal (*NW*) and compression (*CW*) woods of *Pseudotsuga menziesii*. (Redrawn from Latif 1968)

pinoresinol (8–9) > arylglycerol-β-aryl ether (7–8) and also in the order diphenyl (12–13a) > diaryl ether (6–7) > 1,2-diguaiacylpropane (13b–14b). In an end-wise polymer, on the other hand, formed under conditions of low concentration of monomer radicals, the proportion of linkages would be expected to decrease in the order arylglycerol-β-aryl ether > 1,2-diguaiacylpropane > phenylcoumaran. Pinoresinol, diphenyl, and diaryl ether linkages are not found in this type of polymer. Sarkanen (1971) has later stated that the 1,2-diguaiacylpropane linkage does not occur in bulk polymers. Lee argued that since compression wood contains more lignin than normal wood, the rate of lignification is likely to be higher in the former wood and hence also the concentration of coniferyl and p-coumaryl alcohol radicals. As a result, the lignin in compression wood ought to be of the bulk rather than the end-wise polymer type. To prove this hypothesis, milled wood lignins from normal and compression woods of *Pseudotsuga menziesii* were subjected to a detailed structural analysis.

Normal, milled wood lignin contained 13.7% methoxyl and had the formula $C_9H_{9.18}O_{2.87}(OCH_3)_{0.85}$, while milled wood lignin from compression wood contained only 10.9% methoxyl and had the formula $C_9H_{9.75}O_{3.13}(OCH_3)_{0.68}$. The ultraviolet spectra of the two lignins were not identical, as had been claimed by Bland (1958a, b) but differed notably, as can be seen in Fig. 5.18. With the aid of model compounds it could be shown that the difference was due to the pres-

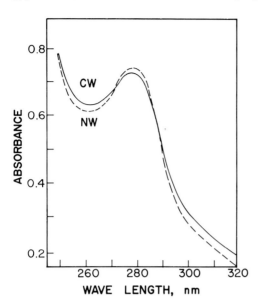

Fig. 5.18. Ultraviolet absorption spectra of milled wood lignins from normal (*NW*) and compression (*CW*) woods of *Pseudotsuga menziesii*. (Lee 1968)

ence of more diphenyl linkages in compression wood than in normal wood lignin.

An alternative explanation of the different ultraviolet spectra has later been offered by Musha and Goring (1975), who pointed out that lignins should exhibit an increasing flattening of the UV absorption peak at 280 nm with an increasing proportion of p-hydroxyphenylpropane units. Such a flattening is evident in the compression wood absorption curve in Fig. 5.18, confirming this prediction. Whiting and Goring (1982) later found that a middle lamella-primary wall lignin containing 40% p-hydroxyphenylpropane units (Chap. 6.4.1) had the same ultraviolet absorbance at 260 nm and 280 nm. The corresponding guaiacyl lignin in the secondary wall, by contrast, showed a minimum at 260 nm and a maximum at 280 nm. A larger number of diphenyl linkages may contribute to the flatness of the UV absorption curve at 280 nm of a compression wood lignin, as suggested by Lee (1968), but the larger number of p-hydroxyphenylpropane units would seem to be a more decisive factor in this respect.

Lignin from the normal wood studied by Lee (1968) contained 1.31 hydroxyl groups per C_9 unit, and compression wood lignin almost the same, or 1.28. On methylation with methanolic hydrogen chloride at room temperature, however, normal wood lignin acquired 0.55 new methoxyl groups per C_9 compared to only 0.43 for compression wood lignin, in both cases at the α-oxygen function in the side chain. Cyclic structures, such as phenylcoumaran and pinoresinol, are relatively resistant to methylation, and linkages of this type were accordingly more numerous in compression wood than in normal wood lignin. Lee (1968) found that normal wood lignin contained 0.20 such units per C_9 compared to 0.16 for compression wood, a result contrary to what would have been expected from their theory.

5.9.2 Lignin in Compression Wood

The arylglycerol-β-aryl ether linkage is cleaved on refluxing with methanolic hydrogen chloride, giving rise to a series of Hibbert's ketones which can be estimated spectrophotometrically. On the basis of lengthy and complex experiments, Lee (1968) came to the conclusion that compression wood lignin contained fewer arylglycerol-β-aryl ether linkages than lignin from normal wood. In another series of experiments evidence was obtained for the occurrence in compression wood lignin of more phenylcoumaran but less 1,2-diguaiacylpropane linkages than in normal wood. There was no difference in carbonyl content between the two lignins, as had also been found by Bland (1961). Some evidence was adduced for the occurrence in lignin from compression wood of more units of the coniferaldehyde type (**16**).

```
    CHO
    |
    CH
    ||
    HC
    |
   [benzene ring]
           OCH₃
    O
    |
   16
```

Summarizing the results of this interesting investigation, it had been shown that, compared to the lignin present in normal softwood, compression wood lignin contained more diphenyl linkages and bonds of the general phenylcoumaran type and fewer arylglycerol-β-aryl ether and 1,2-diguaiacylpropane linkages. A contradiction was that the actual phenylcoumaran bond was less frequent in compression wood than in normal wood lignin, an anomaly that could not be fully explained. On the whole, however, the results agreed with the original hypothesis that the lignin present in compression wood probably is of the bulk rather than the end-wise polymer type. Later results obtained by Kutsuki and Higuchi (1981, 1982) agree with this conclusion (Sect. 9.6.1.4.2).

In the first of a series of extensive investigations on the structure of lignins in normal and compression woods by Sakakibara and his co-workers, Morohoshi and Sakakibara (1971 a, b) studied lignins from *Abies sachalinensis*. Compression wood was isolated from a single band along the periphery of a concentric disc and normal wood from a region immediately inside this band and on the same side. From these facts, as well as from the relatively low lignin content of 35.8% of the compression wood, it would appear that the latter was not of the pronounced type. The yields of milled wood lignin from normal and compression woods were 6.96% and 7.05%, respectively, which is two to three times less than reported by Latif (1968), suggesting that optimum conditions for extraction had not been attained. The milled wood lignin from the compression wood contained 13.2% methoxyl groups compared to 15.3% for that from normal wood, figures very close to those for corresponding Klason lignins. The total hydroxyl contents were 12.9% and 12.5%, and the number of conjugated carbonyl groups per methoxyl group were 0.17 and 0.27, respectively. The lower carbonyl content of the compression wood lignin was confirmed by evaluation of infrared and ultraviolet spectra. Nuclear magnetic resonance spectra indicated that compression wood lignin contained 65% more condensed units than did lignin from normal wood.

On oxidation with nitrobenzene, compression wood lignin afforded 1.9% p-hydroxybenzaldehyde and 12.1% vanillin, corresponding yields for normal wood being 0.5% and 17.1% based on the Klason lignin. When compression wood lignin was methylated and oxidized with potassium permanganate according to Freudenberg's procedure, the major products obtained were p-anisic acid (**17**) (1.73%), veratric acid (**18**) (3.80%), and isohemipinic acid (**22**) (0.21%). Corresponding yields for normal wood were 0.10%, 6.50%, and 0.07%, respectively. Small amounts of other acids were also obtained from both types of wood, namely trimethylgallic acid (**19**), 4-methoxyisophthalic acid (**20**), 4-methoxyphthalic acid (**21**), metahemipinic acid (**23**), and dehydrodiveratric acid (**25**).

17	R, R', R" = H
18	R = OCH$_3$, R', R" = H
19	R, R' = OCH$_3$, R" = H
20	R = COOH, R', R" = H
21	R, R' = H, R" = COOH
22	R = OCH$_3$, R' = COOH, R" = H
23	R = OCH$_3$, R' = H, R" = COOH
24	R, R' = COOH, R" = H

25	R, R' = OCH$_3$
26	R = OCH$_3$, R' = H
27	R, R' = H

The yields of Hibbert's ketones on ethanolysis were 10.4% for compression wood and 12.4% for normal wood. Both tissues afforded as degradation products vanilloyl methyl ketone (**28**), obtained in a yield of 3.34% with compression wood, α-ethoxypropioguaiacone (**29**) (2.88%), guaiacylacetone (**30**) (1.28%), 1-ethoxy-1-(4-hydroxy-3-methoxyphenyl)-2-propanone (**31**) (0.99%), and α-ethoxypropiosyringone (**32**) (0.15%). Corresponding yields of these compounds from normal wood were 3.12%, 5.72%, 2.06%, 1.24%, and 0.24%. Compression wood gave three compounds that were obtained in only trace amounts from normal wood, namely, 4-hydroxy-α-ethoxypropiophenone (**33**) (1.00%),

5.9.2 Lignin in Compression Wood

[Structures 31–35: phenolic compounds with side-chain variations]

- **31**: phenol with OH, OCH$_3$, side chain HCOC$_2$H$_5$ / C=O / CH$_3$
- **32**: phenol with OH, two OCH$_3$ (syringyl), side chain C=O / HCOC$_2$H$_5$ / CH$_3$
- **33**: phenol with OH, side chain C=O / HCOC$_2$H$_5$ / CH$_3$
- **34**: phenol with OH, side chain C=O / C=O / CH$_3$
- **35**: phenol with OH, side chain CH$_2$ / C=O / CH$_3$

4-hydroxybenzoyl methyl ketone (**34**) (0.37%), and 4-hydroxyphenylacetone (**35**) (0.29%).

All these results are in agreement with the conclusions drawn by Morohoshi and Sakakibara (1971 a, b), namely that compression wood lignin contains a higher proportion of p-hydroxyphenylpropane units and is more highly condensed than is the lignin in normal conifer wood. Obviously the lower proportion of guaiacyl units makes a further condensation possible in compression wood.

In the belief that the earlier study of Morohoshi and Sakakibara (1971 a, b) had not contributed to the elucidation of the structure of the lignin present in compression wood, a rather odd notion, Erickson et al. (1973 c) subjected this problem to a renewed investigation (Miksche 1973). Normal and compression wood meals from *Pinus mugo* were heated with sodium sulfide−sodium hydroxide. The soluble kraft lignins thus obtained were subsequently degraded with permanganate-periodate, followed by hydrogen peroxide according to methods previously developed by Miksche and his co-workers (Larsson and Miksche 1969 a, b, Erickson et al. 1973 a, b). The acids resulting from these degradations were methylated with diazomethane, and the methyl esters were separated and identified.

Twelve acids were isolated and identified from both normal and compression woods. The yields of 11 of these acids are summarized in Table 5.40. In addition, compression wood lignin afforded a number of previously unknown acids with two benzene rings, five of which were identified as compounds (**38–42**). Compared to the lignin in normal wood, compression wood lignin gave a much higher proportion of p-anisic acid (**17**), 4-methoxyisophthalic acid (**20**), and the three acids (**26**), (**27**), and (**36**), all entirely or partly derived from p-hydroxyphenylpropane units in the lignin. Acids (**26**), (**36**), and (**38–42**), which consisted of one benzene ring without and one with a methoxyl group ortho to the phenolic oxygen, offered direct evidence that the compression wood lignin had originated by a copolymerization of p-coumaryl alcohol and coniferyl alcohol.

Investigations by Erickson and Miksche have shown that no lignin from any gymnosperm (1974 a) or from *Psilotum, Lycopodium, Selaginella, Isoetes,* and *Equisetum* (1974 b) contains such a high proportion of p-hydroxyphenylpropane units as compression wood lignin. An earlier view that lignin from primitive plants should have a characteristically high content of such units is not correct according to Miksche (1973). It should, however, be mentioned here that

Table 5.40. Yield of methyl esters of 11 acids obtained on oxidative degradation of kraft lignins from normal and compression woods of *Pinus mugo*. All values in mg per g wood. (From data reported by Erickson et al. 1973)

Com-pound	Name	Normal wood	Compression wood
17	p-Anisic acid	3.1	16.5
18	Veratric acid	62.8	58.8
19[a]	Trimethylgallic acid	1.0	0.85
20	4-Methoxyisophthalic acid		2.9
22	Isohemipinic acid	16.9	15.7
23	Metahemipinic acid	3.0	3.2
36		0.7	1.55
37		5.2	4.8
27	Dehydrodianisic acid	0.2	0.45
26		0.75	2.5
25	Dehydrodiveratric acid	13.0	11.9

[a] Includes compound 21

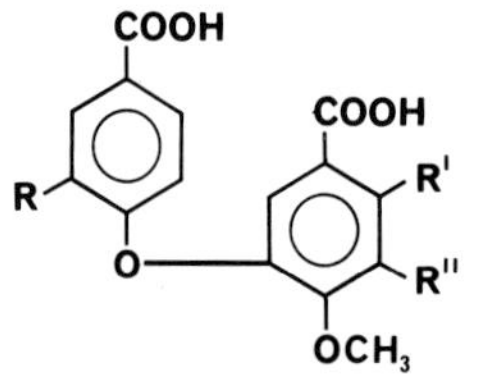

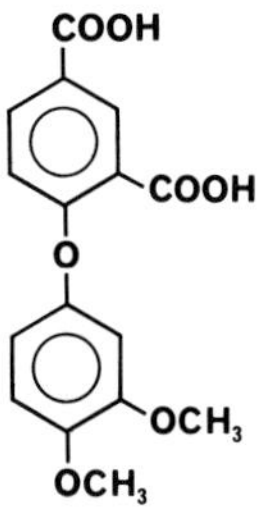

36 R=OCH$_3$, R', R''=H
37 R=OCH$_3$, R'=H, R''=OCH$_3$
38 R, R', R''=H
39 R=COOH, R'=H, R''=OCH$_3$
40 R=OCH$_3$, R'=COOH, R''=H

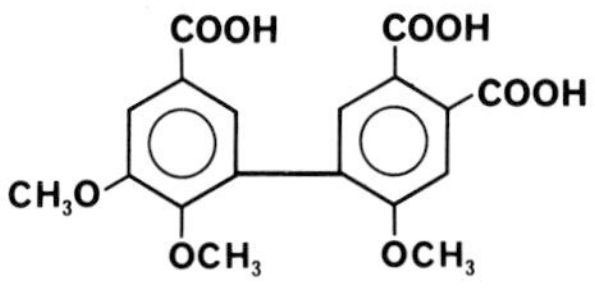

41

42

according to Faix et al. (1977) at least *Psilotum nudum* and the two ferns *Dryopteris felix-mas* and *Pteridium aquilinum* contain a lignin with 16–23% p-hydroxyphenylpropane units. Among the gymnosperms, only three species of *Podocarpus* and the atypical *Tetraclinis articulata* have a higher content of p-hydroxyphenylpropane units. A specimen of *Ginkgo biloba* also had a higher than normal proportion of such units, but this was attributed to the presence of compression wood (Erickson and Miksche 1974a). Higuchi et al. (1967) and Erickson et al. (1973d) have shown that a previous notion that monocotyledons should have a guaiacyl-syringyl lignin with a high proportion of p-hydroxyphenylpropane units may also be erroneous. Lignins isolated from the bark of *Picea abies* and *Taxus baccata*, however, seem to contain more p-hydroxyphenylpropane units than corresponding wood lignins (Andersson et al.

1973). Nimz et al. (1981) have later reported ^{13}C NMR evidence indicating that at least grass lignins may contain a significant proportion of p-hydroxyphenylpropane units.

It is clear from the results obtained by Erickson et al. (1973c) that compression wood lignin is formed by dehydrogenation-polymerization of p-coumaryl alcohol and coniferyl alcohol (Adler 1977). The lignin present in compression wood is unusual in its high content of p-hydroxyphenylpropane units, at least among higher plants. Miksche (1973) has suggested that the lignin in compression wood ought to be referred to as a guaiacyl-p-hydroxyphenyl lignin, which indeed would be a proper designation.

The most extensive investigation so far on the structure of any compression wood lignin is that by Yasuda and Sakakibara on the lignin present in compression wood of *Larix leptolepis*. Most of these results have been reviewed by Sakakibara (1977, 1980). The Klason lignin content and the yield of milled wood lignin were 39.3% and 12.2% for compression wood and 28.1% and 10.5% for normal wood of this species (Yasuda and Sakakibara 1975a). The lignin from compression wood contained 12.6% methoxyl and 18% carbonyl groups compared to 15.2% and 24%, respectively, for normal wood lignin. Compression wood lignin contained 0.76 methoxyl groups per C_9 unit and the lignin in normal wood 0.94. The infrared spectra of the milled wood lignins were very similar, although that from compression wood contained fewer ester groups. Ultraviolet and nuclear magnetic resonance spectra suggested a higher proportion of phenolic hydroxyl groups in the compression wood lignin. The latter had 86% condensed units and normal wood lignin only 56%. On oxidation with nitrobenzene and alkali, compression wood lignin gave 20.0% degradation products and lignin from normal wood 26.1%, based on the Klason lignin content. Compression wood yielded 2.6% p-hydroxybenzaldehyde and 15.1% vanillin, while normal wood afforded only 0.8% of the former and 22.6% of the latter compound.

Ethanolysis of lignins from compression and normal woods gave a mixture of Hibbert's ketones in a total yield of 12.1% and 10.1%, respectively. Three of these ketones were obtained in a much higher yield from compression wood than from normal wood, namely 4-hydroxy-α-ethoxypropiophenone (**33**) (1.01%), 4-hydroxyphenylacetone (**35**) (0.31%), and 1-ethoxy-1-(4-hydroxyphenyl)-2-propanone (**43**) (0.33%), while both woods gave trace amounts of 4-hydroxybenzoyl methyl ketone (**34**). Larger amounts of vanilloyl methyl ketone (**28**), α-ethoxypropioguaiacone (**29**), guaiacylacetone (**30**), and 1-ethoxy-1-(4-hydroxy-3-methoxyphenyl)-2-propanone (**31**) were recovered from normal than from compression wood. The nature and yields of these compounds indicate the presence of fewer arylglycerol-β-aryl ether linkages in compression wood than in normal wood, as had also been found by Lee (1968).

On hydrogenolysis, compression wood lignin gave 26.3% monomeric degradation products and normal wood lignin 31.2%, suggesting a higher proportion of carbon-carbon linkages in the former lignin. The combined yield of p-hydroxyphenylpropane (**44**) and p-hydroxyphenylpropan-3-ol (**45**) was nearly three times higher for compression wood than for normal wood. Degradation with permanganate-periodate gave larger amounts of p-anisic (**17**), 4-methoxyisophthalic (**20**), 4-methoxyphthalic (**21**), isohemipinic (**22**),

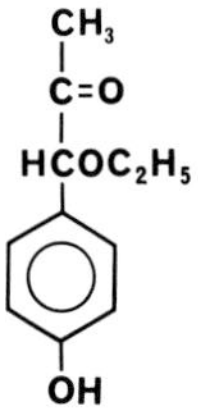

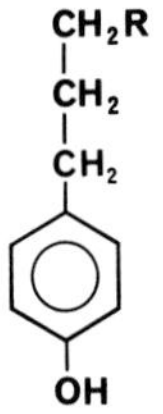

44 R=H
45 R=OH

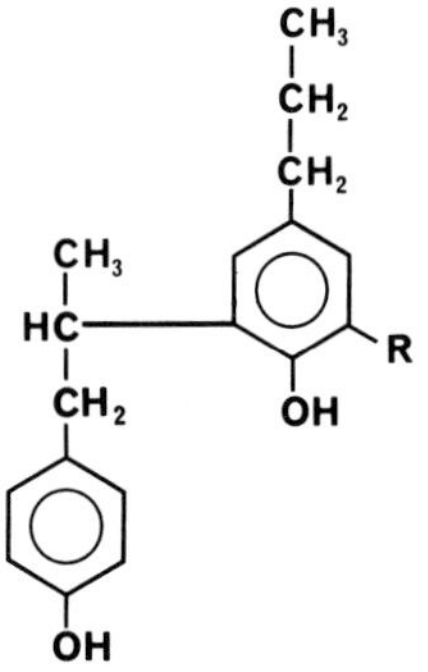

46 R=OCH$_3$
47 R=H

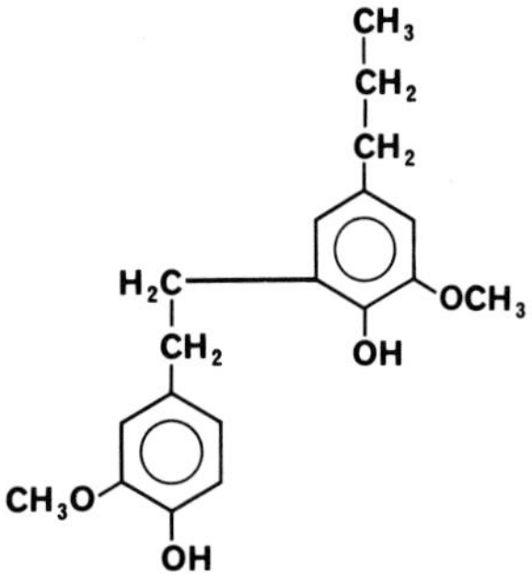

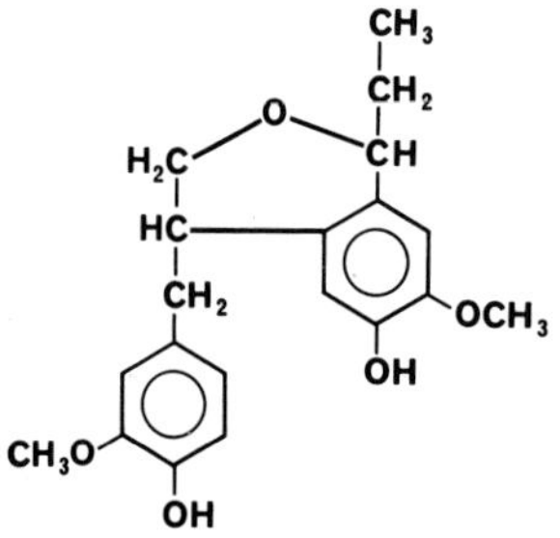

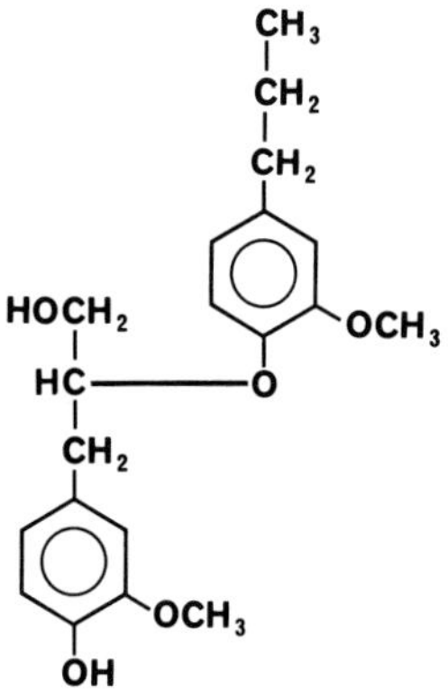

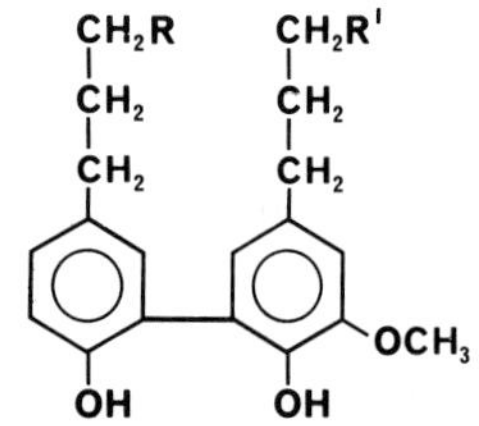

51 R=OH, R'=H
52 R=H, R'=OH

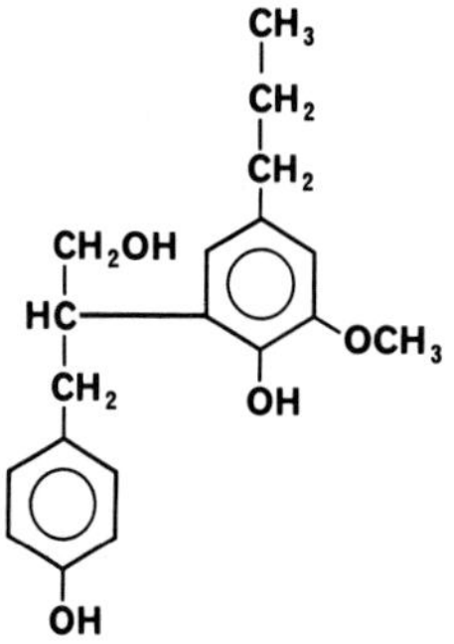

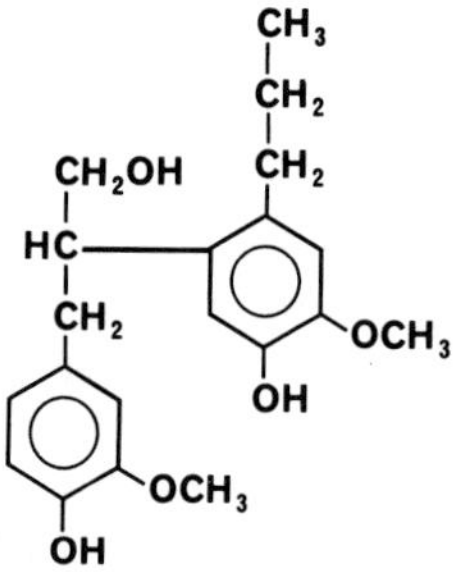

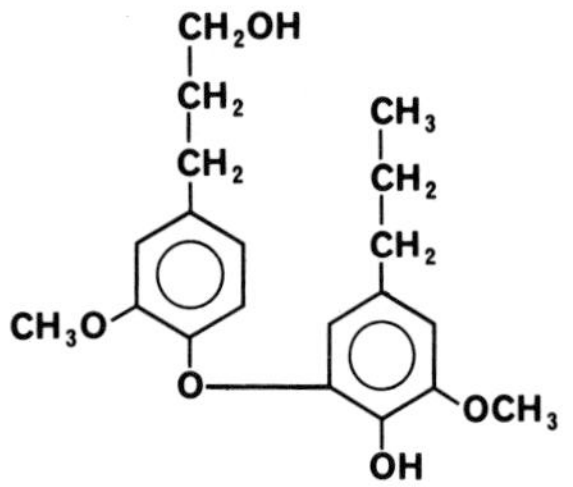

methoxytrimesic (**24**), dehydrodiveratric (**25**), and dehydrodianisic acids (**27**) with compression wood than with normal wood lignins, but the latter afforded more veratric (**18**) and metahemipinic (**23**) acids.

Yasuda and Sakakibara (1975a) concluded from their results, that, in comparison with the lignin in normal softwood, compression wood lignin has a larger number of carbon-carbon linkages, a higher proportion of p-hydroxyphenylpropane units, and contains more condensed residues. The p-hydroxyphenylpropane units are linked to adjacent residues with carbon-carbon and carbon-oxygen linkages and not with ester bonds, such as those known to exist in the lignin present in the monocotyledons. The proportion of arylglycerol-β-arylether linkages is lesser in compression than in normal wood lignin.

In further investigations, Yasuda and Sakakibara subjected compression wood lignin from *Larix leptolepis* to hydrogenolysis in aqueous dioxane solution with copper chromite as a catalyst. The resulting oligomers, largely dimers, were isolated chromatographically and were identified through their mass, infrared and nuclear magnetic resonance spectra. The first three dimers obtained were compounds (**46–48**) (Yasuda and Sakakibara 1975b). Compound (**46**) consisted of a p-hydroxyphenylpropane and a guaiacylpropane unit, linked together by a C_β–C_5 bond, while (**47**) was composed of two p-hydroxyphenylpropane residues. In dimer (**48**) an α-hydroxymethyl group had probably been removed as formaldehyde. The isolation of a guaiacyl dimer containing a 6-membered, cyclic ether link (**49**) is of special interest for it represents a new type of interunit bond in a softwood lignin where most cyclic ether linkages are of the 5-membered, phenylcoumaran type (Yasuda and Sakakibara 1975c, 1976). The same investigation also afforded compound (**50**). The phenylisochroman (**49**) with its C_β–C_6 and C_γ–O–C_α linkages is also present in normal wood lignin (Sakakibara 1980).

In a subsequent study, Yasuda and Sakakibara (1977a) isolated three previously known guaiacyl dimers and four new dimers (**51–54**) from the hydrogenation products of the same compression wood lignin. The nature of compounds (**51**) and (**52**) shows that compression wood lignin contains diphenyl linkages between p-hydroxyphenylpropane and guaiacylpropane units. Compound (**53**), in analogy with compounds (**46**) and (**47**) has a C_β–C_5 bond between these units. Compound (**54**) is another example, in addition to (**49**), of the existence of C_β–C_6 linkages in compression wood lignin, a fact also evident from the formation of metahemipinic acid (**23**) on oxidative degradation. Yasuda and Sakakibara (1977b) also isolated and identified four new guaiacylpropane dimers, one of which was a C_5 aryl ether (**55**), while the other three had C_β–C_5 (**56**) or C_β–C_β (**57, 58**) linkages. Three previously known guaiacyl dimers and one trimer were also obtained. Finally, Yasuda and Sakakibara (1981) have reported the presence among the hydrogenation products from the same compression wood lignin of two trimeric compounds (**59, 60**) with a γ-lactone ring.

Terashima and his co-workers have developed an interesting technique for estimating the proportion of condensed units in lignin (Terashima 1979, Terashima et al. 1979, Tomimura et al. 1979, 1980). Growing shoots were administered ferulic acid, an excellent lignin precursor, randomly labeled with ^{14}C

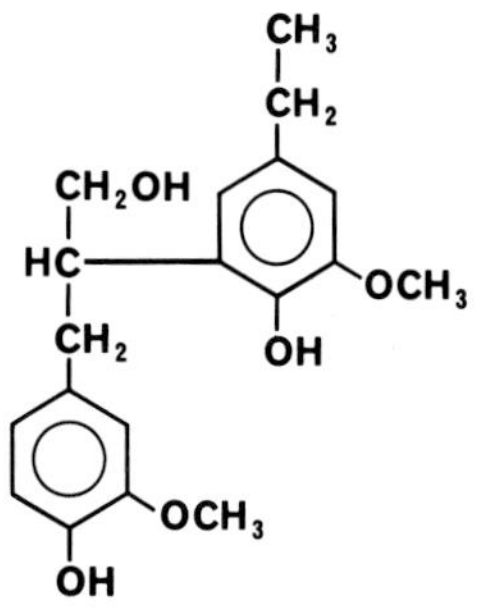

56

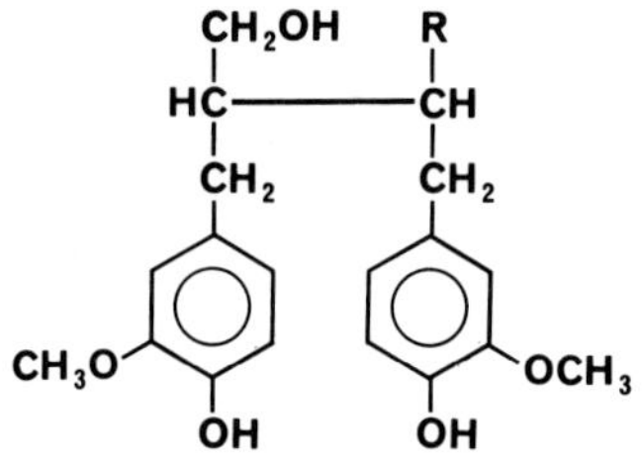

57 R = H
58 R = CH$_3$

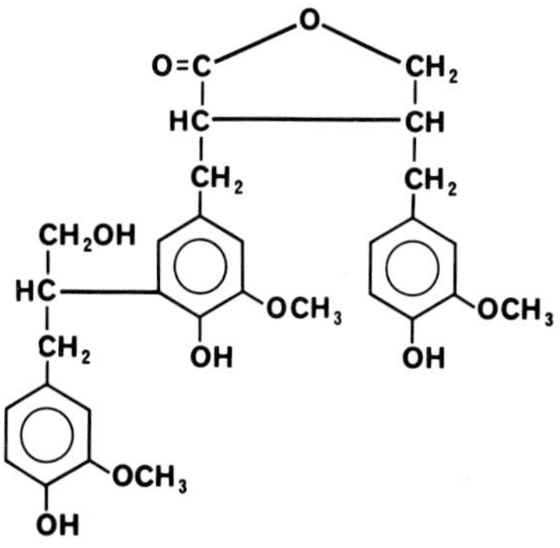

59

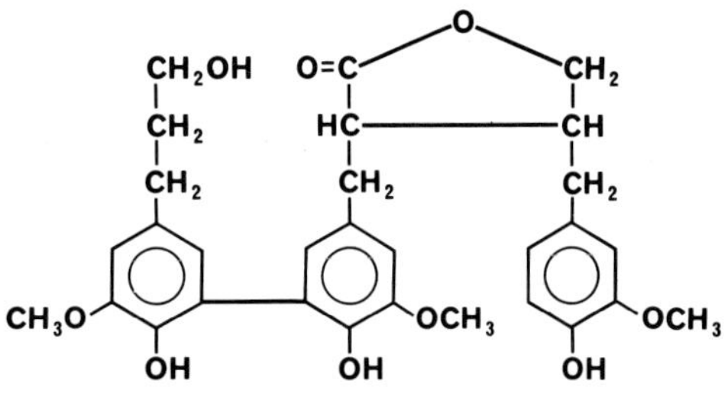

60

over the benzene ring and in addition labeled with tritium-^{3}H at positions 2, 5, or 6 in the ring. The ratio ^{3}H/^{14}C was determined in the differentiating xylem at increasing distances from the cambium. Because condensation involving positions 2, 5, or 6 will result in elimination of tritium, this ratio will decrease with increasing extent of condensation. It was found that the lignin formed early in differentiation was more highly condensed than that produced later. When a *Pinus thunbergii* shoot, growing at an angle of 45°, was administered doubly labeled ferulic acid with the tritium at C-5, the ^{3}H/^{14}C ratio was lower in the lignin located on the lower side of the shoot than in that on the upper. Evidently, the compression wood lignin on the under side contained more units condensed through position 5 than the normal lignin on the upper side. When the same precursor was applied to an area subjected to bending stress, there was no such difference between this lignin and a normal wood lignin, suggesting that stress does not induce formation of compression wood (Chap. 11.5.5) (Terashima et al. 1979).

When the same doubly labeled ferulic acid was administered together with indoleacetic acid, the lignin formed was more condensed than that obtained in a control experiment. A less condensed lignin was obtained after the pine shoot had been treated with abscisic acid (Tomimura et al. 1979). The physiological implications of these results are discussed in Chap. 13.4.

Further experiments with inclined shoots of *Ginkgo biloba* and *Pinus thunbergii* (Tomimura et al. 1980) showed that 52.1% of the lignin units in the normal pine wood were condensed through C-5. In the compression wood, 62.2% of the units were condensed at this position, a figure considerably lower than the 86% reported by Yasuda and Sakakibara (1975a). In Ginkgo, the compression wood lignin was condensed to the same extent as that in the normal wood, as mentioned in Chap. 8.3.1. Only 2% of the units in the pine lignin and 3% in that of *Ginkgo* were condensed through positions 2 and 6, and there was no difference in this regard between normal and compression woods in either of these species.

Carbon-13 NMR studies with various acetylated lignins were carried out by Nimz et al. (1981). The compression wood lignin from *Larix leptolepis* investigated by Yasuda and Sakakibara (1975a) differed in several respects from a normal spruce lignin in its NMR spectrum. Signals for arylglycerol-β-aryl ether linkages were less pronounced in the compression wood spectrum, while other signals indicated many phenylcoumaran bonds. It could be shown that the number of arylglycerol-β-aryl ether structures was negatively correlated with the number of p-hydroxyphenylpropane units. These results confirm those obtained previously by Lee (1968) and by Yasuda and Sakakibara (1975a).

In connection with an investigation of the nature of the lignin present in the middle lamella and in the total tissue of *Picea abies* wood (Chap. 6.4.1), Westermark (personal communication 1984) studied the chemical properties of the corresponding compression wood lignin. Oxidation with nitrobenzene and alkali gave 26 times more p-hydroxybenzaldehyde with compression wood than with normal wood, a result in agreement with the earlier finding of Bland (1961) (Table 5.36).

On acidolysis, compression wood gave five compounds not obtained from normal wood, three of which were identified as p-hydroxybenzaldehyde, 2-hydroxy-1-(4-hydroxyphenyl)-1-propanone, and 3-hydroxy-1-(4-hydroxyphenyl)-2-propanone. The last two are obviously related to compounds **(33)** and **(35)**, respectively, previously recovered after ethanolysis of compression wood lignins by Morohoshi and Sakakibara (1971 a, b) and by Yasuda and Sakakibara (1975a). The compression wood lignin was less brominated than corresponding lignin in normal wood when lignin concentrations were determined by the SEM-EDXA technique. The uptake of bromine was only 0.69 mole per C_9 unit of lignin in compression wood, compared with 1.03 mole per unit of normal wood lignin. Westermark concluded that the structure of the lignin in compression wood is quite different from that of the lignin present in both the entire tissue and the middle lamella of normal wood.

5.9.3 Lignin-Carbohydrate Complexes from Compression Wood

Mukoyoshi et al. (1981) isolated lignin-carbohydrate complexes from compression wood of *Pinus densiflora* in yields of 0.2–0.5%. Two of these complexes were homogeneous on electrophoresis and ultracentrifugation and had molecular weights of 650 000 and 34 000. The former was composed of 40% neutral sugar residues, 2% uronic acids, and 57% lignin, while the latter consisted of 67% neutral sugars, 5% uronic acids, and 30% (sic) lignin. The larger complex had a galactose content of 20.2% and the smaller 21.9%. A lignin-carbohydrate complex from normal wood of the same species contained much less galactan (Azuma et al. 1981). Methylation analysis showed that the galactose residues were derived from a $(1 \rightarrow 4)$-β-D-galactan, slightly branched at C-2, C-3, and C-6, and lacking acetyl groups. The lignin was also linked to a galactoglucomannan of the same composition as that in normal softwood. Both complexes had a high content of arabinose residues, namely 16.4% and 4.8%, respectively. Except for their unique carbohydrate composition, the two lignin-carbohydrate complexes were similar to those obtained from corresponding normal wood. The investigators suggested that most of the galactan in compression wood is linked to lignin.

Minor (1982) studied the chemical linkage between lignin and polysaccharides in compression wood of *Pinus resinosa*. The ball-milled wood was treated with a mixture of cellulolytic enzymes, and the remaining polysaccharides were subjected to a structural analysis by the methylation technique. End groups were analyzed by deuterium labeling. Three quarters of the sugar residues consisted of galactose. The structural analysis showed that the galactan linked to the lignin was of the $(1 \rightarrow 4)$-β-D-type and had an average degree of polymerization of 13. There was probably one bond to the lignin per macromolecule, largely through C_6 in the galactan chain. Small amounts of $(1 \rightarrow 4)$- and $(1 \rightarrow 6)$-linked arabinogalactan were also present.

The investigations reported by Mukoyoshi et al. (1981) and by Minor (1982) strongly suggest that a significant portion of the galactan in compression wood is linked to lignin. It will be recalled (Sect. 5.5.3) that most of the galactan is dissolved when compression wood is delignified with aqueous chlorite. Much of this galactan is located in the S_1 and $S_2(L)$ layers of the tracheid wall (Chap. 6.3.3). Evidently, a chemical bond is established between the galactan and the lignin when the latter is formed in these regions.

5.9.4 Conclusions

The lignin present in compression wood is a true guaiacyl-p-hydroxyphenyl lignin (Eriksson et al. 1973c, Adler 1977, Nimz et al. 1981). It is a copolymer of two p-hydroxycinnamyl alcohols, namely p-coumaryl alcohol and coniferyl alcohol, with about 20% of the units originating from the former. The lignin units are linked together by carbon-carbon and carbon-oxygen bonds. The proportion of ether linkages is lower than in normal wood lignin. Whiting and Goring (1982) have reported that the lignin in the middle lamella-primary wall region of *Picea mariana* tracheids consists of about 40% p-hydroxyphenylpropane and 60% guaiacylpropane units. The similarity between this extracellular lignin and that occurring in compression wood is obvious (Chap. 6.4.1). There are apparently now at least two lignins known that might be of the guaiacyl-p-hydroxyphenyl type. The more recent results obtained by Westermark (personal communication 1984), which suggest that the lignin in the middle lamella region of normal conifer wood might be a guaiacyl lignin, are discussed in Chap. 6.4.1.

All available evidence indicates that compression wood lignin is more highly condensed than a normal softwood lignin, which is not surprising, considering the higher content of p-hydroxyphenylpropane units. The proportion of different functional groups in compression wood lignin is not exactly known. It has been claimed that, compared to a normal softwood lignin, that in compression wood contains more phenolic hydroxyl and coniferaldehyde groups but fewer ester and conjugated carbonyl groups. The total numbers of hydroxyl and of carbonyl groups are stated to be the same in normal and compression woods. Compression wood has a lower frequency of arylglycerol-β-aryl ether bonds, and especially so the p-hydroxyphenylpropane units. The incidence of diphenyl and phenylcoumaran linkages may be higher than in normal wood.

Several investigators (Bland 1958a, Morohoshi and Sakakibara 1971a, b, Terashima et al. 1979, Tomimura et al. 1980) have shown that the lignin present in tension wood is similar to that in corresponding normal wood. Since lignin does not contribute to the tensile strength of wood, this is not unexpected. Unlike tension wood, compression wood always exerts a dynamic pressure along the grain and usually it is also subjected to a considerable static pressure, since it generally occurs on the lower side of a branch or a leaning stem. The more highly condensed nature of the lignin in compression wood is probably an adaptation to these conditions, resulting in a more rigid, three-dimensional polymer with enhanced compressive strength. The lack of ester groups also contributes to this rigidity.

5.10 Extractives

Several phenolic extractives have lately been isolated from both normal and compression woods of *Larix leptolepis* by Sakakibara and his co-workers (Sasaya et al. 1980). Yasuda et al. (1975) extracted normal sapwood and heartwood, side wood, and compression wood of *Larix leptolepis* with methanol. The mixtures obtained were treated in succession with n-hexane, benzene, and ethyl acetate. The yields of the three fractions from the side wood were 0.42%, 0.07%, and 0.30%, respectively, and 0.36%, 0.13%, and 0.78% from compression wood. Six phenolic compounds were isolated from normal and compression woods, namely lignoceryl ferulate (**61**), pinocembrin (**62**), naringenin (**63**), aromadendrin (**64**), taxifolin (**65**), and kaempferol (**66**). One compound was obtained only from compression wood, namely p-hydroxybenzaldehyde.

Takehara et al. (1980) isolated lignans and lignan esters from ethanol extracts of normal and compression woods of *Abies sachalinensis*. Two of the esters were obtained from both normal and compression wood, namely lariciresinol p-coumarate (**67**) and lariciresinol p-ferulate (**68**). A third lignan ester, secoisolariciresinol di-o-coumarate, was obtained only from normal wood. In subsequent investigations, Sasaya et al. 1981) recovered and identified seven additional lignans from both opposite wood and compression wood of the same species, namely conidendrin (**69**), isolariciresinol (**70**), lariciresinol

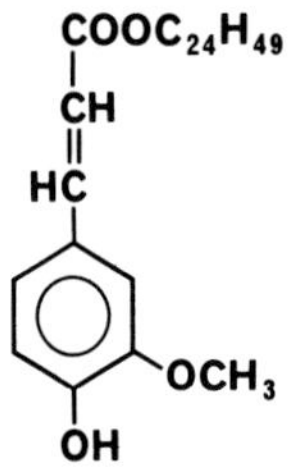

61

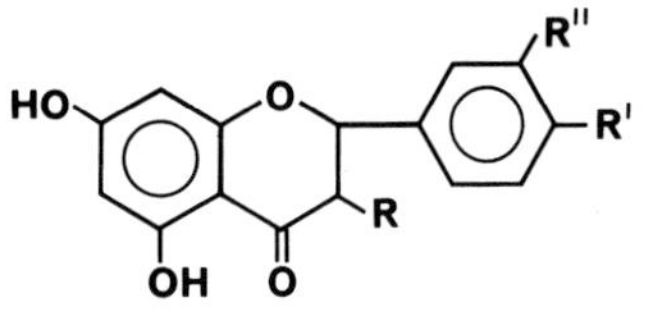

62 R, R', R''=H
63 R, R''=H, R'=OH
64 R, R'=OH, R''=H
65 R, R', R''=OH

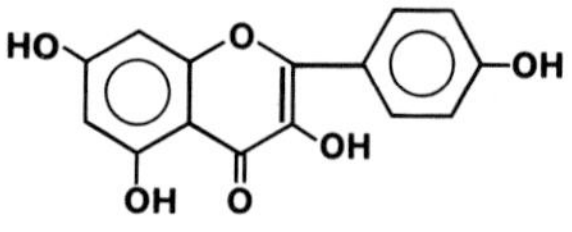

66

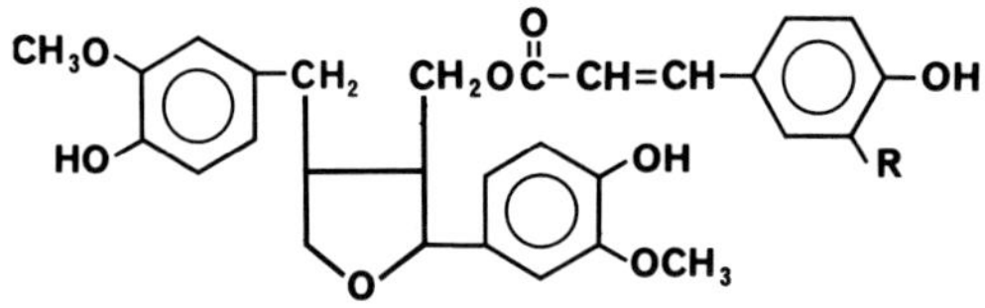

67 R=H
68 R=OCH$_3$

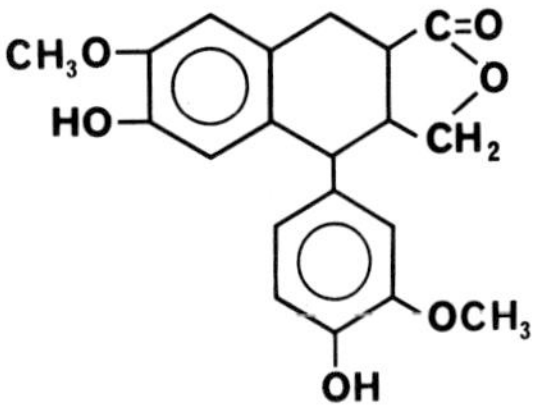

69

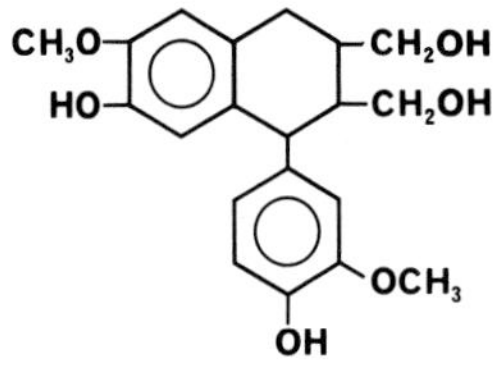

70

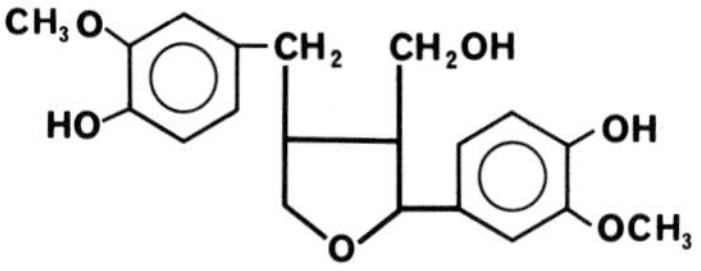

71

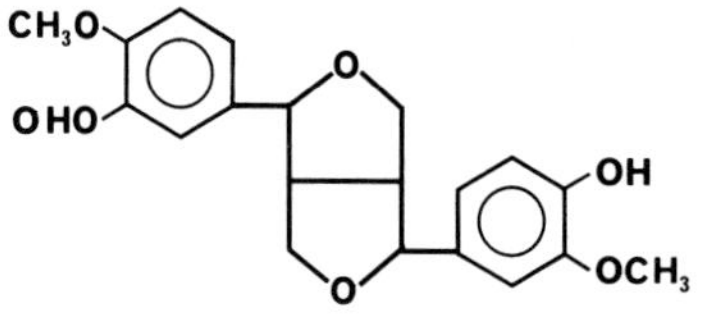

72

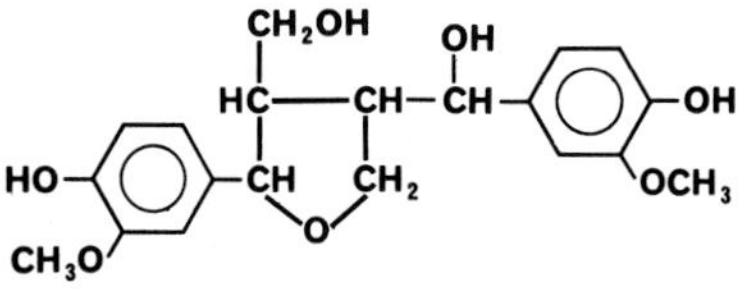

73

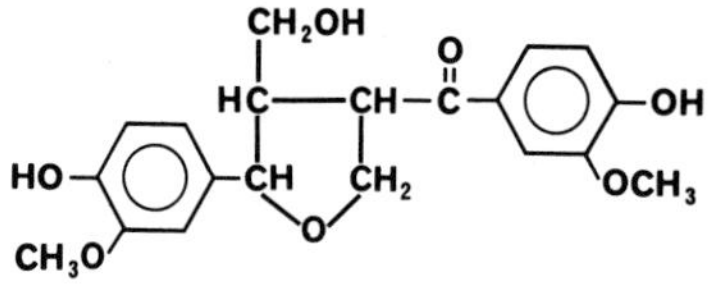

74

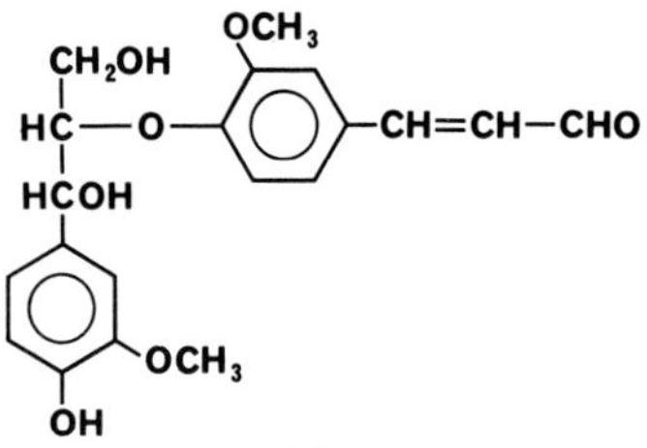

75

(**71**), pinosresinol (**72**), and compounds (**73**), (**74**), and (**75**). Obviously, opposite and compression woods of *Abies sachalinensis* contain the same kinds of lignans. Whether or not this applies to all conifers cannot be decided at present, since only one species has so far been investigated in this respect.

5.11 Microbial Degradation of Compression Wood

5.11.1 Introduction

Wood is decomposed by both bacteria and fungi, and treatment with preservatives is required for protection against this decay. The chemistry of the enzymatic degradation of the principal polymeric wood constituents, cellulose, hemicelluloses, and lignin, has lately been the subject of intensive studies which have served to clarify many hitherto obscure details. Efforts have also been made to clarify the topochemistry of wood decay, but in this respect the results obtained are less spectacular. Recent progress has been summarized in books edited by Nicholas (1973), Liese (1975), Loewus and Runeckles (1977), and Kirk et al. (1980). The biodegradation of lignin, which is of particular interest in this context, has been discussed in reviews by Kirk (1971, 1973, 1975a) and by Kirk et al. (1975, 1977). Wilcox (1968, 1970, 1973) and Liese (1970) have reviewed the ultrastructural aspects of wood decay. Ander and Eriksson (1978) and Eriksson (1981) have considered the degradation and utilization of lignin.

The microbiological degradation of compression wood has so far been studied only with brown, white, and simultaneous rot fungi, and there are no reports on the effect of soft rot fungi or bacteria on this tissue. Brown rot fungi rapidly degrade cellulose and hemicelluloses, especially in softwoods. Lignin is not removed, but it is now known that it becomes considerably modified by the action of these fungi, which cause demethylation, oxidation, and aromatic hydroxylation (Kirk 1975b). White rot fungi, in contrast, are capable of degrading both polysaccharides and lignin. Like that of the brown rots, the action of the white rot fungi is oxidative. The lignin undergoes oxidation of the propane side chain, causing introduction of carbonyl and carboxyl groups, as well as partial demethylation and cleavage of the aromatic ring (Kirk and Chang 1974, 1975). Simultaneous rot fungi destroy all components of the cell wall at approximately the same rate. According to Cowling (1961), depolymerization is immediately followed by utilization of the breakdown products by the fungus.

Brown rot fungi penetrate softwoods via the rays and the lumina of the tracheids. The degradation of the wall polysaccharides proceeds from the lumen toward the middle lamella, evidently under the action of enzymes that diffuse into and through the wall. White rot fungi also erode the cell wall from the lumen, degrading each layer in sequence (Wilcox 1968). Usually the lignin is removed first. White rot fungi tend to attack latewood more rapidly than earlywood, but they generally decompose wood more slowly than brown rots. Eriksson and his co-workers (Eriksson et al. 1980, Ruel et al. 1981) found that the white rot fungus *Sporotrichum pulverulentum* caused a gradual thinning of the cell walls in wood but also bored holes in these walls. Degradation proceed-

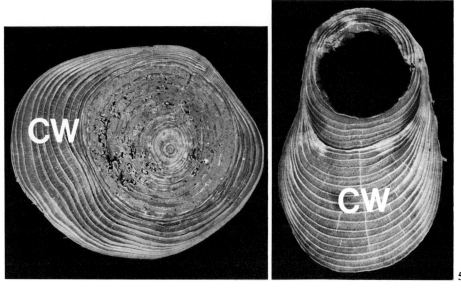

Fig. 5.19. Cross section of a stem of a living *Abies balsamea* tree with compression wood (*CW*) located below several large branches. Only the core of the normal wood had suffered decay. (cf. Fig. 10.51)

Fig. 5.20. Cross section of a stem of a dead *Abies balsamea* tree, long buried in the ground. The erstwhile core had been completely destroyed while the compression wood (*CW*) appears to be intact

ed from the lumen inward, with the S_3 layer most resistant to attack. It could be shown that lignin-degrading enzymes diffused into the walls from the fungal hyphae.

5.11.2 Biodegradation of Compression Wood

Although systematic observations are lacking, it would seem that compression wood is more resistant than normal wood to microbial decay in the forest. The stem section from a still living *Abies balsamea* tree in Fig. 5.19 contains a core of juvenile wood which has suffered visible decomposition, while the surrounding compression wood appears to be intact. A section of a stem from the same species, which had been buried in the ground for a long time, is seen in Fig. 5.20. Here the central core of normal wood had been completely destroyed, whereas the compression wood appeared to be unchanged, having retained its original hardness. These casual observations are fully confirmed by the few reported microscopic examinations of the fungal degradation of compression wood.

Jurášek (1964a, b), using light microscopy, carried out an extensive investigation on the effect of brown and white rot fungi on normal and reaction woods from both softwoods and hardwoods. When compression wood of *Abies alba*

5.11.2 Biodegradation of Compression Wood

Table 5.41. Average weight loss of normal and compression woods from *Picea rubens* after treatment for various periods of time with three brown rot fungi. All values in per cent of original wood. (Côté et al. 1966c)

Time, months	*Poria monticola*		*Lenzites saepiaria*		*Lenzites trabea*	
	Normal wood	Compression wood	Normal wood	Compression wood	Normal wood	Compression wood
2	38	12	34	30	22	8
4	50		64	54	46	10
6		30	68	58	68	17

was subjected to the action of the powerful brown rot fungus *Merulius (Serpula) lacrymans (domesticus)*, a thinning of the cell wall became visible only after a 59% weight loss. The progress of decay was followed with the aid of various staining reagents. Jurášek (1964b) noted that the inner portion of the S_2 layer in the tracheids was decomposed at an advanced stage of decay, a fact that he attributed to the low lignin content of this portion of the cell wall. The white rot fungus *Phellinus (Trametes, Fomes) pini* also caused decay of the inner portion of S_2. There was no defibration at a weight loss of 32%. Maximum weight loss under the action of the simultaneous rot fungus *Trametes gibbosa* was only 25%. At this stage there was no visible thinning of the tracheid walls. Jurášek pointed out that a wood with a high lignin content was more resistant to fungal decay than a less lignified tissue, a conclusion later confirmed by Wilcox (1968). Compression wood is a tissue with an unusually high lignin content, and this fact is therefore at least one of the factors responsible for the resistance of this wood toward degradation by fungi.

In an attempt to visualize the distribution of lignin in the electron microscope by the technique introduced by Meier (1955), Côté et al. (1966c) subjected normal and compression woods of *Picea rubens* to biodegradation for removal of the polysaccharides (Chap. 6.4.2). The three brown rot fungi used for this purpose were *Lenzites (Gleophyllum) saepiaria*, *L. trabea*, and *Poria monticola*. Normal wood gave lignin skeletons similar to those obtained when the polysaccharides were completely removed with hydrofluoric acid (Chap. 6.4).

Average weight losses after 2, 4, and 6 months of decay with the three fungi are summarized in Table 5.41. The weight loss with normal wood after 6 months corresponded approximately to the polysaccharide content of the wood (70%). Degradation of the compression wood proceeded very slowly with both *Poria monticola* and *Lenzites trabea*, and only 17% of material had been removed after incubation for 6 months with the latter fungus. *Lenzites saepiaria*, by contrast, degraded compression wood at a rate that was only slightly lower than that for normal wood. After 3 months, this fungus seemed to have affected only the S_1 layer of the compression wood tracheids, as seen in Fig. 5.21. After 5 months, fungal attack was more general and had proceeded from the intercellular spaces and from the lumina, as shown in Fig. 5.22. At a later stage in the decay, illustrated in Fig. 5.23, both middle lamella and sec-

Fig. 5.21. Compression wood of *Picea rubens* after exposure for 3 months to *Lenzites saepiaria*. Only the S_1 layer shows any signs of decay. Transverse section. TEM. (Côté et al. 1966)

ondary wall had been further eroded. All electron micrographs obtained in this investigation indicated that the hyphae penetrated both through the lumina of the tracheids and through the intercellular spaces of the compression wood.

Jutte and Sachs (1976) studied the interaction between the brown rot fungus *Poria placenta* and normal and compression woods of *Picea abies*, using scanning electron microscopy. In normal wood, the hyphae penetrated through the lumen of the tracheids and were surrounded by a gelatinous sheath. In compression wood, by contrast, no such sheath was observed, and the hyphae grew exclusively inside the intercellular spaces, as shown in Fig. 5.24. From these spaces they could also penetrate into the rays (Fig. 5.25). The tracheid walls were heavily decomposed, and the investigators found it strange that the fungus could degrade the wall without entering the lumina. They suggested that enzymes diffused into the cell wall from the intercellular spaces. Côté et al. (1966c), as already mentioned, had observed fungal hyphae in both lumina and intercellular spaces. The lumen of the tracheids and fibers is also the most common pathway in the invasion of normal wood by various fungi. Both Jurášek (1964a, b) and Côté et al. (1966c) found that brown rot fungi eroded the secondary wall in compression wood tracheids from the lumen.

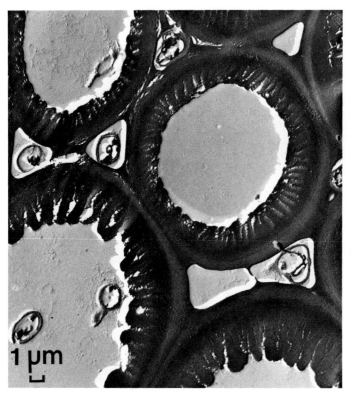

Fig. 5.22. Compression wood of *Picea rubens* partly decomposed by the action of *Lenzites saepiaria* for 5 months. Note the presence of hyphae in both intercellular spaces and lumina. Transverse section. TEM

In a subsequent report, Jutte et al. (1977) briefly described observations in the scanning electron microscope of *Picea* wood that had been in contact with water for over 2 years. Membranes of the half-bordered and simple pit pairs were intact in both normal and compression woods. In the latter wood, neither the margo nor the torus had suffered any degradation.

According to my own observations (Timell, unpublished results), compression wood in *Abies balsamea* is degraded very slowly by *Phellinus pini*. Figure 5.26 shows that the middle lamella and tracheid wall had been only slightly decomposed after incubation for 3 months with this white rot. It is also evident that penetration had occurred via both lumina and intercellular spaces.

It has been observed that there seems to exist a direct relationship between the methoxyl content of a lignin and the susceptibility of a particular wood to degradation by white rot fungi. The lignin in compression wood, because of its lower methoxyl content, is more highly condensed than that in normal softwood, a fact that very likely contributes to its higher resistance toward decay caused by white rots.

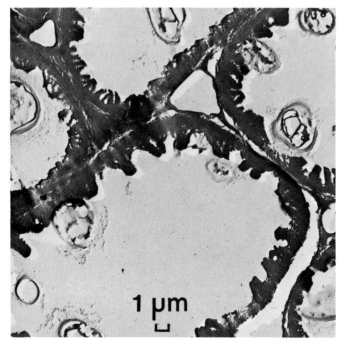

Fig. 5.23. Compression wood of *Picea rubens* at an advanced state of decay by *Lenzites saepiaria*. Transverse section. TEM

5.11.3 Conclusions

The scanty evidence available does not allow many conclusions to be drawn concerning the microbial degradation of compression wood. Nevertheless, a few facts seem to be beyond dispute. It is clear that compression wood is more resistant than corresponding normal wood to decay caused by brown and white rot fungi. The high lignin content of compression wood and the fact that its lignin is more highly condensed (Sect. 5.9.2) are probably responsible for this. Because of its often numerous intercellular spaces, compression wood is presumably more readily penetrated by fungi than normal wood. The observation that the fungal hyphae of a brown rot should grow only in the intercellular spaces does not seem to be of general validity. Other evidence indicates that the hyphae also grow through the lumen of the tracheids, as they do in normal wood.

Both brown and white rot fungi decompose the wall in compression-wood tracheids primarily from the lumen side, another indication of the presence of hyphae there. As degradation proceeds, at least the brown rot fungi also erode the wall from the middle lamella, where they are present in the intercellular spaces. Some brown rots, for example *Lenzites saepiaria*, are able to decompose practically all polysaccharides in compression wood. Unlike the situation in normal wood, the ultrastructure of the tracheids is severely altered in this pro-

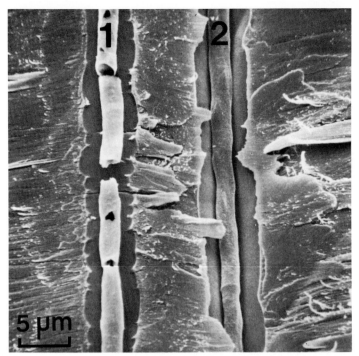

Fig. 5.24. Compression wood of *Picea abies* decomposed by the action of *Poria placenta*. An empty hull of a hypha (*1*) and a normal hypha (*2*) are present in the intercellular spaces. Radial section. SEM. (Jutte and Sachs 1976)

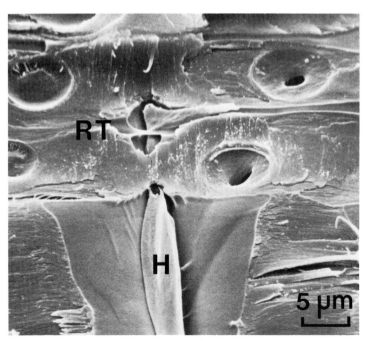

Fig. 5.25. Compression wood of *Picea abies* degraded by *Poria placenta*. A hypha (*H*) in an intercellular space has entered two ray tracheids (*RT*). Radial section. SEM. (Jutte and Sachs 1976)

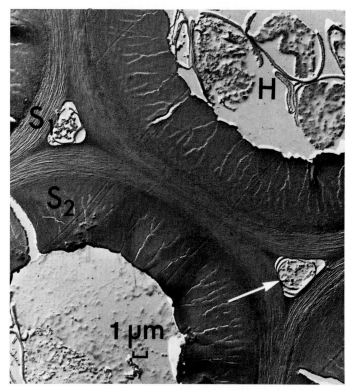

Fig. 5.26. *Abies balsamea* compression wood after exposure for 3 months to *Phellinus pini*. Note the presence of hyphae in both lumina (*H*) and intercellular spaces (*arrow*). Transverse section. TEM

cess, and intact lignin skeletons have so far not been obtained on fungal degradation of compression wood.

The rays in compression wood are also penetrated by the fungal hyphae, but nothing is known regarding the biodegradation of the highly lignified ray cells. The effects of soft rot fungi and bacteria on compression wood also remain to be studied.

5.12 Bark

Bark is often thicker and has a different appearance on the lower side of a leaning stem in comparison with the upper side. Xylem and phloem both arise from the vascular cambium, and it would therefore not be surprising if compression phloem differed from normal phloem. Bark in itself, both the outer bark and the phloem (the inner bark), differs not only in its general anatomy (Chang 1954, Srivastava 1963, den Outer 1967, Esau 1969, Howard 1970) but also in its chemistry (Chang and Mitchell 1955, Painter and Purves 1960, Timell 1961 a, b, 1965 b, c) from wood. Compared to xylem, conifer phloem contains ten times as much inorganic constituents, less acetyl, more uronic acid, galactose, and arabinose, and less glucose, mannose, and xylose residues. The polysaccharide composition of the phloem is not completely known, but it is evident that in conifers, the phloem contains much more pectin and less cellulose and hemicelluloses than does the xylem. Callose, a $(1 \rightarrow 3)$-β-D-glucan, is found only in phloem. Besides lignin, phloem also has a high content of other polyphenolics and especially tannins. The outer bark is rich in suberin.

In their study of the chemical composition of various normal and compression woods, Côté et al. (1966b) subjected bark from seven conifer species to a similar sequence of analyses, comparing either total bark or phloem from the lower (compression wood) side of the stem with corresponding tissue from the upper (opposite wood) side. The results are summarized in Table 5.42. Klason "lignin" in the case of bark includes not only true lignin but also an undetermined amount of tannins. In comparison with xylem (Table 5.22), bark is characterized by its lower content of cellulose, galactoglucomannan, and xylan. Arabinose residues are, however, more numerous in bark than in wood, and most notably so in the genus *Pinus* (Chang and Mitchell 1955, Timell 1961a). Phloem of *Pinus sylvestris* contains significant amounts of a $(1 \rightarrow 5)$-α-L-arabinan (Fu and Timell 1972b), present in only traces in wood.

Comparison between normal and compression phloems in Table 5.42 shows that the two tissues have essentially the same overall chemical composition. The considerable difference in this respect between normal and compression xylems evidently does not extend to the bark. The vascular cambium on the lower side of a branch or a leaning stem is apparently capable of producing a chemically and anatomically (Chap. 4.13) normal phloem toward the outside, while forming at the same time toward the inside a xylem that is entirely different in both these respects.

In some cases compression phloem has been found to contain gelatinous fibers (Liese and Höster 1966), which would probably have a higher than normal cellulose content. Even if present in the tissues analyzed by Côté et al. (1966b), it is doubtful if such gelatinous fibers would have been numerous enough to affect the analytical results.

5.13 General Conclusions

In general, the difference in chemical composition between normal and compression woods of the majority of the conifers is of a quantitative rather than a qualitative nature. All polysaccharides so far isolated from compression wood are also present in normal wood, albeit in very different amounts. An estimate of the lignin and polysaccharide composition of normal and compression woods from a typical conifer is presented in Table 5.43. Compression wood contains 30–40% more lignin and 20–25% less cellulose than normal conifer wood. Laricinan and galactan are present in only trace amounts in normal softwood. The former is a minor and the latter a major constituent of compression wood. The content of galactoglucomannan is twice as high in compression wood, but the xylan content is similar, and the same probably applies to pectin and other

Table 5.42. Summative chemical composition of normal and compression barks from five conifer species. All values in per cent of extractive-free tissues. (Côté et al. 1966 b)

Species and tissue	Type	Lignin	Ash	Acetyl	Uronic anhydride	Galactose	Glucose	Mannose	Arabinose	Xylose
Abies balsamea										
Normal (upper)	Total	32.0	3.2	1.0	12.0	3.3	34.2	5.6	5.1	3.6
Compression (lower)	Total	30.6	3.2	0.8	13.5	3.7	34.8	3.9	5.8	3.7
Normal	Outer	59.9	1.8	0.8	9.6	1.4	17.7	3.3	3.2	2.3
Compression	Outer	53.6	1.7	0.8	7.6	2.3	23.7	4.3	3.4	2.6
Normal	Inner	50.4	2.7	0.9	10.1	3.0	22.5	2.4	3.7	4.3
Compression	Inner	50.3	2.6	0.9	9.8	2.2	20.5	4.8	4.7	4.2
Larix laricina										
Normal	Total	34.7	2.0	0.8	11.4	4.6	33.6	3.4	5.5	4.0
Compression	Total	41.3	2.1	0.7	10.7	4.6	29.7	2.1	5.0	3.8
Picea mariana										
Normal	Total	39.6	3.0	0.6	10.3	1.8	35.3	2.4	3.8	3.2
Compression	Total	39.6	3.2	0.6	11.2	2.5	33.7	2.5	3.7	3.0
Picea rubens										
Normal	Total	26.9	3.5	0.6	14.5	2.4	40.2	2.3	6.2	3.4
Compression	Total	26.9	3.6	0.6	15.6	2.9	37.4	2.0	7.5	3.5
Pinus sylvestris										
Verical stem	Inner	18.3	2.7	1.0	12.5	3.5	45.2	3.7	9.9	3.2
Normal (upper)	Inner	24.2	3.0	1.0	12.2	3.6	38.3	3.3	10.3	4.1
Compression (lower)	Inner	27.3	2.8	0.8	12.0	2.8	41.4	1.8	8.1	3.0

5.13 General Conclusions

Table 5.43. Lignin and polysaccharide composition of normal and compression woods of an average conifer. All values in per cent of extractive-free wood

Constituent	Normal wood	Compression wood
Lignin	30	40
Cellulose	42	30
(1→3)-β-D-Glucan (Laricinan)	Trace	2
Galactoglucomannans	18	9
Galactan	Trace	10
Xylan	8	7
Pectin	1	1
Other polysaccharides	1	1

polysaccharides which include arabinogalactan, arabinan, and xyloglucan. If the amount of polysaccharides present in each tracheid is considered rather than the overall weight percentage, different results are obtained. On this basis, a compression wood tracheid contains as much cellulose, slightly more xylan, and half as much galactoglucomannan as does a tracheid in normal wood.

Qualitatively, the chemical differences between normal and compression woods are less pronounced. The proportions of different galactoglucomannans are probably the same in the two tissues, and their structure is identical, including the presence of acetyl groups. The arabino-4-O-methyl-glucurono-xylan in compression wood carries only half as many arabinose side chains as in normal wood. The exact frequency of the 4-O-methylglucuronic acid side chains is not known. The galactan and laricinan are both slightly branched, acidic polysaccharides. The galactan consists of (1 → 4)-linked β-D-galactose residues which have a few galacturonic acid side chains attached to C-6. Laricinan is a (1 → 3)-linked β-D-glucan which may contain both glucuronic acid and galacturonic acid side chains. Several structural details remain to be established for both polysaccharides. The high lignin content has so far made it impossible to isolate any polysaccharides in representative yields from fully lignified compression wood. All agents used for removing the lignin cause scission of glycosidic bonds, and for this reason the original degree of polymerization of the hemicelluloses in compression wood remains unknown. No attempt has yet been made to isolate an undegraded cellulose from compression wood. It is to be expected that the high lignin content would cause difficulties. All investigators agree that compression-wood cellulose is less crystalline than that present in normal wood.

The lignin in compression wood differs in several respects in its structure from that present in normal softwood. It has a considerably higher proportion of p-hydroxyphenylpropane units and is more condensed than a normal conifer lignin. It also contains fewer arylglycerol-β-arylether and probably more diphenyl and phenylcoumaran interunit bonds and more carbon-carbon bonds. In other respects, it resembles a normal softwood lignin.

Data for inorganic constituents and extractives are few and at times contradictory, but it would appear that compression wood contains slightly more of both than does normal wood. The nature of the polyphenyls and the lignans seem to be the same in the two types of wood. The same applies to the extracellular arabinogalactan which, however, occurs in lesser amounts in compression wood.

References

Adams GA 1960 Structure of an arabinogalactan from tamarack (Larix laricina). Can J Chem 38:280–293

Adams MF, Douglas C 1963 Arabinogalactan – a review of the literature. Tappi 46:544–548

Adler E 1977 Lignin chemistry – past, present and future. Wood Sci Technol 11:169–218

Ahlgren PA, Goring DAI 1971 Removal of wood components during chlorite delignification of black spruce. Can J Chem 49:1272–1275

Ahlgren PA, Yean WQ, Goring DAI 1971 Chlorite delignification of spruce wood. Comparison of the molecular weight of the lignin dissolved with the sizes of pores in the cell wall. Tappi 54:737–740

Ander K, Eriksson KE 1978 Lignin degradation and utilization by micro-organisms. In: Bull MJ (ed) Progress in industrial microtechnology 14, 58 pp

Anderegg RJ, Rowe JW 1974 Lignans, the major component of resin from Araucaria angustifolia knots. Holzforschung 28:171–175

Anderson E, Kesselman J, Bennett EC 1941 Polyuronide hemicelluloses isolated from sapwood and compression wood of white pine, Pinus strobus L. J Biol Chem 140:563–568

Andersson A, Erickson M, Fridh H, Miksche GE 1973 Zur Struktur des Lignins der Rinde von Laub- und Nadelhölzern. Holzforschung 27:189–193

Andrews P, Hough L, Jones JKN 1954 The galactan of Strychnos nux-vomica seeds. J Chem Soc 806–810

Archibald AR, Cunningham WL, Manners DJ, Stark JR, Ryley JF 1963 Studies of the metabolism of the protozoa. 10. The molecular structure of the reserve polysaccharides from Achromonas malhamensis and Peranema trichophorum. Biochem J 88:444–451

Aspinall GO, Begbie R, Hamilton A, Whyte JNC 1967a Polysaccharides of soybeans. III. Extraction and fractionation of polysaccharides from cotyledon meal. J Chem Soc C:1065–1067

Aspinall GO, Begbie R, McKay JE 1962 The hemicelluloses of European larch (Larix decidua). II. The glucomannan component. J Chem Soc 214–219

Aspinall GO, Begbie R, McKay JE 1967b Polysaccharide components of soybeans. Cer Sci Today 12(6):223–229

Aspinall GO, Cottrell IW, Eagan SV, Morrison IM, Whyte JNC 1967c Polysaccharides of soybeans. IV. Partial hydrolysis of the acidic polysaccharide complex from cotyledon meal. J Chem Soc C:1071–1080

Aspinall GO, Fairweather RM 1965 Araucaria bidwilli gum. Carbohyd Res 1:83–92

Aspinall GO, Fairweather RM, Wood TM 1968 Arabinogalactan A from Japanese larch (Larix leptolepis). J Chem Soc C:2174–2179

Aspinall GO, Hunt K, Morrison IM 1967d Polysaccharides of soybeans. V. Acidic polysaccharides from the hulls. J Chem Soc C:1080, 1086

Aspinall GO, Kessler G 1957 The structure of callose from the grape vine. Chem Ind (London), 1296

Aspinall GO, McKenna JP 1968 Araucaria bidwilli gum. II. Further studies on the polysaccharide component. Carbohyd Res 7:244–254

Aspinall GO, Molloy JA, Whitehead CC 1970 Araucaria bidwilli gum: partial acid hydrolysis of the gum. Carbohyd Res 12:143–146

Aspinall GO, Nicolson A 1960 The catalytic oxidation of European larch ε-galactan. J Chem Soc 2503–2507

Aspinall GO, Wood TM 1963 The structure of two water-soluble polysaccharides from Scots pine (Pinus sylvestris). J Chem Soc 1686–1696

Azuma J, Takahashi N, Koshijima T 1981 Isolation and characterization of lignin-carbohydrate complexes from milled wood lignin fraction of Pinus densiflora Sieb et Zucc. Carbohyd Res 93:91–104

Bailey IW, Berkley EE 1942 The significance of x-rays in studying the orientation of cellulose in the secondary wall of tracheids. Am J Bot 29:231–241

Bariska M 1969 Some internal changes in ammonia-treated woody materials. IAWA Bull 1969(2):3–8

Bariska M 1975 Collapse phenomena in beechwood during and after NH_3-impregnation. Wood Sci Technol 9:293–306

Bariska M, Popper R 1971 The behavior of cotton cellulose and beech wood in ammonia atmosphere. Polym Symp 36:199–212

Bariska M, Popper R 1975 Ammonia sorption isotherm of wood and cotton cellulose. Wood Sci Technol 9:153–163

Bariska M, Schuerch C 1977 Wood softening and forming with ammonia. In: Goldstein IS (ed) Wood technology: chemical aspects. ACS Symp Ser 43:326–347

Bariska M, Skaar C, Davidson RW 1969 Studies on the wood-anhydrous ammonia system. Wood Sci 2:65–72

Bariska M, Skaar C, Davidson RW 1970 Water sorption "overshoot" in ammonia-treated wood. Wood Sci 2:232–237

Bariska M, Strasser C 1976 The progressing of plasticization in beech wood as indicated by changes in its mechanical properties. Appl Polym Symp 28:1087–1097

Beakbane AB, Thompson EC 1945 Abnormal lignification in the wood of some apple trees. Nature 156:145–146

Beall FC 1969 Thermogravimetric analysis of wood lignin and hemicelluloses. Wood Fiber 1:215–226

Beattie A, Hirst EL, Percival E 1961 Studies on the metabolism of the Crysophyceae. Comparative structural investigations on leucosin (chrysolaminarin) separated from diatoms and laminarin from the brown algae. Biochem J 79:531–537

Berklee EE, Woodyard OC 1948 Certain variations in the structure of wood fibers. Text Res J 18:519–525

Berzinsh GV, Eglais IJ, Kalninsh AJ 1976 Changes in birch wood structure and strength properties in the process of wood modification with ammonia. Appl Polym Symp 28:1099–1108

Bikales N, Segal L (eds) 1971 Cellulose and cellulose derivatives, Vol 4. Interscience-Wiley, New York, 1411 pp

Björkman A 1956 Studies on finely divided wood. 1. Extraction of lignin with neutral solvents. Svensk Papperstidn 59:477–485

Björkman A 1957 Studies on finely divided wood. 3. Extraction of lignin-carbohydrate complexes with neutral solvents. Svensk Papperstidn 60:243–251

Bland DE 1958a The chemistry of reaction wood. I. The lignins of Eucalyptus goniocalyx and Pinus radiata. Holzforschung 12:36–43

Bland DE 1958b The spectra of reaction wood lignins in relation to wood maturity. Holzforschung 12:115–116

Bland DE 1961 The chemistry of reaction wood. III. The milled wood lignins of Eucalyptus goniocalyx and Pinus radiata. Holzforschung 15:102–106

Bletchley JD, Taylor JM 1964 Investigations on the susceptibility of home-grown Sitka spruce (Picea sitchensis) to attack by the common furniture beetle (Anobium punctatum Deg). J Inst Wood Sci 12:29–43

Bluhm TL, Sarko A 1977a Conformational studies of polysaccharide multiple helices. Carbohyd Res 54:125–138

Bluhm TL, Sarko A 1977b The triple helical structure of lentinan, a linear β-(1 → 3)-D-glucan. Can J Chem 55:293–299

Bose S, Dutta AS 1963 Structure of Salmalia malabarica gum. I. Nature of sugars present and the structure of aldobiouronic acid. J Indian Chem Soc 40:257–262

Boutelje JB 1966 On the antomical structure, moisture content, density, shrinkage, and resin content of the wood in and around knots in Swedish pine (Pinus silvestris L) and Swedish spruce (Picea abies Karst). Svensk Papperstidn 69:1–10

Bouveng HO 1959 Studies on arabogalactans. IV. A methylation study of arabogalactan B from Larix occidentalis Nutt. Acta Chem Scand 13:1877–1883

Bouveng HO 1961 Studies on arabogalactans. V. Barry degradation of the arabogalactans from western hemlock. A kinetic study of the mild acid hydrolysis of arabogalactan A. Acta Chem Scand 15:78–86

Bouveng HO, Lindberg B 1958 Studies on larch arabogalactans. II. Fractionation of the arabogalactan from Larix occidentalis Nutt. A methylation study of one of the components. Acta Chem Scand 12:1977–1984

Bouveng HO, Meier H 1959 Studies on a galactan from Norwegian spruce compression wood. Acta Chem Scand 13:1884–1889

Brax AJ 1936 Undersökningar angående kvistens inflytande på slipmassans egenspaer. (Investigations on the influence of knots on the properties of groundwood pulp.) Svensk Papperstidn 39 (spec issue):7–20

Brodzki P 1972 Callose in compression-wood tracheids. Acta Soc Bot Polon 41:321–327

Brown RG, Lindberg B 1967 Polysaccharides from cell walls of Aureobasidium (Pullularia) pullulans. Acta Chem Scand 21:2379–2382

Browning BL 1967 Methods of wood chemistry. Interscience, New York London Sydney, 882 pp

Browning BL, Bublitz LO 1953 The isolation of holocellulose from wood. Tappi 36:452–458

Caldwell WI, Lork-Horowitz K 1937 X-ray investigations of the structure of wood. Phys Rev 51:998–999

Chang YP 1954 Anatomy of common North American pulpwood barks. TAPPI Monogr Ser 14, 249 pp

Chang YP, Mitchell RL 1955 Chemical composition of common North American pulpwood barks. Tappi 38:315–320

Chochieva MM, Tsvetaeva IP, Yur'eva MK, Zaitseva AF, Petropavlovskii, Nikitin NI 1958 (Distribution of arabinogalactan in Larix dahurica.) Trudy Inst Lesa Akad Nauk SSSR Izuchenie Khim Sostava Drevesiny Daursk Listvennitsy 45:31–49

Chu LC 1972 Comparison of normal wood and first-year compression wood in longleaf pine trees. M S Thesis MS State Univ, 72 pp

Cieslar A 1896 Das Rothholz der Fichte. Cbl. Ges Forstwes 22:149–165

Cieslar A 1897 Über den Ligningehalt einiger Nadelhölzer. Mitt Forstl Versuchsanst Österr 23:1–40

Clarke AE, Stone BA 1960 Structure of the paramylon from Euglena gracilis. Biochim Biophys Acta 44:161–163

Cleve von Euler A 1923 Ligninhaltens storlek och växlingar hos svensk tall och gran. Undersökningar över pappersvedens tekniska egenskaper II. (The lignin content and its variations in Swedish pine and spruce. Investigations on the technical properties of pulpwood. II.) Svensk Skogsvårdsför Tidskr B21

Coles RW 1973 Ultrastructural changes in ammonia-plasticized Corsican pine. IAWA Bull 1973(4):3–8

Correns E 1961 Über anormale Holzfasern. Pap Puu 43:47–62

Côté WA, Jr, Day AC, Kutscha NP, Timell TE 1967a Studies on compression wood. V. Nature of the compression wood formed in the early springwood of conifers. Holzforschung 21:180–186

Côté WA, Jr, Day AC, Simson BW, Timell TE 1966a Studies on larch arabinogalactan. I. The distribution of arabinogalactan in larch wood. Holzforschung 20:178–192

Côté WA, Jr, Kutscha NP, Simson BW, Timell TE 1968 Studies on compression wood. VI. Distribution of polysaccharides in the cell wall of tracheids from compression wood. Tappi 51:33–40

Côté WA, Jr, Pickard PA, Timell TE 1967b Studies on compression wood. IV. Fractional extraction and preliminary characterization of polysaccharides in normal and compression wood of balsam fir. Tappi 50:350–356

Côté WA, Jr, Simson BW, Timell TE 1966b Studies on compression wood. II. The chemical composition of wood and bark from normal and compression regions of fifteen species of gymnosperms. Svensk Papperstidn 69:547–558

Côté WA, Jr, Simson BW, Timell TE 1967c Studies on larch arabinogalactan. II. Degradation of arabinogalactan within the living tree. Holzforschung 21:85–88

Côté WA, Jr, Timell TE 1967 Studies on larch arabinogalactan. III. Distribution of arabinogalactan in tamarack. Tappi 50:285–289

Côté WA, Jr, Timell TE, Zabel RA 1966c Studies on compression wood. I. Distribution of lignin in compression wood of red spruce (Picea rubens Sarg). Holz Roh-Werkst 24:432–438

Cowling EB 1961 Comparative biochemistry of the decay of sweetgum sapwood by white-rot and brown-rot fungi. US Dep Agr Tech Bull 1258, 79 pp

Croon I, Lindberg B, Meier H 1959 Sturcture of a glucomannan from Pinus silvestris L. Acta Chem Scand 13:1299–1304

Curran CE 1936, 1937 Some relations between growth conditions, wood structure, and pulping quality. Pap Trade J 103(11):200–204. Tech Assoc Pap 20:464–468

Dadswell HE, Hawley LF 1929 Chemical composition of wood in relation to physical characteristics. Ind Eng Chem 21:973–975

Dadswell HE, Wardrop AB 1955 Structure and properties of tension wood. Holzforschung 9:97–104

Das A, Rao CVN 1964 Constitution of the glucan from mango (Mangifera indica L). Tappi 47:339–343

Das A, Rao CVN 1965 Constitution of the glucan from mango, Mangifera indica L. Aust J Chem 18:845–850

Das NN, Das A 1977 Structure of the D-galactan isolated from garlic (Allium sativum) bulbs. Carbohyd Res 56:337–343

Davidson RW, Baumgardt WG 1970 Plasticizing wood with ammonia – a progress report. For Prod J 20(3)19–24

Delmer DP 1977 The biosynthesis of cellulose and other plant cell wall polysaccharides. In: Loewus FA, Runeckles VC (eds) The structure, biosynthesis, and degradation of wood. Plenum, New York London, 45–77

Dence CW, Gupta MK, Sarkanen KV 1962 Studies on oxidative delignification mechanisms. II. Reactions of vanillyl alcohol with chlorine dioxide and sodium chlorite. Tappi 45:29–38

den Outer RW 1967 Histological investigations of the secondary phloem of gymnosperms. Medd Landbouwhogeschool Wageningen 67-7, 119 pp

Deslandes Y, Marchessault RH, Sarko A 1980 Triple-helical structure of $(1 \rightarrow 3)$-β-D-glucan. Macromolecules 13:1466–1471

Dietrich G 1973 Untersuchungen über die Astbildung und natürliche Astreinigung der Weißtanne. Beih Forstwiss Cbl 34, 95 pp

Dinwoodie JM 1965 Tensile strength of individual compression wood fibres and its influence on properties of paper. Nature 205:763–764

Dryselius E, Lindberg B 1956 Constitution of resin phenols and their biogenetic relations. XXI. Pinoresinol and its dimethyl ether from Araucaria angustifolia. Acta Chem Scand 10:445–446

Ekman R 1979 Distribution of lignans in Norway spruce. Acta Acad Aboensis 39(3):1–6

Ekman R 1980 Wood extractives of Norway spruce. Publ Inst Wood Chem Pulp Pap Technol Åbo Akad A-330, 42 pp

Erickson M, Larsson S, Miksche GE 1973a Gaschromatographische Analyse von Ligninoxydationsprodukten. VII. Ein verbessertes Verfahren zur Charakterisierung von Ligninen durch Methylierung und oxydativen Abbau. Acta Chem Scand 27:127–140

Erickson M, Larsson S, Miksche GE 1973b Gaschromatographische Analyse von Ligninoxydationsprodukten. VIII. Zur Struktur des Lignins der Fichte. Acta Chem Scand 27:903–914

Erickson M, Larsson S, Miksche GE 1973c Zur Struktur des Lignins des Druckholzes von Pinus mugo. Acta Chem Scand 27:1673–1678

Erickson M, Miksche GE 1974a Charakterisierung der Lignine von Gymnospermen durch oxidativen Abbau. Holzforschung 28:135–138

Erickson M, Miksche GE 1974b Charakterisierung der Lignine von Pteridophyten durch oxidativen Abbau. Holzforschung 28:157–159

Erickson M, Miksche GE, Somfai I 1973d Charakterisierung der Lignine von Angiospermen durch oxidativen Abbau. II. Monokotylen. Holzforschung 27:147–150

Eriksson KE 1981 Fungal degradation of wood components. Pure Appl Chem 53:33–43

Eriksson KE, Grünewald A, Nilsson T, Vallander L 1980 A scanning electron microscopy study of the growth and attack on wood by three white-rot fungi and their cellulase-less mutants. Holzforschung 34:207–213

Eriksson Ö, Goring DAI, Lindgren BO 1980 Structural studies on the chemical bonds between lignins and carbohydrates in spruce wood. Wood Sci Technol 14:267–279

Eriksson Ö, Lindgren BO 1977 About the linkage between lignin and hemicelluloses in wood. Svensk Papperstidn 80:59–63

Esau K 1969 The phloem. Gebrüder Bornträger, Berlin, 505 pp

Eskilsson S, Hartler N 1973 Whole tree pulping. 2. Sulphate pulping. Svensk Papperstidn 76:63–70

Ettling BV, Adams MF 1968 Gel filtration of arabinogalactan from western larch. Tappi 51:116–118

Faix O, Gyzas E, Schweers W 1977 Vergleichende Untersuchungen an Ligninen verschiedener Pteridophyten-Arten. Holzforschung 31:137–144

Fengel D, Grosser D 1975 Chemische Zusammensetzung von Nadel- und Laubhölzern. Holz Roh-Werkst 33:32–34

Fengel D, Wegener G 1983 Wood. Chemistry, ultrastructure, reactions. Walter de Gruyter, Berlin New York, 613 pp

Ford CW, Percival E 1965 The carbohydrates of Phaeodactylum tricornutum. I. Preliminary examination of the organism, and characterization of low molecular weight material and of a glucan. J Chem Soc 7035–7041

Freudenberg K 1964 Entwurf eines Konstitutionsschemas für das Lignin der Fichte. Holzforschung 18:3–9

Freudenberg K 1965 Lignin: its constitution and formation from p-cinnamyl alcohols. Science 148:595–600

Freudenberg K 1968 The constitution and biosynthesis of lignin. In: Freudenberg K, Neish AC, Constitution and biosynthesis of lignin. Springer, New York, 45–122

Freudenberg K, Neish AC 1968 Constitution and biosynthesis of lignin. Springer, New York, 129 pp

Fu YL, Gutmann PJ, Timell TE 1972 Polysaccharides in the secondary phloem of Scots pine (Pinus sylvestris L). I. Isolation and characterization of callose. Cellul Chem Technol 6:507–512

Fu YL, Timell TE 1972a Polysaccharides in compression wood of tamarack (Larix laricina). 5. The constitution of an acidic arabinogalactan. Svensk Papperstidn 75:680–682

Fu YL, Timell TE 1972b Polysaccharides in the secondary phloem of Scots pine (Pinus sylvestris L). II. The structure of an arabinan. Cellul Chem Technol 6:513–515

Fu YL, Timell TE 1972c Polysaccharides in the secondary phloem of Scots pine (Pinus sylvestris L). III. The constitution of a galactoglucomannan. Cellul Chem Technol. 6:517–519

Fukazawa K 1974 The distribution of lignin in compression- and lateral-wood of Abies sachalinensis using ultraviolet microscopy. Res Bull Coll Exp For Hokkaido Univ 31:87–114

Fulcher RG, McCully ME, Setterfield G, Sutherland J 1976 β-1,3-Glucans may be associated with cell plate formation during cytokinesis. Can J Bot 54:539–542

Fuoss RM, Strauss UP 1949 Viscosity of mixtures of polyelectrolytes and simple electrolytes. Ann NY Acad Sci 51:836–851

Gaillard BDE, Thompson NS 1971 Interaction of polysaccharides with iodine. II. The behavior of xylans in different salt solutions. Carbohyd Res 18:137–146

Gaillard BDE, Thompson NS, Morak AJ 1969 The interaction of polysaccharides with iodine. I. Investigation of the general nature of the reaction. Carbohyd Res 11:509–519

Gillham GK, Perlin AS, Timell TE 1958 The configuration of glycosidic linkages in oligosaccharides. VII. 4-O-β-D-galactopyranosyl-D-galactose from white birch wood. Can J Chem 36:1741–1743

Gillham JK, Timell TE 1958 The polysaccharides of white birch (Betula papyrifera). VII. Carbohydrates associated with the alpha-cellulose component. Svensk Papperstidn 61:540–544

Giordano G 1971 Tecnologia del legno. I. La materia prima. Unione Tipographico-Editrice Torinese, Torino, 1086 pp

Glaudemans CPJ, Timell TE 1957 The polysaccharides of white birch (Betula papyrifera). V. The isolation of the hemicellulose. Svensk Papperstidn 60:869−871

Goring DAI, Timell TE 1960 Molecular properties of a native wood cellulose. Svensk Papperstidn 63:524−527

Goring DAI, Timell TE 1962 Molecular weight of native celluloses. Tappi 40:454−460

Goto T, Harada H, Saiki H 1975 Cross-sectional view of microfibrils in gelatinous layer of poplar tension wood (Populus euramericana). Mokzai Gakkaishi 21:537−542

Goto T, Harada H, Saiki H 1978 Fine structure of cellulose microfibrils in poplar gelatinous layer and Valonia. Wood Sci Technol 12:223−231

Gross GG 1977 Biosynthesis of lignin and related monomers. In: Loewus FA, Runeckles VC (eds) The structure, biosynthesis, and degradation of wood. Plenum, New York, London, 141−184

Gross GG 1980 Biochemistry of lignification. Adv Bot Res 8:25−63

Hägglund E 1925 Jämförande undersökningar över utbyte och egenskaper av cellulosa framställd enligt sulfitmetoden av rå och lufttorr granved from olika delar av trädet. (Comparative investigations on the yield and properties of cellulose prepared by the sulfate method from green and air-dry spruce wood from different parts of the tree.) Suom Pap Puutavar 7:38−45

Hägglund E 1951 Chemistry of wood. Academic Press, New York, 631 pp

Hägglund E, Larsson S 1937 Om grankvistens kemiska sammansättning och dess förhållande vid sulfitkokningsprocessen. (On the chemical composition of knots in spruce and their behavior in the sulfite pulping process.) Svensk Papperstidn 40:356−360

Hägglund E, Ljunggren S 1933 Untersuchungen des Rotholzes von Fichte. I. Die chemische Zusammensetzung des Rotholzes. Papier-Fabr 31(27A):35−38. Svensk Kem Tidskr 45:123−129

Hans R 1970 Die Zusammensetzung nichtcellulosischer Polysaccharide in verschiedenen Altersstufen des Holzes von Thuja occidentalis und Abies dahlemensis. Holzforschung 24:60−64

Hans R 1971 Die Polysaccharidzusammensetzung des Holzes verschiedener Altersstufen von Taxodium distichum. Holzforschung 25:15−18

Haq S, Adams GA 1961 Structure of an arabinogalactan from tamarack (Larix laricina). Can J Chem 39:1563−1573

Harada T, Misaki A, Saito H 1968 Curdlan: a bacterial gel-forming β-1,3-glucan. Arch Biochem Biophys 124:292−298

Harris JM 1961 The resinification of knots in radiata pine. NZ For Res Inst Res Note 28, 11 pp

Harris JM 1963 The effect of pruning on the resinification of knots in radiata pine. NZ For Res Inst Res Note 34, 7 pp

Hartig R 1901 Holzuntersuchungen. Altes und Neues. Julius Springer, Berlin, 99 pp

Hasegawa Y, Yamada T, Sumiya K 1970 (Formation of wood under restraint and properties of the wood.) Wood Res (Kyoto) 49:1−17

Hashizume S, Takahashi S 1974 (Variation of arabinogalactan and some extractives in wood, Karamatsu (Larix leptolepis Gordon).) Bull Shinshu Univ For 11:19−45

Hata K 1951 Studies on the pulp of "Akamatsu" (Pinus densiflora S et Z) wood. IV. On the chemical composition of pulp of compression wood. J Jap For Soc 32:257−260

Hay JB, Gray GM 1966 The isolation of O-β-D-galactopyranosyl-(1 → 4)-D-galactose: a product of the partial acid hydrolysis of the galactan of white-lupin (Lupinus albus) seeds. Biochem J 100:33c−35c

Herth W, Franke WW, Bittiger H, Kuppel A, Keilich G 1974 Alkali-resistant fibrils of β-1,3- and β-1,4-glucans: structural polysaccharides in the pollen tube wall of Lilium longiflorum. Cytobiology 9:344−367

Higuchi T 1980 Biochemistry of lignification. Wood Res (Kyoto) 66:1−16

Higuchi T 1981 Biosynthesis of lignin. In: Tanner W, Loewus FA (eds) Encyclopedia of plant physiology. Springer, Berlin Heidelberg New York, (ns) 13B:194−223

Higuchi T, Ito Y, Kawamura I 1967 p-Hydroxyphenylpropane component of grass lignin and the role of tyrosine-ammonia lyase in its formation. Phytochemistry 6:875−881

Hillis WE (ed) 1962 Wood extractives and their significance to the pulp and paper industries. Academic Press, New York London, 513 pp

Hirst EL, Jones JKN, Walder WO 1947 Pectic substances. VII. The constitution of the galactan from Lupinus albus. J Chem Soc 1225–1229

Hoffmann GC, Simson BW, Timell TE 1971 Structure and molecular size of pachyman. Carbohyd Res 20:185–188

Hoffmann GC, Timell TE 1970a Isolation of a β-1,3-glucan (laricinan) from compression wood of Larix laricina. Wood Sci Technol 4:159–162

Hoffmann GC, Timell TE 1970b Isolation and characterization of a galactoglucomannan from red pine (Pinus resinosa) wood. Tappi 53:1896–1899

Hoffmann GC, Timell TE 1972a Polysaccharides in compression wood of tamarack (Larix laricina). I. Isolation and characterization of laricinan, an acidic glucan. Svensk Papperstidn 75:135–142

Hoffmann GC, Timell TE 1972b Polysaccharides in compression wood of tamarack (Larix laricina). II. Isolation and structure of a xylan. Svensk Papperstidn 75:241–242

Hoffmann GC, Timell TE 1972c Polysaccharides in compression wood of tamarack (Larix laricina). III. Constitution of a galactoglucomannan. Svensk Papperstidn 75:297–298

Hoffmann GC, Timell TE 1972d Polysaccharides in ray cells of normal wood of red pine (Pinus resinosa). Tappi 55:733–736

Hoffmann GC, Timell TE 1972e Polysaccharides in ray cells of compression wood of red pine (Pinus resinosa). Tappi 55:871–873

Holzer WF, Lewis HF 1950 The characteristics of unbleached kraft pulp from western hemlock, Douglas-fir, western red cedar, loblolly pine, and black spruce. VII. Comparison of springwood and summerwood fibers of Douglas-fir. Tappi 33:110–112

Hosia M, Lindholm CA, Toivanen P, Nevalainen K 1971 Undersökningar rörande utnyttjandet av barrträdsgrenar som råmaterial för kemisk massa och hård fiberskiva I. (Investigations on the utilization of softwood branches as a raw material for pulp and hardboard I.) Pap Puu 53:49–66

Howard ET 1970 Bark structure of the southern pines. Wood Sci 3:134–148

Howard ET 1973 Physical and chemical properties of slash pine tree parts. Wood Sci 5:312–317

Huwyler HR, Franz G, Meier H 1979 Changes in the composition of cotton fibre cell walls during development. Planta 146:635–642

Jaquet JP, Buchala AJ, Meier H 1982 Changes in the nonstructural carbohydrate content of cotton (Gossypium spp) fibres at different stages of development. Planta 156:481–486

Jarceva NA 1969 (The chemical composition of Pinus sibirica.) Lesn Z Arhangel'sk 12(1):112–116

Jayme G, Blischnok B 1938 Über die chemische Zusammensetzung der verschiedenen Anteile des Kiefernholzes. Holz Roh-Werkst 1:538–543

Jayme G, Harders-Steinhäuser M 1950 Über die chemische Zusammensetzung des Zugholzes in einem Pappelholz. Papier 4:104–113

Jayme G, Harders-Steinhäuser M 1953 Zugholz und seine Auswirkungen in Pappel- und Weidenholz. Holzforschung 7:39–43

Jelsma J 1974 The structure of $(1 \rightarrow 3)$-β-D-glucans. Acta Bot Neerl 23:747–748

Jelsma J, Kreger DR 1975 Ultrastructural observations on $(1 \rightarrow 3)$-β-D-glucans from fungal cell walls. Carbohyd Res 43:200–203

Jiang KS, Timell TE 1972 Polysaccharides in compression wood of tamarack (Larix laricina). 4. Constitution of an acidic galactan. Svensk Papperstidn 75:592–594

Johnsen B, Hovey RW 1918 The determination of cellulose in wood. J Soc Chem Ind 37:132T–137T

Jones DG, Simson BW, Timell TE 1968 Studies on larch arabinogalactan. V. Degradation of arabinogalactan within a 700-year-old western larch tree. Cellul Chem Technol 2:391–399

Jones JKN, Wise LE, Jappe JP 1956 The action of alkali containing metaborate on wood cellulose. Tappi 39:139–141

Jurášek L 1964a Změny v obsahu stavebních složek dřeva při rozkladu způsobeném houbami. (Changes in the content of structural components of wood in decomposition caused by fungi.) Drev Vysk 9:41–56

Jurášek L 1964b Změny v mikroskopické struktuře při rozkladu dřeva dřevokaznými houbami. (Changes in the microstructure at the destruction of wood by wood destroying fungi.) Drev Vysk 9:127–144

Jutte SM, Jongebloed WL, Sachs IB 1977 Influence of water environment on normal and compression wood of a Picea species observed by scanning electron microscopy (SEM). SEM 1977/II Proc Workshop on other biological applications of the SEM/STEM, IIT Res Inst Chicago IL, 683–690

Jutte SM, Sachs IB 1976 SEM observations on brown-rot fungus Poria placenta in normal and compression wood of Picea abies. SEM 1976, VII. Proc Workshop on plant sci appl of the SEM, ITT Res Inst Chicago IL, 535–542

Kantola M 1964 X-ray studies concerning the orientation of crystallites in wood fibers. Faserforsch Textiltech 15:587–590

Kantola M, Kähkönen H 1963 Small-angle x-ray investigation of the orientation of crystallites in Finnish coniferous and deciduous wood fibers. Ann Acad Sci Fenn A-137:1–14

Kantola M, Seitsonen S 1961 X-ray orientation investigations on Finnish conifers. Ann Acad Sci Fenn A-80:1–15

Katz G 1965 The location and significance of the O-acetyl groups in a glucomannan from Parana pine. Tappi 48:34–41

Kenne L, Rosell KG, Svensson S 1975 Studies on the distribution O-acetyl groups in pine glucomannan. Carbohyd Res 44:69–76

Kessler G 1958 Zur Charakterisierung der Siebröhrenkallose. Ber Schweiz Bot Ges 68:5–43

Kibblewhite RP 1973 Effects of beating and wood quality on radiata pine kraft paper properties. NZ J For Sci 3:220–239

Kininmoth JA 1961 Checking of intergrown knots during seasoning of radiata pine sawn timber. NZ For Res Inst Tech Pap 30, 16 pp

Kinnman G 1923 Kvalitetsfordringar på pappersved och skogsvårdsåtgärdernas anpassande därefter. (Quality requirements for pulpwood and the silvical practices required.) Svensk Skogsvårdsför Tidskr 21:201–225

Kirk TK 1971 Effects of microorganisms on lignin. Ann Rev Phytopath 9:185–210

Kirk TK 1973 The chemistry and biochemistry of decay. In: Nicholas DD (ed) Wood deterioration and its prevention by preservative treatments. Vol I. Syracuse Univ Press, Syracuse, 149–177

Kirk TK 1975a Chemistry of lignin degradation by wood-destroying fungi. In: Liese W (ed) Biological transformation of wood by microorganisms. Springer, Berlin Heidelberg New York, 153–164

Kirk TK 1975b Effects of a brown-rot fungus, Lenzites trabaea, on lignin in spruce wood. Holzforschung 29:99–107

Kirk TK, Chang HM 1974 Decomposition of lignin by white-rot fungi. I. Isolation of heavily degraded lignins from decayed spruce. Holzforschung 28:217–222

Kirk TK, Chang HM 1975 Decomposition of lignin by white-rot fungi. II. Characterization of heavily degraded lignins from decayed spruce. Holzforschung 29:56–64

Kirk TK, Chang HM, Lorenz LF 1975 Topochemistry of the fungal degradation of lignin in birch wood as related to the distribution of guiacyl and syringyl lignins. Wood Sci Technol 9:81–86

Kirk TK, Connors WJ, Zeikus JG 1977 Advances in understanding the microbiological degradation of lignin. In: Loewus FA, Runeckles VC (eds) The structure, biosynthesis and degradation of wood. Plenum, New York London, 369–394

Kirk TK, Higuchi T, Chang HM 1980 Lignin biodegradation: microbiology, chemistry and potential applications. CRC Press, Boca Raton, 256 pp (I), 272 pp (II)

Klason P 1923 Om granvedens halt av lignin. (On the lignin content of spruce wood.) Svensk Papperstidn 26:319–321

Klason P 1924 Hur skall man definiera cellulosa? (How should cellulose be defined?) Svensk Papperstidn 27:261–264

Klason P 1931 Untersuchung über den Wechsel des Ligningehaltes des Fichtenholzes als Folge von verschiedenen klimatischen Verhältnissen. Cellulosechemie 12:36–37

Klason P 1932 Om reservligninet i granved. (On the reserve lignin in spruce wood.) Svensk Papperstidn 35:152–155

Klem GG 1933 Undersökelser av granvirkets kvalitet. (Investigations on the quality of spruce wood.) Medd Norsk Skogforsøksves 5(17):197–348

Koch P 1972 Utilization of the southern pines. US Dep Agr Handb 420, 1663 pp

Kooiman P, Adams GA 1961 Constitution of a glucomannan from tamarack (Larix laricina). Can J Chem 39:889–896

Koshijima T 1960 Studies on mannan in wood pulp. III. Acetyl-mannose linkages in the hemicellulose extracted with dimethyl sulfoxide. Mokuzai Gakkaishi 6:194–198

Koshijima T 1962 Studies on mannan in wood pulp. VII. The crystallinity of glucomannan. Agr Biol Chem (Japan) 26:98–105

Koshijima T, Timell TE, Zinbo M 1965 The number-average molecular weight of native hardwood xylans. J Polym Sci C-11:265–279

Köster E 1934 Die Astreinigung der Fichte. Mitt Forstwirtsch Forstwiss 5:393–416

Kowalski J 1972 (Determining certain properties of branch wood of Scots pine from different sites.) Zeszyty Naukowe Szkoly Glownej Gospodarstwa Wiejskiego w Warszawie, Leśnictwo 17:95–122

Kuo CM, Timell TE 1969 Isolation and characterization of a galactan from tension wood of American beech (Fagus grandifolia Ehr). Svensk Papperstidn 72:703–716

Kuo ML, Arganbright DG 1980a Cellular distribution of extractives in redwood and incense cedar. I. Radial variation in cell-wall extractive content. Holzforschung 34:17–22

Kuo ML, Arganbright DG 1980b Cellular distribution of extractives in redwood and incense cedar. II. Microscopic observation of the location of cell wall and cell cavity extractives. Holzforschung 34:41–47

Kuroyanagi S 1953 (Spectrochemical studies of woods and trees. III. On the qualitative analysis of the chemical elements in woods and trees by emission spectroscopy.) Okayama Univ Fac Agr Sci Rep 2:92–106

Kürschner W 1964 Betrachtungen zum gegenwärtigen Stand der chemischen Holzforschung. Holzforschung 18:65–76

Kutsuki H, Higuchi T 1981 Activities of some enzymes of lignin formation in reaction wood of Thuja orientalis, Metasequoia glyptostroboides and Robinia pseudoacacia. Planta 152:365–368

Kutsuki H, Higuchi T 1982 Activities of some enzymes of lignin formation in reaction wood of Thuja orientalis and Metasequoia glyptostroboides. Wood Sci Technol 16:287–291

Ladell JL, Carmichael AJ, Thomas GHS 1968 Current work in Ontario on compression wood in black spruce in relation to pulp yield and quality. Proc 8th Lake States For Tree Improv Conf 1967, US For Serv Res Pap NC-23, 52–59

Lai YZ, Sarkanen KV 1971 Isolation and structural studies. In: Sarkanen KV, Ludwig CH (eds) Lignins. Occurrence, formation, structure and reactions. Wiley-Interscience, New York, 165–240

Larson PR 1969 Incorporation of ^{14}C in the developing walls of Pinus resinosa tracheids: compression wood and opposite wood. Tappi 52:2170–2177

Larsson S, Miksche GE 1969a Gaschromatographische Analyse von Ligninoxydationsprodukten. II. Nachweis eines neuen Verknüpfungsprinzips von Phenylpropaneinheiten. Acta Chem Scand 23:917–923

Larsson S, Miksche GE 1969b Gaschromatographische Analyse von Ligninoxydationsprodukten. III. Oxidativer Abbau von methylierten Björkman-Lignine (Fichte). Acta Chem Scand 23:3337–3351

Latif MA 1968 Comparative study of normal and compression wood lignin of Douglas fir (Pseudotsuga menzeisii Franco). Ph D Thesis Univ WA, Seattle, 69 pp

Lebedev KK 1965a (Ozonization of reaction wood.) Sb Tr Tsentr Nauch-Issled i Proektn Inst Lesokhim Prom 16:248–255

Lebedev KK 1965b (Experimental data on aliphatic double bonds in compression wood.) Sb Tr Tsentr Nauch-Issled i Proektn Inst Lesokhim Prom 16:256–258

Lee CL 1961 Crystallinity of wood cellulose fibers. For Prod J 11:108–112

Lee VPFF 1968 Structural differences in lignin formation between normal and compression wood of Douglas fir. Ph D Thesis Univ WA, Seattle, 52 pp

Liese W 1970 Ultrastructural aspects of woody tissue disintegration. Ann Rev Phytopath 8:231–258

Liese W (ed) 1975 Biological transformation of wood by microorganisms. Springer, Berlin Heidelberg New York, 203 pp

Liese W, Höster HR 1966 Gelatinöse Bastfasern im Phloem einiger Gymnospermen. Planta 69:338–346

Lindberg B 1962 Recent advances in methods of isolating and purifying polysaccharides. Pure Appl Chem. 5:67–75

Lindberg B, Meier H 1957 Studies on glucomannans from Norwegian spruce. Isolation and physical properties. Svensk Papperstidn 60:785–790

Lindberg B, Norin T 1969 Hartsets kemi (Resin chemistry). Svensk Papperstidn 72:143–153

Lindberg B, Rosell KG, Svensson S 1973 Positions of the O-acetyl groups in pine glucomannan. Svensk Papperstidn 76:383–384

Lindgren PH 1958 X-ray orientation investigations on some Swedish cellulose materials. Arkiv Kemi 12(38):437–452

Loewus FA, Runeckles VC (eds) 1977 The structure, biosynthesis, and degradation of wood. Plenum, New York London, 527 pp

Lynch RS, Stillman JE, Timell TE 1968 Studies on larch arabinogalactan. VI. Nature of the galactan framework in an arabinogalactan from western larch. Svensk Papperstidn 71:890–891

Mahadevan PR, Tatum EL 1967 Localization of structural polymers in the cell wall of Neurospora crassa. J Cell Biol 35:295–302

Mahdalík M, Rajčan J, Mlčoušek M, Lábsky O 1971 Changes of some chemical properties of wood treated with liquid ammonia. J Polym Sci C-36:251–263

Manners DJ, Ryley JF, Stark JR 1966 Studies on the metabolism of the protozoa. The molecular structure of the reserve polysaccharide from Astasia ocellata. Biochem J 101:323–327

Marchessault RH, Deslandes Y 1979 Fine structure of $(1 \rightarrow 3)$-D-glucans: curdlan and paramylon. Carbohyd Res 75:231–242

Marchessault RH, Deslandes Y, Ogawa O, Sundararajan PR 1977 X-ray diffraction data for β-$(1 \rightarrow 3)$-D-glucan. Can J Chem 55:300–303

Marchessault RH, Settineri W, Winter W 1967 Crystallization of xylan in the presence of cellulose. Tappi 50:55–59

Marchessault RH, Timell TE 1960 The x-ray pattern of crystalline xylans. J Phys Chem 64:704

Marton R, Rushton P, Sacco JS, Sumiya K 1972 Dimensions and ultrastructure of growing fibers. Tappi 55:1499–1504

Meier H 1955 Über den Zellwandabbau durch Holzvermorschungspilze und die submikroskopische Struktur von Fichtenholztracheiden und Birkenholzfasern. Holz Roh-Werkst 13:323–338

Meier H 1958a Barium hydroxide as a selective precipitating agent for hemicelluloses. Acta Chem Scand 12:144–146

Meier H 1958b Studies on hemicelluloses from pine (Pinus sylvestris L). Acta Chem Scand 12:1911–1918

Meier H 1961 Isolation and characterization of an acetylated glucomannan from pine (Pinus silvestris L). Acta Chem Scand 15:1381–1385

Meier H 1962 Studies on a galactan from tension wood of beech (Fagus silvatica L). Acta Chem Scand 16:2275–2283

Meier H 1965 On the chemistry of reaction wood. In: Chimie et biochimie de la lignine, de la cellulose et des hémicelluloses. Actes Symp Int Grenoble 1964, 405–412

Meier H, Buchs I, Buchala AJ, Homewood T 1981 $(1 \rightarrow 3)$-β-D-glucan (callose) is a possible intermediate in the biosynthesis of cellulose in cotton fibres. Nature 289:821–822

Meinert MC, Delmer DP 1977 Changes in biochemical composition of the cell wall of the cotton fiber during development. Plant Physiol 59:1088–1097

Meshitsuka G, Lee ZA, Nakano J, Eda S 1982 Studies on the nature of lignin-carbohydrate bonding. J Wood Chem Technol 2:251–267

Miksche GE 1973 Studies on the structure of gymnosperm and angiosperm lignins. Abstr Gothenburg Diss Sci 31, 24 pp

Minor JL 1982 Chemical linkage of pine polysaccharides to lignin. J Wood Chem Technol 2:1–16

Miyazaki K, Smelstorius JA, Harwood BJ, Stewart CM 1971 The formation of uronic acid during holocellulose preparation. Appita 24:452–454

Mork E 1928 Om tennar. (On compression wood.) Tidsskr Skogbr 36(suppl):1−41

Morohoshi N, Sakakibara A 1971a (The chemical composition of reaction wood I.) Mokuzai Gakkaishi 17:393−399

Morohoshi N, Sakakibara A 1971b (The chemical composition of reaction wood II.) Mokuzai Gakkaishi 17:400−404

Mosimann H, Svedberg T 1942 Sedimentations- und Diffusionsmessungen am wasserlöslichen Polysaccharid aus Lärchenholz. Koll-Z 100:99−105

Mukoyoshi SI, Azuma JI, Koshijima T 1981 Lignin-carbohydrate complexes from compression wood of Pinus densiflora Sieb et Zucc. Holzforschung 35:233−240

Musha Y, Goring DAI 1975 Distribution of syringyl and guaiacyl moieties in hardwoods as determined by ultraviolet microscopy. Wood Sci Technol 9:45−58

Nacu A, Constantinescu O 1973 Utilizarea crengilor de răşinoase la fabricarea celulozei sulfat. (Utilization of coniferous branch wood in the manufacture of sulfate pulp.) Celul Hirt 22(2):49−58

Nakajima T, Tamari K, Matsuda K, Tanaka H, Ogasawara N 1972 Studies on the cell wall of Piricularia oryzae. III. The chemical structure of the β-D-glucan. Agr Biol Chem 36:11−17

Nečesaný V, Lábsky O 1973 Changes in structure of ammonia treated beech wood. Drev Vysk 18:225−234

Nelmes BJ, Preston RD 1968 Cellulose microfibril orientation in rubbery wood. J Exp Bot 19:515−525

Nelmes BJ, Preston RD, Ashworth D 1973 A possible function of microtubules suggested by their abnormal distribution in rubbery wood. J Cell Sci 13:741−751

Nevo A, Sharon N 1969 The cell wall of Peridinium westii, a non-cellulosic glucan. Biochim Biophys Acta 183:161−175

Nicholas DD (ed) 1973 Wood deterioration and its prevention by preservative treatments. Vol. I. Degradation and protection of wood. Syracuse Univ Press, Syracuse, 380 pp

Nimz HH, Robert D, Faix O, Nemr M 1981 Carbon-13 NMR spectra of lignins. 8. Structural differences between lignins of hardwoods, softwoods, grasses, and compression wood. Holzforschung 35:16−26

Onaka F 1949 (Studies on compression and tension wood.) Mokuzai Kenkyo, Wood Res Inst Kyoto Univ 1, 88 pp Transl For Prod Lab Can 93 (1956), 99 pp

Ott E, Spurlin HM, Grafflin MW 1954 Cellulose and cellulose derivatives, Vols I−III, Interscience, New York, 1601 pp

Painter TJ, Purves CB 1960 Polysaccharides in the inner bark of white spruce. Tappi 43:729−736

Panshin AJ, de Zeeuw CH 1980 Textbook of wood technology, 4th ed. McGraw-Hill, New York, 722 pp

Parham RA 1970 Structural effects of ammonia treatment of the wood of Pinus taeda L. Ph D Thesis SUNY Coll Environ Sci For Syracuse, 248 pp

Parham RA 1971a Crystallinity and ultrastructure of ammoniated wood. Wood Fiber 2:311−320

Parham RA 1971b Crystallinity and structure of ammoniated wood. II. Ultrastructure. Wood Fiber 3:22−34

Parham RA, Davidson RW, de Zeeuw CH 1972 Radial-tangential shrinkage of ammonia-treated loblolly pine wood. Wood Sci 4:129−136

Phillip B, Jacopian V, Casperson G 1970 Chemische und morphologische Untersuchungen zum zeitlichen Verlauf eines Calciumbisulfitaufschlusses von Lärchenholz. Faserforsch Textiltech 21:153−164

Pillow MY, Bray MW 1935 Properties and sulphate pulping characteristics of compression wood. Pap Trade J 101(26):361−364

Poller S 1978 Studie über die chemische Zusammensetzung von Wurzel-, Stamm- und Astholz zweier Kiefern (Pinus silvestris L) unterschiedlichen Alters. Holztechnologie 19:22−25

Preston RD, Ranganatan V 1948 The fine structure of the fibres of normal and tension wood in beech (Fagus sylvatica L). Forestry 21:92−98

Rickey RJ, Hamilton JK, Hergert HL 1974 Chemical and physical properties of tumor-affected Sitka spruce. Wood Fiber 6:200−210

Rocensh KA 1976 Rheological features of wood plasticized with ammonia. Appl Polym Symp 28:1109−1116

Roudier AJ, Eberhard L 1965 Recherches sur les hémicelluloses du bois de pin maritime des Landes (Pinus pinaster Solander subsp P maritima (Poiret) Fieschi et Gaussen). IV. Polyosides extraits de ce bois par l'eau bouillante. Constitution d'une arabinane parme ceux-ci. Bull Soc Chim (France) 460–464

Roudier AJ, Galzin J 1965 Recherches sur les hémicelluloses du bois de pin maritime des Landes (Pinus pinaster Solander subsp P maritima (Poiret) Fieschi et Gaussen). V. Étude d'une fraction riche en substances pectiques extraite par la soude à 0.5% d'une holocellulose au chlorite et précipitée par l'alcohol à 50%. Bull Soc Chim (France) 2480–2484

Ruel K, Barnoud F, Eriksson KE 1981 Micromorphological and ultrastructural aspects of spruce wood degradation by wild-type Sporotrichum pulverulentum and its cellulase-less mutant Cel 44. Holzforschung 35:158–171

Saito H, Misaki A, Harada T 1968 A comparison of the structure of curdlan and pachyman. Agr Biol Chem 32:1261–1269

Sakakibara A 1977 Degradation products of protolignin and the structure of lignin. In: Loewus FA, Runeckles VC (eds) The structure, biosynthesis, and degradation of wood. Plenum, New York London, 117–139

Sakakibara A 1980 A structural model of softwood lignin. Wood Sci Technol 14:89–100

Sarkanen KV 1971 Precursors and their polymerization. In: Sarkanen KV, Ludwig CH (eds) Lignins. Occurrence, formation, structure and reactions. Wiley-Interscience, New York, 95–163

Sarkanen KV, Chang HM, Allen GG 1967a Species variations in lignin. II. Conifer lignins. Tappi 50:583–587

Sarkanen KV, Chang HM, Ericsson B 1967b Species variations in lignins. I. Infrared spectra of guaiacyl and syringyl models. Tappi 50:572–575

Sarkanen KV, Kakehi K, Murphy RA, White H 1962 Studies on oxidative delignification mechanisms. I. Oxidation of vanillin with chlorine dioxide. Tappi 45:24–29

Sarkanen KV, Ludwig CH (eds) 1971 Lignins. Occurrence, formation, structure and reactions. Interscience-Wiley, New York, 916 pp

Sasaya T, Takehara T, Kobayashi T 1980 Extractives of Todomatsu Abies sachalinensis Masters. II. Lignans in the compression and opposite woods from leaning stems. Mokuzai Gakkaishi 26:759–764

Sasaya T, Takehara T, Miki K, Sakakibara A 1980 (Phenolic constituents of Larix leptolepis Gord.). Res Bull Coll Exp For Hokkaido Univ 37:837–860

Schreuder HR, Côté WA, Jr, Timell TE 1966 Studies on Compression wood. III. Isolation and characterization of a galactan from compression wood of red spruce. Svensk Papperstidn 69:641–657

Schuerch C 1963 Plasticizing wood with liquid ammonia. Ind Eng Chem 55:39

Schuerch C 1964 Wood plasticization. For Prod J 14:377–384

Schuerch C 1966 Method of forming wood and formed wood product. US Pat 3, 282, 313

Schuerch C, Davidson RW 1971 Plasticizing wood with ammonia – control of color change. J Polym Sci C-36:231–239

Schwerin G 1958 The chemistry of reaction wood. II. The polysaccharides of Eucalyptus goniocalyx and Pinus radiata. Holzforschung 12:43–48

Scurfield G, Bland DE 1963 The anatomy and chemistry of "rubbery" wood in apple var Lord Lambourne. J Hortic Sci 38:297–306

Shelbourne CJA, Ritchie KS 1968 Relationship between degree of compression wood development and specific gravity and tracheid characteristics in loblolly pine (Pinus taeda L). Holzforschung 22:185–190

Sherrard EC, Kurth EF 1933 Distribution of extractives in redwood. Ind Eng Chem 25:300–302

Siau JF 1984 Transport processes in wood. Springer, Berlin Heidelberg New York Tokyo, 245 pp

Simson BW, Côté WA, Jr, Timell TE 1968 Studies on larch arabinogalactan. IV. Molecular properties. Svensk Papperstidn 71:699–710

Simson BW, Timell TE 1977 Polysaccharides in cambial tissues of Populus tremuloides and Tilia americana. III. Isolation and constitution of an arabinogalactan. Cellul Chem Technol 12:63–67

Sjöström E 1981 Wood Chemistry. Fundamentals and applications. Academic Press, New York, 223 pp
Sondheimer E, Simpson WG 1962 Lignin abnormalities of "rubbery apple wood". Can J Biochem Physiol 40:841 – 846
Sonntag P 1904 Über die mechanischen Eigenschaften des Roth- und Weißholzes der Fichte und anderer Nadelhölzer. Jahrb Wiss Bot 39:71 – 105
Srivastava LM 1963 Secondary phloem in the Pinaceae. Univ CA Press, Berkeley Los Angeles, 69 pp
Stockman L, Hägglund E 1948 Om granvedens träpolyoser och deras förhållande vid hydrolys. (On the wood polyoses in spruce wood and their behavior on hydrolysis.) Svensk Papperstidn 51:269 – 274
Sumiya K, Hasegawa Y, Yamada T 1969 Formation of Sugi trees (Cryptomeria japonica D Don) under constant deflection and their infrared spectra. Wood Res (Kyoto) 46:10 – 18
Swenson HA, Kaustinen HM, Bachhuber JJ, Carlson JA 1969 Fractionation and characterization of larchwood arabinogalactan polymers A and B. Macromolecules 2:142 – 145
Takashima T, Akabane F, Miyasaka M 1974 The chemical and physicochemical properties of Amamatsu compression wood and some properties of the pulp. Bull Exp For Tokyo Univ Agr Technol 11:73 – 78
Takehara T, Kobayashi T, Sasya T 1980 Extractives of Todomatsu Abies sachalinensis Masters. I. Lignan esters in the compression and opposite woods from leaning stem. Mokuzai Gakkaishi 26:274 – 279
Tanaka F, Koshijima T 1984 Orientation distributions of cellulose crystallites in Pinus densiflora woods. Wood Sci Technol 18:177 – 186
Tanaka F, Koshijima T, Okamura K 1981 Characterization of cellulose in compression and opposite woods of a Pinus densiflora tree grown under the influence of strong wind. Wood Sci Technol 15:265 – 273
Tarkow H, Krueger J 1961 Distribution of hot-water soluble material in cell walls and cavities of redwood. For Prod J 11:228 – 229
Taylor FW 1979 Variation of specific gravity and fiber length in loblolly pine branches. J Inst Wood Sci (46) 8 (4):171 – 175
Terashima N 1979 (Gravitational stimuli and wood lignin formation.) Kagaku te Seibutsu (Chem and Biol) 17(2):124 – 127
Terashima N, Tomimura Y, Araki H 1979 Heterogeneity in formation of lignin. III. Formation of condensed type structure with bond at position 5 of guaiacyl nucleus. Mokuzai Gakkaishi 25:595 – 599
Thomas BB 1945 An improved method for large-scale laboratory preparation of holocellulose. Pap Ind 26:1281 – 1284
Thompson NS, Heller HH, Hankey JD, Smith O 1966 The isolation and the carbohydrate composition of the epithelial cells of longleaf pine (Pinus palustris Mill). Tappi 49:401 – 405
Thompson NS, Kaustinen 1964 Noncellulosic polysaccharides of spruce holocellulose. Pap Puu 47:637 – 650
Thompson NS, Kaustinen OA 1966 Orienting studies of the less abundant polysaccharides of conifers. Tappi 49:83 – 90
Thompson NS, Kremers RO, Kaustinen OA 1968a Effects of alkali on mature and immature jack pine holocelluloses. I. Separation of tracheids. Tappi 51:123 – 127
Thompson NS, Kremers RE, Kaustinen OA 1968b Effects of alkali on mature and immature jack pine holocelluloses. II. Comparison of certain polysaccharide components. Tappi 51:127 – 131
Thunell B 1945 Trä. Dess byggnad och felaktigheter. (Wood. Its structure and defects.) Byggnadsstandardiseringen, Stockholm 103 pp
Timell TE 1955 Chain length and chain-length distribution of native white spruce cellulose. Pulp Pap Mag Can 56(7):102 – 112
Timell TE 1957a Nitration as a means of isolating the alpha-cellulose component of wood. Tappi 40:30 – 33
Timell TE 1957b Carbohydrate composition of ten North American species of wood. Tappi 40:569 – 572
Timell TE 1957c The acyl groups of wood. 1. The formyl groups. Svensk Papperstidn 60:762 – 766

Timell TE 1957d Molecular weight of native celluloses. Svensk Papperstidn 60:836–842

Timell TE 1957e Molecular properties of seven native wood celluloses. Tappi 40:25–29

Timell TE 1961a Isolation of polysaccharides from the bark of gymnosperms. Svensk Papperstidn 64:651–661

Timell TE 1961b Characterization of four celluloses from the bark of gymnosperms. Svensk Papperstidn 64:685–688

Timell TE 1961c Isolation of galactoglucomannans from the wood of gymnosperms. Tappi 44:88–96

Timell TE 1964 Wood hemicelluloses I. Adv Carbohyd Chem 19:247–302

Timell TE 1965a Wood hemicelluloses II. Adv Carbohyd Chem 20:409–483

Timell TE 1965b Wood and bark polysaccharides. In: Côté WA, Jr (ed) Cellular ultrastructure of woody plants. Syracuse Univ Press, Syracuse, 127–156

Timell TE 1965c Bark polysaccharides. In: Chimie et biochimie de la lignine, de la cellulose et des hémicelluloses. Actes Symp Int Grenoble 1964, 99–111

Timell TE 1967 Recent progress in the chemistry of wood hemicelluloses. Wood Sci Technol 1:45–70

Timell TE 1981 Recent progress in the chemistry, structure, and formation of compression wood. The Ekman Days (1981), Stockholm SPCI Rep 38, Vol 1:99–147

Timell TE 1982 Recent progress in the chemistry and topochemistry of compression wood. Wood Sci Technol 16:83–122

Timell TE, Jahn EC 1951 A study of the isolation and polymolecularity of paper birch holocellulose. Svensk Papperstidn 54:831–846

Tomada M, Murayama K 1965 The polysaccharides from Lupinus luteus seed. I. Identification of the sugar component and isolation of 2,3,6-tri-O-methyl-D-galactose from the methylated substance. Yakugaku Zasshi 85:511–514

Tomimura Y, Yokoi T, Terashima N 1979 Heterogeneity in formation of lignin. IV. Various factors which influence the degree of condensation at position 5 of guaiacyl ring. Mokuzai Gakkaishi 25:743–748

Tomimura Y, Yokoi T, Terashima N 1980 Heterogeneity in formation of lignin. V. Degree of condensation in guaiacyl nucleus. Mokuzai Gakkaishi 26:37–41

Ulfsparre S 1928 Något om tjurved och därav framställd sulfat- och sulfitcellulosa. (On compression wood and sulfate and sulfite pulps prepared therefrom.) Svensk Papperstidn 31:642–644

Urbas B, Bishop CT, Adams GA 1963 Occurrence of D-glucuronic acid in tamarack arabinogalactan. Can J Chem 41:1522–1524

Wang PY, Bolker HI, Purves CB 1964 Ammonolysis of uronic ester groups in birch xylan. Can J Chem 42:2434–2439

Wardrop AB, Dadswell HE 1948 The nature of reaction wood. I. The structure and properties of tension wood fibres. Aust J Sci Res B-1:3–16

Wardrop AB, Dadswell HE 1950 Nature of reaction wood. II. Cell wall organization of compression-wood tracheids. Aust J Sci Res B-3:1–13

Wardrop AB, Dadswell HE 1955 Nature of reaction wood. IV. Variations in cell wall organization of tension-wood fibres. Aust J Bot 3:177–189

Warsi SA, Whelan WJ 1957 Structure of pachyman, the polysaccharide component of Poria cocos Wolf. Chem Ind (London) 1573

Watanabe H, Tsutsumi J, Kanagawa H 1962 Properties of branch wood; especially on specific gravity, tracheid length, and appearance of compression wood. Bull Kyushu Univ For 35:91–96

Waterkeyn L 1981 Cytochemical localization and function of the 3-linked glucan callose in the developing cotton fibre cell wall. Protoplasma 106:49–67

Waterkeyn L, Caeymaex S, Decamps E 1982 La callose des trachéides du bois de compression chez Pinus silvestris et Larix decidua. Bull Soc R Bot Belg 115:149–155

Watson AJ, Dadswell HE 1957 Paper making properties of compression wood from Pinus radiata. Appita 11(3):56–70

Wegelius T 1939 The presence and properties of knots in Finnish spruce. Acta For Fenn 48:1–191

Wegelius T 1946 Det finska granvirkets egenskaper och kvalitetsvariationer. (Properties and variations in quality of Finnish spruce wood.) Svensk Papperstidn 49:51–61

Whiting P, Goring DAI 1981 The morphological origin of milled wood lignin. Svensk Papperstidn 84:R120–R122

Whiting P, Goring DAI 1982 Chemical characterization of tissue fractions from the middle lamella and secondary wall of black spruce tracheids. Wood Sci Technol 16:261–267

Wilcox WW 1968 Changes in wood microstructure through progressive stages of decay. US For Serv Res Pap FPL-70, 46 pp

Wilcox WW 1970 Anatomical changes in wood cell wall attacked by fungi and bacteria. Bot Rev 36:1–28

Wilcox WW 1973 Degradation in relation to wood structure. In: Nicholas DD (ed) Wood deterioration and its prevention by preservative treatments, Vol I. Syracuse Univ Press, Syracuse, 107–148

Wise LE, Hamer PL, Peterson FC 1933 The chemistry of wood. IV. Water-soluble polysaccharide of eastern larch. Ind Eng Chem 25:184–187

Wloch W, Hejnowicz Z 1983 Location of laricinan in compression wood tracheids. Acta Soc Bot Polon 52:201–203

Wood JR, Goring DAI 1971 The distribution of lignin in stem wood and branch wood of Douglas fir. Pulp Pap Mag Can 72(3):T95–T102

Wood PJ 1980 Specificity in the interaction of direct dyes with polysaccharides. Carbohyd Res 85:271–287

Wu WT, Wilson JW 1967 Lignification within coniferous growth zones. Pulp Pap Mag Can 68(4):T159–T164, T171

Yasuda S, Sakakibara A 1975a The chemical composition of lignin from compression wood. Mokuzai Gakkaishi 21:363–369

Yasuda S, Sakakibara A 1975b Hydrogenolysis of protolignin in compression wood. I. Isolation of two dimers with $C_\beta-C_5$ and $C_\beta-C_3$ composed of p-hydroxyphenyl and guaiacyl nuclei and two p-hydroxyphenyl nuclei, respectively. Mokuzai Gakkaishi 21:370–375

Yasuda S, Sakakibara A 1975c Isolation of a new dimeric "condensed" type compound from hydrogenolysis products of compression wood lignin. Mokuzai Gakkaishi 21:639–640

Yasuda S, Sakakibara A 1976 Hydrogenolysis of protolignin in compression wood. II. Isolation of two dimers with β-aryl ether linkage and phenylisochroman structure. Mokuzai Gakkaishi 22:606–612

Yasuda S, Sakakibara A 1977a Hydrogenolysis of protolignin in compression wood. III. Isolation of four dimeric compounds with carbon to carbon linkage. Mokuzai Gakkaishi 23:114–119

Yasuda S, Sakakibara A 1977b Hydrogenolysis of protolignin in compression wood. IV. Isolation of a diphenyl ether and three dimeric compounds with carbon to carbon linkage. Mokuzai Gakkaishi 23:383–387

Yasuda S, Sakakibara A 1981 Hydrogenolysis of protolignin in compression wood. V. Isolation of two trimeric compounds with γ-lactone ring. Holzforschung 35:183–187

Yasuda A, Tahara S, Sakakibara A 1975 The phenolic constituents of normal and reaction woods of Karamatsu, Larix leptolepis' Gord. Res Bull Coll Exp For Hokkaido Univ 32:55–62

Yean WQ, Goring DAI 1965 The molecular weight of lignosulphonates from morphologically different subdivisions of wood structure. Svensk Papperstidn 68:787–790

Young RA, Sarkanen KV, Johnson PG, Allen GG 1972 Marine plant polymers. III. A kinetic analysis of the alkaline degradation of polysaccharides with specific reference to $(1 \rightarrow 3)$-β-D-glucans. Carbohyd Res 21:111–122

Young WD, Laidlaw RA, Packman DF 1970 Pulping of British-grown softwoods. VI. The pulping properties of Sitka spruce compression wood. Holzforschung 24:86–98

Yundt A 1951 Crystalline hemicelluloses. I. Crystalline and amorphus xylan from barley straw. II. Crystalline xylan from paper birch. III. Acid and enzymatic hydrolysis of xylans. IV. Crystalline mannan. Tappi 34:89–95

Zevenhuizen LPTM, Bartnicki-Garcia S 1969 Chemical structure of the insoluble hyphal wall glucan of Phytophthora cinnamomi. Biochem J 8:1496–1502

Zherebov LP 1946 (Mechanical functions of the chemical constituents of wood.) Bumazh Prom 21(3/4):14–26.

Zobel BJ, Thorbjornsen E, Henson F 1960 Geographic, site, and individual tree variation in wood properties of loblolly pine. Silvae Genet 9:149–158

Chapter 6 Distribution of Chemical Constituents

CONTENTS

6.1	Introduction	409
6.2	Macroscopic Distribution	410
6.2.1	Juvenile Wood and Mature Wood	410
6.2.2	Sapwood and Heartwood	414
6.2.3	Earlywood and Latewood	415
6.3	Microscopic Distribution of Polysaccharides	416
6.3.1	Introduction	416
6.3.2	Tracheids and Ray Cells	418
6.3.3	Distribution of Polysaccharides in the Tracheids	422
6.3.4	Distribution of Arabinogalactan	432
6.4	Microscopic Distribution of Lignin	433
6.4.1	Introduction	433
6.4.2	Distribution of Lignin in the Tracheids	437
6.4.3	Distribution of Lignin in the Ray Cells	455
6.5	Distribution of Extractives	456
6.6	General Conclusions	457
References		459

6.1 Introduction

Wood is an extremely heterogeneous material of great anatomical, physical, and chemical complexity, a fact that was not always fully appreciated in the past. In later years, attention has been directed toward this lack of chemical uniformity, for example, by Meier (1957, 1962, 1964), de Zeeuw (1965), Wu and Wilson (1967), and Panshin and de Zeeuw (1980), and studies have been reported on the chemistry of separate wood regions, the different types of wood cells, and the location of the various wood components within the cell wall (Parameswaran and Liese 1982).

Reaction wood has as yet been studied far less than normal wood in this respect. With tension wood, difficulties arise, because vessels and rays, which always constitute a considerable proportion of the entire wood, usually remain the same as in normal wood, and because tension wood fibers often occur scattered among normal fibers and tend to develop only in earlywood. Compression wood is easier to study on a macroscopic scale. Pronounced compression wood is frequently formed over large areas containing no normal tracheids and it can occupy an entire growth ring. The first few rows of tracheids in the earlywood are usually different in structure from those deposited later, as shown in Chap. 4.5, but they do not seem to differ in any chemical respect from the latter (Côté et al. 1967a). In mild or moderate compression woods, by contrast, compression-wood tracheids can be absent in the earlywood while present in the latewood (Fukazawa 1974). The ray cells in compression wood do not have the same polysaccharide composition as the tracheids.

The macroscopic or gross distribution of the chemical constituents in compression wood has so far received little attention, although sapwood and heartwood and also earlywood and latewood have been subjected to comparative chemical analyses. The ray cells in compression wood have been separated from the tracheids and examined for their polysaccharide composition. The distribution of the polysaccharides within the cell wall of the tracheids in compression wood has been studied to some extent, but the picture that has emerged is far from complete. Several excellent methods are available for determining the location of lignin in wood on a microscopic scale. As a result, the distribution of lignin in compression wood is fairly well known. There are very few reports on the distribution of the extractives in compression wood.

The ultrastructural localization of the components in wood cell walls was recently reviewed by Parameswaran and Liese (1981, 1982). The distribution of the chemical constituents in compression wood has been the subject of two recent reviews by Timell (1981, 1982).

6.2 Macroscopic Distribution

6.2.1 Juvenile Wood and Mature Wood

Juvenile wood, sometimes referred to as core wood, is formed by a young vascular cambium and is deposited within the first 5−20 growth rings. It differs anatomically, physically, and chemically from the outer, mature (adult) wood. The growth rings are often wide in juvenile wood, but this is by no means always the case. Irrespective of the rate of growth (Larson 1972), juvenile wood consists largely of earlywood and has a correspondingly low density. Juvenile wood resembles compression wood in several respects (Zobel 1975). The incidence of compression wood is often quite high in the juvenile region of a conifer stem (Zobel 1975, 1981, 1984, Bendtsen 1978, Harris 1981, Kellison 1981) (Chaps. 10.3.6.2 and 19.2).

Juvenile wood has a lower content of cellulose than mature wood, a fact noted by many investigators, such as Zobel and McElwee (1958), Zobel et al. (1959, 1966), Kennedy and Jaworsky (1960), Schütt and Augustin (1961), Hale and Clermont (1963), Packman and Laidlaw (1967), Zenker and Poller (1968), Burkart and Watterston (1968) and Uprichard (1971). Zobel and McElwee (1958) claimed that the proportion of cellulose in juvenile and mature woods are positively correlated, but this correlation has later been found to be at best weak (Zobel et al. 1966) or nonexistent (Burckart and Watterston 1968). Environmental effects are also of importance. Burckart and Watterston (1968) found that juvenile wood contained relatively more cellulose in suppressed than in dominant trees. It is also clear that comparisons should be made on the basis of earlywood and latewood and not between the entire growth rings.

Several investigators have reported higher lignin and pentosan contents for juvenile wood than for mature wood (Packman and Laidlaw 1967, Zenker and

6.2.1 Juvenile Wood and Mature Wood

Poller 1968, Uprichard 1971, Uprichard and Lloyd 1980). Both earlywood and latewood in juvenile wood of *Pseudotsuga menziesii* were found by Thompson et al. (1966 a, b) to contain more xylose and less mannose residues than did mature wood. Unexpectedly, the galactose and arabinose contents were lower in juvenile wood, and there seems to have been no difference in the glucose content between the two tissues. In *Pinus radiata*, juvenile wood has a higher content of lignin, xylan, and glucomannan than mature wood, whereas the cellulose content is lower (Harwood 1971). Larson (1966) determined the chemical composition of earlywood and latewood from juvenile and mature woods of *Pinus resinosa*. Two trees were used, one 10 years and the other 58 years old. Earlywood and latewood were carefully separated from a number of annual increments, and the wood was analyzed for lignin content and relative sugar composition. Variations were found to occur between earlywood and latewood, to be discussed later, but there were also changes associated with increasing age of the trees. During the first 10 years of growth, there was a slight, gradual decline in lignin content of the earlywood with increasing age, from 29.3% to 27.3%. At the same time, the relative proportion of galactose, arabinose, and xylose residues decreased considerably, in the case of xylose from 16.9% to 12.9%. The proportion of glucose residues increased from 54.3% to 64.2% and that of mannose from 15.5% to 17.0%. Similar changes occurred in the latewood except for the lignin content, which remained constant. Larson (1966) concluded that rapid, fluctuating changes in cell wall composition take place during the juvenile period of the first 15–20 years. Thereafter, changes are more steady and less conspicuous. Part of the changes in the proportion of lignin and polysaccharides may be a consequence of an increase in tracheid wall thickness, but superimposed upon this there must be changes associated with the increasing age of the wood. Uprichard and Lloyd (1980) have more recently argued that the variations in chemical composition only reflect changes in fiber morphology. The tracheids in juvenile wood have a thin wall throughout most of the growth rings and therefore a thinner S_2 layer than the mature wood. A low cellulose content in juvenile wood is accordingly to be expected.

Larson (1966) also examined a young *Pinus resinosa* tree that leaned 30° from the vertical and as a result had formed compression wood. Data obtained for the first nine growth rings are summarized in Table 6.1 and also shown graphically in Fig. 6.1. In this case only latewood was examined, obviously because compression wood does not form earlywood in the accepted meaning of this term. The trends in Table 6.1 for the side wood are the same as those in normal wood. The changes in the opposite wood, which are referred to in Chap. 21.4, are similar to those in the side wood. Compared to the latter, opposite wood contained more cellulose and less xylan.

A comparison between the side wood and the compression wood data in Table 6.1 reveals several interesting facts. In the second growth ring, the only chemical difference between the two types of wood is the slightly higher lignin and galactose and the somewhat lower xylose content of the compression wood. In the third ring, the differences are still very modest, but after that, the two types of wood become more and more dissimilar. While the glucose content of the normal wood increases with increasing age, it remains, except for minor

Table 6.1. Lignin and carbohydrate contents of nine growth rings of compression, normal (side) and opposite woods in *Pinus resinosa*. (Larson 1966)

Age of ring, years	Lignin per cent	Sugar residues in relative per cent				
		Galactose	Glucose	Mannose	Arabinose	Xylose
Compression wood						
2	31.2	8.2	55.4	16.1	5.6	14.7
3	30.6	9.3	56.9	16.0	3.6	14.2
4	35.8	16.3	57.7	12.0	2.2	11.8
5	36.2	14.7	57.4	12.6	2.8	12.5
6	36.3	13.7	58.7	13.3	2.7	11.6
7	38.7	18.4	56.9	11.4	2.5	10.8
8	35.8	17.7	56.4	12.6	2.5	10.8
9	35.1	17.7	55.5	13.0	2.7	11.1
Normal (side) wood						
2	29.2	5.2	55.9	16.9	5.6	16.4
3	28.2	3.1	57.7	17.5	4.5	17.2
4	27.4	2.1	60.7	17.4	3.8	16.0
5	28.2	2.2	60.6	18.0	3.7	15.5
6	27.1	1.5	63.4	17.3	2.5	15.3
7	26.9	2.0	61.1	18.4	3.5	15.0
8	27.8	2.0	63.6	19.8	3.0	11.6
9	26.3	2.3	62.5	20.8	2.9	11.5
Opposite wood						
2	30.9	6.4	55.8	15.2	8.6	14.0
3	26.2	3.1	57.4	17.1	4.5	17.9
4	26.0	2.5	60.3	17.8	3.6	15.8
5	26.1	2.8	60.2	17.9	3.8	15.3
6	26.4	2.2	65.2	16.4	3.0	13.2
7	26.0	2.3	64.6	17.4	3.2	12.5
8	27.5	1.9	66.5	18.2	2.6	10.8
9	26.3	2.2	64.4	20.8	2.9	9.7

fluctuations, on the same level in the compression wood. For lignin, galactose, and mannose, the changes with age are opposite in the two tissues. In normal wood, the proportion of lignin exhibits a modest but continuous decline with age, whereas in compression wood it increases from a minimum value of 30.6% to a maximum of 38.7%. The reduction in lignin content with increasing age of normal wood has usually been attributed to the fact that cell wall thickness also increases with age, thus reducing the effect of the high lignin concentration in the middle lamella, but this explanation is open to criticism on several points (Wu and Wilson 1967). Instead of decreasing, as it does in normal wood, the proportion of galactose more than doubles over 7 years in the compression wood. The proportion of mannose is gradually reduced in the latter but increases in the former wood. Arabinose shows a decline in both tissues, and so does xylose, but the decrease is less pronounced in compression wood so that, in the ninth ring, the xylose content has become the same in the two woods.

6.2.1 Juvenile Wood and Mature Wood

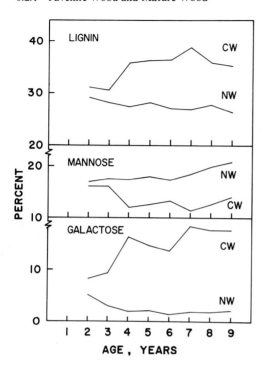

Fig. 6.1. Changes in lignin, mannose, and galactose contents with age of normal (*NW*) and compression (*CW*) woods in the juvenile zone of a *Pinus resinosa* stem. (Drawn from data reported by Larson 1966)

Larson (1966) mentions that the compression wood was not of the most extreme type. It is likely that the compression wood produced in the seventh growth ring was fully developed, whereas that present in the eighth and ninth rings was probably of the moderate type. Judging from its chemical composition alone, the latewood in the second and third increments could hardly be classified as compression wood, and it would have been interesting to have some information concerning its anatomy. The changes that occurred within the later rings were much less drastic than those distinguishing the third and fourth increments from each other.

Larson's results indicate that severe compression wood is not formed in extremely young wood, but this conclusion is based on the chemical composition of only two growth rings, namely the second and third and would require the presence of pronounced compression wood in the latewood portion of these increments. Obviously, additional information is required before any firm conclusions can be drawn. Ideally, a tree known to have grown at a considerable constant lean throughout its life and for at least 20 years should be examined, and not only chemically but also anatomically. Compression wood is very common at the center of forest trees, since it is an important regulatory factor in the life of young seedlings (Chap. 10.3.6.2). It would be surprising if pronounced compression wood were not formed during these early years of a tree when it serves a vital function.

6.2.2 Sapwood and Heartwood

Heartwood is defined as that center portion of a stem where no living ray cells remain. In most cases this wood has a darker color than the sapwood because of its high content of extractives, especially polyphenols. The cytological, chemical, and physiological changes associated with the transition of sapwood into heartwood have been reviewed by Hugentobler (1965), Stewart (1966), Ziegler (1968), Hillis (1968a, b, 1977), Higuchi (1976), and Fukazawa et al. (1980). When comparing the chemical composition of sapwood and heartwood, some investigators have failed to notice that the latter often consists of juvenile wood, whereas the former may be mature wood, and that this might be the reason for their different lignin and polysaccharide contents. It is very likely that once a polysaccharide has been deposited in the secondary wall, it suffers little further change, such as a metabolic breakdown. Claims that the transition of sapwood into heartwood should be associated with changes in the pectin or hemicellulose contents must be regarded as unfounded.

Data for the chemical composition of sapwood and heartwood of compression wood are meager. Dadswell and Hawley (1929) examined sapwood and heartwood of *Pseudotsuga menziesii*, a species known for its high content of cell wall extractives in the heartwood. As can be seen from Table 5.1, heartwood of both normal and compression woods had a slightly higher lignin content than the sapwood and contained less cellulose and pentosans. It is difficult to assess these results, since the extractive content was not reported for the heartwood.

Côté et al. (1966a) determined the chemical composition of normal and compression woods in the heartwood and sapwood regions of an eccentric stem of *Larix laricina*, shown in cross section in Fig. 5.3. The summative values obtained are listed in Table 5.25. The heartwood portion of the normal core undoubtedly consisted of juvenile wood, and this is also reflected in its lower glucose content in comparison with the sapwood. Both sapwood and heartwood of the compression wood were mature wood. If a correction is applied for the 7% of arabinogalactan present in the heartwood, it is found that the sapwood and heartwood were very similar in chemical composition, except for the lignin content, which was higher in the heartwood. It is quite possible that tannins and other polyphenols were included in the heartwood lignin value. There is accordingly no reason to believe that heartwood compression wood should differ chemically from sapwood in any respect other than in its higher content of extractives.

Côté et al. (1966a) found the heartwood zone of normal wood to contain 10.5% arabinogalactan compared to 7.03% for the same region in compression wood. The lower arabinogalactan content of compression wood was later confirmed for the same species by Côté and Timell (1967). Hillis (1960) noted that bands of tension wood in *Acacia* sp. had a lower content of polyphenols than did adjacent bands of normal wood. Hillis et al. (1962) later found that formation of tension wood in *Angophora costata* was associated with low levels of sugars and starch in the sapwood and of methanol-soluble extractives in the heartwood. They suggested that formation of the thick, gelatinous layer in the tension wood fibers might have left the tree with less photosynthetic material

6.2.3 Earlywood and Latewood

available for biosynthesis of sugars, starch, and extractives. It is possible that a similar explanation could be adduced for the lesser amounts of the extracellular arabinogalactan in the compression wood. In this case additional photosynthate would have been required for synthesis of the thick tracheid wall. It should, however, be noted that the fact that compression wood usually contains more extractives than corresponding normal wood militates against this interpretation (Chap. 5.2.8).

6.2.3 Earlywood and Latewood

Earlywood and latewood, formerly referred to as springwood and summerwood, each have their characteristic anatomical features, described in Chap. 4.2. Earlywood is formed during the early part of the growing season when the supply of auxin is abundant and growth is rapid, while latewood is produced at a later stage, when the auxin level is lower but supply of photosynthate larger. That chemical differences exist between earlywood and latewood has been known for a long time. Ritter and Fleck (1926) were the first to notice that earlywood consistently has a higher lignin content than latewood, both in softwoods and hardwoods, and this observation has since been confirmed by many investigators, such as Bailey (1936a, b), Curran (1936), Hata (1950), Leopold (1961), Hale and Clermont (1963), Wilson and Wellwood (1965), Larson (1966), Stamm and Sanders (1966), Wu and Wilson (1967), Siddiqui (1976), and Fukazawa and Imagawa (1981). Ritter and Fleck (1926) advanced an explanation for the higher lignin concentration in earlywood, pointing out that this must be interpreted as a dilution effect, in that the highly lignified middle lamella constitutes a larger part of the total tracheid in the thin-walled earlywood than in the thick-walled latewood. Most later investigators, such as Stamm and Sanders (1966) and Siddiqui (1976), have accepted this explanation with the exception of Wu and Wilson (1967), who have drawn attention to the fact that most lignin is not located in the middle lamella as was believed by Ritter. Since the concentration of lignin is about three times higher in the middle lamella than in the secondary wall (Sect. 6.4.1), it would nevertheless seem very probable that a dilution effect is really operating in this case.

Several investigators, for example, Ritter and Fleck (1926), Leopold (1961), Hale and Clermont (1963), and Siddiqui (1976), have reported that earlywood contains less cellulose than latewood, a fact also borne out by the results of many pulping studies. Others, however, such as Meier (1957, 1962) and Thompson et al. (1966a), have observed no difference between the two woods in this respect.

Meier (1962, 1964) found 20.3% glucomannan in earlywood and 24.8% in latewood of *Pinus sylvestris*. Corresponding figures for xylan were 18.6% and 14.1% respectively. Larson (1966) and Thompson et al. (1966a) also noted a higher concentration of glucomannan in the latewood, and the latter investigators confirmed the lower concentration of xylan in this zone. According to Meier (1962, 1964), glucomannan is located in the center of S_2 and this could explain its higher concentration in the thick-walled latewood tracheids. Surpris-

Table 6.2. Chemical composition of normal wood and first-formed and later-formed compression woods of *Abies balsamea*. (Côté et al. 1967a)

Wood	Lignin, per cent	Sugar residues in relative per cent				
		Galactose	Glucose	Mannose	Arabinose	Xylose
Normal wood	31.7	1.6	71.8	16.2	1.8	8.4
First-formed compression wood	37.1	22.2	50.9	14.7	2.8	9.4
Later-formed compression wood	38.8	24.5	49.5	11.9	3.3	10.8

ingly, Ifgu and Labosky (1972) found the chemical composition to be almost constant across each growth ring in *Pinus taeda*.

As ist emphasized in Chap. 4.2, a distinction between earlywood and latewood in the accepted sense of these terms is of doubtful value in the case of pronounced compression wood, since all tracheids in such wood have the same appearance except for the few rows of cells formed at the very beginning and at the very end of the growing season. It must also be noted that cases are known where severe compression wood occurred uniformly throughout an entire growth ring (Phillips 1940, Watson and Dadswall 1957, Core et al. 1961, Yoshizawa et al. 1981, 1982) (Chap. 4.5). This view concerning the earlywood-latewood concept as applied to compression wood has also been expressed by Casperson (1962), Shelbourne and Ritchie (1968) and Brazier (1969). It is not surprising, therefore, to find contradictory statements concerning the chemical composition of the earlywood and latewood zones of compression wood. Dadswell and Hawley (1929) found more lignin in latewood than in earlywood of *Sequoia sempervirens* compression wood, but the lignin content of the latter tissue was not determined directly. In *Pinus taeda* compression wood, Watson and Hodder (1954) obtained the same lignin content for earlywood and latewood. Côté et al. (1967a) determined the lignin and carbohydrate composition of first-formed compression wood and the wood formed subsequently in *Abies balsamea*. The analytical data in Table 6.2 show that the two tissues were very similar in chemical composition, and that the first-formed wood was entirely different from that of normal fir wood. The high content of galactan and the low cellulose and galactoglucomannan contents in similar first-formed compression wood was later confirmed by Côté et al. (1968b).

6.3 Microscopic Distribution of Polysaccharides

6.3.1 Introduction

Determination of the microscopic distribution of the polysaccharides in wood is still an enterprise fraught with great difficulties. Usually, the aim is to establish the concentration of cellulose, hemicelluloses, and pectin in the middle lamella and in the different types of wood cells and, within each cell, in the different layers of the secondary wall. Tracheids, ray cells, fibers, vessels, and epithelial cells have all been separated and analyzed for their chemical consituents. The middle lamella of softwood tracheids and the gelatinous layer (G-layer) of tension wood fibers have also been isolated and analyzed, but nobody has as yet been able to separate the S_1, S_2, and S_3 layers from any wood cell, and the same applies to the warty layer. Instead, several indirect methods have been developed. Most recently, the middle lamella together with the layer S_1, and the S_2 layer together with S_3 have been isolated from tracheids and fibers. Attempts have also been made to stain cell-wall polysaccharides by various means, thus rendering them visible in the light or electron microscopes.

Histochemical techniques (Hayat 1975, Berlyn and Miksche 1976) are generally nonspecific. Ruthenium red, for example, long used as a supposedly specific stain for pectin, actually reacts with all polysaccharides containing uronic acid groups, as first pointed out by Kerr and Bailey (1934). Asunmaa and Lange (1953, 1954a, b) tried to determine the distribution of cellulose and hemicellulose in delignified wood by esterifying the polysaccharides with p-phenylazobenzoyl chloride, followed by spectrophotometric determination, but the procedure has been criticized on several accounts (Ruch and Hengartner 1960, Hengartner 1961, Meier 1962). More recently, Hoffmann and Parameswaran (1976) delignified and oxidized wood with acid chlorite. The resulting carboxylic acid groups at the reducing end of the polysaccharides were converted to salts of heavy metals, such as silver or lead. These cations would be too few to render the long cellulose chains visible in the electron microscope, but the much shorter hemicellulose molecules could be located in this way. This is a promising method, but it unfortunately does not distinguish between the different types of hemicelluloses, and it probably also overemphasizes acidic polysaccharides, such as pectin or xylan, at the expense of the neutral ones.

Almost 50 years ago, A. J. Bailey (1936a, b) was able to isolate minute amounts of middle lamella from thin, transverse sections of *Pseudotsuga menziesii* wood. The material obtained could be analyzed for lignin, cellulose, and pentosan. Using ultrasonic treatment, Norberg and Meier (1966) 30 years later succeeded in separating the G-layer from the remainder of the fiber wall in tension wood.

The cambial zone in trees consists of middle lamella and primary wall material. In the spring, the gelatinous, highly swollen, active cambial tissue can be removed after the loose bark has been peeled from the wood. If isolated with great care, this material consists only of undifferentiated tissues and lends itself to studies of its chemical composition. The technique was first applied by Allsopp and Misra (1940) to trees and was later used by several investigators, such as Sultze (1957) and Kremers (1963, 1964, 1965). Simson and Timell (1978) improved the method further in their investigations on the polysaccharides constituents in the primary wall of *Populus tremuloides* and *Tilia americana*. In another approach, suspension-cultured cells with only primary walls have been used as a source of such walls. Using cells of *Acer pseudoplatanus* and other species,

Albersheim and his co-workers were able to establish the detailed chemical composition of these primary walls in a series of now classical investigations (McNeil et al. 1979, Darvil et al.1980). Similar studies of primary walls in suspension-cultured *Rosa glauca* cells were most recently reported by Chambat et al. (1984) and by Joseleau and Chambat (1984).

A general method for assessing the polysaccharide distribution in wood tracheids or fibers was developed 25 years ago by Meier (1961, 1962, 1964) (Meier and Wilkie 1959). Thin radial sections of the cambial zone and differentiating xylem were examined in polarized light. Because the cellulose microfibrils have a different orientation in the different cell-wall layers, the latter can be distinguished from one another, isolated, and analyzed for constituent neutral sugars. The tissues obtained are in different stages of development and consist in succession of middle lamella and primary wall (M + P), $M + P + S_1$, $M + P + S_1 + S_{2\text{ outer}}$, $M + P + S_1 + S_2$, and $M + P + S_1 + S_2 + S_3$. From the content of sugar residues in the different sections, the thickness of the cell-wall layers, and the known composition of the major wood polysaccharides, Meier was able to establish the distribution of these polysaccharide over the entire cell wall. The method has the disadvantages that it is summative and that it does not lend itself to analysis for acidic polysaccharides, such as pectin.

Larson (1969a) studied the incorporation of $^{14}CO_2$ in developing tracheids of normal earlywood and latewood of 5-year-old *Pinus resinosa* trees. The many interesting results obtained did not provide information concerning the final distribution of polysaccharides and lignin within a particular tracheid. Instead, they allowed conclusions to be drawn concerning the extent of metabolic activity of a tracheid at a particular stage of its development and the rate of incorporation of cellulose, hemicelluloses, and lignin into the tracheid wall. The data pertaining to the cambial cells for these reasons did not agree with the known polysaccharide composition of the M + P layer.

Hardell and Westermark (1981 a, b) were recently able to isolate the M + P, $M + P + S_1$, and $S_2 + S_3$ layers from partly delignified *Picea abies* fibers by carefully removing the first two wall regions with special tweezers under a stereomicroscope. This is a valuable extension of Meier's technique, which is not applicable to fibers, but it still does not allow the isolation of separate S_1, S_2, and S_3 layers.

6.3.2 Tracheids and Ray Cells

In conifers, the ray tracheids and ray parenchyma cells comprise only 1 – 2% by weight of the wood. Where whole wood is analyzed or resolved into its individual polymer constituents, the results therefore by and large apply to the tracheids. For a chemical examination of the ray cells, the latter have to be separated from the tracheids (Harlow and Wise 1928). A physical separation of ray tracheids and ray parenchyma can at present only be achieved by micromanipulation and does not appear to have been attempted. Ray cells in conifers have long been known to differ in chemical composition from the longitudinal tracheids. Bailey (1936a, b) found that such cells in *Pseudotsuga menziesii* had

6.3.2 Tracheids and Ray Cells

a lignin content of 41.1% and a pentosan content of 6.6%. The entire wood contained 35.2% (earlywood) and 31.6% (latewood) lignin and 5.5% pentosans. Using an optical method of measurement, Harada and Wardrop (1960) obtained a value of 43.5% for the lignin content of the ray parenchyma cells in *Cryptomeria japonica*. All available microscopic evidence indicates that ray cells are more highly lignified than tracheids in normal softwood.

Perilä (1961, 1962) and his coworkers studied the polysaccharide composition of the ray cells in normal wood of *Pinus sylvestris* (Perilä and Heitto 1959) and *Picea abies* (Perilä and Seppä 1960). The wood was delignified by heating with 2-aminoethanol, followed by chlorination, and the longitudinal tracheids and the short ray cells were separated by sieving. The ray tracheids and ray parenchyma were analyzed for constituent sugars, and a mixture of hemicelluloses was isolated by extraction with alkaline borate. The insoluble residue was a cellulose, still containing 20–30% hemicelluloses. The most important result of these investigations was the discovery that conifer ray cells contain more xylan and less galactoglucomannan than the tracheids. The hemicelluloses obtained from the pine tracheids consisted of 70% mannose and 20% xylose residues, whereas the ray cell hemicelluloses were composed of 30% mannose and 50% xylose. Usind Perilä's results, Meier (1964) calculated that delignified tracheids in *Pinus sylvestris* contain 25% galactoglucomannan and 17% arabino-4-*O*-methylglucurono-xylan, corresponding figures for the ray cells being 20% and 28%, respectively. In the pulping industry, the so-called fines contain a large proportion of ray cells, and this material has long been known to have a high content of pentosan (Roschier et al. 1959).

Borzakovskaya and Sharkov (1974) separated under the microscope resin canal tissue, ray cells, and young parenchyma cells in branches of *Pinus sylvestris* and compared their chemical compositiosn with that of the entire wood. The parenchyma cells contained four times less cellulose than did the wood. Mannose was the major sugar constituent of the wood hemicelluloses, while those present in the parenchyma cells were largely based on arabinose, xylose, and galactose. In a later investigation with stem wood of *Picea abies*, earlywood and latewood were found to have the same chemical composition (Borzakovskaya et al. 1975). The parenchyma cells had a low cellulose content. Whereas mannose was the dominating sugar residue in the tracheids, the ray cells contained mostly xylose. These results confirm the data reported by Perilä (1961, 1962). Studies by Meier and Welck (1965) and by Mullis et al.(1976) also indicate a high xylan content in hardwood ray cells.

Hoffmann und Timell (1972a, b) determined the polysaccharide composition of ray cells from normal and compression woods of *Pinus resinosa*. Thin, tangential shavings were delignified with acid chlorite, the resulting holocellulose was defibered mechanically, and the longitudinal tracheids were separated by screening from the short ray tracheids and ray parenchyma cells. Some hemicelluloses, 3–5% of the wood, were lost in the chlorite liquor, a drawback that could not be avoided. The yield of ray cells from normal and compression woods were 1.8 and 1.0%, respectively. The sugar composition of the tracheids and ray cells are shown in Table 6.3. The difference between the tracheids from normal wood and from compression wood was the usual, with more galactose

Table 6.3. Carbohydrate composition of tracheids and ray cells in normal and compression woods of *Pinus resinosa*. All values in relative per cent. (Hoffmann and Timell 1972a, b)

Cell type and wood	Galactose	Glucose	Mannose	Arabinose	Xylose
Tracheids					
Normal wood	0.4	73.8	14.3	1.8	9.7
Compression wood	4.7	73.5	9.5	3.2	9.1
Ray cells					
Normal wood	2.4	72.9	9.2	3.1	12.4
Compression wood	2.6	72.0	11.6	1.4	12.4

and less mannose residues in the compression-wood tracheids. Within the limits of accuracy attainable, the ray cells from the two woods had the same carbohydrate composition. In comparison with the tracheids, the ray cells contained more xylose residues, as had already been noted by Perilä and Heitto (1959).

The ray cells were subjected to a series of successive fractional extractions of the type used previously by Schreuder et al. (1966), and by Côté et al. (1967b). Each fraction was analyzed for constituent sugars, and attempts were made to isolate pure polysaccharides by various techniques of which only a few, such as complexing with barium hydroxide, proved successful. Arabino-4-*O*-methylglucurono-xylan and galactoglucomannan were obtained pure and were subjected to a structural analysis. Both polysaccharides had the same constitution in normal and compression wood ray cells. A $(1 \rightarrow 3)$-linked β-D-glucan was also obtained from some fractions. It was obviously the same hemicellulose that had previously been found to occur in compression wood tracheids (Hoffmann and Timell 1970, 1972) (Chap. 5.4). The glucan was present in ray cells from both normal and compression woods but could not be detected in normal wood tracheids. Some of the fractions also contained much pectin, as indicated by their high proportion of galacturonic acid residues. The polysaccharide composition of the seven different fractions is summarized in Table 6.4.

A comparison is made in Table 6.5 between the chemical composition of ray cells and tracheids in normal and compression woods from *Pinus resinosa*, assuming a lignin content of 40% for both types of ray cells. The ray cells in compression wood probably contain as much lignin as do the tracheids in this type of wood, but they lack almost completely the galactan present in the tracheids and contain more cellulose. Inspection of the data reveals that the ray cells in compression wood were chemically indistinguishable from those in normal wood. Evidently, the chemical changes associated with the transition from normal to compression wood formation involve only the tracheids, while the ray cells remain unchanged in lignin content, polysaccharide composition, and structure of their hemicelluloses. Anatomically, the tracheids suffer profound changes on initiation of compression wood formation, but the ray parenchyma cells are only slightly enlarged in some species and remain unaltered in others (Timell 1972) (Chap. 4.7).

6.3.2 Tracheids and Ray Cells

Table 6.4. Polysaccharide composition of extracts from delignified ray cells of normal and compression woods of *Pinus resinosa*. All values in relative per cent. (Hoffmann and Timell 1972 a, b)

Extract and wood	Galacto-glucomannans	1, 3-Glucan	Xylan	Pectin
Water				
Normal wood	50		50	Trace
Compression wood	60		40	
Ammonium oxalate, 0.5%				
Normal wood	25		35	40
Compression wood	30		50	20
Sodium carbonate, 10%				
Normal wood	10	3	55	32
Compression wood	4	30	45	21
Potassium hydroxide, 1%				
Normal wood	14	5	50	31
Compression wood	6	14	70	10
Potassium hydroxide, 10%				
Normal wood	15	14	41	Trace
Compression wood	18	18	60	4
Potassium hydroxide, 24%				
Normal wood	38	22	40	
Compression wood	47	20	33	Trace
Sodium hydroxide, 25%-Borate, 5%				
Normal wood	76		7	
Compression wood	83		4	

Table 6.5. Chemical composition of ray cells annd tracheids in normal and compression woods of *Pinus resinosa*. All values in per cent of extractive-free wood. (Hoffmann and Timell 1972 a, b)

Constituent	Normal wood		Compression wood	
	Ray cells	Tracheids	Ray cells	Tracheids
Lignin	40	28	40	40
Cellulose	35	42	35	30
Galactoglucomannans	9	20	11	9
1,3-Glucan (laricinan)	2		2	2
Galactan	Trace	Trace	Trace	10
Arabino-4-*O*-methylglucurono-xylan	11	8	10	7
Pectin	2	1	1	1
Other polysaccharides	1	1	1	1

The xylan in compression-wood tracheids contains one arabinose side chain per 16–18 xylose residues (Chap. 5.7), whereas that in both tracheids and ray cells of normal wood has one arabinose per 8–9 xylose units. The xylan in the compression wood ray cells had one arabinose per 8–9 xylose residues, again demonstrating that the ray cells have the same chemical composition in normal and compression woods.

Topochemical features of compression wood other than those involving the tracheids and ray cells are not known. If ray tracheids and ray parenchyma cells were to be separated, it would have to be done by micromanipulation, a most laborious task if enough material were to be collected for a chemical analysis. Thompson and his co-workers (Thompson et al. 1966b, Thompson and Hankey 1968) have developed an ingenious method for isolation of epithelial cells from carefully delignified conifer wood. Interestingly, these cells, which line the resin canals, contain higher proportions of arabinose, galactose, and uronic acid residues than the tracheids, features typical of cambial tissues or immature tracheids. No attempt to isolate epithelial cells from compression wood has been made so far, although such an isolation would probably be no more difficult to achieve than with normal wood.

6.3.3 Distribution of Polysaccharides in the Tracheids

The two polysaccharides most characteristic of compression wood are the galactan and the laricinan, both of which are acidic (Chaps. 5.4 and 5.5). It has been observed by several investigators (Bouveng and Meier 1959, Schreuder et al. 1966, Côté et al. 1967b) that the galactan can be removed from a compression wood holocellulose with great ease, and it was suggested that it is probably located in the outer portion of the cell wall. Jiang and Timell (1972) found that practically all of the galactan was removed from the wood on prolonged delignification of compression wood. Although the glucan cannot be extracted quite as readily, it is nevertheless completely removed with dilute alkali, whereas the glucomannan requires extraction with strong alkali and borate.

Only one attempt has as yet been made to determine the complete polysaccharide distribution in compression wood tracheids, namely by Côté et al. (1968b), who adopted Meier's method for this purpose. Specimens of normal and compression woods of *Abies balsamea* were collected at a time when latewood was being formed. Immediate embedding in celloidin had several advantages and did not affect the carbohydrate composition of the wood. Tracheids at different stages of development were collected under a polarizing microscope from thin radial sections, such as that shown in Fig. 6.2. The microfibrillar orientation in the S_2 layer differed enough from the almost transverse orientation in S_1 to make it possible to ascertain the point at which S_2 began to be deposited onto S_1. The contrast between the different zones, was, however, somewhat less than with normal wood.

The volume ratio between the middle lamella and the cell-wall layers was estimated from electron micrographs of untreated wood. The ratio of the cell-wall thickness of the layers M + P, S_1, and S_2 was 1:8:41 for normal wood, in

6.3.3 Distribution of Polysaccharides in the Tracheids

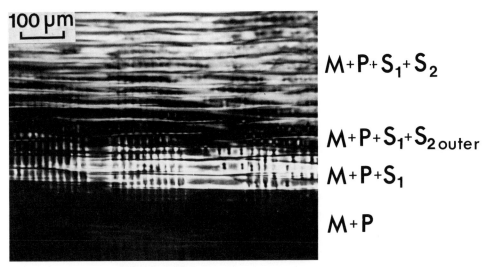

Fig. 6.2. A thin radial section of developing compression wood of *Abies balsamea* viewed in polarized light. LM. (Kutscha 1968, Timell 1981)

good agreement with the ratio of 1:5:44 reported by Meier and Wilkie (1959) for latewood tracheids from pine. The ratio for compression wood was 1:12:37, reflecting the thicker S_1 layer in this tissue. Volume ratios of the unlignified cell wall layers were known to equal weight ratios (Chap. 7.2.3). The proportions of neutral sugars present in each section were determined by the method of Simson and Timell (1967), which had been developed for this purpose. These data and the known ratio between the cell wall layers gave the relative amounts of neutral sugar residues in each layer. From the known composition of cellulose and the major hemicelluloses in normal and compression woods, the proportion of polysaccharides present in each region could be calculated.

The sections from normal balsam fir wood were found to contain an abnormally large number of galactose residues after initiation of secondary wall formation. The reason for this anomaly could not be ascertained, but it was believed that it might be associated with the fact that the tree used had previously been sampled regularly over a long period of time, a procedure that could have affected its metabolism (Evert and Kozlowski 1967, Evert et al. 1973). Comparison between normal and compression woods, therefore, had to be restricted to the cambial layer. The results obtained are summarized in Table 6.6 and shown graphically in Fig. 6.3. Comparison with the data of Meier and Wilkie (1959) and with results obtained with xylem scrapings of *Abies balsamea* and *Pinus sylvestris* by Haas and Kremers (1964), revealed that the M+P layer had the same polysaccharide composition in normal and compression woods, suggesting that the changes that result in formation of compression wood are initiated after the primary wall has been formed (Chaps. 9.6.1 and 12.5.2).

The galactan content was very high in S_1 and in the outer portion of S_2, regions where little or no galactan is present in normal wood. More than 80% of

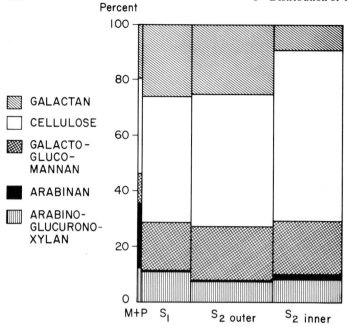

Fig. 6.3. Distribution of polysaccharides in the different cell-wall layers of a compression-wood tracheid in *Abies balsamea*. (Kutscha 1968, Timell 1981)

Table 6.6. Polysaccharide composition of the middle lamella and cell-wall layers of tracheids in normal and compression woods of *Abies balsamea*. All values in per cent of total polysaccharide content. (Côté et al. 1968)

Polysaccharide	Normal wood	Compression wood			
	M+P	M+P	S_l	$S_{2\,outer}$	$S_{2\,inner}$
Galactan	17.0	19.6	26.0	25.0	8.9
Cellulose	40.8	34.5	45.2	47.7	61.6
Galactoglucomannan	7.4	10.9	17.9	19.2	19.3
Arabinan	20.9	23.3	0.4	0.6	1.9
Arabino-4-*O*-methyl-glucurono-xylan	14.1	11.7	10.5	7.5	8.4

the total amount of galactan was present in these zones. The relatively high concentration of galactan in the M + P layer had little effect because of the small size of this layer. One third of the arabinan, however, was located here, and perhaps actually much more, since the figures for the arabinan content of the secondary wall layers must be deemed very uncertain. It is clear that most of the galactan in compression wood is located in the outer region of the cell wall, as had been inferred earlier from its great accessibility. It will be recalled that 80%

6.3.3 Distribution of Polysaccharides in the Tracheids

of the total galactan could be removed from *Picea rubens* holocellulose by extraction with mild reagents, whereas 20% required strong swelling agents (Schreuder et al. 1966) (Chap. 5.5.3).

Slightly less than 80% of the total cellulose and galactoglucomannan occurred in the S_2 layer, a proportion also found in normal wood. The distribution of xylan was uniform in both types of wood. The outer part of S_2 contained less cellulose and galactoglucomannan in compression wood than in normal wood, largely because of the much higher content of galactan in compression wood. The outer S_2 is the zone in a compression wood tracheid that has more lignin and less carbohydrates than any other region.

In addition to neutral sugar residues, cambial tissue in conifers contains 5–20% uronic acid residues (Alsopp and Misra 1940, Thornber and Northcote 1961a, b, 1962, Haas and Kremers 1964), probably largely galacturonic acid derived from pectin. Cambial tissue of *Populus tremuloides* according to Sultze (1957) and Simson and Timell (1978a) contains 38% and 25% uronic acid residues, respectively. Meier and Wilkie (1959) quote a figure of 20% for *Pinus sylvestris*. If all uronic acid residues are assumed to be derived from pectin, and if all the galactan and arabinan are regarded as constituents of the so-called pectic triad, the M + P layer should contain 40–50% of pectic material. The suspension-cultured tissues of *Acer pseudoplatanus* investigated by Talmadge et al. (1973) contained 16% rhamnogalacturonan, 8% of galactan, and 10% arabinan, all presumably members of the so-called pectic triad. Simson and Timell (1978a) found their cambial tissue from *Populus tremuloides* to contain 20% rhamnogalacturonan, 18% galactan, and 9% arabinan. The M + P layer in compression wood very likely has the same amount of pectic material. The exact quantity remains unknown, but it probably accounts for 30–40% of this layer in both normal and compression woods.

In a continuation of the earlier investigation (Larson 1969a) referred to previously in this chapter (Sect. 6.2.1), Larson (1969b) studied the incorporation of $^{14}CO_2$ in developing tracheids of compression and opposite woods of *Pinus resinosa*. The trees, which were 5 years old and accordingly still producing juvenile wood, were exposed to $^{14}CO_2$ for 2 h, after which they were allowed to grow under controlled conditions in a greenhouse for 1, 3, 5, 7, and 9 days, respectively. Five of them were tilted at an angle of 45° 3 weeks before the treatment, while another five specimens were inclined immediately after exposure to the $^{14}CO_2$ had been terminated. Data obtained with normal latewood in the earlier investigation (Larson 1969a) were used for comparison. The first tracheid fraction comprised cambial cells with only a primary wall. The second consisted partly of cells with an expanding primary wall and partly of cells in which deposition of S_1 had begun. All other fractions represented successive stages of secondary wall formation.

When the pulse of ^{14}C photosynthate became available, maximum activity would obviously be incorporated into the wall of the most actively metabolizing tracheids. In the trees that had been bent prior to the exposure to $^{14}CO_2$, this peak in ^{14}C activity occurred in fraction 2 after one day, and then shifted gradually inward as the time was extended to 9 days. The cambium was much more active on the lower than on the upper side of the inclined stem. In agree-

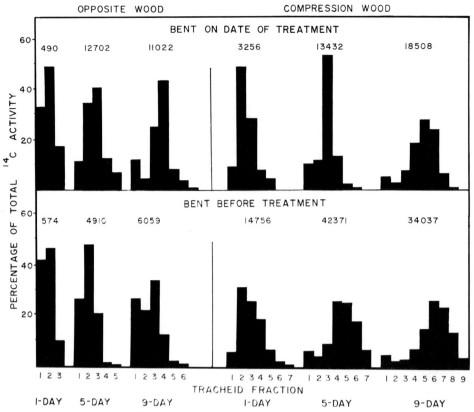

Fig. 6.4. Distribution of ^{14}C activity among tracheid fractions of *Pinus resinosa* trees harvested after 1, 5, and 9 days. The total ^{14}C activity for all fractions is shown above each histogram. (Larson 1969b)

ment with the results of Kennedy and Farrar (1965) (Chap. 12.5.2), it was found that one row of tracheids was added to the compression wood every day, whereas one new row was added to the opposite wood every fourth day. The detailed distribution of the ^{14}C activities in the two series of experiments is summarized graphically in Fig. 6.4. The metabolic activity on the upper side was low, and the cells matured close to the cambium. In the trees tilted after treatment with $^{14}CO_2$, one tracheid was at first formed every second day, but later, as formation of compression wood was initiated, one tracheid was added every day. At the same time, there was a reduction in cambial activity in the opposite wood. Evidently, photosynthate was transferred from the upper to the lower side of the stem when the latter was inclined. An example of the distribution of the ^{14}C activity among the different wall constituents is shown in Fig. 6.5.

Glucose reached its peak activity in fraction 3 of the compression wood, that is, at the time when the S_1 layer was initiated. As new tracheid rows were added, this peak moved inward. The distribution of mannose and xylose activi-

6.3.3 Distribution of Polysaccharides in the Tracheids

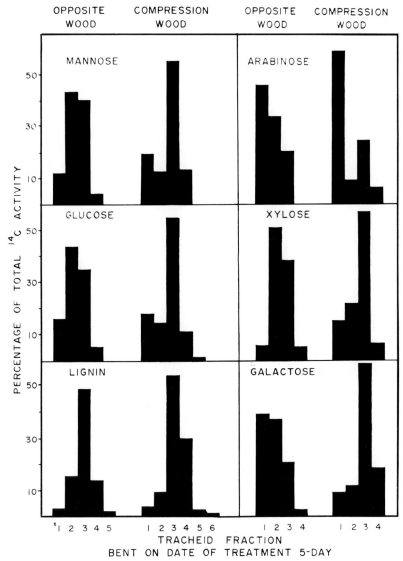

Fig. 6.5. Distribution of total ^{14}C activity among wall constituents in a 5-day tree. (Larson 1969b)

ties was similar to that of glucose, whereas the highest activity of arabinose occurred in fraction 1, that is, in the primary wall. The distribution patterns for these sugar residues in the opposite wood were basically the same, although they were affected by the lower cambial activity on this side. Galactose, which is abundant in all primary plant cell walls, reached its highest activity in fractions 1 and 2 of the opposite wood. In the compression wood, however, most of the galactose activity was at first concentrated in fraction 2 and subsequently in fractions 3–5 as more tracheids were formed.

Table 6.7. Percentage distribution of total ^{14}C recovered in cell-wall components of third-year internodes from 5-year-old *Pinus resinosa* stems that had been bent before treatment with $^{14}CO_2$. (Larson 1969b)

Constituent	Compression wood	Opposite wood
Glucose	29.1	44.6
Mannose	4.7	10.2
Xylose	4.7	9.1
Galactose	16.3	6.6
Arabinose	3.0	6.0
Lignin	42.2	23.5

In these experiments, the composition of a single, mature tracheid should result, if all fractions are added together. Data for the 3-, 5-, 7-, and 9-day trees are averaged in Table 6.7. Compared to normal wood, the distribution of ^{14}C activity reflects the higher lignin and galactose and lower glucose and mannose contents of compression wood. Opposite wood consistently contained more cellulose (glucose) and less lignin than normal wood. While this may apply to juvenile wood, it does not appear to be true for mature opposite wood (Timell 1973a) (Chap. 21.4.1).

The distribution of ^{14}C activity in each tracheid fraction is summarized in Fig. 6.6. It is clear that in every fraction, glucose residues predominated. The ratio of glucose to mannose was 6.3:1 in compression wood compared to 3.9:1 for normal wood in accordance with the lesser galactoglucomannan content of the former tissue. There was less incorporation of xylose residues in the compression wood. Recovery of arabinose residues was also lower, probably reflecting the lower incidence of arabinose side chains in the compression wood xylan (Chap. 5.7). The ratio of mannose to galactose was 4.5:1 in normal latewood but 0.4:1 in compression wood. The increase in galactose residues on tilting was not only pronounced but was also very rapid and was detectable only one day after inclination. It obviously represented the incorporation of the acidic, $(1 \rightarrow 4)$-β-D-galactan typical of compression wood (Chap. 5.5).

If the data for fraction 1 are averaged, the values in Table 6.8 are obtained. These ^{14}C recovery values indicate the number of different sugar residues added to the primary walls at the stage of cell enlargement. It is clear that the incorporation of the different polysaccharides into the primary wall was independent of the position of the stem. This observation is in agreement with the earlier results of Côté et al. (1968b), who found that primary cambial walls of normal and compression woods in *Abies balsamea* were identical in chemical composition and thickness.

Larson's (1969b) results indicate that the first cell wall entity to be chemically affected when a tree is displaced in the S_1 layer, and in this case the first noticeable change was a rapid and pronounced increase in the galactose content, caused by the deposition of the acidic $(1 \rightarrow 4)$-β-D-galactan. His data also

6.3.3 Distribution of Polysaccharides in the Tracheids

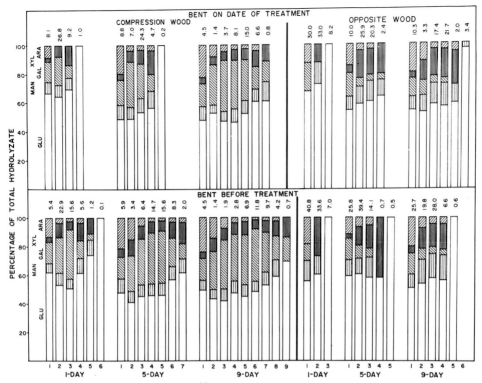

Fig. 6.6. Percentage distribution of ^{14}C activity in wood sugars within each tracheid fraction for 1-, 5-, and 9-day trees. The total ^{14}C activity recovered is indicated above each bar. (Larson 1969b)

Table 6.8. Percentage recovery of ^{14}C-labeled sugar residues in the cambial zone of compression and opposite woods and in normal latewood. (Larson 1969b)

Sugar residue	Cambial zone		Normal latewood
	Compression wood	Opposite wood	
Galactose	15.3	15.1	15.9
Glucose	51.5	54.5	53.2
Mannose	8.7	9.7	10.0
Arabinose	20.2	16.7	17.8
Xylose	4.7	4.0	3.1

show that most of this galactan is synthesized during the formation of the S_1 and the outer S_2 layers, as had been found by Côté et al. (1968b). Larson concluded that several features are typical of the outer portion of the secondary wall in compression wood tracheids, namely a loose compaction of the cellulose microfibrils, a high rate of wall formation, a high metabolic activity, and a high content of galactan. The ratio between mannose and galactose activities in the compression wood remained the same throughout the formation of the secondary wall, suggesting that perhaps the galactan was taking the place of the galactoglucomannan in normal wood.

In investigations also discussed in Chaps. 5.4.1 and 7.3.3.2, Brodzki (1972) and Wloch (1975) observed, within the helical cavities of compression wood that had been treated with aniline blue, a fluorescent material, referred to by them as "callose". The $(1 \rightarrow 3)$-β-D-glucan present in the sieve elements of the secondary phloem and for which the term *callose* is usually reserved, has long been known to show such fluorescence. As recently pointed out by Wloch and Hejnowicz (1983), there can be no doubt that the "callose" observed by Brodzki (1972) and Wloch (1975) is identical with the $(1 \rightarrow 3)$-β-D-glucan first shown to be present in compression wood by Hoffmann and Timell (1970, 1972c) and designated by them as *laricinan*. Brodzki (1972) found that some growth rings contained little or no callose, and often only small amounts were present in the tracheids formed at the beginning and at the end of the growing season. None occurred in the epithelial cells surrounding the resin canals. Occasionally, the intercellular spaces were lined with callose.

Boyd (1978) (Chap. 7.3.3.2) has pointed out that the evidence originally presented by Brodzki (1972) for the presence of a 1,3-glucan in the helical cavities is inconclusive. He suggested that the polysaccharide might instead be deposited as a thin layer on the surface of the helical ribs. Callose is known to be formed in sieve elements of the phloem in response to wounding. If the helical cavities were of schizogenous origin, as suggested by Boyd (1973), the larician could serve the function of sealing the faces of the helical cavities (Boyd 1978).

Recently, Wloch and Hejnowicz (1983) again examined the location of the 1,3-glucan, which they now referred to as *laricinan*. Compression wood specimens of *Pinus strobus* were treated either with aniline blue or with phloroglucinol-hydrochloric acid, a lignin stain of universal applicability. An example of their results is shown in Fig. 6.7. According to the investigators, those portions of the inner secondary wall that showed fluorescence did not respond to the lignin stain, and the lignified regions did not fluoresce. Since the lignin is known to occur in the helical ribs, it follows that the fluorescent material, namely the 1,3-glucan, must be located within the helical cavities between the ribs. Inspection of the original photomicrographs indicates that this interpretation is not impossible, but also that the resolution attained is still not sufficiently good to eliminate all doubts, at least not in my mind.

Waterkeyn et al. (1982) subjected the topochemistry of the 1,3-glucan in compression wood to a renewed examination. The two species studied were *Larix decidua* and *Pinus sylvestris*. The polysaccharide, here referred to as callose, was located by fluorescence microscopy after staining with aniline blue. The callose was present only in the S_2 layer of the tracheids, where it formed a thin helical band, as shown in Fig. 6.8. There was no callose on the inner surface of the cell wall or between the helical ribs, as suggested by Brodzki (1972). Waterkeyn et al. (1982) believe that the 1,3-glucan is formed at the same time as the cellulose and is incorporated into the cell wall by apposition. They suggested that the helical cavities develop by a contraction of the callose bands, a hypothesis referred to in Chap. 9.6.1.3. In an earlier study, Waterkeyn (1981)

6.3.3 Distribution of Polysaccharides in the Tracheids

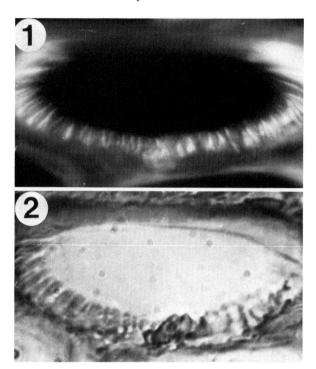

Fig. 6.7. Compression wood of *Pinus strobus* treated either with aniline blue (**1**) or with phloroglucinol-hydrochloric acid (**2**) in order the show the location of the fluorescent 1,3-glucan and the dark-staining lignin, respectively. Transverse section. LM. (Wloch and Hejnowicz 1983)

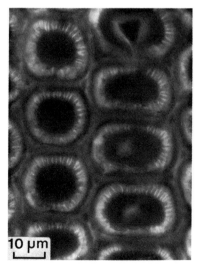

Fig. 6.8. Compression wood of *Pinus sylvestris*, treated with aniline blue. The location of the 1,3-glucan is indicated by the radial pattern of the white fluorescence. (Waterkeyn et al. 1982)

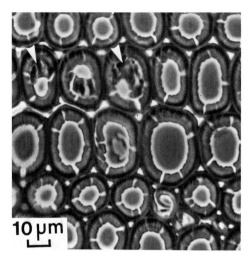

Fig. 6.9. Compression wood in the heartwood zone of *Larix laricina*, showing the presence of arabinogalactan (*arrowheads*) in the lumen of a few tracheids. Transverse section. LM. (Simson et al. 1968)

had shown that a transitory 1,3-glucan is deposited as a thin layer on the inner surface of the secondary wall in differentiating cotton fibers (Chap. 5.4.7).

In summary, the tracheids in compression wood differ in their polysaccharide distribution from those in normal wood largely in their high concentration of galactan in the S_1 and the outer portion of the S_2 layer. The low concentration of polysaccharides in the latter layer is unique to compression wood. The distribution of cellulose, galactoglucomannan, and xylan is probably the same in the two tissues. The laricinan is present in the helical ribs in the inner S_2 layer. The middle lamella and the primary wall very probably have the same chemical composition in normal and compression woods. A study of the chemistry of the active cambial zone in compression wood should not present undue difficulties and would be desirable.

6.3.4 Distribution of Arabinogalactan

Like its counterpart in normal wood, arabinogalactan in compression wood occurs only in the heartwood. Côté et al. (1966a) found that in normal wood most of the arabinogalactan in larch was located in the lumen of the tracheids. Compression wood contains less of this polysaccharide (Chap. 5.8). As shown in Fig. 6.9, it is deposited in the lumen of the tracheids also in compression wood. The entire lumen was seldom filled, and there seemed to be no arabinogalactan in the intercellular spaces (Côté and Timell 1967).

More recently, Teratani et al. (1976) studied the radial distribution of the extractives, including arabinogalactan, in a stem of *Larix gmelinii*. The disc used for sampling "contained some compression wood", but how much is not mentioned in the report. Judging from the only moderately eccentric cross section, the wider portion of the growth rings could hardly have consisted of pure, pronounced compression wood. Each annual increment was subdivided into earlywood and latewood. The sapwood contained only 0.5% carbohydrates, largely arabinogalactan, and the heartwood 2–12%. The concentration of arabinogalactan in the latter region was consistently three times higher in the earlywood than in the latewood. Near the pith, this pattern of intra-increment variation disappeared.

6.4 Microscopic Distribution of Lignin

6.4.1 Introduction

The microscopic distribution of lignin in wood is of considerable interest, both from a fundamental and from a practical point of view. It has not been studied as extensively as the chemistry of lignin, but it has, nevertheless, attracted interest for a long time and also, until recently, involved much controversy. Our present information on the distribution of lignin in wood we owe largely to the brilliant investigations carried out in the course of the last 15 years by Goring and his co-workers, who have also contributed much to our knowledge of how lignin is removed from wood in the technical pulping and bleaching processes.

Selective staining followed by examination under the light microscope is one of the oldest methods for locating lignin in wood (von Aufsess 1973). Besides the classical phoroglucinol and safranin-fast green reagents, many others have been used, for example iodine-malachite green (Bucher 1968). Kutscha and Gray (1972) tested several of the available staining reagents and applied the four best to normal and compression woods of *Abies balsamea*. Staining with potassium permanganate (Crocker 1921) has in later years become an excellent tool for assessing the distribution of lignin in wood and pulps by electron microscopy (Hepler and Newcomb 1963, Helper et al. 1970, Bland et al. 1971, Parham 1974, Kutscha and Schwarzmann 1975, Saka et al. 1979a). In this reaction, permanaganate is reduced to manganese oxide by the lignin. Although it has been claimed that acidic polysaccharides should also respond to this stain (Hoffmann and Parameswaran 1976), the effect of these is probably negligent. Kutscha (1968) used permanganate to study the lignification of *Abies balsamea* compression wood (Chap. 9.6.1.4.2).

The distribution of lignin in the wood of conifers can also be determined by microscopic examination of so-called lignin skeletons that remain after the polysaccharides have been removed with concentrated acid (Bailey and Kerr 1937, Staudinger 1942, Jayme and Fengel 1961). Because it does not swell or otherwise distort the cells in wood, concentrated hydrofluoric acid has in later years been the preferred reagent for hydrolyzing the cellulose and hemicelluloses (Sachs et al. 1963, Sachs 1965, Côté and Day 1969, Norberg 1969, Parham and Côté 1971, Parham 1974, Fujii et al. 1981). Timell made extensive use of this technique in his studies on the distribution of lignin in compression and opposite woods (Côté et al. 1967a, 1968a, Timell 1972, 1973b, 1978a, b). Brown rot fungi have also been used with good results for removing the polysaccharides from wood (Meier 1955, Sachs et al. 1963, Côté et al. 1966a). Collapse of enzymically produced lignin skeletons has been observed by Cowling

(1965) and by Chou and Levi (1971), but this does not occur under proper experimental conditions. The location of lignin in wood can be determined by examination in the electron microscope of holocelluloses after the lignin has been eliminated with acid chlorite (Jayme and Fengel 1961), an indirect method that lends itself particularly well to compression wood (Wergin and Casperson 1961, Casperson 1963a, Côté et al. 1968a, Timell 1972, 1978a, b).

A drawback common to all techniques now mentioned is that they do not offer any quantitative data for the distribution of lignin in wood. While some information of this kind can be obtained from the overall lignin content of a specimen and the volume fraction of its morphological entities (Berlyn and Mark 1965, Stamm and Sanders 1966), this approach has its obvious limitations. One of the best methods for establishing the quantitative distribution of lignin in wood entails examination of thin sections in ultraviolet light, an almost ideal approach, since lignin is the only component of extractive-free wood that absorbs in the ultraviolet region. This technique was first applied by Lange in his pioneering investigations on the chemical nature and distribution of lignin in normal, compression, and tension woods (Lange 1944, 1945, 1954a, b, c, 1957, 1958, Asunmaa and Lange 1954a, b). Some of Lange's results were later questioned by Frey (1959) and by Ruch and Hengartner (1960).

When it became possible to prepare ultrathin wood sections, Goring and his co-workers were able to develop ultraviolet microscopy into a powerful tool for quantitative determination of the distribution of lignin in wood, including its exact concentration, its chemical nature, and even its content of functional groups in different types of cells and cell-wall layers. Important phases of these investigations were, first, the development of reliable microscopic techniques (Scott et al. 1969, Scott and Goring 1970b, Wood and Goring 1974) and the determination of the exact absorbance of the guaiacyl lignin in softwoods and the much lower absorbance of the guaiacyl-syringyl lignin in hardwoods. The quantitative distribution of lignin could be determined for normal softwood (Fergus et al. 1969, Scott and Goring 1970a), compression wood (Wood and Goring 1971), and hardwood (Fergus and Goring 1970b). It was found that different types of lignin occurred in the fibers, vessels, and ray cells of hardwoods (Fergus and Goring 1970a, Musha and Goring 1975a, b), a result later challenged by Obst (1982, 1983). In softwood, the lignin in the secondary wall contained twice as many phenolic hydroxyl groups as that in the middle lamella (Yang and Goring 1978, 1980). The technique has since been applied by several other investigators, for example Imagawa et al. (1976), Gromov et al. (1977), and Boutelje and Eriksson (1982). One of the major results of the studies carried out by Goring and his co-workers is that it has finally become clear that only 20–30% of the total lignin in normal softwood and hardwood is present in the M + P layer, and that the remainder is located within the secondary wall, as originally predicted by Berlyn and Mark (1965). The concentration of lignin in M + P varies from 70% to 100%, depending on where the measurement is made. It is 20–25% in the secondary wall of tracheids and fibers, but higher in the walls of ray cells and vessels. Several of the results obtained by Goring and his associates were later confirmed by Hardell et al. (1980a, b). A direct photometric scanning method using an ultraviolet microscopic image analyzer has

6.4.1 Introduction

been developed by Fukazawa and Imagawa (1981), who consider it to be superior to the indirect technique used by Goring and co-workers (Takano et al. 1983).

Other methods have also been used for establishing the quantitative distribution of lignin wood. Lange and Kjaer (1957) introduced the use of interference microscopy for this purpose, a technique that was later further developed by Boutelje (1972). Kutscha and McOrmond (1972) used fluorescence microscopy for studying the distribution of lignin in wood. They pointed out that, while this method is capable of yielding quantitative information, it is probably inferior to ultraviolet microscopy.

In later years, application of scanning or transmission electron microscopy in conjunction with energy-dispersive x-ray analysis (EDXA) has been developed into a valuable tool for determining the quantitative distribution of lignin in wood (Saka et al. 1978, 1979b, 1981, 1982a, b, Saka and Thomas 1982a, b, Kolar et al. 1982). In this procedure, lignin is brominated selectively, and the concentration of bromine in different regions of the wood is determined by x-ray analysis. Most recently, it has been found that the lignin in the secondary wall of *Picea mariana* tracheids was 1.70 times more reactive in bromination than the middle lamella lignin. When a correction was introduced for this fact, the results obtained by the EDXA and the UV methods agreed fairly well (Saka et al. 1982b). The TEM-EDXA method was recently used in studies of the chlorination (Kuang et al. 1983) and the sulfonation (Beatson et al. 1984) of the middle lamella and the wall lignins in *Picea mariana*.

Hardell et al. (1980a, b) and Hardell and Westermark (1980b) disintegrated wood of *Picea abies* and *Betula verrucosa* and resolved the material by a series of sievings and sedimentations into three main fractions, namely fibers, ray cells, and a finely divided product consisting of middle lamella and primary wall. The lignin present in these fractions was characterized by various methods.

An ingenious method for separating middle lamella and secondary wall tissues was recently developed by Goring and his co-workers (Whiting et al. 1981). It is based on the facts that lignin has a lower density that the polysaccharides in wood (Chap. 7.2.1), and that the concentration of the lignin in the middle lamella-primary wall is about three times higher than in the secondary wall. Finely ground wood meal of *Picea mariana* could be resolved into fractions with a lignin content ranging from 20% to 60%. The former consisted largely of secondary wall material and the latter of M + P. Experiments with these two fractions afforded some exceedingly interesting results. Whiting and Goring (1981b) could demonstrate that most of the milled-wood lignin originates from the secondary wall of spruce tracheids, and that the lignin in this wall is more reactive toward both pulping (Whiting and Goring 1981a, Whiting et al. 1982) and bleaching (Whiting and Goring 1982c) reagents. They could also determine the carbohydrate composition of the middle lamella (Whiting and Goring 1983). The higher free phenolic hydroxyl content of the lignin in the secondary wall has been confirmed by both Whiting and Goring (1982b) and Sorvari et al. (1983). About 60% of the lignin units in the middle lamella were unmethylated, suggesting the presence of a guaiacyl-p-hydroxyphenyl lignin in this region, a lignin that could be expected to be more highly condensed and less reactive than the lignin in the secondary wall. According to Whiting and Goring (1982a), the middle lamella lignin in normal softwood, like the lignin in compression wood, is a copolymer of p-coumaryl alcohol and coniferyl alcohol (Chap. 5.9.2).

Very recently, this view has been challenged by Westermark (personal communication, 1984) in an investigation of a middle lamella fraction from normal wood of *Picea abies*, isolated by the sieving technique developed by Hardell et al. (1980b) and Hardell and Westermark (1981b). The M + P gave the same lignin degradation products as the whole wood on oxidation with nitrobenzene and alkali and on acidolysis, and it was also brominated to the same extent. Westermark concluded from her findings that the lignin present in the middle lamella of conifers is a guaiacyl lignin and does not contain many p-hydroxyphenylpropane units. She suggested that the earlier, different results obtained by Whiting and Goring (1982a) could have been caused by an accidental inclusion of compression wood in the *Picea mariana* specimen used by these investigators. If this were the case, the density gradient technique developed by Whiting et al. (1981) would not have been able to distinguish between the M + P of the entire wood and the $S_2(L)$ layer of the compression wood tracheids, since both have the same, high concentration of lignin. To this can be added that other portions of the lignin-rich secondary wall of the compression wood tracheids could also have been included in the middle lamella fraction. The presence of mild compression wood is easily overlooked, and, unfortunately, even this grade has been found to contain more lignin than normal wood and at times also an $S_2(L)$ layer, its lack of other compression wood characteristics notwithstanding (Yumoto et al. 1983) (Chap. 9.8). The subject needs to be studied further. One disadvantage of the sieving method used by Westermark seems to be that at least in this case it did not allow as sharp a separation of middle lamella and secondary wall material as did the sedimentation technique applied by Whiting and Goring (1982a).

Sjöström (personal communication 1984) and his co-workers recently resolved finely divided *Picea abies* wood into two main fractions of different density, representing middle lamella and secondary wall materials, which were further subdivided into subfractions by extraction with various solvents. A thorough characterization of the middle lamella and secondary wall lignins by several different methods indicated only minor differences between the two lignins, thus corroborating the results obtained by Westermark and her co-workers with the same species. In agreement with Whiting and Goring, the middle lamella lignin was found to contain more p-hydroxyphenyl groups than the wall lignin, and it was also more highly condensed.

Several of the methods referred to above have been applied in studies on the topochemistry of lignin in compression wood. Qualitative information has been obtained by observing stained sections under the light microscope or by examining sections treated with permanganate in the transmission electron microscope. Lignin and, less frequently, holocellulose skeletons have also been used for the same purpose. Ultraviolet light microscopy was adopted at an early date for estimating the qualitative distribution of lignin in compression wood

6.4.2 Distribution of Lignin in the Tracheids

tracheids, but it was only much later that quantitative information was obtained by Wood and Goring (1971). The differential sedimentation technique of Whiting et al. (1981) has not yet been applied to compression wood. It is undoubtedly less suited for this tissue which, in comparison with normal softwood, has a somewhat lower concentration of lignin in the middle lamella and a much higher lignin concentration in the secondary wall, and especially so in the $S_2(L)$ layer. It should perhaps be added here that Obst (1983) recently has questioned the accuracy of the pyrolytic gas chromatography technique used by Whiting and Goring (1982b).

6.4.2 Distribution of Lignin in the Tracheids

As mentioned in Chap. 2, Sanio as early as 1860 reported that one of the layers in the tracheids of compression wood seemed to be as highly lignified as the primary wall. His more detailed description of this layer in a later communication (Sanio 1873) makes it almost certain that he was referring to the highly lignified, isotropic outer portion of the S_2 layer, now usually called the $S_2(L)$ layer (Côté et al. 1968a, Timell 1973c). Timell (1980) has drawn attention to the fact that these observations by Sanio have been overlooked by all later investigators.

Hartig (1901), on treating compression wood with concentrated sulfuric acid, observed that the primary wall was strongly lignified. Sonntag (1904) removed lignin from compression wood of *Picea abies* with acid chlorite and examined the remaining holocellulose with the aim of establishing the location of the lignin. He also attempted to locate the lignin directly by staining the wood with phloroglucinol. Sonntag made the important observation that both "lamellae" in the compression-wood tracheids were strongly lignified, but the outer more so than the inner. It is very likely that what Sonntag was observing was the $S_2(L)$ and the inner part of the S_2 layer. He also mentions the occurrence of an intercellular substance but, unlike Sanio (1873), makes no reference to what is now known as the S_1 layer.

More than a quarter of a century later, Dadswell (1931) noted that, after compression wood had been treated with concentrated sulfuric acid, the remaining lignin appeared to be organized in a radial pattern. Bailey and Kerr (1937), treated compression wood of *Pinus roxburghii* with 72% sulfuric acid and observed the presence of radial discontinuities in the secondary wall. They also noted that the first-formed layer of the secondary wall was separated from the thick, inner layer by "an isotropic layer of noncellulosic composition". When viewing transverse sections of compression wood of *Taxodium distichum* under polarized light, Bailey and Berkley (1942) found both the outer and the inner portions of the secondary wall to be conspicuously birefringent. There was, however, also an intermediate region that was isotropic or only feebly birefringent, a property that the investigators attributed, not to the occurrence in this zone of longitudinally oriented fibrils, but to the largely noncellulosic nature of this zone.

Dadswell and Ellis (1940) treated sections of compression wood from *Araucaria cunninghamii, Pinus radiata,* and *Tsuga heterophylla* with 72% sulfric

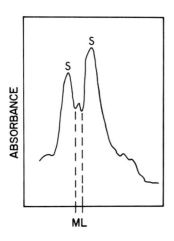

Fig. 6.10. Densitometer tracing in ultraviolet light across a double tracheid wall of compression wood in *Picea abies*. The concentration of lignin is higher in the outer part of the secondary wall (*S*) than in the middle lamella (*ML*). (Lange 1958)

acid. The residual lignin on microscopic examination appeared to be organized in a coarse, radial pattern, very different from the radioconcentric arrangement in normal wood. In less pronounced compression wood, the radial pattern was finer and less distinct. Similar results were obtained 10 years later by Wardrop and Dadswell (1950) in a study of the ultrastructure and lignin distribution of compression wood. No evidence was obtained for the occurrence of an isotropic layer inside S_1 in *Taxodium distichum* as observed by Bailey and Kerr (1937), but there were some indications of such a layer in *Pinus pinaster*, *P. radiata*, and *Pseudotsuga menziesii*. Wardrop and Dadswell concluded that the occurrence of this isotropic layer could not be regarded as typical of compression-wood tracheids.

Wardrop and Dadswell (1950) observed a radial lignin distribution in transverse and a helical pattern in longitudinal sections of compression-wood tracheids, when specimens of *Araucaria cunninghamii*, *Pinus pinaster*, *P. radiata*, and several other species were heated with 72% sulfuric acid. In *Pinus*, which had coarse spiral striations in the cell wall, the lignin pattern was coarse, whereas in *Araucaria*, where the striations were very fine, the lignin pattern was finer. Evidently, the organization of the lignin merely reflected that of the original S_2 layer with its helical cavities. Similar results were reported later by Wardrop (1954) for compression wood of *Pinus pinaster*. Matsumoto (1950) removed the polysaccharides with 72% sulfuric acid from compression wood of *Cryptomeria japonica* and *Thujopsis dolabrata*. The lignin skeletons thus revealed were very similar to those reported by later investigators, such as Wergin (1965).

In his extensive investigations on the lignin distribution in various woods, Lange (1954a, 1957, 1958) examined compression wood of *Picea abies*. The pattern of ultraviolet light transmission across two cell walls is shown in Fig. 6.10. This is a distribution curve very different from that found for normal wood. According to Lange (1954a), a narrow band containing relatively large amounts of lignin is present between S_1 and S_2. This region forms a network, described as a system of adjacent, hollow tubes, which are highly lignified. An

6.4.2 Distribution of Lignin in the Tracheids

ultraviolet micrograph showed a dark band inside a brighter layer, presumably S_1. The inner part of the S_2 layer had no absorption and no helical cavities. Lange concluded that this region of the cell wall contained very little, if any, lignin, a finding that has not been supported by later investigators. Wergin (1965) has pointed out that the cells studied by Lange were not typical of those usually found in compression wood, since they lacked a rounded outline, intercellular spaces, and helical cavities in S_2. Inspection of the best reproduction of Lange's photomicrograph (Lange 1958) reveals that this is correct. The photomicrographs obtained by Wood and Goring (1971) (Fig. 6.17) indicate that, in ultraviolet light and at Lange's magnification, helical cavities, if present, should have been distinguishable. Côté et al. (1967a) found that the tracheids formed at the beginning of the growing season often lacked a rounded outline, intercellular spaces, and helical cavities, but that they still differed from normal tracheids in their chemical composition and distribution of lignin (Chap. 4.5). That the tissue studied by Lange should be of this earlywood type seems precluded, however, by the thickness of the cell walls.

Casperson (1959a, b, 1962, 1963a, b, c, Casperson and Zinsser 1965) in several communications emphasized that the helical cavities in compression wood are not filled with lignin. Instead, the lignin is located, together with cellulose, in the helical ribs, and this fact causes the apparent radial distribution of lignin (Wardrop 1954).

On examination of compression wood tracheids of *Pinus radiata*, Wardrop and Bland (1961) and Wardrop and Davis (1964) observed between S_1 and S_2 a region with very low birefringence, thus confirming the original observations of Bailey and Kerr (1937) and Bailey and Berkley (1942). This layer also had a high ultraviolet absorption. It could be shown by electron microscopy that this zone did not contain any axially or steeply helically oriented microfibrils that could have caused extinction. Instead, it was concluded that the weakly birefringent band, while containing some lamellae of cellulose, consisted largely of lignin.

A different pattern of lignin distribution in compression wood was advocated by Casperson (1963a, b, 1964, 1965), who contended that most of the lignin was located in the middle lamella, the primary wall, and S_1. On treatment with acid chlorite, the lignin in these regions, and especially that in S_1, was rapidly removed, while that in S_2 was least attacked. Elimination of the lignin left microfibrils of cellulose in the S_1 layer which, according to Casperson (1965), is more highly lignified than in normal wood. A cellulose skeleton, obtained by Casperson (1962, 1963a, c) (Correns 1961, Wergin and Casperson 1961) from compression wood of *Picea pungens* is shown in Fig. 6.11.

Wergin (1965) studied the distribution of lignin in compression wood of *Picea abies* by ultraviolet microscopy. The primary wall, which was found to be unusually wide, showed only slight absorption in ultraviolet light. A thin, strongly absorbing layer was identified as S_1. The remainder of the secondary wall exhibited fairly strong absorption. Extinction curves, similar to those reported previously by Lange (1954a), were obtained with both normal and compression woods. Wergin concluded that the primary wall is less highly lignified in compression wood than in normal wood, that the S_1 layer is highly lignified,

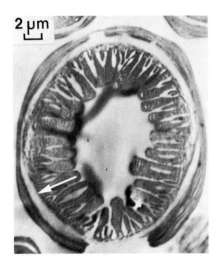

Fig. 6.11. Cellulose skeleton in compression wood of *Picea pungens* after removal of hemicelluloses and lignin. Note the low concentration of cellulose in the S_2 (L) layer (*arrow*). Transverse section. TEM. (Casperson 1962, 1963a, c)

and that the remainder of the secondary wall, contrary to the opinion of Lange (1954a), contains a considerable amount of lignin.

In a study devoted to the decomposition of wood by various fungi, Jurásek (1964a, b) made some observations on the distribution of lignin in compression wood of *Abies alba*. No significant changes in cell-wall structure could be observed at a weight loss as high as 33% when the wood was exposed to the brown rot fungus *Merulius (Serpula) lacrymans* which selectively attacks the lignin in wood. At a weight loss of 59%, a part of the secondary wall had decomposed. Jurasek noted that most of the lignin was located in the compound middle lamella and in the outer part of the secondary wall. The primary wall, surprisingly, was said to be only slightly lignified.

In a study on the distribution of lignin in normal and compression wood tracheids of *Abies balsamea* and *Picea rubens*, Côté et al. (1966b) arrived at conclusions quite different from those of Casperson (1963a, c) and Wergin (1965). Figure 6.12 shows a section of *Larix laricina* compression wood which had been stained with hematoxylin, a reagent which imparts a dark color to lignin. Immediately evident from the micrograph is the relatively low degree of lignification of S_1, the presence of a highly lignified layer inside S_1, and the moderate concentration of lignin in the inner, fissured part of S_2. Attempts to use *p*-(acetoxymercuri)aniline (Sachs et al. 1963) for visualizing the lignin were not successful.

As mentioned in Chap. 5.11, Côté et al. (1966b) were not able to produce satisfactory lignin skeletons by treating compression wood of *Abies balsamea* and *Picea rubens* with brown rot fungi. It appeared that most of the carbohydrate portion of the wood was at first inaccessible to the fungi, and that the polysaccharides later became accessible only after the encrusting lignin had been partly removed. The investigators noted that a highly lignified region seemed to be located immediately inside S_1, and that this layer resisted breakdown by the brown rot fungi used. Better results were obtained when the poly-

6.4.2 Distribution of Lignin in the Tracheids

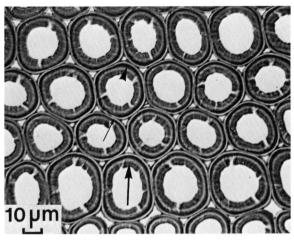

Fig. 6.12. Compression wood of *Larix laricina* stained with hematoxylin to indicate the extent of lignification of the S_1 (*small arrow*), S_2 (L) (*arrowhead*) and inner S_2 (*large arrow*) layers. The S_2 (L) region is most highly lignified. Transverse section. LM. (Côté et al. 1968a)

saccharides were hydrolyzed with concentrated hydrofluoric acid, a technique later applied to compression wood also by Fujii et al. (1981).

In an attempt to improve the method for obtaining lignin skeletons from compression wood, Côté et al. (1968a) gradually increased the concentration of the hydrofluoric acid from 30% to 80%, extended the time of hydrolysis with the 80% acid, reduced the concentration stepwise to 30%, and finally heated with 4% acid to complete the hydrolysis of the glycosidic bonds. Normal wood containing 27% lignin and compression wood with 38% lignin from *Larix laricina* were used in this study. A transverse section of lignin skeletons of compression wood tracheids is shown in Fig. 6.13. The proportion of the total lignin present in the M + P layer of the central tracheid is unusually small and probably less than 5% because of the six intercellular spaces. The S_1 and S_2(L) layers, both of which are about 1 µm thick, differ strikingly from each other in their lignin concentration. The S_1 layer appears to contain very little lignin, whereas the S_2(L) seems to be as highly lignified as the middle lamella. The remainder of S_2 consists of a uniform and relatively dense lignin meshwork. The helical cavities extend up to, but do not enter, the S_2(L) layer.

In a communication to A. H. Westing, reported by him in 1968 and later cited by Wood and Goring (1971), I attempted to estimate the quantitative distribution of lignin in compression wood from a visual inspection of electron micrographs such as that shown in Fig. 6.12. Similar electron micrographs of lignin skeletons from normal wood, where the distribution of lignin was better known at that time, were used for reference. I estimated that only 5–10% of the total lignin in compression wood could be located in the middle lamella-primary wall region. The S_1 layer was assumed to contain 5–10% of the total lignin and have a lignin concentration of only 10%. The concentration of lignin in the S_2(L) layer was probably the same as in M + P, that is 60–70%. About 40%

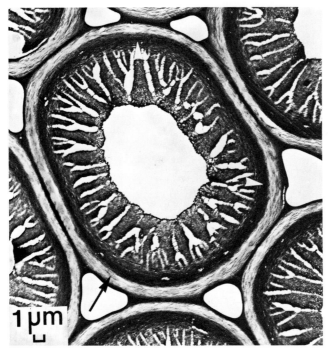

Fig. 6.13. Lignin skeleton of compression wood tracheids in *Larix laricina*, suggesting a spuriously low concentration of lignin in the S_1 layer (*arrow*). Transverse section. TEM. (Côté et al. 1968a)

of the total lignin was located there. The same proportion, 40%, was assumed to be present in the inner S_2 layer, which probably had a lignin concentration of 40%. As pointed out later by Wood and Goring (1971), these estimates are, with one notable exception, remarkably close to their own, measured values. The exception is the S_1 layer, where later research has shown that the real concentration of lignin is within the range of 30–40% (Wood and Goring 1971, Fukazawa 1974) and not 10%. There are two reasons for this discrepancy. First, the apparent transverse orientation of the lignin lamellae in the S_1 layer tends to minimize its concentration in transverse sections of lignin skeletons. In holocellulose, the transverse orientation of the cellulose microfibrils similarly gives a spurious impression of a high density of polysaccharides in transverse sections (Figs. 6.15, 6.16, and 6.25). Second, lignin skeletons are prone to disintegrate in regions of relatively low lignin concentration, as first observed by Ritter (1925). There can be little doubt that a sizeable portion, perhaps two thirds, of the lignin originally present in the thin section seen in Fig. 6.13 had been accidentally removed at some stage of the acid treatment. In other words, the extremely low lignin density in this section, which was evident from several electron micrographs, was a preparation artifact.

That my earlier figure of 10% had seriously underestimated the lignin concentration in S_1 later became evident from an inspection of a large number of

6.4.2 Distribution of Lignin in the Tracheids

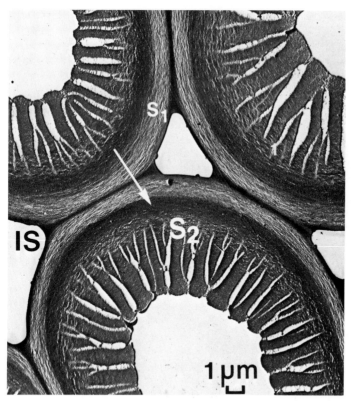

Fig. 6.14. A lignin skeleton of the same origin (*Larix laricina*) as that in Fig. 6.13 but showing a higher lignin concentration in S_1. *Arrow* indicates S_2 (L). Transverse section. TEM

electron micrographs obtained with acid-treated compression wood sections other than that shown in Fig. 6.13. The lignin skeleton of tracheids in *Larix laricina* compression wood seen in Fig. 6.14 undoubtedly gives a more correct picture of the distribution of lignin in this wood. It should perhaps also be noted that the helical ridges are better preserved in this electron micrograph. This lignin skeleton indicates that the lignin concentrations are quite similar in S_1 and the inner S_2, that is, 30–40%.

Chlorite holocelluloses at various stages of delignification were prepared by Côté et al. (1968a) from compression wood of *Larix laricina* and were examined in the electron microscope. An early stage of delignification is shown in Fig. 6.15 where only the S_2(L) layer exhibits visible signs of lignin removal. Parham and Côté (1971) later made the same observation with *Pinus taeda* compression wood. A portion of a completely delignified compression wood tracheid is shown in Fig. 6.16. Remnants of the primary wall can be seen outside S_1. These microfibrils have no preferred orientation and are finer than those in the secondary wall. The S_1 layer consists of lamellae of transversely oriented microfibrils. The low polysaccharide content of the S_2(L) layer is clearly

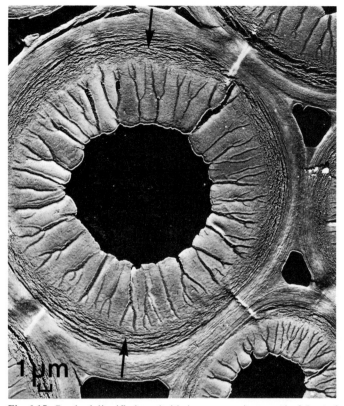

Fig. 6.15. Partly delignified tracheids in compression wood of *Larix laricina*. Some lignin has been removed from the S_2 (L) layer (*arrows*). Transverse section. TEM. (Côté et al. 1968a)

discernible. Most of the short microfibrils have fallen over for lack of support. Those that had not been displaced seem to have the same orientation as the microfibrills in the inner S_2. The helical ribs in S_2 are remarkably well preserved and consist of cellulose microfibrils that have been sectioned almost transversely. Individual microfibrils can be readily detected.

These results confirm the conclusion of Wardrop and Davies (1964) that the S_2(L) layer contains microfibrils of cellulose, and they also show the correctness of the earlier opinion of Bailey and Berkley (1942), that this layer is isotropic because of its high lignin content. Also attested is the observation of Wergin and Casperson (1961) and Casperson (1964) that the inner part of S_2 contains lamellae of cellulose. This layer has a high concentration of lignin, which is uniformly distributed. Lange's (1954a) statement that this region should be unlignified is not correct. Côté et al. (1966b) suggested that the orientation of the cellulose microfibrils in the S_2(L) layer might be the same as in S_1, that is, almost transverse. Inspection of the holocelluloses from *Larix laricina*, especially that in Fig. 6.14, shows that this suggestion is incorrect. The microfibril orientation in S_2(L) is the same as in the remainder of the S_2 layer.

6.4.2 Distribution of Lignin in the Tracheids

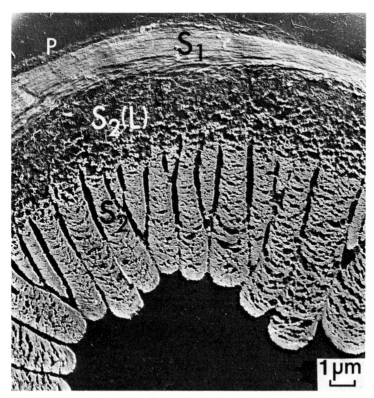

Fig. 6.16. Completely delignified tracheids (holocellulose skeleton) of a portion of a compression wood tracheid in *Larix laricina*. The primary wall (*P*) can be seen outside S_1, and short (sectioned) microfibrils can be discerned in S_2 (L). The helical ribs contain transversely sectioned microfibrils (cf. Fig. 4.55). Transverse section. TEM. (Côté et al. 1968a)

The interpretation offered by Casperson, Correns, Wergin, and their co-workers of the lignin distribution in compression wood was briefly referred to above. As explained in Chap. 4.4.3, the conclusions drawn by these investigators must be due to a misinterpretation of their light and electron micrographs. In a later report on compression wood of *Larix decidua*, Casperson (1968) corrected his earlier assignments of the different cell wall layers in compression wood tracheids. With this new interpretation, the lignin distributions observed by Casperson and by Timell become identical.

A truly quantitative determination of the distribution of lignin in compression wood was first achieved by Wood and Goring (1971). The method used previously with normal wood was applied to compression wood from the lower side of a large branch of *Pseudotsuga menziesii*, 13 cm in diameter and 108 years old. Ultraviolet micrographs of transverse sections from four successive positions across a growth ring are shown in Fig. 6.17. They represent first-formed earlywood (A), transitional earlywood (B), fully developed, early compression wood (C), and late compression wood with thick cell walls (D). The helical cavities are revealed with a clarity never attained with visible light

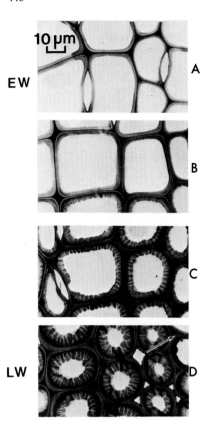

Fig. 6.17 A–D. Four sections from a growth ring on the lower side of a branch of *Pseudotsuga menziesii*. **A** First-formed earlywood (normal). **B** Transitional earlywood. **C** Early compression wood. **D** Late compression wood. Transverse sections. UVLM. (Wood and Goring 1971)

thanks to the improved resolution possible with ultraviolet light. Visual inspection of the photomicrographs shows two dark regions with a high lignin density, namely the M + P layer and the $S_2(L)$ layer, and two zones with a lesser lignin concentration, which are the S_1 and the inner, fissured portion of the S_2 layer.

The variation in absorbance of the S_2 layer at 280 nm with cell position within a growth ring is shown in Fig. 6.18. The transition, marking the first appearance of helical cavities, occurred in ring 6. At this point, there was a sharp increase in absorbance and accordingly in lignin concentration. The lignin density of the S_1 and $S_2(L)$ layers also increased rapidly from the 5th to the 6th cell, as shown in Fig. 6.19. It remained more or less constant thereafter except for the S_1 layer in the tangential walls, where the lignin density decreased somewhat in the transition from earlywood to latewood. As could be expected, the ultraviolet absorbance of the $S_2(L)$ was throughout higher than that of the S_1 layer. Interestingly, both these layers, but not the inner part of the S_2, had a higher lignin concentration at the cell corners than at the tangential walls.

The relative volume of the different morphological regions was carefully determined by a statistical count-point method. From these data and the values obtained on evaluation of the ultraviolet photometer tracings, the data in

6.4.2 Distribution of Lignin in the Tracheids

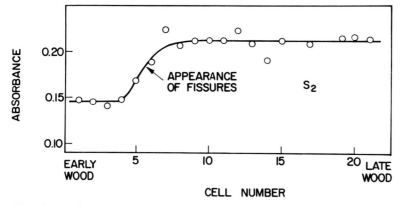

Fig. 6.18. Variation across a growth ring of ultraviolet absorbance of the S_2 layer in the tracheids shown in Fig. 6.17. The absorbance increases rapidly at the point where the helical cavities begin to appear (**B** in Fig. 6.17). (Wood and Goring 1971)

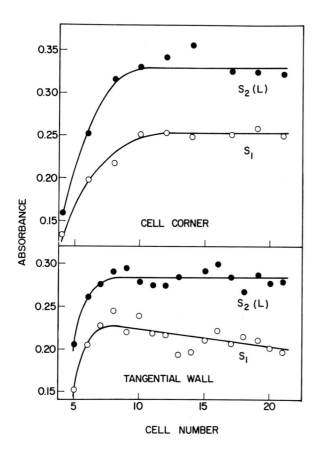

Fig. 6.19. Variation across a growth ring in the ultraviolet absorbance of the S_1 and S_2 (L) layers at the cell corner and at the tangential wall in the tracheids shown in Fig. 6.17. (Wood and Goring 1971)

Table 6.9. The distribution of lignin in tracheids of branch compression wood in *Pseudotsuga menziesii*. (Wood and Goring 1971)

Morphological entity	Volume, percent	Lignin, percent of total	Lignin concentration, weight percent
S_1	19.7	18.5	40
$S_2(L)$	19.3	25.0	54
S_2 inner	53.9	46.8	36.3
Middle lamella, along wall	4.5	5.2	49
Middle lamella, at cell corner	2.5	4.5	75
Middle lamella, total	7.0	9.7	58

Table 6.9 were obtained. As had been found with normal wood, the concentration of lignin in the intercellular region was highest at the cell corners, namely 75%, a concentration assumed to represent the lignin content of the true middle lamella. Along the radial and tangential walls, between the cell corners, the combined lignin content of the primary wall and the intercellular substance was 49%, giving an average lignin content of the entire M + P layer of 59%. The S_1 layer had a lignin concentration of 40%, which is higher than the 36.3% lignin content of the inner S_2. The concentration of lignin in the $S_2(L)$ layer was 54%, which is only slightly lower than the average value for the M + P layer.

The branch compression wood contained 42.0% lignin. Only 9.7% of the total lignin was located in the middle lamella–primary wall region. The S_1 layer contained 18.1% of the total amount of lignin, the $S_2(L)$ layer 25.0%, and the inner part of S_2 46.8%. Compared to earlywood and latewood in normal wood, a higher proportion of the total lignin was located within the secondary wall.

For normal earlywood and latewood of *Picea mariana*, Fergus et al. (1969) found that the concentration of lignin was 50% and 60% along the wall and 85% and 100% at the cell corners, respectively (Table 6.10). According to Wood and Goring (1971), the corresponding lignin concentrations in their compression wood were considerably lower, namely 49% and 75%. In his study of the dormant cambial zone in compression wood of *Picea abies*, Timell (1980a) observed that the primary walls were often quite thick. He pointed out that, because of the way in which cambial cells divide, with a new primary wall being deposited after each division, parental primary walls could be expected to traverse intercellular regions with a variable frequency. Timell found, as had previously been observed by both Wardrop and Dadswell (1952) and Casperson and Zinsser (1965), that primary walls often crossed intercellular corner areas in compression wood. He suggested that this might be the reason for the relatively low concentration of lignin in the true middle lamella of compression wood, and that the unusually thick primary walls might be responsible for the low lignin concentration along the wall (Chap. 9.2.2).

6.4.2 Distribution of Lignin in the Tracheids

Table 6.10. Distribution of lignin in normal earlywood and latewood of *Picea mariana* and in branch compression wood of *Pseudotsuga menziesii*. (Fergus et al. 1969, Wood and Goring 1971)

Proportion and concentration	Early-wood	Late-wood	Compression wood
Percentage of total lignin in the middle lamella	28	18	10
Concentration of lignin in the middle lamella, weight percent			
Along wall	50	60	49
At cell corners	85	100	75
Concentration of lignin in the secondary wall, weight percent	23	22	42

Wood and Goring (1971), in conclusion, pointed out that all regions of the secondary wall of their compression wood had higher than normal concentrations of lignin, and that this "extra" lignin was not confined to the $S_2(L)$ layer, as has sometimes been assumed. They also noted that the concentration of lignin in the S_1 layer was rather variable. One further fact, not mentioned by Wood and Goring, should be added, namely that in the earlywood tracheids, the $S_2(L)$ layer was present only at the cell corners (Fig. 6.17).

Table 6.10 shows a comparison between the distribution of lignin in normal and compression woods, based upon the results obtained by Goring and his co-workers. The proportion of the total lignin located outside the secondary wall is two to three times larger in normal wood than in compression wood. The frequency of intercellular spaces appears to have been unusually low in the branch compression wood examined by Wood and Goring (1971). In other compression wood specimens, only 5% of the lignin might be located in the M + P layer, as suggested by Timell (Westing 1968). The concentration of lignin along the cell walls in the middle lamella of compression wood is the same as that in normal earlywood but lower than in latewood. At the cell corners, the intercellular region of compression wood has a remarkably low lignin concentration in comparison with normal wood. When compared with this wood, the average lignin concentration in the secondary wall of compression wood is almost twice as high.

The techniques developed by Goring and his co-workers were a few years later adopted by Fukazawa (1974) for determining the distribution of lignin in compression and normal woods in *Abies sachalinensis*. Fukazawa studied the ultrastructure and lignin distribution of the cells across an entire growth ring in zones containing less and less compression wood, an approach that made it possible for him to follow the gradual transition of fully developed compression wood into normal wood. The location of his samples is shown in Fig. 6.20. On

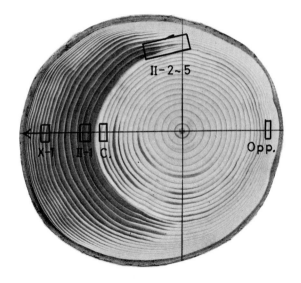

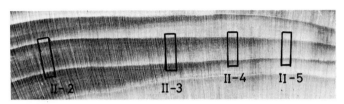

Fig. 6.20. Cross section of a stem of *Abies sachalinensis*, showing the location of the compression-wood areas examined. (Fukazawa 1974)

going from the under side to the flank of the leaning stem, compression wood as usual first ceased to be formed at the beginning of the growing season and disappeared last from the material deposited at the very end of the season.

Ultraviolet micrographs were obtained at 280 nm, representing all stages in the transition of compression wood into normal wood. Eighteen of them are shown in Fig. 6.21 and 6.22. The tracheids were classified into seven different types, the first four of which comprised all cells except those formed at the very beginning and at the very end of the growing season, while the remaining three consisted of the last-formed cells. Type A, seen in micrographs 2 and 3, consisted of typical, pronounced compression wood tracheids with a rounded outline, large intercellular spaces, a thick S_1 layer, an $S_2(L)$ layer that formed a continuous ring around the tracheid, and helical cavities in the inner portion of the S_2. Micrographs 1 and 12 include cells of type B. Here the tracheids were more square in outline and rounded only at the cell corners, S_1 was still thick, but the $S_2(L)$ layer was present only at the corners. The helical cavities were either absent or poorly developed. In type C tracheids, seen in micrographs 6 and 16, the outline was also angular. Here the S_1 layer thickened only at the cell corners

6.4.2 Distribution of Lignin in the Tracheids

Table 6.11. Distribution of lignin in compression wood tracheids of *Abies sachalinensis*. (Fukazawa 1974)

Morphological entity	Relative absorbance	Lignin concentration, weight per cent
S_1	1.1	29
$S_2(L)$	1.6	42
$S_{2\,inner}$	1.0	26
Middle lamella, along wall	1.9	49
Middle lamella, at cell corners	2.5	65

and the $S_2(L)$ layer was entirely restricted to this region. Type D, finally, not shown here, was normal wood.

A different classification had to be adopted for the last-formed, flattened tracheids. Type a cells, shown in micrograph 4, were round at the corners and possessed thick S_1 and $S_2(L)$ layers. Tracheids of type b were more angular in outline, as shown in micrograph 13, and the thick portion of S_1 and the $S_2(L)$ layers were restricted to the cell corners. Normal wood was referred to as type c. The first-formed earlywood cells were of type B throughout.

The silver density of the negatives of the ultraviolet micrographs was determined with a microdensitometer. Four of the density profiles across a tracheid double wall are shown in Fig. 6.23, two for normal wood and two for compression wood. Relative lignin concentrations in the various regions were calculated from these density data by the equation developed by Scott et al. (1969). As had been done by these investigators, lignin concentrations were also computed from absorbance values obtained with transverse sections in ultraviolet light. The spectra were all those of a typical guaiacyl lignin, with no difference between normal and compression woods. From these values and the proportions of the different morphological entities, the distribution of lignin within the compression wood tracheids could be determined. The final results are summarized in Table 6.11.

A comparison with the data obtained by Wood and Goring (1971) indicates for *Abies sacchalinensis* consistently lower lignin concentrations in all regions except for the radial-tangential part of the middle lamella, where they are the same. Possibly, the reason for this difference is that the branch compression wood studied by Wood and Goring was of a more severe type. The relative lignin contents of the middle lamella regions and the cell wall layers, on the other hand, were strikingly similar in the two specimens. In both cases the S_1 layer was found to have a slightly higher lignin concentration than the inner portion of the S_2 layer, while the $S_2(L)$ contained about 50% more lignin than the latter. The highest lignin concentration occurred at the cell corners. Fukazawa estimated on the basis of his absorbance measurement that his compression wood contained 46.0% lignin, the highest value reported for the lignin content of any

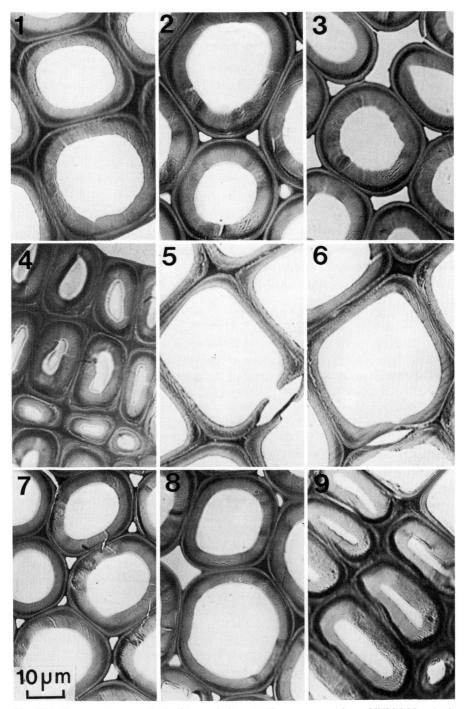

Fig. 6.21. Compression wood of *Abies sachalinensis*. Transverse sections. UVLM Nos. 1–9. (Fukazawa 1974)

6.4.2 Distribution of Lignin in the Tracheids

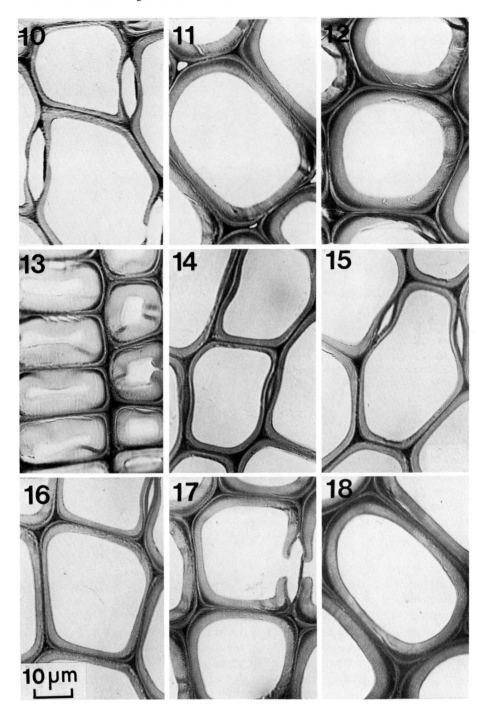

Fig. 6.22. Compression wood of *Abies sachalinensis*. Transverse section. UVLM Nos. 10–18. (Fukazawa 1974)

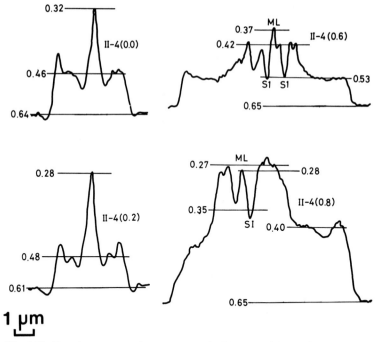

Fig. 6.23. Densitometer tracings across a double tracheid wall in sample II-4 in Fig. 6.20 at four different positions (*numbers in parentheses*). *Other numbers* indicate silver density of the negative. (Fukazawa 1974)

compression wood. Comparison with values obtained by direct chemical or spectrophotometric analysis would have been of interest.

The distribution of lignin in the tracheids of compression wood is perhaps the most characteristic feature of this tissue. It is found in all gymnosperm species known to form compression wood, and it occurs not only in the severe but also in the mild and moderate grades of this wood. The $S_2(L)$ layer is usually highly lignified in the first-formed earlywood tracheids (Côté et al. 1967a, Yoshizawa et al. 1982), and it is always so in the last-formed, latewood cells (Chaps. 4.5 and 4.6).

In mild and moderate compression woods the lignin in the $S_2(L)$ layer tends to be restricted to the cell corners instead of extending along the entire circumference. This is obvious from the photomicrographs published by Wood and Goring (1971) and has since been confirmed in more detail by Fukazawa (1974) and by Yumoto et al. (1983). In their investigations of the transition of normal wood into compression wood, Yumoto et al. (1983) observed some tracheids located within a region between normal wood and compression wood with a trace of an $S_2(L)$ layer. The S_2 of these cells had a somewhat higher lignin content, as manifested by its ultraviolet absorption, than in normal wood. No $S_2(L)$ layer was present, the lignin being uniformly distributed throughout S_2. It is interesting to note in this connection that according to Kucera and

Philipson (1977) lignification is uniform across the secondary wall in compression wood tracheids in *Phyllocladus alpinus.* Yumoto et al. (1982) observed, in the transition from normal wood to compression wood, intermediate tracheids where the concentration of lignin was higher in the inner than in the outer S_2 layer, an unusual situation (Fig. 9.53). Similar observations were recently reported by Yoshizawa et al. (1985) for *Taxus cuspidata* (Chap. 9.8).

In their investigation of the lignification of compression wood tracheids in *Cryptomeria japonica,* which is dealt with in Chap. 9.6.1.4.2, Fujita et al.(1978) recorded ultraviolet densitometer tracings across double cells walls. These tracings, some of which are reproduced in Fig. 9.44, indicate that in the mature tracheids the concentration of lignin in the $S_2(L)$ layer was notably higher than in the middle lamella along the wall. This agrees with the earlier results obtained by Wood and Goring (1971), but not with the data reported by Fukazawa (1974).

In summary, the feature in the distribution of lignin that distinguishes compression wood tracheids most patently from those in normal wood is the occurrence inside S_1 of the $S_2(L)$ layer, where the concentration of lignin is at least as high as in the M + P. Its high lignin content renders this layer as seemingly isotropic as the intercellular substance, so that it extinguishes polarized light. Unlike the true middle lamella at the cell corners, however, the $S_2(L)$ layer is not devoid of cellulose microfibrils, and it probably also contains some galactan. The S_1 and the inner, fissured portion of S_2 are also highly lignified. Their lignin content is 30–40% compared to only 20–25% in normal wood. The middle lamella–primary wall region, by contrast, has a lower lignin concentration than in normal wood. Another important difference is, of course, the fact that in compression wood the space between the cell corners often fails to lignify.

Approximately 90% of the lignin in compression wood tracheids is located within the secondary wall, a higher proportion than the 75–80% in normal wood. The inner S_2 contains slightly less than half of the total lignin, the $S_2(L)$ layer one quarter, and the S_1 layer one fifth. The fact that the entire secondary wall is so highly encrusted by lignin is undoubtedly one of the major factors determining the physical properties of compression wood, such as the hard, brash nature of this wood and its high compressive and low tensile strength.

6.4.3 Distribution of Lignin in the Ray Cells

The distribution of lignin in the ray cells of compression wood has been discussed in only two reports. Côté et al. (1968a) removed the polysaccharides from compression wood of *Larix laricina* and examined lignin skeletons of the ray cells in tangential sections of the wood. A typical example is shown in Fig. 6.24. The ray cell wall appears to be lignified to the same extent as the inner S_2 layer of the adjacent tracheid.

A transverse section of delignified compression wood of *Larix laricina* is shown in Fig. 6.25 (Timell 1972). Remnants of primary walls are seen outside both the longitudinal tracheids and the ray cells. The polysaccharide density of

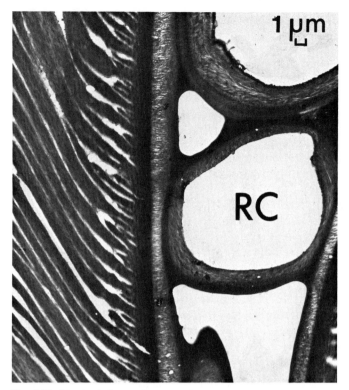

Fig. 6.24. Lignin skeleton of compression wood of *Larix laricina*, showing the high lignin content of the wall in three ray cells (*RC*). Tangential section. TEM. (Côté et al. 1968a)

the secondary wall in the ray cells and in the S_1 and inner part of S_2 layers in the tracheids appear to be the same. Originally, the three zones probably all contained about 40% lignin, suggesting that the apparent, high polysaccharide density of these delignified cell walls is an artifact, caused by swelling of the polysaccharides.

The similarity between the lignin skeletons of the ray cells and the inner S_2 layer of the tracheids (Fig. 6.24) indicates that the former probably are lignified to the same extent as the latter and that accordingly both have a lignin content of 35–40% (Table 6.10). Ray parenchyma cells in normal softwood have been reported to contain 35–40% lignin (Bailey 1936a, b, Harada and Wardrop 1960, Fergus et al. 1969, Yamamoto 1982). Evidently, the ray parenchyma cell walls are lignified to the same degree in normal and compression woods. It will be recalled that they also have the same polysaccharide composition (Hoffmann and Timell 1972a, b).

6.5 Distribution of Extractives

The distribution of the extractives in compression wood has so far been investigated only by Teratani et al. (1976), who examined a disc from a *Larix gmelinii*

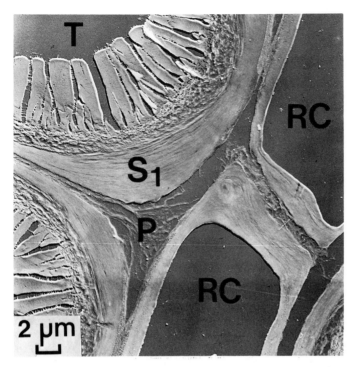

Fig. 6.25. Delignified compression wood (holocellulose) of *Larix laricina* with portions of two ray parenchyma cells (*RC*) and three adjacent tracheids (*T*). The polysaccharide density appears to be the same in the wall of the rays as in the S_1 layer of the tracheids. Remnants of primary wall (*P*) are located between the cells. Transverse section. TEM

stem that "contained some compression wood". The concentration of extractives soluble in ethanol-benzene was higher in the heartwood than in the sapwood. The radial distribution was irregular, and there was a middle region in the heartwood with a low concentration of extractives. Within the outer heartwood, latewood contained more extractives but less arabinogalactan than the earlywood. The investigators concluded that the formation of arabinogalactan and of extractives soluble in ethanol-benzene are not correlated with each other.

6.6 General Conclusions

In comparing normal and compression woods with respect to the distribution of their chemical constituents, both similarities and differences become apparent. As with normal wood, juvenile compression wood differs chemically from the mature tissue, namely in lacking the high galactan and lignin contents of the latter. The earlywood and latewood zones of normal conifer wood are less distinct in compression wood. The atypical, first-formed earlywood tracheids do not seem to differ significantly in chemical composition from those formed later.

The ray cells in compression wood have the same chemical composition as in normal wood and apparently also the same high lignin content. Both tracheids and ray cells in compression wood contain 35–40% lignin and 10% galactoglucomannan. The galactan forming 10% of the tracheids is entirely missing in the rays, which instead contain somewhat more cellulose and xylan than the tracheids. Obviously, neither the anatomy nor the chemical composition of the ray cells is modified in conifer reaction wood. In tension wood, both vessels and rays appear to remain the same as in normal wood, and only the fibers are altered, anatomically as well as chemically. It has so far not been possible to examine ray tracheids and ray parenchyma of compression wood separately, nor have the epithelial cells surrounding the resin canals been studied chemically, although methods are available for their isolation (Thompson et al. 1966a, b, Kibblewhite et al. 1971, Kibblewhite and Thompson 1973).

The distribution of the polysaccharides across the tracheid wall in compression wood is only incompletely known. The middle lamella and primary wall evidently have the same composition as in normal wood, with a relatively low cellulose and a high pectin content. To date, no cambial tissue from the compression wood side of actively growing trees has been collected and analyzed, nor have the different polysaccharides in this tissue been characterized. Evidence obtained by the summative microdissection method developed by Meier shows that the $(1 \rightarrow 4)$-β-D-galactan is largely located in the S_1 and outer S_2 layers. The $(1 \rightarrow 3)$-β-D-glucan, referred to by some as laricinan and by others as callose, occurs together with cellulose in the inner portion of the S_2 layer. The zone with the lowest concentration of polysaccharides in the secondary wall is the $S_2(L)$ layer.

If different hemicelluloses could be preferentially removed from thin holocellulose sections of compression wood, examination in the electron microscope might furnish more information as to their approximate location. A more promising approach would be to make use of the oxidation-staining technique developed by Hoffmann and Parameswaran (1976) for establishing the location of hemicelluloses in plant cell walls. Best results would, however, probably be obtained by application of the recent peeling technique developed by Hardell and Westermark (1981a, b). This would allow a direct chemical analysis of the $M + P$, S_1, and S_2 layers in a compression wood fiber, which would mean a complete analysis, since the S_3 layer, which cannot be isolated by this method, is absent in severe compression wood tracheids.

The distribution of lignin in the tracheids of compression wood is highly characteristic and sets this tissue apart from normal wood. The unusually high lignin and low polysaccharide contents of the $S_2(L)$ layer are unique among wood cells and render this layer as isotropic in polarized light as the middle lamella. Its presence undoubtedly influences the properties of compression wood. The fact that also the S_1 and the inner portion of the S_2 layer are highly lignified is, together with the large microfibril angle in S_2, the major reason for the high compressive strength of compression wood (Chap. 7.4.2.10). Both the high lignin content and the distribution of this lignin contribute to the poor pulping and bleaching characteristics of compression wood (Chaps. 19.4 – 19.6).

Additional insight into the distribution of both polysaccharides and lignin is gained from studies on the way in which these polymers are deposited in formation of the cell wall. Advantageous in this respect is the fact that the cell-wall polysaccharides are formed earlier than the lignin, so that each layer of the wall can be examined while it still consists of only polysaccharides. The gradual progress of lignification can be followed in detail by examination in the electron microscope of tissues stained with permanganate, by ultraviolet microscopy, or by the SEM-EDXA technique. Information obtained in this way on the formation of the cell walls in compression wood is discussed in Chap. 9.

References

Allsopp A, Misra P 1940 The constitution of the cambium, the new wood, and the mature sapwood of the common ash, the common elm, and the common Scotch pine. Biochem J 34:1078–1084

Asunmaa S, Lange PW 1953 Esterification of the carbohydrates in the plant cell wall with p-phenylazobenzoyl chloride for microspectrographic investigation. Svensk Papperstidn 56:85–94

Asunmaa S, Lange P 1954a A note on the change in cell wall dimensions of carbohydrate fibres after esterification with p-phenylazobenzoyl chloride and subsequent swelling in pyridine. Svensk Papperstidn 57:498–500

Asunmaa S, Lange PW 1954b The distribution of "cellulose" and "hemicellulose" in the cell wall of spruce, birch and cotton. Svensk Papperstidn 57:501–516

Bailey AJ 1936a Lignin in Douglas-fir. Composition of the middle lamella. Ind Eng Chem Anal Ed 8:52–55

Bailey AJ 1936b Lignin in Douglas-fir. The pentosan content of the middle lamella. Ind Eng Chem Anal Ed 8:389–391

Bailey IW, Berkley EE 1942 The significance of x-rays in studying the orientation of cellulose in the secondary wall of tracheids. Am J Bot 29:231–241

Bailey IW, Kerr T 1937 The structural variability of the secondary wall as revealed by "lignin" residues. J Arnold Arbor 18:261–272

Beatson RP, Gancet C, Heitner C 1984 The topochemistry of black spruce sulfonation. Tappi J 67(3):82–85

Bendtsen BA 1978 Properties of wood from improved and intensively managed trees. For Prod J 28(10):61–72

Berlyn GP, Mark RE 1965 Lignin distribution in wood cell walls. For Prod J 16:140–141

Berlyn GP, Miksche JP 1976 Botanical microtechnique and cytochemistry. IA State Univ Press, Ames, 326 pp

Bland DE, Foster RC, Logan AF 1971 The mechanism of permanganate and osmium tetroxide fixation and the distribution of lignin in the cell wall of Pinus radiata. Holzforschung 25:137–168

Borzakovskaya VS, Sharkov VI 1974 (Investigation of the carbohydrate composition of parenchyma cells of Scots pine wood by paper chromatography.) Khim Drev 2:3–5

Boud JD 1973 Helical fissures in compression wood cells: causative factors and mechanics of development. Wood Sci Technol 7:92–111

Boutelje J 1972 Calculation of lignin concentration and porosity of cell wall regions by interference microscopy. Svensk Papperstidn 75:683–686

Boutelje JB, Eriksson I 1982 An UV microscopic study of lignin in middle lamella fragments from fibers and mechanical pulp of spruce. Svensk Papperstidn 85:R39–R42

Bouveng HO, Meier H 1959 Studies on a galactan from Norwegian spruce compression wood (Picea abies Karst). Acta Chem Scand 13:1884–1889

Boyd JD 1978 Significance of laricinan in compression wood tracheids. Wood Sci Technol 12:25–35

Brazier JD 1969 Some considerations in appraising within-ring density. Proc FAO/IUFRO World Consult For Tree Breeding, Washington DC, FO-FTB-69-4/2, 9 pp

Brodzki P 1972 Callose in compression wood tracheids. Acta Soc Bot Polon 41:321–327

Bucher H 1968 Der mikroskopische Nachweis von Lignin in verholzten Fasern durch Jodmalachitgrün. Papier 22:390–396

Burkart LF, Watterston KG 1968 Effect of environment on ratio of cellulose to lignin in shortleaf pine. For Prod J 18(5):25–28

Casperson G 1959a Elektronenmikroskopische Untersuchungen des Zellwandaufbaus beim Reaktionsholz der Coniferen. Ber Deutsch Bot Ges 72:230–235

Casperson G 1959b Mikroskopischer und submikroskopischer Zellwandaufbau beim Druckholz. Faserforsch Textiltech 10:536–540

Casperson G 1962 Über die Bildung der Zellwand beim Reaktionsholz. Holztechnologie 3:217–223

Casperson G 1963a Reaktionsholz. Seine Struktur und Bildung. Habilitationschr Humboldt-Univ Berlin DDR, 116 pp

Casperson G 1963b Die Bildung der Zellwand als morphologisch-physiologisches Problem. Symp Perspektiven der Grundlagenforschung des Holzes (Medzinárodné kolokvium), Bratislava, 49–54

Casperson G 1963c Chemische, anatomische und physiologische Eigenheiten des Reaktionsholzes. Chemiefaser-Symp, 1962, Akademie-Verlag, Berlin DDR, 39–52

Casperson G 1964 Über den lamellaren Aufbau der verholzten Zellwand. 3rd Eur Reg Conf Electron Microscopy. Publ House Czech Acad Sci, Praha, 169–170

Casperson G 1965 Zur Anatomie des Reaktionsholzes. Svensk Papperstidn 68:534–544

Casperson G 1968 Anatomische Untersuchungen an Lärchenholz (Larix decidua Miller). Faserforsch Textiltech 19:467–476

Casperson G, Zinsser A 1965 Über die Bildung der Zellwand beim Reaktionsholz. 3. Zur Spaltenbildung im Druckholz von Pinus sylvestris L. Holz Roh-Werkst 23:49–55

Chambat G, Barnoud F, Joseleau JP 1984 Structure of the primary cell walls of suspension-cultured Rosa glauca cells. I. Polysaccharides associated with cellulose. Plant Physiol 74:687–693

Chou CK, Levi MP 1971 An electron microscopical study of the penetration and decomposition of tracheid walls of Pinus sylvestris by Poria vaillantii. Holzforschung 25:107–112

Core HA, Côté WA, Jr, Day AC 1961 Characteristics of compression wood in some native conifers. For Prod J 11:356–362

Correns E 1961 Über anormale Holzfasern. Pap Puu 43:47–62

Côté WA, Jr, Day AC 1969 Wood ultrastructure of the southern pines. NY Coll For Syracuse Tech Publ 95, 70 pp

Côté WA, Jr, Day AC, Kutscha NP, Timell TE 1967a Studies on compression wood. V. Nature of the compression wood formed in the early springwood of conifers. Holzforsch 21:180–186

Côté WA, Jr, Day AC, Simson BW, Timell TE 1966a Studies on larch arabinogalactan. I. The distribution of arabinogalactan in wood. Holzforschung 20:178–192

Côté WA, Jr, Day AC, Timell TE 1968a Studies on compression wood. VII. Distribution of lignin in normal and compression wood of Tamarack (Larix laricina (Du Roi) K Koch). Wood Sci Technol 2:13–37

Côté WA, Jr, Kutscha NP, Simson BW, Timell TE 1968b Studies on compression wood. VI. Distribution of polysaccharides in the cell wall of tracheids from compression wood of balsam fir (Abies balsamea (L) Mill). Tappi 51:33–40

Côté WA, Jr, Pickard PA, Timell TE 1967b Studies on compression wood. IV. Fractional extraction and preliminary characterization of polysaccharides in normal and compression wood of balsam fir. Tappi 50:350–356

Côté WA, Jr, Timell TE 1967 Studies on larch arabinogalactan. III. Distribution of arabinogalactan in tamarack. Tappi 50:285–289

Côté WA, Jr, Timell TE, Zabel RA 1966b Studies on compression wood. I. Distribution of lignin in compression wood of red spruce (Picea rubens Sarg). Holz Roh-Werkst 24:432–438

Cowling EB 1965 Microorganisms and microbial enzyme systems as selective tools in wood anatomy. In: Côté WA Jr (ed) Cellular ultrastructure of woody plants. Syracuse Univ Press, Syracuse, 341–368

Crocker EC 1921 An experimental study of the significance of "lignin" color reactions. Ind Eng Chem 13:625–627
Curran CE 1936 Some relations between growth conditions, wood structure and pulping quality. Pap Trade J 103(11):200–204
Dadswell HE 1931 Distribution of lignin in the cell wall of wood. J Coun Sci Ind Res Aust 4:185–186
Dadswell HE, Ellis J 1940 Study of the cell wall. I. Methods of demonstrating lignin distribution in wood. J Coun Sci Ind Res Aust 13:44–54
Dadswell HE, Hawley LF 1929 Chemical composition of wood in relation to physical characteristics: preliminary study. Ind Eng Chem 21:973–975
Darvill A, McNeil M, Albersheim P, Delmer DP 1980 The primary cell wall of flowering plants. In: Tolbert NE (ed) The plant cell. Academic Press, New York, 91–162
de Zeeuw CH 1965 Variability in wood. In: Côté WA, Jr (ed) Cellular ultrastructure of woody plants. Syracuse Univ Press, Syracuse, 457–471
Evert RF, Kozlowski TT 1967 Effect of isolation of bark on cambial activity and development of xylem and phloem in trembling aspen. Am J Bot 54:1045–1054
Evert RF, Kozlowski TT, Davis JD 1973 Influence of phloem blockage on cambial growth of sugar maple. Am J Bot 59:632–641
Fergus BJ, Goring DAI 1970a The location of guaiacyl and syringyl lignins in birch xylem tissue. Holzforschung 24:113–117
Fergus BJ, Goring DAI 1970b The distribution of lignin in birch wood as determined by ultraviolet microscopy. Holzforschung 24:118–124
Fergus BJ, Procter AR, Scott JAN, Goring DAI 1969 The distribution of lignin in sprucewood as determined by ultraviolet microscopy. Wood Sci Technol 3:117–138
Frey HP 1959 Über die Einlagerung des Lignins in der Zellwand. Holz Roh-Werkst 17:313–318
Fujii T, Harada H, Saiki H 1981 Ultrastructure of "amorphous layer" layer in xylem parenchyma cell wall of angiosperm species. Mokuzai Gakkaishi 27:149–156
Fujita M, Saiki H, Harada H 1978 The secondary wall formation of compression wood tracheids. II. Cell wall thickening and lignification. Mokuzai Gakkaishi 24:158–163
Fukazawa K 1974 The distribution of lignin in compression- and lateral-wood of Abies sachalinensis using ultraviolet microscopy. Res Bull Coll Exp For Hokkaido Univ 31:87–114
Fukazawa K, Imagawa H 1981 Quantitative analysis of lignin using an UV microscopic image analyser. Variation within one growth increment. Wood Sci Technol 15:45–55
Fukazawa K, Yamamoto K, Ishida S 1980 The season of heartwood formation in genus Pinus. In: Bauch J (ed) Natural variation of wood properties. Mitt Bundesforschungsanst Forst-Holzwirtsch, Hamburg-Reinbek 131, 113–131
Gromov VS, Evdokimov AM, Abramovich TL 1977 (Distribution of lignin in wood and topochemistry of its delignification studied by ultraviolet microspectrophotometry. 1. Lignin distribution in the cell walls of birchwood tracheary element.) Khim Drev 6:73–79
Haas BR, Kremers RE 1964 The pectic substances as an index to the chemistry of wood formation. Tappi 47:568–573
Hale JD, Clermont LP 1963 Influence of prosenchyma cell-wall morphology on basic physical and chemical characteristics of wood. J Polym Sci C-2:253–261
Harada H, Wardrop AB 1960 Cell wall structure of ray parenchyma cells of a softwood (Cryptomeria japonica). Mokuzai Gakkaishi 6:34–41
Hardell HL, Leary GJ, Stoll M, Westermark U 1980a Variations in lignin structure in defined morphological parts of spruce. Svensk Papperstidn 83:44–49
Hardell HL, Leary GJ, Stoll M, Westermark U 1980b Variations in lignin structure in defined morphological parts of birch. Svensk Papperstidn 83:71–74
Hardell HL, Westermark U 1981a The carbohydrate composition of the outer cell wall of spruce fibers. The Ekman-Days 1981 (Stockholm), SPCI Rep 38 Vol 1:32–34
Hardell HL, Westermark U 1981b Techniques for separation of wood elements and cell wall layers. The Ekman-Days 1981 (Stockholm), SPCI Rep 38 Vol 5:17–19
Harlow WM, Wise LE 1928 The chemistry of wood. I. Analysis of wood rays in two hardwoods. Ind Eng Chem 20:720–727

Harris JM 1981 Effect of rapid growth on wood processing. Proc 17th IUFRO World Congr Japan, Div 5:117–125

Hartig R 1901 Holzuntersuchungen. Altes und Neues. Julius Springer, Berlin, 99 pp

Harwood VD 1971 Variation in carbohydrate analyses in relation to wood age in Pinus radiata. Holzforschung 25:73–77

Hata K 1950 (Studies on the pulp of "Akumatsu" (Pinus densiflora S et Z) wood. III. On the chemical composition of spring wood and summer wood in "Akumatsu" (Pinus densiflora S et Z) wood.) J Jap For Soc 32:257–261

Hayat MA 1975 Positive staining for electron microscopy. Van Nostrand Reinhold, New York, 361 pp

Hengartner H 1961 Die Fluoreszenzpolarisation der verholzten Zellwand. Holz Roh-Werkst 19:303–309

Hepler PK, Fosket DE, Newcomb EH 1970 Lignification during secondary wall formation in Coleus: an electron microscopic study. Am J Bot 57:85–96

Hepler PK, Newcomb EH 1963 The fine structure of young tracheary xylem elements arising by redifferentiation of parenchyma in wounded Coleus stem. J Exp Bot 14:496–503

Higuchi T 1976 Biochemical aspects of lignification and heartwood formation. Wood Res (Kyoto) 59/60:180–199

Hillis WE 1960 Factors influencing the formation of phloem and heartwood polyphenols. Holzforschung 14:105–110

Hillis WE 1968a Chemical aspects of heartwood formation. Wood Sci Technol 2:241–259

Hillis WE 1968b Heartwood formation and its influence on utilization. Wood Sci Technol 2:260–267

Hillis WE 1977 Secondary changes in wood. In: Loewus FA, Runeckles VC (eds) The structure, biosynthesis, and degradation of wood. Plenum, New York London, 247–309

Hillis WE, Humphreys FR, Bamber RK, Carle A 1962 Factors influencing the formation of phloem and heartwood polyphenols. Holzforschung 16:114–121

Hoffmann GC, Timell TE 1970 Isolation of a β-1,3-glucan (laricinan) from compression wood of Larix laricina. Wood Sci Technol 4:159–162

Hoffmann GC, Timell TE 1972a Polysaccharides in ray cells of normal wood of red pine (Pinus resinosa). Tappi 55:733–736

Hoffmann GC, Timell TE 1972b Polysaccharides in ray cells of compression wood of red pine (Pinus resinosa). Tappi 55:871–873

Hoffmann GC, Timell TE 1972c Polysaccharides in compression wood of tamarack (Larix laricina). I. Isolation and characterization of laricinan, an acidic glucan. Svensk Papperstidn 75:135–142

Hoffmann P, Parameswaran N 1976 On the ultrastructural localization of hemicelluloses within delignified tracheids of spruce. Holzforschung 30:62–70

Hugentobler UH 1965 Zur Cytologie der Kernholzbildung. Vierteljahrschr Naturforsch Ges Zürich 110:321–342

Ifgu G, Labosky R Jr 1972 A study of loblolly pine growth increments. I. Wood and tracheid properties. Tappi 55:524–529

Imagawa H, Fukazawa K, Ishida S 1976 Study of the lignification of Japanese larch, Larix leptolepis Gord. Res Bull Coll Exp For Hokkaido Univ 33:127–138

Jayme G, Fengel D 1961 Beitrag zur Kenntnis des Feinbaus der Fichtenholztracheiden. II. Beobachtungen an Ultradünnschnitten von delignifiziertem Holz und Ligningerüsten. Holzforschung 15:98–102

Jiang KS, Timell TE 1972 Polysaccharides in compression wood of tamarack (Larix laricina). IV. Constitution of an acidic galactan. Svensk Papperstidn 75:592–594

Joseleau JP, Chambat G 1984 Structure of the primary cell walls of suspension-cultured Rosa glauca cells. II. Multiple forms of xyloglucan. Plant Physiol 74:694–700

Jurásek L 1964a Změny v obsahu stavebních složek dřeva při rozkladu způsobeném houbami. (Changes in the content of the structural components of wood on decomposition caused by fungi.) Drev Vysk 9:41–56

Jurásek L 1964b Změny v mikroskopické struktuře při rozkladu dřeva dřevokazynými houbami. (Changes in the microstructure on decomposition of wood by wood-destroying fungi.) Drev Vysk 9:127–144

Kellison RC 1981 Characteristics affecting quality of timber from plantations, their determination and scope for modification. Proc 17th IUFRO World Congr Japan, Div 5:77−88

Kennedy RW, Farrar JL 1965 Tracheid development in tilted seedlings. In: Côté WA, Jr (ed) Cellular ultrastructure of woody plants. Syracuse Univ Press, Syracuse, 419−453

Kennedy RW, Jaworsky JM 1960 Variation in cellulose content in Douglas-fir. Tappi 43:25−28

Kerr T, Bailey IW 1934 The cambium and its derivative tissues. IV. Structure, optical properties and chemical composition of the so-called middle lamella. J Arnold Arbor 15:327−349

Kibblewhite RP, Thompson NS 1973 The ultrastructure of the middle lamella region in resin canal tissue isolated from slash pine holocellulose. Wood Sci Technol 7:112−126

Kibblewhite RP, Thompson NS, Williams DG 1971 A study of the bonding forces between the epithelial cells surrounding the resin canals of slash pine (Pinus elliottii Engelm) holocellulose. Wood Sci Technol 5:101−120

Kolar JJ, Lindgren BO, Treiber E 1982 The distribution of lignin between fiber wall and middle lamella. Svensk Papperstidn 85:R21−R26

Kremers RE 1963 The chemistry of developing wood. In: Browning BL (ed) The chemistry of wood. Interscience, New York, 369−404

Kremers RE 1964 Cambial chemistry. A way of looking at a stick of pulpwood. Tappi 47:292−296

Kremers RE 1965 The chemistry of wood formation. In: Chimie et biochimie de la lignine, de la cellulose et des hémicelluloses. Actes Symp Int Grenoble 1964, 469−480

Kuang SJ, Saka S, Goring DAI 1983 The distribution of chlorine in chlorinated spruce and birch wood as determined by TEM-EDXA. Appl Polym Symp 37:483−490

Kucera LJ, Philipson WR 1977 Growth eccentricity and reaction anatomy in branchwood of Drimys winteri and five native New Zealand trees. NZ J Bot 15:517−524

Kutscha NP 1968 Cell wall development in normal and compression wood of balsam fir, Abies balsamea (L) Mill. Ph D Thesis NY State Coll For Syracuse, 231 pp

Kutscha NP, Gray JR 1972 The suitability of certain stains for studying lignification in balsam fir, Abies balsamea (L) Mill. Life Sci Agr Exp Sta Univ ME, Orono, Tech Bull 53, 50 pp

Kutscha NP, McOrmond RR 1972 The suitability of using fluorescent microscopy for studying lignification in balsam fir. Life Sci Agr Exp Sta Univ ME, Orono, Tech Bull 62, 15 pp

Kutscha NP, Schwarzmann JM 1975 The lignification sequence in normal wood of balsam fir. Holzforschung 29:79−84

Lange PW 1944 Om ligninets natur och fördelning i granved. (On the nature of the lignin and its distribution in spruce wood.) Svensk Papperstidn 47:262−265

Lange PW 1945 Ultraviolettabsorbtionen av fast lignin. (The ultraviolet absorption of solid lignin.) Svensk Papperstidn 48:241−245

Lange PW 1950a The distribution of lignin in the cell wall of normal and reaction wood from spruce and a few hardwoods. Svensk Papperstidn 57:525−532

Lange PW 1950b Mass distribution in the cell walls of Swedish spruce and birch. Svensk Papperstidn 57:533−536

Lange PW 1954c The distribution of the components in the plant cell wall. Review of earlier work and future possibilities. Svensk Papperstidn 57:563−566

Lange PW 1957 Verteilung der Zellwandkomponenten in der Zellwand. In: Treiber E (ed) Die Chemie der Pflanzenzellwand. Springer, Berlin, 259−280

Lange PW 1958 The distribution of the chemical constituents throughout the cell wall. In: Bolam F (ed) Fundamentals of papermaking fibres. Tech Sect Brit Pap Board Makers' Assoc, London, 147−185

Lange PW and Kjaer A 1957 Kvantitativ kemisk analys av cellväggens delar i ved- och cellulosafibrer med använding av interferensmikroskopi. (Quantitative chemical analysis of the different parts of the cell wall in wood and cellulose fibers by interference microscopy.) Norsk Skogind 11:425−432

Larson PR 1966 Changes in chemical composition of wood cell walls associated with age in Pinus resinosa. For Prod J 16(4):37−45

Larson PR 1969a Incorporation of ^{14}C in the developing walls of Pinus resinosa tracheids (earlywood and latewood). Holzforschung 23:17−26

Larson PR 1969b Incorporation of ^{14}C in the developing walls of Pinus resinosa tracheids: compression wood and opposite wood. Tappi 52:2170−2177

Larson PR 1972 Evaluating the quality of fast-grown coniferous wood. Proc 1972 Ann Meet West Stand Manag Comm, Seattle WA, 7 pp

Leopold B 1961 Chemical composition and physical properties of wood fibers. II. Alkali extraction of holocellulose fibers from loblolly pine. Tappi 44:232–235

Matsumoto T 1950 (Studies on compression wood. 1. Spiral cracks in the tracheid wall.) J Jap For Soc 32:16–20

McNeil M, Darvill AG, Albersheim P 1979 The structural polymers of the primary cell walls of dicots. Progress in the chemistry of organic natural products. Springer, Wien New York, 37:192–249

Meier H 1955 Über den Zellwandabbau durch Holzvermorschungspilze und die submikroskopische Struktur von Fichtentracheiden und Birkenholzfasern. Holz Roh-Werkst 13:323–338

Meier H 1957 Discussion of the cell wall organization of tracheids and fibres. Holzforschung 11:41–66

Meier H 1961 The distribution of polysaccharides in wood fibers. J Polym Sci 51:11–18

Meier H 1962 Chemical and morphological aspects of the fine structure of wood. Pure Appl Chem 5:37–52

Meier H 1964 General chemistry of cell walls and distribution of the chemical constituents across the walls. In: Zimmermann MH (ed) The formation of wood in forest trees, Academic Press, New York London, 137–151

Meier H, Welck A 1965 Über den Ordnungszustand der Hemicellulosen in den Zellwänden. Svensk Papperstidn 68:878–881

Meier H, Wilkie KCB 1959 The distribution of polysaccharides in the cell-wall of tracheids of pine (Pinus silvestris L). Holzforschung 13:177–182

Mullis RH, Thompson NS, Parham RA 1976 The localization of pentosans within the cell wall of aspen (Populus tremuloides Michx) by high resolution autoradiography. Planta 132:241–248

Musha Y, Goring DAI 1975a Distribution of guiacyl moieties in hardwoods as indicated by ultraviolet microscopy. Wood Sci Technol 9:45–58

Musha Y, Goring DAI 1975b Cell dimensions and their relationship to the chemical nature of the lignin from the wood of broad-leaved trees. Can J For Res 5:259–268

Norberg PH 1969 Electron microscope studies of the morphology of fibers subjected to different chemical pulping conditions. Svensk Papperstidn 72:575–582

Norberg PH, Meier H 1966 Physical and chemical properties of the gelatinous layer in tension wood fibres of aspen (Populus tremula L). Holzforschung 20:174–178

Obst JR 1982 Guaiacyl and syringyl lignin composition in hardwood cell components. Holzforschung 36:143–152

Obst Jr 1983 Analytical pyrolysis of hardwood and softwood lignins and its use in lignin-type determination of hardwood vessel elements. J Wood Chem Technol 3:377–397

Packman DF, Laidlaw RA 1967 Pulping of British-grown softwoods. IV. A study of juvenile, mature and top wood in a large Sitka spruce tree. Holzforschung 21:38–45

Panshin AJ, de Zeeuw CH 1980 Textbook of wood technology. 4th ed. McGraw-Hill, New York, 722 pp

Parameswaran N, Liese W 1981 Ultrastructural localization of wall components in wood cells. The Ekman-Days 1981 (Stockholm), SPCI Rep 38 Vol 1:16–31

Parameswaran N, Liese W 1982 Ultrastructural localization of wall components in wood cells. Holz Roh-Werkst 40:145–155

Parham RA 1974 Distribution of lignin in kraft pulp as determined by electron microscopy. Wood Sci 6:305–315

Parham RA, Côté WA, Jr 1971 Distribution of lignin in normal and compression wood of Pinus taeda L. Wood Sci Technol 5:49–62

Perilä O 1961 The chemical composition of carbohydrates of wood cells. J Polym Sci 51:19–26

Perilä O 1962 The chemical compositions of wood cells. III. Carbohydrates of birch cells. Suom Kemistil B-35:176–178

Perilä O, Heitto P 1959 The chemical compositions of wood cells. I. Carbohydrates of pine cells. Suom Kemistil B-32:76–80.

Perilä O, Seppä T 1960 The chemical compositions of wood cells. II. Carbohydrates of spruce cells. Suom Kemistil B-33:114–116

Phillips EWJ 1940 A comparison of forest- and plantation-grown African pencil cedar (Juniperus procera Hochst) with special reference to the occurrence of compression wood. Emp For 19:282–288

Ritter GJ 1925 Distribution of lignin in wood. Microscopical study of changes in wood structure upon subjection to standard methods of isolating cellulose and lignin. Ind Eng Chem 17:1194–1197

Ritter GJ, Fleck LC 1926 Chemistry of wood. IX. Springwood and summerwood. Ind Eng Chem 18:608–609

Roschier RH, Ahava J, Ilvonen U 1959 Investigation of the fines (short fibre fraction) removed from sulphite cellulose. Pap Puu 41:133–139

Ruch F, Hengartner H 1960 Quantitative Bestimmung der Ligninverteilung in der pflanzlichen Zellwand. Beih Z Schweiz Forstver 30:75–92

Sachs IB 1965 Evidence of lignin in the tertiary wall of certain wood cells. In: Côté WA, Jr (ed) Cellular ultrastructure of woody plants. Syracuse Univ Press, Syracuse, 335–339

Sachs IB, Clark IT, Pew JC 1963 Investigation of lignin distribution in the cell wall of certain woods. J Polym Sci C-2:203–212

Saka S, Thomas RJ 1982a Evaluation of the quantitative assay of lignin distribution by SEM-EDXA-technique. Wood Sci Technol 16:1–8

Saka S, Thomas RJ 1982b A study of lignification in loblolly pine tracheids by the SEM-EDXA technique. Wood Sci Technol 16:168–179

Saka S, Thomas RJ, Gratzl JS 1978 Lignin distribution determined by energy-dispersive analysis of x-rays. Tappi 61(1):73–76

Saka S, Thomas RJ, Gratzl JS 1979a Lignin distribution in soda-oxygen and kraft fibers as determined by conventional electron microscopy. Wood Fiber 11:99–108

Saka S, Thomas RJ, Gratzl JS 1979b Lignin distribution by energy dispersive x-ray analysis. In: Inglet GE, Falkehag SI (eds) Dietary fibers: chemistry and nutrition. Academic Press, New York, 15–29

Saka S, Thomas RJ, Gratzl JS 1981 Lignin distribution in Douglas-fir and loblolly pine as determined by energy dispersive x-ray analysis. The Ekman-Days 1981 (Stockholm), SPCI Rep 38 Vol 1:35–42

Saka S, Thomas RJ, Gratzl JS, Abson D 1982a Topochemistry of delignification in Douglas-fir wood with soda-anthraquinone and kraft pulping as determined by SEM-EDXA. Wood Sci Technol 16:139–153

Schreuder HR, Côté WA, Jr, Timell TE 1966 Studies on compression wood. III. Isolation and characterization of a galactan from compression wood of red spruce. Svensk Papperstidn 69:641–657

Schütt P, Augustin H 1961 Die Verteilung des Cellulosegehaltes im Stamm. Untersuchungen über eine Methode der züchterischen Probenahme an 30jährigen Murraykiefern. Papier 15:661–665

Scott JAN, Goring DAI 1970a Lignin concentration in the S_3 layer of softwoods. Cellul Chem Technol 4:83–93

Scott JAN, Goring DAI 1970b Photolysis of wood microsections in the ultraviolet microscope. Wood Sci Technol 4:237–239

Scott JAN, Procter AR, Fergus BJ, Goring DAI 1969 The application of ultraviolet microscopy to the distribution of lignin in wood. Description and validity of the technique. Wood Sci Technol 3:73–92

Shelbourne CJA, Ritchie KS 1968 Relationships between degree of compression wood development and specific gravity and tracheid characteristics in loblolly pine (Pinus taeda L). Holzforschung 22:185–190

Siddiqui KM 1976 Relationship between cell wall morphology and chemical composition of earlywood and latewood in two coniferous species. Pak J For 26:21–34

Simson BW, Timell TE 1967 A method for determination of the carbohydrate composition of microgram quantities of plant tissues. Tappi 50:437–477

Simson BW, Timell TE 1978 Polysaccharides in cambial tissues of Populus tremuloides and Tilia americana. I. Isolation, fractionation, and chemical composition of the cambial tissues. Cellul Chem Technol 12:39–50

Sonntag P 1904 Über die mechanischen Eigenschaften des Roth- und Weissholzes der Fichte und anderer Nadelhölzer. Jahrb Wiss Bot 39:71–105
Sorvari J, Pietarila V, Nygren-Konttinen A, Klemola A, Laine JE, Sjöström E 1983 Attempts at isolating and characterizing secondary wall and middle lamella from spruce wood (Picea abies). Pap Puu 66:117–121
Stamm AJ, Sanders HT 1966 Specific gravity of the wood substance of loblolly pine as affected by chemical composition. Tappi 49:397–400
Staudinger M 1942 Chemische Anatomie des Holzes. Holz Roh-Werkst 5:193–201
Stewart CM 1966 Excretion and heartwood formation in living trees. Science 153:1068–1074
Sultze RF Jr 1957 A study of the developing tissues of aspenwood. Tappi 40:985–994
Takano T, Fukazawa K, Ishida S 1983 Within-ring variation of lignin in Picea glehnii by UV microscopic image analysis. Res Bull Coll Exp For Hokkaido Univ 40:709–722
Talmadge KW, Keegstra K, Bauer WD, Albersheim P 1973 The structure of plant cell wall. I. The macromolecular components of the walls of suspension-cultured sycamore cells with a detailed analysis of the pectic polysaccharides. Plant Physiol 51:158–173
Teratani F, Haraguchi S, Kai Y 1976 Radial distribution of the extractives in the compression wood of Guimatsu, Larix gmelinii Gord. Mokuzai Gakkaishi 22:133–136
Thompson NS, Hankey JD 1968 The isolation of the ephithelia of conifers. Tappi 51:88–93
Thompson NS, Heller HH, Hankey JD, Smith O 1966a Investigations on the isolation and carbohydrate composition of different components of trees and wood. Pulp Pap Mag Can 67(2):T541–T551
Thompson NA, Heller HH, Hankey JD, Smith O 1966b The isolation and the carbohydrate composition of the epithelial cells of longleaf pine (Pinus palustris Mill). Tappi 49:401–405
Thornber JP, Northcote DH 1961a Changes in the chemical composition of a cambial cell during its differentiation into xylem and phloem tissue in trees. I. Main components. Biochem J 81:449–455
Thornberg JP, Northcote DH 1961b Changes in the chemical composition of a cambial cell during its differentiation into xylem and phloem tissue in trees. II. Carbohydrate constituents of each main component. Biochem J 81:455–464
Thornberg JP, Northcote DH 1962 Changes in the chemical composition of a cambial cell during its differentiation into xylem and phloem tissue in trees. 3. Xylan, glucomannan and α-cellulose fractions. Biochem J 92:340–346
Timell TE 1972 Beobachtungen an Holzstrahlen im Druckholz. Holz Roh-Werkst 30:267–273
Timell TE 1973a Studies on opposite wood in conifers. I. Chemical composition. Wood Sci Technol 7:1–5
Timell TE 1973b Studies on opposite wood in conifers. III. Distribution of lignin. Wood Sci Technol 7:163–172
Timell TE 1973c Ultrastructure of the dormant and active cambial zones and the dormant phloem associated with formation of normal and compression woods in Picea abies (L) Karst. SUNY Coll Environ Sci For, Syracuse, Tech Publ 96, 94 pp
Timell TE 1978a Helical thickenings and helical cavities in normal and compression woods of Taxus baccata. Wood Sci Technol 12:1–15
Timell TE 1978b Ultrastructure of compression wood in Ginkgo biloba. Wood Sci Technol 12:89–103
Timell TE 1980 Karl Gustav Sanio and the first scientific description of compression wood. IAWA Bull (ns) 1:147–153
Timell TE 1981 Recent progress in the chemistry, ultrastructure, and formation of compression wood. The Ekman-Days 1981 (Stockholm), SPCI Rep 38 Vol 1:99–147
Timell TE 1982 Recent progress in the chemistry and topochemistry of compression wood. Wood Sci Technol 16:83–112
Uprichard JM 1971 Cellulose and lignin content in Pinus radiata D Don. Within-tree variation in chemical composition, density and tracheid length. Holzforschung 25:97–105
Uprichard JM, Lloyd JA 1980 Influence of tree age on the chemical composition of radiata pine. NZ J For Sci 10:551–557
von Aufsess H 1973 Mikroskopische Darstellung des Verholzungsgrades durch Färbemethoden. Holz Roh-Werkst 31:24–33
Wardrop AB 1954 The fine structure of the conifer tracheid. Holzforschung 8:12–29

Wardrop AB, Bland DE 1961 Lignification in reaction wood. 140th Meet Am Chem Soc Chicago IL 1961 Abstr Pap 5E
Wardrop AB, Dadswell HE 1950 The nature of reaction wood. II. The cell wall organization of compression wood tracheids. Aust J Sci Res B-3:1−13
Wardrop AB, Dadswell HE 1952 The cell wall structure of xylem parenchyma. Aust J Sci Res B-5:223−236
Wardrop AB, Davies GW 1964 The nature of reaction wood. VIII. The structure and differentiation of compression wood. Aust J Bot 12:24−38
Waterkeyn L 1981 Cytochemical localization and function of the 3-linked glucan callose in the developing cotton fibre cell wall. Protoplasma 106:49−67
Waterkeyn L, Caeymaex S, Decamps E 1982 La callose des trachéides du bois de compression chez Pinus silvestris et Larix decidua. Bull Soc R Bot Belg 115:149−155
Watson AJ, Dadswell HE 1957 Paper making properties of compression wood from Pinus radiata. Appita 11:56−70
Watson AJ, Hodder IG 1954 Relationship between fibre structure and handsheet properties in Pinus radiata. Appita 8:290−310
Wergin W 1965 Über Entstehung und Aufbau von Reaktionsholzzellen. 4. Nachweis der Ligninverteilung in den Zellwänden des Druckholzes durch Untersuchungen im UV-Licht. Flora 156:322−331
Wergin W, Casperson G 1961 Über Entstehung und Aufbau von Reaktionsholzzellen. 2. Morphologie der Druckholzzellen von Taxus baccata L. Holzforschung 15:44−50
Westing AH 1968 Formation and function of compression wood in gymnosperms. Bot Rev 34:51−78
Whiting P, Abbot J, Yean WQ, Goring DAI 1982 Topochemical differences in the kinetics of kraft pulping of spruce wood. CPPA Trans 83(12):TR109−TR112
Whiting P, Favis BD, St-Germain FGT, Goring DAI 1981 Fractional separation of middle lamella and secondary wall tissue from spruce wood. J Wood Chem Technol 1:29−42
Whiting P, Goring DAI 1981a The topochemistry of delignification shown by pulping middle lamella and secondary wall tissue from black spruce wood. J Wood Chem Technol 1:111−122
Whiting P, Goring DAI 1981b The morphological origin of milled wood lignin. Svensk Papperstidn 84:R120−R122
Whiting P, Goring DAI 1982a Chemical characterization of tissue fractions from the middle lamella and secondary wall of black spruce tracheids. Wood Sci Technol 16:261−267
Whiting P, Goring DAI 1982b Phenolic hydroxyl analysis of lignin by pyrolytic gas chromatography. Pap Puu 65:592−594, 599
Whiting P, Goring DAI 1982c Relative reactivities of middle lamella and secondary wall lignin of black spruce wood. Holzforschung 36:303−306
Whiting P, Goring DAI 1983 The composition of the carbohydrates in the middle lamella and the secondary wall of tracheids from black spruce wood. Can J Chem 61:506−508
Wilson JW, Wellwood RW 1965 Intra-increment chemical properties of certain western Canadian coniferous species. In: Côté WA, Jr (ed) Cellulose ultrastructure of woody plants. Syracuse Univ Press, Syracuse, 551−559
Wloch W 1975 Longitudinal shrinkage of compression wood in dependence on water content and cell wall structure. Acta Soc Bot Polon 44:217−229
Wloch W, Hejnowicz Z 1983 Location of laricinan in compression-wood tracheids. Acta Soc Bot Polon 52:201−203
Wood JR, Goring DAI 1971 The distribution of lignin in stem wood and branch wood of Douglas fir. Pulp Pap Mag Can 72(3):T95−T102
Wood JR, Goring DAI 1974 Ultraviolet microscopy at wavelengths below 240 nm. J Microsc 100:105−111
Wu YT, Wilson JW 1967 Lignification within coniferous growth zones. Pulp Pap Mag Can 68(4):T159−T164
Yamamoto K 1982 (Yearly and seasonal process of maturation of ray parenchyma cells in Pinus species.) Res Bull Coll Exp For Hokkaido Univ 39:245−296
Yang JM, Goring DAI 1978 A comparison of the concentration of free phenolic hydroxyl groups in the secondary wall and middle lamella regions of softwoods. CPPA Trans 4(1):2−5

Yang JM, Goring DAI 1980 The phenolic hydroxyl content of lignin in spruce wood. Can J Chem 58:2411–2414

Yoshizawa N, Idei T, Okamoto K 1981 Structure of inclined grown Japanese black pine (Pinus thunbergii Parl). Distribution of compression wood and cell wall structure of tracheids. Bull Utsunomiya Univ For 17:89–105

Yoshizawa N, Itoh T, Shimaji K 1982 Variation in features of compression wood among gymnosperms. Bull Utsunomiya Univ For 18:45–64

Yoshizawa N, Koike S, Idei T 1985 Formation and structure of compression wood tracheids induced by repeated inclination in Taxus cuspidata. Mokuzai Gakkaishi 31:325–333

Yumoto M, Ishida S, Fukazawa K 1982 Studies on the formation and structure of the compression-wood cells induced by artificial inclination in young trees of Picea glauca. II. Transition from normal to compression wood revealed by SEM-UVN combination method. J Fac Agr Hokkaido Univ 60:312–335

Yumoto M, Ishida S, Fukazawa K 1983 Studies on the formation and structure of the compression-wood cells induced by artificial inclination in young trees of Picea glauca. IV. Gradation of severity of compression wood tracheids. Res Bull Coll Exp For Hokkaido Univ 40:409–454

Zenker R, Poller S 1968 Über die unterschiedliche Beschaffenheit von dünnem und starkem Kiefernholz. Arch Forstwes 17:501–511

Ziegler H 1968 Biologische Aspekte der Kernholzbildung. Holz Roh-Werkst 26:61–68

Zobel BJ 1975 Using the juvenile wood concept in the southern pines. South Pulp Pap Manuf 38(9):14–16

Zobel BJ 1981 Wood quality from fast-grown plantations. Tappi 64(1):71–74

Zobel BJ 1984 The changing quality of world wood supply. Wood Sci Technol 18:1–17

Zobel BJ, McElwee RL 1958 Natural variation in wood specific gravity of loblolly pine, and an analysis of contributing factors. Tappi 41:158–161

Zobel BJ, Stonecypher R, Browne C, Kellison RC 1966 Variation and inheritance of cellulose in the southern pines. Tappi 49:383–387

Zobel BJ, Webb C, Henson F 1959 Core or juvenile wood of loblolly and slashpine trees. Tappi 42:345–356

Chapter 7 Physical Properties of Compression Wood

CONTENTS

7.1	Introduction	470
7.2	Specific Gravity	470
7.2.1	Introduction	470
7.2.2	Specific Gravity and Its Variation in Normal Wood	471
7.2.3	Specific Gravity of Compression Wood	476
7.2.4	Specific Gravity of Branch Wood	488
7.2.5	Specific Gravity of Knot Wood and Associated Stem Wood	496
7.2.6	Conclusions	497
7.3	Wood–Water Relationships	498
7.3.1	Moisture Content	498
7.3.1.1	Introduction	498
7.3.1.2	Compression Wood	499
7.3.1.3	Branch Wood and Knot Wood	505
7.3.2	Permeability to Water	506
7.3.3	Shrinkage and Swelling	507
7.3.3.1	Normal Wood	507
7.3.3.2	Compression Wood	513
7.3.3.3	Branch Wood	527
7.3.3.4	Conclusions	528
7.4	Mechanical Properties	530
7.4.1	Introduction	530
7.4.2	Compression Wood	531
7.4.2.1	Early Contributions and General Review	531
7.4.2.2	Compression Parallel to the Grain	534
7.4.2.2.1	Maximum Crushing Strength	534
7.4.2.2.2	Stress at Proportional Limit	542
7.4.2.2.3	Young's Modulus of Elasticity	543
7.4.2.2.4	Creep and Failure in Longitudinal Compression	545
7.4.2.3	Tension Parallel to the Grain	549
7.4.2.3.1	Tensile Strength	549
7.4.2.3.2	Young's Modulus of Elasticity	551
7.4.2.4	Static Bending	552
7.4.2.4.1	Introduction	552
7.4.2.4.2	Modulus of Rupture	552
7.4.2.4.3	Stress at Proportional Limit	554
7.4.2.4.4	Modulus of Elasticity	555
7.4.2.4.5	Work to Maximum Load	558
7.4.2.4.6	Work to Proportional Limit	558
7.4.2.4.7	Total Work	558
7.4.2.5	Toughness	559
7.4.2.6	Shearing Strength	559
7.4.2.7	Hardness	561
7.4.2.8	Fracture	561
7.4.2.9	Strength Properties and Extent of Compression Wood Development	568
7.4.2.10	Factors Determining the Strength Properties of Compression Wood	570

7.4.3	Branch Wood	574
7.4.4	Conclusions	577
7.5	General Conclusions	578
References		579

7.1 Introduction

The chemical composition and anatomy of wood together determine its physical characteristics, including color, density, wood−water relationships, and all the different mechanical properties. Chemical and morphological factors directly affect the pulping characteristics of wood and, indirectly, by their effect on strength properties, the utilization of wood for lumber. All strength characteristics, in their turn, are influenced by the density and moisture content of the wood. For this reason, the mechanical properties of compression wood will be treated here together with density and wood−water relationships. Traditionally, the mechanical properties of compression wood have been considered separately from other physical characteristics but are, of course, physical phenomena. Color is discussed in connection with detection of compression wood in Chap. 3.3.

The most characteristic properties of fully developed compression wood are of a physical nature: its dark color, its high density, its exceptionally high longitudinal shrinkage, its hardness, its high compressive strength and elasticity, and its low tensile strength and rigidity. These are all properties that profoundly influence the behavior of compression wood when it is subjected to various physical forces. For the living tree, they on the whole make compression wood a highly valuable and often vital tissue. When wood is to be utilized to maximum advantage, the same properties render compression wood an undesirable material from almost any point of view.

The mechanical characteristics of the cell wall in wood tracheids and fibers have been the subject of many excellent contributions, especially in later years, such as those by Schniewind (1962a, c, 1981), Mark (1967, 1972, 1981), Cave (1968), Kollmann and Côté (1968), Schiewind und Barrett (1969), Mark and Gillis (1970, 1973), Barrett et al. (1972), and Dinwoodie (1975). The most recent contribution is the fine monograph by Bodig and Jayne (1982) on the mechanics of wood and wood composites. Transport processes in wood and wood−water relationships are ably dealt with in books by Siau (1971, 1984), Skaar (1972), and Zimmermann (1983). The subject of wood and moisture has also been reviewed by Tarkow (1981).

7.2 Specific Gravity

7.2.1 Introduction

Specific gravity is a rather misleading term, since it has nothing to do with gravity, and it should long ago have been replaced by another term, such as *density index* (Panshin and de Zeeuw 1980). Unfortunately, it is now probably too well established to be changed. Specific gravity is the ratio of the mass of a substance to the mass of an equal volume of water. It is expressed without units. *Density* is the ratio of a mass of a quantity of a substance to the volume of that quantity and is expressed in mass per unit substance. In this book density is expressed throughout in grams per milliliter (cm^3), and the terms specific gravity and density are considered to be equivalent as long as the metric system is used. According to the SI convention, density is expressed in kilograms per cubic meter. Like so many other units in this system (Zimmermann 1983), it has no practical advantage over the older convention. A disadvantage is that in the SI system density is 1000 times larger than specific gravity. Because wood shrinks on drying, the highest specific gravity is obtained when the volume is that of oven-dry wood. The minimum value is associated with green (water-saturated) wood. It is sometimes referred to as *basic specific gravity*.

Specific gravity is the single most important property of wood because it is the best single wood quality indicator (Mitchell 1961, Olson and Arganbright 1977, Gonzales and Kellogg 1978, Desch and Dinwoodie 1980). Not surprisingly, it has received more attention than any other wood characteristic. In pulp manufacture, the density of the wood determines the pulp yield and, partly because density is closely correlated with latewood content, many of the properties of the final paper (Zobel and McElwee 1958, van Buijtenen 1964, Kellogg and Gonzales 1976). All strength characteristics of lumber are related to the density of the wood. Despite the fact that it is not a simple, independent parameter, specific gravity is strongly inherited (Chap. 14.3.2.2). The density of wood can be determined with considerable accuracy and reasonable ease by any of several methods now available, including the classical water displacement procedures, the maximum moisture content method, and the mercury immersion (Ericson 1959, 1966) techniques. Most convenient are the β-ray and, especially, the x-ray densitometry techniques referred to in Chap. 3.4. Specific gravity and its variation in the tree have been discussed in several books concerned with wood, for example those authored by Kollmann (1951), Knuchel (1954), Trendelenburg and Mayer-Wegelin (1955), Paul (1963), Stamm (1964), Knigge and Schulz (1966), Kollmann and Côté (1968), Giordano (1971), Koch (1972), Bosshard (1974), Desch and Dinwoodie (1980), Panshin and de Zeeuw (1980), and Dinwoodie (1981). Physiological factors influencing the specific gravity of wood have been discussed by Brown (1970) and by Larson (1962, 1963, 1973). Recent research on the density and other properties of wood from the southern pines has been reviewed by Manwiller (1978).

7.2.2 Specific Gravity and Its Variation in Normal Wood

The density of wood must not be confused with the density of the wood substance itself, which is fairly constant and depends on the composition of the cell wall and on its porosity. Extractives can be present in the cell wall, for example, in *Sequoia sempervirens* (Tarkow and Krueger 1961, Kuo and Arganbright 1980a, b), thus increasing the apparent density of the latter (Keith 1969), but they can usually be readily removed and can therefore be disregarded in this context. Whether dehydrated cell walls are porous or not has been a moot question (Berlyn 1964, 1968). As a result of recent investigations, the original contention of Stamm (1946) that the dry cell wall in wood contains only a few or no voids, has been generally accepted (Stone and Scallan 1965, Stone et al. 1966, Wilfong 1966, Berlyn 1969, Petty 1971, 1981). In a study of 18 species, Kellogg and Wangaard (1969, 1970) found the void volume to vary from 1.64 to 4.76% of the dry cell wall (Wangaard 1969). Using a different technique, Weatherwax and Tarkow (1968a,b) arrived at a void volume of about 4%.

The density of the dry, nonporous cell wall can be calculated from the relative amounts of cellulose, hemicelluloses, and lignin and from the density of each constituent (Vorreiter 1955). The methods available for a direct determination of the density of dry cell wall material have long been a subject of much controversy, centering on the question which displacement liquid should be used in measuring the volume. Water has been most frequently used but suffers from the disadvantage that it is compacted in the cell wall. Other agents include helium, nonswelling organic liquids, such as hexane or benzene, or silicones, all of which afford lower density values than does water. Optical methods give lower values than those determined by displacement.

Experimental density values vary according to the displacement media used. Compaction of water is now believed to be less than was previously assumed, and the difference between values obtained with water and organic liquids has been attributed to the presence of about 4% void volume in the dehydrated cell wall, voids that are penetrated by the water but not by the nonswelling organic media (Kellogg and Wangaard 1969). Other explanations have been offered by Hermans (1946), Stamm (1964) and Ramaiah and Goring (1965). It is possible that both compression of water and the ability of the latter to penetrate voids play a role here (Mark 1967). Christensen and Hergt (1968), using various nonpolar liquids, arrived at densities of 1.434 and 1.437 g ml^{-1} for the cell wall material in two softwoods. In their study of 13 hardwoods and 5 softwoods, Kellogg and Wangaard (1969) found cell wall densities ranging from 1.433 to 1.497 g ml^{-1} when measurements were carried out pycnometrically in toluene. When water was used instead, densities varied from 1.517 to 1.529 for softwoods. The porosity of wood is determined by anatomical parameters such as cell wall thickness, size of the cells, occurrence of cavities or micropores in the cell wall, and presence of intercellular spaces. Extraneous substances contribute to the density of wood (Hartwig 1973). Heartwood for this reason is generally heavier than sapwood.

Specific gravity is strongly correlated with the proportion of latewood, a fact demonstrated in a large number of investigations, such as the recent ones by Voorhies (1972), Petrić (1974), Barnes et al. (1977), Warren (1979), and Taylor and Burton (1982). Latewood, with its thick tracheid wall and small lumen, has a density that is two to five times as high as that of the earlywood with its large, thin-walled tracheids. Ifgu and Labosky (1972) found that in *Pinus taeda* the specific gravity increased more than 100% from the beginning toward the end of each growth ring. As a result, species with a wide zone of latewood, such as the southern pines, have a heavier wood than species characterized by a

7.2.2 Specific Gravity and Its Variation in Normal Wood

relatively narrow latewood zone, such as *Pinus ponderosa*. The mean specific gravity of earlywood varies in conifers from 0.25 to 0.35 and in latewood from 0.60 to 0.90.

Basic specific gravity of common softwoods in the United States ranges from 0.29 for *Thuja occidentalis* to 0.56 for *Pinus elliottii*. Firs have a lighter wood than spruces, and larch wood is heavier than spruce wood. Among the pines, basic specific gravity varies considerably between different species. It is only 0.34 for *Pinus strobus* and 0.54−0.56 for *P. palustris* and *P. elliottii*. Each species exhibits its own, characteristic overall variation in specific gravity among a population of trees. *Pinus strobus*, like all hapoxylon pines, has a very narrow distribution, ranging from 0.31 to 0.46 (dry volume). The diploxylon pines, in contrast, have an unusually wide density distribution, extending for *Pinus sylvestris* from 0.30 to 0.86 (Trendelenburg and Meyer-Wegelin 1955, Knigge and Schulz 1966). Fir, larch, and spruce also have fairly wide distribution curves.

The variation in wood density within an individual tree is often as great or even greater than that between different trees. Branches usually contain much compression wood, and branch wood as a result is on the average heavier than normal stem wood. Root wood, in contrast, is usually lighter than stem wood.

The variation of specific gravity within the bole of a tree is of great practical importance and can occur in either the horizontal or the vertical direction. At a given height, changes occur from the pith toward the bark, a variation that is often correlated with rate of growth and age. Within a given growth ring, specific gravity varies from the base to the top of the tree. Brief accounts of the variation of density in the stem of trees are given in the textbooks and monographs referred to above. Other reviews have been presented by Trendelenburg (1934, 1935, 1937a,b, 1939a,b), Paul and Smith (1950), Goggans (1961), Mitchell (1963, 1964, 1965), Scaramuzzi (1965), Hakkila (1966), Hakkila and Uusvaara (1968), and Schalck (1967). The most comprehensive of the earlier surveys is that by Spurr and Hsiung (1954). An extensive review of the entire literature was prepared by Elliott (1970). Investigations on the density of wood from southern pines have been reported by Saucier (1972), Zobel et al. (1972), Wahlgren and Schumann (1972), and by Bendsen and Ethington (1972). The density of wood from *Pseudotsuga menziesii* has been studied by Erickson and Harrison (1974). Age, proportions of juvenile wood and top wood, and geographic and species variations were all found to affect the specific gravity. Ledig et al. (1975) noted that in *Pinus rigida* wood density and tracheid length increased from north to south, and that both properties were closely related to climatic variables.

Taras (1965) found earlywood specific gravity for *Pinus elliottii* to decrease from the pith until the 8th−12th growth ring, after which it remained constant. Latewood, on the other hand, increased rapidly in density within this period, after which it decreased gradually toward the cambium. Total wood followed the trend of the latewood. Both earlywood and latewood decreased in density until breast height and then increased toward the top. In other cases earlywood has generally been found to remain more or less constant in density from pith to cambium. The specific gravity of latewood, on the other hand, increases or decreases in this direction in the same manner as the overall specific gravity of the wood (Harris and Birt 1972).

Specific gravity in the softwoods can increase from the pith toward the bark, first decrease and then increase, or decrease continuously from the pith outward (Panshin and de Zeeuw 1980). Schniewind (1962 b, c) has shown that a gradual increase in specific gravity from pith to bark entails the best use of available stem material from a structural point of view. As a tree grows older, its wood tends to become lighter. The decrease in specific gravity from pith to bark shown by *Juniperus virginiana* may be a consequence of the greater proportion of intercellular spaces in the sapwood and in the heartwood of this species (McGinnes and Dingeldein 1969, McGinnes and Phelps 1972). The transition from juvenile wood (Chap. 6.2.1) at the center of a stem toward mature wood is usually accompanied by an increase in density.

In *Pinus contorta* grown in Alberta, Taylor et al. (1982) found the radial variation of specific gravity to depend on stem height. At breast height, it first decreased in the first few increments near the pith and then increased toward the outside. Above a height of 7 m, density decreased from pith to bark. Björklund (1982), by contrast, has reported that in *Pinus contorta* var. *latifolia* planted in Finland the radial change in density was the same at all stem heights, first decreasing and subsequently increasing from pith to bark.

Few aspects of wood science have aroused more controversy or caused more heated discussions than the question whether or not the rate of radial growth influences the density of wood. Many of the early discussions are now less relevant since they failed to take into account the occurrence of juvenile wood. The most common earlier view was that the wider the growth ring, the lower was the density of the wood. This opinion was challenged by Turnbull (1947, 1948) (Turnbull and du Plessis 1946), who maintained that no correlation existed between rate of growth and specific gravity of wood. A sometimes caustic controversy gradually ensued in the forestry literature, the contributions of Aldrich and Hudson (1955a, b, 1958, 1959) being especially mordant. These authors drew attention to the fact that Turnbull never mentions the possible occurrence of compression wood, although this wood is quite common in South African conifers and could have affected his results. In later years the position taken by Spurr and Hsiung (1954) has been widely, albeit not universally, accepted. Spurr and Hsiung pointed out that any tree will tend to produce more narrow growth rings with advancing age, and they concluded that age was probably a more decisive factor than ring width in determining specific gravity. Obviously, the position of the wood in relation to the center of the tree has to be taken into account when assessing the correlation between growth rate and specific gravity.

In later years, many investigators have come to the conclusion that there is no or, at best, only a weak correlation between rate of growth and specific gravity in conifers, and especially in pines (Chalk 1953, Zobel and Rhodes 1955, Banks and Schwegmann 1957, Larson 1957, Rendle 1958, Rendle and Phillips 1958, Zobel and McElwee 1958, Wahlgren and Fassnacht 1959, Mozina 1960, Risi and Zeller 1960, Zobel et al. 1960, Goggans 1962, Thor and Brown 1962,

7.2.2 Specific Gravity and Its Variation in Normal Wood

Thor 1964, 1965, Ralston and McGinnes 1964, Choong et al. 1970, Lubardić and Nicolić 1970, Taylor and Burton 1982). It should be pointed out, however, that density has been found to be negatively correlated with growth rate, especially in younger trees by several investigators, such as McKimmy (1959), Pearson and Fielding (1961), Wellwood and Smith (1962), Bernhart (1964), Krahmer (1966), Chiang and Kennedy (1967), Hakkila and Uusvaara (1968), Wellwood and Jurasz (1968), Kennedy and Swann (1969), Brazier (1970), and Nepveu and Birot (1979). In many other cases, such as those reported by Knigge (1961, 1962), rate of growth and age have both been found to influence specific gravity. Actually there is considerable evidence that density is negatively correlated with growth rate, provided that age is constant. Schalck (1967), who has presented a stimulating review of this complex question, concludes that there is evidence both for and against an effect of age and rate of growth on density. He has suggested that age and growth rate, independently of each other and indirectly, both may influence wood density. He himself found, for example, that two environmental factors could have the same effect on ring width but an opposite effect on the structure, and thus also on the density, of the wood.

More than 200 years ago, Duhamel du Monceau noted that wood at the base of a tree was heavier than that in the crown. Almost 100 years later, Sanio (1873) showed that the proportion of latewood decreases with height, an observation later corroborated on many occasions, for example by Turnbull (1937). In the annual increment, the latewood part accordingly tapers toward the top, whereas the earlywood tapers toward the stem base, which is what would be expected from the general physiology of tree growth. Superimposed on the earlywood-latewood variation is the variability in density with height of these two tissues themselves. In earlywood, density tends to decrease to a height of 2–5 m and then to increase. The specific gravity of latewood first increases to this height and then decreases (Knigge 1958b, Taras 1965).

A majority of the conifers, including all *Araucaria*, *Larix*, and *Pseudotsuga*, and almost all *Abies* species investigated tend to develop a lighter wood with increasing stem height, albeit more slowly within the living crown than below. This is a typical trend in two thirds of the pine species examined, for example in *Pinus caribaea* var. *hondurensis* (Cown 1981), *P. contorta* (Taylor et al. 1982), *P. densiflora* (Yasawa et al. 1951), *P. ponderosa* (Cockrell 1943, Conway and Minor 1961, Markstrom and Yerkes 1972, Okkonen et al. 1972, Markstrom et al. 1983), and *P. taeda* (Lenhart et al. 1977). In spruce, by contrast, the density of the wood first decreases until a point ½ to ¾ of the total stem length, after which it begins to increase toward the top. This type of variation has been demonstrated for *Picea abies* by Nylinder (1953) and for *P. engelmannii*, *P. glauca*, and *P. mariana* by Okkonen et al. (1972). Hakkila (1966) found that the crown of the *Picea abies* trees used in his density survey occupied 68% of the total length of the stem, whereas the corresponding figure for *Pinus sylvestris* was only 43%. The difference between the two species in proportion of crown-formed wood was probably responsible for the different variation of density with height. It should be noted, however, that firs (*Abies* spp.), which also have a long crown, resemble pines in forming a wood that becomes lighter toward the top.

Specific gravity of larch wood decreases in the clear bole, but begins to increase in the crown portion.

A thorough study of the relationship between specific gravity and tree height has been reported by Okkonen et al. (1972). In 17 of the 27 conifer species investigated, density decreased with increase in height, in five it increased, in two species the density first decreased but later increased with height, and in three species there was no apparent trend. The first case was attributed to an increase in specific gravity with age, caused by an increase in the proportion and density of the latewood. Species included in this group were *Larix occidentalis, Pinus ponderosa, P. resinosa,* the southern pines, and *Pseudotsuga menziesii.* An increase in specific gravity with increase in height was noted for species such as *Abies balsamea* and *Thuja plicata,* both species with a low proportion of latewood. This trend was attributed to an increase in the density of the earlywood. Farr (1973), by contrast, found that in *Pseudotsuga menziesii* stem specific gravity first increased to stem mid height and then decreased. Heger (1974a) observed that in *Abies balsamea* and *Picea mariana* minimum density occurred at mid stem but at 70% total height in *Pinus contorta.* In a subsequent investigation, Heger (1974b) found that density reached a minimum at 30 – 40% of total height in *Abies balsamea* trees from another location. The minimum value was not related to the base of the live crown.

In summary, the pattern of density variation in the axial direction of the stem obviously varies considerably. Even within the same species, different trends are observed for forest and plantation-grown trees, indicating that genetic and environmental factors influence the variation in specific gravity. Krahmer (1966) has reported that the density of the wood in *Tsuga heterophylla* increases toward the top in vigorous trees but decreases in slow-growing ones. With *Pinus resinosa,* Baker (1967) observed a decrease in specific gravity from the base toward the top in forest trees, while the trend was opposite in plantation trees. Kärkkäinen (1984) has recently reported that in *Picea abies* growing in Finland smaller trees had a lower density than taller specimens.

The effect of lean and eccentric growth within areas free of compression wood has attracted little attention until most recently. Pawsey and Brown (1970) found that in slightly leaning stems of *Pinus radiata,* sampling only on the side opposite to that of the inclination resulted in an underestimation of the mean density of an entire diametric sample by 2 – 3%. Tracheid length, on the other hand, did not differ between sampling directions.

7.2.3 Specific Gravity of Compression Wood

Compression wood is almost always heavier than corresponding normal wood. Accidental inclusion of compression wood among normal wood will therefore result in spuriously high density values for the latter. In studies on wood properties, such as specific gravity or tracheid length (Chap. 4.3.2), the possible presence of compression wood has to be taken into account (Hale and Prince 1936, 1940, Hale et al. 1961, Haasemann 1967). It is best, of course, to avoid any inclusion of compression wood, an approach used by many investigators, such

7.2.3 Specific Gravity of Compression Wood

as Greenhill and Dadswell (1940), Klem et al. (1945), Pillow (1949), Kennedy and Wilson (1954), Wellwood (1960), Einspahr et al. (1964), Nicholls and Fielding (1965), Hakkila (1966), Nicholls (1967), Kennedy et al. (1968), Pawsey and Brown (1970), Zobel et al. (1972), Olesen (1973), Ward and Gardiner (1976), Barnes et al. (1977), Gonzalez and Kellogg (1978), Seth (1979), and Talbert and Jett (1981). In a recent progeny trial in *Pinus radiata* grown in New Zealand, Bannister and Vine (1981) collected specimens for density determinations so as to minimize the incidence of compression wood, taking increment borings on the upper side of any leaning stem. The results obtained nevertheless suggested that some compression wood had been included in the samples.

Nördlinger (1878) seems to have been the first to determine the specific gravity of what must have been compression wood. A leaning *Larix decidua* tree was found to have an eccentric cross section, with the widest part of the growth rings on the lower side. The specific gravity of this wood (compression wood) was 0.710, whereas that formed on the opposite, upper side had a density of 0.565. Similar results were obtained, but not reported, with fir, pine, and spruce trees (Chap. 2.2).

In 1896, Cieslar (1896) and Hartig (1896) both reported density data for compression wood, apparently the first time such information had been gathered in a systematic way. In the horizontally growing part of a stem of *Picea abies*, Cieslar in one case found compression wood formed on the lower side to have the extremely high specific gravity (dry volume) of 0.83, a value 84% higher than that for wood on the upper side, namely 0.45. Hartig determined the specific gravity of wood formed on the lower and upper sides in different positions along several branches of *Picea abies* (Table 7.6). Commenting on these results 5 years later, Hartig (1901) pointed out that the highest density he had observed for any compression wood, namely 0.87, corresponded to the density of the heaviest oakwood. Schwappach (1897, 1898) determined the specific gravity of opposite and compression woods from stems of *Pinus sylvestris*. His values defy any rational interpretation, and it is probable, as remarked by Rothe (1930), that his specimens were far from representative of either type of wood.

The specific gravity of compression wood from conifer stems has since been determined by many investigators, albeit often only casually and in connection with other studies. More systematic determinations were made by Verrall (1928), Trendelenburg (1931, 1932), Pillow and Luxford (1937), Onaka (1949), Wardrop (1951), Kaburagi (1952), Harris (1977), Seth und Jain (1978), Seth (1979), Jain and Seth (1979, 1980), and Timell (1981, 1982). Comparisons have usually been made with normal wood or with opposite wood. The latter wood, as is discussed in Chap. 21, is not identical with normal wood in either anatomy or physical properties, a fact that should be kept in mind when considering these values. A summary of some of the available data is presented in Table 7.1. When studying wood from branches or knots, investigators have generally not used pure tissues, but been satisfied with the lower part of the branch or even with an entire branch. In these cases, comparison has usually been with opposite wood, as the amount of normal wood in a conifer branch tends to be small.

Table 7.1. Specific gravity of compression and normal woods with the volume measured in green, air-dry, or oven-dry conditions

Species	Condition	Specific gravity			References
		Compression wood	Normal wood	Ratio	
Abies alba					
Earlywood		0.36 – 0.40	0.22 – 0.28		Constantinescu 1955
Latewood		0.65 – 0.70	0.67 – 0.75		
	Air-dry	0.583	0.442		Giordano 1971
Abies concolor	Green	0.470	0.346	1.36	Pillow and Luxford 1937
	Air-dry	0.509	0.375	1.36	
Abies procera	Green	0.388	0.355	1.09	Paul et al. 1959
	Oven-dry	0.435	0.405	1.07	
Abies sachalinensis	Air-dry	0.639	0.384	1.66	Ueda et al. 1972
	Air-dry	0.566	0.396	1.43	Ueda 1973
Chamaecyparis obtusa	Air-dry	0.79	0.53	1.49	Onaka 1949
Ginkgo biloba	Air-dry	0.489	0.450	1.09	Ueda 1973
Larix decidua	Air-dry	0.710	0.565	1.26	Nördlinger 1878
	Air-dry	0.681	0.381	1.79	Giordano 1971
Larix laricina	Oven-dry	0.778	0.617	1.26	Verrall 1928
Picea abies	Oven-dry	0.788	0.562	1.40	Cieslar 1896
		0.669	0.452	1.48	
	Green	0.522	0.459	1.14	Hartig 1901
	Air-dry	0.487	0.469	1.04	Janka 1909
	Oven-dry	0.450	0.439	1.03	
	Green	0.53	0.36	1.47	Rothe 1930
	Green	0.707 – 0.733	0.359 – 0.389		Rak 1957
	Oven-dry	0.766 – 0.795	0.405 – 0.439		
	Oven-dry	0.81	0.47	1.72	Ollinmaa 1959
	Air-dry	0.527	0.446	1.18	Giordano 1971
Picea glauca	Green	0.387	0.316	1.22	Perem 1958
	Air-dry	0.392	0.332	1.18	
Picea mariana	Oven-dry	0.749	0.407	1.84	Verrall 1928
	Air-dry	0.528	0.396	1.33	Ladell et al. 1970
Picea sitchensis		0.516	0.316	1.63	Young et al. 1970
		0.86	0.46	1.87	Rickey et al. 1974
Pinus densiflora	Green	0.57	0.47	1.21	Onaka 1949
	Air-dry	0.84	0.64	1.31	
	Oven-dry	0.62	0.54	1.15	
Pinus palustris	Oven-dry	0.788	0.513	1.54	Verrall 1928
	Oven-dry	0.757	0.549	1.38	Chu 1972
Pinus ponderosa	Green	0.467	0.354	1.32	Pillow and Luxford 1937
	Air-dry	0.499	0.372	1.34	

7.2.3 Specific Gravity of Compression Wood

Table 7.1 (continued)

Species	Condition	Specific gravity			References
		Compression wood	Normal wood	Ratio	
Pinus radiata		0.470	0.401	1.17	du Toit 1963
		0.73	0.46	1.59	Kibblewhite 1973
	Air-dry	0.358	0.357	1.00	Harris 1977
		0.361	0.356	1.01	
		0.598	0.445	1.34	
		0.633	0.409	1.55	
		0.697	0.485	1.44	Nicholls 1982
Pinus resinosa	Green	0.415	0.376	1.10	Perem 1958
	Air-dry	0.448	0.400	1.12	
Pinus sylvestris	Air-dry	0.477	0.436	1.10	Schwappach 1897, 1898
Pinus taeda	Green	0.584	0.519	1.13	Pillow and Luxford 1937
	Air dry	0.619	0.586	1.06	
Pinus sp. (Southern pine)	Oven-dry	0.66	0.57	1.16	Heck 1919, Markwardt and Wilson 1935
Pseudotsuga menziesii	Green	0.451	0.439	1.03	Trendelenburg 1931, 1932
	Green	0.513	0.428	1.20	Pillow and Luxford 1937
	Air-dry	0.527	0.459	1.15	
	Green	0.494	0.354	1.40	
	Air-dry	0.527	0.400	1.32	
Sequoia sempervirens	Green	0.506	0.380	1.33	Markwardt and Wilson 1935, Pillow and Luxford 1937
	Air-dry	0.510	0.380	1.34	
Sequoiadendron giganteum	Green	0.55	0.32	1.72	Cockrell and Knudson 1973, Cockrell 1974
		0.53	0.38	1.39	
Thuja occidentalis	Oven-dry	0.560	0.251	2.23	Verrall 1928
Thuja plicata	Oven-dry	0.672	0.431	1.56	

The specific gravity of the cell wall in compression wood tracheids can be computed from the density of each cell wall constituent and the relative composition of the wall. Vorreiter (1955) has reported such calculations for compression wood, branch wood, and normal wood of *Picea abies*, assuming the cellulose, hemicellulose, and lignin to have densities of 1.58, 1.50, and 1.40, respectively. If extractive-free compression wood tracheids consist of about 29% cellulose, 31% hemicelluloses, and 40% lignin, its density should be 1.482. For normal spruce wood with 44% cellulose, 26% hemicelluloses, and 30% lignin, the value is 1.505.

According to more recent investigations by Stamm and Sanders (1966) and by Stamm (1969), the density of softwood lignin varies from 1.33 to 1.35. The density of fully crystalline cellulose is 1.59–1.60, but since wood cellulose is only 60–70% crystalline, its density is probably closer to 1.56. The hemicelluloses, which are amorphous in their native state in wood, probably have a density of about 1.50. Normal wood of *Picea abies* contains, on the average, 42% cellulose, 28% hemicelluloses, and 30% lignin. Corresponding figures for fully developed compression wood are 30%, 30%, and 40% (Table 5.43). Using these figures, the specific gravity of the wall material in normal spruce wood tracheids should be 1.480 and in compression wood tracheids 1.457.

Beall (1972) determined the density of individual wood hemicelluloses, obtaining values of 1.52–1.53 for a 4-*O*-methylglucuronoxylan, an arabinogalactan, and a compression wood galactan and a value of 1.80 for an arabino-4-*O*-methylglucuronoxylan. Unfortunately, the degree of crystallinity of the polysaccharides was unknown. All wood hemicelluloses are amorphous in their native state but, once isolated, often in a modified condition, they have a tendency to crystallize to an extent that depends on how they were obtained. In comparing calculated and experimentally determined values for wood from various species, Beall (1972) nevertheless noted a good agreement. For three conifers the density values were all within the range of 1.48–1.51.

The density of the wall material in normal and compression woods from *Abies balsamea* was determined by Côté et al. (1968). Measurements were carried out in water with thin, radial sections of wood according to the flotation technique of Wilfong (1966). Normal and compression woods were found to have a cell wall density of 1.474 and 1.427, respectively. Tissues from incompletely lignified S_1 and S_2 layers also had a lower density in compression wood than in normal wood. Paulson (1971) determined the density of prehydrolysis-kraft pulps from *Pinus elliottii*. Fibers from compression wood had a density varying from 1.52 to 1.54, compared to a narrow range around 1.54 for normal wood. He attributed the lower density of the former fibers to their higher lignin content. The statement by Neel (1967) that the cells in compression wood have a higher density than those in normal wood because of their higher lignin content is incorrect.

It is clear that the cell wall substance is lighter in compression wood than in normal wood, and obviously because compression wood contains less cellulose and more lignin than does normal wood, and probably also because compression wood cellulose is less crystalline than that in normal wood (Chap. 5.3.2). The difference in cell wall density is significant and cannot be neglected when attempts are made to calculate the absolute amount of material within a given volume of wood. This was first emphasized by Mayer-Wegelin (1931) in a critical review of the investigations reported by Rothe (1930), and was later reiterated by Trendelenburg (1932) and Onaka (1949).

Taking as an example *Picea abies*, the maximum specific gravity (oven-dry volume) reported for compression wood of this species is on the average 0.80 (Cieslar 1896, Rak 1957, Ollinmmaa 1959). For normal wood, the average value is 0.43 (Trendelenburg and Mayer-Wegelin 1955, Knigge and Schulz 1966). Oven-dry compression wood accordingly contains 0.80/1.457 or 55.0% of solid material per unit volume. The corresponding value for normal spruce wood is 0.43/1.480 or 29.1%. Rak (1957) has reported specific gravities of 0.73 and 0.39 for compression and normal woods, respectively, of *Picea abies*, when measured

7.2.3 Specific Gravity of Compression Wood

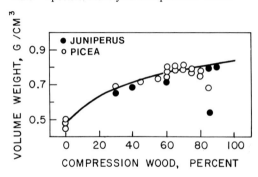

Fig. 7.1. Variation of volume weight (dry) with the proportion of compression wood in *Juniperus communis* and *Picea abies*. (Redrawn from Ollinmaa 1959)

with green volume. In their natural state, the two tissues accordingly contain 50.1% and 26.4% material per unit volume. The maximum values computed by Rothe (1930) for compression and normal woods of the same species are 43.6% and 27.6%, respectively. Fully developed compression wood of *Picea abies* evidently contains almost twice as much cell wall material per unit volume as corresponding normal wood.

The maximum ratio between the specific gravities of dry compression wood and normal wood is probably 1.9–2.0. With one exception (Verrall 1928), the ratios listed in Table 7.1 are all considerably lower. In some cases compression wood appears to be no heavier than normal wood. It is impossible to decide from the values in this table whether or not certain species form heavier compression wood than others. Specific gravity values of 0.7–0.8 have been reported for species such as *Larix laricina* (Verrall 1928), *Picea abies* (Cieslar 1896, Rak 1957, Ollinmaa 1959), *Pinus sylvestris* (Nördlinger 1878), *P. densiflora* and *P. palustris* (Onaka 1949), and *Thuja plicata* (Verrall 1928). The reason for the large differences between the reported density values is, of course, that compression wood can occur in different forms, ranging from nearly normal via mild and moderate to severe types with a gradual increase in specific gravity in this direction. How density increases with increasing percentage of compression wood is shown in Fig. 7.1 for *Juniperus communis* and *Picea abies* (Ollinmaa 1959).

According to Pillow and Luxford (1937), compression wood is 40% heavier than normal wood in species with relatively light wood, such as *Pinus ponderosa*, while in species with heavy wood such as *P. taeda*, the difference is only 15%. For *Abies balsamea* and *Picea mariana* Hale and Prince (1940) found compression wood to have an average density that was 50% higher than that of normal wood at the same height. According to Vanin (1949), compression wood is on the average 43% heavier than normal wood. Dadswell and Wardrop (1949) believed that extreme compression wood is only 33% heavier than corresponding normal wood. Even lesser differences between the density of normal and compression wood have been reported by Paul et al. (1959). In *Pinus pinaster* trees exposed to westerly winds, Polge and Illy (1967) found the mild compression wood formed on the eastern, leeward side to be only 7% heavier than that on the windward side. Values reported by Keith (1974) indicate practically the same density for normal and compression woods in *Picea glauca*.

Table 7.2. Specific gravity of compression wood and adjacent normal wood in nine conifer species. Dry weight and green volume. (Timell 1981, 1982)

Species	Specific gravity		Ratio
	Compression wood	Normal wood	
Abies balsamea	0.63	0.29	2.17
Picea abies, P. glauca, P. mariana, P. rubens	0.64	0.36	1.78
Pinus sylvestris	0.61	0.32	1.91
Pseudotsuga menziesii	0.61	0.36	1.69
Thuja occidentalis	0.59	0.28	2.11
Tsuga canadensis	0.58	0.35	1.66
Average	0.61	0.33	1.85

Timell (1981, 1982) determined the specific gravity of normal and compression woods from nine conifer species. Normal side wood and compression wood were obtained from the same disc and from the same growth rings. The compression woods were of the pronounced type, and the few rows of angular, thin-walled tracheids formed at the beginning of the growing season were excluded. The results obtained are summarized in Table 7.2. The average specific gravity of the compression woods was 0.61, a value 85% higher than the average density of the normal woods, 0.33.

If specific gravity determinations of fully developed, severe compression wood are going to have much relevance, sampling must be carried out with care. In most conifer species, severe compression wood occupies the entire growth ring with the exception of the first few tangential cell rows, and this is the type of wood that should be selected. Alternatively, intra-increment sampling should be used, and the first-formed cell rows should be discarded.

Separate determinations of the specific gravity of earlywood and latewood from compression wood have also been reported. Pillow and Luxford (1937) determined the specific gravity of composite wood, earlywood and latewood in normal and compression woods of three conifer species, obtaining the values shown in Table 7.3. Compression wood was 60–110% heavier in the latewood than in the earlywood zone. The difference for normal wood, however, was 150–220% and, as a result, the ratio between the specific gravities of compression and normal woods was much larger in earlywood than in latewood and especially in *Sequoia sempervirens*. Similar data, shown in Table 7.4, extending over eight growth rings were obtained by Wardrop (1951) for *Pinus radiata*. In this case, however, the difference in density between compression and normal woods was much less than in the species investigated by Pillow and Luxford (1937). It is possible that only moderate compression wood had developed in some of the growth rings examined by Wardrop (1951).

7.2.3 Specific Gravity of Compression Wood

Table 7.3. Specific gravity of compression and normal woods in several annual rings and within earlywood and latewood zones in the same tree stem. (Courtesy of US Forest Service). (Pillow and Luxford 1937)

Species	Several annual rings			Earlywood			Latewood		
	Compression wood	Normal wood	Ratio	Compression wood	Normal wood	Ratio	Compression wood	Normal wood	Ratio
Pinus taeda	0.60	0.52	1.15	0.41	0.32	1.28	0.65	0.85	0.76
Pseudotsuga menziesii	0.59	0.46	1.28	0.35	0.29	1.21	0.73	0.82	0.89
Sequoia sempervirens	0.51	0.38	1.34	0.43	0.21	2.05	0.70	0.67	1.04

Table 7.4. Tracheid length in the latewood and basic specific gravity in the earlywood and latewood zones of compression and normal woods from nine growth rings of *Pinus radiata*. (Wardrop 1951)

Growth ring, No.	Tracheid length in latewood, mm		Basic specific gravity			
			Earlywood		Latewood	
	Compression wood	Normal wood	Compression wood	Normal wood	Compression wood	Normal wood
4	2.18	2.30	0.437	0.406	0.531	0.452
5	2.49	2.67	0.475	0.358	0.522	0.505
6	2.48	3.02	0.374	0.352	0.568	0.545
7	2.61	3.07	0.476		0.568	0.515
8	2.90	3.32	0.458		0.617	0.545
9	3.10	3.36	0.477		0.694	0.630
10	3.24	3.63	0.588		0.633	0.660
11	3.41	3.58	0.457	0.406	0.606	0.600
12	3.32	3.57	0.550	0.431	0.712	0.550

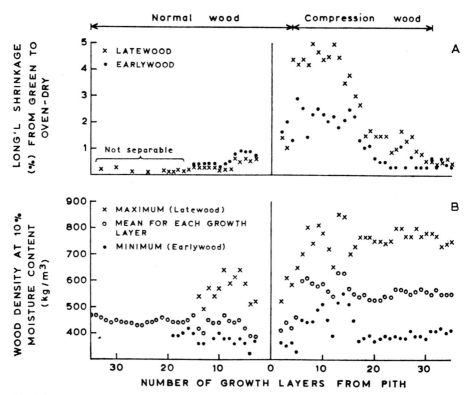

Fig. 7.2. Longitudinal shrinkage (**A**) and density (**B**) of normal and compression woods in the earlywood and latewood zones in a disc with 35 growth rings from a *Pinus radiata* stem. (Harris 1977)

7.2.3 Specific Gravity of Compression Wood

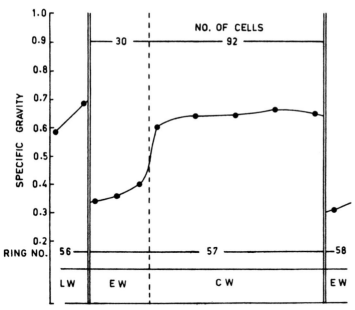

Fig. 7.3. Variation of specific gravity across a growth ring in stem wood of *Pinus wallichiana*. *Double vertical lines* indicate ring boundary. *LW* normal latewood; *EW* normal earlywood; *CW* compression wood. (Jain and Seth 1979)

Harris (1977) determined the specific gravity of the earlywood and latewood portions of opposite wood, considered by the investigator as normal, and compression wood in a leaning *Pinus radiata* stem. Some of the results are shown graphically in Fig. 7.2. In the compression wood region, the density of the latewood was very high, exceeding that of the normal latewood. Earlywood, by contrast, had a density that was often no higher than that of the normal earlywood. There was little variation in density of the compression wood with distance from the pith.

Seth and Jain (1978) and Jain and Seth (1979) investigated the variation in specific gravity in normal and compression woods of *Pinus wallichiana*. In one growth ring, composed of 76% compression wood, the first-formed earlywood had a density of only 0.324 and the last-formed compression wood 0.725. An example of the density variation across such an increment is shown in Fig. 7.3. Within the compression wood zone, the density remained constant at a level similar to that of the normal latewood.

Surprisingly, the relationship between specific gravity and extent of compression wood development has been a subject of some controversy. Shelbourne and Ritchie (1968) were the first to study this, using juvenile wood of *Pinus taeda*. They could not find any difference in specific gravity between normal wood and compression wood growth rings, even when only earlywood was examined, and in latewood the density actually *decreased* from normal through mild to severe compression wood. Their results are presented graphically in Fig. 7.4. Earlywood, in this investigation, was rigorously defined, but latewood

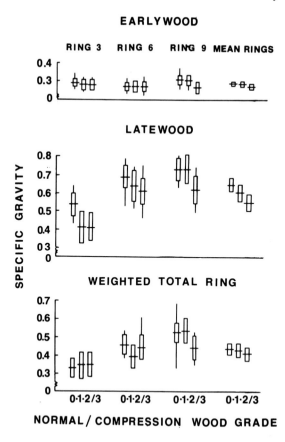

Fig. 7.4. Variation of specific gravity with extent of compression wood development in three growth rings of a *Pinus taeda* stem. *0* normal wood; *1* mild compression wood; *2/3* moderate and severe compression woods. *Horizontal lines* indicate mean, *boxes* signify confidence belt, and *vertical lines* indicate range. Cf. Fig. 4.31. (Modified from Shelbourne and Ritchie 1968)

in the compression wood zone could not be strictly delineated. It is probable, as mentioned by the investigators, that the low specific gravity of the total growth rings in compression wood was caused by the presence of increments no wider than those in normal wood and containing an unusually low proportion of latewood. This, however, does not explain the exceptionally low density of the earlywood portion of the compression wood. It would actually appear as if this earlywood were closer in properties to normal wood than to typical compression wood, a suspicion that is strengthened by the fact that the compression wood tracheids in this zone contained few helical cavities and were only slightly shorter than normal earlywood tracheids. This explanation probably also applies to the results reported by Mackney and Mathieson (1948) and by Keith (1974), who found normal and compression woods to have the same specific gravity. Seth and Jain (1978) have pointed out that Shelbourne and Ritchie characterized their compression woods only on the basis of relative opacity and not by microscopic observations, adding that no allowance seems to have been made in this investigation for the influence of age, height, or tree to tree variations on wood density.

7.2.3 Specific Gravity of Compression Wood

Howland and Paterson (1971) observed a weak, negative correlation between specific gravity and content of compression wood in seven tropical pine species growing in Malawi, for example *Pinus kesiya* (Chaps. 10.2.6.3 and 14.3.3). The extent of stress defects and specific gravity were also poorly correlated. Boone and Chudnoff (1972) likewise noted a poor correlation between density and proportion of compression wood in *Pinus caribeaea*.

Unlike previous investigators, Seth and Jain (1978) could establish a linear, positive correlation between specific gravity and proportion of compression wood. A statistical analysis gave the relationship

$$Y = 0.3123 + 0.0030 \, X$$

where Y is the average specific gravity and X is the percentage of compression wood. Obviously, normal wood of this species (*Pinus wallichiana*) had a density of 0.312 and pure compression wood 0.612, values almost the same as those obtained by Timell (1981, 1982) with *Pinus sylvestris*. The compression wood was accordingly almost twice as heavy as corresponding normal wood. The presence of compression wood in a growth ring had no influence on the first-formed earlywood in the next ring, as can be seen from Fig. 7.3. The density of the wood in this region was also unaffected by stem inclination (Jain and Seth 1980). The investigators recommended that this part of each increment be used when collecting samples for specific gravity determinations of wood in both vertical and leaning trees (Jain and Seth 1979, 1980). All evidence indicates that there should exist a direct, positive relationship between average specific gravity of wood and the proportion of that wood which consists of compression wood, as found by Seth and Jain (1978). Earlier investigators (Shelbourne and Ritchie 1968, Howland and Paterson 1971) might not have assessed the proportion of compression wood with sufficient accuracy.

The reason why compression wood, notwithstanding its lighter cell-wall substance, always is much heavier than normal wood, is, of course, that its tracheids differ from those in normal softwood. Compared to earlywood tracheids, those in compression wood have a much thicker cell wall (Chap. 4.3.4). They are, as pointed out by Rak (1957) also shorter (Chap. 4.3.2), which means greater wood substance per unit length. Normal latewood tracheids, on the other hand, often have a smaller lumen than those in compression wood, and as a result compression wood sometimes exhibits a lower density than normal latewood (Koehler 1933, Pillow et al. 1936, 1959, Pillow and Luxford 1937, Shelbourne and Ritchie 1968), a fact to which attention was first drawn by Jaccard and Frey (1928a). Two examples of this are shown in Table 7.3. The fact that short, thick-walled tracheids occupy almost the entire growth ring in severe compression wood is evidently decisive enough to outweigh factors that would otherwise render compression wood less heavy than normal wood, namely the lighter cell wall substance, the intercellular spaces, and the helical cavities in the S_2 layer. In contrast to normal wood, no negative correlation has ever been observed between specific gravity and rate of growth of compression wood (Perem 1960). On the contrary, vigorous, rapid growth is generally associated with formation of severe, and accordingly dense, compression wood.

Table 7.5. Variation of specific gravity with height above the ground of compression and normal woods of *Pinus palustris*. (Chu 1972)

Verticl position	Specific gravity	
	Compression wood	Normal wood
Base	0.744	0.615
Middle	0.752	0.538
Top	0.771	0.499

In a study of the variation in density in wood-plastics composites, Kandeel (1979) impregnated normal, compression, and opposite woods of *Pinus halepensis* with polystyrene. The density variation between the growth rings was least in the composite material made from compression wood. In this composite, the change in density was greater in the earlywood than in the latewood, while the opposite was observed with normal wood. Impregnation with polystyrene resulted in a higher maximum crushing strength with normal wood than with compression or opposite woods.

The distribution of the specific gravity of compression wood in radial and axial directions of a stem has rarely been studied. Shelbourne and Ritchie (1968), in the investigation referred to above, noted the same trends in specific gravity variation at sampling heights of 6, 20, and 40 feet. Kaburagi (1952) determined the distribution of compression wood in the stem of a 15.5 m high *Abies mayriana* tree, which leaned in one direction over the lower two thirds of the bole and then leaned slightly in the opposite direction over a short distance (overcorrection), finally achieving vertical growth in the top portion. For comparison, the density distribution was also established for the same tree, as shown in Fig. 10.57. The results indicate a fair correlation between the density distribution and the distribution of the compression wood. An even closer agreement would probably have been realized if a distinction had been made between compression woods of different severity.

Chu (1972) found that in *Pinus palustris* compression wood had a higher specific gravity than normal wood. As shown in Table 7.5, the density of the normal wood decreased with increasing height above the ground, as is usually the case, whereas that of compression wood increased slightly. The reason for this was probably that the trees had been exposed to a hurricane a few years earlier. The wind had affected the upper portion of the stem more than the lower, causing formation of a more severe compression wood in the upper part of the stem (Chap. 15.2.1.4).

7.2.4 Specific Gravity of Branch Wood

Branch wood and knot wood are heavier, while root wood is generally lighter than stem wood. Branch wood always has a high density because of its high content of compression wood. Unlike stem wood and branch wood, knot wood has a resin content that can reach values of 30–40% by weight, a fact that

7.2.4 Specific Gravity of Branch Wood

further contributes to the often exceedingly high density of this type of wood (Chap. 10.6.3).

Specific gravity values for whole branches represent the average of the densities of compression wood, opposite wood, and normal side wood in the branch. If only the lower part of a branch is considered, the proportion of compression wood will be high, but some normal wood will usually be included. Unless the resin is first removed by extraction from knot wood, density data for knots have little significance. The wood surrounding the knot in the stem frequently also has a higher density than normal stem wood (Chap. 10.6.4).

Nördlinger (1878) appears to have been the first to determine the specific gravity of branch wood. The density of the upper portion of a branch of *Pinus sylvestris* was 0.715 and that of the lower 0.867, unusually high values for both normal and compression woods of this species. The average density of the entire branch wood was considerably higher than that of the stem wood.

Hartig (1896) determined the specific gravity of the upper and lower portions of several conifer branches. Of the large material amassed, Table 7.6 shows data obtained for a heavy branch, located 8 m above the ground in a 100-year-old *Picea abies* tree. The wood on the lower side had a consistently higher specific gravity than that on the upper, obviously a consequence of its high content of compression wood. Other density data obtained by Hartig (1896) for branches are shown in Table 7.7 together with values reported by later investigators. Density and other physical properties of stem wood and various parts of branches in *Pinus sylvestris* as determined by Götze et al. (1972) are summarized in Table 7.8.

The data in Table 7.6, as well as other results obtained by Hartig (1896) for *Picea abies* and *Pinus sylvestris*, indicate that the specific gravity of both lower and upper branch wood decreases toward the branch tip. This observation has since been made by several other investigators with these species (Encev 1962, Fellegi et al. 1962, Boutelje 1965, 1966, Götze 1969, Hakkila 1969, 1971, Eskilsson 1972, Götze et al. 1972, Atmer and Thörnqvist 1982) and also with *Abies alba* (Michels 1941), *Cryptomeria japonica* (Ando and Kataoka 1961), and *Pinus taeda* (Taylor 1979). For *Abies alba, Picea abies,* and *Pinus sylvestris* Encev (1962) found densities of 0.922, 0.813, and 0.613, respectively, in branches close to the stem and 0.630, 0.611, and 0.481 at a distance of 1 m from the bole.

Yasawa et al. (1951) determined the specific gravity, moisture content, and volumetric shrinkage of stem wood and branch wood in a 63-year-old *Pinus densiflora* tree. Their results with sapwood, summarized in Table 7.9, show a higher density, a lower moisture content, and a lesser volumetric shrinkage for the branch wood. In a study of branch wood of the same species, Watanabe et al. (1962) found the under side of the branches to contain compression wood and have a higher specific gravity than the upper side.

Ando and Kataoka (1961) investigated the distribution of density in branches of *Cryptomeria japonica*. No clear trend can be discerned in their summarizing figure. Some of the branches had a higher density at the base, some at the tip, and a few in the middle portion. The average specific gravity for all branches in two trees was 0.63 at the branch base, 0.56 at the middle portion, and 0.54 at the top.

Table 7.6. Growth characteristics, specific gravity, moisture content, and volumetric shrinkage of lower and upper portions of a *Picea abies* branch. (Hartig 1896)

Distance from stem, m	Number of growth rings	Vertical diameter, cm	Horizontal diameter, cm	Amount of wood, g		Specific gravity		Moisture, per cent		Volumetric shrinkage, per cent	
				Lower	Upper	Lower	Upper	Lower	Upper	Lower	Upper
0 – 0.2	66	7.1	6.5	77.4	68.6	0.853	0.769	27.0	30.0	9.0	11.0
1.0 – 1.2	63	6.2	5.7	63.2	56.4	0.686	0.642	34.7	38.9	7.9	12.1
2.0 – 2.2	54	4.6	4.2	70.5	62.4	0.806	0.677	35.9	42.4	7.9[a]	12.6[a]
3.0 – 3.2	44	3.6	3.1	65.4	54.3	0.731	0.622	42.1	47.9	10.4	12.7

[a] In the original article these two figures are interchanged. This is probably a misprint

7.2.4 Specific Gravity of Branch Wood

Table 7.7. Specific gravity (dry weight and volume) of lower and upper portions of branches

Species	Specific gravity			References
	Lower	Upper	Ratio	
Abies alba	0.693	0.497	1.39	Michels 1941
Picea abies	0.769	0.677	1.14	Hartig 1896
	0.633	0.519	1.22	
	0.630	0.524	1.20	
	1.00[a]	1.01[a]	0.99	Boutelje 1966
	1.08[a]	1.07[a]	1.01	
	0.8	0.5	1.6	Eskilsson 1972
Pinus densiflora	0.67	0.56	1.20	Watanabe et al. 1962
Pinus sylvestris	0.862	0.715	1.21	Nördlinger 1878
	0.552	0.436	1.27	Hartig 1896
	0.957	0.870	1.10	Götze et al. 1972
	0.99[a]	0.98[a]	1.01	Boutelje 1966
	0.96[a]	0.89[a]	1.08	
Pinus taeda	0.52	0.37	1.41	Taylor 1979

[a] Refers to knot with full resin content

Table 7.8. Physical properties of stem and branch woods of *Pinus sylvestris*. (Götze et al. 1972)

Property	Stem wood	Branch wood		
		Normal wood	Compression wood	Total wood
Density (dry volume)	0.49	0.421	0.561	0.466
Moisture content, per cent	32[a], 142[b]	72.9	64.0	
Volumetric swelling, per cent	14.8	12.7	10.7	11.2
Compression strength (green wood), kg cm^{-2}		144	278	
Compression strength (dry wood), kg cm^{-2}	550	580	802	635
Bending strength (green wood), kg cm^{-2}	486	437	578	461
Impact strength (dry wood), cm kg cm^{-2}	50	40	33	38

[a] Heartwood
[b] Sapwood

Table 7.9. Some physical properties of stem and branch woods (sapwood) of *Pinus densiflora*. (Yasawa et al. 1951)

Property	Stem wood	Branch wood
Specific gravity		
Green condition	0.933	0.984
Oven-dry weight, green volume	0.384	0.487
Oven-dry weight, oven-dry volume	0.437	0.534
Moisture content (green), per cent	145	98
Volumetric shrinkage, per cent	13.8	9.8

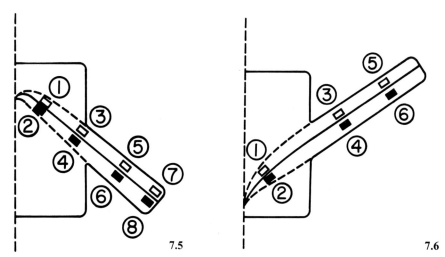

Fig. 7.5. Sample locations in knot wood and branch wood of *Picea abies* (Table 7.10). (Redrawn from Boutelje 1966)

Fig. 7.6. Sample locations in knot wood and branch wood of *Pinus sylvestris* (Table 7.11). (Redrawn from Boutelje 1966)

Table 7.10. Resin and moisture contents, radial and tangential shrinkage, and specific gravity (dry volume) of branch and knot woods of *Picea abies*. All values except those for specific gravity in per cent. Locations are shown in Fig. 7.5. (Boutelje 1966)

Location	Resin	Moisture	Radial shrinkage	Tangential shrinkage	Specific gravity
1	5.8	38	9.4	7.8	0.85
2	2.7	28	2.0	2.6	1.01
3	6.4	78	7.9	8.1	0.80
4	2.1	45	1.2	2.9	1.03
5	2.5	104	7.7	7.0	0.66
6	1.6	47	2.2	2.6	0.99
7	2.1	83	6.0	6.7	0.65
8	1.0	52	2.4	3.0	0.94

In his comprehensive study of branch wood and knot wood of *Picea abies* and *Pinus sylvestris*, Boutelje (1965, 1966) examined the physical properties of branches and associated knots at certain selected points, indicated in Figs. 7.5 and 7.6. The results obtained are summarized in Tables 7.10 and 7.11. As could be expected, densities were throughout higher on the lower side of both knots and branches. Upper and lower branch woods became progressively lighter with increasing distance from the center of the stem.

In a large *Picea abies* tree Hakkila (1969) found the branches to have a very high basic specific gravity at the base of the branch, ranging from 0.70 to 0.89 (Fig. 7.7). At a distance of 1 m from the stem, the density had dropped to 0.50,

7.2.4 Specific Gravity of Branch Wood

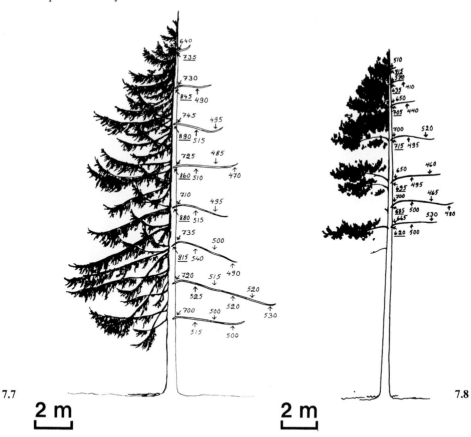

Fig. 7.7. Variation of basic density within and between the branches in a *Picea abies* tree. *Underlined numbers* refer to knot wood. (Hakkila 1969, 1971)

Fig. 7.8. Variation of basic density within and between the branches in a *Pinus sylvestris* tree. *Underlined numbers* refer to knot wood. (Hakkila 1971)

Table 7.11. Resin and moisture contents, radial and tangential shrinkage, and specific gravity (dry volume) of branch and knot woods of *Pinus sylvestris*. All values except those for specific gravity in per cent. Locations are shown in Fig. 7.6. (Boutelje 1966)

Location	Resin	Moisture		Radial shrinkage	Tangential shrinkage	Specific gravity
		Original wood	Resin-free wood			
1	43.1	15	26	2.8	3.3	1.07
2	27.5	18	25	3.2	3.2	1.15
3	10.5	44	49	4.0	6.4	0.53
4	27.3	20	28	2.4	2.9	0.92
5	4.3	40	42	4.0	6.2	0.51
6	12.9	23	27	2.4	3.0	0.80

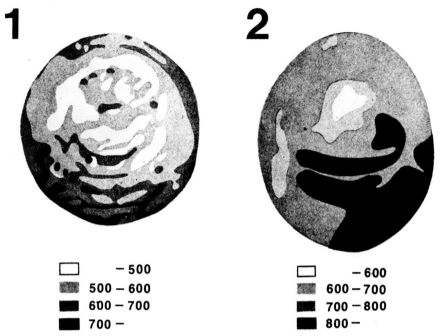

Fig. 7.9. Density distribution within a cross section at the base of a branch in *Pinus sylvestris* (**1**) and *Picea abies* (**2**). (Hakkila 1971)

a value that remained almost constant from this point and to the tip of the branch. Larger branches tended to have a higher density than smaller ones. Branches with a diameter of 1−2 cm had a specific gravity of 0.531 and those with a diameter of 4−6 cm, 0.621. Similar results were later obtained with *Pinus sylvestris*, as shown in Fig. 7.8 (Hakkila 1971). Also in this species, density values were especially high close to the stem. The density distribution over branch cross sections seen in Fig. 7.9 suggests a basic difference in this respect between spruce and pine.

Normal stem woods of *Picea abies* and *Pinus sylvestris* trees growing in Finland have been reported to have specific gravities of 0.383 and 0.398 (Hakkila 1969), 0.399 and 0.415 (Hakkila 1971), and 0.405 and 0.400 (Lehtonen 1980), respectively. For branch wood, by contrast, the average specific gravity is much higher for spruce than for pine. This has been reported by many investigators, including Hartig (1896), Enčev (1962, 1972), Boutelje (1966), Alestalo and Hentola (1966), Hakkila (1969, 1971), and Kärkkäinen (1976). The same appears to apply to other *Picea* and *Pinus* species. Fegel (1938, 1941), as can be seen in Table 7.12, found that *Picea mariana* and *P. rubens* had a heavier branch wood than did *Pinus resinosa* and *P. strobus*. According to my own observations, much of the compression wood present on the lower side of pine branches is not of the heavy, severe type, while this is the rule in spruce branches. The reason for this difference is probably that the branches in spruce have to support a larger mass of needles in relation to their diameter than do the much thicker

7.2.4 Specific Gravity of Branch Wood

Table 7.12. Specific gravity, moisture content in green condition, and volumetric shrinkage of branch and stem woods from seven conifer species. All values except those for specific gravity in per cent. (Fegel 1938)

Species	Specific gravity		Moisture content		Volumetric shrinkage	
	Branch	Stem	Branch	Stem	Branch	Stem
Abies balsamea	0.48	0.35	35.2	74.3	11.7	10.3
Picea mariana	0.60	0.42	37.5	41.9	13.9	12.1
Picea rubens	0.51	0.33	47.5	57.7	12.8	11.0
Pinus resinosa	0.43	0.35	48.2	114.1	11.5	10.3
Pinus strobus	0.40	0.33	120.4	76.2	9.8	8.2
Thuja occidentalis	0.46	0.27	42.4	34.2	11.2	7.2
Tsuga canadensis	0.55	0.34	41.8	65.1	10.7	9.5

pine branches. The accumulation of snow and ice also tends to be larger on spruce than on pine branches.

The data reported by Fegel (1938) (Table 7.12) indicate that in the conifers native to northeastern United States branch wood is on the average 25–70% heavier than normal stem wood. Several other investigators have made similar comparisons. According to Janka (1904) branch wood of *Picea abies* has a specific gravity (dry volume) varying from 0.67 to 0.88, values twice as high as those valid for normal stem wood of the same species. Densities of branch wood and stem wood have been reported to be 0.45 and 0.30, respectively, in *Abies balsamea* (Young and Chase 1965), 0.55 and 0.44 in *Picea rubens* (Kurrle 1963), and 0.57 and 0.42 in *Tsuga heterophylla* (Worster and Vinje 1968). Benić (1970) found that branch wood of *Abies alba* had a density of 0.666, which was twice the density of normal stem wood. Gava (1974) has reported that branch wood in *Picea abies* is much denser than stem wood.

The branches of *Pinus taeda* used by Taylor (1979) in a study of density and tracheid length variability (Chap. 4.3.2.3) had wider growth rings on their lower side, where they contained compression wood of the moderate type. At one quarter of total branch length from the stem, the lower branch wood had a specific gravity of 0.52, while that of the upper was only 0.37. The density of the former had decreased to 0.50 at three quarters branch length. In the radial direction, the specific gravity of the upper branch wood first remained constant and later increased slowly toward the outside. The density of the wood on the lower side was 0.48 in the second growth ring. It increased rapidly to a maximum of 0.54 in ring 15 and then decreased at the same rapid rate, reaching a value of 0.45 in ring 30.

Lehtonen (1980) has reported results of an extensive investigation of the incidence (Chap. 10.6.2) and density of knots and associated branch stumps in *Picea abies* and *Pinus sylvestris* trees growing in Finland. The cross section of the branches was eccentric, with the average horizontal diameter 5.3% smaller than the vertical diameter in the spruce and 7.1% smaller in the pine. In *Picea abies* the eccentric growth was most pronounced in branches growing on the southern side of the stem. There was no such variation in the pine branches. Stem wood of both spruce and pine had a basic density of 0.40. Branch wood had a much higher density, namely 0.810 in the spruce and 0.620 in the pine. In *Pinus sylvestris* the specific gravity of the branch stumps decreased markedly with increasing stem height, but it remained constant in *Picea abies*.

Present evidence regarding the relationship between average branch wood density and branch size is contradictory. Boutelje (1966) and Lehtonen (1980) found that in *Picea abies* the average specific gravity of the branch wood decreased with increasing branch size, but Enčev (1962) and Kärkkäinen (1976) found that it increased. In *Pinus sylvestris* wood density has been reported to be higher in large than in small branches by Enčev (1962), Kärkkäinen (1976) and Lehtonen (1980), but Boutelje (1966) and Poller (1978) observed an opposite trend. Phillips et al. (1976) found branch wood density to increase with increasing branch size in four species of southern pines. Very probably, these contradictory results are due to the non-uniform nature of branch wood with its lighter side wood and opposite wood and its heavy compression wood, the proportion of which strongly influences the average density of the total branch wood. The occurrence of compression wood in conifer branches is only partly under genetic control (Chap. 14). It is strongly influenced by many and variable ecological factors, such as exposure to wind, snow, and ice (Chap. 15.2).

Compression wood occurs more regularly and frequently in larger amounts in branches than in stems, and it is therefore to be expected that branch wood should normally be heavier than stem wood. Surprisingly, several exceptions to such a rule have been reported for the genus *Pinus*. Lee (1971a, b) found that in *Pinus nigra* var. *nigra* the mean specific gravity of stem wood was 0.393, whereas that of branch wood was only 0.352. Eskilsson and Hartler (1973) obtained a density of 0.41 for stem wood and only 0.34 for the wood in coarse branches in *Pinus sylvestris*. Corresponding values for *Picea abies* were 0.43 and 0.66. According to Howard (1973), stem wood in three *Pinus elliottii* var. *elliottii* trees had an average density of 0.42 and branch wood only 0.35. In four species of southern pines, Phillips et al. (1976) found branch wood to have a lower specific gravity (0.45) than stem wood (0.47) in *Pinus echinata* and *P. taeda*. The difference was greater in *Pinus elliottii* (0.49 compared to 0.56) and *P. palustris* (0.43 compared to 0.52). The reason for this unexpected trend is not clear. Possibly the branches examined contained an unusually small amount of compression wood and a high proportion of juvenile wood. According to my own observations, this is not a rare phenomenon in pines, but it is seldom found in spruces.

7.2.5 Specific Gravity of Knot Wood and Associated Stem Wood

Although knots are branches encased in a stem, knot wood is generally denser than branch wood. One reason for this is that knots, unlike branches, always have a high resin content, sometimes 30% or more (Chap. 10.6.3). According to Wegelius (1939), knot wood is three times as heavy as stem wood and half as heavy as branch wood. Its density can vary between 0.5 and 1.4 and is usually above 1.0. Boutelje (1966) has reported densities ranging from 0.85 to 1.15 for knot wood in *Picea abies* and *Pinus sylvestris* (Tables 7.10 and 7.11). According to Levčenko (1969), knot wood in the latter species had a density twice that of stem wood. Enčev (1972) compared the density of stem wood and the wood of live knots in trunks of *Abies alba, Picea abies,* and *Pinus sylvestris.* In all three species, the knot wood was 2 to 2.5 times denser than the stem wood. Enčev attributed the high density of the knots to their narrow growth rings and high resin content. Compression wood present on the lower side of the branches must also have contributed to the high density. Lehtonen (1980) found knot wood in *Picea abies* and *Pinus sylvestris* to have a specific gravity of 0.895 and 0.750, respectively, values that were both higher than those obtained for corresponding branch woods (0.810 and 0.620). The density of the spruce knots decreased only in the upper part of the stem, whereas that of the pine knots decreased continuously with increasing height. A very large knot of *Tsuga canadensis* according to my own measurements had a density of 1.103, obviously a result of its high content of both compression wood and resin (Chap. 10.6.2).

Several investigators have concluded that the stem wood adjacent to a knot (Chap. 10.6.4) has properties intermediate to those typical of these two types of wood (Kininmoth 1961, Boutelje 1966, Baker 1968, von Wedel et al. 1968, Götze 1969). Schultze-Dewitz and Götze (1973) determined the density of stem wood surrounding knots in *Picea abies* and *Pinus sylvestris.* The transition from the lighter stemwood to the heavier knot wood was abrupt in the pine and above the embedded branch in the spruce, but was gradual on the lower side of the latter species. Lehtonen (1980) found that in these two species stem wood closest to a knot had the highest density. The decrease in density above a branch whorl in *Pinus sylvestris* is more rapid than that below the whorl, where it is both lesser and more gradual (Götze 1969, Lehtonen 1980). The reason for this is probably that there is usually more compression wood below than above a knot (Chap. 10.3.8.2).

7.2.6 Conclusions

When fully developed, compression wood is without any exceptions much heavier than normal wood. The density of its cell wall substance is somewhat lower than that of normal wood, but this fact and the presence of the helical cavities and intercellular spaces are more than compensated for by the occurrence over almost the entire growth ring of thick-walled, short tracheids. Values reported in the literature for the density of compression wood vary considerably. The reason for this is probably to be found in variations in collection techniques. Ideally, the compression wood to be measured should be of the severe type throughout. The specific gravity of softwood is positively correlated with the proportion of compression wood present. When compared with normal side wood from the same growth ring and at the same stem height, compression wood is twice as heavy as normal wood. Mild and moderate compression woods have a lower density. Because of its slightly larger lumen, compression wood is generally somewhat lighter than normal latewood.

Branches have a denser wood than the stem because of their more narrow growth rings and because of the almost universal presence on their lower side of compression wood. The average specific gravity of the entire branch wood is strongly influenced by the proportion of compression wood present. The density of branch wood is higher in *Picea* than in *Pinus*. The high resin content of knot wood makes this wood even denser than branch wood. The density of branch wood decreases toward the branch tip. Because branches contain three different kinds of wood, namely compression, opposite, and side woods, the density of total branch wood tends to vary in an irregular and often unpredictable manner.

The strength properties of wood are positively correlated with its specific gravity. When comparison is made between the strength characteristics of normal and compression woods, this fact has to be taken into account. In itself, compression wood is in several respects stronger than normal wood. When its higher specific gravity is taken into consideration, however, much of this apparent superiority disappears.

7.3 Wood−Water Relationships

7.3.1 Moisture Content

7.3.1.1 Introduction

Wood is strongly hygroscopic, and when dry, it absorbs water avidly. Green wood in living trees contains a high proportion of water which varies with the species and also with the season. In conifers, the upper portion of the stem contains more water than the lower. Moisture contents of 50−100% are common in softwoods. The absorption of water by wood has been reviewed by Stamm (1964) who has also contributed much to this subject, while practical aspects have been discussed by Besley (1960). The sorption of aqueous and nonaqueous liquids by wood has been thoroughly reviewed by Venkatsevaran (1970). Wood−moisture relations in connection with kiln drying of lumber are discussed by Bramhall and Wellwood (1976). Skaar's (1972) excellent book *Water in Wood* offers a comprehensive treatment of the subject.

All cell wall components in wood have some affinity for water. Cellulose, because of its numerous hydroxyl groups, is strongly hygroscopic. The larger part of wood cellulose, however, is crystalline, and in the crystalline areas the hydroxyl groups are involved in formation of hydrogen bonds. In native cellulose, these regions are inaccessible to water. The hemicelluloses possess not only hydroxyl groups but frequently also carboxyl groups and are probably entirely amorphous in their native state. As a result, wood hemicelluloses are even more hydroscopic than cellulose (Ziegler 1974).

Lignin has a lesser affinity for water. It is, however, also hydrophilic thanks to its numerous aliphatic and phenolic hydroxyl groups. In a thoughtful discussion of wood reactions, Schuerch (1965) has emphasized that the behavior of lignin is that of a polar, aromatic polymer containing numerous hydroxyl groups. Christensen and Kelsey (1958, 1959a, b) (Kelsey and Christensen 1959) found that lignin, on sorption of water, like cellulose, behaved as a swelling gel. According to Runkel and Lüthgens (1956) (Runkel 1954), cellulose is far less hygroscopic than the hemicelluloses, while lignin has a low affinity for water. Ziegler (1974) believes that lignin controls the sorption of wood by restraining the swelling of the polymeric system in the cell wall.

As a result of an extensive investigation on the sorption of water by *Eucalyptus regnans* and its polymeric constituents, Christensen and Kelsey (1958, 1959a, b), concluded that cellulose contributed 47%, the hemicelluloses 37% and lignin 16% of the total sorption of water by this wood. Most woods, because of the high sorptive capacity of the hemicelluloses, are more hygroscopic than cellulose alone, but the sorption isotherms of cellulose and wood are quite similar.

The extractives in wood also influence its sorption of water, especially at high relative humidities. According to Wangaard and Granados (1967), the fiber saturation point is enhanced on removal of the extraneous substances. Only polymolecular sorption is affected, however. Heartwood absorbs water less readily than does sapwood because of its higher content of hydrophobic extractives (Bailey and Preston 1969). Latewood wall substance has a greater capacity to bind water than earlywood (Boutelje 1962a).

7.3.1.2 Compression Wood

Mer (1887) was the first to report observations on the water sorption of compression wood. He found that this tissue absorbed more water than normal wood, and that it also lost more water when dried in the air. Actually, the opposite is true, as was demonstrated simultaneously by Hartig (1896) and by Cieslar (1896). In one case Cieslar found compression wood of *Picea abies* to contain 33.1% water compared to 40.8% for normal wood.

The fact that green compression wood contains less water than normal wood under the same conditions has been confirmed by all later investigators. A summary of reported values is presented in Table 7.13. The absolute values vary, depending on external conditions, but the ratios of the water content of compression wood to that of normal wood are almost all in the range of 0.6−0.8, that is, green compression wood contains ⅔ to ¾ less water than normal wood. According to Kaburagi (1952) normal wood of *Abies mayeriana* holds twice as much water as does compression wood in the sapwood zone. Burger (1941) mentions that in *Pinus sylvestris,* green compression wood contains considerably less water than normal wood.

Jaccard and Frey (1928b) placed oven-dry normal and compression woods of *Pinus nigra* and *Pseudotsuga menziesii* in a moist chamber and followed the sorption of water. The two species gave practically identical results. The sorption curves for the latter are shown in Fig. 7.10. Compression wood absorbed water at a slightly lower rate than did normal wood and also attained a lower level of equilibrium moisture content, that is, it had a lower fiber saturation point. Verrall (1928), in the same year, carried out similar experiments at 90%

Table 7.13. Moisture content of green compression and normal woods. All values in per cent of oven-dry wood

Species	Compression wood	Normal wood	Ratio	References
Abies alba	80	185	0.43	Giordano 1971
Abies concolor	109.0	187.1	0.58	Pillow and Luxford 1937
Chamaecyparis obtusa	66.8	95.6	0.70	Yazawa 1944
Chamaecyparis pisifera	53.2	95.4	0.56	Yazawa 1944
Ginkgo biloba	91.7	103.5	0.89	Onaka 1949
Picea abies	32.5	40.0	0.81	Cieslar 1896
Picea mariana	38.2	58.2	0.66	Verrall 1928
Pinus densiflora	105.7	175.3	0.60	Onaka 1949
	89.0	146.0	0.61	
	75.0	93.0	0.82	
Pinus nigra	76.2	114.7	0.66	Jaccard and Frey 1928 b
Pinus ponderosa	88	133	0.66	Markwardt and Wilson 1935
	85	121	0.70	
	78	138	0.57	
Pinus strobus	42.3	54.2	0.78	Burns 1920
	35	48	0.73	
	73.2	98.9	0.74	Pillow and Luxford 1937
Pinus taeda	77.2	101.9	0.76	Pillow and Luxford 1937
Pseudotsuga menziesii	44.7	52.4	0.85	Jaccard and Frey 1928 b
	27.4	42.5	0.64	Pillow and Luxford 1937
	43.3	58.3	0.74	
Sequoia sempervirens	102	114	0.89	Markwardt and Wilson 1935
	89	129	0.69	
	106	126	0.84	
Sequoiadendron giganteum	135	180	0.75	Cockrell and Knudson 1973

Table 7.14. Moisture adsorbed by oven-dry normal and compression woods at 90% (Verrall 1928) and at 100% (all other investigators) relative humidity. All values in per cent of oven-dry wood

Species	Compression wood	Normal wood	Ratio	References
Larix laricina	23.2	24.1	0.96	Verrall 1928
Picea abies	10.4 – 11.1	30.7 – 37.4		Rak 1957
Picea glauca	29.5	33.5	0.88	Perem 1958, 1960
Picea mariana	19.4	22.0	0.88	Verrall 1928
Picea sp.	19.5	21.9	0.89	Verrall 1928
Pinus nigra	25.5	27.1	0.94	Jaccard and Frey 1928 b
Pinus resinosa	27	31	0.87	Perem 1958, 1960
Pseudotsuga menziesii	25.4	27.0	0.94	Jaccard and Frey 1928 b
Thuja occidentalis	20.1	20.6	0.98	Verrall 1928
Thujopsis dolabrata	23.2	24.7	0.94	Mori 1933

7.3.1.2 Compression Wood

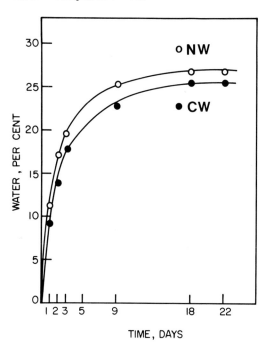

Fig. 7.10. Sorption of water by normal (*NW*) and compression (*CW*) woods of *Pseudotsuga menziesii*. (Redrawn from Jaccard and Frey 1928b)

relative humidity, observing almost the same rate of absorption. Additional data have later been reported by Mori (1933), Rak (1957), and Perem (1958, 1960). The results are summarized in Table 7.14. If the unusually low value for *Picea abies* compression wood is disregarded, the fiber saturation point of compression wood is 5–10% lower than that of corresponding normal wood and varies within a range of 20–30%. Verrall (1928), albeit unable to carry out direct measurements, concluded that the fiber saturation point must be lower for compression wood than for normal wood. Schulz and Bellmann (1982) found that the fiber saturation point of their *Picea abies* specimens decreased with increasing proportions of compression wood, as shown in Fig. 7.11. In other experiments, normal wood had a fiber saturation point of 34.5% compared to only 24.8% for compression wood.

In later years, several investigators have reported values for the fiber saturation point of gymnosperm woods which are significantly higher than those estimated earlier. Stone and Scallan (1967) obtained a value of 40% for *Picea mariana*. Ahlgren et al. (1972) used a solute exclusion method to determine the fiber saturation point of solvent-extracted normal and compression woods of *Pseudotsuga menziesii*. The values obtained were 46% and 39%, respectively. The investigators attributed the lower value of compression wood to the denser and more rigid structure of this tissue. The fiber saturation point of tension wood was also lower than that of normal wood. Boutelje (1972), using interference microscopy, measured the refractive index and volume of the cell walls in *Picea abies* and *Pinus sylvestris*. The calculated fiber saturation point was

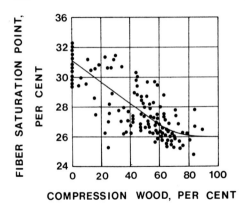

Fig. 7.11. Variation of the fiber saturation point with the proportion of compression wood present in *Picea abies*. (Modified from Schulz and Bellmann 1982)

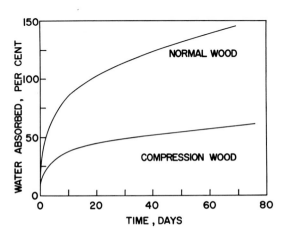

Fig. 7.12. Sorption of water by air-dry normal and compression woods from four conifer species. (Redrawn from Verrall 1928)

35–37%. From these results, which were obtained by entirely different methods, it would appear that the true fiber saturation point of normal conifer wood should be 40 ± 5%, or about 10% higher than was believed previously. It is 5–20% lower for compression wood.

Siau (1984) has recently pointed out that in these investigations use was made of thin wood sections, and that the results obtained cannot be directly applied to gross wood. He prefers to define as the fiber saturation point the moisture content at which the physical properties of wood cease to vary with the water content. The level at which this occurs coincides closely with the previously accepted, lower values for the fiber saturation point of wood.

Verrall (1928) submerged air-dried compression and normal woods in water for various lengths of time and followed the uptake of water for 73 days. Figure 7.12 shows the average rate of imbibition of water of four species. It was much faster for normal wood than for compression wood. The fiber saturation point was reached after 1 or 2 days and henceforth only free water was taken up. At the end of the experiment, compression wood had imbibed only 60% water compared to 148% for normal wood. Rak (1957) found that air-dry compression

7.3.1.2 Compression Wood

Table 7.15. Moisture content of air-dry compression and normal woods. All values in per cent of oven-dry wood

Species	Compression wood	Normal wood	Ratio	References
Abies alba	11.7	11.3	1.04	Giordano 1971
Abies concolor	11.9	11.7	1.02	Pillow and Luxford 1937
Picea abies	14.4	14.0	1.03	Janka 1909
	14.1	13.9	1.01	
	7.8	6.2	1.26	Ulfsparre 1928
	15.4	13.5	1.14	
Picea mariana	9.2	8.7	1.06	Ladell et al. 1968
Pinus ponderosa	12.6	12.0	1.05	Pillow and Luxford 1937
Pinus resinosa	12.6	12.0	1.05	Markwardt and Wilson 1935
	10.6	10.0	1.06	
	12.7	12.1	1.05	
Pinus taeda	11.7	11.1	1.05	Pillow and Luxford 1947
Pinus sp.	12.4	11.6	1.07	Heck 1919
	10.8	11.7	0.92	Markwardt and Wilson 1935
Pseudotsuga menziesii	8.9	10.2	0.87	Trendelenburg 1931, 1932
	11.3	11.1	1.02	
	14.7	12.0	1.23	
	13.0	14.3	0.91	
	11.9	13.3	0.89	
	12.6	11.6	1.09	Pillow and Luxford 1937
	12.1	11.5	1.05	
Sequoia sempervirens	10.5	9.9	1.06	Markwardt and Wilson 1935
	9.7	8.8	1.10	
	10.0	8.6	1.16	
Sequoiadendron giganteum	14.0	13.0	1.08	Cockrell and Knudson 1973

wood of *Picea abies* imbibed 70–75% water, while normal sapwood took up as much as 192–217%. These are much larger differences than those shown in Table 7.13 between compression and normal woods in their green state. Obviously, drying impairs the sorptive capacity of compression wood much more than normal wood. For the same species Schulz and Bellmann (1982) found normal wood to imbibe 190% water and compression wood only 100%. They attributed this difference to the higher lignin content of compression wood.

When brought from a higher to a lower moisture content by drying in the air, compression wood loses less water than does normal wood. As can be seen in Table 7.15, with only three exceptions, compression wood has been found to have a higher equilibrium moisture content than normal wood in the air-dry condition, usually 5–15%.

In a study of the sorption of water vapor by normal and compression woods of *Tsuga canadensis*, Ziegler (1974) found that sorption took place in two cycles. The first was controlled by diffusion and the second by molecular rearrangement of the wood polymers. The sorption rate was negatively correlated with the square of the wall density. The activation energy for diffusion in the radial direction was 12 kcal per mole for normal wood and 18 kcal for compression wood.

Popper and Bosshard (1976) investigated wood–water relationships from calorimetric measurements, comparing normal wood and branch compression wood of *Abies alba*. The heat of sorption of normal wood was found to be 1428 cal per mole, and that of compression wood 1257 cal. Normal and compression woods had approximately the same absorptive capacity. Compression wood, however, had an 8% larger active surface and absorbed somewhat more water than normal wood as a result of its greater mass. Normal wood had a distinctly greater affinity for water than compression wood. The investigators attributed this to the higher lignin content of the latter wood.

In considering the compression wood–water relationship, it must be remembered that fully developed compression wood contains twice as much cell wall material per unit volume as does normal wood. This fact probably accounts for the lower rate of absorption of compression wood, but it does not explain the lower fiber saturation point or the lower moisture content at 90% relative humidity (Verrall 1928). Vorreiter (1963) and Raczkowski and Stempień (1967) claim that the sorptive capacity of wood decreases considerably with increasing density of the cell wall substance. If this were generally true, compression wood, whose tracheid walls are slightly less dense than those of normal wood, should have a higher fiber saturation point, which it does not.

Verrall (1928) and Onaka (1949) have suggested that the lower fiber saturation point of compression wood could possibly be a result of its high content of lignin (Bosshard 1956, Knigge 1958a), which has a lower sorptive capacity than cellulose and hemicelluloses. The hygroscopic polysaccharides are heavily encrusted by lignin in compression wood, a fact that would contribute to the low sorptive capacity of the tracheid walls in compression wood. Vorreiter (1963) has drawn attention to the fact that hydrophobic wood constituents tend to impair absorption of water. Jaccard and Frey (1928b) suggested that the lower sorptive capacity of compression wood could perhaps be caused by a higher content of inorganic salts.

The high lignin content of compression wood is probably also responsible for the higher moisture content attained by this wood when dried in the air, compared to normal wood. Very likely, the lignin surrounding the cellulose and hemicelluloses renders it difficult for the water present in these polysaccharides to escape. The larger amount of cell wall material may also contribute. Why green compression wood or compression wood saturated with water contains so little free water compared to normal wood, is not difficult to explain. The small lumen and the shortness of the compression wood tracheids must be responsible for this, the helical cavities and the intercellular spaces notwithstanding.

The fact that compression wood both absorbs and loses less water than does normal wood is bound to affect its strength properties. Perem (1958, 1960) has pointed out that a wood with a low fiber saturation point may show strength characteristics that can be attained by a wood with a high fiber saturation point only after it has lost some water in drying.

7.3.1.3 Branch Wood and Knot Wood

Because of their high content of compression wood, branch wood and knot wood should have a lower moisture content than normal stem wood. The extremely high resin content of knot wood further impairs the sorptive capacity of this type of wood.

Hartig (1896), as can be seen from Table 7.6, found the lower portion of *Picea abies* branches to contain noticeably less water than the upper part. The specimens investigated by Fegel (1938) could not be examined immediately after cutting, and the figures for moisture content listed in Table 7.12 are accordingly of limited significance. In five of the seven species studied, branch wood had a lower moisture content than stem wood. In *Pinus densiflora* the moisture content of stem wood is considerably higher than that of branch wood, as seen from Table 7.9 (Yazawa et al. 1951).

Boutelje (1965, 1966) measured the moisture content in and around knots in *Picea abies* and *Pinus sylvestris*, obtaining the results shown in Table 7.16. Knot wood in the sapwood zone of both species had a three to four times lower moisture content than the surrounding stem wood. In the heartwood the differences were much less, especially for the spruce. As could be expected, pine branches contained much more resin than did those from spruce, and in both species the resin content of the branch wood was appreciably higher on the upper side than on the lower (Table 7.17). Moisture contents differed surprisingly little between the two portions. However, as pointed out by the investigator, the lower, compression wood side clearly contained less water if the calculations were based on resin-free wood. The low moisture content of the opposite wood was undoubtedly a result of its exceptionally high resin content. The fact that this wood contained more resin than the compression wood on the lower side can presumably be attributed to the narrow lumina of the compression wood tracheids. Boutelje also determined the moisture content of upper and lower knot wood and branch wood in the locations shown in Figs. 7.5 and 7.6. As can be seen from Tables 7.10 and 7.11, the wood on the lower side, with only one exception, contained less water than did that on the upper. In *Picea abies*, where only small amounts of resin were present, the compression wood on the under side contained less resin than did the opposite wood above. In *Pinus sylvestris*, this was the case only with the knot wood closest to the stem. Further out along

Table 7.16. Average moisture content in and around knots in *Picea abies* and *Pinus sylvestris*. All values in per cent. (Boutelje 1966)

Nature of wood	Picea abies	Pinus sylvestris
Knotwood in sapwood part of the stem	21	31
Knotwood in heartwood part of the stem	18	23
Wood around knots in sapwood part of the stem	88	88
Wood around knots in heartwood part of the stem	20	33

Table 7.17. Resin and moisture contents and shrinkage characteristics of lower and upper branch woods of *Picea abies* and *Pinus sylvestris*. All values in per cent. (Boutelje 1966)

Property	*Picea abies*		*Pinus sylvestris*	
	Lower	Upper	Lower	Upper
Resin content	10.8	16.0	27.1	40.5
Moisture content, original wood	27	29	22	19
Moisture content, resin-free wood	30	35	30	32
Longitudinal shrinkage	6.1	0.5	4.6	1.7
Radial shrinkage	3.1	7.8	2.3	2.8
Tangential shrinkage	3.7	8.8	2.9	3.3

the branch, compression wood had three times as much resin as did opposite wood. There seems to be no explanation for this peculiar fact. If moisture contents are based on resin-free wood, the three compression-wood specimens contained 25–28% water, close to the fiber saturation point of this tissue.

7.3.2 Permeability to Water

The permeability of wood to water and aqueous and organic solutions is of great technical importance and has been the subject of much research. Comprehensive reviews of the movements of fluids in wood have been authored by Stamm (1967a, b). More recent are the two excellent monographs by Zimmermann (1983) and Siau (1984).

There have been very few investigations on the permeability of compression wood. Jaccard and Frey (1928b) prepared thin tangential sections of normal and compression woods of *Pinus nigra* and *Pseudotsuga menziesii*. Placing the membranes at the end of a glass tube, they first measured the rate of filtration of water. The membrane from Douglas-fir compression wood gave a rate of filtration of 0.91 mm h^{-1} and that from opposite wood 1.67 mm h^{-1}. Similar results were obtained when the tubes were filled with a 2-M sucrose solution and inserted into water. Water was imbibed at a lower rate through compression wood. The rate of diffusion of the sugar molecules from the tube into the water was the same in the two cases, indicating that the permeability of compression and opposite woods were the same. The semipermeability of the former was, however, slightly larger than that of the latter. Compression wood contains fewer pits on the radial walls of the tracheids than does normal wood (Chap. 4.10.1). The larger resistance to filtration of compression wood could be a result of the scarcity of pitting (Trendelenburg 1932), but it is likely that the small lumen of the tracheids also contributes (Onaka 1949).

7.3.3.1 Normal Wood

The experiments of Jaccard and Frey (1928b) only relate to the movement of fluid in the radial direction. No measurements seem to have been reported concerning the flow of liquid in the axial direction of compression wood. Onaka (1949) mentions that if a small branch is inserted into a staining solution, the solution rises more slowly in compression wood than in opposite wood. According to Mitchell (1967), the abnormal wood, very similar to compression wood, which is produced in fir trees infested by the balsam woolly aphid, is inferior to normal wood in conducting sap (Chap. 20.8.4.3).

Although more experiments are obviously needed, it would appear that compression wood offers more resistance to liquid flow than normal wood in both the longitudinal and the radial directions. The role of the pits in the axial conduction of liquid in compression wood is obscure and needs further investigation. It seems evident, however, that the resistance to flow to a considerable part must be caused by the narrow lumen of the compression-wood tracheids. According to Petty and Puritch (1970), the lumen may be responsible for more than half of the resistance to liquid flow in the first-formed earlywood, which contains many pits.

7.3.3 Shrinkage and Swelling

7.3.3.1 Normal Wood

As wood takes up and loses water with changes in the surrounding atmosphere, it swells and shrinks in a complex fashion that has attracted much interest and also generated much controversy over the years. The factors that determine these changes in wood dimensions are as yet not fully understood, although in recent years considerable progress has been made. Shrinkage of wood, which has been more widely studied than swelling, is governed by the chemical composition, ultrastructure, and gross anatomy of wood. Several reviews have been published discussing the relationship between wood structure and shrinkage, such as those by Pentoney (1953), Bosshard (1956, 1961, 1974), Hale (1957), Boutelje (1958, 1962a,b, 1972, 1973), Frey-Wyssling (1968), Preston (1968), Stamm and Smith (1969) and Boyd (1974b). An excellent review, both critical and comprehensive, has been presented by Kelsey (1963).

Normal wood shrinks to different extents in longitudinal, radial, and tangential directions. Longitudinal shrinkage, along the grain, is usually quite small, only 0.1–0.3%. Radial shrinkage is generally much larger, 2–8% and sometimes even more on drying from green to oven-dry condition. Tangential shrinkage is as a rule twice as large as that in the radial direction, and total volumetric shrinkage is accordingly considerable. Less shrinkage occurs on drying from the green to the air-dry condition. The small longitudinal shrinkage of normal wood is a great advantage in wood utilization. Conversely, the large transverse shrinkage is a drawback which is compounded by the fact that tangential shrinkage is so much larger than that in the radial direction. As a result logs tend to split on drying and lumber becomes distorted.

When wood is dried from a green to an air-dry state it will either expand slightly in length or, more commonly, contract. Further drying to an oven-dry condition always involves shrinkage. The denser the wood, the lesser the longitudinal shrinkage, but this relationship seems to apply only within each species.

Koehler (1931) suggested that the longitudinal shrinkage of wood was largely governed by the microfibrillar angle in the S_2 layer, increasing with an increase in this angle. He also noted that earlywood, where the microfibrils are more steeply oriented than in latewood, shrinks more than the latter. To this can be added, as first pointed out by Hale (1957), that the S_2 layer is more dominating in latewood, whereas in earlywood, where S_2 is much thinner, the S_1 and S_3 layers would be expected to play a larger role. With their transverse or almost transverse orientation of microfibrils, these two layers would, of course, tend to enhance any longitudinal shrinkage.

In many cases the correlation between microfibril angle and shrinkage has been found to be weak, and several investigators have therefore attempted to find other explanations. Frey-Wyssling (1940a, b, 1943) suggested that radial and tangential shrinkage depends on the amount of middle lamella present in each direction. According to this theory, the middle lamella contributes directly to the dimensional changes, and the longitudinal shrinkage is unrelated to the orientation of the microfibrils in the S_2 layer. Similar views were later expressed by Bosshard (1956, 1961) and by Nečesaný (1966). Actually, as pointed out by Kelsey (1963), Frey-Wyssling's theory lacks experimental verification. Matsumoto (1950), repeating Frey-Wyssling's experiments, came to the conclusion that shrinkage and swelling of wood does not take place in the middle lamella but in the secondary wall.

The anisotropy of transverse shrinkage has been even more difficult to explain than the difference between longitudinal and transverse shrinkage. Many investigators have believed that rays are able to restrain radial shrinkage (Schniewind 1959), and while this appears to be true in certain cases, it cannot be the only explanation, for the anisotropy persists in woods free of large rays. Boutelje (1962a, b) came to the conclusion that rays have no effect on the anisotropy of transverse shrinkage. He also rejected the idea that longitudinal shrinkage should be positively correlated with the microfibril angle. Instead, he suggested that most water in the cell wall is located between the lamellae in the cell wall layers and that this is the reason why longitudinal shrinkage is small and transverse shrinkage is large. More recent theories, to be discussed later, do not agree with this concept.

According to Frey-Wyssling (1940b, 1943), the difference between radial and tangential shrinkage is caused by the middle lamella, a view also shared by Bosshard (1956, 1961), Nečesaný (1966) and Kato and Nakato (1968). While it is true that the middle lamella contains some pectin, which swells in water, 80–100% of this region consists of lignin, a polymer with a lesser swelling ability than either cellulose or hemicelluloses. According to Kelsey (1963), transverse shrinkage is restrained by both the primary wall and the middle lamella, facts which have long been recognized in wood pulping and in beating of pulp fibers. Bosshard (1956, 1961), also noted that transverse shrinkage increases on delignification of wood.

7.3.3.1 Normal Wood

It has long been known that in conifers the helical pitch is steeper in the tangential than in the radial walls of the tracheids, a fact which has been attributed to the higher frequency of pits in the radial walls. Trendelenburg (1939 b) proposed that the transverse shrinkage anisotropy perhaps could be due to this difference. This concept was rejected by Pentoney (1953), and Hale (1957) later pointed out that species with only moderate radial pitting nevertheless exhibit strong anisotropy in transverse shrinkage.

In those woods that have clearly distinguishable earlywood and latewood, the shrinkage behavior of each of these tissues offers an explanation for the fact that tangential shrinkage is larger than radial shrinkage for whole wood. Latewood always shows a larger transverse shrinkage than earlywood. The reason for this is that S_2 is more dominating in latewood than in earlywood tracheids, where the S_1 and S_3 layers reduce the transverse shrinkage (Vintila 1939, Pentoney 1953, Nakato and Kadita 1955). In the radial direction, the shrinkage of the whole wood is the average of the large contribution of the latewood and the lesser contribution of the earlywood. In the tangential direction, on the other hand, the thick-walled latewood is able to carry the thin-walled earlywood with it. As a result, tangential shrinkage always exceeds that in the radial direction (Hale 1957). While this theory certainly is attractive and has found considerable acceptance (Panshin and de Zeeuw 1980), it does not seem to be applicable to all woods (Kelsey 1963).

In later years the swelling and shrinkage of wood has been subjected to new theoretical treatments by Barber, Cave, Harris, and Meylan in important contributions which also have a direct bearing on the shrinkage of compression wood. In the initial approach, Barber and Meylan (1964) assumed that only the S_2 layer influenced the shrinkage, and that no restraint was offered by the other two layers. It was further assumed that the microfibrils in S_2 were oriented at a constant angle to the fiber axis, and that the cell wall could be treated as if it were flat. One part of the cell wall consists of an amorphous, isotropic matrix. When this matrix imbibes water, it swells and shows an isotropic strain, thus becoming uniformly larger. The wall also contains long, partly crystalline microfibrils of cellulose, embedded in the matrix at a certain angle. Water cannot enter the crystalline regions of the microfibrils, and when it is removed from wood, the microfibrils do not change in length. The matrix, however, shrinks in volume, a change that is resisted by the microfibrils, and as a consequence the matrix is deformed. The extent of this deformation depends on the extensibility of the microfibrils and the elasticity of the matrix. The net result is an anisotropic shrinkage or, if water is imbibed, an anisotropic expansion.

Strains are produced in the longitudinal, tangential, and radial directions of the cell wall. If E is the elastic modulus of the cellulose microfibrils, and S is the modulus of rigidity of the matrix, three different strain ratios can be obtained for different microfibril angles and different ratios of E/S. Longitudinal shrinkage is small for microfibril angles less than 20° and can become negative around 30°. After a minimum at this angle, shrinkage increases rapidly with increasing microfibril angle. At no stage is the relationship between longitudinal shrinkage and this angle linear. Tangential shrinkage decreases continuously with increasing microfibril angle, but the reduction becomes significant

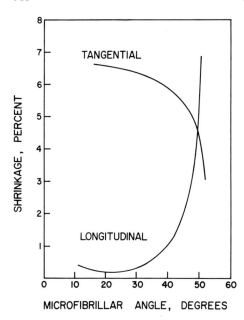

Fig. 7.13. Relationship between microfibril angle and longitudinal and tangential shrinkage of normal wood in *Pinus jeffreyi*. (Redrawn from Meylan 1968 and Meylan and Probine 1969)

only at angles larger than 15°. The radial shrinkage increases slowly with increasing angle up to about 45° and then decreases.

The value of the ratio between the elastic modulus of the microfibrils and the shear modulus of the matrix also has a pronounced effect on the shrinkage. At high ratios, longitudinal expansion can occur on desorption at angles between 10° and 40°, and tangential expansion can take place if the angle is 50° or larger. The radial shrinkage assumes a sharp maximum at 45°. The curves for longitudinal and tangential shrinkage cross each other at 45–58°. As long as the cell wall contains much water, the modulus of the matrix is low and E/S accordingly large. In drying from the green to the air-dry state, longitudinal shrinkage should accordingly be slight or even negative. On further drying, the matrix becomes more rigid and the modulus of shear increases, causing an increase in E/S. On drying from the air-dry to the oven-dry state, wood should always contract in length. Both these predictions agree with experimental results (Cockrell 1947).

The theory does not offer any explanation for the anisotropy of transverse shrinkage in wood. In an attempt to explain this phenomenon, Barber and Meylan (1964) introduced two assumptions, namely that the microfibrils are oriented at different angles in the tangential and radial walls, and that the ray cells reinforce the microfibrils in the direction of the rays. The theoretical treatment showed that if the angle in the radial wall is 1.5 times larger than that in the tangential, the ratio between tangential and radial shrinkage should be 1.75:1. Similarly, it was found that if wood contains 5% ray cells and thus 5% microfibrils oriented in the radial direction, the ratio between tangential and radial shrinkage should be 1.45:1.

7.3.3.1 Normal Wood

Harris and Meylan (1965) compared the reinforced matrix theory with experimental results obtained using wood of *Pinus radiata* with microfibril angles ranging from 15° to 45°. The predicted nonlinear relationship between microfibril angle and shrinkage could be confirmed and also the pronounced increase in longitudinal shrinkage at angles larger than 30°. As could be expected, the tangential shrinkage of earlywood was less than that of latewood, whereas the shrinkage in the longitudinal direction was larger. Since these results are valid at identical microfibrillar angles, the difference between the two tissues must be attributed to the greater thickness of the S_2 layer in the latewood. In earlywood, the S_1 and S_3 layers evidently influence the shrinkage behavior to a larger extent than in the latewood. A contributing factor could be the presence of a higher porportion of a more crystalline cellulose in latewood than in earlywood.

Additional verification of the reinforced matrix theory was provided by Meylan (1968) (Meylan and Probine 1969) in a study of the shrinkage behavior of *Pinus jeffreyi* wood. Juvenile wood of this species was found to contain an S_2 layer with a microfibril angle as high at 50°. Figure 7.13 shows the curvilinear relationships obtained with *Pinus radiata*. Longitudinal shrinkage was small for angles less than 25°. At larger angles there was a rapid increase in longitudinal shrinkage, associated with an almost as rapid decrease in tangential shrinkage. The two curves crossed each other at 49°, as predicted by the theory.

Barber (1968) extended the reinforced matrix theory, applying his calculations to a curved cell wall forming a cylinder and taking into account the restraining effect of the S_1 layer. In one case S_1 was assumed to contain a horizontally oriented helix of microfibrils, rendering this layer rigid in the transverse direction but with no strength in the axial direction. In a second case the S_1 layer was assumed to be isotropic. When no sheath was restraining the movement of the S_2 layer, the results were essentially the same as those obtained for a flat cell wall. Differences apparently become noticeable only when the ratio of the outer and inner diameters of the cylindrical cell exceeds 2.0, that is, when the cell wall is unusually thick. If the shrinkage of the S_2 layer is restricted by a sheath containing transversely oriented microfibrils, such as those in the S_1 layer, the outer diameter shrinks slightly less and the inner diameter much less. In the longitudinal direction there is little or no expansion on drying at high E/S ratios. The behavior of a cell enclosed in an isotropic sheath is not much different, at least not at angles less than 40°.

A sheath consisting of transversely oriented microfibrils will enhance the longitudinal shrinkage only slightly. It is in the transverse direction that the restraining influence of the sheath is most decisive. Compared to the earlier theory, the new deductions predict less shrinkage of the lumen and shrinkage rather than swelling in the longitudinal direction when water is lost, both facts that agree with most experimental results. The modified theory also predicts that cells which differ only in the thickness of their wall could exhibit different shrinkage behavior.

The reinforced matrix theory of Barber and Meylan was a highly significant contribution to our understanding of the shrinkage and swelling behavior of wood. Although received with little enthusiasm in some quarters, the theory was soon endorsed by others (Frey-Wyssling 1968). Hann (1969) found that the shrinkage of several hardwoods and softwoods agreed with the predictions of Barber and Meylan. Gaby (1971) studied the longitudinal shrinkage of juvenile wood from *Pinus echinata* and *Pinus taeda* as a function of the microfibril angle, which in this case varied from 32° to 53°. Maximum shrinkage observed was 2.2%. The relationship between shrinkage and microfibril angle was curvilinear and similar to that observed with juvenile wood by Meylan (1968).

Barrett et al. (1972) have pointed out that the theory of Barber and Meylan does not treat cellulose as hygroscopic and orthotropic. They also question whether it can actually be assumed that the volume of the cellulose microfibrils in the cell wall is small compared to that of the lignin. Instead, they developed a cell wall model where the wall is composed of many orthotropic layers, oriented at different angles to the cell axis. Composite shrinkage was calculated from the elastic and shrinkage characteristics of each component. Three primary variables were considered within each layer, namely thickness, microfibril angle, and chemical composition.

Meylan (1972) measured the longitudinal shrinkage of a large number of wood specimens of *Pinus radiata* with microfibril angles varying from 10° to 40°. The relationships between microfibril angle and incremental longitudinal swelling and shrinkage during adsorption and desorption were in agreement with those predicted by Barber's (1968) theory. Shrinkage increased rapidly above an angle of 28°, and no minimum could be detected. Meylan pointed out that 40% of the matrix consists of hemicelluloses, which are oriented in the same direction as the cellulose microfibrils, and that it therefore could be expected to swell anisotropically.

Cave (1972a,b, 1975) generalized the theories developed by Barber and Meylan by taking into account all three layers in the secondary wall, the proportion of cellulose, and the effect of moisture content on the stiffness of the matrix. In this treatment, wood is considered as a fiber composite, consisting of cellulose microfibrils embedded in a matrix that swells in water and whose shear modulus is a function of the moisture content. It could be shown that the relative thickness of the M + P, S_1, and S_3 layers markedly influence longitudinal shrinkage (Cave 1976). Cave (1978a,b) later developed a model that predicts the variation in the longitudinal Young's modulus as well as the longitudinal shrinkage with changes in the wood moisture content. Cave's theory represents a refinement of the earlier treatments that has so far not been surpassed.

Boyd (1974b), after reviewing the available evidence, has come to the conclusion that variations in density, distribution of hemicelluloses, cellulose content, microfibril angle, and location of rays, all fail to account for the anisotropic shrinkage of wood. Instead, he has suggested that the shrinkage anisotropy of both earlywood and latewood in gymnosperms is proportional to differences in the extent of lignification between the radial and tangential tracheid walls. The difference in shrinkage between earlywood and latewood is caused by variations in cell form and wall thickness. The different microfibril angles in the radial and tangential walls of the tracheids (Boyd 1974a) and the content of extractives in the heartwood are also factors of some, albeit less, importance. In a subsequent investigation of the shrinkage of normal and reaction woods of *Pinus radiata* and *Eucalyptus* spp., Boyd (1977) came to the conclusion that the microfibril angle in S_2 has a decisive influence on both longitudinal and volumetric shrinkage. Other factors are, however, also involved, namely the thickness of the cell walls, the thickness of the S_1 layer relative to S_2, and the lignin content. All these factors are additive.

Tang and Smith (1975) studied the longitudinal shrinkage of delignified, isolated conifer tracheids, using scanning electron microscopy. They concluded that the anisotropic shrinkage of single fibers is influenced by a number of factors, including the microfibril angle, the size and texture of the microfibrils, and the thickness of the cell wall.

7.3.3.2 Compression Wood

One of the most characteristic properties of compression wood is its exceptionally high longitudinal shrinkage, which is occasionally 10 (Paul and Smith 1950), 20 (Verrall 1928), or even 40–50 (King 1954, Nicholls 1982) times larger than that of normal wood. Compression wood also swells much more than normal wood, but this phenomenon has been studied far less than its shrinkage. For *Pinus sylvestris* compression wood, the swelling pressure has been reported to be larger by 460% in the longitudinal and by 40% in the transverse direction in comparison with normal wood (Raczkowski and Krauss 1979).

The excessive shrinkage in length is one of the most undesirable characteristics of compression wood from a utilization point of view. The shrinkage behavior of compression wood has also attracted more interest than any other physical property of this wood (Koehler 1931, 1946, Koehler and Luxford 1951, Anonymous 1960). Following the first mention of compression wood, 40 years elapsed before the true nature of its shrinkage properties were realized. Cieslar (1896) pointed out that compression wood is hard and brittle, and that swelling and shrinkage of such tissues are small. In his first investigation on compression wood, Hartig (1896) measured only its volumetric shrinkage. Normal wood from a stem of *Picea abies* had a volumetric shrinkage of 16.0%, but compression wood only 9.5%. Similar results were obtained with branch wood from *Picea abies*, shown in Table 7.6, and from *Pinus sylvestris*, although in these cases the differences between the upper and lower portions of the branch were far less.

In the course of the following 5 years, Hartig (1901) carried out further experiments, which were reported in his last, summarizing publication. On the basis of his earlier, extensive investigations, Hartig had come to the conclusion that wood shrinks more volumetrically the more cell wall substance it contains, that is, the higher its density. Making certain assumptions, he calculated that a certain *Picea abies* compression wood should shrink approximately 20%. The actual volumetric shrinkage, however, was only 8.8%. Hartig (1901) now measured the longitudinal as well as the volumetric shrinkage of opposite and compression woods from the stem and branches of *Picea abies*. Opposite wood from branches had a longitudinal shrinkage of only 0.09% while that of compression wood was 1.29%. Corresponding values for the volumetric shrinkage were 10.5

Table 7.18. Volumetric shrinkage of compression and normal woods of five conifer species in per cent of oven-dry volume and shrinkage values adjusted for differences in specific gravity. (Verrall 1928)

Species	Shrinkage			Shrinkage/specific gravity		
	Compression wood	Normal wood	Ratio	Compression wood	Normal wood	Ratio
Larix laricina	10.4	21.0	0.50	13.4	34.4	0.39
	8.9	20.5	0.43	12.6	29.4	0.43
Picea mariana	9.2	12.3	0.75	12.3	30.2	0.41
	8.1	13.1	0.62	11.4	30.3	0.38
	9.9	16.9	0.59	13.4	30.8	0.44
	8.5	13.3	0.64	12.0	29.5	0.41
Pinus palustris	9.9	14.4	0.69	12.5	31.0	0.40
Thuja occidentalis	7.2	8.1	0.89	12.8	32.3	0.40
	7.4	8.6	0.86	13.5	29.7	0.45
Thuja plicata	7.3	12.7	0.57	10.9	29.5	0.37

and 6.4%, respectively. It was clear that compression wood shrinks much more longitudinally than normal wood, but that it shrinks less in the radial and tangential directions. The composite, volumetric shrinkage of compression wood is less than for normal wood, especially for the latewood portion (Vintila 1939).

Following the early studies of Hartig, many investigators have examined the shrinkage properties of compression wood and compared them to those of normal wood, but more extensive data have been reported only by Verrall (1928), Pillow and Luxford (1937), Onaka (1949), Kaburagi (1952, 1962), Rak (1957) and Harris (1977). Verrall (1928) measured the volumetric shrinkage of compression and normal woods from six conifer species, obtaining the values shown in Table 7.18. In many of these cases the difference between compression and normal woods was larger than had been reported up to that time, probably because Verrall used specimens consisting of fully developed compression wood. He also divided each shrinkage value with the specific gravity of the wood, thus eliminating the effect of density on the shrinkage. As can be seen from Table 7.18, the values for compression and normal woods were now almost the same for all test specimens and the ratio between them remained constant at 0.40 ± 0.03. Perem (1960), many years later, obtained similar results with compression and normal woods of *Picea glauca* and *Pinus resinosa*, as shown in Table 7.19. Normal wood shrinks volumetrically in direct proportion to its density (Newlin and Wilson 1917, 1919). Evidently, the higher density of compression wood does not cause this wood to shrink more than normal wood. When the effect of the specific gravity is eliminated, compression wood shrinks 2.5 times less in volume than normal wood.

Values reported for the longitudinal, radial, and tangential shrinkages of compression and normal woods are summarized in Table 7.20. These values vary considerably within each group and for both types of wood, but especially

7.3.3.2 Compression Wood

Table 7.19. Increase in specific maximum crushing strength from green to air-dry condition and ratio of volumetric shrinkage to basic density of normal wood and mild, moderate, and severe compression woods of *Picea glauca* and *Pinus resinosa*. (Perem 1960)

Type of wood	*Picea glauca*		*Pinus resinosa*	
	Increase in strength, per cent	Volumetric shrinkage/ basic density	Increase in strength per cent	Volumetric shrinkage/ basic density
Normal wood	139.2	34.8	143.4	30.5
Mild compression wood	126.6	28.9	130.1	27.2
Moderate compression wood	115.8	23.3	108.1	19.6
Severe compression wood	104.1	13.9	100.9	14.8

for compression wood. Obviously, different investigators used wood specimens containing different proportions of compression wood or compression wood of different degrees of severity. This situation, of course, renders many of these data of limited usefulness.

Values stated in the literature to represent either the average or the highest longitudinal shrinkage observed for compression wood also vary considerably. Koehler (1931) quotes a maximum shrinkage for compression wood of 5.4%, observed with stem wood of *Pinus ponderosa*. According to an anonymous (1960) report, the highest longitudinal shrinkage recorded at the Forest Products Laboratory at Madison was 5.78%, which was measured with wood from the lower side of a branch of the same species, a statement later repeated by Hallock (1965). Verrall (1928) observed values up to 4.1%. A longitudinal shrinkage of 1.6% from green to oven-dry condition has been reported for compression wood of *Sequoia sempervirens*, compared to an average value of only 0.18% for corresponding normal wood (Koehler and Luxford 1931, Luxford and Markwardt 1932). According to Pillow and Luxford (1937) severe sompression wood can shrink as much as 5% along the grain, but values in the range of 0.3–2.5% are more common. In their review of compression wood, Dadswell and Wardrop (1949) mention a shrinkage of 3–4%. Dadswell (1958, 1963), in later reviews, claims that compression wood on drying from green to oven-dry conditions can shrink up to 10% in length, a statement also made by Paul and Smith (1950).[1] Côté and Day (1965) mention that compression wood in extreme cases can shrink 6–7% longitudinally, but they do not state whether these figures are based on actual observations. Wooten et al. (1967) observed a maximum shrinkage along the grain of 8.6% for compression wood of *Pinus taeda*. Microtome sections were used in these measurements, and as pointed out by the investigators, the results might not have been representative of bulk wood. Wooten (1967) mentions that compression wood from *Pinus taeda* trees leaning

---

1 Not 20% as quoted by White (1965) in her review of reaction wood.

Table 7.20. Longitudinal, radial, and tangential shrinkage of compression and normal woods. Shrinkage values are from green to oven-dry state in per cent of green condition

Species	Longitudinal shrinkage			Radial shrinkage			Tangential shrinkage			Tangential/Radial		References
	Compression wood	Normal wood	Ratio	Compression wood	Normal wood	Ratio	Compression wood	Normal wood	Ratio	Compression wood	Normal wood	
Abies alba	2.8	0.16	17.65	2.3	3.89	0.59	2.9	8.14	0.36	1.26	2.09	Michels 1941
	3.6	0.6	6.00	1.6	4.5	0.36	3.6	7.0	0.51	2.25	1.56	Giordano 1971
	0.20–1.93	0.015–1.45		0.95–2.90	3.13–3.50		3.45–6.30	4.90–7.85				Constantinescu 1955
Abies concolor	0.54	0.12	4.50									Pillow and Luxford 1937
Abies mayriana	2.36			3.13			4.37			1.40		Kaburagi 1952
Abies procera	0.35	0.15	2.33									Paul et al. 1959
Chamaecyparis obtusa				2.97			3.51			1.18		Yazawa 1944
				2.20			2.50			1.14		
Larix laricina	2.22	0.12	18.5	2.14	6.92	0.31	4.35	10.6	0.41	2.03	4.95	Verrall 1928
Picea abies	0.71	0.18	3.94									Hartig 1901
	1.29	0.09	14.3									
	1.78–2.41	0.29–0.47		2.51–2.72	3.84–4.56	0.44	3.18–3.40	7.64–9.30	0.48	1.26	2.02	Rak 1957
	3.58	0.32	11.2	2.11	4.78		3.15	6.53		1.49	1.37	Ollinmaa 1959
Picea mariana	2.90	0.26	11.2	2.24	3.99	0.56	3.15	7.14	0.44	1.41	1.79	Verrall 1928
Picea sp.	1.10	0.21	5.24	2.64	6.25	0.42	4.60	6.70	0.69	1.74	1.07	
	1.87	0.24	7.80	2.21	4.92	0.45	3.31	8.05	0.41	1.50	1.64	
	2.25	0.18	12.5	2.39	4.51	0.53	3.14	7.50	0.42	1.31	1.66	
Pinus densiflora	2.03	0.24	8.45	2.58	5.57	0.46	3.94	8.38	0.47	1.53	1.50	Onaka 1949
	2.83	0.19	14.9	1.98	5.40	0.37	2.98	7.90	0.38	1.51	1.46	
Pinus palustris	2.13	0.18	11.8									Verrall 1928

7.3.3.2 Compression Wood

Species												Reference
Pinus ponderosa	0.80	0.21	3.81	2.2	3.9	0.56	5.1	6.4	0.80	2.32	1.64	Pillow and Luxford 1937
Pinus radiata	4.55[a] 6.93[a]	0.24[b] 0.17[b]	19.0 40.8				3.26	7.12	0.46			Nicholls 1982
Pinus taeda	1.60	0.14	11.4									Pillow and Luxford 1937
Pinus sp. (Southern pine)	2.46	0.41	6.0	2.15	4.63	0.46	2.61	6.23	0.42	1.21	1.35	Heck 1919, Markwardt and Wilson 1935
Pseudotsuga menziesii	0.20–1.10	0.10–0.25										Trendelenburg 1932
	0.67	0.17	3.94	2.5	3.4	0.74	4.2	5.9	0.71	1.68	1.74	Pillow and Luxford 1937
Sequoia sempervirens	1.19 0.49	0.14 0.17	8.50 2.88	1.4	1.5	0.93	2.4	3.5	0.69	1.71	2.33	Cockrell 1974
Sequoiadendron giganteum	0.65 0.67	0.12 0.07	5.41 9.57	3.9 3.9	3.7 4.9	1.05 0.80	4.3 3.9	7.9 7.9	0.54 0.49	1.10 1.00	2.11 1.61	
Thuja occidentalis	2.32	0.19	12.2	1.45	2.41	0.60	2.75	5.17	0.53	1.90	2.15	Verrall 1928
Thuja plicata	0.94	0.25	3.76									

[a] Earlywood
[b] Latewood

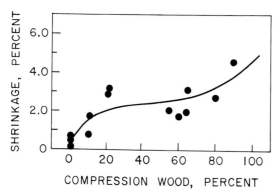

Fig. 7.14. Variation of longitudinal shrinkage with the proportion of compression wood in *Picea abies*. (Ollinmaa 1959)

30° from the vertical had a longitudinal shrinkage of 5%. According to Boutelje (1966, 1972) compression wood from the lower side of knots in *Picea abies* had a longitudinal shrinkage not exceeding 6%. Boone and Chudnoff (1972) state that longitudinal shrinkage of lumber from *Pinus caribaea* increases with increasing proportion of compression wood. A maximum shrinkage of 7.16% has been reported by Wloch (1975) for compression wood of *Pinus sylvestris*, but Harris (1977) found severe compression wood of *P. radiata* to have a maximum longitudinal shrinkage of only 4.49%.

If facts rather than hearsay are considered, it appears that maximum longitudinal shrinkage of compression wood is in the range of 6–7%, very large values indeed, when it is remembered that normal wood shrinks only 0.1–0.3% along the grain. In a 10-foot board, it would mean a shrinkage of 7–8 inches. Lumber containing both types of wood, as a result will warp on drying. The data in Table 7.20 indicate that compression wood can shrink about 20 times as much as corresponding normal wood, but that values in the range of three to ten times are more common (Paul et al. 1959).

The radial shrinkage of compression wood is generally 2–3%, although figures as low as 1% have also been recorded. Values for normal wood fall in the range of 4–6%, about twice as high as for compression wood. Tangentially, compression wood shrinks 3–4% and normal wood 6–8%. For normal wood, the ratio between tangential and radial shrinkage is usually 1.5–2, but for compression wood this ratio is generally only 1.5. According to Dadswell and Wardrop (1949) there is a high, positive correlation between longitudinal shrinkage and extent of compression wood development in native, Australian conifers. Quantitative data reported by Ollinmaa (1959), summarized in Fig. 7.14, indicate the same.

Several explanations have been offered for the high longitudinal shrinkage of compression wood. Sonntag (1904) seems to have been the first to suggest that it could be caused by the large microfibril angle (the helical striations were actually observed) in the secondary wall. However, he also pointed out that this could not be the only factor involved but that chemical composition, and especially extent of lignification ("incrustation") probably also played a role.

7.3.3.2 Compression Wood

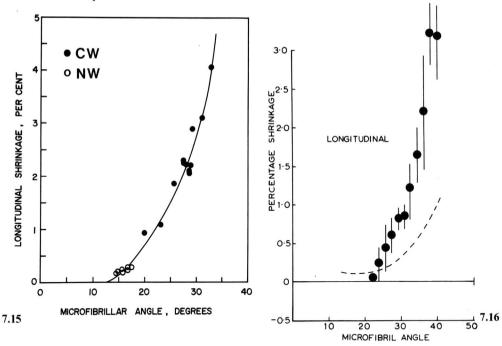

Fig. 7.15. Relationship between microfibril angle and longitudinal shrinkage of normal (*NW*) and compression (*CW*) woods from 12 conifer species. (Drawn from data reported by Verrall 1928)

Fig. 7.16. Relationship between microfibril angle and longitudinal shrinkage (oven-dry) of compression wood in *Pinus radiata*. *Broken line* is based on data for normal latewood (Fig. 7.13). *Vertical lines* represent two standard errors about mean values (*black dots*). (Harris and Meylan 1965)

Koehler (1924, 1946) attributed the large shrinkage to the flat orientation of the microfibrils in the S_2 layer. In this, as in several other respects, compression wood accordingly resembles earlywood rather than latewood. Jaccard and Frey (1928b) developed a qualitative theory dealing with the relationship between microfibril angle and swelling anisotropy of wood. Having observed that this angle was the same in opposite and compression woods, they rejected any influence of this angle and instead attributed the higher longitudinal shrinkage of compression wood to the fact that this wood had a lesser ability than normal wood to imbibe water. Since the orientation of the microfibrils in the S_2 layer actually is much steeper in *normal* wood than in compression wood, this theory is open to doubt, as was already indicated by Trendelenburg (1932).

Verrall (1928) measured the longitudinal shrinkage of compression wood from several species as a function of the pitch of the striations, that is, as a function of the microfibril angle in the S_2 layer. The results of these experiments, which were obviously carried out with great care, are shown graphically in Fig. 7.15. Values for normal and compression woods can be fitted to the same curve and indicate that shrinkage increases rapidly with increasing microfibril angle. The curve is nonlinear, as required by the reinforced matrix theory of Barber and Meylan (1964). The continuous increase in longitudinal shrinkage with increasing angle begins, however, at 10–15° rather than at 30° as predicted by the later theory.

The large longitudinal shrinkage of compression wood has been attributed to the large microfibril angle in S_2 by many later investigators, such as Trendelenburg (1932), Kollmann (1935), Pillow and Luxford (1937), Onaka (1949), Anonymous (1952), Hale (1957), Rak (1957), Boutelje (1966), Bernhart (1965), Foulger (1966) and Preston (1968). It has, however, been pointed out that factors other than the microfibril angle may also influence the swelling and shrinkage behavior of compression wood (Preston 1968). Such factors could include the high lignin content and the peculiar distribution of lignin, the relatively low crystallinity of the cellulose, and several anatomical features.

Onaka (1949) rejected Frey-Wyssling's (1940b, 1943) theory concerning the effect of the middle lamella on shrinkage, correctly pointing out that the pectin which was supposed to cause this region to swell in water, actually is present in only small amounts. He instead sought other explanations for the high longitudinal shrinkage of compression wood. Referring to the earlier treatment presented by Kollmann (1935), a much simplified version of the later theory of Barber and Meylan (1964), he agreed that the large shrinkage of compression wood along the grain could be a result of the 45° inclination of the microfibrils in the S_2 layer. However, he also considered other factors to affect shrinkage, and especially the thick cell wall and the high lignin content of compression wood. He came to this conclusion by the difficulty of explaining how normal wood can show such a small longitudinal shrinkage despite the occasional occurrence of microfibril angles as large as 20–35°. The reinforced matrix theory (Barber and Meylan 1964) predicts minimum shrinkage exactly within this range, and this aspect of Onaka's (1949) contributions is therefore now of less relevance.

Dadswell et al. (1958) (Dadswell 1963) offered an entirely different explanation of the longitudinal shrinkage of compression wood. Pointing out that the S_1 layer with its transverse orientation of the microfibrils is larger in compression wood than in normal wood and apparently also only slightly lignified, they suggested that this layer could influence the longitudinal shrinkage. They were of the opinion that such a shrinkage on the part of S_1 would not be noticeably restrained by the S_2 layer because of the presence of helical cavities in the latter. In her review of the relationship between wood shrinkage and structure, Kelsey (1963) pointed out that cases are known where compression and normal woods exhibit the same microfibril angle in S_2, yet the former wood invariably shrinks much more along the grain than the latter. She did not offer any explanation of this behavior, but felt that factors other than the microfibril angle must also be involved.

In comparing the results predicted by the reinforced matrix theory with experimental data, Harris and Meylan (1965) discussed the shrinkage characteristics of compression wood. As can be seen in Fig. 7.16, the longitudinal shrinkage of compression wood from *Pinus radiata* is much larger than that of normal

7.3.3.2 Compression Wood

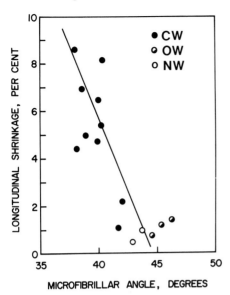

Fig. 7.17. Variation of longitudinal shrinkage with microfibril angle of normal (*NW*), opposite (*OW*), and compression (*CW*) woods in 2-year-old seedlings of *Pinus taeda*. (Drawn from data reported by Wooten et al. 1967)

latewood and also larger than that of normal earlywood at identical microfibril angles. Citing Dadswell (1958), Harris and Meylan (1965) stated that the anomalous shrinkage of compression wood in relation to its microfibril angle must be a consequence of its ultrastructure. Besides the influence of the thick S_1 layer, they also considered the possibility that the high longitudinal shrinkage could be caused by a twisting of the helical ridges in S_2, a twisting that would be restrained by the compound middle lamella. It is also possible that the helical cavities are able to close on shrinking. They drew attention to the fact that compression wood contains less cellulose than normal wood and accordingly should have a lower E/S ratio, which should result in higher longitudinal shrinkage.

Wooten et al. (1967) measured the longitudinal shrinkage and microfibril angles in compression, opposite, and normal woods from 2-year-old seedlings of *Pinus taeda*. Since the original diagram is somewhat misleading (Westing 1968), the data obtained have been replotted in Fig. 7.17. While it is doubtful if any straight line could be drawn through these experimental points, it is nevertheless clear that the unusually high longitudinal shrinkage of the compression wood was not positively correlated with the microfibrillar angle in S_2. In fact, if there were any correlation, it would actually be negative, since the normal and opposite woods, which showed only slight shrinkage, had larger microfibril angles than the compression wood specimens. It should be kept in mind, however, as mentioned by Panshin and de Zeeuw (1980), that all samples consisted of juvenile wood, and that different results may have been obtained if mature wood had been examined. Wooten et al. (1967) concluded, as had several investigators before them, for example Schniewind (1962b), that some other factors, in addition to the microfibril angle in S_2, must determine the longitudinal shrinkage of compression wood. The possible influence of the S_1 layer, already emphasized by other investigators (Dadswell et al. 1958, Dadswell 1963), was mentioned, and it was suggested that both the S_1 and the S_2 layers might contribute to the high longitudinal shrinkage of compression wood. In tracheids with exceptionally thin S_2 layers, the concomitant large longitudinal shrinkage could be caused by the influence of the S_1 and the S_3 layers (Koehler 1931, Hale and Clermont 1963).

Barrett et al. (1972) developed a theoretical shrinkage model for wood cell walls which was used for calculating the longitudinal and transverse shrinkage strains of normal earlywood and latewood and also for compression wood. The compression-wood model was assumed to have the composition previously suggested by Mark and Gillis (1970). Here, the thickness ratios of the P, S_1, and S_2 layers were 7:28:65 with a microfibril angle of 90° in the primary wall, ±80° in S_1, and 40° in S_2. The authors pointed out that the chemical composition of compression wood differs considerably from that of normal wood and ought to be taken into account, although this was not done in the present analysis. According to these computations, latewood should shrink much less than earlywood in the longitudinal direction. The shrinkage of compression wood should be large, albeit not as large as that of earlywood.

In his study of the effect of the microfibril angle on the relationship between longitudinal shrinkage and moisture content, Meylan (1972) examined some specimens consisting of pronounced compression wood. Unlike the case with normal wood, the relation between shrinkage and moisture content was linear, and there was no hysteresis. According to von Pechmann (1973) anatomical factors contributing to the high longitudinal shrinkage of compression wood are the helical cavities and the absence of a restraining S_3 layer.

Cockrell (1974) studied the shrinkage of normal and compression woods of *Sequoiadendron giganteum*, obtaining the values listed in Table 7.20. According to Cockrell, the longitudinal shrinkage of 0.6–0.7% must be regarded as modest for what appeared to be well-developed compression wood. It is, however, by no means unique, for all of Cockrell's data are astonishingly similar to those reported by Pillow and Luxford (1937) for normal and compression woods of *Pseudotsuga menziesii*, and they are not very different from those found by these investigators for *Sequoia sempervirens* (Table 7.20). The relatively small microfibril angles in the giant sequoia compression wood, 25–32°, were probably responsible for the small longitudinal shrinkage. All shrinkage values, longitudinal, radial, and tangential, were larger when the wood was dried quickly to 100 °C than when it was first air-dried.

According to Brodzki (1972) a 1,3-glucan, which he referred to as "callose" is deposited within the helical cavities of compression-wood tracheids. It is obviously the same polysaccharide as the $(1 \rightarrow 3)$-β-D-glucan first identified in compression wood by Hoffmann and Timell (1970, 1972) and given the name laricinan (Chap. 5.4). In a later investigation, Wloch (1975) found that compression wood of *Pinus sylvestris* varied in longitudinal shrinkage from 1% to 8%. The shrinkage was directly related to the thickness of the tracheid wall and especially to that of the S_2 layer. Inspection in fluorescent light of sections stained with aniline blue revealed the presence of more "callose" the thicker the tracheid wall. From these observations Wloch concluded that the glucan was responsible for the high longitudinal shrinkage of compression wood. He could

7.3.3.2 Compression Wood

find no relationship between the shrinkage and the microfibril angle in S_2. He also postulated that callose interacts with lignin in differentiating compression wood, generating a compressive stress along the grain (Chap. 17).

Boyd (1978) objected to Wloch's conclusions and convincingly demonstrated that neither growth stresses nor longitudinal shrinkage can be attributed to the presence of a $(1 \rightarrow 3)$-β-D-glucan in compression wood. While this might seem fairly obvious (Wilson 1981), Boyd's arguments have undoubtedly served a valuable purpose in laying to rest an unfounded theory. Boyd (1978) also pointed out that the callose observed by Brodzki (1972) and Wloch (1975) probably is located on the surface of the helical ribs and not in the cavities (Chap. 6.3.3). Wloch's observation that the longitudinal shrinkage of compression wood is positively correlated with tracheid-wall thickness is correct and was later confirmed by Boyd (1977) and Harris (1977).

Boyd (1977) studied the relationship between shrinkage and wood structure in normal and reaction woods of *Pinus radiata* and *Eucalyptus regnans*. The microfibril angle in S_2 was always the major factor determining both longitudinal and volumetric shrinkage, but it was never the only factor involved. The thickness of the cell wall, the relative proportions of the S_1 and S_2 layers, and the extent of lignification also influenced the shrinkage. With regard to compression wood, Boyd drew attention to several additive factors responsible for its large longitudinal shrinkage, namely the thick cell wall, the S_1 layer, the large microfibrillar angle in S_2, and the presence of the helical cavities. It is interesting to note in this connection that, according to Cave (1972b), the proportions of S_1 and S_2 can be expected to influence strongly the longitudinal shrinkage.

Harris (1977) investigated the influence of microfibril angle, density, and anatomical properties on the longitudinal shrinkage of different grades of compression wood in *Pinus radiata*. It should be noted that the polarized light method used (Cousins 1972) measured the average microfibril angle in all layers of the tracheid wall in both opposite and compression woods. Three grades of compression woods were examined, probably corresponding to mild (grade 1), moderate (grade 2), and pronounced (grade 3) compression woods. Their average longitudinal shrinkage was 1.62%, 2.46%, and 2.67%, respectively, compared to 0.60% for normal side wood. The influence of microfibril angle on longitudinal shrinkage is shown graphically in Fig. 7.18. The shrinkage increased with increasing microfibril angle in both compression and opposite woods. In the latter tissue it was independent of the severity of the compression wood present. Shrinkage of the compression wood increased more rapidly with increasing microfibril angle from grade 1 to grade 3. Obviously, this angle was not the only factor determining the longitudinal shrinkage of compression wood. Harris estimated that it was responsible for only 30% of the variance in shrinkage. Among anatomical properties, such as thickness of the cell wall, intercellular spaces, helical cavities, or presence of an S_3 layer, only tracheid wall thickness was consistently correlated with a large longitudinal shrinkage. This was also clear from the fact that at a constant microfibril angle, shrinkage of the compression wood was directly related to the difference in density between opposite and compression woods. Harris (1977) concluded that both microfibril

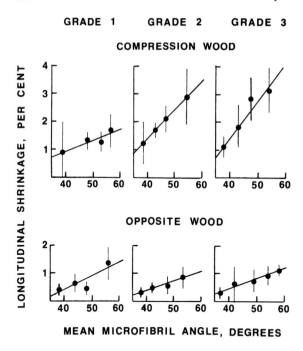

Fig. 7.18. Relationship between microfibril angle and longitudinal shrinkage in three grades of compression wood, ranging from mild to pronounced, and the corresponding opposite woods. *Pinus radiata*. (Redrawn from Harris 1977)

angle and thickness of the tracheid wall are responsible for the large longitudinal shrinkage of compression wood.

Meylan (1981) has reported that in the primitive, vessel-less dicotyledon *Pseudowintera colorata* (Chap. 8.4) longitudinal shrinkage was positively correlated with the microfibril angle in tissues from the lower and upper sides of a branch. The only difference between the tracheids in these two locations was the larger microfibril angle on the upper side. This angle seemed to be the only factor influencing the shrinkage behavior of the two woods.

In their studies of the physical properties of specimens from a board of *Picea abies* that contained much compression wood, Schulz and Bellmann (1982) (Schulz et al. 1984) found that the longitudinal swelling and samples rich in compression wood was ten times larger than that of normal wood. It increased slowly at low concentrations of compression wood, more rapidly at medium concentrations, and again more slowly at higher concentrations, as shown in Fig. 7.19. An attempt to calculate the shrinkage of the entire, original board from the data obtained for the separate specimens was not entirely successful.

The relatively small transverse and volumetric shrinkage of compression wood has also elicited several attempts at an explanation. Hartig (1901) thought that the small transverse shrinkage could easily be explained by the structure of compression wood. He pointed out that the cell wall substance contracts to the same extent as in normal wood, but that the outer volume of the tracheids changes less. Instead, the helical, lamellar ridges in the inner secondary wall

7.3.3.2 Compression Wood

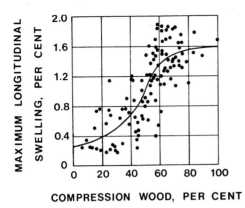

Fig. 7.19. Variation of maximum longitudinal swelling with the proportion of compression wood present in *Picea abies*. (Modified from Schulz and Bellmann 1982 and Schulz et al. 1984)

contract, and the lumen and the helical cavities increase in volume. Sonntag (1904) objected to Hartig's explanation and instead attributed the small transverse shrinkage to the high lignin content of compression wood. When wood was treated with chlorine, the delignified material was found to possess a greater capacity for both swelling and shrinkage, an observation later confirmed by Bosshard (1956). However, normal wood from certain conifers, for example *Tsuga canadensis*, is almost as highly lignified as severe compression wood, yet such wood has a high transverse shrinkage. It seems unlikely, therefore, that a high lignin content per se should be a decisive factor in restricting swelling and shrinking.

Verrall (1928) found that neither the high specific gravity nor the large microfibril angle of compression wood could fully account for the small tangential shrinkage. He suggested that the checks formed on drying and the low fiber saturation point might be responsible. Rak (1957) likewise attributed the small volumetric shrinkage to the low fiber saturation point but also to the high lignin content. Considering that the fiber saturation point of compression wood differs only slightly from that of normal wood, however, this could only be a minor factor. Kelsey (1963) found that on drying, the lumen expands more in compression wood than in normal wood, as had been postulated by Hartig (1901). This, according to Kelsey, could be attributed to the fact that the S_3 layer is missing in compression-wood tracheids, a circumstance also mentioned by Onaka (1949). In addition, Kelsey (1963) refers to the spoke-like arrangement of the microfibrils in S_2.

Harris and Meylan (1965) determined the tangential shrinkage of *Pinus radiata* compression wood. As can be seen in Fig. 7.20, the shrinkage of compression wood was less than that of normal latewood for the same microfibril angle, a fact recognized earlier by Verrall (1928). The cross-over point of the transverse and longitudinal curves also occurred at a smaller angle than the 45–48° valid for normal wood. Harris and Meylan pointed out that on shrinking the change in the cross section of the cell wall substance will not be translated into a decrease in the outer cell diameter, but rather in an opening of the

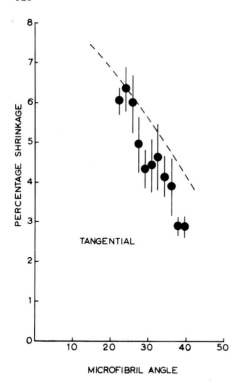

Fig. 7.20. Relationship between microfibril angle and tangential shrinkage (oven-dry) of compression wood in *Pinus radiata*. Broken line is based on data for normal latewood (Fig. 7.13). *Vertical lines represent two standard errors about mean values (black dots).* (Harris and Meylan 1965)

helical cavities on the inside, an explanation similar to that offered by Hartig (1901) and later also advanced by Kelsey (1963). The data in Table 7.20 indicate that the transverse anisotropy of compression wood is less than that of normal wood. This is in agreement with the general trend in normal wood, where this anisotropy decreases with increasing density of the wood. It will be recalled that latewood exhibits much less anisotropy than does earlywood (Pentoney 1953).

Cockrell (1974) found the ratio between tangential and radial shrinkage to be only 1.0 and 1.1 for compression woods, compared to 2.1 and 1.6 for corresponding normal woods, a situation caused by the lesser tangential shrinkage of the compression wood. Inspection of the data in Table 7.20 reveals that, while a value of 1.1 – 1.2 for the ratio of tangential to radial shrinkage has been reported on a few occasions (Heck 1919, Michels 1941, Yazawa et al. 1951, Rak 1957), in most instances this ratio has been found to be significantly larger. Its average value, based on all data in Table 7.20, is 1.51 for compression wood and 1.71 for normal wood. That the transverse anisotropy of compression wood is less than that of normal wood is evident, although by how much is not quite clear.

Compression wood is not the only xylem tissue that exhibits a large longitudinal shrinkage. Juvenile wood, which has a large microfibrillar angle in S_2, also shrinks much more than normal wood along the grain, often ten times as much (Koehler 1938, Paul 1930, 1957a). In some conifers, second-growth, rapidly growing trees at times have a wood which likewise has a high longitudinal shrinkage (Anonymous 1960, Paul 1957b). An exceptionally large shrinkage is found in sclerenchyma cells of pine cones, where the orientation of the microfibrils is transverse, an unusual arrangement. These fibers, as a result, have a longitudinal shrinkage of 10–36%, which serves the purpose of opening the cones (Allen and Wardrop 1964, Harlow et al. 1964).

7.3.3.3 Branch Wood

In the first study devoted to the shrinkage of compression wood, Hartig (1896) measured the volumetric shrinkage of the lower and upper portions of numerous branches from *Picea abies* and *Pinus sylvestris*. Some of his results are listed in Table 7.6. With only few exceptions, the lower part of the branch, where the compression wood was located, contracted somewhat less in volume on drying. This is, of course, what would be expected, and the reason why the difference between upper and lower portions was not larger must have been that neither consisted of pure opposite and compression woods, respectively. Sonntag (1904) noted that the longitudinal shrinkage of opposite wood was exceptionally low despite a microfibrillar angle of 20°. He found this fact difficult to understand. It is actually at this angle that minimal shrinkage along the grain should occur according to the reinforced matrix theory (Sect. 7.3.3.1).

Fegel (1938) found that branch wood of seven conifer species exhibited a slightly larger longitudinal shrinkage than did corresponding stem wood, as seen in Table 7.12. Branch wood shrinks less volumetrically than stem wood in *Pinus densiflora* (Table 7.9) (Yazawa et al. 1951). More extensive data have been reported by Boutelje (1965, 1966, 1973) in his investigations on knot wood and branch wood in *Picea abies* and *Pinus sylvestris*. The results summarized in Tables 7.10 and 7.11 show that the lower portion of the branches invariably had a smaller radial and tangential shrinkage than the upper part. The large differences between upper and lower sections, especially in the case of *Picea abies*, indicate that the former must have consisted largely of opposite and normal woods and the lower of compression wood. The same conclusion applies to the data in Table 7.17, although in this case it is probable that the upper part of the pine branch contained some compression wood.

As can be seen from Table 7.8, branch compression wood in *Pinus sylvestris* swells less volumetrically than either normal branch wood or stem wood. In the longitudinal direction, on the other hand, branch compression wood swells more than normal wood (Götze et al. 1972). In a study of checking of intergrown knots in *Pinus radiata*, Kininmonth (1961) found that these knots, their higher density notwithstanding, shrank less in volume than the adjacent wood. He attributed this to the high compression wood content of conifer branches.

The complex shrinkage of branches has lent itself to a quaint but rather practical use which undoubtedly goes far back in time. It was first mentioned by Hartig (1901), who states that in several rural areas of Germany, for example in Berchtesgaden, dead spruce branches are used for weather forcasting. Such branches bend downward in dry, good weather, but start to bend upward when rain is approaching and humidity begins to increase. The explanation is, of course, simple. With decreasing moisture in the air, the compression wood on the under side of the branches contracts far more along the grain than the opposite wood on the upper, and the branch is pulled down. When humidity rises, the compression wood swells more longitudinally and forces the branch upward. Engler (1924) has also commented on this use of branch compression wood, which apparently must have been widespread. According to an anonymous article, a spruce branch was once used as a hygrometer at the Forest Products Laboratory (U.S. Forest Service) (Anonymous 1924). Its lower part was nailed to a door, and the tip described an arc of 80° as the humidity of the indoors changed from extremely dry in the winter, when the limb was bent down, to the disagreeably high humidity characteristic of the summer months, when it became almost upright.

Mork (1928) mentions that a forester in Norway has a spruce branch on his house wall as a weather indicator and that he always inspects it before he sets out into the forest, superstitious though this may appear to his less knowledgeable neighbors. The scientific aspects were discussed in the same year by Jaccard and Frey (1928b).

Köster (1934) relates how he was once surprised in a spruce forest by a sudden downpour which came after a long drought. Standing in the stillness after the rain had stopped, he could hear a brisk and clear crackling in all the dead spruce branches as they began to bend upward. He also mentions that as branches begin to decay, the effect disappears. The phenomenon has been referred to also by others (Trendelenburg 1932, Pillow and Luxford 1937, Hartmann 1943, Vihrov 1960). Little (1967) used the extent of bending of a shoot on drying to measure the total amount of compression wood present (Chap. 3.5).

7.3.3.4 Conclusions

The interaction between compression wood and water is complex, and it is likely that many factors are involved. The fact that compression wood absorbs less water than normal wood is probably a result of its higher lignin content, since lignin is less hygroscopic than either cellulose or hemicelluloses. Once imbibed, the water is trapped more effectively inside compression wood, which therefore loses less water than normal wood on drying in the air. The fiber saturation point of compression wood is only slightly lower than that of normal wood, but it is possible that the strength properties of compression wood are affected by this fact.

The low permeability of compression wood is probably not to be attributed to any special anatomical feature of the pits. The pits, however, are fewer in number than in normal wood (Chap. 4.10). Flow of liquid is largely impeded by the narrow lumen of the thick-walled tracheids.

The conspicuously high longitudinal shrinkage of compression wood can be explained in terms of its chemical composition and structure. The first, and

7.3.3.4 Conclusions

most obvious factor, is the large microfibril angle in the S_2 layer which is associated with a large shrinkage along the grain as soon as it exceeds about 30°. The low cellulose and high lignin content of compression wood result in a low E/S ratio, and according to the reinforced matrix theory this leads to higher longitudinal shrinkage at a given microfibril angle. It is evident, however, that these two factors alone are not sufficient to account for the entire extent of shrinkage, which in extreme cases has been found to be as high as 6–7%.

Additional structural factors are undoubtedly involved, and prime among them the thick tracheid wall in compression wood. The influence of the S_1 layer probably deserves more attention than it has received in the past. This layer is, both on an absolute and a relative basis, much thicker in compression wood than in normal wood tracheids (Chap. 4.4.4). It is probably not enough to consider only the restraining influence of this layer; its active contribution to swelling and shrinkage along the grain must also be taken into account. The orientation of the microfibrillar helix in S_1 is almost transverse to the fiber axis. Judging from the shrinkage taking place in pine cone scales, which show a similar arrangement of the microfibrils, the S_1 layer, were it completely free from restraint, would probably shrink over 30% in the longitudinal direction. In tension wood fibers, the S_2 layer is much reduced in thickness compared to normal wood, leaving the S_1 layer to play a greater role. In all likelihood, this is the main reason for the high longitudinal shrinkage of tension wood (Norberg and Meier 1966). Other factors which probably contribute to the large longitudinal shrinkage of compression wood, and which have been mentioned in the literature, are the helical cavities and the absence of a restraining S_3 layer. The influence of the intercellular spaces has been found to be minor (Harris 1977).

Despite its high density, compression wood exhibits less transverse shrinkage than normal wood. The lower fiber saturation point and the higher lignin content are probably unimportant in this context. More decisive is the large microfibrillar angle, but even this factor cannot account for the entire phenomenon. Again, the peculiar anatomy of the compression wood tracheids is most likely involved. The absence of the S_3 layer will make it possible for the S_2 to contract on drying, resulting in an expansion of the lumen which has been observed by several investigators. The helical cavities in S_2 would also tend to close. The net result would be that the outer dimensions of the tracheids would change less than in normal wood, where the S_3 layer impedes internal movements.

Radial and tangential shrinkage do not differ as much in compression wood as they do in normal wood. The virtual lack of true earlywood in severe compression wood is probably responsible for this. There are no indications that the microfibril angle should be different in the radial and tangential walls of the rounded compression wood tracheids, nor do the rays seem to play any role. The relatively small radial, tangential, and volumetric shrinkage, and the modest transverse shrinkage anisotropy are all characteristics that should make compression wood superior in utilization to normal wood. Unfortunately, all these desirable properties are more than offset by the remarkably large shrinkage and swelling in the longitudinal direction. It is this enormous shrinkage along the grain which, more than any other property, renders compression wood unsuited for lumber, where it almost invariably is present together with normal wood. The difference in longitudinal shrinkage between the two types of wood causes warping, twisting, and bowing (Chap. 18.4.2).

The large shrinkage of compression wood along the grain is not the only property of this wood that makes it detrimental in lumber. Next in importance to the shrinkage on drying is the fact that compression wood, especially in the air-dry state, in many respects is inferior in strength to normal wood.

7.4 Mechanical Properties

7.4.1 Introduction

The mechanical, or strength, properties of compression wood, because of their practical importance, have been the subject of several extensive investigations and have been mentioned or referred to briefly on many more occasions. Although the number of detailed reports is limited, the material on hand is quite large. A considerable number of conifer species have been studied and both green and air-dry woods have been examined. Most strength properties of wood increase with increasing specific gravity, with the exact relationship varying for different strength characteristics. The density of compression wood is always higher than that of normal wood. Since it is of interest to know the intrinsic strength properties of the material itself, the effect of density has often been eliminated in comparisons between compression and normal woods, that is, specific strength properties have been used. In many investigations, compression wood has not been compared with normal wood but with opposite wood from the same growth ring. Opposite wood, however, differs in several respects from normal wood, having, for example, a higher tensile strength.

When green wood dries, its strength properties undergo a considerable change, and normal wood as a rule becomes stronger with increasing loss of water. These changes are not the same for normal and compression woods, and comparisons between the two tissues must be restricted to values obtained under equal conditions. The mechanical properties of green wood are of decisive importance to the living tree. Strength characteristics of air-dry wood are important in the utilization of wood as lumber. The mechanical properties of juvenile wood are in general inferior to those of mature wood (Pearson and Gilmore 1971, 1980).

The strength properties of branch wood are determined not only by the compression wood on the lower side, but also by the opposite wood on the upper and the normal wood located on the flanks (side wood). Because of this complex situation, the overall strength characteristics of branch wood are generally intermediate between those of normal and compression woods and depend on the relative amount of the three different tissues present.

Excellent reviews of the mechanical properties of wood in general, some of them mentioned in Sect. 7.1, have been authored by Newlin and Wilson (1917), Koehler (1924), Markwardt and Wilson (1935), Wangaard (1950), Brown et al.

1952, Silvester 1967, Kollmann 1968, Dinwoodie 1981, and Schniewind (1981). The *Wood Handbook*, published by the Forest Products laboratory at Madison, Wisconsin (1974) contains a wealth of information. The cell-wall mechanics of tracheids has been admirably treated by Mark (1967). The relationship between structure and mechanical properties of wood has been discussed in an excellent review by Dinwoodie (1975).

7.4.2 Compression Wood

7.4.2.1 Early Contributions and General Review

Nördlinger (1890) mentions that when compression wood is formed in *Picea abies*, the compressive strength of the wood increases, albeit not in proportion to the increase in specific gravity. The first quantitative determinations of the strength properties of compression wood were carried out by Hartig (1896, 1901), who measured the modulus of elasticity in static bending and in compression and tension parallel to the grain of compression and normal woods of *Picea abies*. The values obtained for compression wood were 58 900, 68 650, and 63 900 kg cm^{-2}, while those for normal wood were 108 000, 121 800, and 116 000 kg cm^{-2}, respectively. All three moduli were only about half as large for compression wood as for normal wood, a result confirmed by all later investigators except Rothe (1930). It is likely (Trendelenburg 1932) that green wood was used in all these tests. Schwappach (1897, 1898), who carried out extensive investigations on the density and strength properties of several conifers, found that compression wood, formed on the leeward side of *Pinus sylvestris* trees, was weaker in compression parallel to the grain than opposite wood from the windward side of the stem. His values, however, are highly erratic (Rothe 1930) and in almost 40% of the cases, compression wood was actually found to be the stronger wood.

Sonntag (1904), in an excellent treatise which deserves more attention than it has received by later investigators, discussed several aspects of compression wood in *Picea abies*, including its mechanical properties. Suffering from a lack of proper equipment, he used very small test samples, only 1 mm^2 in cross section, an unsatisfactory technique according to Rothe (1930). Sonntag, however, was evidently a careful investigator, for his results agree well with those obtained by later scientists using more sophisticated methods. He found that air-dry *Picea abies* compression wood had a maximum crushing strength which was 44% greater than that of corresponding opposite wood, a fact which he attributed to the thick cell walls in the former. Confirming Hartig's (1896, 1901) earlier results, Sonntag found the modulus of elasticity in tension to be much greater for normal wood than for compression wood. Extensibility, however, was the same, or 1.5–2% for both types of wood. Sonntag also studied the bending behavior of branches, as referred to later (Chap. 7.4.3). In this connection, he stressed the importance of distinguishing between results obtained within and beyond the limit of proportionality, a warning not heeded by all later investigators.

Sonntag (1909) also made an attempt to correlate the total extensibility at rupture of a large number of plant materials with their microfibril angle. Normal wood of *Picea abies* with a microfibril angle of 20.5 ° had an extensibility of 1.6—1.9%. Corresponding figures for compression wood from the same species were 40.5 ° and 1.4—2%, or approximately the same extensibility, a somewhat unexpected result. Compression wood from the lower side of branches from *Pseudotsuga menziesii*, on the other hand, had microfibril angles within the range of 48—71 ° and in this case the extensibility at rupture was 4.7—7.0%. Sonntag attributed the high extensibility of this compression wood to the large microfibril angle in the S_2 layer and considered the chemical composition of the wood to be of less importance.

Among major investigations devoted to the strength properties of compression wood since 1904 are those of Rothe (1930), Trendelenburg (1931, 1932), Luxford and Markwardt (1932), Markwardt and Wilson (1935), Pillow and Luxford (1937), Onaka (1949), Perem (1958, 1960), and Ollinmaa (1959). Later contributions have been made by Bernhart (1966), von Pechmann and Courtois (1970), Reinhardt (1972), Cockrell and Knudson (1973), von Pechmann (1973), Ohsako et al. (1973), Ohsako (1975, 1976, 1978), Ohsako and Kato (1977), and Ohsako and Tsutsumi (1977). Some of these investigations have been summarized by Kärkkäinen and Raivonen (1977). Data and discussions have also been contributed by others. Janka (1904, 1909, 1915) reported on the hardness and compressive strength of compression and normal woods from *Picea abies*. A beam of southern yellow pine in a factory roof which had failed because of the presence of compression wood was studied by Heck (1919), representing the first report of this type in the English-language literature.[2] A year later, Burns (1920) published results of a few strength tests with normal and compression woods of *Pinus strobus*.

Verrall (1928), in an unpublished report, discussed compressive and tensile strength and hardness of compression wood from two species. More extensive data were reported by Mori (1933) for wood of *Thujopsis dolabrata*, unfortunately in a paper with a too brief English summary. Mayer-Wegelin (1931) presented a critical review of Rothe's (1930) investigation, which is referred to in Sect. 7.4.2.2.3. Münch (1937, 1938), under whose direction this work had been carried out, later offered additional comments. In a report on the causes of brashness in wood, Koehler (1933) briefly referred to the strength properties of compression wood. Even briefer is the reference made by Peck (1933) in the same year. Berkley (1934) in an extensive investigation of the properties of wood from *Pinus echinata*, *P. palustris*, and *P. taeda*, studied the mechanical properties of compression wood. Unfortunately, his concept of what constitutes

---

2 Heck's contribution, published in an American engineering journal, has been misquoted twice in the later literature. Trendelenburg (1932), referring only to the Forest Products Laboratory in Madison, Wisconsin, correctly transcribed Heck's data into the metric system but reported the species as *Pinus ponderosa*. Onaka (1949), who apparently based his information on Trendelenburg's publication, also refers to the species as *Pinus ponderosa*. In addition, he quotes Heck under the name of *Fleck*.

7.4.2.1 Early Contributions and General Review

compression wood differs from that of other investigators, and it is accordingly difficult, if not impossible, to evaluate his results.

Banks (1954) studied the mechanical properties of wood from trees growing in South Africa. He pointed out that compression wood is not as strong as normal wood of the same density, a fact that he attributed to the large microfibril angle and the high lignin content of this wood. Banks (1957) also investigated the mechanical properties of wood from the major (compression wood) and minor (opposite wood) radius in stems of *Pinus montezumae*. The former wood was stronger in compression and harder but weaker in modulus of elasticity and toughness than the latter. According to Sakata and Saiki (1961), compression wood of *Cryptomeria japonica* is weaker than normal wood. Suzuki (1968, 1969) carried out comprehensive investigations on the relationship between strength properties and cell structure of several conifers, among them *Abies firma*, *Cryptomeria japonica*, *Larix leptolepis*, and *Pinus densiflora*. Extensive data were recorded for the tensile strength characteristics of compression wood of these species. Tensile properties of compression wood tracheids of *Picea jezoensis* were investigated at the same time by Furuno et al. (1969). Pearson and Gilmore (1971) studied the effect of compression wood on the moduli of rupture and elasticity in static bending of wood from *Pinus taeda* (Sect. 7.4.2.9).

Many investigators have found the mechanical properties of wood to vary because of the erratic occurrence of compression wood in the stem. Campbell (1968), for example, attributed part of the abnormally high variability of the strength properties of *Pinus radiata* and *P. patula*, grown in Kenya, to the presence of compression wood. Bernhart (1966) reported strength data for normal wood and "wood containing compression wood" from *Picea abies*. In agreement with other investigators, he found that wherever compression wood occurred in the stem, it lowered the strength of the wood when the latter was tested in the air-dry condition, and not only in tension and bending but also in compression. Green compression wood was, however, stronger than normal wood in compression. The density of the compression wood samples tested was practically the same as for the normal wood, suggesting that these specimens probably contained little fully developed compression wood. Bernhart's data for this reason are not representative of such wood. They are, nevertheless, of considerable interest in demonstrating how even occurrence of a moderate amount of compression wood impairs many strength properties of stem wood. Similar investigations were later carried out by von Pechmann and Courtois (1970) with stem wood from *Pseudotsuga menziesii* and by Reinhardt (1972) with *Larix leptolepis*, both grown in Germany. Stem zones containing compression wood had a reduced compressive and bending strength and a very low modulus of elasticity in bending. The density of the wood was, however, relatively unaffected (von Pechmann 1969, 1973).

Branch wood was studied by several of the early investigators, such as Hartig (1896, 1901), Ursprung (1906), Janka (1904), and Sonntag (1904). Fegel (1938, 1941) compared certain strength properties of stem wood and branch wood from several conifer species. A detailed discussion, but few data, were reported by Wegelius (1939).

Knots, which have a high content of compression wood, generally impair the strength characteristics of lumber, especially in compression and bending. Information on this subject is available in several publications dealing with wood defects in general (Chap. 1.2).

In the following, each strength property and the mode of fracture of compression wood will be treated in relation to normal or opposite woods. The variation of the mechanical properties with extent of compression wood development and the factors which probably determine these characteristics will also be discussed. Branch wood is dealt with separately. Most of the data reported in the literature are included, but no attempt has been made to review all information available. For the sake of convenience and ease of survey, some duplication of the tabulated material has been unavoidable. In a few cases, strength values are presented both in metric and in English units.

7.4.2.2 Compression Parallel to the Grain

7.4.2.2.1 Maximum Crushing Strength. Maximum crushing strength measures the ability of wood to sustain a slowly applied end load over a short period. It is a property that is easy to determine, but it is not very sensitive to minor changes in the wood tested. It is positively correlated with the specific gravity of the wood (Yamai 1957). Only measurements in compression parallel to the grain have been reported for compression wood.

More extensive data for the maximum crushing strength of compression wood were first published by Trendelenburg (1931, 1932). His results, obtained with *Pseudotsuga menziesii*, are listed in Table 7.21. They also include moisture content, ring width, and specific gravity of the wood, as well as Young's modulus of elasticity and stress at proportional limit. Similar data were reported 3 years later by Markwardt and Wilson (1935) for *Pinus echinata, P. ponderosa,* and *Sequoia sempervirens.* This study is one of the most comprehensive that has so far been devoted to the mechanical properties of compression wood and particularly with respect to static bending. Data for the last two species, which are the most complete, are shown in Table 7.22 and include values for both green and air-dry wood. Specific strength properties not reported in the original publication have been calculated from the values for specific gravity, using the relationships between this parameter and the different strength properties adopted by Panshin and de Zeeuw (1980).

Further measurements at the Forest Products Laboratory were at the same time undertaken by Pillow and Luxford (1937) and reported in their review of the properties of compression wood. These included values obtained previously by Markwardt and Wilson (1935) for *Pinus ponderosa* and *Sequoia sempervirens,* as well as new data for *Abies concolor, Pinus taeda,* and *Pseudotsuga menziesii.* To facilitate comparison, all five species have been included in Table 7.23, which also contains data obtained later by Perem (1958, 1960) with *Picea glauca* and *Pinus resinosa* and by Cockrell and Knudson (1973) with *Sequoiadendron giganteum.* Certain results obtained with green and air-dry woods are compared in Table 7.24. Finally, a summary is given in Table 7.25 of reported

7.4.2.2 Compression Parallel to the Grain

Table 7.21. Strength in compression parallel to the grain and related properties of compression and normal woods of *Pseudotsuga menziesii*. All strength properties in kg cm^{-2}. (Trendelenburg 1931, 1932)

Wood	Moisture, per cent	Ring width, mm	Specific gravity G	Maximum crushing strength C		Modulus of elasticity Y		Stress at proportional limit σ_{PL}	
				C	C/G	Y	Y/G	σ_{PL}	$\sigma_{PL/G}$
Compression	Green	8.8	0.452	243	538	77 700	171 900	154	341
Normal	Green	4.4	0.423	208	492	107 000	253 000	155	366
Compression	Green	7.9	0.448	219	489	91 600	204 500	68	152
Normal	Green	4.9	0.429	223	520	94 900	221 200	97	226
Compression	Green	7.7	0.503	268	533	77 400	153 900	144	286
Normal	Green	4.9	0.461	247	536	119 300	258 800	162	351
Compression	11.3	4.2	0.570	455	798	81 500	143 000	214	375
Normal	11.1	2.9	0.604	630	1043	158 900	263 100	333	551
Compression	8.9	6.1	0.523	372	711	63 300	121 000	176	337
Normal	10.2	2.7	0.603	657	1090	159 200	264 000	473	784

Table 7.22. Strength and related properties of compression and normal woods of *Pinus ponderosa* and *Sequoia sempervirens*. (Courtesy of US Forest Service). (Markwardt and Wilson 1935)

Property	Pinus ponderosa				Sequoia sempervirens			
	Green		Air-dry		Green		Air-dry	
	Compression wood	Normal wood	Compression wood	Normal wood	Compression wood	Normal wood	Compression wood	Normal wood
Static bending								
Moisture content, per cent	88	133	12.6	12.0	102	114	10.5	9.9
Specific gravity, G	0.47	0.35	0.50	0.37	0.51	0.38	0.51	0.38
Modulus of rupture, psi, R	6 120	4 640	11 710	9 840	7 470	7 310	8 890	10 210
$R/G^{1.5}$	18 990	22 410	33 120	43 720	20 510	31 210	24 410	43 590
Modulus of elasticity, psi $\times 10^{-3}$, E	842	1 074	1 019	1 345	685	1 110	788	1 253
E/G^2	3 812	8 767	4 080	9 825	2 634	7 687	3 030	8 675
Work to proportional limit, inch pounds inch^{-3}, W_{PL}	0.94	0.47	0.63	2.19				
W_{PL}/G^2	4.26	3.84	2.52	16.0				
Work to maximum load, inch pounds inch^{-3}, W_{ML}	8.8	4.0	15.7	7.6	6.9	7.5	6.5	6.0
W_{ML}/G^2	39.8	32.7	62.8	55.5	26.5	51.9	25.0	41.6
Total work, inch pounds inch^{-3}, W_T	45.6	14.4	16.2	10.8				
W_T/G^2	206.4	117.6	64.8	78.9				
Compression parallel to the grain								
Moisture content, per cent	78	138	12.7	12.1	106	126	10.0	8.6
Specific gravity, G	0.47	0.35	0.50	0.37	0.51	0.37	0.51	0.38
Modulus of elasticity, (pounds inch^{-2}) $\times 10^{-3}$, Y	996	1 476						
Y/G	2 119	4 217						
Stress at proportional limit, psi, σ_{PL}	2 090	2 140	5 970	5 220	4 640	3 950	7 250	7 160
σ_{PL}/G	4 447	6 114	11 940	14 110	9 098	10 680	14 220	18 840
Maximum crushing strength, psi, C_{max}	3 300	2 340						
C_{max}/G	7 021	6 686						
Toughness								
Moisture content, per cent	85	121	10.6	10.0	89	129	9.7	8.8
Specific gravity	0.49	0.37	0.53	0.38	0.52	0.37	0.49	0.37
Toughness per specimen, inch-pounds	173.4	100.7	100.4	79.2	69.5	83.0	64.4	64.5

7.4.2.2 Compression Parallel to the Grain

values of maximum crushing strength, here expressed in the metric system as kg cm^{-2} rather than as pounds per square inch (psi).

Sonntag (1904), as already noted, had come to the conclusion that green compression wood is much stronger than normal wood in compression parallel to the grain. Rothe (1930) found that compression wood of *Picea abies* was 61% stronger than normal wood in this respect. Verrall (1928) concluded that compressive strength of air-dry compression wood was about 1.8 times larger than that of normal wood, the species tested being *Picea* sp. and *Thuja occidentalis*. Burns (1920), on the other hand, found compression wood of *Pinus strobus* to be three times stronger in longitudinal compression than normal wood. It is possible, as suggested by Verrall (1928), that this anomalous result was due to an especially low moisture content of the compression wood. As shown in Table 7.25, maximum crushing strength is sensitive to the water content of the wood, approximately doubling on drying from the green to the air-dry condition. Compression woods of *Abies alba* (Giordano 1971), *A. sachalinensis* (Ueda et al. 1972), *Larix decidua*, and *Picea abies* (Giordano 1971) have also been found to be stronger in compression parallel to the grain than corresponding normal woods.

Schulz and Bellmann (1982) (Schulz et al. 1984) found that with air-dry *Picea abies* wood compressive strength per unit density along the grain first increased slowly with increasing proportions of compression wood until a value of 27% had been reached, after which it decreased relatively rapidly (Fig. 7.21.1). With water-saturated wood, there was a slow increase in compressive strength with increasing content of compression wood (Fig. 7.21.2).

The data reported by Trendelenburg (1931, 1932) should be accepted with reservation. As pointed out by the investigator himself, these specimens of *Pseudotsuga menziesii* compression wood had originally been collected for another study (Trendelenburg 1931), and it is clear from the reported density values that few, if any, of the samples tested consisted of severe compression wood. It will be noted, for example, that in the last two cases, normal wood was denser than corresponding compression wood (Table 7.21).

Inspection of the data in Tables 7.23 and 7.24 reveals that in all cases so far reported green compression wood has been found to be stronger in maximum crushing strength than corresponding normal wood. In some cases, the ratio is as low as 1.1, as seen from Table 7.25, in others it varies from 1.3 to 1.7. When the higher density of compression wood is taken into account, compression wood is surprisingly similar to normal wood in compressive strength, being sometimes slightly stronger and sometimes slightly weaker. The rate of increase in maximum crushing strength with increasing specific gravity of the wood is higher for normal wood than for compression wood, as can be seen in Fig. 7.22 (Perem 1960).

Most strength properties of wood improve on drying, but compression wood generally improves less than does normal wood. The data in Table 7.24 show that the maximum crushing strength of normal wood approximately doubles on going from the green to the air-dry state. For compression wood, the change is less. Air-dry compression wood is weaker than corresponding normal wood in only three of the 11 species listed in Table 7.25, but the differences are, with

Table 7.23. Strength and related properties of green and air-dry compression and normal woods of eight conifer species. (Pillow and Luxford 1937, Perem 1960, Cockrell and Knudson 1973)

Species	Type of wood	Moisture, per cent	Specific gravity, G	Static bending			
				Modulus of rupture, psi, R		Modulus of Elasticity, psi $\times 10^{-3}$ E	
				R	$R/G^{1.5}$	E	E/G^2
Abies concolor	Compression	109.0	0.470	7 570	23 500	984	4 450
		11.9	0.509	12 700	34 970	1108	4 280
	Normal	187.1	0.346	6 040	29 680	1180	9 860
		11.7	0.375	10 460	45 550	1327	9 440
Picea glauca	Compression	Green	0.387	5 290	21 970		
		10	0.392	9 620	39 200		
	Normal	Green	0.316	4 610	25 950		
		10	0.332	9 270	48 460		
Pinus ponderosa	Compression	87.5	0.467	6 120	19 180	842	3 860
		12.6	0.499	11 710	33 220	1019	4 090
	Normal	133.3	0.354	4 640	22 030	1074	8 570
		12.0	0.372	9 840	43 370	1345	9 720
Pinus resinosa	Compression	Green	0.415	5 355	20 030		
		10	0.448	10 280	34 280		
	Normal	Green	0.376	4 830	20 900		
		10	0.400	10 530	41 620		
Pinus taeda	Compression	77.2	0.584	8 490	19 020	919	2 700
		11.7	0.619	13 960	28 660	1156	3 020
	Normal	101.9	0.519	8 170	21 850	1016	3 770
		12.1	0.586	15 870	35 380	2364	6 880
Pseudotsuga menziesii	Compression	43.3	0.513	8 010	21 800	1016	3 860
		12.1	0.527	12 500	32 670	1188	4 280
	Normal	58.3	0.428	6 780	24 210	1369	7 470
		10.5	0.459	12 950	41 640	1666	7 910
Sequoia sempervirens	Compression	102.0	0.506	7 470	20 750	685	2 680
		10.5	0.510	8 890	24 410	788	3 030
	Normal	113.7	0.380	7 310	31 210	1110	7 690
		9.9	0.380	10 210	43 590	1253	8 680
Sequoiadendron giganteum	Compression	135	0.54	10 750	27 090	1090	3 740
		14	0.59	13 930	30 740	1140	3 270
	Normal	180	0.34	6 520	32 890	1100	9 520
		13	0.35	9 820	47 430	1240	10 120
	Opposite	246	0.28	5 840	39 420	930	11 860
		15	0.29	8 950	57 310	1080	12 840

7.4.2.2 Compression Parallel to the Grain

Work, inch-pounds inch^{-3}		Maximum crushing strength, psi, C		Tension parallel to the grain, psi, T		Toughness, inch-pound	Deflection at maximum load, inch
To maximum load	Total	C	C/G	T	T/G		
14.42	46.63	3580	7 620			141	0.70
18.24	27.88	5900	11 590			114	0.63
6.46	19.00	2780	8 030			130	0.40
9.68	17.96	5220	13 920			116	0.40
		2395	6 190			271	
		5270	13 440			121	
		1980	6 270			208	
		4990	15 030			186	
8.76	41.60	3300	7 070	9 690	20 750	173	0.55
15.74	16.17	5970	11 960			100	0.59
4.02	14.39	2340	6 610	11 780	33 280	101	0.34
7.56	10.82	5210	14 000			79	0.35
		2390	5 760			385	
		5590	12 480			161	
		2130	5 660			351	
		5400	13 500			186	
24.5	56.76	4380	7 500			196	1.07
21.21	22.02	7200	11 630			109	0.65
10.59	35.69	3760	7 245			251	0.46
10.90	33.00	8530	14 560			173	0.31
14.88	34.44	4150	8 090	10 880	21 210	182	0.68
12.29	12.36	7140	13 550	12 800	24 290	89	0.47
7.77	18.79	3280	7 660	13 850	32 360	185	0.45
11.23	22.73	7230	15 750	13 200	28 760	204	0.39
6.88	9.99	4640	9 170	5 910	11 680	83	0.41
6.51	8.11	7250	14 220	7 560	14 820	64	0.39
7.52	15.71	3950	10 400	10 140	26 680	83	0.41
6.04	8.24	7160	18 840	8 850	23 290	65	0.29
24.2	30.5	5260	9 740	12 150	22 500	250	
20.7	23.6	7030	11 920	17 210	29 170	280	
5.6	12.8	3660	10 760	10 670	31 380	180	
6.3	11.8	5380	15 370	9 490	27 110	120	
7.39	12.4	2780	9 930	9 600	34 290	150	
7.13	12.6	4900	16 900	10 280	35 450	190	

Table 7.24. Ratio of strength properties of compression and normal woods in air-dry and green conditions. Calculated from data reported by Pillow and Luxford (1937), Perem (1958), and Cockrell and Knudson (1973)

Species	Modulus of rupture in static bending		Modulus of elasticity in static bending		Maximum crushing strength		Tension parallel to grain	
	Compression wood	Normal wood	Compression wood	Normal wood	Compression wood	Normal wood	Compression wood	Normal wood
Abies concolor	1.68	1.73	1.12	1.12	1.65	1.88		
Picea glauca	1.82	2.01			2.20	2.52		
Pinus ponderosa	1.91	2.12	1.21	1.25	1.81	2.23		
Pinus resinosa	1.92	2.18			2.34	2.54		
Pinus taeda	1.65	1.94	1.26	1.31	1.64	2.27		
Pseudotsuga menziesii	1.56	1.91	1.17	1.22	1.72	2.21	1.18	0.95
Sequoia sempervirens	1.19	1.40	1.15	1.13	1.56	1.81	1.28	0.87
Sequoiadendron giganteum	1.30	1.51	1.04	1.13	1.34	1.47	1.42	0.89

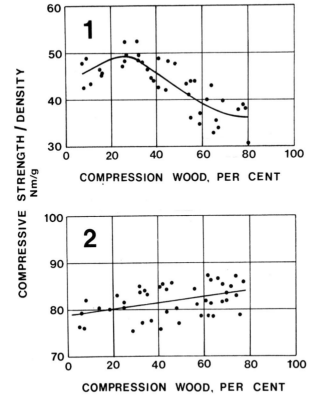

Fig. 7.21. Variation of compressive strength per unit density with the proportion of compression wood present in *Picea abies* for air-dry (**1**) and water-saturated (**2**) wood. (Modified from Schulz and Bellmann 1982)

7.4.2.2 Compression Parallel to the Grain

Table 7.25. Maximum crushing strength parallel to the grain of green and air-dry compression and normal woods. All strength values in kg cm^{-2}

Species	Moisture	Maximum crushing strength			Maximum crushing strength/ Specific gravity			References
		Compression wood	Normal wood	Ratio	Compression wood	Normal wood	Ratio	
Abies concolor	Green	252	195	1.29	536	565	0.95	Pillow and Luxford 1937
	Air-dry	415	367	1.13	815	979	0.83	
Abies sachalinensis	Air-dry	488	384	1.27	760	977	0.78	Ueda et al. 1972
Picea abies	Green	403	241	1.67				Rothe 1930
	Air-dry	358	344	1.04				Janka 1909
Picea glauca	Green	168	139	1.21	940	950	0.99	Perem 1958
	Air-dry	370	351	1.05	434	440	0.99	
					945	1057	0.89	
Pinus densiflora	Green	300	210	1.43	526	436	1.21	Onaka 1949
	Green	312	233	1.34	546	495	1.10	
Pinus ponderosa	Green	232	165	1.41	497	465	1.07	Pillow and Luxford 1937
	Air-dry	420	366	1.15	841	985	0.85	Markwardt and Wilson 1935
Pinus resinosa	Green	168	150	1.03	405	400	1.01	Perem 1958
	Air-dry	393	380	1.24	876	950	0.92	
Pinus taeda	Green	308	264	1.17	527	509	1.04	Pillow and Luxford 1937
	Air-dry	506	600	0.84	818	1023	0.80	
Pseudotsuga menziesii	Green	243	226	1.08	520	517	1.01	Trendelenburg 1931, 1932
	Air-dry	414	644	0.64	755	1065	0.71	
	Green	292	231	1.26	569	539	1.06	Pillow and Luxford 1937
	Air-dry	502	508	0.99	952	1107	0.86	
Sequoia sempervirens	Green	326	278	1.17	644	731	0.88	
	Air-dry	510	503	1.01	999	1325	0.75	
Sequoiadendron giganteum	Green	370	257	1.44	685	756	0.91	Cockrell and Knudson 1973
	Air-dry	494	378	1.31	838	1080	0.78	
Thujopsis dolabrata	Air-dry	415	348	1.19	825	916	0.90	Mori 1933

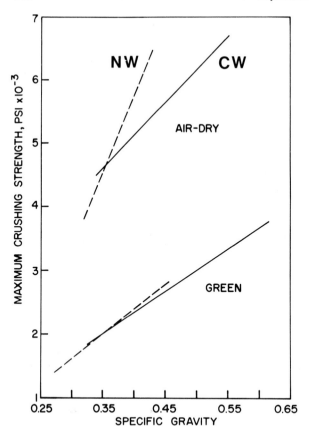

Fig. 7.22. Variation of maximum crushing strength with specific gravity for green and air-dry normal (*NW*) and compression (*CW*) woods of *Pinus resinosa*. Note the increase in maximum crushing strength caused by drying. (Courtesy of Canadian Forestry Service). (Redrawn from Perem 1960)

one single exception, very slight in the other cases. Actually, the variations are so minor that compression wood must be considered to have about the same maximum crushing strength as normal wood in the air-dry condition. The fact that air-dry compression wood has a slightly higher moisture content than normal wood is probably of no significance in this case.

When the difference in density of the two types of wood is considered, this picture changes considerably. Except for *Pinus densiflora* (Onaka 1949), all air-dry compression wood specimens referred to in Table 7.25 are weaker than corresponding normal wood in compression on an equal density basis, the ratios varying from 0.71 to 0.99.

7.4.2.2.2 Stress at Proportional Limit. Stress at proportional limit has been measured for compression wood only in compression parallel to the grain. It denotes the maximum stress at which load and deformation are still proportional and increases with the specific gravity of the wood. Values obtained by Trendelenburg (1932) with *Pseudotsuga menziesii* and by Markwardt and Wilson (1935) with *Pinus ponderosa* and *Sequoia sempervirens* are shown in Tables 7.21 and 7.22.

7.4.2.2 Compression Parallel to the Grain

According to Trendelenburg (1932), green and even more so air-dry compression wood is weaker than normal wood in stress at proportional limit in static bending. The data of Markwardt and Wilson (1935), which were probably obtained with more severe compression wood, indicate that both green and air-dry compression wood can be equal to or stronger than normal wood in this respect. In *Abies sachalinensis*, stress at proportional limit of air-dry compression wood has been reported to be 219 kg cm^{-2} compared to only 253 kg cm^{-2} for corresponding normal wood (Ueda et al. 1972). All investigators agree that compression wood is weaker than normal wood in stress at proportional limit when comparison is made on the basis of unit weight.

7.4.2.2.3 Young's Modulus of Elasticity. Young's modulus of elasticity in compression is a measure of the ability of a piece of wood to resume its original size after the stress has been removed. It is the slope of the linear portion (elastic line) of the stress−strain relationship. The larger the modulus, the greater the stiffness of the material.

The modulus of elasticity in compression was the mechanical property of compression wood that was most extensively studied by the early investigators. Hartig (1896, 1901) and Sonntag (1904), as already mentioned, found it to be much less for compression wood than for normal wood of *Picea abies*. Rothe (1930), who also used compression wood of *Picea abies*, arrived at an entirely different conclusion. A co-worker of Münch, he was primarily interested in the function of compression wood in the living tree. It was natural, therefore, that Rothe carried out all his strength tests with green compression wood which he obtained from the stem of a large, leaning spruce tree which exhibited overcorrection. Rothe (1930) noted that compression wood was stronger than normal wood in longitudinal compression, especially in regions where the stem was curved. He also found that compression wood had a modulus of elasticity of 45 330 kg cm^{-2} and corresponding opposite wood only 10 770 kg cm^{-2}. He accordingly concluded, adopting Münch's teleologic attitude, that not only the high compressive strength, but also the high modulus of elasticity of compression wood served the function of protecting the tree against rupture. This conclusion was soon challenged from two sides.

Trendelenburg (1931, 1932), who had earlier obtained entirely different results with wood from *Pseudotsuga menziesii*, pointed out that it could hardly be correct to compare compression wood in this respect with opposite wood. He drew attention to the fact that Janka (1909) had found normal wood of *Picea abies* to have a Young's modulus of 96 800, a value not too different from that reported earlier by Hartig (1901) and also more in agreement with Trendelenburg's own results, shown in Table 7.21. Compared to normal wood, compression wood accordingly has a smaller modulus of elasticity in compression, and this is true of both green and air-dry woods, as shown in Table 7.26. This conclusion was completely accepted by Münch (1938) a few years later.

Further criticism of Rothe's investigation was offered by Mayer-Wegelin (1931) in a review of recent publications. He pointed out, as had actually been admitted by Rothe (1930) himself, that practically all measurements had been performed above the limit of proportionality of the wood, and that what had

Table 7.26. Young's modulus in compression of green and air-dry compression and normal woods of *Abies sachalinensis, Picea abies* and *Pseudotsuga menziesii.* All strength values in kg cm^{-2}

Species	State	Young's modulus			References
		Compression wood	Normal wood	Ratio	
Abies sachalinensis	Air-dry	63 300	101 800	0.62	Ueda et al. 1972
Picea abies	Green	68 650	128 100	0.52	Hartig 1896, 1901
	Green	45 330	107 700	0.42	Rothe 1930
	Air-dry	97 600	113 600	0.86	Janka 1909
Pseudotsuga menziesii	Green	77 233	107 066	0.72	Trendelenburg 1931, 1932
	Air-dry	72 400	159 050	0.46	

really been measured was the elasticity of the wood between the limits of proportionality and rupture. Mayer-Wegelin (1931) accordingly expressed doubt that any reliable conclusions could be drawn from these values concerning the role of compression wood in protecting a tree mechanically.

This criticism is formally correct, yet it would be rendering Rothe an injustice to deny his contribution any merit. As pointed out by Trendelenburg (1932), a determination of the modulus of elasticity alone is not enough in this case. What Rothe's data show particularly clearly is instead, first, how much less compression wood is deformed than opposite wood when subjected to a stress beyond the limit of proportionality and, second, how superior it is in elastic recovery. Both these aspects are well brought out in the stress – strain diagrams reported by Rothe, one of which is shown in Fig. 7.23. Thanks to the low modulus of elasticity in bending of compression wood, leaning stems and branches bend readily at moderate loads, thus protecting the tree from rupture.

The discussion so far has been limited to wood in the green state. According to Trendelenburg (1932), Young's modulus is much smaller for air-dry compression wood than for normal wood. The values reported by Markwardt and Wilson (1935) for *Pinus ponderosa* indicate that the difference between the two types of wood is larger in the air-dry than in the green state, as seen in Table 7.22. Based on unit density, this difference, as could be expected, is even larger (Tables 7.21 – 7.23). In a study of the properties of wood from *Pseudotsuga menziesii*, von Pechmann and Courtois (1970) noted that low values for the modulus of elasticity were associated with the presence of compression wood and high values with the occurrence of normal wood.

Ueda et al. (1972) have reported values for the modulus of elasticity in compression (E) of different woods from *Abies sachalinensis*. Values for normal wood ranged from 90 000 to 110 000 kg cm^{-2} in the air-dry condition. For compression wood they were only 60 000 – 80 000 kg cm^{-2}, thus again demonstrating the invariably low modulus of elasticity in compression of this tissue. The same investigators found that the torsional modulus of rigidity (modulus of

7.4.2.2 Compression Parallel to the Grain

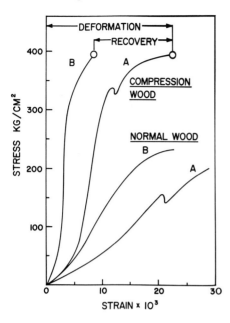

Fig. 7.23. Stress-strain curves for normal and compression woods of *Picea abies* before (*A*) and after (*B*) stress had been removed. (Redrawn from Rothe 1930)

elasticity in shear) (G) was strongly and positively correlated with the specific gravity of the wood. The mean value for normal wood was 6000 kg cm^{-2}, while pronounced compression wood had a value of 17 000 kg cm^{-2}. They suggested that the E/G ratio could serve as a useful diagnostic index for compression wood, as this ratio is larger than 13 for normal wood but only about 4 for compression wood. Similar results were obtained with compression wood of *Ginkgo biloba* (Ueda 1973) (Table 7.33).

7.4.2.2.4 Creep and Failure in Longitudinal Compression. Keith and Côté (1968) and Keith (1971, 1972) studied the deformation and failure of wood when stressed in compression. Keith (1974) later extended these investigations to include compression wood from the same *Picea glauca* tree that had been used previously. Values obtained for maximum crushing strength and stress at proportional limit with test specimens containing 9% and 18% moisture are summarized in Table 7.27. The specific gravity values also listed here indicate that the former specimen consisted of more severe compression wood than did the latter. The relatively low density of both samples suggests that they either were mixtures of normal and compression woods or consisted of only mild compression wood. Pure, pronounced compression wood of *Picea glauca* has a density of 0.64 (Table 7.2).

As has often been noted by others, the specific compressive strength of compression wood was less than that of normal wood. The stress at proportional limit as a fraction of the maximum crushing strength was also less for compression wood. In creep tests, carried out over a 24-h period, it was found that all components of strain were abnormally high in compression wood, as shown in

Table 7.27. Data for normal and compression-wood specimens of *Picea glauca* compressed to failure in short-term tests. (Keith 1974)

Characteristic	Moisture content: 9%		Moisture content: 18%	
	Normal wood	Compression wood	Normal wood	Compression wood
Maximum crushing strength, psi	6362	6955	4320	4153
Basic specific gravity	0.345	0.451	0.352	0.392
Specific strength, psi	18 441	15 421	12 273	10 594
Stress at proportional limit in per cent of maximum crushing strength	83.9	72.0	81.5	75.1

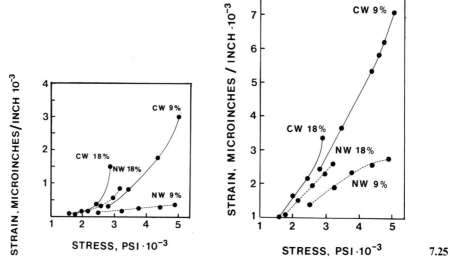

Fig. 7.24. Creep strain after 24 h for compression (*CW*) and normal (*NW*) woods at moisture contents of 9% and 18% in *Picea glauca*. (Modified from Keith 1974)

Fig. 7.25. Total strain after 24 h for compression (*CW*) and normal (*NW*) woods at moisture contents of 9% and 18% in *Picea glauca*. (Modified from Keith 1974)

Figs. 7.24 and 7.25. In one case, creep strain was, for example, ten times larger for compression wood than for normal wood. In their studies on the tensile strength of compression wood, Suzuki (1968) and Furuno et al. (1969) obtained low values for stress and strain at the proportional limit and very high values at the point of failure. Ueda et al. (1972) also found compression wood to exhibit an abnormally low stress at the proportional limit when tested in axial compression. Keith (1974) noted that at high values of stress the deviations from linear viscoelastic flow, previously observed for normal wood (Keith 1972), were even greater for compression wood. Grossman and Kingston (1954) had earlier not-

7.4.2.2 Compression Parallel to the Grain

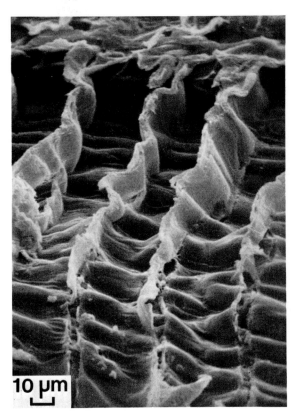

Fig. 7.26. Severe deformation (buckling) of compression-wood tracheids in *Picea glauca*, in a region of macroscopic compression failure. Earlywood zone. Longitudinal surface. SEM. (Keith 1974)

ed that compression wood on bending exhibited more creep and nonlinear behavior.

Several earlier investigators, such as Wardrop and Dadswell (1947), Côté and Day (1965), and Keith and Côté (1968), had noted a characteristic absence of slip planes and compression failures in compression wood. Keith (1974) likewise could not observe any microscopic compression failures, even in specimens tested close to failure. Some samples developed macroscopic compression failures in the earlywood region of the compression wood. Probably as a result of excessive strain in this region, the earlywood tracheids tended to buckle under load, as shown in Fig. 7.26. Microscopic creases and folds were finer and more sparse in the latewood area, probably the only region containing fully developed compression wood. Figure 7.27 shows the distortion of the microfibrils in the S_2 layer. Failures were especially frequent in areas where the tracheids were in contact with rays, a region where macroscopic failure is also common in normal wood.

The fact that compression wood, when stressed in compression, does not develop slip planes and microscopic failures prior to gross failure was attributed by Keith (1974) to the large microfibril angle of the tracheids in this tissue. Any compressive stress along the grain would be only partly translated into a stress

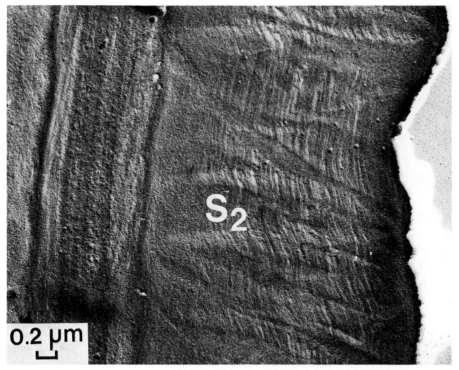

Fig. 7.27. Section of a compression-wood tracheid in *Picea glauca*, showing distortions of the microfibrils in S_2 associated with compression creases and folds. Lines of microscopic compression failures are absent. Longitudinal section. TEM. (Keith 1974)

Table 7.28. Tensile strength parallel to the grain of green and air-dry compression and normal woods. All strength values in kg cm^{-2}

Species	Tensile strength			Tensile strength/specific gravity			References
	Compression wood	Normal wood	Ratio	Compression wood	Normal wood	Ratio	
Picea abies							
Earlywood	645	1180	0.55				Sonntag 1904
Latewood	1590	2360	0.67				
Pinus densiflora	311	925	0.34	540	1970	0.27	Onaka 1949
Pinus ponderosa	681	828	0.82	1459	2340	0.62	Pillow and Luxford 1937
Pseudotsuga menziesii	765	974	0.79	1491	2275	0.66	
	900	928	0.97	1708	2222	0.77	
Sequoia sempervirens	415	713	0.58	821	2061	0.40	
	532	622	0.86	1042	1637	0.64	

7.4.2.3 Tension Parallel to the Grain

Table 7.29. Tensile strength properties of normal latewood and compression wood of *Picea jezoensis*. (Furuno et al. 1969)

Type of wood	Condition	Ultimate stress, kg cm^{-2}	Strain at failure $\times 10^3$	Stress at proportional limit, kg cm^{-2}	Strain at proportional limit, kg cm^{-2}	Young's modulus, kg cm^{-2} $\times 10^3$
Normal wood	Wet	1340	25.3	610	9.6	64
	Dry	2380	20.8	1470	12.2	121
Compression wood	Wet	510	58.3	170	7.1	24
	Dry	810	36.1	410	9.1	5.8

parallel to the microfibrils. This is probably a correct explanation and is also in agreement with earlier observations reported by Wardrop and Addo-Ashong (1964).

7.4.2.3 Tension Parallel to the Grain

7.4.2.3.1 Tensile Strength. Normal wood is usually two to three times stronger in tension than in compression. Compression wood also has a higher tensile than compressive strength, as can be seen from the data in Tables 7.23 and 7.28. Only in two cases has the tensile strength of compression wood been found to be so low that it approached the strength in compression, namely with *Sequoia sempervirens* (Pillow and Luxford 1937) and *Pinus densiflora* (Onaka 1949). Compared to normal wood, green compression wood is weak in tension in these two species as well as in *Pseudotsuga menziesii*. Contrary to the situation with respect to compressive strength, the difference between the two woods is less in the air-dry than in the green state. The reason for this is that strength in tension decreases on drying with normal wood, but increases with compression wood. If differences in density are eliminated, compression wood is much weaker in tension than normal wood. Burns (1920) claimed that compression wood of *Pinus strobus* was stronger in tension than normal wood, but Verrall (1928) found the opposite with *Picea mariana* and *Thuja plicata*. Furuno et al. (1969) have reported detailed measurements of the behavior in tension of compression and normal woods from *Picea jezoensis*. The data in Table 7.29 and the stress–strain curves in Fig. 7.28 show the inferiority of compression wood in tensile strength. Similar results were reported at the same time by Suzuki (1968, 1969) with wood from other Japanese conifers.

Wardrop (1951), in a study of the breaking load in tension of *Pinus radiata* wood, noted the same linear increase in breaking load with increasing basic density for both compression and normal woods. At any given density, however, compression wood had a lower value of breaking load compared to normal wood, as is shown in Fig. 7.29. The relationship between breaking load in tension and tracheid length of normal and compression woods, on the other

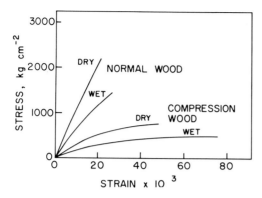

Fig. 7.28. Tensile stress-strain curves for normal latewood and compression wood of *Picea jezoensis* in dry and wet conditions. (Redrawn from Furuno et al. 1969)

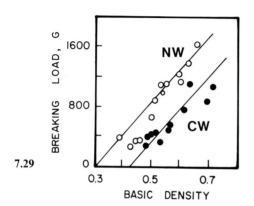
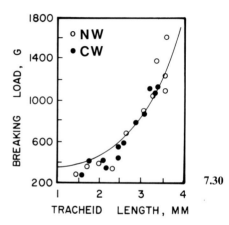

Fig. 7.29. Relationship between breaking load in tension and basic density of the xylem in a specimen of *Pinus radiata* containing normal (*NW*) and compression (*CW*) woods. (Redrawn from Wardrop 1951)

Fig. 7.30. Relationship between breaking load in tension and tracheid length in a specimen of *Pinus radiata* containing normal (*NW*) and compression (*CW*) woods. (Drawn from data reported by Wardrop 1951)

hand, could be described by a single, nonlinear relationship, shown in Fig. 7.30. Nordman and Quickström (1970) tested the tensile strength of normal and compression wood sections of *Pinus sylvestris*. Normal latewood was stronger in tension than earlywood. Compression wood, its thick tracheid wall notwithstanding, was much weaker than both, and had a tensile strength only a quarter of that of normal latewood.

Cockrell and Knudson (1973) have reported extensive strength data for compression, normal, and opposite woods of *Sequoiadendron giganteum*, obtaining the values for ultimate stress in tension parallel to the grain listed in Table 7.23. Contrary to all other reported measurements, these results indicate a greater tensile strength for compression wood than for normal wood and not only in the air-dry but also in the green state. The reason for this discrepancy is not known. Undoubtedly, the wood specimens tested were somewhat unusual in

7.4.2.3 Tension Parallel to the Grain

having been obtained from a leaning, strongly suppressed tree without a top. The resulting, extremely slow growth is reflected in the narrow growth rings, namely 24 per inch in the compression and 36 in the opposite wood zones, compared to 8 in the normal wood. The data for stress at proportional limit in tension show compression wood to be much weaker than normal wood in this respect, the values in the green conditions being 4670 psi and 7530 psi, respectively. This fact, in conjunction with the greater breaking strength, agrees well with the large plasticity of compression wood, a characteristic attributed by Cockrell and Knudson to the large microfibril angle and the helical cavities.

7.4.2.3.2 Young's Modulus of Elasticity. All investigators agree that the modulus of elasticity in tension of compression wood is much lower than that of either opposite or normal woods. Hartig (1896, 1901) found it to be slightly more than half of that for normal wood of *Picea abies*. The data in Table 7.29 indicate that for compression wood of *Picea jezoensis*, values for Young's modulus were half or less of those valid for normal wood (Furuno et al. 1969). Suzuki (1968, 1969) obtained similar results. According to Nordman and Quickström (1970) compression wood of *Pinus sylvestris* has a modulus of elasticity in tension which is less than half of that of normal latewood.

Mark and Gillis (1973) have analyzed the relationship between Young's modulus of elasticity in axial tension and the microfibrillar angle in the S_2 layer, treating wood fibers as anisotropic, composite lamellates. They found that the critical microfibril angle is 15°. At angles larger than this value, which is typical of both juvenile and compression woods, the axial fiber stiffness will be substantially reduced. At very small angles, the hemicellulose-lignin matrix has almost no influence on the modulus. When the microfibril angle in S_2 is 40°, on the other hand, the matrix alone determines the modulus in tension, which is now approximately 1000 kg mm^{-2}, irrespective of the characteristics of the cellulose microfibrils. Angles of 40°–50° are typical of compression-wood tracheids. The unusually low modulus of elasticity in tension of these tracheids is accordingly attributable to their matrix, which is not only relatively larger (70%) than in normal tracheids (58%), but also contains somewhat more lignin. Mark and Gillis also found that the absence of the layer S_1 will not affect the modulus of elasticity in tension at large microfibril angles in S_2. Whether this holds for compression-wood tracheids with their thick S_1 layer has not been established.

Tang and Hsu (1973) studied the effect of fiber ultrastructure on the elastic properties of the cell wall, using models that included earlywood, latewood, and compression wood fibers. They came to the general conclusion that the elastic properties of wood fibers depend not only on the microfibril angle in the S_2 layer and the size of each layer, but also on the crossed structure of the S_3 layer and on the spacing between the microfibrils in each cell wall layer. In certain cases the spacing can become the most important factor, influencing elasticity and other tensile properties of the fibers.

7.4.2.4 Static Bending

7.4.2.4.1 Introduction. Many of the tests concerned with the strength of compression wood in static bending have been carried out with branches when the properties of compression, opposite, and side woods all have to be considered. In this section, discussion will be limited to results obtained with compression wood alone.

7.4.2.4.2 Modulus of Rupture. The modulus of rupture in bending represents the stress at the top and bottom of a beam at maximum load. It measures the ability of a beam to accept a slowly applied load for a short time. When tested in the green condition, compression wood has a higher modulus of rupture than normal wood, as can be seen from the data in Tables 7.22, 7.23, and 7.30, although in some cases the difference is quite small. Normal wood, which, as already mentioned, is much stronger in tension than in compression along the grain, exhibits a modulus of rupture in bending which is determined largely by its strength in longitudinal compression. Failure in bending of normal wood usually begins on the side of compression. Compression wood, by contrast, has a high maximum crushing strength but is relatively weak in tension. Its larger modulus of rupture, compared to normal wood, is a result of its higher strength in compression. On bending, compression wood tends to fail at about the same time on the compression and tension sides (Trendelenburg 1932, Onaka 1949). According to Parem (1958, 1960), failure in tension is more common with compression wood than with normal wood when testing for toughness.

In the air-dry condition, compression wood is comparable to normal wood in modulus of rupture, being sometimes slightly stronger and sometimes slightly weaker, as seen from Tables 7.23 and 7.30. The ratios in Table 7.24 show that the modulus of rupture increases less rapidly on drying than that of normal wood which is probably a reflection of the similar behavior of the maximum crushing strength. Janka (1909) noted that compression wood of *Picea abies*, when air-dry, was weaker in all strength properties in bending than normal wood. The failed factory roof beam studied by Heck (1919) contained compression wood with a much lower modulus of rupture than that of the normal wood, as seen in Table 7.31.

When based on unit density, the modulus of rupture of green compression wood is generally somewhat and occasionally considerably lower than that of normal wood (Table 7.23). When tested air-dry, compression wood is always much weaker than normal wood on this basis. According to Perem (1958, 1960), no test showed more clearly the inferiority of compression wood than did the static bending tests when specific strength values were used for comparison. Mori (1933) had earlier arrived at a similar conclusion. The relationships between modulus of rupture and specific gravity for normal and compression woods of *Pinus resinosa* (Perem 1960) are shown in Fig. 7.31.

Kučera (1970, 1971) has reported extensive data for the bending strength of normal and compression woods of *Picea abies*. In most cases, compression wood had approximately the same strength as normal wood but it was considerably weaker in two instances. In a later study, by contrast, it was found that bending

7.4.2.4 Static Bending

Table 7.30. Strength in static bending and related properties of compression and normal woods of *Pseudotsuga menziesii*. (Trendelenburg 1931, 1932)

Wood	Moisture, per cent	Ring width, mm	Specific gravity G	Modulus of rupture, kg cm^{-2}		Modulus of elasticity, kg cm^{-2}		Stress at proportional limit kg cm^{-2}		Work to proportional limit, cm –kg cm^{-3}	
				R	R/G$^{1.5}$	E	E/G^2	σ_{PL}	$\sigma_{PL}/G^{1.5}$	W_{PL}	W_{PL}/G^2
Compression	Green	8.6	0.453	506	1660	81 400	396 700	269	882	0.63	3.07
Normal	Green	4.6	0.418	470	1740	100 100	572 900	273	1010	0.39	2.23
Compression	Green	7.4	0.434	515	1800	77 400	410 900	204	714	0.67	3.56
Normal	Green	5.2	0.436	500	1740	89 200	469 200	254	882	0.61	3.21
Compression	Green	6.6	0.487	540	1590	84 200	355 000	243	715	0.40	1.69
Normal	Green	4.8	0.468	538	1680	100 300	457 900	215	672	0.46	2.10
Compression	14.7	8.6	0.472	737	2270	75 100	337 100	378	1170	0.77	3.46
Normal	12.0	7.2	0.465	716	2260	105 200	486 530	454	1430	0.50	2.31
Compression	13.0	4.4	0.526	884	2320	94 400	341 190	462	1210	0.96	3.47
Normal	14.3	2.9	0.581	890	2010	139 900	414 440	555	1250	0.72	2.13
Compression	11.9	6.0	0.568	746	1743	86 500	268 110	270	631	0.64	1.98
Normal	13.3	2.8	0.599	882	1903	144 700	403 290	453	977	0.62	1.73

Table 7.31. Strength properties in static bending of air-dry compression and normal southern pine woods. (Heck 1919)

Property	Compression wood	Normal wood	Ratio
Modulus of rupture, psi	9000	11 730	0.77
Modulus of elasticity, 1000 psi	994	1 495	0.66
Stress at proportional limit, psi	6520	8 550	0.76
Work to maximum load, inch-pound per cubic inch	5.50	8.20	0.67
Deflection at maximum load, inches	0.62	0.67	0.93

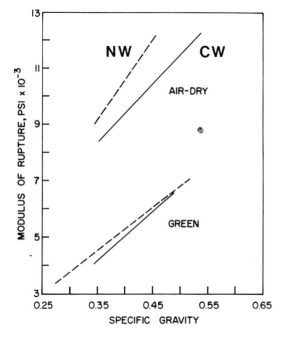

Fig. 7.31. Variation of modulus of rupture with specific gravity for green and air-dry normal (*NW*) and compression (*CW*) woods of *Pinus resinosa*. Note the increase in modulus of rupture caused by drying. (Courtesy of Canadian Forestry Service). (Redrawn from Perem 1960)

strength increased somewhat with increasing content of compression wood in *Picea abies* and *Pinus sylvestris* (Kučera 1973). In *Pinus caribaea* trees containing up to 50% compression wood, bending strength and modulus of rupture did not vary with the proportion of compression wood (Boone and Chudnoff 1972).

7.4.2.4.3 Stress at Proportional Limit. Stress at proportional limit in bending is, of course, always less than the modulus of rupture, which is the computed stress at maximum load. It has been measured for compression wood by Heck (1919), Trendelenburg (1932), Markwardt and Wilson (1935), and Cockrell and Knudson (1973), whose results are summarized in Tables 7.30–7.32. Markwardt and Wilson and Cockrell and Knudson found green compression

7.4.2.4 Static Bending

Table 7.32. Stress at proportional limit in static bending for compression and normal woods of *Pinus ponderosa* (Markwardt and Wilson 1935)) and for compression, normal, and opposite woods of *Sequoiadendron giganteum* (Cockrell and Knudson 1973). All values in psi

Species	Type of wood	Moisture, per cent	Specific gravity (G)	Stress at proportional limit σ_{PL}	$\sigma_{PL}/G^{1.5}$
Pinus ponderosa	Compression wood	88	0.47	3730	11 580
		12.6	0.50	6620	18 720
	Normal wood	133	0.35	3010	14 540
		12.0	0.35	7250	35 010
Sequoiadendron giganteum	Compression wood	135	0.54	6550	16 510
		14	0.59	6540	14 430
	Normal wood	180	0.34	4300	21 690
		13	0.35	7400	35 740
	Opposite wood	246	0.28	3460	23 350
		15	0.39	6630	27 220

wood to be stronger than normal wood in stress at proportional limit in bending, but Heck and Trendelenburg found it to be consistently weaker. In the air-dry condition, compression wood is always weaker than normal wood in this respect.

7.4.2.4.4 Modulus of Elasticity. When a beam is subjected to a load, its deflection is inversely proportional to its modulus of elasticity, which measures the rigidity of a material under a stress that does not exceed the limit of proportionality. The fact that compression wood has an unusually low modulus of elasticity in bending has been recognized since the pioneering investigations of Hartig (1896, 1901), who found that, to cause a certain deflection, only half the force was required for green compression wood in comparison with normal wood.

Inspection of the data in Tables 7.22, 7.23, and 7.30 reveals that green compression wood always is less rigid than normal wood, although the difference is not always as large as that observed by Hartig. The increase in modulus of elasticity on drying is modest and is the same for normal and compression woods, a rare situation. As usual, the difference between normal and compression woods is larger when comparison is made on the basis of unit density. Data obtained by Ueda et al. (1972) and by Ueda (1973), summarized in Table 7.33, indicate that compression wood in *Abies sachalinensis* and *Ginkgo biloba* also have smaller moduli of elasticity in bending than similar normal wood. Levčenko (1974) found that compression and opposite woods from a pine branch had a larger modulus of elasticity than normal stem wood. This is one of the few instances where compression wood has been claimed to be more rigid in bending than normal wood.

The modulus of elasticity in bending tends to increase with increasing wood density. In some *Pseudotsuga menziesii* trees, von Pechmann (1969) found that the values were poorly correlated with the density. A closer inspection indicated that normally the values for the modulus of elasticity were concentrically distributed over the cross section of a stem. In another tree with a one-sided crown the radial growth had been eccentric, and the wood within the wider portion of the growth rings had a lower modulus of elasticity than the remainder. Microscopic examination revealed that this abnormally flexible tissue consisted of compression wood. Similar results obtained with *Larix leptolepis* and *Pseudotsuga menziesii* were later reported by von Pechmann and Courtois (1970) and by von Pechmann (1973), as already mentioned. Measurements carried out by Reinhardt (1972) indicated that regions containing compression wood in the stems were inferior to normal wood, not only in modulus of elasticity in bending (80 000 kg cm^{-2}) and tensile strength, which is to be expected, but also in compressive strength which is more unusual but is explained by the fact that testing was carried out with air-dry wood.

Relationships between load and deflection (stress–strain curves) for normal and compression woods in green or air-dry conditions are shown in Fig. 7.32 for *Pseudotsuga menziesii* (Trendelenburg 1932), in Fig. 7.33 for *Pinus taeda* (Pillow and Luxford 1937) and in Fig. 7.34 for *Sequoiadendron giganteum* (Cockrell and Knudson 1973). The first two diagrams clearly bring out the much greater deflection of compression wood at equal load in comparison with normal wood.

As can be seen from the data in Table 7.23 and in Fig. 7.34, the results obtained by Cockrell and Knudson (1973) differ from those reported by the other investigators. Cockrell and Knudson found the modulus of elasticity in static bending to be only slightly smaller for compression wood than for normal wood in both green and air-dry conditions. Green compression and normal woods had identical stress–strain curves until the proportional limit was reached, and the two woods differed very little in this respect in the air-dry condition. The investigators considered their normal and compression woods to be essentially equivalent with respect to mechanical properties, which they obviously were. This is a rather unusual situation, which is probably related to the likewise un-

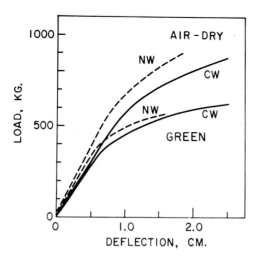

Fig. 7.32. Relationship between load and deflection (stress-strain curves) in static bending for green and air-dry normal (*NW*) and compression (*CW*) woods of *Pseudotsuga menziesii*. (Redrawn from Trendelenburg 1932)

7.4.2.4 Static Bending

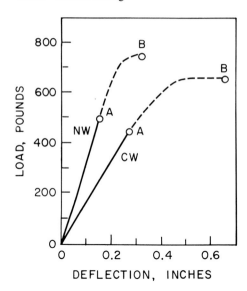

Fig. 7.33. Relationship between load and deflection in static bending for air-dry normal (*NW*) and compression (*CW*) woods of *Pinus taeda*. *A* load at proportional limit; *B* maximum load. (Courtesy of the US Forest Service). (Redrawn from Pillow and Luxford 1937)

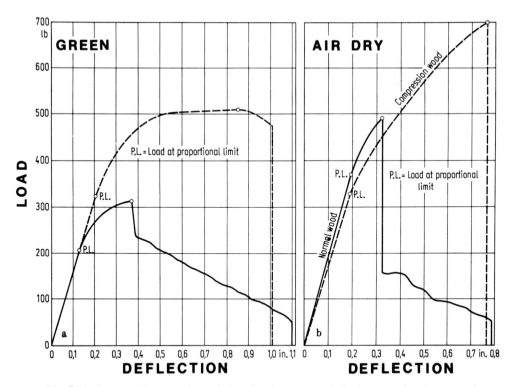

Fig. 7.34. Stress-strain curves in static bending for green and air-dry normal and compression woods of *Sequoiadendron giganteum*. (Cockrell and Knudson 1973)

usual nature of the compression wood tested, as mentioned previously. The giant sequoia tree from which the test materials were obtained had become displaced 33 years earlier, so that it was now leaning against a larger tree. At the time of sampling, it had a dead top, and it had obviously been suppressed at the time when the compression wood had been formed, but not earlier, when the normal wood was laid down. As a result, and contrary to the usual situation, the growth rings were more narrow in the compression wood than in the normal wood. It can be questioned, therefore, if the material tested really consisted of typical compression wood.

Boone and Chudnoff (1972) found that the modulus of elasticity in bending of lumber from *Pinus caribaea* decreased with increasing content of compression wood, whereas work to maximum load was not affected. Ohsaka et al. (1973) studied the properties of the wood in some 10-year-old *Pinus thunbergii* trees, where the stem was leaning and the branches had become pendulous. Wide growth rings and compression wood had been formed on the lower side of the stem. Tensile strength and modulus of elasticity in bending increased basipetally in both stem and branches. The investigators suggested that this decrease in strength together with a similar increase in bending stress were responsible for the pendulous branch habit.

7.4.2.4.5 Work to Maximum Load. Work to maximum load in static bending measures the ability of a beam to absorb shocks that cause stress beyond the limit of proportionality. Heck (1919) reported it to be less for compression wood than for normal wood of southern yellow pine, but other investigators (Markwardt and Wilson 1935, Pillow and Luxford 1937) have found it to be larger for both green and air-dry compression woods (Tables 7.22 and 7.23).

7.4.2.4.6 Work to Proportional Limit. Work to proportional limit measures the toughness of a material when stressed to the limit of proportionality. Trendelenburg (1932), as seen from Table 7.30, found it to be larger for both green and air-dry compression woods in comparison with corresponding normal woods. Based on unit density, green compression wood was either stronger or weaker than normal wood, whereas air-dry compression wood was always weaker. Markwardt and Wilson (1932) (Table 7.22) came to the conclusion that in *Pinus ponderosa*, compared with normal wood, work at proportional limit was twice as large for green but only one third as large for air-dry compression wood. With *Sequoiadendron giganteum* wood, Cockrell and Knudson (1973) found green compression wood and both green and air-dry normal woods to be equally strong, whereas air-dry compression wood was less than half as strong.

7.4.2.4.7 Total Work. Total work in static bending measures the toughness of a material under a stress that results in complete failure. Compression wood has a high modulus of rupture and also exhibits a large deflection under maximum load. Since work is the product of force and distance moved, it is to be expected that total work for compression wood should be large. This is indeed the case, as can be seen from Tables 7.22 and 7.23. When measured in the green condition, total work is generally much larger for compression wood than for normal wood. In the air-dry state it has sometimes been found to be larger for compression wood, but at other times to be smaller (Table 7.23), usually because complete failure often takes place in compression wood once maximum load has been exceeded (Pillow and Luxford 1937).

7.4.2.5 Toughness

Toughness indicates the ability of a material to resist shock. Mori (1933) found it to be lower for compression wood than for normal wood of *Thujopsis dolabrata* in the air-dry condition. Markwardt and Wilson (1935) (Table 7.22), on the other hand, observed a higher toughness for both green and air-dry compression woods of *Pinus ponderosa*. In *Sequoia sempervirens*, normal wood was tougher than compression wood in the green state and equally tough when air-dry. Pillow and Luxford (1937) (Table 7.23) found green and air-dry compression woods to be superior, equal, or inferior to corresponding normal woods in toughness.

The most thorough study on the toughness of compression wood was carried out by Perem (1958, 1960) (Table 7.23). Although measured values varied within a wide range, the mean value for all samples tested was significantly higher for green compression wood than for normal wood. When tested in the air-dry condition, compression wood of *Picea glauca* had a lower toughness than normal wood. With *Pinus resinosa*, the difference was not statistically significant. It will be noted from Table 7.23, that toughness, unlike most strength characteristics of wood, is considerably reduced on drying from green to air-dry condition. Toughness, as can be seen in Fig. 7.35, increased much more slowly with increase in specific gravity for compression wood than for normal wood, and the increase was slightly less for air-dry than for green compression wood. Dinwoodie (1971) has mentioned an example of low toughness in *Tsuga* compression wood. Cockrell and Knudson (1973), in contrast to other investigators, found compression wood from *Sequoiadendron giganteum* to be not only stronger than corresponding normal wood under all conditions, but also to increase, in toughness on drying from the green to the air-dry state (Table 7.23).

7.4.2.6 Shearing Strength

Maximum shearing strength parallel to the grain is the stress required to shear off from the test specimen a lip with a length of 2 inches in the direction of the grain. Mori (1933) (Onaka 1949) appears to be the first to have measured it for compression wood. Air-dry compression wood of *Thujopsis dolabrata* had a strength of shear of 8.5 kg cm^{-2} on the radial and 10.5 kg cm^{-2} on the tangential surfaces, corresponding values for normal wood being 8.2 and 10.8 kg cm^{-2}, respectively, or essentially the same. According to Sawada (1951), compression wood from *Abies mayriana* had a greater maximum shearing strength than normal wood, but was weaker when its higher density was taken into account.

The modulus of elasticity in shear, usually referred to as the modulus of rigidity, is associated with shear deformation in one plane and with shear

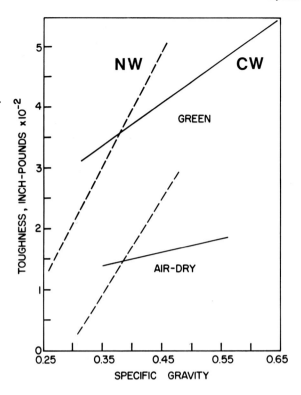

Fig. 7.35. Variation of toughness with specific gravity for green and air-dry normal (*NW*) and compression (*CW*) woods in *Pinus resinosa*. Note the lesser increase in toughness with increasing specific gravity of the compression wood. (Courtesy of Canadian Forestry Service.) (Perem 1960)

Table 7.33. Specific gravity, modulus of elasticity in bending, and modulus of elasticity in shear (modulus of rigidity) of air-dry normal and compression woods of *Abies sachalinensis* and *Ginkgo biloba*. (Ueda 1973)

Species	Type of wood	Specific gravity	Modulus of elasticity, kg cm^{-2} E	Modulus of rigidity, kg cm^{-2} G	E/G
Abies sachalinensis	Compression wood	0.566	79 300	12 900	6.15
	Normal wood	0.396	104 300	6 200	16.8
Ginkgo biloba	Compression wood	0.489	54 600	11 100	4.92
	Normal wood	0.450	92 200	9 000	10.2

stresses in the remaining two of the longitudinal, radial, and tangential planes. Values reported by Ueda and his co-workers (Ueda et al. 1972, Ueda 1973) (Table 7.33) indicate that compression wood from *Abies sachalinensis* and *Ginkgo biloba* had a larger modulus of rigidity than corresponding normal wood. Characteristically, the ratio of the modulus of elasticity in bending to the modulus of rigidity was much lower for compression wood (4–11) than for normal wood (9–22) (Sect. 7.4.2.2.3). Levčenko (1974) found the modulus of

Table 7.34. Hardness of air-dry compression and normal woods of *Thujopsis dolabrata*. (Mori 1933, Onaka 1949)

Section	Hardness, kg cm^{-2}			Specific hardness		
	Compression wood	Normal wood	Ratio	Compression wood	Normal wood	Ratio
Transverse	344.8	200.4	1.72	695	529	1.31
Radial	251.4	107.0	2.35	507	288	1.76
Tangential	242.2	124.8	1.94	488	335	1.46

elasticity in shear to be larger for compression and opposite woods from a pine branch than for normal stem wood.

7.4.2.7 Hardness

Hardness, by definition, is the load required to embed an 0.444-inch steel sphere to one half of its diameter into the wood. That compression wood is unusually hard has probably always been known. Many of its common names reflect this fact, such as "glassy wood", "hard streak", "Nagelhart", and "Eichiges Holz" (Chap. 3.2). Janka (1915), as well as previous investigators referred to by him, all found compression wood to be exceedingly hard.

Verrall (1928) measured the hardness of normal and compression woods of *Thuja occidentalis*.[3] Hardness values measured on transverse, radial, and tangential surfaces were 250, 234, and 520 pounds, respectively, for normal wood and 931, 1140, and 1175 pounds for compression wood. On the average, compression wood was 3.2 times harder than normal wood. Data reported by Mori (1933) are shown in Table 7.34. Apparently compression wood of *Thujopsis dolabrata* was about twice as hard as normal wood. When the effect of differences in density is eliminated, compression wood was still considerably harder. Similar results have been reported for *Abies mayriana* by Sawada (1951). According to Banks (1957), compression wood from eccentric *Pinus montezumae* trees is harder than normal wood.

7.4.2.8 Fracture

When compression wood fails in compression, tension, or bending, it almost invariably fractures in a manner characteristic of brash wood. Excellent reviews of brashness in lumber have been contributed by Koehler (1933) and by Dinwoodie (1971, 1976). Brittleness should not be equated with brashness. The

---

[3] Verrall's (1928) hardness values for *Picea* sp. are too low because the wood split during the test.

former is a normal property of materials such as cast iron, ceramics, chalk, or glass, whereas brashness is an abnormal property of wood. Brittle materials can be strong, but brash wood is always weak. A tough wood has a high resistance to propagation of cracks, being able to absorb large amounts of energy. A brash wood has a poor resistance to shock and a low ability to absorb energy. Failing load, deflection, and work to maximum load are also low. The ratio of tensile to compressive axial strength is much lower than in tough wood. Brash wood breaks abruptly with a smooth fracture and with little deflection. Tough wood breaks more slowly and gradually, splinters in its fracture, and deflects prior to rupture. Brashness in wood can be a consequence of several factors, such as low density, large microfibril angle, and compression failures (Dinwoodie 1976). Juvenile wood is often brash.

In conifers, wood with a high proportion of earlywood tends to be brash and rupture suddenly without splintering. The reasons for this behavior are several. First, compared to latewood, earlywood tracheids are thin-walled and weak. The microfibril angle in the S_2 layer of earlywood is usually larger than in latewood, and it is well known that the greater this angle, the greater will be the tendency for rupture between, rather than across the microfibrils, which, of course, requires less energy (Mark 1967). A third reason for the brashness of earlywood is probably the greater role in this wood of the relatively weaker middle lamella and S_1 layer compared to latewood, where the S_2 layer is thicker (Koehler 1933). The slightly higher lignin and lower cellulose contents of earlywood are probably of less importance.

The mechanism of wood fracture on a microscopic level has been studied by several investigators, such as Garland (1939), Wardrop (1951), Stone (1955), Kollmann (1963), Wardrop and Addo-Ashong (1964), Porter (1964), Debaise et al. (1966), Mark (1967), Koran (1967, 1968), Davies (1968), Grozdits and Ifju (1969), Borgin (1971, 1973), and Côté and Hanna (1983). It has been generally agreed that the S_1 layer is the weakest of the layers in a conifer tracheid, and that failure ordinarily occurs between the primary wall and S_1 or within S_1 (Mark 1967, Koran 1967, 1968, Davies 1968, Keith and Côté 1968) and between the S_1 and S_2 layers (Garland 1939, Wardrop and Addo-Ashong 1963, Davies 1968). Mark (1967) concluded that failure should never begin in the middle lamella, as this region is stronger in tension than are the different cell-wall layers. Grozdits and Ifju (1969) found that, in *Tsuga canadensis* earlywood under tensile stress, rupture took place across the cell wall. In latewood, on the other hand, failure occurred in the middle lamella and in the S_1 layer.

Saiki et al. (1972) observed that, in tension parallel to the grain, the S_2 layer of the tracheids broke either in a splintering fracture or so that the fracture crossed the wall perpendicular to the fiber axis, producing two types of faces, one smooth and the other uneven. Tracheids were also separated from one another, especially in the latewood. In a later investigation (Furukawa et al. 1973), it was observed that earlywood tracheids tended to fracture in the region of cross field pitting.

Kibblewhite (1973) and Kibblewhite and Harwood (1973) studied the fracture in tension of *Pinus radiata* wood which had been delignified with acid chlorite to various degrees. Tracheids containing more than 2% lignin failed across their wall (40%) or within the S_1 and S_2 layers (25% each). At lignin levels less than 2%, fracture occurred within the compound middle lamella. When the partly delignified wood was treated with alkali, fracture took place within this region even at a lignin content of 13.3%, apparently because the alkali had penetrated the middle lamella, swelling and softening its lignin.

Borgin et al. (1975) studied the effect of aging on the ultrastructure of wood varying in age from 900 to 4400 years. They found that fracture had occurred especially in the middle lamella-S_1 region, but also in the matrix between the microfibrils in the secondary wall. Complete disintegration of the wood apparently did not occur as long as the cellulose microfibrils remained intact.

Comparing the behavior of brash and tough woods under tension, Mark (1967) postulated that failure first occurs in the S_1 layer in both cases. In brash wood, rupture of the remainder of the cell wall probably follows very soon, whereas in tough wood shear in S_1 takes place prior to cell wall failure. Schniewind and Barrett (1969) improved Mark's theory and also predicted failure of S_1. The model developed by Mark and Gillis (1970) predicts maximal stresses in the S_1 layer for both normal and compression wood tracheids. Compared to the earlier theory (Mark 1967), the transverse stress in S_1 is 65% greater.

External conditions also influence the mode of rupture. Debaise et al. (1966) noted that slow fracture involved separation within the compound middle lamella and the S_1 layer. By contrast, rapid fracture occurred across the entire cell wall. Increase in temperature, which plasticizes both lignin and hemicelluloses, especially in the presence of water, increases interfiber fracture by reducing the strength of fiber to fiber bonding by as much as 80% (Koran 1967, 1968). Similarly, when the middle lamella is gradually broken down in pulping of wood, interfiber rupture also becomes more common, first between S_1 and S_2 and later in the middle lamella (Lagergren et al. 1957). Grozdits and Ifju (1969) observed that tracheids could be pulled apart almost intact at a stage in cell wall development when only the middle lamella, the primary wall and the S_1 layer had been formed, that is, before the tracheids had been cemented together with lignin. At a later stage, rupture occurred between S_2 and S_3, probably because these two layers had not yet become lignified.

Aiuchi and Ishida (1978, 1979, 1981) studied the effects of compression perpendicular to the grain in *Abies sachalinensis* wood, using scanning electron microscopy. When radially compressed, tracheids developed cracks that followed the orientation of the microfibrils in the S_2 layer. These fissures could continue through the S_3 layer. Checks sometimes began to form at the pit apertures, extending on the inner wall surface in the same direction as the microfibrils. Similar checks are very common in compression wood, as discussed in Chap. 4.4.5.4.8. The deformation of transversely compressed wood has also been studied by Kisser and Schuster (1977) and by Grozdits and Chauret (1978).

Compression wood, despite its thick tracheid walls and resulting high density, nevertheless must be considered a very brash type of wood. It differs from normal, brash wood in deflecting considerably before breaking, especially in bending. As in all brash woods, rupture in compression wood is abrupt. Factors that possibly contribute to the brashness of compression wood are its high con-

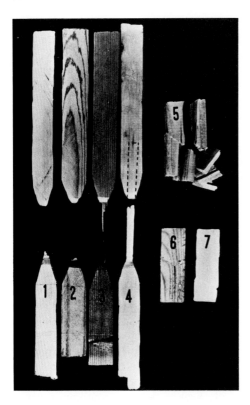

Fig. 7.36. Failure of normal and compression wood specimens in tension and compression. **Failure in tension** *1* clean break in compression wood of *Thuja plicata;* *2* clean break in compression wood of *Picea mariana;* *3* splintery break in normal wood of *Thuja plicata;* *4* failure of vertical shear in normal wood of *Picea mariana.* Break across the grain occurred at pin knot. **Failure in compression** *5* failure in compression wood of *Thuja plicata;* *6* oblique shear in normal wood of *Thuja plicata;* *7* oblique shear in normal wood of *Picea mariana.* (Verrall 1928)

Fig. 7.37. Typical toughness failures in green test specimens of *Picea glauca* and *Pinus resinosa.* Specimens *1–4* strong in toughness; Specimens *5–8* weak in toughness; Specimens *1* and *5* normal wood; *2* and *6* mild compression wood; *3* and *7* moderate compression wood; *4* and *8* pronounced compression wood. (Courtesy of Canadian Forestry Service.) (Perem 1960)

7.4.2.8 Fracture

tent of lignin, large microfibril angle, and short tracheids (Dinwoodie 1971, 1976).

Verrall (1928) tested compression wood of *Thuja plicata* in compression to failure. Corresponding normal wood failed by oblique shear, as shown in Fig. 7.36. Compression wood in all cases broke suddenly into several pieces with great force. In tension, also shown in Fig. 7.36, normal wood of *Thuja plicata* broke with splintering, whereas compression wood of both this species and of *Picea mariana* ruptured straight across the grain. Wardrop (1951) found that when *Pinus radiata* compression wood was ruptured in tensile stress, the line of failure followed the direction of the microfibrillar helix in S_2. Furuno et al. (1969), who subjected wood of *Picea jezoensis* to tensile stress, observed that both normal and compression woods ruptured with a splintering failure in the green state, as did also dry, normal wood. Dry compression wood, in contrast, failed directly across the grain.

When a beam of compression wood is broken in bending, the fracture is always brash, although straight rupture across the grain is rare. Sometimes the fracture forms a zigzag, but more frequently it assumes the shape of a wide Y, as shown in Fig. 7.37 (Koehler 1933, Pillow and Luxford 1937, Kraemer 1950, von Pechmann and Courtois 1970, von Pechmann 1973, Cockrell and Knudson 1973). Normal wood, if not brash, ruptures in bending with splintering. Perem (1958, 1960) observed different types of failure during testing for toughness of green woods from *Picea glauca* and *Pinus resinosa* as shown in Fig. 7.37. Samples strong in toughness exhibited compression fractures on the compression side of the tested beam, and this occurred with both normal and compression woods. Specimens weak in toughness failed in tension only.

Excellent examples of brash fractures caused by presence of compression wood in lumber have been furnished by Dinwoodie (1971). Figure 7.38 shows an instance of an extreme, brash fracture, where the concentric bands of dark compression wood are clearly visible. The *Tsuga* sp. lumber in Fig. 7.39, which

Fig. 7.38. Extremely brittle fracture in a beam caused by the presence of severe compression wood (*CW*). (Dinwoodie 1971)

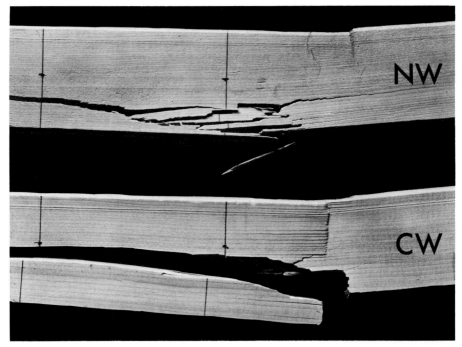

Fig. 7.39. Failure in bending in ladder stiles of *Tsuga* sp. The upper stile consisted of normal wood *(NW)* and failed with splintering. The lower stile, which contained compression wood *(CW)*, failed with a brittle fracture. (Dinwoodie 1971)

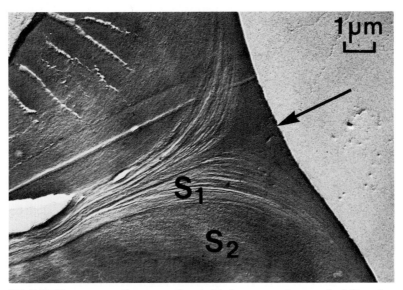

Fig. 7.40. Fracture at cell corners of tracheids in compression wood of *Picea jezoensis*, tested to tensile failure. *Arrow* indicates direction of stress. Transverse section. TEM. (Furuno et al. 1969)

7.4.2.8 Fracture

Fig. 7.41. Tension fracture of compression wood tracheids in *Cryptomeria japonica*. Fracture occurred along the helical cavities in S_2 but also across the helical ribs (*arrow*). Transverse surface. SEM. (Saiki et al. 1972)

failed in bending with a brash fracture, contained compression wood. It had a toughness value which was only 77% of that of normal wood.

Harada and his co-workers studied the tensile fracture of compression wood, using light and electron microscopy. When compression wood of *Picea jezoensis* was stressed in tension in a water-saturated condition, the failure was of the splintering type, with the tracheids usually separating along the middle lamella, as shown in Fig. 7.40 (Furuno et al. 1969). In the dry state, on the other hand, failure occurred across the tracheid wall, almost perpendicularly to the fiber axis. Saiki et al. (1972) found that, when stressed in tension parallel to the grain, compression wood of *Cryptomeria japonica* fractured along the helical cavities in the tracheid walls and also across the helical ribs, as shown in Fig. 7.41. It was later observed that the initial failure occurred in several locations along the helical ribs. Final fracture took place when the ribs were broken perpendicularly to the direction of the microfibrils (Furukawa et al. 1973).

In the course of the preparation of ultrathin sections for electron microscopy, fracture frequently is observed in compression wood. Typically, failure occurs between the compound middle lamella and S_1, between S_1 and $S_2(L)$, and between $S_2(L)$ and the remainder of S_2. Wood and Goring (1971), in preparing ultrathin sections of normal wood of *Pseudotsuga menziesii*, noted that failure occurred at the boundary between the primary wall and the S_1 layer, but also between S_1 and S_2 and between S_2 and S_3. In latewood, failure also took place in the middle lamella. The results obtained by Keith in his studies on the deformation and failure that occurred when normal (Keith 1971, 1972) and compression woods (Keith 1974) of *Picea glauca* were stressed in compression parallel to the grain are discussed in Sect. 7.4.2.2.4. Macroscopic compression failures tended to develop in the earlywood region of the compression wood, where the tracheids buckled under the load. Microscopic compression failures were not observed.

7.4.2.9 Strength Properties and Extent of Compression Wood Development

It is evident from the tabulated strength and other physical data that great variations exist throughout. It is probably futile to attempt a comparison of different species. Instead, one has to assume, at least at the present time, that fully developed compression wood has approximately the same mechanical properties irrespective of the species where it occurs. It is true that helical cavities do not occur in all compression woods, and that their presence or absence must have some effect on the mechanical properties of the wood. At this time, however, the exact influence of this feature remains uncertain.

The reason for the variability of the published data very likely resides in the fact that often no clear distinction was made between mild, moderate, and severe compression woods. In some cases, the test material may have included opposite or normal wood. Actually, it is fairly obvious from some reports that tests were carried out with whatever material happened to be on hand. It is not surprising then that strength data for the "same" compression wood can be so different in different investigations.

The variation of strength properties with extent of compression wood development is of considerable practical importance. Results obtained by Pillow and Luxford (1937) with mild and severe compression woods, as well as normal woods of *Pinus taeda* and *Pseudotsuga menziesii*, are shown in Table 7.35. Mild compression wood is usually intermediate between normal and severe compression woods in mechanical properties, although several exceptions do exist. In the green condition, severe compression wood is equal or superior to mild compression wood except in modulus of elasticity. When dry, it is inferior in modulus of rupture, work to maximum load, total work, and modulus of elasticity. Both green and air-dry, pronounced compression woods are superior to the mild type in maximum crushing strength. Based on unit weight, air-dry severe compression wood is inferior to mild compression wood in all strength properties, and the same applies, with only a few exceptions, to green wood.

Ollinmaa (1959) measured the compression strength of *Juniperus communis* and *Picea abies* specimens containing different proportions of compression woods. The results, summarized in Fig. 7.42 show that compressive strength increased linearly with increasing percentage of compression wood.

Extensive tests have been reported by Perem (1958, 1960) for normal wood and for mild, moderate, and severe compression woods of *Picea glauca* and

7.4.2.9 Strength Properties and Extent of Compression Wood Development

Table 7.35. Strength and related properties of normal wood and mild and severe compression woods of *Pinus taeda* and *Pseudotsuga menziesii*. (Courtesy of US Forest Service). (Pillow and Luxford 1937)

Species	Type of wood	Moisture, per cent	Specific gravity G	Static bending		Modulus of elasticity, psi $\times 10^{-3}$, E		Work, inch-pounds inch^{-3}		Maximum crushing strength, psi, C	
				Modulus of rupture, psi, R				To maximum load	Total		
				R	R/G$^{1.5}$	E	E/G^2			C	C/G
Pinus taeda	Normal wood	98.9	0.527	8190	21 410	1806	6500	10.6	36.0	3820	7 250
		11.1	0.583	15 920	35 760	2380	7000	10.9	32.1	8480	14 550
	Mild compression wood	92.9	0.527	8 560	22 370	1169	4210	12.8	54.4	4210	7 990
		11.5	0.593	15 540	34 030	1290	3670	25.7	25.7	6990	11 790
	Severe compression wood	73.2	0.596	8 540	18 560	817	2300	26.9	60.1	4470	7 500
		11.8	0.648	13 160	25 230	945	2250	22.7	22.7	7320	11 300
Pseudotsuga menziesii	Normal wood	42.5	0.460	7 990	25 610	1562	7380	8.4	27.3	3950	8 590
		11.6	0.510	15 140	41 570	1945	7460	15.0	33.8	8230	16 140
	Mild compression wood	79.0	0.480	6 830	20 540	999	4340	15.2	38.4	3520	7 330
		12.7	0.498	13 700	38 980	1180	4760	16.8	16.8	6640	13 330
	Severe compression wood	27.4	0.590	9 800	21 620	1001	2880	21.1	45.4	4730	8 020
		12.6	0.604	13 170	28 060	1212	3320	12.4	12.4	8330	13 790

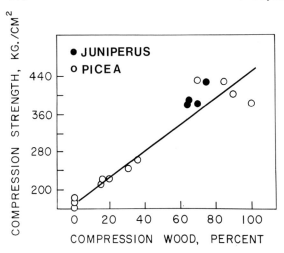

Fig. 7.42. Relationship between compressive strength parallel to the grain and proportion of compression wood in *Juniperus communis* and *Picea abies*. (Ollinmaa 1959)

Pinus resinosa (Table 7.36). Both modulus of rupture and maximum crushing strength increased in the above order for green and air-dry woods. The specific strength properties, on the other hand, tended to decrease in this order, and were highest for normal and lowest for severe compression woods. When tested green, severe compression wood had the highest toughness and normal wood the lowest, while mild and moderate compression woods were intermediate. In the air-dry condition, toughness was lowest for moderate compression wood and highest for normal wood. Cross sections of the four different test materials are shown in Fig. 7.37. Obviously, the classification of mild and moderate compression woods must have been somewhat arbitrary.

Another example has been reported by Pearson and Gilmore (1971), who determined the moduli of rupture and elasticity in bending of juvenile and mature woods of *Pinus taeda* containing varying proportions of compression wood. These investigators were unable to observe any correlation between the amount of compression wood present and the strength values, in spite of the fact that the contents of compression wood varied from zero to 61%. Two specimens, both containing 13% compression wood, had moduli of elasticity that differed by more than 100%. A sample with 9% compression wood had a modulus of rupture that was almost twice as large as that of a sample with 8% of this tissue. Another two specimens had the same modulus of rupture, although one consisted of normal wood and the other contained 40% compression wood. The results recently obtained by Schulz and Bellmann (1972) are referred to in Sect. 7.4.2.2.1 (Fig. 7.21).

7.4.2.10 Factors Determining the Strength Properties of Compression Wood

The different strength properties of compression wood are ultimately determined by factors on the molecular and ultrastructural level. In most cases, the

7.4.2.10 Factors Determining the Strength Properties of Compression Wood

Table 7.36. Strength and related properties of normal wood and mild, moderate, and severe compression woods of *Picea glauca* and *Pinus resinosa*. (Perem 1958, 1960)

Species	Type of wood	Moisture	Modulus of rupture, psi, R		Maximum crushing strength, psi, C		Toughness, inch-pound, T	
			R	R/G$^{1.5}$	C	C/G	T	T/G
Picea glauca	Normal wood	Green	4 610	25 500	1980	6 380	208	648
		Air-dry	9 270	47 800	4990	15 260	186	555
	Mild compression wood	Green	4 675	24 200	2005	6 130	256	773
		Air-dry	8 865	44 800	4750	13 890	125	367
	Moderate compression wood	Green	5 345	23 900	2160	6 135	229	609
		Air-dry	9 610	40 550	5020	13 250	132	295
	Severe compression wood	Green	5 830	18 300	3165	6 580	338	697
		Air-dry	10 310	34 500	6070	13 430	132	295
Pinus resinosa	Normal wood	Green	4 830	21 000	2130	5 695	351	926
		Air-dry	10 530	41 450	5400	13 860	186	463
	Mild compression wood	Green	5 490	20 300	2290	5 740	389	958
		Air-dry	10 020	38 200	5565	13 250	172	409
	Moderate compression wood	Green	5 245	19 700	2455	5 915	383	880
		Air-dry	10 260	34 800	5560	12 320	151	321
	Severe compression wood	Green	4 890	15 900	2840	5 635	434	833
		Air-dry	10 900	21 500	5740	11 320	157	298

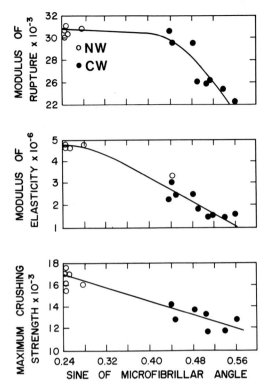

Fig. 7.43. Relationship between the sine of the average microfibril angle and three strength properties per unit density in normal (*NW*) and compression (*CW*) woods of *Pinus taeda*. (Courtesy of US Forest Service). (Redrawn from Pillow and Luxford 1937)

relationships are fairly clear. The high compressive strength of compression wood has been attributed to the thick tracheid walls ever since Sonntag (1904) expressed this view (Trendelenburg 1932, Onaka 1949). Undoubtedly this is correct, but other factors must also play a role. Trendelenburg (1932) has mentioned one, namely the fact that compression wood tracheids contain relatively few pits, but Koehler (1933) feels, first, that the number of pits has little or no effect on the strength of a fiber and, second, that it is not even certain that the pits cause a weakening of the adjacent cell wall. The high lignin content and large microfibril angle of compression wood, on the other hand, must be factors that contribute to its compressive strength.

Pillow and Luxford (1937) perceived that the microfibril angle in the S_2 layer of the tracheids may have an influence on the mechanical properties of wood, noticing that this angle increases in the order normal wood, mild compression wood, severe compression wood. Relationships between three strength parameters and the sine of the microfibrillar angle in S_2 are shown in Fig. 7.43. Maximum crushing strength decreases nearly linearly with the sine of increasing angle, but it would appear that for compression wood alone the relationship is not statistically significant, a conclusion that applies also to normal wood.

Schniewind (1962c) has pointed out that a high concentration of lignin in the secondary wall could lead to a higher lateral stability of the microfibrils. This positive correlation between lignin content and strength in compression

7.4.2.10 Factors Determining the Strength Properties of Compression Wood

along the grain is undoubtedly partly responsible for the high strength in compression of green compression wood. Mark and Gillis (1973) have shown that the stiffness of cellulosic fibers with a microfibril angle in S_2 exceeding 25° is insensitive to the properties of the cellulose. Instead, it depends largely on the characteristics of the matrix, that is, the hemicelluloses and the lignin. There can be little doubt that in compression wood the large microfibril angle of its tracheids also contributes decisively to the high compressive strength of this wood (Mark 1981).

The low tensile strength of compression wood is probably a consequence of several factors, namely the large microfibril angle in S_2, which has been shown to be associated with a large total creep in tension parallel to the grain (El-Osta and Wellwood 1972), the presence of helical cavities in S_2, and the low cellulose and high lignin contents (Knigge 1958a, Wardrop 1951, Schniewind 1962c). The numerous intercellular spaces probably weaken the cohesion between the tracheids. Frey-Wyssling (1968) has briefly reviewed the strength characteristics of wood, considering the cell wall as a system of rodlets (cellulose) embedded in an isotropic matrix (lignin and hemicelluloses). It is evident that tensile strength derives from the cellulose. However, as long as cellulose remains stronger in tension than the matrix, the strength of the entire, composite material will depend largely on the length of the elementary fibrils. Lignin and probably also the hemicelluloses contribute in themselves little or nothing to tensile strength.

Münch (1938) discussed in great detail the effect of the helical orientation of the microfibrils on the compressive and tensile strength of wood. One conclusion from his deductions was that tensile strength decreases with increasing microfibril angle in S_2. The presence of the helical cavities probably has the same effect (Onaka 1949), conferring upon the tracheid the characteristics of a spring.

The more recent calculations reported by Mark and Gillis (1973) suggest that the arguments advanced by Frey-Wyssling (1968) must be modified. It is only in fibers with a microfibril angle less than 10° that fiber stiffness depends largely on the properties of the cellulose. In compression wood tracheids with their large angles, the characteristics of the hemicellulose and lignin strongly influence the fiber stiffness. Mark and Gillis (1970, 1973) could show that for a microfibril angle of 40°, which is common in compression wood tracheids, the axial fiber modulus was 1000 kg cm^{-2}, irrespective of the elastic constants assigned to the cellulose. If the matrix constants were reduced by 90%, the matrix value decreased to only 200 kg cm^{-2}. It is obvious, as pointed out by the authors, that tensile tests of fibers with large microfibril angles in S_2 would furnish much information about the properties of the hemicelluloses and lignin.

Strength properties in static bending are a function of the strength in compression and tension. The modulus of rupture of compression wood is relatively low, largely because the tensile strength of this tissue is so low. Fig. 7.43 shows that the modulus of rupture decreases rapidly with increasing microfibril angle in S_2 but only at angles larger than 25°–30°.

It is characteristic of compression wood that the moduli of elasticity in compression, tension, and bending are all about the same and only 50–60% of those valid for normal wood. Modulus of elasticity in bending decreases with increasing microfibril angle (Fig. 7.43). Yamai (1951), in an attempt to explain the low modulus of elasticity of compression wood, compared a compression wood tracheid with two helical steel springs, with one (S_2) inside the other (S_1). By treating this model mathematically, he came to the conclusion that the orientation of the microfibrils in the S_1 and S_2 layers could account for the low modulus of elasticity. This is an approach very similar to that used earlier by Münch (1938). The fact that compression wood can be compressed, drawn out, and bent so much more readily than normal wood must be attributed to the flat microfibril helix in S_2 and also to the high lignin and low cellulose contents.

The brash nature of compression wood, which at first might appear surprising in view of the thick tracheid walls, is due to several factors, one of which is the high lignin content. Another is the ultrastructure of the compression wood tracheids with their thick S_1 layer, where the orientation of the microfibrils is almost transverse and the wall can offer little resistance to tensile stress. In the S_2 layer, the large microfibril angle must have the same effect as it has in normal earlywood (Sect. 7.4.2.8), favoring rupture between the microfibrils rather than across the cell wall (Mark 1967). The presence of the helical cavities must also weaken the tracheid wall and cause it to rupture abruptly.

With the exception of work in static bending, most strength properties of wood increase on going from green to air-dry conditions. Compression wood improves far less in strength on drying than does normal wood, one exception being the modulus of elasticity in bending which changes equally for both woods. Sonntag (1904), referring to an earlier suggestion by Hartig (1901), believed that a reason for this behavior could be that air entered the helical cavities on drying. A more likely explanation, however, is that proposed by Trendelenburg (1932), namely that the high longitudinal shrinkage of compression wood causes formation of cracks in the tracheid walls and also creates strong stresses in the wood. The deleterious effect of microscopic compression failures on the strength of wood is a related phenomenon (Chap. 15.2.1.5). Cracks develop especially readily in the secondary wall of compression-wood tracheids, both because of the absence of an S_3 layer and because of the already existing helical cavities.

7.4.3 Branch Wood

Studies on the strength properties of branch wood have largely been concerned with static bending. Branches have to be able to support not only their own weight but also occasional additional loads, such as snow and ice. The function of opposite and compression woods in branch wood is discussed in Chap. 10.4.

Hartig (1901) measured the deflection of *Picea abies* branches which were loaded down with weights. First, the loads were applied with the branch in its natural position, that is, with the compression wood on the lower side. In a second series of experiments the weights were applied to the branch after it had been inverted, so that the opposite wood was on the lower side. The deflections observed in the two cases were almost identical, a rather surprising result.

7.4.3 Branch Wood

Table 7.37. Results of bending experiments with branches of *Pinus thunbergii* in their natural and in the reversed positions. Green branches with bark. Span: 30 cm, center loading. (Onaka 1949)

Diameter of branch mm	Deflection at a load of 5 kg, cm		Limit of proportionality, kg		Load at failure, kg	
	Natural position	Reversed position	Natural position	Reversed position	Natural position	Reversed position
14.2	1.07	1.37	5.2	4.2	8.7	6.2
15.5	0.79	0.90	7.2	5.0	10.3	8.0
16.0	0.91	1.41	3.5	3.0	10.0	7.0
17.0	0.51	0.63	12.0	6.3	13.0	9.2

A few years later, Sonntag (1904) carried out similar tests with branches, also of *Picea abies*. In a typical experiment, a branch, in its natural orientation, thick as a thumb and 46 cm long, was deflected 7.9 cm when it was loaded with 1.5 kg. On removal of the weight, it returned to within 0.65 cm of its original position where it remained permanently. When the same load was applied with the branch in the reverse position, the deflection was 8.9 cm and the permanent displacement was 1.0 – 1.2 cm. Similar results have been reported by all subsequent investigators. In both Hartig's (1901) and Sonntag's (1904) experiments, the limit of elasticity must have been exceeded by far. Realizing this, Sonntag (1904) also undertook measurements within this limit. In this case, the deflection remained the same, regardless of the orientation of the branch. Evidently, the difference in deflection between branches in their natural and opposite orientations appears only when the limit of proportionality has been exceeded.

Similar experiments were later carried out by Onaka (1949) with branches of *Pinus thunbergii*, *P. densiflora*, and *Chamaecyparis obtusa*. Results obtained with the first species are shown in Table 7.37. Here too, the deflection was less when the branch was in its natural rather than in the reverse position, the proportional limit was higher, and the load at failure was larger. The general conclusion to be drawn from these experiments is, of course, that the location of opposite wood on the upper and compression wood on the under side best serves to support the branch.

The mechanical properties of branch wood have been measured by several investigators, albeit not as extensively as those of isolated compression wood. Janka (1904) found a maximum crushing strength of 650 kg cm^{-2} for air-dry branch wood of *Picea abies*, considered by him to be a very high value. More extensive data were obtained by Fegel (1938, 1941) for branch and stem wood of seven American conifers. Some of his results are shown in Tables 7.38 and 7.39. Green branch wood was superior and air-dry branch wood equal to corresponding stem wood in compressive strength. In static bending, air-dry branch wood was inferior to stem wood in all properties except work to maximum load. Branch wood was tougher than stem wood.

Table 7.38. Average strength properties in compression parallel to the grain of green and air-dry stem and branch woods from various conifer species. All values in psi. (Fegel 1938, 1941)

Property	Stem wood		Branch wood	
	Green	Air-dry	Green	Air-dry
Maximum crushing strength	2633	5520	3660	5628
Stress at proportional limit	1958	3838	2111	3446

Table 7.39. Average strength properties in static bending of air-dry stem and branch woods from various conifer species. (Fegel 1941)

Property	Stem wood	Branch wood
Modulus of rupture, psi	5483	4850
Modulus of elasticity, psi $\times 10^{-3}$	1014	912
Work to proportional limit, inch-pounds inch^{-3}	0.81	0.60
Work to maximum load, inch-pounds inch^{-3}	5.08	10.64
Fiber stress at proportional limit, psi	3515	2530

As could be expected from the high content of compression wood, the tensile strength of branch wood is low. Ursprung (1906) found that wood on the upper side of a branch was stronger in tension than that on the lower side. Opposite wood appears to have an especially high tensile strength. Sonntag (1904) found such wood from a branch of *Picea abies* to be twice as strong as the compression wood.

Wegelius (1939) discussed the mechanical properties of branch and knot wood, including his own results, in the form of various graphs. In agreement with earlier investigators, he observed that a branch will support a larger load when in its natural rather than in the reversed position. Compressive strength was directly proportional to the percentage of compression wood in the branch, but tensile strength perpendicular to the grain and cleavability were independent. Confirming earlier measurements by Brax (1936), Wegelius found branch wood and knot wood to be very hard. That knot wood is harder than normal wood can readily be observed on old, wooden floors where the knots always have withstood wear and tear much better than the surrounding stem wood and as a result protrude above the latter.[4]

Enčev (1972) has reported data for the compressive strength along the grain of stem wood and knot wood of *Abies alba, Picea abies,* and *Pinus sylvestris.* The

---

[4] A good example of this is found in the hall on the third floor of the Beethovenhaus in Bonn where millions must have trodden to visit the small room where the greatest of all composers was born.

values, which were obtained with the woods in the green conditions, varied more for stem wood than for knot wood, which on the average was 42% stronger than the former. The difference in density, on the other hand, was more than 100%, and, according to Enčev, one could have expected a greater strength in compression for knot wood. The possibility that the high density of the knot wood could have been due to a high content of resin does not seem to have been taken into consideration. Enčev's suggestion that the presence of compression wood should lower the compressive strength of branch wood is difficult to accept for green compression wood is generally superior to normal wood in this respect. Knot wood was 2.2–4.6 times harder than stem wood.

Götze et al. (1972) determined some strength properties of stem wood and various parts of branches in *Pinus sylvestris*. The data in Table 7.8 show that the compressive strength of branch compression wood was noticeably higher than of normal stem or branch woods in both the green and the dry states. Unlike the values obtained with normal wood, those for compression wood varied considerably. Bending strength was 32% greater for compression wood, but values for impact strength were 15–20% less. It is not clear whether the "normal" branch wood consisted only of side wood or if opposite wood was included.

7.4.4 Conclusions

It is difficult to draw definitive conclusions concerning the mechanical properties of compression wood. Published data are sometimes contradictory, tests for all strength properties have seldom been carried out, and in some cases it is clear that the specimens tested did not consist entirely of fully developed compression wood. As pointed out by Westing (1965), reliable information on green, severe compression wood is especially scarce. This is unfortunate, particularly as it is not difficult to locate pure, large-sized specimens of such wood. A few data for *Pinus ponderosa* in Table 7.40, based on calculations reported by Westing, summarize the more important strength characteristics of compression wood in comparison with normal wood.

Table 7.40. Comparison of some strength properties of green normal and compression woods of *Pinus ponderosa*. Values calculated on the basis of data reported by Markwardt and Wilson (1935) and Pillow and Luxford (1937), all in kg cm^{-2}. (Westing 1965)

Force required to cause	Compression wood	Normal wood	Ratio
Longitudinal compression, 0.1%	70	104	0.67
Stretching, 0.1%	59	76	0.78
Failure in longitudinal compression	232	165	1.41
Failure in longitudinal tension	681	828	0.82
Failure in bending	430	326	1.32

Green compression wood, when tested for compression parallel to the grain, is superior to corresponding normal wood in maximum crushing strength and stress at proportional limit. It invariably has a lower modulus of elasticity. In tension parallel to the grain, green compression wood is always inferior to normal wood with respect to tensile strength and Young's modulus of elasticity. When subjected to static bending, green compression wood is superior to normal wood in modulus of rupture and stress at proportional limit. Work to proportional limit, work to maximum load, and total work are all very large for compression wood. The modulus of elasticity in bending is low compared to that of normal wood. When loaded beyond the limit of proportionality, compression wood deflects less than normal wood and has a greater ability to return to its original position. Within the limit of proportionality, it requires about half the force to deflect compression wood a certain distance in comparison with normal wood. Toughness is a highly erratic property. Often, it is higher for green compression wood than for normal wood, but in other cases it is lower. Green compression wood is always harder than normal wood, especially the compression wood present in knots. Compression wood is invariably brash. It breaks abruptly with a clean fracture, but, unlike brash, normal wood, with considerable prior deflection.

Compression wood improves less in strength on drying than normal wood. As a result, it is inferior to corresponding normal wood in more respects in the air-dry than in the green condition. Air-dry compression wood is superior to normal wood only in maximum crushing strength and work to maximum load. It is inferior in tensile strength, stress at proportional limit in bending, work at proportional limit, total work, all three moduli of elasticity, and usually, albeit not always, in toughness.

When mechanical properties are considered on the basis of unit weight it becomes evident that much of the strength of compression wood derives from its high specific gravity, that is, its thick tracheid walls. The cell wall material per se is weaker in most respects than that in normal wood. When comparison is made on the basis of equal density, it is found that compression wood is inferior to normal wood in all strength properties except work to maximum load and total work. In maximum crushing strength it seems to be equal to normal wood. Air-dry compression wood, when considered on a unit weight basis, is weaker than normal wood in all aspects except, occasionally, work to maximum load and total work.

In branches, opposite wood on the upper side has a high tensile strength, whereas compression wood on the under side is superior in compression. Because of this, branches deflect less when loaded down in their natural than in the reverse position and are also better able to return to their original position on removal of a load that exceeded the limit of proportionality. Compression wood in branches has the same mechanical properties as that in the stem.

7.5 General Conclusions

The chemical composition and ultrastructure of compression wood are, directly or indirectly, responsible for all the physical properties of this remarkable tissue. The intercellular spaces, the thickness of the tracheid wall, the narrow lumen, the orientation of the microfibrils in the S_1 and S_2 layers, the presence of the helical cavities, the highly lignified secondary wall, and its peculiar distribution of lignin all have their influence on each property: the high specific gravity, the lower fiber saturation point, the reduced permeability, the low radial and tangential but exceedingly large longitudinal shrinkage, the high compressive but low tensile strength, the considerable ability to deflect in bending, the high elasticity, the brashness in fracture, and the hardness.

It may at first seem surprising that compression wood, which in the living tree has to support often tremendous loads and actually is in a state of compression, is not constructed better for this purpose. It must, however, be remembered that the chief function of compression wood in nature is not to withstand compressive stress but to exert an active, dynamic pressure by expanding longitudinally (Chap. 17.6). The eccentric cross section of branches and leaning stems offers a good protection against rupture. Green compression wood actually has a relatively high compressive strength and the ability to bend rather than break when subjected to stress, and opposite wood has a high tensile strength. Green compression wood is also more elastic than normal wood. These aspects are discussed in more detail in Chap. 17.

From a wood utilization point of view, compression wood has many drawbacks. Most serious is its high longitudinal shrinkage which causes warping and other distortions in lumber containing both normal and compression woods. Its second most serious disadvantage is its low ability to increase in strength on drying, which renders it weaker than normal wood not only in tension, which it is under all circumstances, but also in many properties related to compression and bending. A third drawback is the brash rupture characteristic of compression wood which, where it occurs in construction lumber, can render it a danger to life and property. As a fourth disadvantage must be listed the great hardness of compression wood, which makes it difficult, if not impossible, to nail or to work with ordinary tools. Here, as in so many other cases, what is beneficial and useful to the living tree, is not always of advantage to man in his attempts at maximum utilization of a natural material.

References

Ahlgren PA, Wood JR, Goring DAI 1972 The fiber saturation point of various morphological subdivisions of Douglas-fir and aspen wood. Wood Sci Technol 6:81–84

Aiuchi T, Ishida S 1978 An observation of the failure process of softwood under compression perpendicular to the grain in the scanning electron microscope. Mokuzai Gakkaishi 24:507–510

Aiuchi T, Ishida S 1979 An observation of the failure process of softwood under compression perpendicular to the grain in the scanning electron microscope. II. Morphology of the failure as observed in longitudinal section of tracheid wall. Res Bull Coll Exp For Hokkaido Univ 36:623–632

Aiuchi T, Ishida S 1981 An observation of the failure process of softwood under compression perpendicular to the grain in the scanning electron microscope. III. On the radial compression. Res Bull Coll Exp For Hokkaido Univ 38:73–82

Aldridge F, Hudson RH 1955a Growing quality softwoods. Q J For 49:109–114

Aldridge F, Hudson RH 1955 Growing quality softwoods. A critical examination of the Turnbull hypothesis. Q J For 49:260–270

Aldridge F, Hudson RH 1958 Growing quality softwoods. Variation in strength and density of Picea abies-specimens taken from a commercial assignment. Q J For 52:107–114
Aldridge F, Hudson RH 1959 Growing quality softwoods. Q J For 53:210–219
Alestalo A, Hentola Y 1966 Sulfaattisellua havupuiden kuorellisista latvuksista, okista ja kannoista. (Sulfate pulps from unbarked softwoods tops, branches, and stumps.) Pap Puu 48: 737–742
Allen R, Wardrop AB 1964 The opening and shedding mechanism of the female cones of Pinus radiata. Aust J Bot 12:125–134
Ando T, Kataoka H 1961 (Bulk density of branches of Cryptomeria japonica D Don.) J Jap For Soc 43:280–283
Anonymous 1924 Spruce limb serves as humidity indicator. West Coast Lumberm 25 (536):26b (Febr 1)
Anonymous 1952 Shrinkage and swelling of wood in use. US For Serv FPL Rep R736, 10 pp
Anonymous 1960 (1930) Longitudinal shrinkage of wood. US For Serv FPL Rep 1093, 10 pp
Atmer B, Thörnqvist T 1982 Fiberegenskaper i gran (Picea abies Karst) och tall (Pinus silvestris L). (The properties of tracheids in spruce (Picea abies Karst) and pine (Pinus silvestris L).) Swed Univ Agr Sci Dep For Prod Rep 134, 78 pp
Bailey PJ, Preston RD 1969 Some aspects of softwood permeability. I. Structural studies with Douglas fir sapwood and heartwood. Holzforschung 23:114–120
Baker G 1967 Estimating specific gravity in plantation-grown red pine. For Prod J 17(8):21–24
Baker G 1968 Effect of proximity to limb whorls on wood density. For Prod J 18(9):112–113
Banks CH 1954 The mechanical properties of timbers with particular reference to those grown in the Union of South Africa. J S Afr For Assoc 24:44–65
Banks CH 1957 A comparison of the strength of timber from the major and minor radius of eccentric trees. Brit Commonw 7th For Conf, Gov Printer Pretoria, 10 pp
Banks CH, Schwegmann LM 1957 The physical properties of fast- and slow-grown Pinus patula and P. taeda from South African sources. J S Afr For Assoc 30:44–59
Bannister MH, Vine MH 1981 An early progeny trial in Pinus radiata. 4. Wood density. NZ J For Sci 11:221–243
Barber NF 1968 A theoretical model of shrinking wood. Holzforschung 22:97–103
Barber NF, Meylan BA 1964 The anisotropic shrinkage of wood. Holzforschung 18:146–156
Barnes RD, Woodend JJ, Schweppenhauser MA, Mullin LJ 1977 Variation in diameter growth and wood density in six-year-old provenance trials of Pinus caribaea Morelet on five sites in Rhodesia. Silvae Genet 26:163–167
Barrett JD, Schiewind AP, Taylor RL 1972 Theoretical shrinkage model for wood cell walls. Wood Sci 4:178–192
Beall FC 1972 Density of hemicelluloses and their relationship to wood substance density. Wood Fiber 4:114–116
Bendtsen BA, Ethington RL 1972 Properties of major southern pines. II. Structural properties and specific gravity. US For Serv Res Pap FPL-177, 10 pp
Benić R 1970 Neki elementi koji utiču na mogućnost iskorišćenja drva jelovih grana. (Some factors which influence the utilization of fir branch wood.) Drvna Ind 21(11/12):199–201
Berkley EE 1934 Certain physical and structural properties of three species of southern yellow pine correlated with the compression strength of their wood. Ann MO Bot Gard 21:241–328
Berlyn GP 1964 Recent advances in wood anatomy: the cell walls in secondary xylem. For Prod J 14:467–476
Berlyn GP 1968 A hypothesis for cell-wall density. For Prod J 18(2):34–36
Berlyn GP 1969 Microspectrophotometric investigations of free space in plant cell walls. Am J Bot 56:498–506
Bernhart A 1964 Über die Rohdichte von Fichtenholz. Holz Roh-Werkst 22:215–228
Bernhart A 1965 Frischfeuchtigkeit und Schwindverhalten von Fichtenholz. Forstwiss Cbl 84:347–356
Bernhart A 1966 Über die statische und dynamische Kurzzeitfestigkeit von Fichtenholz – absolut, rohdichtebezogen und unter Druckholzeinfluß. Forstwiss Cbl 85:275–295
Besley L 1960 Relationship between wood fibre properties and paper quality. Pulp Pap Mag Can 61(7):136–146

Björklund T 1982 Kontortamännyn puutekniset ominaisuudet. (Technical properties of lodgepole pine wood.) Fol For 522, 25 pp

Bodig J, Jayne BA 1982 Mechanics of wood and wood composites. Van Nostrand Reinhold, New York, 736 pp

Boone RS, Chudnoff M 1972 Compression-wood formation and other characteristics of plantation-grown Pinus caribaea. US For Serv Res Pap ITF-13, 16 pp

Borgin K 1971 The cohesive failure of wood studied with the scanning electron microscope. J Microsc 94:1−11

Borgin K 1973 The failure of the ultrastructure of wood under mechanical stresses. Meet IUFRO-5 S Afr 2:58−65

Borgin K, Parameswaran N, Liese W 1975 The effect of aging on the ultrastructure of wood. Wood Sci Technol 9:87−98

Bosshard HH 1956 Über die Anisotropie der Holzschwindung. Holz Roh-Werkst 14:285−295

Bosshard HH 1961 Influence of the microscopic and submicroscopic structure on the anisotropic shrinkage of wood. Recent Adv Bot II., Univ Toronto Press, Toronto, 1714−1720

Bosshard HH 1974 Holzkunde. Band 2, Birkhäuser, Basel Stuttgart, 312 pp

Boutelje JB 1958 Rechnerische Betrachtungen über die Schwankung der maximalen Raumquellung bei der Kiefer (Pinus silvestris L). Holz Roh-Werkst 16:413−418

Boutelje JB 1962a On shrinkage and change in microscopic void volume during drying, as calculated from measurements on microtome cross sections of Swedish pine (Pinus sylvestris L). Svensk Papperstidn 65:209−215

Boutelje JB 1962b The relationship of structure to transverse anisotropy in wood with reference to shrinkage and elasticity. Holzforschung 16:33−46

Boutelje JB 1965 The anatomical structure, moisture content, density, shrinkage and resin content of the wood in and around knots of Swedish redwood (Pinus silvestris L) and Swedish whitewood (Picea abies Karst). Proc Meet Sect 41 IUFRO, Melbourne Vol 2, 13 pp

Boutelje JB 1966 On the anatomical structure, moisture content, density, shrinkage and resin content of the wood in and around knots in Swedish pine (Pinus silvestris L) and Swedish spruce (Picea abies Karst). Svensk Papperstidn 69:1−10

Boutelje JB 1972 Change in refractive index and volume of cellulosic cell walls with moisture content, as measured by interference microscopy. Svensk Papperstidn 75:361−367

Boutelje JB 1973 On the relationship between structure and the shrinkage and swelling of the wood in Swedish pine (Pinus silvestris) and spruce (Picea abies). Svensk Papperstidn 75:78−83

Bramhall G, Wellwood RW 1976 Kiln drying of western Canadian lumber. Can For Serv Rep VP-X-159, 112 pp

Boyd JD 1974a Relating delignification to microfibril angle differences between tangential and radial faces of all wall layers in wood cells. Drev Vysk 19:41−54

Boyd JD 1974b Anisotropic shrinkage of wood: identification of the dominant determinants. Mokuzai Gakkaishi 20:473−482

Boyd JD 1977 Relationship between fibre morphology and shrinkage of wood. Wood Sci Technol 11:3−22

Boyd JD 1978 Significance of larcinan in compression wood tracheids. Wood Sci Technol 12:25−35

Brax AJ 1936 Undersökningar angående kvistens inflytande på slipmassans egenskaper. (Investigations on the influence of knots on the properties of groundwood pulp.) Svensk Papperstidn 39 (spec issue):7−20

Brazier JD 1970 Timber improvement. II. The effect of vigour on young-growth Sitka spruce. Forestry 43:135−150

Brodzki P 1972 Callose in compression wood tracheids. Acta Soc Bot Polon 41:321−327

Brown CL 1970 Physiology of wood formation in conifers. Wood Sci 3:8−22

Brown HP, Panshin AJ, Forsaith CC 1952 Textbook of wood technology. Vol II. McGraw-Hill, New York, 783 pp

Burger H 1941 Holz, Blattmenge und Zuwachs. V. Fichten und Föhren verschiedener Herkunft auf verschiedenen Kulturen. Mitt Schweiz Centralanst Forstl Versuchswes 22:10−62

Burns GP 1920 Eccentric growth and the formation of redwood in the main stem of conifers. VT Agr Exp Sta Bull 219, 16 pp

Campbell PA 1968 Variability of the strength properties of the Kenya pines. 2nd Sess 3rd East Afr Timber Symp Nairobi, 4 pp

Cave ID 1968 The anisotropic elasticity of the plant cell wall. Wood Sci Technol 2:268–278

Cave ID 1972a Swelling of a fibre reinforced composite. Wood Sci Technol 6:157–161

Cave ID 1972b A theory of the shrinkage of wood. Wood Sci Technol 6:284–292

Cave ID 1975 Wood substances as a water-reactive fibre-reinforced composite. J Microsc 104:47–52

Cave ID 1976 Modelling the structure of the softwood cell wall for computation of mechanical properties. Wood Sci Technol 10:19–28

Cave ID 1978a Modelling moisture-related mechanical properties of wood. I. Properties of the wood constituents. Wood Sci Technol 12:75–86

Cave ID 1978b Modelling moisture-related mechanical properties of wood. II. Computation of properties of a model of wood and comparison with experimental data. Wood Sci Technol 12:127–139

Chalk L 1953 Variation of density in stems of Douglas fir. Forestry 26:33–36

Chiang CI, Kennedy RW 1967 Influence of specific gravity and growth rate on dry wood production in plantation-grown white spruce. For Chron 43:165–173

Choong ET, Box BH, Fogg PJ 1970 Effects of intensive cultural management on growth and certain wood properties of young loblolly pine. Wood Fiber 2:105–112

Christensen GN, Hergt HFA 1968 The apparent density of wood in non-swelling liquids. Holzforschung 22:165–170

Christensen GN, Kelsey KE 1958 The sorption of water vapour by the constituents of wood: determination of sorption isotherms. Aust J Appl Sci 9:265–282

Christensen GN, Kelsey KE 1959a The sorption of water vapour by the constituents of wood. II. The swelling of lignin. Aust J Appl Sci 10:284–293

Christensen GN, Kelsey KE 1959b Die Sorption von Wasserdampf durch die chemischen Bestandteile des Holzes. Holz Roh-Werkst 17:189–203

Chu LC 1972 Comparison of normal wood and first-year compression wood in longleaf pine trees. M S Thesis MS State Univ, 72 pp

Cieslar A 1896 Das Rothholz der Fichte. Cbl Ges Forstwes 22:149–165

Cockrell RA 1943 Some observations on density and shrinkage of ponderosa pine wood. Am Soc Mech Eng Trans 65:729–739

Cockrell RA 1947 Explanation of longitudinal shrinkage of wood, based on interconnected chain molecule concept of cell wall structure. Am Soc Mech Eng Trans 69:931–935

Cockrell RA 1974 A comparison of latewood pits, fibril orientation and shrinkage of normal and compression wood of giant sequoia. Wood Sci Technol 8:197–206

Cockrell RA, Knudson RM 1973 A comparison of static bending, compression and tension parallel to grain and toughness properties of compression wood and normal wood of a giant sequoia. Wood Sci Technol 7:241–250

Constantinescu A 1956 Cercetări preliminare asupra formatiunii de lemn de compresiune la bradul (Abies alba) din pădurea Chilerei-Valea Timisului. (Preliminary investigations of the formation of compression wood in Abies alba in Chilerei forest in the Timis valley.) Ind Lemn 5:455–457

Conway EM, Minor CO 1961 Specific gravity of Arizona ponderosa pine pulpwood. US For Serv Res Note RM-54, 3 pp

Côté WA, Jr, Day AC 1965 Anatomy and ultrastructure of reaction wood. In: Côté WA, Jr (ed) Cellular ultrastructure of woody plants. Syracuse Univ Press, Syracuse, 391–418

Côté WA, Jr, Hanna RB 1983 Ultrastructural characterization of wood fracture surfaces. Wood Fiber Sci 15:135–163

Côté WA, Jr, Kutscha NP, Simson BW, Timell TE 1968 Studies on compression wood. VI. Distribution of polysaccharides in the cell wall of tracheids from compression wood. Tappi 51:33–40

Cousins WJ 1972 Measurement of mean microfibril angles of wood tracheids. Wood Sci Technol 6:58

Cown DJ 1981 Wood density of Pinus caribaea var hondurensis grown in Fiji. NZ J For Sci 11:244–253

Dadswell HE 1958 Wood structure variations occurring during tree growth and their influence on properties. J Inst Wood Sci 1:11–33

References

Dadswell HE 1963 Tree growth–wood property inter-relationships. VIII. Variations in structure and properties in wood grown under abnormal conditions. In: Maki TE (ed) A special field institute in forest biology, NC State Univ School For, Raleigh 1960, 55–66

Dadswell HE, Wardrop AB 1949 What is reaction wood? Aust For 13:22–33

Dadswell HE, Wardrop AB, Watson AJ 1958 The morphology, chemistry and pulping characteristics of reaction wood. In: Bolam F (ed) Fundamentals of papermaking fibres. Tech Sect Brit Pap Board Makers' Assoc London, 187–219

Davies GW 1968 Microscopic observations of wood fracture. Holzforschung 22:177–180

Debaise GR, Porter AW, Pentoney RE 1966 Morphology and mechanics of wood fracture. Mater Res Stand 6:493–499

Desch HE, Dinwoodie JM 1980 Timber. Its structure, properties and utilisation, 6th ed. Timber Press, Forest Grove, 410 pp

Dinwoodie JM 1971 Brashness in timber and its significance. J Inst Wood Sci (28) 5 (4):3–11

Dinwoodie JM 1975 Timber – review of the structure-mechanical property relationship. J Microsc 104:3–32

Dinwoodie JM 1976 Causes of brashness in timber. In: Baas P, Bolton AJ, Catling DM (eds) Wood structure in biological and technological research. Leiden Bot Ser 2, Leiden Univ Press, Leiden, 238–252

Dinwoodie JM 1981 Timber. Its nature and behaviour. Van Nostrand Reinhold, New York, 190 pp

du Toit AJ 1963 A study of the influence of compression wood on the warping of Pinus radiata D Don timber. S Afr For J 44:11–15

Einspahr DW, Goddard RE, Gardner HS 1964 Slash pine, wood and fiber property heritability study. Silvae Genet 13:103–109

Elliott GK 1970 Wood density in conifers. For Bur Oxford Tech Commun 8, 44 pp

El-Osta MLM, Wellwood RW 1972 Short-term creep as related to microfibril angle. Wood Fiber 4:26–32

Enčev EA 1962 (Density and compression strength of branchwood of Scots pine, Norway spruce and silver fir.) Nauč Trud Lesoteh Inst Sofija 10:177–190

Enčev E 1972 Untersuchung einiger physikalisch-mechanischer Eigenschaften der Äste von Nadelhölzern. Holztechnologie 13:232–236

Engler A 1924 Heliotropismus und Geotropismus der Bäume und deren waldbauliche Bedeutung. Mitt Schweiz Centralanst Forstl Versuchswes 13:225–283

Erickson HD, Harrison AT 1974 Douglas-fir wood quality studies. I. Effects of age and stimulated growth on wood density and anatomy. Wood Sci Technol 8:207–226

Ericson B 1959 A mercury immersion method for determining the wood density of increment core sections. Stat Skogsforskningsinst Stockholm, Avd Skogsprod Rapp 1, 31 pp

Ericson B 1966 Determination of basic density in small wood samples. Skogshögskolan Stockholm Inst Skogsprod Rapp Upps 9, 34 pp

Eskilsson S 1972 Whole tree pulping. 1. Fibre properties. Svensk Papperstidn 75:397–402

Eskilsson S, Hartler N 1973 Whole tree pulping. 2. Sulphate pulping. Svensk Papperstidn 76:63–70

Farr WA 1973 Specific gravity of western hemlock and Sitka spruce in southeast Alaska. Wood Sci 6:9–12

Fegel AC 1938 A comparison of the mechanical and physical properties and the structural features of root-, stem- and branch-wood. Ph D Thesis State Univ Coll For Syracuse, 66 pp

Fegel AC 1941 Comparative anatomy and varying physical properties of trunk, branch, and root wood in certain northeastern trees. NY State Coll For Syracuse Bull 55, 19 pp

Fellegi J, Janči J, Kubelka V, Zemánek R 1962 (Manufacture of pulp from small-dimensioned wood.) Bumazh Prom 37(11):13–15

Forest Products Laboratory 1974 Wood handbook: wood as an engineering material. US Dept Agr, Agr Handb 72, 216 pp

Foulger AN 1966 Longitudinal shrinkage pattern in eastern white pine stems. For Prod J 16(12):45–47

Frey-Wyssling 1940a Die Anisotropie des Schwindmaßes auf Holzquerschnitt. Holz Roh-Werkst 3:43–45

Frey-Wyssling A 1940b Die Ursachen der anisotropen Schwindung des Holzes. Holz Roh-Werkst 3:349–353

Frey-Wyssling A 1943 Weitere Untersuchungen über die Schwindungsanisotropie des Holzes. Holz Roh-Werkst 6:197–198

Frey-Wyssling A 1968 The ultrastructure of wood. Wood Sci Technol 2:73–83

Furukawa I, Saiki H, Harada H 1973 Continous observation of tensile fracture process of single tracheid by scanning electron microscope. Mokuzai Gakkaishi 19:399–402

Furuno T, Saiki H, Harada H 1969 Ultrastructural feature of compression wood tracheids stressed to tensile failure. Mokuzai Gakkaishi 15:104–108

Gaby LI 1971 Longitudinal shrinkage and fibril angle variation in corewood of southern pine. GA For Res Counc Res Pap 70, 4 pp

Garland H 1939 A microscopic study of coniferous wood in relation to its strength properties. Ann MO Bot Gard 26(1):1–78

Gava M 1974 (Factors that influence the shedding of dead branches in Picea abies.) Ind Lemn Silvicult Exploat Pădur 89(1):8–12

Giordano G 1971 Tecnologia del legno: la materia prima I. Unione Tipografico-Editrice Torinese, Torino, 1068 pp

Goggans JF 1961 The interplay of environment and heredity as factors controlling properties in conifers with special emphasis on their effects on specific gravity. NC State Coll School For, Raleigh, Tech Rep 11, 56 pp

Goggans JF 1962 The correlation, variation and inheritance of wood properties in loblolly pine (Pinus taeda L). NC State Coll School For, Raleigh, Tech Rep 14, 155 pp

Gonzales JS, Kellogg RM 1978 Evaluating wood specific gravity in a tree-improvement program. Can For Serv Rep VP-X-183, 14 pp

Götze H 1969 Untersuchungen zum physikalisch-technischen Eigenschaftsbild dünner Durchforstungshölzer der Baumarten Kiefer und Fichte. Arch Forstwes 18:1155–1161

Götze H, Günther B, Luthard H, Schultze-Dewitz G 1972 Eigenschaften und Verwendung des Astholzes von Kiefer (Pinus silvestris L). 2. Physikalische und physikalisch-technische Eigenschaften. Holztechnologie 13:20–27

Greenhill WI, Dadswell HE 1940 The density of Australian timbers. 2. Air-dry and basic density for 172 timbers. CSIRO Aust Pamphl 92, 75 pp

Grossman PUA, Kingston RST 1954 Creep and stress relaxation in wood during bending. Aust J Appl Sci 5:403–417

Grozdits GA, Chauret G 1978 The distribution and form of deformation in transversely compressed wood. Can For Serv Rep OP-X-204-E, 47 pp

Grozdits RA, Ifgu G 1969 Development of tensile strength and related properties in differentiating coniferous xylem. Wood Sci 1:137–147

Haasemann W 1967 Untersuchungen zur genetisch bedingten Variation der Dichte des Lärchenholzes (Larix europaea D C). Arch Forstwes 16:479–486

Hakkila P 1966 Investigations on the basic density of Finish pine, spruce and birch wood. Commun Inst For Fenn 61.5, 98 pp

Hakkila P 1969 Weight and composition of the branches of large Scots pine and Norway spruce trees. Commun Inst For Fenn 67.6, 37 pp

Hakkila P 1971 Coniferous branches as a raw material source. Commun Inst For Fenn 75.1, 60 pp

Hakkila P, Uusvaara O 1968 On the basic density of plantation-grown Norway spruce. Commun Inst For Fenn 66.6, 22 pp

Hale JD 1957 The anatomical basis of dimensional changes of wood in response to changes in moisture content. For Prod J 7:140–144

Hale JD, Clermont LP 1963 Influence of cell-wall structure on basic physical and chemical characteristics of wood. Proc Meet IUFRO Sect 41, Madison WI, 27 pp

Hale JD, Perem E, Clermont LP 1961 Importance of compression wood and tension wood in appraising wood quality. 13th IUFRO Congr, Wien, 0-186, 23 pp

Hale JD, Prince JB 1936 A study of variation in density of pulpwood. Pulp Pap Mag Can 37(8):458–459

Hale JD, Prince JB 1940 Density and rate of growth in the spuces and balsam firs of eastern Canada. Dom For Serv Can Bull 94, 43 pp

Hallock H 1965 Sawing to reduce warp of loblolly pine studs. US For Serv Res Pap FPL-51, 52 pp

References

Hann RA 1969 Longitudinal shrinkage of seven species of wood. US For Serv Res Note FPL-0203, 13 pp

Harlow WM, Côté WA, Jr, Day AC 1964 The opening mechanism of pine cone scales. J For 62:538–540

Harris JM 1977 Shrinkage and density of radiata pine compression wood in relation to its anatomy and mode of formation. NZ J For Sci 7:91–106

Harris JM, Birt DV 1972 Use of beta rays for early assessment of wood density development in provenance trial. Silvae Genet 21:21–25

Harris JM, Meylan BA 1965 The influence of microfibril angle on longitudinal and tangential shrinkage in Pinus radiata. Holzforschung 19:144–153

Hartig R 1896 Das Rothholz der Fichte. Forstl Naturwiss Z 5:96–105, 157–169

Hartig R 1901 Holzuntersuchungen. Altes und Neues. Julius Springer, Berlin, 99 pp

Hartmann F 1943 Die Frage der Gleichgewichtsreaktion von Stamm und Wurzel heimischer Waldbäume. Biol Generalis 17:367–418

Hartwig GLF 1973 Radial specific gravity, moisture content, equilibrium moisture content, and shrinkage/swelling in Pinus pinaster and Pinus radiata. Meet IUFRO-5 S Afr 2:445–458

Heck GE 1919 "Compression" wood and failure of factory roof-beam. Eng News-Rec 83:508–509

Heger L 1974a Longitudinal variation of specific gravity in stems of black spruce, balsam fir, and lodgepole pine. Can J For Res 4:321–326

Heger L 1974b Relationship between specific gravity and height in the stem of open- and forest-grown balsam fir. Can J For Res 4:477–481

Hermans PH 1946 Contributions to the physics of cellulose fibres. Elsevier, Amsterdam, 221 pp

Hoffmann GC, Timell TE 1970 Isolation of a β-1,3-glucan (laricinan) from compression wood of Larix laricina. Wood Sci Technol 4:159–162

Hoffmann GC, Timell TE 1972 Polysaccharides in compression wood of tamarack (Larix laricina). 1. Isolation and characterization of laricinan, an acidic glucan. Svensk Papperstidn 75:135–141

Howard ET 1973 Physical and chemical properties of slash pine tree parts. Wood Sci 5:312–317

Howland P, Paterson DN 1979 Variation in morphology, grade turn-out and physical properties in seven pine and one cypress grown on an average Malawi site. Res Rec Malawi For Res Inst 39, 55 pp

Ifgu G, Labosky P, Jr 1972 A study of loblolly pine growth increments. I. Wood and tracheid properties. Tappi 55:524–529

Jaccard P, Frey A 1928a Einfluß von mechanischen Beanspruchungen auf die Micellarstruktur, Verholzung und Lebensdauer der Zug- und Druckholzelemente beim Dickenwachstum der Bäume. Jahrb Wiss Bot 68:844–866

Jaccard P, Frey A 1928b Quellung, Permeabilität und Filtrationswiderstand des Zug- und Druckholzes von Laub- und Nadelbäumen. Jahrb Wiss Bot 69:549–571

Jain KK, Seth MK 1979 Intra-increment variation in specific gravity of wood of blue pine. Wood Sci Technol 13:239–248

Jain KK, Seth MK 1980 Effect of bole inclination on ring width, tracheid length and specific gravity of wood at breast height in blue pine. Holzforschung 34:52–60

Janka G 1904 Untersuchungen über die Elastizität und Festigkeit der österreichischen Bauhölzer. II. Fichten von Nord-Tirol, vom Wienerwald und Erzgebirge. Mitt Forstl Versuchswes Österr 28, 313 pp

Janka G 1909 Untersuchungen über die Elastizität und Festigkeit der österreichischen Bauhölzer. III. Fichte aus den Karpaten, aus dem Böhmerwalde, Ternovanerwalde und den Zentralalpen. Technische Qualität des Fichtenholzes im Allgemeinen. Mitt Forstl Versuchswes Österr 15, 127 pp

Janka G 1915 Die Härte der Hölzer. Mitt Forstl Versuchswes Österr 39, 114 pp

Kaburagi Z 1952 Forest-biological studies on the wood quality. 4. On the moisture content, the bulk-density in green lumber and the shrinkage of the compression wood of Todo-fir. Bull Tokyo For Exp Sta 52:53–78

Kaburagi Z 1962 Forest-biological studies of wood quality. 15. On the volumetric shrinkage of Todo-fir − the apparent values of its relationship to the bulk-density. Gov For Exp Sta Bull 144, 53−96

Kandeel SAE 1979 Beta-ray determination of interincremental density variation in wood-plastic composites. Wood Sci 12:59−64

Kärkkäinen M 1976 Puu ja kuoren tiheys ja kosteus sekä kuoren osus koivun, kuusen ja männyn oksissa. (Density and moisture content of wood and bark, and bark percentage in the branches of birch, Norway spruce, and Scots pine.) Silva Fenn 10:212−236

Kärkkäinen M 1984 Effect of tree social status on basic density of Norway spruce. Silva Fenn 18:115−120

Kärkkäinen M, Raivonen M 1977 Reaktiopuun mekaaninen lujuus. (Mechanical strength of reaction wood.) Silva Fenn 11:87−96

Kato H, Nakato K 1968 The transverse anisotropic shrinkage of wood and its relation to the cell wall structure. I. The lignin distribution in the radial and tangential walls of coniferous wood tracheids. Kyoto Univ For Bull 40:284−292

Keith CT 1969 Resin content of red pine wood and its effect on specific gravity determinations. For Chron 45:338−343

Keith CT 1971 The anatomy of compression failure in relation to creep-inducing stresses. Wood Sci 4:71−82

Keith CT 1972 The mechanical behavior of wood in longitudinal compression. Wood Sci 5:234−244

Keith CT 1974 Longitudinal compressive creep and failure development in white spruce compression wood. Wood Sci 7:1−12

Keith CT, Côté WA, Jr 1968 Microscopic characterization of slip lines and compression failures in wood cell walls. For Prod J 18(3):67−74

Kellogg RM, Gonzales JS 1976 Relationships between anatomical and sheet properties in western hemlock kraft pulp. 1. Anatomical relationships. 2. Sheet property relationships. CPPA Trans Tech Sect 2(3):69−77

Kellogg RM, Wangaard FF 1969 Variation in the cell-wall density of wood. Wood Fiber 1:180−204

Kellogg RM, Wangaard FF 1970 Variations in cell wall and wood substance density. In: Page DH (ed) The physics and chemistry of wood pulp fibers. STAP 8, TAPPI New York, 100−106

Kelsey KE 1963 A critical review of the relationship between the shrinkage and structure of wood. CSIRO Div For Prod Aust Technol Pap 28, 35 pp

Kelsey KE, Christensen GN 1959 The sorption of water vapour by the constituents of wood. II. Heats of sorption. Aust J Appl Sci 10:269−283

Kennedy IE, Jessome AP, Tetro FJ 1968 Specific gravity survey of eastern Canadian woods. For Branch Can Publ 1221, 40 pp

Kennedy RW, Swann GW 1969 Comparative specific gravity and strength of amabilis fir and western hemlock grown in British Columbia. Can For Serv Rep VP-X-50, 16 pp

Kennedy RW, Wilson JW 1954 Studies on smooth and cork-bark Abies lasiocarpa. I. Fibre length comparisons. Pulp Pap Mag Can 55(7):13−132

Kibblewhite RP 1973 Effects of chlorite delignification on the structure and chemistry of Pinus radiata wood. Cellul Chem Technol 7:659−668

Kibblewhite RP, Harwood VD 1973 Effects of alkaline extraction on the structure and chemistry of lignified and delignified Pinus radiata wood. Cellul Chem Technol 7:669−678

King WW 1954 Cause and remedy for warped southern pine 2×2's. South Lumberm 189(2361)31−34

Kininmoth JA 1961 Checking of intergrown knots during seasoning of radiata pine sawn timber. NZ For Serv For Res Inst Tech Pap 30, 16 pp

Kisser J, Schuster R 1977 Makro- und mikroskopische Untersuchungen an querdruckbelastetem Holz. Holzforsch Holzverwert 29:45−53

Klem GG, Løschbrandt F, Bade O 1945 Undersøkelser av granvirke i forbindelse med slipe- og sulfitkokeforsøk. (Investigations of spruce wood in connection with experiments in groundwood and sulfite pulp manufacture.) Medd Norsk Skogforsøksves 9(31):1−127

Knigge W 1958a Das Phänomen der Reaktionsholzbildung und seine Bedeutung für die Holzverwertung. Forstarchiv 29:4−10

References

Knigge 1958b Untersuchungen über die Beziehungen zwischen Holzeigenschaften und Wuchs der Gastbaumart Douglasie (Pseudotsuga taxifolia Britt). Schriftenr Forstl Fak Univ Göttingen 20, 101 pp

Knigge W 1961 Der Einfluß verschiedener Wuchsbedingungen auf Eigenschaften und Verwertbarkeit des Nadelholzes. Allg Forst-Jagdztg 132:149–158

Knigge W 1962 Untersuchungen über die Abhängigkeit der mittleren Rohdichte nordamerikanischer Douglasienstämme von unterschiedlichen Wuchsbedingungen. Holz Roh-Werkst 20:352–360

Knigge W, Schulz H 1966 Grundriß der Forstbenutzung. Paul Parey, Hamburg, Berlin, 584 pp

Knuchel H 1954 Das Holz. Sauerländer, Aarau Frankfurt, 472 pp

Koch P 1972 Utilization of the southern pines. US Dep Agr, Agr Handb 420, 1663 pp

Koehler A 1924 The properties and uses of wood. Mc-Graw-Hill, New York, 354 pp

Koehler A 1931 Longitudinal shrinkage of wood. Am Soc Mech Eng Trans WDI-53-2:17–20

Koehler A 1933 Causes of brashness in wood. US Dep Agr Tech Bull 342, 39 pp

Koehler A 1938 Rapid growth hazards usefulness of southern pine. J For 36:153–159

Koehler A 1946 Longitudinal shrinkage of wood. Timb Can 7(4):60–66

Koehler A, Luxford RF 1931, 1951 The longitudinal shrinkage of redwood. US For Serv Rep FPL-R1297, 3 pp

Kollmann FFP 1935 Die Bedeutung der Physik für die moderne Holzforschung. Z Forst-Jagdwes 67:75–85

Kollmann FFP 1951 Technologie des Holzes und der Holzwerkstoffe. Springer, Berlin Göttingen Heidelberg Vol I, 1050 pp

Kollmann FFP 1963 Phenomena of fracture in wood. Holzforschung 17:65–71

Kollmann FFP 1968 Physics of wood: mechanics and rheology of wood. In: Kollmann FFP, Côté WA, Jr Principles of wood science and technology. I. Solid wood. Springer, Berlin Heidelberg New York, 160–419

Kollmann FPP, Côté WA, Jr 1968 Principles of wood science and technology. I. Solid wood. Springer, Berlin Heidelberg New York, 592 pp

Korán Z 1967 Electron microscopy of radial tracheid surfaces of black spruce separated by tensile failure at various temperatures. Tappi 50:60–67

Korán Z 1968 Electron microscopy of tangential tracheid surfaces of black spruce produced by tensile failure at various temperatures. Svensk Papperstidn 71:567–576

Köster E 1934 Die Astreinigung der Fichte. Mitt Forstwirtsch Forstwiss 5:393–416

Kraemer JH 1950 Growth–strength relations of red pine. J For 48:842–849

Krahmer RL 1966 Variation of specific gravity in western hemlock trees. Tappi 49:227–232

Kučera B 1970 Einfluß einiger Fehler auf die Biegefestigkeit von Fichtenholz. Holztechnologie 11:219–224

Kučera B 1971 Virkesfeilenes innflytelse på bøyefasthet hos granved. (The influence of wood defects on the bending strength of spruce wood.) Norsk Skogind 25:295–302

Kučera B 1973 Holzfehler und ihr Einfluß auf die mechanischen Eigenschaften der Fichte und Kiefer. Holztechnologie 14:8–17

Kuo ML, Arganbright DG 1980a Cellular distribution of extractives in redwood and incense cedar. I. Radial variation in cell-wall extractive content. Holzforschung 34:17–22

Kuo ML, Arganbright DG 1980b Cellular distribution of extractives in redwood and incense cedar. II. Microscopic observation of the location of cell wall and cell cavity extractives. Holzforschung 34:41–47

Kurrle FL 1963 Nitric acid digestion of logging residues of red pine. Pulp yield and physical properties. Tappi 46:267–272

Ladell JL, Carmichael AJ, Thomas GHS 1968 Current work in Ontario on compression wood in black spruce in relation to pulp yield and quality. Proc 8th Lake States For Tree Improv Conf, US For Serv Res Pap NC-33, 52–60

Lagergren S, Rydholm S, Stockman L 1957 Studies on the interfibre bonds in wood. 1. Tensile strength of wood after heating, swelling and delignification. Svensk Papperstidn 60:632–644

Larson PR 1957 Effect of environment on the percentage of summerwood and specific gravity of slash pine. Yale Univ School For Bull 63, 82 pp

Larson PR 1962 Auxin gradients and the regulation of cambial activity. In: Kozlowski TT (ed) Tree growth. Ronald, New York, 97–117

Larson PR 1963 Microscopic wood characteristics and their variation with tree growth. Proc IUFRO Sect 41 Meet Madison WI, 28 pp

Larson PR 1973 The physiological basis for wood specific gravity in conifers. IUFRO-5 Meet, S Afr, 2:672–680

Ledig FT, Zobel BJ, Matthias MF 1975 Geoclimatic patterns in specific gravity and tracheid length in wood of pitch pine. Can J For Res 5:318–329

Lee CH 1971 Trunkwood–branchwood specific gravity and tracheid length relationships in Pinus nigra. For Sci 17:62–63

Lehtonen I 1980 Knots in Scots pine (Pinus sylvestris L) and Norway spruce (Picea abies (L) Karst) and their effect on the basic density of stemwood. Commun Inst For Fenn 95.1, 34 pp

Lenhart JD, Shinn KH, Cutter BE 1977 Specific gravity at various positions along the stem of planted loblolly pine trees. For Prod J 27(9):43–44

Levčenko VP 1969 (Physical and mechanical properties of the wood of pine knots.) Lesn Zh Arhangel'sk 12(1):93–96

Levčenko VP 1974 (Elastic deformations of pine branch wood.) Izv Vuz Lesn Zh 17(4):153–154

Little CHA 1967 Some aspects of apical dominance in Pinus strobus L. Ph D Thesis Yale Univ, New Haven, 234 pp

Lumbardić S, Nicolić M 1970 (The effect of ring width and latewood percent on some physical and mechanical properties of pine from Mount Goć.) Sumarstvo 23:31–40

Luxford RG, Markwardt LJ 1932 The strength and related properties of redwood. US Dep Agr Tech Bull 305, 48 pp

Mackney AW, Mathieson CJ 1948 Sulphate pulping characteristics of pronounced compression wood occurring in New Zealand grown Monterey pine, P radiata D Don. Proc 9th Ann Pulp Pap Coop Res Conf, Burnie Tasmania, 111–125

Manwiller FG 1978 Southern pine properties related to complete-tree utilization – a review of the literature from 1971–1977. In: McMillin CW (ed) Complete tree utilization of southern pine. For Prod Res Soc, Madison WI, 29–40

Mark RE 1967 Cell wall mechanics of tracheids. Yale Univ Press, New Haven, London, 310 pp

Mark RE 1972 Mechanical behavior of the molecular components of fibers. In: Jayne BA (ed) Theory and design of wood and composite materials. Syracuse Univ Press, Syracuse, 49–82

Mark RE 1981 Molecular and cell wall structure of wood. In: Wangaard FF (ed) Wood: its structure and properties. EMMSE Project, PA State Univ, University Park, 43–100

Mark RE, Gillis PP 1970 New models in cell wall mechanics. Wood Fiber 2:79–95

Mark RE, Gillis PP 1973 The relationship between fiber modulus and S2 angle. Tappi 56(4):164–167

Markstrom DC, Troxell HE, Boldt CE 1983 Wood properties of immature ponderosa pine after thinning. For Prod J 33(4):33–36

Markstrom DC, Yerkes VP 1972 Specific gravity variation with height in Black Hills ponderosa pine. US For Serv Res Note RM-213, 4 pp

Markwardt LJ, Wilson TRC 1935 Strength and related properties of woods grown in the United States. US Dep Agr Tech Bull 479, 99 pp

Matsumoto T 1950 (The anisotropic shrinkage of wood.) Morioka Coll Agr For Iwate Univ Bull 26:81–88

Mayer-Wegelin H 1931 Neue Arbeiten über die Eigenschaften des Holzes. Forstarchiv 7:229–234

McGinnes EA, Jr, Dingeldein TW 1969 Selected wood properties of eastern redcedar (Juniperus virginiana L) grown in Missouri. Univ MO Agr Exp Sta Res Bull 960, 19 pp

McGinnes EA, Jr, Phelps JE 1972 Intercellular spaces in eastern redcedar (Juniperus virginiana L). Wood Sci 4:225–229

McKimmy MD 1959 Factors related to variation of specific gravity in young-growth Douglas fir. OR For Prod Res Cent Bull 8, 15 pp

Mér E 1887 De la formation du bois rouge dans le sapin et l'épicea. CR Acad Sci 104:376–378

Meylan BA 1968 Cause of high longitudinal shrinkage in wood. For Prod J 18(4):75–78

Meylan BA 1972 The influence of microfibril angle on the longitudinal shrinkage–moisture content relationship. Wood Sci Technol 6:292–301

Meylan BA 1981 Reaction wood in Pseudowintera colorata – a vesselless dicotyledon. Wood Sci Technol 15:81–92
Meylan BA, Probine MC 1969 Microfibril angle as parameter in timber quality assessment. For Prod J 19(4):30–34
Michels P 1941 Feuchtigkeitsverteilung im Holz des Weißtannenstammes, Gewicht und Schwindmaß des Weißtannenholzes. Mitt Forstwirtsch Forstwiss 12:295–329
Mitchell HL 1961 A concept of intrinsic wood quality, and nondestructive methods for determining quality in standing timbers. US For Serv FPL Rep 2233, 24 pp
Mitchell HL 1963 Specific gravity variation in North American conifers. Proc Meet IUFRO Sect 41, Madison WI, 32 pp
Mitchell HL 1964 Patterns of variation in specific gravity of southern pines and other coniferous species. Tappi 47:276–283
Mitchell HL 1965 Patterns of specific gravity variation in North American conifers. Proc Soc Am For Meet Denver CO, 169–179
Mitchell RG 1967 Translocation of dye in grand and subalpine fir infested by the balsam woolly aphid. US For Serv Res Note PNW-46, 17 pp
Mori S 1933 On the physical properties of "compression wood." Imp For Exp Sta Bull 33:35–48
Mork E 1928 Om tennar. (On compression wood.) Tidsskr Skogbr 36 (spec issue):1–41
Mozina I 1960 Über den Zusammenhang zwischen Jahrringbreite und Raumdichte bei Douglasienholz. Holz Roh-Werkst 18:409–413
Münch E 1937 Entstehungsursachen und Wirkung des Druck- und Zugholzes der Bäume. Forstl Wochenschr Silva 25:337–341, 345–350
Münch E 1938 Statik und Dynamik des schraubigen Baues der Zellwand, besonders des Druck- und Zugholzes. Flora 32:357–424
Nakato K, Kadita S 1955 (On the cause of the anisotropic swelling of wood. VI. On the relationships between the annual ring and the anisotropic shrinkage.) J Jap For Soc 37:22–25
Nečesaný V 1966 Der Anteil von Zellwand und Mittellamelle am Schwind- und Quellmaß des Holzes. Holz Roh-Werkst 24:470–473
Neel PL 1967 Factors influencing tree trunk development. Proc Int Tree Shade Conf 43:293–303
Nepveu G, Birot Y 1979 Les corrélations phénotypiques juvénile-adulte pour la densité du bois et la vigueur chez l'épicea. Ann Soc For 36:125–149
Newlin JA, Wilson TRC 1917 Mechanical properties of woods grown in the United States. US Dept Agr Bull 556, 47 pp
Newlin JA, Wilson TRC 1919 The relation of the shrinkage and strength properties of wood to its specific gravity. US Dept Agr Bull 676, 35 pp
Nicholls JWP 1967 Assessment of wood quality for tree breeding. IV. Pinus pinaster Ait grown in western Australia. Silvae Genet 16:21–28
Nicholls JWP 1982 Wind action, leaning trees and compression wood in Pinus radiata D Don. Aust For Res 12:75–91
Nicholls JWP, Fielding JM 1965 The effect of growth rate on wood characteristics. Appita 19:24–30
Norberg PH, Meier H 1966 Physical and chemical properties of the gelatinous layer in tension wood fibres of aspen (Populus tremula L). Holzforschung 20:174–178
Nördlinger H 1878 Liegt an schiefen Bäumen das bessere Holz auf der dem Himmel zugekehrten oder auf der unteren Seite? Cbl Ges Forstwes 4:246–247, 494–495
Nördlinger H 1890 Die gewerblichen Eigenschaften der Hölzer. Cotta'sche Buchhandlung, Stuttgart, 92 pp
Nordman LS, Quickström B 1970 Variability of the mechanical properties of fibers within a growth period. In: Page DH (ed) The physics and chemistry of wood pulp fibers. STAP 8, TAPPI, New York, 177–200
Nylinder P 1953 Volymviktsvariationer hos planterad gran. (Variations in density of planted spruce.) Medd Stat Skogsforskningsinst 43.3, 43 pp
Ohsako Y 1975 (A study of the physical and mechanical properties of newly formed compression wood.) J Soc Mat Sci Jap 24:849–854
Ohsako Y 1976 (A study of some properties at the early stage of the xylem formation of Japanese red pine (Pinus densiflora Sieb et Zucc).) Mem Fac Educ Kumamoto Univ 25:5–15

Ohsako Y 1978 (Mechanical properties of xylem in newly formed branches of Japanese black pine (Pinus thunbergii Parl).) Mokuzai Gakkaishi 24:778–783

Ohsako Y, Kato H 1977 (A study of the material properties and the formation of wood. IV. On newly formed branches of Japanese red pine (Pinus densiflora Sieb et Zucc).) Mokuzai Gakkaishi 23:521–527

Ohsako Y, Kato H, Nobuchi T 1973 Studies on the properties of natural bending wood formed in tree growth. 1. On Taiwan Akamatsu (Pinus massoniana Lamb) planted in Kyoto University Experimental Forest in Kamigamo. Bull Kyoto Univ For 45:238–251

Ohsako Y, Tsutsumi Y 1977 (Material properties and formation of wood. V. Formation of wood under restraint.) Mem Fac Educ Kumamoto Univ 26:53–61

Okkonen EA, Wahlgren HE, Maeglin RR 1972 Relationships of specific gravity to tree height in commercially important species. For Prod J 22(7):37–42

Olesen PO 1973 Forest tree improvement. 6. The influence of the compass direction on the basic density of Norway spruce (Picea abies L) and its importance for sampling for estimating the genetic value of plus trees. Arboretet, Hørsholm, Akademisk Vorlag, København, 58 pp

Ollinmaa PJ 1959 Rektiopuututkimuksia. (Study on reaction wood.) Acta For Fenn 72, 54 pp

Olson JR, Arganbright DG 1977 The uniformity factor – a proposed method for expressing variations in specific gravity. Wood Fiber 9:202–210

Onaka F 1949 (Studies on compression and tension wood.) Mokuzai Kenkyo, Wood Res Inst Kyoto Univ 1, 88 pp. Transl For Prod Lab Can 93 (1956), 99 pp

Panshin AJ, de Zeeuw CH 1980 Textbook of wood technology, 4th ed. McGraw-Hill, New York, 722 pp

Paul BH 1930 The application of silviculture in controlling the specific gravity of wood. US Dep Agr Tech Bull 168, 20 pp

Paul BH 1957a Juvenile wood in conifers. US For Serv FPL Rep 2094, 5 pp

Paul BH 1957b Lengthwise shrinkage in ponderosa pine. J For 7:408–410

Paul BH 1963 The application of silviculture in controlling the specific gravity of wood. US Dep Agr Tech Bull 1288, 97 pp

Paul BH, Dohr AW, Drew JT 1959 Some physical and mechanical properties of Noble fir. US For Serv FPL Rep 2168, 14 pp

Paul BH, Smith DM 1950 Summary of growth in relation to quality of southern yellow pine. US For Serv FPL Rep 1751, 19 pp

Paulson JC 1971 Unbleached pulp uniformity in prehydrolysis-kraft pulp from slash pine. Svensk Papperstidn 74:397–401

Pawsey CK, Brown AG 1970 Variation in properties of breast height wood samples of trees of Pinus radiata. Aust For Res 4(3):15–25

Pearson FGO, Fielding HA 1961 Some properties of individual growth rings in European larch and Japanese pine and their influence upon specific gravity. Holzforschung 15:82–89

Pearson RG, Gilmore RC 1971 Characterization of the strength of juvenile wood of loblolly pine (Pinus taeda L). For Prod J 21(1):23–31

Pearson RG, Gilmore RC 1980 Effect of fast growth rate on the mechanical properties of loblolly pine. For Prod J 30(5):47–54

Peck EC 1933 Specific gravity and related properties of softwood lumber. US Dep Agr Tech Bull 343, 24 pp

Pentoney RE 1953 Mechanisms affecting tangential vs radial shrinkage. J For Prod Res Soc 3(2):27–32, 86

Perem E 1958 The effect of compression wood on the mechanical properties of white spruce and red pine. For Prod J 8:235–240

Perem E 1960 The effect of compression wood on the mechanical properties of white spruce and red pine. For Prod Lab Can Tech Note 13, 22 pp

Petrić B 1974 Utjecaj starosti i širine goda na strukturu i volumnu težinu bijele borovine. (Influence of the age and width of the annual ring on the structure and density of Scots pine wood (Pinus sylvestris L).) Ann Exp For (Zagreb) 17:157–228

Petty JA 1971 The determination of fractional void volume in conifer wood by microphotometry. Holzforschung 25:24–29

Petty JA 1981 Determination of percent void volume in conifer wood. Holzforschung 35:95–96

Petty JA, Puritch GS 1970 The effect of drying on the structure and permeability of the wood of Abies grandis. Wood Sci Technol 4:140–154

Phillips DR, Clark A, III, Taras MA 1976 Wood and bark properties of southern pine branches. Wood Sci 8:164–169

Pillow MY 1949 Variations of longitudinal shrinkage of second-growth Douglas fir. J For 47:383–391

Pillow MY, Luxford RF 1937 Structure, occurrence, and properties of compression wood. US Dep Agr Tech Bull 546, 32 pp

Pillow MY, Schafer ER, Pew JG 1936, 1959 Occurrence of compression wood in black spruce and its effect on properties of ground wood pulp. Pap Trade J 102(16):36–38, US For Serv FPL Rep 1288, 3 pp

Polge H, Illy G 1967 Observations sur l'anisotropie du pin maritime des Landes. Ann Sci For 24:205–231

Poller S 1978 Studie über die chemische Zusammensetzung von Wurzel-, Stamm- und Astholz zweier Kiefern (Pinus silvestris L) unterschiedlichen Alters. Holztechnologie 19:22–25

Popper R, Bosshard HH 1976 Kalorimetrische Messungen zur Darstellung von Unterschieden im Cellulose-Lignin-System von Druck- und Normalholz. Holz Roh-Werkst 34:281–288

Porter AW 1964 The mechanics of fracture in wood. For Prod J 13:325–331

Preston RD 1968 Note on differential or anisotropic swelling and shrinkage of wood. In: Desch HE (ed) Timber, its structure and properties. 4th ed, Macmillan, London, 350–355

Raczkowski J, Krauss A 1979 (Swelling pressure of pine compression wood.) Drev Vysk 24(2):1–11

Raczkowski J, Stempień C 1967 Zur Beziehung zwischen der Rohdichte und der Reindichte von Holz. Holz Roh-Werkst 25:380–383

Rak J 1957 Fysikální vlastnosti reakčního dřeva smrku. (Physical properties of spruce reaction wood.) Drev Vysk 2(1):27–52

Ralston RA, McGinnes EA, Jr 1964 Shortleaf pine wood density unaffected by ring growth. South Lumberm 208(2592):17–19

Ramiah MV, Goring DAI 1965 The thermal expansion of cellulose, hemicellulose, and lignin. J Polym Sci C-11:27–48

Reinhardt A 1972 Untersuchungen über die Festigkeitseigenschaften und die Schnittholzqualität der japanischen Lärche. Diss, Ludwig-Maximilians-Univ, München, 141 pp

Rendle BJ 1958 Some recent work on factors affecting the quality of softwoods. Q J For 52:308–311

Rendle BJ, Phillips EWJ 1958 The effect of rate of growth (ring width) on the density of softwoods. Forestry 31:113–120

Rickey RG, Hamilton JK, Hergert HL 1974 Chemical and physical properties of tumor-affected Sitka spruce. Wood Fiber 6:200–210

Risi J, Zeller E 1960 Specific gravity of the wood of black spruce (Picea mariana Mill B S P) grown on a Hylocomium-Cornus site type. Fond Recher For Univ Laval PQ Contr 6, 70 pp

Rothe G 1930 Druckfestigkeit und Druckelastizität des Rot- und Weißholzes der Fichte. Thar Forstl Jahrb 81:204–231

Runkel ROH 1954 Die Sorption der Holzfaser in morphologisch-chemischer Betrachtung. Holz Roh-Werkst 12:226–232

Runkel ROH, Lüthgens M 1956 Untersuchungen über die Heterogenität der Wassersorption der chemischen und morphologischen Komponenten verholzter Zellwände. Holz Roh-Werkst 14:424–441

Saiki H, Furukawa I, Harada H 1972 An observation on tensile fracture of wood by scanning electron microscope. Bull Kyoto Univ For 43:309–319

Sakata K, Saeki H 1961 (Wood properties of Chizu-Sugi (Cryptomeria japonica D Don grown in Chizu district in Tottori prefecture). III. Compression test parallel to the grain on air-dried wood.) Bull Tottori Univ For 2:47–63

Sanio K 1873 Anatomie der gemeinen Kiefer (Pinus silvestris L). Jahrb Wiss Bot 9:50–126

Saucier JR 1972 Wood specific gravity of eleven species of pine. For Prod J 22(3):32–33

Sawasa M 1951 Tests on the shear and hardness of compression wood of Todo-fir (Abies mayriana Miyabe et Kudo). J Jap For Soc 33:379–383

Scaramuzzi G 1965 The relationships of fibre wall thickness, fibre diameter and percentage of summer wood (late wood) to specific gravity. Proc Meet IUFRO Sect 41, Melbourne, Vol 2, 11 pp

Schalck J 1967 Über die Rohdichte und Festigkeit des Schwarzkiefernholzes (Pinus nigra Arnold) und den Zusammenhang zwischen Rohdichte und Holzstruktur, untersucht an belgischen Aufforstungsbeständen. Beih Forstwiss Cbl 24, 86 pp

Schniewind AP 1959 Transverse anisotropy of wood. For Prod J 9:350–359

Schniewind AP 1962a Research on the mechanical properties of wood as it relates to structural utilization – prospects and new dimensions. J Inst Wood Sci 9:12–26

Schniewind AP 1962b Horizontal specific gravity variation in tree stems in relation to their support function. For Sci 8:111–118

Schniewind AP 1962c Mechanical behavior of wood in the light of its anatomic structure. In: Schniewind AP (ed) The mechanical behavior of wood. Conf Univ CA, Berkeley, 136–146

Schniewind AP 1981 Mechanical behavior and properties of wood. In: Wangaard FF (ed) Wood: its structure and properties. EMMSE Project, PA State Univ, University Park, 233–270

Schniewind AP, Barrett JD 1969 Cell-wall model with complete shear restraint. Wood Fiber 1:205–214

Schuerch C 1965 Physical restrictions to homogeneous reactions on wood. Ind Eng Chem Prod Res Develop 4:51–66

Schultze-Dewitz G, Götze H 1973 Untersuchungen zur Faserlänge, Raumdichte und Druckfestigkeit inter- und circumnodalien Holzes der Baumarten Kiefer (Pinus sylvestris L), Fichte (Picea abies Karst) und Douglasie (Pseudotsuga menziesii Franco). Drev Vysk 18:33–44

Schulz H, Bellmann B 1982 Untersuchungen an einem durch Druckholz stark verkrümmten Fichtenbrett. Inst Holzforsch Univ München, 85 pp

Schulz H, Bellmann B, Wagner L 1984 Druckholzanalyse in einem stark verkrümmten Fichtenbrett. Holz Roh-Werkst 42:109

Schwappach A 1897 Untersuchungen über Raumgewicht und Druckfestigkeit des Holzes wichtiger Waldbäume. I. Kiefer. Julius Springer, Berlin, 130 pp

Schwappach A 1898 Untersuchungen über Raumgewicht und Druckfestigkeit des Holzes wichtiger Waldbäume. II. Fichte, Weißtanne, Weymouthskiefer und Rotbuche. Julius Springer, Berlin, 136 pp

Seth MK 1979 Studies on the variation and correlation among some wood characteristics in blue pine (Pinus wallichiana A B Jackson). Ph D Thesis Himachal Pradesch Univ Simla, 168 pp

Seth MK, Jain KK 1978 Percentage of compression wood and specific gravity in blue pine (Pinus wallichiana A B Jackson). Wood Sci Technol 12:17–24

Shelbourne CJA, Ritchie KS 1968 Relationships between degree of compression wood development and specific gravity and tracheid characteristics in loblolly pine (Pinus taeda L). Holzforschung 22:185–190

Siau JF 1971 Flow in wood. Syracuse Univ Press, Syracuse, 131 pp

Siau JF 1984 Transport processes in wood. Springer, Berlin Heidelberg New York Tokyo, 245 pp

Silvester FD 1967 Timber. Its mechanical properties and factors affecting its structural use. Pergamon, Oxford, 152 pp

Skaar C 1972 Flow in wood. Syracuse Univ Press, Syracuse, 218 pp

Sonntag P 1904 Über die mechanischen Eigenschaften des Roth- und Weißholzes der Fichte und anderer Nadelhölzer. Jahrb Wiss Bot 39:71–105

Sonntag P 1909 Die duktilen Pflanzenfasern, der Bau ihrer mechanischen Zellen und die etwaigen Ursachen der Duktilität. Flora 99:203–259

Spurr SH, Hsiung WY 1954 Growth rate and specific gravity in conifers. For Prod J 52:191–200

Stamm AJ 1946 Passage of liquids, vapors and dissolved materials through softwoods. US Dep Agr Tech Bull 929, 80 pp

Stamm AJ 1964 Wood and cellulose science. Ronald, New York, 549 pp
Stamm AJ 1967a Movement of fluids in wood. I. Flow of fluids in wood. Wood Sci Technol 1:122–141
Stamm AJ 1967b Movement of fluids in wood. II. Diffusion. Wood Sci Technol 1:205–230
Stamm AJ 1969 Correlation of structural variations of lignins with their specific gravity. Tappi 52:1498–1502
Stamm AJ, Sanders HT 1966 Specific gravity of the wood substance of loblolly pine as affected by chemical composition. Tappi 49:397–400
Stamm AJ, Smith WE 1968 Laminar sorption and swelling theory for wood and cellulose. Wood Sci Technol 3:301–323
Stone JE 1955 The rheology of cooked wood. I. Introduction and discussion. II. Effect of temperature. Tappi 38:449–459
Stone JE, Scallan AM 1965 A study of cell wall structure by nitrogen absorption. Pulp Pap Mag Can 66(8):T407–T414
Stone JE, Scallan AM 1967 The effect of component removal upon the porous structure of the cell wall of wood. II. Swelling in water and the fiber saturation point. Tappi 50:496–501
Stone JE, Scallan AM, Aberson GMA 1966 The wall density of native cellulose fibres. Pulp Pap Mag Can 67(5):T263–T269
Suzuki M 1968 The relationship between elasticity and strength properties and cell structure of coniferous wood. Gov For Exp Sta Bull 212, 149 pp
Suzuki M 1969 Relation between Young's modulus and the cell-wall structures of Sugi (Cryptomeria japonica D Don). Mokuzai Gakkaishi 15:278–284
Talbert JT, Jett JB 1981 Regional specific gravity values for plantation grown loblolly pine in the southeastern United States. For Sci 27:801–807
Tang RC, Hsu NN 1973 Analysis of the relationship between the microstructure and elastic properties of the cell wall. Wood Fiber 5:139–151
Tang RC, Smith ND 1975 Investigation of anisotropic shrinkage of isolated softwood tracheids with scanning electron microscope. I. Longitudinal shrinkage. Wood Sci 8:415–424
Taras MA 1965 Some wood properties of slash pine (Pinus elliottii Engelm) and their relationship to age and height within the stem. Ph D Thesis NC State Univ Raleigh, 166 pp
Tarkow H 1981 Wood and moisture. In: Wangaard FF (ed) Wood: its structure and properties. EMMSE Proj PA State Univ, University Place, 155–194
Tarkow H, Krueger J 1961 Distribution of hot-water soluble material in cell walls and cavities of redwood. For Prod J 11:228–229
Taylor FW 1979 Variation of specific gravity and tracheid length in loblolly pine branches. J Inst Wood Sci (46) 8(4):171–175
Taylor FW, Bruton JD 1982 Growth ring characteristics, specific gravity, and fiber length of rapidly grown loblolly pine. Wood Fiber 14:204–210
Taylor FW, Wang EIC, Micko MM 1982 Differences in the wood of loblolly pine in Alberta. Wood Fiber 14:296–309
Thor E 1964 Variation in Virginia pine. I. Natural variation in wood properties. J For 62:258–262
Thor E 1965 Variation in some wood properties of eastern white pine. For Sci 11:451–455
Thor E, Brown SJ 1962 Variation among six loblolly pine provenances tested in Tennessee. J For 60:476–480
Timell TE 1981 Recent progress in the chemistry, ultrastructure, and formation of compression wood. The Ekman-Days 1981 (Stockholm), SPCI Rep 38 Vol 1:99–147
Timell TE 1982 Recent progress in the chemistry and topochemistry of compression wood. Wood Sci Technol 16:83–122
Trendelenburg R 1931 Festigkeitsuntersuchungen an Douglasienholz. Mitt Forstwirt Forstwiss 2(1):132–203
Trendelenburg R 1932 Über die Eigenschaften des Rot- oder Druckholzes der Nadelhölzer. Allg Forst-Jagdztg 132:1–14
Trendelenburg R 1934 Untersuchungen über das Raumgewicht der Nadelhölzer. I. Grundlagen und vergleichende Auswertung bisheriger Forschungen. Thar Forstl Jahrb 85:649–747

Trendelenburg R 1935 Schwankungen des Raumgewichts wichtiger Nadelhölzer nach Wuchsgebiet, Standort und Einzelstamm. Z Ver Deutsch Ing 79, Ser 4:85–89

Trendelenburg R 1937a Über Stammwuchsuntersuchungen und ihre Auswertung in der Holzforschung. Holz Roh-Werkst 1:3–13

Trendelenburg R 1937b Neuere Erkenntnisse über den Zusammenhang zwischen Wuchsbedingungen und Raumgewicht der Nadelhölzer. Mitt Fachausschusses Holzfragen 17:3, Ver Deutsch Ing-Verlag

Trendelenburg R 1939a Das Holz als Rohstoff. Lehmann, Berlin, 435 pp

Trendelenburg R 1939b Über Fasersättigungsfeuchtigkeit, Schwindmaß und Raumdichtezahl wichtiger Holzarten. Holz Roh-Werkst 2:12–17

Trendelenburg R, Mayer-Wegelin H 1955 Das Holz als Rohstoff, 2nd ed. Carl Hanser, München, 541 pp

Turnbull JM 1937 Variation in strength of pine timbers. J S Afr For Assoc 1:52–59

Turnbull JM 1947 Some factors affecting wood density in pine stems. 5th Brit Emp For Conf, Dep For Pretoria S Afr, 22 pp

Turnbull JM 1948 Some factors affecting wood density in pine stems. J S Afr For Assoc 16:22–43

Turnbull JM, du Plessis CP 1946 Some sidelights on the rate of growth bogey. J S Afr For Assoc 14:29–46

Ueda K 1973 Studies on the mechanical properties of reaction woods. 2. The elastic constant of Icho (Ginkgo biloba L), Yamanarashi (Populus sieboldii Miq) and Yachidamo (Fraxinus mandshurica var japonica Maxim). Res Bull Coll Exp For Hokkaido Univ 30:379–388

Ueda K, Iijima Y, Yokoyama T 1972 Studies on the mechanical properties of reaction woods. 1. The elastic constants of Todomatsu Abies sp. Res Bull Coll Exp For Hokkaido Univ 29:327–334

Ulfsparre S 1928 Något om tjurved och därav framställd sulfat- och sulfitcellulosa. (On compression wood and sulfate and sulfite pulps prepared therefrom.) Svensk Papperstidn 31:642–644

Ursprung A 1906 Untersuchungen über die Festigkeitsverhältnisse an exzentrischen Organen und ihre Bedeutung für die Erklärung des exzentrischen Dickenwachstums. Bot Cbl Beih 19(1):393–408

van Buijtenen JP 1964 Anatomical factors influencing wood-specific gravity of slash pines and the implications for the development of high-quality pulpwood. Tappi 47:401–404

Vanin SI 1949 Drevesinovedenie, Moskva. (Quoted by Kučera B 1971)

Venkatsewaran A 1970 Sorption of aqueous and nonaqueous media by wood and cellulose. Chem Rev 70:619–637

Verrall AF 1928 A comparative study of the structure and physical properties of compression wood and normal wood. M S Thesis, Univ MN, St Paul, 37 pp

Vihrov VE 1960 (Movement of the branches of trees.) Priroda, Moskva 49(1):127–128

Vintila E 1939 Untersuchungen über Raumgewicht und Schwindmaß von Früh- und Spätholz bei Nadelhölzern. Holz Roh-Werkst 2:345–357

von Pechmann H 1969 Der Einfluß von Erbgut und Umwelt auf die Bildung von Reaktionsholz. Beih Z Schweiz Forstver (Festschr Hans Leibundgut) 46:159–169

von Pechmann H 1973 Beobachtungen über Druckholz und seine Auswirkungen auf die mechanischen Eigenschaften von Nadelholz. Meet IUFRO-5 S Afr 2:1114–1122

von Pechmann H, Courtois H 1970 Untersuchungen über die Holzeigenschaften von Douglasien aus linksrheinischen Anbaugebieten. Forstwiss Cbl 89:88–112

von Wedel KW, Zobel BJ, Shelbourne CJA 1968 Prevalence and effects of knots in young loblolly pine. For Prod J 18(19):97–103

Voorhies G 1972 A profile of young-growth southwestern ponderosa pine wood. North AZ Univ AZ For Notes 7, 16 pp

Vorreiter L 1955 Rechnungsmäßige Bestimmung der Zellwanddichte aus den Holzkonstituenten. Holz Roh-Werkst 13:185–188

Vorreiter L 1963 Fasersättigungsfeuchte und höchste Wasseraufnahme der Hölzer. Holzforschung 17:139–146

Wahlgren HE, Fassnacht DL 1959 Estimating tree specific gravity from a single increment core. US For Serv FPL Rep 2146, 9 pp

Wahlgren HE, Schumann R 1972 Properties of major southern pines. I. Wood density survey. US For Serv FPL Rep 176, 58 pp

Wangaard FF 1950 The mechanical properties of wood. Wiley, New York, 377 pp

Wangaard FF 1969 Cell-wall density of wood with particular reference to the southern pines. Wood Sci 1:222–226

Wangaard FF, Granados LA 1967 The effect of extractives on water-vapor sorption by wood. Wood Sci Technol 1:253–277

Ward D, Gardiner JJ 1976 The influence of spacing on tracheid length and density in Sitka spruce. Irish For 33:39–56

Wardrop AB 1951 Cell-wall organization and the properties of the xylem. I. Cell-wall organization and the variation of breaking load in tension of the xylem in conifer stems. Aust J Sci Res B-4:391–414

Wardrop AB, Addo-Ashong FW 1964 The anatomy and fine structure of wood in relation to its mechanical failure. Proc Tewkbury Symp on Fracture 1962, CSIRO Aust Dep For Prod Rep 560:169–200

Wardrop AB, Dadswell HE 1947 Contributions to the study of the cell wall. The occurrence, structure, and properties of certain cell wall deformations. Counc Sci Ind Res Aust Bull 221:14–32

Warren WG 1979 The contribution of earlywood and latewood specific gravities to overall wood specific gravity. Wood Fiber 11:127–135

Watanabe H, Tsutsumi J, Kanagawa H 1962 Properties of branchwood: especially on specific gravity, tracheid length, and appearance of compression wood. Kyushu Univ For Bull 35:91–96

Weatherwax RC, Tarkow H 1968a Cell-wall density of dry wood. For Prod J 18(2):83–85

Weatherwax RC, Tarkow H 1968b Density of wood substance: Importance of penetration and adsorption compression of the displacement fluid. For Prod J 18(7):44–46

Wegelius T 1939 The presence and properties of knots in Finnish spruce. Acta For Fenn 48, 191 pp

Wellwood RW 1960 Specific gravity and tracheid length variations in second-growth western hemlock. J For 58:361–368

Wellwood RW, Jurasz PE 1968 Variation in sapwood thickness, specific gravity, and tracheid length in western red cedar. For Prod J 18(12):37–46

Wellwood RW, Smith JGH 1962 Variation in some important qualities of wood from young Douglas fir and hemlock trees. Univ BC Fac For, Vancouver, Res Pap 50, 15 pp

Westing AH 1965 Formation and function of compression wood in gymnosperms. Bot Rev 31:381–480

Westing AH 1968 Formation and function of compression wood in gymnosperms. Bot Rev 34:51–78

White DJB 1965 The anatomy of reaction tissues in plants. In: Carthy JD, Duddington CL (eds) Viewpoints in biology IV, Butterworth, London, 54–82

Wilfong JG 1966 Specific gravity of wood substance. For Prod J 16(1):55–61

Wilson BF 1981 The development of growth strains and stresses in reaction wood. In: Barnett JR (ed) Xylem cell development. Castle House, Tunbridge Wells, 275–289

Wloch W 1975 Longitudinal shrinkage of compression wood in dependence on water content and cell-wall structure. Acta Soc Bot Polon 44:217–229

Wood JR, Goring DAI 1971 The distribution of lignin in stem wood and branch wood of Douglas fir. Pulp Pap Mag Can 72(3):T95–T102

Wooten TE 1967 Compression wood – a common defect in southern pines. For Farmer 28(3):14–15

Wooten TE, Barefoot AC, Nicholas DD 1967 The longitudinal shrinkage of compression wood. Holzforschung 21:168–171

Worster HE, Vinje MG 1968 Kraft pulping of western hemlock tree tops and branches. Pulp Pap Mag Can 69(7):T308–T311

Yamai E 1957 On the orthotropic properties of wood in compression. J Jap For Soc 39:328–338

Yamai R 1951 (A mechanical model of the cell wall of compression wood.) Kyushu Univ Fac Agr Sci Bull 13:234–237

Yazawa K 1944 Report of Gifu Agr Coll. (Cited by Onaka 1949)

Yazawa K, Tate S, Iwata H 1951 Research on specific gravity, water content in the green condition, volumetric shrinkage for trunk and branches and the specific gravity, water content in the green conditions, volumetric shrinkage of the springwood and summerwood of Japanese red pine (Pinus densiflora S et Z). J Jap For Soc 33:34–39

Young HE, Chase AJ 1965 Fiber weight and pulping characteristics of the logging residue of seven tree species in Maine. ME Agr Exp Sta Tech Bull 17, 43 pp

Young WD, Laidlaw RA, Packman DF 1970 Pulping of British-grown softwoods. VI. The pulping properties of Sitka spruce compression wood. Holzforschung 24:86–98

Ziegler GA 1974 Water vapor sorption of softwood cell-wall constituents. Ph D Thesis PA State Univ, 161 pp

Zimmermann MH 1983 Xylem structure and the ascent of sap. Springer, Berlin Heidelberg New York Tokyo 143 pp

Zobel BJ, Kellison RC, Matthias MF, Hatcher AV 1972 Wood density of the southern pines. NC Agr Exp Sta Tech Bull 208, 56 pp

Zobel BJ, McElwee RL 1958 Natural variation in wood-specific gravity of loblolly pine, and an analysis of contributing factors. Tappi 41:158–161

Zobel BJ, Rhodes RR 1955 Relationship of wood-specific gravity in loblolly pine (Pinus taeda L) to growth and environmental factors. TX For Serv Tech Rep 11, 32 pp

Zobel BJ, Thorbjornsen E, Henson F 1960 Geographic, site, and individual tree variation in wood properties of loblolly pine. Silvae Genet 9:149–158

Chapter 8 Origin and Evolution of Compression Wood

CONTENTS

8.1	Introduction	597
8.2	The Fossil Record of Compression Wood	598
8.3	Compression Wood in Primitive Gymnosperms	604
8.3.1	*Ginkgo biloba*	604
8.3.2	Primitive Conifers and Taxads	613
8.4	Compression Wood in Primitive Angiosperms	614
8.5	General Conclusions	616
References		617

8.1 Introduction

Westing (1965) has pointed out that the evolutionary history of compression wood has not yet been investigated, although information could readily be obtained by examination of available specimens of fossil gymnosperm wood. How long has compression wood existed? We only know that at the present time it occurs in *Ginkgo biloba*, the Coniferales, and the Taxales but not in the Cycadales or the Gnetales. Tissues resembling compression wood have also been observed in a few primitive angiosperms with a xylem consisting largely of tracheids.

Clues to the origin of compression wood can be obtained by a search of the fossil record or by an examination of this tissue among ancient members of the gymnosperms. Hints as to the time when compression wood may have first appeared in the course of plant evolution have gradually accumulated in this manner. At the same time it has become evident that compression wood may have undergone a certain evolutionary development.

No attempt will be made here to discuss the evolution of land plants or their secondary xylem. Excellent texts dealing with this subject and with paleobotany in general have been published by Seward (1933), Chamberlain (1935), Andrews (1947, 1961, 1974), Arnold (1947), Müller-Stoll (1951), Walton (1953), Darrah (1960), Delevoryas (1962, 1966), Barghoorn (1964), Gothan and Weyland (1965), Sporne (1965), Mägdefrau (1968), and Banks (1970a,b). The ecological factors that might have influenced the evolution of the xylem have been discussed by Carlquist (1975). Taylor (1981) and Stewart (1983) have summarized the present status of the rapidly changing field of paleobotany in two recent, excellent monographs. The origin and evolution of compression wood have been reviewed by Timell (1981, 1983).

8.2 The Fossil Record of Compression Wood

No systematic search for compression wood in fossil gymnosperms has so far been made. Bailey (1933, 1953) pointed out the dangers inherent in neglecting the occurrence of root wood or compression wood in fossils, a neglect that in his opinion has resulted in an unfortunate multiplication of superfluous form genera. Patel (1968) has drawn attention to one such mistake involving compression wood.

Most fossilized or petrified woods have undergone some degradation. Almost invariably cellulose and hemicelluloses are decomposed more rapidly than the more resistant lignin, although instances when cellulose had survived surprisingly intact have been reported (Crook et al. 1965, Morey and Morey 1969a,b, 1971a,b, Fengel 1974). Compounds from degraded lignin have been found to occur in silicified wood from Arizona 200 million years old (Sigleo 1978) and also in early Silurian plant fossils (Niklas and Pratt 1980). Gorczynski et al. (1969) found that the S_2 and S_3 layers in fossilized tracheids and fibers were destroyed most rapidly, while the S_1 layer persisted longer, and the primary wall was still structurally intact after the entire secondary wall had decomposed. Schmid (1967) could not observe any S_3 layer in fossilized xylem of the paleozoic genera *Callixylon* and *Cordaites*, probably because this layer had been degraded. According to Barghoorn (1952a,b) the lamellae of cellulose within a single cell wall can be decomposed at different rates. If the S_3 has a higher lignin content than the S_1 and S_2, this layer tends to be better preserved in ancient wood than the remainder of the secondary wall, as reported, for example, by Sachs (1965).

The tracheids in compression wood are highly lignified and are more resistant to microbial degradation than normal wood tracheids (Chap. 5.11.2). For these reasons, compression wood could be expected to be equally well or better preserved in comparison with corresponding normal wood. Rounded tracheids, intercellular spaces, and helical cavities or spiral striations are generally considered reliable diagnostic features of compression wood. Unfortunately, rounded tracheids and intercellular spaces are also present in normal wood of several genera, such as *Ginkgo*, *Agathis*, *Araucaria*, *Cupressus*, and *Juniperus* (Chap. 4.4.2). Helical cavities are often missing in first-formed compression wood. They also seem to be lacking in root wood as well as in stem and branch compression wood of some gymnosperm genera. The most reliable diagnostic feature of compression wood should therefore be the absence of an S_3 layer in the tracheids. Unfortunately, this layer, as already mentioned, might be decomposed in fossilized wood, and it may be present in mild or moderate types of compression wood. Because it is such a thin layer it is, moreover, not easily distinguished in sometimes poorly preserved fossil specimens. For all these reasons, utmost care must be exercized when attempts are made to identify compression wood in fossilized or petrified specimens.

Between the appearance of the first, primitive land plants and the extensive and highly developed flora of the Carboniferous there occurred a period of rapid plant evolution from the Upper Silurean to the end of the Lower Devonian, a time interval of about 30 million years (Chaloner 1970). The Lycophyta, which had appeared during the Upper Devonian, were probably the dominant plants during the entire Carboniferous period and were among the first to assume an arborescent habit. The stem structure of these trees is fairly well known (Barghoorn and Scott 1958). Trees such as *Lepidodendron* or *Sigillaria* possessed a unifacial vascular cambium which produced a secondary xylem to the inside but no secondary phloem to the outside (Eggert and Kanemoto 1977). The secondary xylem consisted of ray cells and large, scalariform tracheids, which were long and had very thin walls. The angular tracheids were arranged in regular files. This wood, termed manoxylic by Seward, must have been soft and weak and probably served largely for conduction of water. It formed a band no wider than a few centimeters that decreased in width with increasing height. These large trees obviously depended on other means for support, namely an extensive cortex and periderm. Inspection of available photomicrographs gives no indication of the presence of any rounded, thick-walled tracheids in the secondary xylem. It is, actually, unlikely that compression wood, even if present, could have effected any righting, considering the limited amount of xylem present.

Among the Sphenophyta, which developed simultaneously with the Lycophyta, the *Calamites* (Equisetales) were large trees. They contained a larger proportion of secondary xylem, sometimes 13 cm wide and, unlike the Lycophyta, seem to have produced a secondary phloem. The tracheids had thick walls. Although these trees contained more secondary xylem than any other fossil cryptogams (Taylor 1981), neither their stem nor their branches seem to have formed any compression wood. The Pteridophyta or ferns, which originated in the Lower Carboniferous, may have had a vascular cambium at this time and some were arboreal. Most of their vascular tissue, however, was probably primary.

In 1960 Beck (1960a, b) reported his important discovery that the foliage genus *Archaeopteris* was organically connected with the stem genus *Callixylon*, both from the Upper Devonian time, 360–370 million years ago. *Archaeopteris* (the older name) was a large tree with the appearance and reproduction of a fern, yet with a stem anatomy remarkably similar to that of the extant gymnosperms. These and related trees are now grouped within the Progymnospermophyta (Beck 1962a, b, 1964a, b, 1966, 1970, 1971, 1976, 1981, Namboodiri and Beck 1968, Bonamo 1975), a division which at present is considered to comprise three orders (Scheckler and Banks 1971a, b, Scheckler 1975, 1976, 1978). It is likely that the gymnosperms originated from these progymnosperms. The oldest progymnosperms were the Aneurophytales, which appeared 370 million years ago, comprising genera such as *Aneurophyton*, *Tetraxylopteris* (Beck 1957), and *Triloboxylon* (Scheckler and Banks 1971a, b, Scheckler 1976). These trees produced not only a massive secondary xylem but also a secondary phloem, a feature observed also in the later Archaeopteridales (Arnold 1930, Scheckler 1978) and the Protopityales (Walton 1969). The wood was clearly pycnoxylic and was composed of both tracheids and ray cells.

The structure of the secondary xylem of *Archaeopteris* (*Callixylon*) is now known in considerable detail thanks to the studies of Arnold (1930), Beck (1960b, 1962b, 1970, 1979), and Beck et al. (1982). Schmid (1967) examined the xylem in the electron microscope. Distinct growth rings with earlywood and

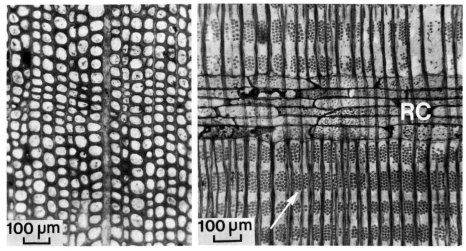

Fig. 8.1. Tracheids and ray cells in *Callixylon arnoldii*. The tracheids have an angular outline, but in many the lumen is rounded. Transverse section. LM. (Beck 1962b)

Fig. 8.2. Tracheids and ray cells (*RC*) in *Callixylon newberryi*. Note the radial alignment of the multiseriate pits in the tracheid walls (*arrow*). Radial section. LM. (Beck 1970)

latewood have been observed only rarely. The tracheids were regularly arranged, thick-walled, and usually angular in outline. Several of the specimens examined by Arnold (1930), as well as the wood of *Callixylon arnoldii* studied by Beck (1962b), contained tracheids with a rounded lumen outline, as shown in Fig. 8.1. Circular, bordered pits were present in the radial walls in groups of 6–20 and in two or three radially aligned, vertical files, as seen in Fig. 8.2, a striking feature of this wood. Tangential pitting occurred only in the latewood. The tracheid wall consisted of S_1 and S_2 layers, but whether or not an S_3 layer was originally present cannot be decided, for, albeit never observed, it could have been removed by decay (Schmid 1967). Rays were uniseriate, biseriate, or multiseriate and consisted of both parenchyma and tracheids, a remarkedly advanced feature. The similarity between this type of xylem and that found in extant gymnosperms is striking. Apparently, as pointed out by Schmid (1967), the basic features of the xylem cell wall evolved very early in the land plants and have not changed significantly since the Devonian period.

Whether the progymnosperms were capable of forming compression wood cannot be decided on the basis of the fossil evidence currently available. The presence of rounded, thick-walled tracheids, such as those shown in Fig. 8.1, would suggest a possible presence of compression wood, but the fact that these tracheids occurred among others with an angular outline argues against such a conclusion. For reasons stated above, the absence of an S_3 layer cannot be interpreted as a feature of compression wood in this case. On the other hand, in the progymnosperms, compression wood could have served a righting function.

Among the first true gymnosperms, the Pteridospermophyta or seed ferns produced a manoxylic secondary xylem with thin-walled tracheids. Although these trees often developed extensive wood, it has been characterized as very unspecialized (Andrews 1961). In the Cycodophyta, the stem consists of relatively little xylem and a large cortex. Both the pit membrane (Eicke and Metzner-Küster 1961, Eicke 1963) and the ray cells (Greguss 1958, 1968) are considered to be primitive. No compression wood has been observed in the extant cycads, although it should be noted that these trees may represent reduced, primitive forms whose Mesozoic ancestors were more arboreal in nature (Zimmermann 1930).

The Cordaitales, Coniferales, and Taxales, which belong to the Coniferophyta, are believed to have evolved from some progymnosperm and are all characterized by the presence of an extensive secondary xylem. The *Cordaites* were large, dominant trees during the Carboniferous period, but they became extinct in the Lower Triassic. Their wood was simple, consisting only of long, thick-walled, and narrow tracheids and uniseriate rays. In many respects the xylem resembled that of the extant *Araucaria*. A *Cordaites* species examined by Schmid (1967) had narrow, diagonal pit apertures and no torus. Although the xylem of the Cordaitales in some respects is primitive, compression wood, if present, could probably have functioned as a righting tissue.

The most successful of all gymnosperms, the Coniferales, date back to the Upper Carboniferous and reached their highest development in the Jurassic and Cretaceous periods. They probably originated from the Voltziales, trees that had a compact wood containing few parenchyma cells and no resin canals. The Taxales evidently existed as a separate order as early as during the Jurassic. Although some consider the conifers to be a dying race, it should be remembered that in terms of the land area they have dominated and the time they have existed (300 million years), the conifers have been more successful than any other group of land plants (Banks 1970a). The major reason for this success has probably been the unique structure of their longitudinal tracheids, which makes it possible for these cells to contribute to mechanical support, while at the same time conducting water over large distances (Carlquist 1975).

It has long been a moot question which of the extant conifer families should be considered as most primitive. Some paleobotanists favor the Araucariaceae, a family whose fossil record goes back to the Paleozoic (Florin 1963). According to Eicke (1958), their pit structure must be regarded as primitive. Greguss (1955) believes that *Pinus* has the most advanced xylem among the conifers.

There are few reports of the occurrence of compression wood in gymnosperm fossils. The first such observation was made by Lämmermayr (1901), who claimed to have observed it in *Cupressinoxylon fissum*. *Cupressinoxylon* is a genus to which many fossil coniferous stems have been assigned, especially from the Jurassic and Cretaceous. According to Lämmermayr, the fossil wood, which had been described in 1850 by Goeppert, was reddish-brown in color, had slit-like openings over the pits, and contained spiral striations in the cell wall. Lämmermayr also observed oblique cavities in the thick cell walls of fossilized wood of *Pinus thunbergii*. Available paleobotanical literature indicates that either normal, secondary xylem of *Cupressinoxylon* has features characteristic of compression wood, or fossil wood from this large genus frequently consists of compression wood.

Barber (1898) described in great detail stem, root, and branch woods of *Cupressinoxylon vectense*. Rounded tracheids were present in the branch wood, which might have contained compression wood. They occurred, however, in narrow rather than in wide growth rings, and a cross section of the branch, which was unusually well preserved, showed no eccentricity. Doubt therefore exists whether this particular wood really came from a branch, a fact also mentioned by Barber. No spiral striations can be observed in the tracheid walls in longitudinal sections.

Lutz (1930) carried out a thorough investigation of an exceptionally well preserved, silicified specimen of *Cupressinoxylon jurassica* from the Upper Jurassic, collected by Wieland in South Dakota. Although about 150 million years old, the wood resembled extant *Cupressus* and *Chamaecyparis* rather closely, even in specific characteristics, such as ray cell pitting. Ray tracheids were not present. The longitudinal tracheids were angular in outline, but where the cell wall was thick, the lumen was rounded. Faint, spiral markings could be distinguished in the tangential walls of some tracheids, a feature indicating the possible presence of mild compression wood. At least one of the many fossil gymnosperm woods from Hungary which have been described by Greguss (1967) probably consisted of compression wood, namely a specimen of *Cupressinoxylon secretiferum*. Characteristically, the earlywood comprised only 5–10 tracheid rows while the latewood had 35–40 rows of tracheids which all had equally thick walls. In longitudinal section, all tracheids displayed spiral striations, which were oriented at an angle of about 45° with respect to the fiber axis and were clearly discernible, as shown in Fig. 8.3. Greguss (1968) found the striated tracheid walls to be "highly reminiscent of the so-called red species of *Picea* and *Larix*." Considering the entire anatomical evidence, the presence of fully developed compression wood must be deemed likely.

Greguss (1967) has also described a fossilized specimen of *Araucarioxylon* from the Permian, with many of the features typical of compression wood, such as intercellular spaces and tracheids with a rounded outline, a thick cell wall, and a uniform size. The apparent lack of helical cavities is probably not decisive, for such cavities are also missing or rare in compression wood of the extant Araucariaceae. Very probably, this represents an authentic specimen of fossilized compression wood, and if its assignment to the Permian period is correct, compression wood must have existed for at least 250 million years.

A possible example of a fossil compression wood from the Mesozoic is found in a silicified log of *Cedrus alaskensis*, described by Barghoorn (1952a). This Jurassic specimen, collected in northern Alaska, had been subjected to microbial decomposition, so that only the residual lignin skeleton had become petrified. The transverse section in Fig. 8.4 shows some angular and some round cells, all with a rounded lumen. Fine, helical cavities and also larger drying checks are present throughout the cell wall of every tracheid. Barghoorn (1952b) has also described fossilized compression wood of *Pinus strobus* from postglacial sediments approximately 4000 years old.

Morey and Morey (1969a) examined several fossilized woods, including an unusually well-preserved Pleistocene specimen of *Picea* compression wood from Wisconsin, dated at 11 000 years. The transverse section in Fig. 8.5 shows

8.2 The Fossil Record of Compression Wood

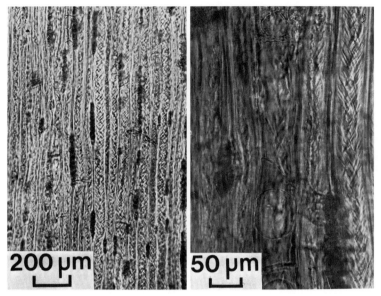

Fig. 8.3. Two specimens of *Cupressinoxylon secretiferum* with tracheids containing spiral striations (helical cavities). Longitudinal section. LM. (Greguss 1967)

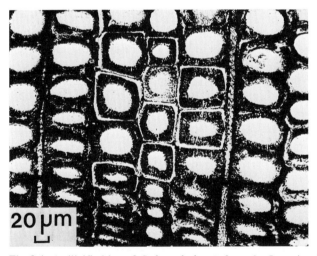

Fig. 8.4. A silicified log of *Cedrus alaskensis* from the Jurassic with lignin skeletons. Note the rounded outline of the lumen and the radial cavities in the inner part of the secondary wall. Transverse section. LM. (Barghoorn 1952a)

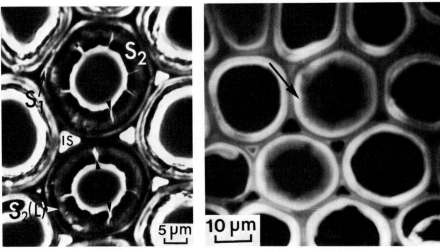

Fig. 8.5. Fossilized compression wood of *Picea* sp. from two Creeks, Wisconsin (Pleistocene), with round tracheids and intercellular spaces (*IS*). *Arrowheads* indicate checks terminating at the S_2 (L) layer. Transverse section. LM (phase contrast). (Morey and Morey 1969a)

Fig. 8.6. Fluorescence photomicrograph of a portion of the specimen in Fig. 8.5, showing the high concentration of lignin in the S_2 (L) layer (*arrow*). Transverse section. (Morey and Morey 1969a)

typical compression wood tracheids with a rounded outline and intercellular spaces. Several of the tracheids are unusually well-preserved and have thick walls. In others, the decomposition appears to have progressed from the lumen toward the outside of the tracheid. Occasionally, the S_2 layer has become detached from S_1, obviously as a result of degradation of the S_2(L) layer. Helical cavities cannot be distinguished but radial checks are present. Spiral striations, oriented at an angle of 35–45° to the fiber axis, could be readily observed in longitudinal sections. The fluorescence micrograph in Fig. 8.6 indicates a high concentration of lignin in the outer S_2. Morey and Morey (1969b, 1971a,b) have also reported anatomical studies of a lignified wood from Senftenberg in Germany, dating from the Miocene and similar to extant *Cryptomeria* and *Taxodium*. Whether or not this specimen contained an S_2(L) layer, as claimed by the investigators, cannot be decided from the micrographs. The wood lacked intercellular spaces, and the tracheids were angular in outline and contained no helical cavities. It would therefore seem unlikely that this specimen could have consisted of compression wood.

8.3 Compression Wood in Primitive Gymnosperms

8.3.1 *Ginkgo biloba*

Ginkgo biloba is one of the oldest still existing trees on earth. The Ginkgophytes date back to the Lower Permian or Upper Carboniferous periods, 280

8.3.1 *Ginkgo biloba*

million years from the present, and the genus *Ginkgo* to the Triassic, 210 million years ago. The Ginkgophytes developed rapidly during the Triassic, flourished and reached their zenith in the Jurassic, and were still dominant trees during the Cretaceous and the early Tertiary. Once represented by several genera and many species, *Ginkgo biloba* is now the sole survivor of its race, having found a last home in China, where it may still exist in the wild state. It was probably saved from extinction by ancient Buddhist priests, who began to plant it in their temple gardens. At the present time, ginkgo is widely planted as an ornamental and has adapted itself well even to the harsh environment of big cities. A "living fossil" in the words of Charles Darwin, ginkgo has probably come closer to extinction than any other tree except *Metasequoia glyptostroboides*. It is truly a unique tree and not only because it seems to have remained unchanged since the early Mesozoic, but also because its foliage is that of a fern, its fertilization that of a cycad, its seed that of a taxad, and its stem that of a conifer. *Ginkgo biloba* has long captured the imagination of botanists and paleobotanists and has been the subject of many excellent reviews. Among the more extensive ones are those by Seward and Gowan (1900), Seward (1938), Andrews (1947), Kobenza (1957), Sporne (1965), and Major (1967). A complete compilation of the literature on ginkgo until 1959 has been prepared by Franklin (1959).

The vascular tissues in *Ginkgo biloba* are very similar to those in modern conifers. The stem is largely composed of secondary xylem, which, except for certain details in its mode of formation, is very similar to coniferous wood in general (Eicke 1964, Eicke and Ehling 1965). The long tracheids are irregular in size, and considerably more so than in most conifers, ranging from 30 to 55 µm in diameter (Greguss 1955, 1958, Eicke 1964). Two distinct types of tracheids seem to be present, one wide and one narrow, the latter having a thicker cell wall than the former (Srivastava 1963). The tracheids carry one, two, or very seldom, three rows of circular, bordered pits in their radial walls. The wide earlywood tracheids have a few pits in their tangential wall, while those in the latewood have many. The pits resemble those in *Araucaria*. The membrane is supported by radially arranged microfibrils and lacks a torus (Eicke 1964, Eicke and Ehling 1965). The parenchyma cells in the xylem contain druces of calcium oxalate crystals, which are also frequent in the phloem, both features typical of the cycads (Seward and Gowan 1900). On the whole, the wood of ginkgo resembles structurally that of the conifers. The irregular size of the tracheids and the abundance of calcium oxalate crystals are, nevertheless, notable characteristics which distinguish ginkgo wood from that of the Coniferales.

Except for some of the extraneous substances, chemically the wood of ginkgo is that of a typical conifer. The cellulose content is 40–42% (Timell 1960), compared to an average of 42% for most conifers. It was shown at an early date that the acetyl groups in the wood are associated with the holocellulose portion (Tang et al. 1935). Orcel's (1958) contention that ginkgo should be closer to the ferns than to the conifers in hemicellulose composition is of doubtful validity. Timell (1960) subjected wood and bark from male and female ginkgo trees to a summative chemical analysis. Except for its high content of calcium oxalate, the wood was in all respects chemically identical with a conifer wood. Jabbar

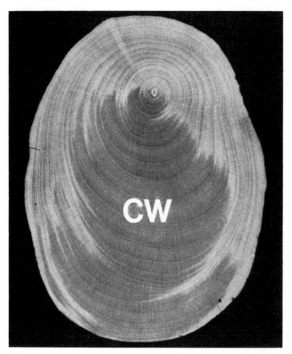

Fig. 8.7. Cross section of a large *Ginkgo biloba* branch at its base, close to the stem. The branch, having escaped apical control, had bent upward with radial growth promotion and formation of dark compression wood (*CW*) on the lower side

Mian and Timell (1960a, b) isolated an arabino-4-*O*-methylglucuronoxylan and a galactoglucomannan from ginkgo wood. These hemicelluloses had the same structure as corresponding polysaccharides in conifer woods. The chemical similarity between ginkgo and the conifers was later confirmed by Eicke and Ehling (1965). Interestingly, the shell of the ginkgo nut contains more arabino-4-*O*-methylglucurono-xylan than galactoglucomannan, whereas the latter polysaccharide predominates in the wood (Maekawa and Kitao 1975). The lignin content of ginkgo wood is 30–34%, a range found in many conifers. Chemical composition, reactions (Traynard and Eymery 1956), color tests (Gibbs 1958), ultraviolet, infrared, and other characteristics (Kawamura and Higuchi 1963, 1964, 1965a, b) all show that the lignin is of the guaiacyl type present in most plants preceding ginkgo in evolution and also in all but a few of the conifers. Some of the low molecular weight extractives in ginkgo are typical of this tree, for example the bisflavonoid ginkgetin (Kawano 1959).

Because *Ginkgo biloba* has existed for such a long time, this species is of great interest in any attempt to trace the origin and evolution of compression wood in the gymnosperms. Unlike the cycads, ginkgo develops compression wood when displaced from its equilibrium position in space or when one of its laterals has escaped from the apical control of the leader. An example of the latter case is shown in Fig. 8.7. As in the conifers, the compression wood in

8.3.1 Ginkgo biloba

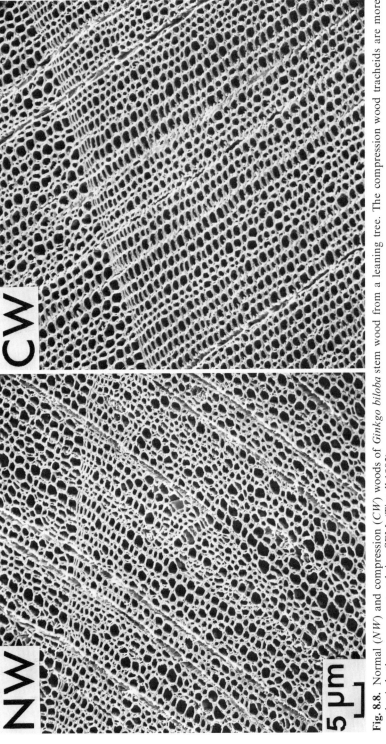

Fig. 8.8. Normal (*NW*) and compression (*CW*) woods of *Ginkgo biloba* stem wood from a leaning tree. The compression wood tracheids are more regular in both arrangement and size. SEM. (Timell 1983)

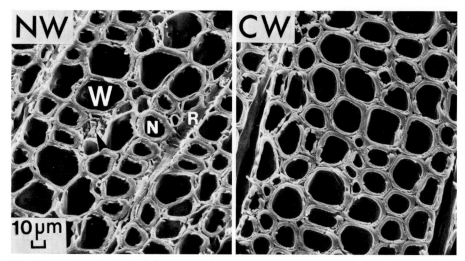

Fig. 8.9. Normal and compression woods of *Ginkgo biloba* with tracheids and rays (*R*). The normal tracheids are either wide (*W*) or narrow (*N*), and some of the latter have a thicker wall and more rounded outline than the wide ones. *Arrowhead* indicates crystals. Many of the compression wood tracheids are angular, and all have the same, relatively thin wall. Intercellular spaces are numerous. Transverse surface. SEM. (Timell 1978, 1981, 1983)

ginkgo is brown in color and is most frequently located within the wide growth rings on the lower side of a branch or an inclined stem. Figure 8.8 shows a comparison between normal and compression woods in a ginkgo stem. In the normal wood, the tracheids vary more in cross sectional dimension and are arranged more irregularly than in coniferous woods. The arrangement of the tracheids is noticeably more regular in the compression wood and they also vary less in size than do the normal tracheids. As can be better seen in Fig. 8.9, the difference between normal and compression woods is less conspicuous than in most of the conifers. The normal wood consists of large, thin-walled and smaller, thick-walled tracheids. The latter are sometimes rounded. In the compression wood, many of the tracheids have an angular rather than a round outline, and all of them have a rather thin wall. Even the angular tracheids are, however, surrounded by intercellular spaces.

Lämmermayr (1901) was the first to notice the presence of compression wood in *Ginkgo biloba*, an observation later confirmed by Böning (1925) and Onaka (1949). He could not discern any helical cavities, but Onaka claims that they are definitely present. Timell (1978b, 1981, 1983) subjected the histology and ultrastructure of ginkgo compression wood to a renewed investigation, using material both from a leaning stem and from an upturned branch (Fig. 8.7). Examination in the transmission electron microscope gave a clear picture of the organization of the tracheid wall. As shown in Fig. 8.10, the secondary wall is composed of a relatively thick S_1 layer with transversely oriented cellulose microfibrils and a moderately thick S_2 layer, lined on the lumen side with a warty layer. There is no S_3 layer. The S_2 is solid throughout with a smooth inner surface, that is, there are no helical cavities. These observations were later con-

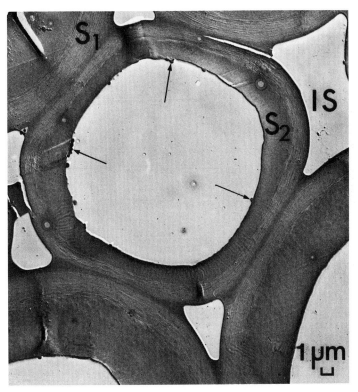

Fig. 8.10. Compression wood tracheids in *Ginkgo biloba*. The central tracheid borders on one large (*IS*) and four smaller intercellular spaces. Neither helical cavities nor checks are present. A warty layer terminates S_2 on the lumen side (*arrows*). Transverse section. TEM. (Timell 1978)

firmed by scanning electron microscopy. Figures 8.11 and 8.12 show a solid and smooth inner surface of the secondary wall with microfibrils oriented at an angle of 40–50° to the tracheid axis. The numerous openings in the wall, all oriented in the same direction as the microfibrils, are spiral drying checks. It is very likely that Onaka (1949), as first suggested by Westing (1965), had been observing such checks which are artifacts (Chap. 4.4.5.4.8). The organization of the bordered pits of compression wood is the same in Ginkgo as in the conifers (Fig. 8.13). The membrane lacks a torus. The lignin skeleton in Fig. 8.14 reveals the same distribution of lignin as in conifer compression wood. The central tracheid has an unusually thick, highly lignified $S_2(L)$ layer and an almost perfectly circular outline. The complete lack of any cavities is again clearly indicated.

Values for the summative chemical composition of normal and compression woods from *Ginkgo biloba* are listed in Table 8.1. Timell (1981, 1983) has recently drawn attention to the fact that the difference in this respect between normal and compression woods is less pronounced in ginkgo than in the conifers. Selected data, obtained on renewed analysis of normal and compression

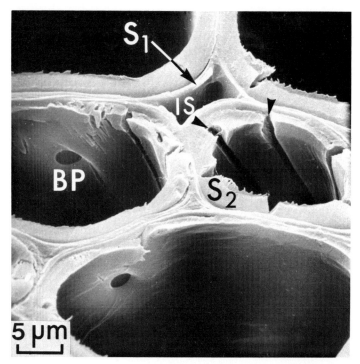

Fig. 8.11. Compression wood tracheids in *Ginkgo biloba*. Note the smooth inner surface of the S_2 layer. *IS* intercellular space. *Arrowheads* indicate drying checks. *BP* bordered pit. Transverse surface. SEM. (Timell 1981, 1983)

Table 8.1. Summative chemical composition of normal and compression woods of *Ginkgo biloba*. All values in per cent of extractive-free, dry wood

Component	Normal wood	Compression wood
Lignin	33.4	37.2
Ash	1.1	0.6
Acetyl	1.3	0.8
Uronic anhydride	4.6	5.5
Residues of:		
Galactose	3.5	6.8
Glucose	40.1	36.1
Mannose	9.5	5.5
Arabinose	1.6	1.2
Xylose	4.9	6.3

8.3.1 Ginkgo biloba

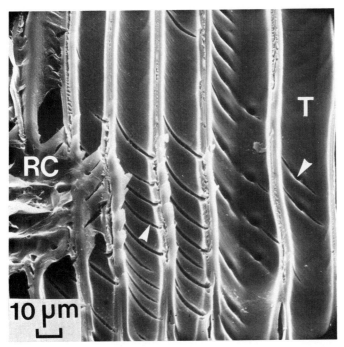

Fig. 8.12. Tracheids (*T*) and ray cells (*RC*) in compression wood of *Ginkgo biloba*. The tracheids have numerous drying checks in their wall (*arrowheads*), some of them crossing pits. Note the absence of helical cavities. Radial surface. SEM. (Timell 1978)

Table 8.2. Comparison between the contents of lignin and galactose, glucose, and mannose residues in normal and compression woods of *Ginkgo biloba* and *Pinus strobus*. All values in per cent of extractive-free, dry wood

Component	*Ginkgo biloba*		*Pinus strobus*	
	Normal wood	Compression wood	Normal wood	Compression wood
Lignin	33	37	29	40
Galactose	4	7	4	11
Glucose	40	36	44	32
Mannose	10	6	9	4

woods from a ginkgo stem, are compared in Table 8.2 with corresponding values for *Pinus strobus*, a typical conifer. Compared to the conifer compression wood, that in ginkgo contains less galactan and lignin and more cellulose and galactoglucomannan. Evidently, the compression wood in ginkgo is closer in chemical composition to the corresponding normal wood than is that in white pine. Brodzki (1972) found that Ginkgo compression wood lacked the $(1 \rightarrow 3)$-

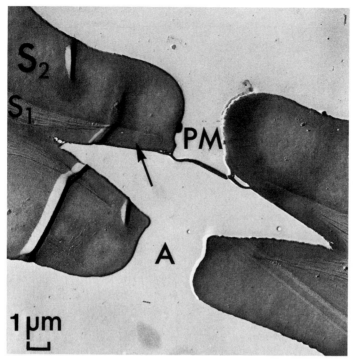

Fig. 8.13. Aspirated bordered pit in a compression wood tracheid in *Ginkgo biloba* with the initial pit border (*arrow*) and the layers S_1 and S_2. The pit membrane (*PM*) has no torus. *A* aperture. Transverse section. TEM

β-D-glucan typical of conifer compression woods. Tomimura et al. (1980) have reported that, in comparison with the conifers, ginkgo has a compression wood lignin that is less condensed at C-5 (Chap. 5.9.2).

In a study of some physical properties of ginkgo compression wood, Ueda (1973) found this wood to have a specific gravity of 0.489, only 9% higher than that of the normal wood. Compression wood of *Abies sachalinensis*, on the other hand, had a density that was 43% higher than that of the normal wood. The low density of the compression wood of *Ginkgo biloba* is probably a consequence of its relatively thin tracheid walls. The difference in modulus of rigidity between normal and compression woods was also less in ginkgo than in the fir.

It is evident that *Ginkgo biloba* forms a compression wood that in several respects differs less from normal wood than is the case in the conifers. Compared with the tracheids in the latter, those in ginkgo have a more angular outline and a thinner wall. Their most conspicuous feature is the absence of any helical cavities and ribs. Ginkgo compression wood is also closer to normal wood in its chemical composition. Other ultrastructural characteristics of the tracheids and their distribution of lignin, on the other hand, are the same in ginkgo and conifer compression woods.

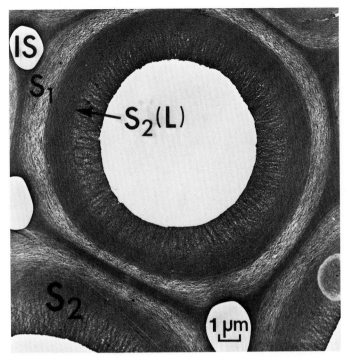

Fig. 8.14. Lignin skeleton of tracheids in compression wood of *Ginkgo biloba*. The S_2 (L) layer has a high lignin concentration. *IS* intercellular space. Note the absence of helical cavities. Transverse section. TEM. (Timell 1978, 1981, 1983)

8.3.2 Primitive Conifers and Taxads

Among the Coniferales, the Araucariaceae, which originated in the Permian period, are generally regarded as more primitive than the other five families (Delevoryas 1962). Wardrop and Dadswell (1950) observed helical cavities in compression woods of *Araucaria bidwillii* and *A. cunninghamii*, and Brodzki (1972) later discerned fine striations in the former species as well as in *Agathis alba*. Lämmermayr (1901), by contrast, could not detect any striations in compression woods of these two genera, nor could Höster (1971), who, however, might not have examined severe compression wood. My own observations with pronounced compression wood confirm Lämmermayr's as regards *Agathis robusta* and *Araucaria bidwillii*, neither of which possessed any helical cavities (Fig. 8.15). More recently, Yoshizawa et al. (1982) have found that while helical cavities obviously occur very seldom in *Araucaria*, such cavities can develop in the last-formed tracheids of *Araucaria angustifolia* (Chap. 4.4.5.4.4).

In the Taxales, which probably date back to the Jurassic, the compression wood tracheids do not develop any helical cavities but instead retain the helical thickenings of the normal wood, as described in Chap. 4.4.5.5. The younger *Pseudotsuga menziesii* (Timell 1981) and *P. japonica* (Yoshizawa et al. 1982), by

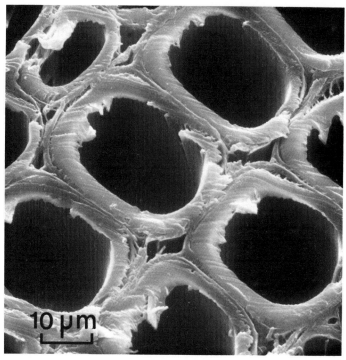

Fig. 8.15. Tracheids in compression wood of *Agathis robusta* with neither helical cavities nor checks. (Cf. Fig. 4.80.) Transverse surface. SEM. (Timell 1981, 1983)

contrast, replace their characteristic helical thickenings with cavities and ribs when changing from normal- to compression-wood formation, as illustrated for Douglas-fir in Figs. 4.85 and 4.99. Yoshizawa et al. (1982) have suggested that *Pseudotsuga* must be at a stage of evolution when helical thickenings are disappearing from compression-wood tracheids.

8.4 Compression Wood in Primitive Angiosperms

Höster and Liese (1966) have pointed out that it is not the taxonomic classification of a certain species that decides whether it is likely to form compression wood or tension wood but instead the nature of its predominant cell type, a view previously expressed also by Nečesaný (1955 a, b). Trees and shrubs with a xylem consisting mostly of axial tracheids tend to develop compression wood, regardless of whether they are gymnosperms or angiosperms. Plants with a xylem and phloem composed largely of fibers form tension wood.

The Ranales are considered to include some of the most primitive angiosperm families. Some of them have a xylem consisting only of tracheids and parenchyma cells with no vessels, namely the Amborellaceae, Tetracentraceae, Trochodendraceae, Winteraceae, and *Sacandra* of the Chloranthaceae (Patel

8.4 Compression Wood in Primitive Angiosperms

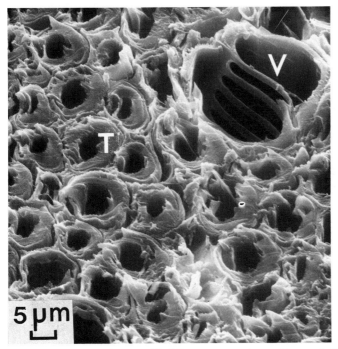

Fig. 8.16. "Compression wood" of *Buxus sempervirens* with round, thick-walled tracheids (*T*) and vessels (*V*). Transverse surface. SEM. (Timell 1981, 1983)

1974, Meylan and Butterfield 1982, Carlquist 1983). Dadswell and Wardrop (1949) observed compression wood on the upper side of branches of *Drimys aromatica* (Winteraceae). Doubts concerning the validity of this finding have been expressed by Kucera and Philipson (1977) who found that inclined branches of *Drimys winteri* developed neither tension wood nor compression wood (Chap. 3.1).

Kucera and Philipson (1978) and Meylan (1981) studied the anatomy and other properties of branch reaction wood in *Pseudowintera colorata* (Winteraceae). The secondary xylem of this primitive angiosperm contains no vessels, and almost the entire tissue is composed of tracheids. The rays are much higher and wider than in the gymnosperms. There was a pronounced radial growth promotion on the lower side of the branches which decreased with increasing distance from the stem. Compared to those on the upper side, the tracheids on the lower were longer and wider and had a thicker tangential wall (Kucera and Philipson 1978). The microfibril angle was large, 35–36°, a feature typical of compression-wood tracheids, but it was also large on the upper side, namely 26–29° (Meylan 1981). Staining failed to indicate any increase in lignification in the lower side tracheids. Both rays and ray cells were larger on the under side of the laterals.

It is obvious that the xylem present on the lower side of the branches in *Pseudowintera colorata* cannot be regarded as compression wood. Apparently, this primitive angiosperm species forms neither compression wood nor tension wood. The hypotrophic growth of its branches is characteristic of the gymnosperms, but is also found in a large number of angiosperm species. Nečesaný (1955 a, b) has suggested that in angiosperms increasing xylem specialization is associated with an increasing tendency to develop tension wood, a view later supported and further attested by Höster and Liese (1966) and Kucera and Philipson (1978). Tension wood clearly cannot be expected to develop in a xylem composed of tracheids and lacking both fibers and vessels. The reason why no compression wood is formed remains obscure. It should perhaps be mentioned at this point that the relationship between the vessel-less angiosperm trees and bushes and the gymnosperms is far from clear (Carlquist 1975). Bailey (1953) has emphasized that these dicotyledons cannot be derived from the conifers.

Members of the genus *Buxus* (Carlquist 1982) have a secondary xylem composed of tracheids but with vessels also present (Chap. 3.1). Lämmermayr (1901) noted the occurrence of reddish brown, sickle-shaped zones within the wide growth rings on the lower side of *Buxus sempervirens* branches, but he could not detect any anatomical difference between these tracheids and those on the upper side. Similar observations were later reported by Onaka (1949) with *Buxus japonica* and *B. microphylla* var. *sinica,* and by Nečesaný (1955a,b) and Höster (1966) with *B. sempervirens* (Fig. 3.1). According to Höster and Liese (1966), the tracheids on the lower side of *Buxus sempervirens* branches resemble those in compression wood in their rounded outline but have no intercellular spaces. The proportion of vessels is conspicuously reduced. As can be seen in Fig. 8.16, these thick-walled cells also lack helical cavities (Timell 1981, 1983). It would be advisable for the time being not to refer to these cells as compression-wood tracheids, their rounded outline and thick wall notwithstanding. More information regarding their ultrastructure and chemical composition is required before their nature can be fully assessed.

8.5 General Conclusions

It is clear, as pointed out by Westing (1965), that compression wood plays a crucial role in the regulation of tree form in gymnosperms. It is also true that the arboreal habit of the gymnosperms probably depends on their ability to control their orientation in space by forming compression wood. A gymnosperm tree that lacked the capacity to produce compression wood could probably not survive very long. For this reason it is likely that in the gymnosperms the ability to form compression wood probably is as old as the arborescent habit itself (Westing 1965).

We do not know when or where compression wood first appeared in the long evolution of land plants. It might have occurred in the Progymnospermophyta 360–370 million years ago but very likely not in the Lycophyta, Sphenophyta, or Pteridophyta. The Cycodophyta probably never developed compression wood since their extant members do not form it. The secondary xylem is a very conservative part of a tree, and the fact that *Ginkgo biloba* produces compression wood makes it highly probable that all Ginkgophyta did so. If this is correct, compression wood must have existed at the time of the Lower Permian, 250–280 million years ago. Within the Coniferophyta, the extinct Cordaitales might have contained compression wood. The presence of compression wood in the Araucariaceae also places this tissue in the Permian period.

The present fossil evidence is both sparse and uncertain and allows no firm conclusions. Compression wood tracheids might have been present in *Cupressinoxylon* from the Jurassic and in *Araucarioxylon* from the Permian. If compression wood did occur in the latter fossil, this tissue must have existed 250–280 million years ago, which agrees with its presence in *Ginkgo biloba*. The only fully authentic fossil compression wood so far observed is from the recent Pleistocene.

It would appear that compression wood has undergone a certain evolutionary development in the course of its long existence. The compression wood in ginkgo is both structurally and chemically closer to normal wood than that formed in the Coniferales and Taxales. The helical cavities would seem to have developed somewhat later than the other structural characteristics of the tracheids in compression wood (Timell 1978a, 1981, 1983, Yoshizawa et al. 1982). They are absent in *Ginkgo biloba* and apparently formed only rarely in the primitive Araucariaceae. In the Taxales the helical thickenings of their normal wood are retained in the compression wood but not in the more recent *Pseudotsuga*. The helical cavities and ribs are probably best developed in *Pinus*, which by many is considered to be the most advanced genus among the conifers.

Interestingly, when the formation of compression wood gradually ceases in a conifer, the helical cavities are the first to disappear (Chap. 9.8). In many genera, but not in *Pinus*, the first rows of tracheids deposited at the beginning of the growing season have an angular outline, are wide and thin-walled, and lack helical cavities, almost exactly as in *Ginkgo biloba* (Chap. 4.5). As has been observed so often, ontogeny seems to recapitulate phylogeny.

In conclusion, it is likely that compression wood has existed for a very long time, even on a geologic time scale, perhaps for as long as 300 million years. Throughout this period it has served the function of making possible the necessary movements of orientation in the Ginkgoales, Coniferales, and Taxales. With the advent of the arboreal angiosperms in the Mesozoic, a new type of xylem appeared, quite different from that of the gymnosperms. These angiosperms also had to perform movements, but in these trees a different mechanism was developed for this purpose, entailing the formation of tension wood.

References

Andrews HN, Jr 1947 Ancient plants and the world they lived in. Comstock, Ithaca, 279 pp
Andrews HN, Jr 1961 Studies in paleobotany. Wiley, New York, 487 pp
Andrews HN 1974 Paleobotany 1947–1972. Ann Miss Bot Gard 61:179–202
Arnold CA 1930 The genus Callixylon from the Upper Devonian of central and western New York. MI Acad Sci Arts Lett 11:1–50

Arnold CA 1947 An introduction to paleobotany. McGraw-Hill, New York, 433 pp
Bailey IW 1933 The cambium and its derivative tissues. VII. Problems in identifying the wood of Mesozoic Coniferae. Ann Bot 47:145–157
Bailey IW 1953 Evolution of the tracheary tissue of land plants. Am J Bot 40:4–8
Banks HP 1970a Evolution and plants of the past. Wadsworth, Belmont, 170 pp
Banks HP 1970b Major evolutionary events and the geological record of plants. Biol Rev 45:317–454
Barber CA 1898 Cupressinoxylon vectense; a fossil conifer from the Lower Greensand of Shanklin in the Isle of Wight. Ann Bot 12:329–361
Barghoorn ES 1952a Degradation of plant material and its relation to the origin of coal. Sec Conf Orgin Constit Coal, Crystal Cliffs NS, 181–203
Barghoorn ES 1952b Degradation of plant tissues in organic sediments. J Sedim Petrol 22:34–41
Barghoorn ES 1964 Evolution of cambium in geological time. In: Zimmermann MH (ed) The formation of wood in forest trees. Academic Press, New York, 3–17
Barghoorn ES, Scott RA 1958 Degradation of the plant cell wall and its relation to certain tracheary features of the Lepidodendrales. Am J Bot 45:222–227
Beck CB 1957 Tetraxylopteris schmidtii ge et sp nov, a probable pteridosperm precursor from the Devonian of New York. Am J Bot 44:35–367
Beck CB 1960a Connection between Archaeopteris and Callixylon. Science 131:1524–1525
Beck CB 1960b The identity of Archaeopteris and Callixylon. Brittonia 12:351–368
Beck CB 1962a Reconstructions of Archaeopteris and further consideration of its phylogenetic position. Am J Bot 49:373–382
Beck CB 1962b Plants of the New Albany shale. II. Callixylon arnoldii sp nov. Brittonia 14:322–327
Beck CB 1964a The woody, fern-like trees of the Devonian. Mem Torrey Bot Club 21:26–37
Beck CB 1964b Predominance of Archaeopteris in Upper Devonian flora of western Catskills and adjacent Pennsylvania. Bot Gaz 125:126–128
Beck CB 1966 On the origin of gymnosperms. Taxon 15:337–339
Beck CB 1970 The appearance of gymnospermous structure. Biol Rev 45:379–400
Beck CB 1971 On the anatomy and morphology of lateral branch systems of Archaeopteris. Am J Bot 58:758–784
Beck CB 1976 Current status of the Progymnospermopsida. Rev Palaeobot Palynol 21:5–23
Beck CB 1979 The primary vascular system of Callixylon. Rev Palaeobot Palynol 28:103–115
Beck CB 1981 Arachaeopteris and its role in vascular plant evolution. In: Niklas K (ed) Paleobotany, paleoecology, and evolution. Praeger, New York, 193–230
Beck CB, Coy K, Schmid R 1982 Observations on the fine structure of Callixylon wood. Am J Bot 69:54–76
Bonamo PM 1975 The Progymnospermopsida: building a concept. Taxon 24:569–579
Böning K 1925 Über den inneren Bau horizontaler und geneigter Sprosse und seine Ursachen. Mitt Deutsch Dendrol Ges 35:86–102
Brodzki P 1972 Callose in compression wood tracheids. Acta Soc Bot Polon 41:321–327
Carlquist S 1975 Ecological strategies of xylem evolution. University of California Press, Berkeley Los Angeles London, 259 pp
Carlquist S 1982 Wood anatomy of Buxaceae: correlations with ecology and phylogeny. Flora 172:463–491
Carlquist S 1983 Wood anatomy of Bubbia (Winteraceae) with comments on origin of vessels in dicotyledons. Am J Bot 70:578–590
Chaloner WG 1970 The rise of the first land plants. Biol Rev 45:353–377
Chamberlain CJ 1935 Gymnosperms. Structure and evolution. Chicago University Press, Chicago, 484 pp
Crook FM, Nelson PF, Sharp DW 1965 An examination of ancient Victorian woods. Holzforschung 19:153–156
Dadswell HE, Wardrop AB 1949 What is reaction wood? Aust For 13:22–33
Darrah WC 1960 Principles of paleobotany. Ronald Press, New York, 295 pp
Delevoryas T 1962 Morphology and evolution of fossil plants. Holt, Rinehart, Winston, New York, 189 pp
Delevoryas T 1966 Plant diversification. Holt, Rinehart, Winston, New York, 145 pp

References

Eggert DA, Kanemoto NY 1977 Stem phloem of a Middle Pennsylvanian lepidodendron. Bot Gaz 138:102–111

Eicke R 1958 Beitrag zur Kenntnis der submikroskopischen Struktur der Araucariaceenhölzer. Ber Deutsch Bot Ges 71:231–240

Eicke R 1963 Feinbauuntersuchungen an den Tracheiden von Cycadeen. II. Cycas media and Cycas pectinata. Ber Deutsch Bot Ges 76:229–234

Eicke R 1964 Ginkgo biloba, Entwicklungsstand des Holzes und seines Feinbaus. Ber Deutsch Bot Ges 77:379–384

Eicke R, Ehling E 1965 Die Ausbildung der jungen Tracheiden im Holz von Ginkgo biloba L. Ber Deutsch Bot Ges 78:326–337

Eicke R, Metzner-Küster, I 1961 Feinbauuntersuchungen an den Tracheiden von Cycadeen. I. Cycas revoluta und Encephalartos spec. Ber Deutsch Bot Ges 74:99–104

Fengel D 1974 Polysaccharide in fossilen Hölzern. Naturwissenschaften 61:450

Florin R 1963 The distribution of conifer and taxad genera in time and space. Acta Horti Berg 20(4):121–212

Franklin AH 1959 Ginkgo biloba L: historical summary and bibliography. Virg J Sci 10:131–176

Gibbs RD 1958 The Mäule reaction, lignins, and the relationships between woody plants. In: Thimann KV (ed) The physiology of forest trees. Ronald, New York, 269–312

Gorczyński T, Molski B, Pogorzelska I 1969 Struktura drewna z wykopaliska "Rynek Warzywny" w Szczecinie. (Wood structure from the archaeological excavation "Vegetable Market" in Szczecin in Poland.) Rocz Dendrol 23:5–38

Gothan W, Weyland H 1964 Lehrbuch der Paläobotanik. Akademie-Verlag, Berlin, 594 pp

Greguss P 1955 Identification of living gymnosperms on the basis of xylotomy. Akadémiai Kiadó, Budapest.

Greguss P 1958 Some recent data on the xylotomy of the Cycas, Zamia and Ginkgo. Acta Biol Szeged 4:143–147

Greguss P 1967 Fossil gymnosperm woods in Hungary. Akadémiai Kiadó, Budapest, 136 pp

Greguss P 1968 Xylotomy of the living cycads. Akadémiai Kiadó, Budapest, 260 pp

Höster HR 1966 Über das Vorkommen von Reaktionsgewebe in Wurzeln und Ästen der Dikotyledonen. Ber Deutsch Bot Ges 79:211–212

Höster HR 1971 Das Vorkommen von Reaktionsholz bei Tropenhölzern. Mitt Bundesforschungsanst Forst-Holzwirtsch 82:225–231

Höster HR, Liese W 1966 Über das Vorkommen von Reaktionsgewebe in Wurzeln und Ästen der Dikotyledonen. Holzforschung 20:80–90

Jabbar Mian A, Timell TE 1960a Studies on Ginkgo biloba L. II. The constitution of an arabino-4-O-methylglucurono-xylan from the wood. Svensk Papperstidn 63:769–774

Jabbar Mian A, Timell TE 1960b Studies on Ginkgo biloba L. III. Constitution of a glucomannan from the wood. Svensk Papperstidn 63:884–888

Kawamura I, Higuchi T 1963 Studies on the properties of lignins of the plants in various taxonomical positions. I. On the UV absorption spectra of lignins. Mokuzai Gakkaishi 9:182–188

Kawamura I, Higuchi T 1964 Studies on the properties of lignins of plants in various taxonomical positions. II. On the IR absorption spectra of lignins. Mokuzai Gakkaishi 10:200–206

Kawamura I, Higuchi T 1965a Studies on the properties of lignins in the plants in various taxonomical positions. III. On the color reactions, methoxyl contents and aldehydes which are yielded by nitrobenzene oxidation. Mokuzai Gakkaishi 11:19–22

Kawamura I, Higuchi T 1965b Comparative studies of milled wood lignins from different taxonomical origins by infrared spectroscopy. In: Chimie et biochimie de la lignine, de la cellulose et des hémicelluloses. Actes Symp Int Grenoble 1964, 439–456

Kawano N 1959 Structure of scidopitysin and ginkgetin. Chem Ind (London) 368–369

Kobendza R 1957 Milorząb dwudzielny. (Ginkgo biloba L, Syn Salisburia adantifolia Sm) Roczn Dendrol Polsk Tow Bot 12:39–65

Kucera LJ, Philipson WR 1977 Growth eccentricity and reaction anatomy in branchwood of Drimys winteri and five native New Zealand trees. NZ J Bot 15:517–524

Kucera LJ, Philipson WR 1978 Growth eccentricity and reaction anatomy in branchwood of Pseudowintera colorata. Am J Bot 65:601–607

Lämmermayr L 1901 Beiträge zur Kenntnis der Heterotrophie von Holz und Rinde. Sitzungsber Kaiserl Akad Wiss Mat-Naturwiss Cl, Wien, Pt 1 110:29–62

Lutz HJ 1930 A new species of Cupressinoxylon (Goeppert) Gothan from the Jurassic of South Dakota. Bot Gaz 90:92–107

Maekawa E, Kitao K 1975 Isolation and characterization of hemicellulose from a Ginkgo nut shell. Wood Res (Kyoto Univ) 58:33–44

Mägdefrau K 1968 Paläobiologie der Pflanzen. VEB Gustav Fischer, Jena, 549 pp

Major RT 1967 The Ginkgo, the most ancient living tree. Science 157:1270–1273

Meylan BA 1981 Reaction wood in Pseudowintera colorata – a vessel-less dicotyledon. Wood Sci Technol 15:81–92

Meylan BA, Butterfield BG 1982 Pit membrane structure in the vessel-less woods of Pseudowintera dandy (Winteraceae). IAWA Bull ns 3:167–175

Morey PR, Morey ED 1969a Observations on Epon embedded Griffin Hill peat (Massachusetts), Two Creeks Picea (Wisconsin), Cedrus penhallowii (Sierra Nevada, California) and Callixylon (Delaware, Ohio). Palaeontographica 125:73–80

Morey PR, Morey ED 1969b Senckenberg lignite: a lignitized wood with apparently original cellulose and lignin. Science 164:836–838

Morey PR, Morey ED 1971a Anatomy of a lignitized wood from Senftenberg. Am J Bot 58:621–626

Morey PR, Morey ED 1971b The cell-wall residue of fossil wood from Senftenberg. Am J Bot 58:627–633

Müller-Stoll R 1951 Mikroskopie des zersetzten und fossilisierten Holzes. In: Freund H (ed) Handbuch der Mikroskopie in der Technik. V. Mikroskopie des Holzes und des Papiers. Umschau, Frankfurt am Main, 726–816

Namboodiri KK, Beck CB 1968 A comparative study of the primary vascular system of conifers. III. Stelar evolution in gymnosperms. Am J Bot 55:464–472

Nečesaný V 1955a Výskyt reačkního dřeva s hlediska taxonomického. (Occurrence of reaction wood from the taxonomic point of view.) Sborník vysoké školy zemědělské a lesnické fakulty, Brno Sec C, 3:131–149

Nečesaný V 1955b Vztah mezi reačkním dřevem listnatých a jehličnatých dřevin. (The relationship between the reaction wood in gymnosperms and angiosperms.) Biológia 10:642–647

Niklas KJ, Pratt LM 1980 Evidence for lignin-like constituents in early Silurian (Llandoverian) plant fossils. Science 209:396–397

Onaka F 1949 (Studies on compression and tension wood.) Mokuzai Kenkyo, Wood Res Inst Kyoto Univ 1, 88 pp. Transl For Prod Lab Can 93 (1956), 99 pp

Orcel M 1958 Études sur les polyholosides des bois de quelques Préphanérogames. C R Acad Sci 246:2402–2404

Patel RN 1968 Wood anatomy of Podocarpaceae indigenous to New Zealand. 3. Phyllocladus. NZ J Bot 6:3–8

Patel RN 1974 Wood anatomy of the dicotyledons indigenous to New Zealand. NZ J Bot 12:19–32

Sachs IB 1965 Evidence of lignin in the tertiary wall of certain wood cells. In: Côté WA, Jr (ed) Cellular ultrastructure of woody plants. Syracuse University Press, Syracuse, 335–339

Scheckler SE 1975 A fertile axis of Triloboxylon ashlandicum, a progymnosperm from the Upper Devonian of New York. Am J Bot 62:923–934

Scheckler SE 1976 Ontogeny of progymnosperms. I. Shoots of Upper Devonian Aneurophytales. Can J Bot 54:202–219

Scheckler SE 1978 Ontogeny of progymnosperms. II. Shoots of Upper Devonian Archaeopteridales. Can J Bot 56:3136–3170

Scheckler SE, Banks HP 1971a Anatomy and relationships of some Devonian progymnosperms from New York. Am J Bot 58:737–751

Scheckler SE, Banks HP 1971b Proteokalon. A new genus of progymnosperm from the Devonian of New York State and its bearing on phylogenetic trends in the group. Am J Bot 58:874–884

Schmid R 1967 Electron microscopy of wood of Callixylon and Cordaites. Am J Bot 54:720–729

References

Seward AC 1933 Plant life through the ages. Cambridge University Press, Cambridge, 603 pp
Seward AC 1938 The story of the maidenhair tree. Sci Prog 32:420–440
Seward AC, Gowan J 1900 The maidenhair tree (Ginkgo biloba L). Ann Bot 14:109–154
Sigleo AC 1978 Degraded lignin compounds isolated in silicified wood 200 million years old. Science 200:1054–1055
Sporne KR 1965 Morphology of gymnosperms: structure and evolution of primitive seedplants. Hutchinson Univ Library, London, 216 pp
Srivastava LM 1963 Cambium and vascular derivatives of Ginkgo biloba. J Arnold Arbor 44:165–192
Stewart WN 1983 Palaeobotany and the evolution of plants. Cambridge Univ Press, Cambridge, 405 pp
Tang YC, Wang YW, Wang HL 1935 Über die Bindungsweise der Essigsäure im Ginkgoholz (Ginkgo biloba). Cellulosechemie 16:90–92
Taylor TN 1981 Paleobotany. An introduction to fossil plant biology. McGraw-Hill, New York, 589 pp
Timell TE 1960 Studies on Ginkgo biloba L. 1. General characteristics and chemical composition. Svensk Papperstidn 63:652–657
Timell TE 1978a Helical thickenings and helical cavities in normal and compression woods of Taxus baccata. Wood Sci Technol 12:1–15
Timell TE 1978b Ultrastructure of compression wood in Ginkgo biloba. Wood Sci Technol 12:89–103
Timell TE 1981 Recent progress in the chemistry, ultrastructure, and formation of compression wood. Int Symp Wood Pulp Chem, Stockholm 1981, SPCI Rep 38(1):99–147
Timell TE 1983 Origin and evolution of compression wood. Holzforschung 37:1–10
Tomimura Y, Yokoi T, Terashima N 1980 Heterogeneity in formation of lignin. V. Degree of condensation of guaiacyl nucleus. Mokuzai Gakkaishi 26:37–41
Traynard P, Eymery A 1956 Délignifications par les solutions hydrotropiques. III. Lignines hydrotropiques et taxonomie des végétaux dont elles ont été extraites. Holzforschung 10:43–45
Ueda K 1973 Studies on the mechanical properties of reaction wood. II. The elastic constants of Icho (Ginkgo biloba L), Yamanarashi (Populus sieboldii Miq) and Yachidamo (Fraxinus manchurica var japonica Maxim). Res Bull Coll Exp For Hokkaido Univ 30:379–388
Walton J 1953 An introduction to the study of fossil plants. A & C Black, London, 201 pp
Walton J 1969 On the structure of the silicified stem of Protopitys and roots associated with it from the Carboniferous limestone, Lower Carboniferous (Mississippian) of Yorkshire, England. Am J Bot 56:808–813
Wardrop AB, Dadswell HE 1950 The nature of reaction wood. II. The cell-wall organization of compression-wood tracheids. Aust J Sci Res B3:1–13
Westing AG 1965 Formation and function of compression wood in gymnosperms. Bot Rev 31:381–480
Yoshizawa N, Itoh T, Shimaji K 1982 Variation in features of compression wood among gymnosperms. Bull Utsunomiya Univ For 18:45–64
Zimmermann W 1930 Die Phylogenie der Pflanzen. Gustav Fischer, Jena

Chapter 9 Formation of Compression Wood

CONTENTS

9.1	Introduction	623
9.2	The Dormant Cambial Zone	625
9.2.1	Introduction	625
9.2.2	Size and Organization of the Cambial Zone	627
9.2.3	Ultrastructure of the Cambial Zone	629
9.3	The Dormant Secondary Phloem	632
9.4	The Active Cambial Zone	636
9.5	The Differentiating Secondary Phloem	642
9.6	The Differentiating Secondary Xylem	645
9.6.1	The Differentiating Tracheids	645
9.6.1.1	Cell Enlargement	645
9.6.1.2	Formation of the S_1 Layer	646
9.6.1.3	Formation of the S_2 Layer	652
9.6.1.4	Lignification	667
9.6.1.4.1	Lignification of Normal Wood	667
9.6.1.4.2	Lignification of Compression Wood	669
9.6.1.5	Formation of the Warty Layer	676
9.6.1.6	Formation of the Bordered Pits	676
9.6.1.7	Formation of the First-Formed Tracheids	676
9.6.2	The Differentiating Ray Cells	679
9.7	The Rate of Formation of Compression Wood	680
9.7.1	Introduction	680
9.7.2	The Rate of Formation of Compression Wood	682
9.7.3	Conclusions	685
9.8	The Transition Between Normal Wood and Compression Wood	686
9.9	General Conclusions	695
References		698

9.1 Introduction

Although the formation of compression wood has been studied for more than 30 years, the number of investigations devoted to this subject have until quite recently been very few. The first study on cell division and cell wall formation in compression wood was reported in 1952 by Wardrop and Dadswell. It was followed 12 years later by a more detailed cytological investigation by Wardrop and Davies (1964). Results of a similar study were reported shortly afterward by Casperson and Zinsser (1965) (Casperson 1965). Of the many observations made by Kutscha (1968) on the formation of compression wood in *Abies balsamea*, only some were published (Côté et al. 1968 b). Timell (1973) studied the ultrastructural cytology of the active and dormant cambial zone and the dormant phloem in normal and compression woods of *Picea abies*. Further observations on the ultrastructure of the dormant cambial zone of the same compression wood were reported in a later publication (Timell 1980b).

In the first of a series of important investigations on the formation of compression wood in *Cryptomeria japonica* by Harada and his co-workers, Fujita et al. (1973) adduced new evidence for the origin of the helical cavities and ridges in the secondary wall. Subsequent studies were concerned with the formation of this wall (Fujita et al. 1978a) and the cell organelles involved (Fujita et al. 1978b). Autoradiographic studies were later reported on the deposition of cellulose, hemicelluloses (Fujita and Harada 1978) and lignin (Fujita and Harada 1979). In an investiagation on the formation of compression wood in *Abies balsamea,* Timell studied the ultrastructure of the active cambial zone (Timell 1979a) and the differentiation of the xylem and phloem cells (Timell 1979b, 1980a). Data for the rate of formation of compression wood have been reported by Kennedy and Farrar (1965) and Larson (1969b). The seasonal variations of this rate have been studied by Kutscha et al. (1975) and Yumoto et al. (1982a, b).

For obvious reasons, attention in these investigations has largely been focused on the tracheids in the xylem, while the xylem ray cells have been examined only by Kutscha (1968) and Timell (1973, 1979a, b). The dormant and active compression phloem has as yet only been investigated by Timell (1973, 1980a). Most investigations on compression xylem have been restricted to the division of the cambial fusiform initials, cell enlargement, cell wall thickening, and the final lignification of the tracheids. Formation of the warty layer and of the bordered pits have been examined only twice (Kutscha 1968, Timell 1979b). The origin of the so-called first-formed tracheids has attracted considerable interest (Côté et al. 1967, Höster 1974, Yoshizawa et al. 1982). Observations on the transition between normal wood and compression wood have been reported by Fukazawa (1973), Fujita et al. (1979), Yumoto et al. (1982a, b, 1983), and Yoshizawa et al. (1984, 1985).

There is general agreement regarding several aspects of compression wood formation, for example the stage at which the tracheids begin to assume their final, rounded outline and the manner in which the entire tissue becomes lignified. One aspect is still the subject of considerable controversy, namely the formation of the helical cavities. Several investigators believe that these peculiar fissures are the result of a tangential contraction of the inner portion of the secondary wall, at least initially (Wardrop and Davies 1964, Kutscha 1968, Côté et al. 1968b, Timell 1979b, Waterkeyn et al. 1982). Casperson and Zinsser (1965), by contrast, concluded that the helical cavities and ridges are the product of an interrupted deposition of wall material effected by the cytoplasm. Fujita et al. (1973) have presented evidence for the correctness of this view. The final answer to the question which of these two interpretations is the correct one must still be held in abeyance.

9.2 The Dormant Cambial Zone

9.2.1 Introduction

According to the nomenclature now most frequently used (Schmid 1976), the *cambium* consists of a single layer of initial cells, capable of undergoing both periclinal and anticlinal divisions, while the initials and their dividing xylem and phloem derivatives together constitute the *cambial zone*. The uniseriate concept was first proposed by Sanio (1873) who was also the first to observe the organization and function of the vascular cambium. The presence of a single initial in each radial cambial file is now well established for the conifers (Mahmood 1968), but whether the same applies to the arboreal angiosperms is less certain (Catesson 1964, 1974, 1980). The mode of division proposed by Sanio (1873) more than 100 years ago has since been confirmed by several later investigators, including Mischke (1890), Newman (1956), Mahmood (1968, 1971), Murmanis (1970, 1971), Timell (1980b), and Imagawa and Ishida (1981). According to this concept, an original initial in the xylem divides to form a new initial and a mother cell. The new initial divides again, giving a third initial and a second mother cell, while the original mother cell divides into two daughter cells. These four cells, which are arranged in pairs, form a group now known as *Sanio's four* (Mahmood 1968). The two daughter cells divide once more, producing two pairs of xylem tissue cells, referred to by Mahmood (1968) as *the enlarging four*. When phloem is being produced, the mother cell divides directly into a pair of phloem tissue cells that undergo no further division. After division has ceased, the differentiating xylem and phloem tissue cells enlarge and subsequently mature. In the xylem, the latter phase always entails deposition of a secondary cell wall.

After the cell plate has been formed, each of the two new cells encloses itself with a new primary wall, as was first observed by Sanio (1873) and most recently confirmed by Mahmood (1968). Because of this fact and as a result of the entire pattern of division, the tangential walls vary in thickness but are always thinner than the radial walls. Tangential walls that abut on the radial walls at a rounded angle are older than those that meet these walls at a sharp angle (Sanio 1873). This and the variable thickness of the tangential walls made it possible for Sanio (1873) to locate the initial cells in the vascular cambium of *Pinus sylvestris*. Both methods are still used for this purpose (Timell 1980b, Imagawa and Ishida 1981, Catesson and Roland 1981).

When cell division has ceased at the end of the growing season, the cambial zone enters a period of endogenously imposed dormancy, referred to as *rest* (Romberger 1963, Perry 1971, Little and Bonga 1974). This is followed by a period of *quiescence*, which is a dormancy induced by external factors, usually a low temperature. In moderate climates the cambial region sometimes does not enter the period of rest (Barnett 1971a, Worrall 1971, Little and Bonga 1974). Albeit wider than in the angiosperms, the dormant cambial zone in the gymnosperms usually comprises only four to eight rows of cells. A few phloem tissue cells often overwinter in a partly differentiated state, while xylem cells do so

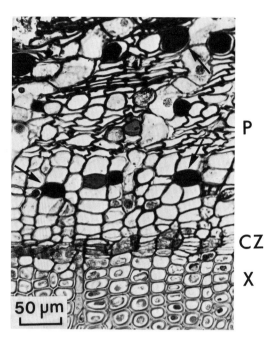

Fig. 9.1. Compression wood of *Abies balsamea* on September 13. Note the discontinuous bands of tannin cells in the phloem (*arrows*). The sieve cells in the outer bark are crushed. *X* xylem; *CZ* cambial zone; *P* phloem. LM. (Kutscha et al. 1975)

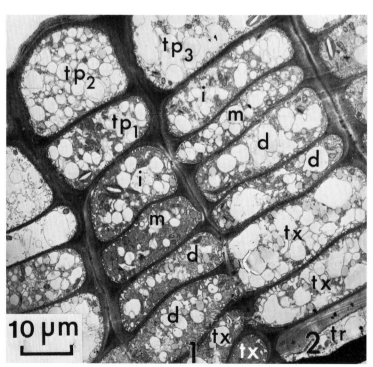

Fig. 9.2. Dormant cambial zone in compression wood of *Picea abies* containing initial (*i*), mother (*m*), and two daughter (*d*) cells, which form two groups of Sanio's four, xylem (*tx*) and phloem (*tp*) tissue cells, and a tracheid (*tr*). Transverse section. TEM. (Timell 1980b)

less frequently. Neither of these radially enlarged cells are a part of the cambial zone proper.

The organization and function of the vascular cambium are dealt with in a book by Philipson et al. (1971) and have also been the subject of several review articles, such as those by Bannan (1962, 1968), Philipson and Ward (1965), Catesson (1974, 1980), and Jacquiot (1984). The terminology in this field, which is still under debate, has been discussed by Wilson et al. (1966), Butterfield (1975), Schmid (1976), and Timell (1980b).

9.2.2 Size and Organization of the Cambial Zone

A transverse section of the compression xylem, cambial zone, and phloem in *Abies balsamea* at the onset of cambial deactivation is shown in Fig. 9.1 (Kutscha et al. 1975). Cytoplasm is still present in the lumen of the tracheids, and those next to the cambial region do not seem to be fully differentiated. The cambial zone is very narrow and consists of only four cells, probably Sanio's four. There are discontinuous rows of dark-staining tannin cells among the sieve cells in the phloem.

In his study of the cambial region and the phloem associated with formation of compression wood in *Picea abies,* Timell (1973) found the dormant cambial zone to comprise six to eight cell rows. A renewed examination (Timell 1980b) revealed the presence of four to eight rows with an average number of 6.0 cells in each cambial file. Höster (1974) observed five to six cell rows in the dormant cambium of compression wood in the same species, and Kutscha et al. (1975) noted the occurrence of five to seven rows in *Abies balsamea.* It is well known that the number of cells can vary considerably also within adjacent files of the dormant cambium. An example of this is shown in Fig. 9.2, where file 1 contains seven cambial cells, whereas file 2 has only four. A survey of the reported width of the cambial zone in normal wood of 11 conifer species (Timell 1980b) indicates that the average number of cell rows is four to eight.

The organization of the dormant cambial zone in compression wood has only been reported by Timell, first in 1973 and later, after a more thorough examination of the material available, in 1980 (b). The compression tissues were collected in late October from the lower side of a *Picea abies* tree with a pronounced basal sweep. In the transverse section shown in Fig. 9.2, file 1 contains a pair of phloem tissue cells and a group of Sanio's four, comprising an initial and a mother cell together with a pair of daughter cells, and finally two xylem tissue cells. File 2 consists of a phloem cell, a group of Sanio's four and two xylem cells. The primary wall around each pair and the parental wall of the initial enclosing the group of Sanio's four can be readily distinguished. The tangential end walls of the group of four are thicker than the walls between or within the two pairs.

It has been noted that division and differentiation often are not synchronous events even in adjacent files and that cambial activity appears to cease cell by cell (Wardrop 1957, Mahmood 1968). An example of this is seen in Fig. 9.3. Groups of Sanio's four are present in all four files. In file 4 this group is located

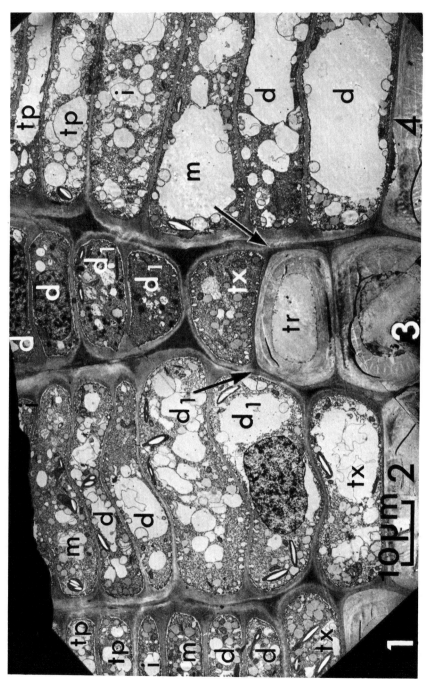

Fig. 9.3. Dormant cambial zone in *Picea abies* compression wood with one file (*3*) further developed than the other two. Groups of Sanio's four are present in files *1*, *2*, and *4* and a pair of daughter cells (d_1) from a previous division in file 2. A parental primary wall encloses the xylem tissue cell and the last-formed tracheid in file 3 (*arrows*). Transverse section. TEM. (Timell 1980b)

next to the xylem. A pair of daughter cells from a previous sequence of divisions is present in both files 2 and 3. The xylem tissue cell and the last-formed tracheids in the latter file are enclosed by a common primary wall.

A region with a more synchronous cessation of cambial growth is shown in Fig. 9.4. A group of Sanio's four and another that is destined to become an enlarging four are present in files 1 and 4. The parental wall of the original initial can still be seen enclosing the two groups of four in each file. A less common arrangement of the cambial cells is shown in Fig. 9.5, where file 1 contains the usual group of Sanio's four but also a group of three, which consists of a pair of xylem tissue cells and an earlier daughter cell that had failed to divide. A more common group of three is one composed of an undivided mother cell and a pair of xylem tissue cells. Undivided mother cells have also been observed by Murmanis (1971) in the active cambial zone of *Pinus strobus*. The group of Sanio's four in file 2 is followed by a pair of radially enlarged xylem cells.

Timell (1980b) concluded from his observations that the group of Sanio's four always overwinters intact, while the group of expanding four does not always do so. The number of cells in this group that remain in the cambial zone can vary from zero to four. Which of the active cambial cells that will overwinter depends on several, partly fortuitous circumstances. A comparison with the organization of the dormant cambial zone in normal coniferous wood (Mahmood 1968, 1971, Murmanis 1970, 1971, Timell 1973, Tsuda 1975a, Imagawa and Ishida 1981) with that in compression wood indicates that there are no differences between the two tissues in this respect. Timell (1980b) noted that parental primary walls frequently traversed intercellular regions at cell corners, as observed in compression wood also by Wardrop and Dadswell (1952) and by Casperson and Zinsser (1965). It is possible that the relatively low lignin concentration of 0.75 g/g in this region, reported by Wood and Goring (1971) can be attributed to this fact, as mentioned in Chap. 6.4.2.

9.2.3 Ultrastructure of the Cambial Zone

The ultrastructural cytology of the dormant cambial zone in compression wood of *Picea abies* has been studied by Timell (1973, 1980b). Initial, mother, and daughter cells could not be distinguished from one another on the basis of their ultrastructure, a well-known fact (Srivastava and O'Brien 1966a, Evert and Deshpande 1970). As is typical of dormant cambial regions in general, the fusiform cells had a dense cell content and were filled with many small vacuoles (Figs. 9.2–9.5). The plasma membrane was frequently infolded. The nucleus was centrally situated and elongated. Golgi cisternae and endoplasmic reticulum were sparse, and mitochondria were never abundant, but free ribosomes were frequent. Plastids and oblong amyloplasts were common, although they were not as numerous as lipid droplets. Cambial cells destined to become ray parenchyma cells or ray tracheids were filled with lipid material (Fig. 9.4). The protoplasm was similar in all respects to that present in the dormant cambial region in normal wood of conifer species such as *Abies sachalinensis* (Tsuda 1975a, Hirakawa et al. 1979), *Cryptomeria japonica* (Itoh

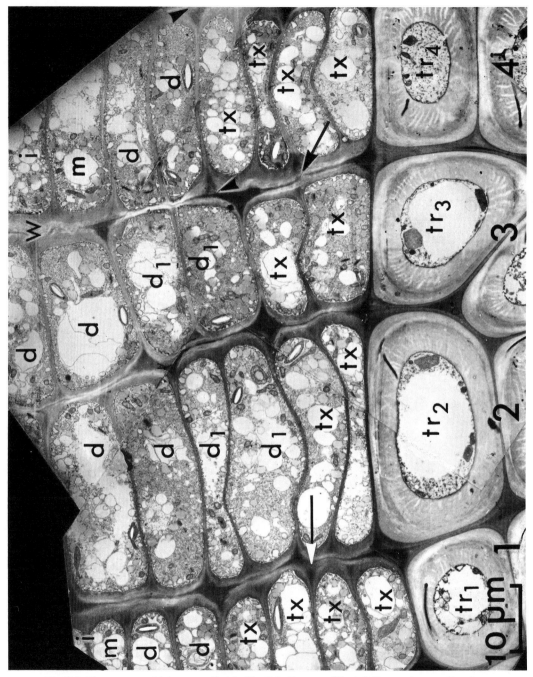

Fig. 9.4. The same cambial zone as that in Fig. 9.3. Groups of Sanio's four and expanding four are present in files *1* and *4*. Files *2* and *3* contain a pair of daughter cells from a previous division (d_1) and a pair of xylem tissue cells (*tx*). The two pairs of the expanding four are enclosed by a common primary wall (*arrows*). The parental wall of the original initial can be seen between the two groups of four in file 4 (*arrowheads*). Transverse section. TEM. (Timell 1980b)

9.2.3 Ultrastructure of the Cambial Zone

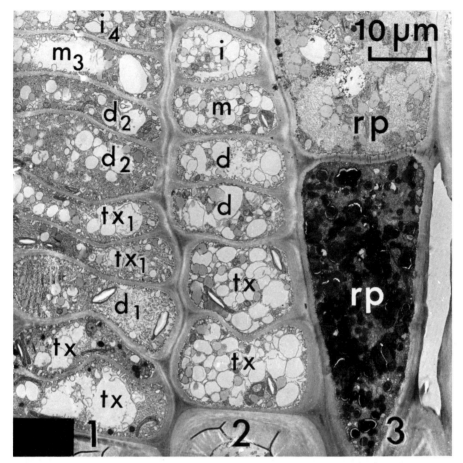

Fig. 9.5. The same cambial zone as that in Figs. 9.3 and 9.4 with two files of fusiform cells (*1* and *2*) and one file (*3*) of cambial ray parenchyma cells (*rp*). File 1 contains a group of Sanio's four and a group of three which consists of a pair of xylem tissue cells (*tx*) and an earlier daughter cell (*tx*) that had failed to divide. A group of Sanio's four is present in file 2. There is an abundance of dark-staining lipid droplets in the inner part of the two parenchyma cells. Transverse section. TEM. (Timell 1980b)

1971), *Picea abies* (Timell 1973), and *Pinus strobus* (Srivastava and O'Brien 1966a, Murmanis 1971). As can be seen from Fig. 9.4, the last-formed compression-wood tracheids usually contained remnants of cytoplasm, often as a parietal layer. The mitochondria seemed to be the last cell organelles to disappear.

Timell (1980b) concluded that the dormant cambial zone is the same regardless of whether the cambial cells are going to develop into normal or compression wood cells in the following spring. He also added that this is hardly surprising in view of the fact that the compression wood stimulus is received, not in the active cambial zone but in the enlarging xylem (Chap. 12.5.2).

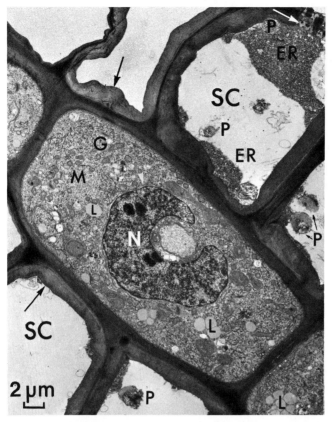

Fig. 9.6. Dormant compression phloem in *Picea abies.* The central ray parenchyma cell has an infolded nucleus (*N*). Its cytoplasm contains endoplasmic reticulum, a few Golgi cisternae (*G*), many mitochondria (*M*), numerous lipid droplets (*L*), and plastids. The sieve cells (*SC*) have thick, lamellated, and folded secondary walls (large, *black arrows*) and contain endoplasmic reticulum (ER) and plastids (P). The *white arrow* indicates a sieve area with pores lined with callose. Transverse section. TEM. (Timell 1973)

9.3 The Dormant Secondary Phloem

There has been only one report on the ultrastructure of the dormant secondary phloem associated with formation of compression wood, namely by Timell (1973), who compared normal and compression tissues in *Picea abies*. A transverse section of several ray parenchyma and sieve cells is shown in Fig. 9.6. The mature sieve cells have the nacreous walls typical of the sieve cells in the Pinaceae, where they are considered to be secondary (Abbe and Crafts 1939, Esau 1969, Evert 1977). These walls are relatively thick, occasionally folded, distinctly lamellar, and unlignified (Srivastava and O'Brien 1966b, Miyakawa et al. 1973, Neuberger and Evert 1974, 1975, 1976). There has been some controversy regarding the orientation of the cellulose microfibrils in the secondary wall of the sieve cells in the gymnosperms (Srivastava 1969), but the most

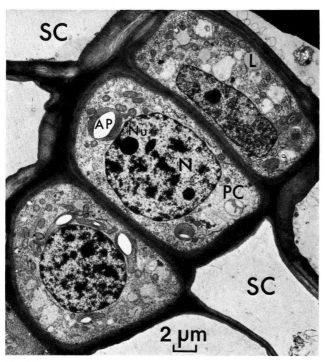

Fig. 9.7. Dormant compression phloem in *Picea abies* with sieve cells (*SC*) and three axial parenchyma cells (*PC*). The two lower sieve cells are still immature. The nuclei in the ray cells have one or two nucleoli (*Nu*). The ergastic material includes lipid droplets (*L*) and amyloplasts (*AP*) of various sizes. Transverse section. TEM. (Timell 1973)

common view at present (Evert 1977) seems to be that these microfibrils are located in the plane of the wall (Chafe and Doohan 1972). One of the sieve cells in Fig. 9.6 still contains aggregates of smooth, tubular endoplasmic reticulum, known to occur in large amounts near the sieve areas of mature cells (Neuberger and Evert 1976). P-type plastids with a circular outline and containing protein crystals are also present. Callose is deposited within the sieve pores.

The axial parenchyma cells in the gymnosperms often store starch, lipid material, tannin, or crystals. Examples of cells that do not serve such functions are shown in Fig. 9.7. These cells contain a round nucleus that tends to become elongated as the cell matures. Amyloplasts and lipid droplets are present in the cytoplasm, but they are not abundant. Murmanis and Evert (1967) observed no starch in the dormant phloem of *Pinus strobus*, but Barnett (1974b) found it to be present in *Pinus radiata*.

Axial parenchyma cells containing tannin were frequent in the dormant compression phloem, where they occurred very close to the cambial zone, as has also been observed in normal phloem (Kollmann and Schumacher 1961, Barnett 1974b). Always strongly osmiophilic, the tannin sometimes appeared as a black, solid mass but more often it was present as a dark, granular material.

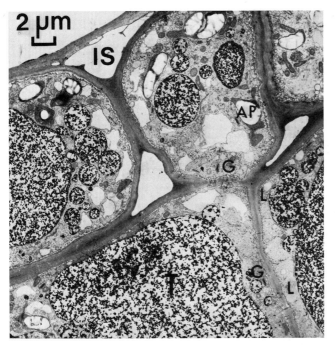

Fig. 9.8. Tannin cells in dormant compression phloem of *Picea abies*. The cells are separated from one another by intercellular spaces (*IS*) and contain tannin vacuoles of widely differing sizes. There are Golgi bodies (*G*), amyloplasts (*AP*), and lipid droplets (*L*). Transverse section. TEM. (Timell 1973)

Typical tannin cells are shown in Fig. 9.8. The tannin is present here in one large vacuole as well as in several small ones, as has also been observed more recently with suspension cultures of various gymnosperms (Chafe and Durzan 1973, Baur and Walkinshaw 1974, Parham and Kaustinen 1976). Sometimes the large, central tannin vacuole occupied almost the entire cell, although the nucleus was never affected, an observation also made by Barnett (1974b). Because of this situation, the cell organelles were often located close to the plasma membrane. Cells with massive deposits of tannin usually had only a few amyloplasts, but the two components were by no means mutually exclusive, as has been pointed out by Chafe and Durzan (1973). Lipid material was sparse, a fact also noted in *Pinus radiata* by Barnett (1974b). Axial parenchyma cells with crystals were not observed in the compression phloem examined by Timell (1973). Although stated to be present in the secondary phloem of *Picea abies* (Holdheide 1951, Srivastava 1963), they are evidently rare (den Outer 1967).

A ray parenchyma cell near the cambial zone is seen in Fig. 9.6. This cell has a central, deeply infolded nucleus. The dense cytoplasm contains numerous mitochondria, some endoplasmic reticulum, Golgi bodies, and a few plastids. Lipid droplets are relatively common, but there are no amyloplasts. The ray parenchyma cells were considerably longer than the axial parenchyma cells (Murmanis and Evert 1967) and always several times longer than wide. As in all

9.3 The Dormant Secondary Phloem

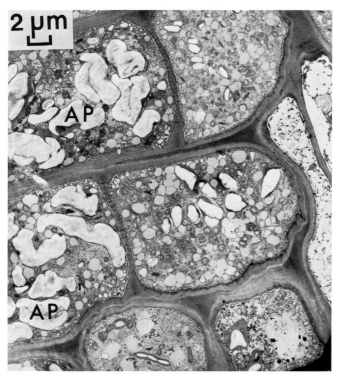

Fig. 9.9. Ray parenchyma cells in dormant compression phloem of *Picea abies* close to the cambial zone. Two of the cells contain numerous amyloplasts (*AP*) completely filled with starch. Radial section. TEM

Pinaceae, the phloem rays were uniseriate and several cells high. They very seldom contained any tannin material.

Some ray cells contained large amounts of starch and also much lipid material, as shown in Fig. 9.9. Typically, all amyloplasts were completely filled with starch, an observation also made by Barnett (1974b). Timell (1973) found the ray cells in the dormant, normal phloem in *Picea abies* to contain less starch than did these cells in the compression phloem. This observation is probably not significant, however, for the time of collection, the nature of the tree, and its location were all quite different in the two cases. Starch has long been known to undergo often drastic seasonal variations in both gymnosperm (Murmanis and Evert 1967, Little 1970a, b, Tsuda and Shimaji 1971) and angiosperm (Sauter 1966a, b, Pomeroy and Siminovitch 1971) trees.

Timell (1973) concluded from his direct comparison between the two dormant phloems that these tissues were the same, irrespective of whether they were associated with the formation of normal or compression wood. A comparison with information obtained by others with normal, dormant phloem leads to the same conclusion. This is, of course, not unexpected when it is recalled (Chap. 12.5.2) that the compression wood stimulus is received, not in the differentiating phloem but in the differentiating xylem.

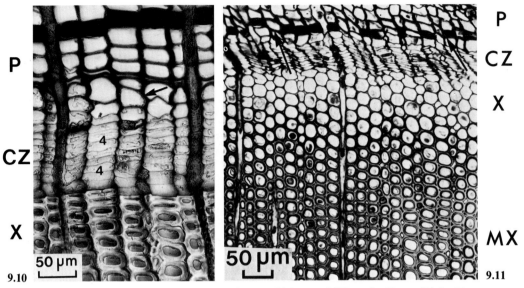

Fig. 9.10. Compression xylem (*X*), an active cambial zone (*CZ*), and phloem (*P*) in *Picea abies* (in May). The angular and radially flattened cambial cells are arranged in groups of two, three, and four (*4*). A secondary wall is being deposited in the phloem (*arrow*) but not in the xylem cells which are in the process of enlarging. Transverse section. LM. (Modified from Höster 1974)

Fig. 9.11. Differentiating phloem (*P*), an active cambial zone (*CZ*), differentiating xylem (*X*), and mature xylem (*MX*) in compression wood of *Abies balsamea* (in July). Most cambial files contain eight cells. The enlarging cambial cells rapidly assume a rounded outline at this time of the year. Note the interrupted row of tannin cells (*arrows*). Transverse section. LM. (Modified from Kutscha and Gray 1972)

9.4 The Active Cambial Zone

Initiation of cambial activity associated with formation of compression wood has only been studied on two occasions. In *Picea abies* growing in northern Germany, Höster (1974) found the first activity to involve overwintering phloem mother cells which enlarged radially, following which they either developed into a sieve cell or divided into two phloem cells. Shortly thereafter, on April 14, the initial cell divided to produce phloem tissue cells. A few days later, xylem daughter cells commenced to enlarge and divide. The cambial zone and the adjacent xylem and phloem, as they appeared on May 19, are shown in Fig. 9.10. The dividing and enlarging cells, as pointed out by Höster, are arranged in groups of two, three, or four cells. Two of the files contain a group of Sanio's four together with a group of expanding four. It should be noted that at this stage no xylem cells had yet begun to deposit a secondary wall, although such a wall is being formed in some of the sieve cells. The angular, radially flattened, and obviously vacuolated cambial cells are indistinguishable from those associated with formation of normal wood in conifers (Srivastava and O'Brien

9.4 The Active Cambial Zone

1966a, Itoh et al. 1968, Murmanis 1970, 1971, Barnett 1973, Tsuda 1975a, Catesson and Roland 1981, Imagawa and Ishida 1981).

In *Abies balsamea* forming normal wood, Kutscha et al. (1975) observed the first cambial activity on April 12 in central New York in the form of enlargement of an overwintering sieve cell. Enlarging phloem tissue cells were also the first to signify a renewed cambial activity in trees producing compression wood, but they did not appear until April 26. At this time there was an increase in the number of cambial cells from 6 to 10 in both normal and compression woods. New xylem and phloem cells appeared on May 3 in the compression wood. Other investigators, such as Alfieri and Evert (1968a) and Murmanis and Sachs (1969b), have reported that division in the phloem begins 6 weeks prior to that in the xylem, but this is probably not a universal pattern, since others, for example Bannan (1955), have found the xylem to become active earlier than the phloem. Kutscha et al. (1975) noted that activation of the cambial zone preceded bud opening by 4–5 weeks.

The cambial zone reached its maximum width of 16 cell rows on May 30 in compression wood, 1 week earlier than in normal wood. Both time and maximum width of the cambial zone agree with other observations on normal wood in *Picea glauca* (Gregory 1971) and in *Pinus* spp. (Alfieri and Evert 1968, Gregory 1971). In a study of the active cambial zone in compression wood in *Picea abies* trees growing in Zürich, Timell (1973) noted that the cambial zone became active during April but that, as could be expected, the exact date varied with the location of the tree. A section collected in July is shown in Fig. 9.11 (Kutscha and Gray 1972). In tilted *Picea glauca* trees, Yumoto et al. (1982a) observed 14–20 cambial cells in the compression wood region, a number that did not vary with the time of inclination.

As is mentioned in Chap. 4.3.2.4, Wardrop and Dadswell 1952 found that in *Pinus pinaster* there was a higher rate of anticlinal divisions in the active cambial zone of compression wood in comparison with normal wood. In one case, illustrated in Fig. 4.36, only two anticlinal divisions had occurred within 100 tangential sections of normal wood compared to 14 such divisions in the subsequent 200 sections of compression wood. Wide growth rings, reflecting a high rate of radial growth, were associated with an increased number of anticlinal divisions in both normal and compression woods. The higher frequency of anticlinal divisions in the cambial zone was held to be responsible for the short tracheids typical of compression wood (Chap. 4.3.2.4).

In his investigation of the active cambial zone in compression wood of *Abies balsamea* growing in central New York, Timell (1979a) observed that this zone consisted of eight to ten rows of cells in early May. In tissues collected on July 22 there were eight rows of dividing cambial cells and four to six rows of enlarging xylem and phloem tissue cells, as shown in Fig. 9.12. This is in agreement with the results reported by Kutscha et al. (1975), who also observed eight rows of cambial cells in July in compression wood of the same species.

Light micrographs of the active cambial zone in trees forming compression wood have also been reported by Wardrop and Dadswell (1952) for *Pinus radiata*, by Kutscha and Gray (1972) for *Abies balsamea*, and by Fujita et al. (1978a) for *Cryptomeria japonica* (Fig. 9.13). In the latter two species, there ap-

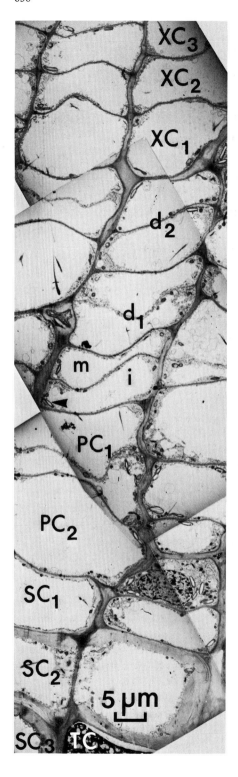

Fig. 9.12. Three radial rows of cambial cells and their derivatives in compression wood of *Abies balsamea*. *Xylem side:* Initial cell (i), mother cell (m), daughter cells (d_1 and d_2), and enlarging cells ($XC_1 - XC_3$). *Phloem side* phloem cambial cells (PC_1 and PC_2), differentiating sieve cells ($SC_1 - SC_3$), and a tannin cell (TC). Transverse section. TEM. (Timell 1979a)

9.4 The Active Cambial Zone

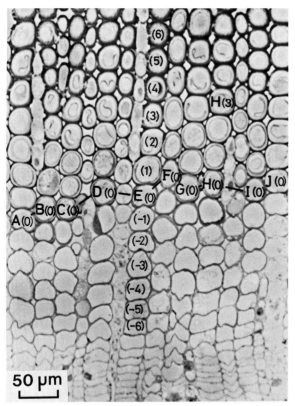

Fig. 9.13. Active cambial zone and differentiating compression wood in *Cryptomeria japonica*. The originally angular and radially flattened cells are gradually assuming a rounded outline. Cells (–6) to (–3) are enlarging. S_1 is being deposited in cells (–2) and (–1). The S_2 layer is forming in cell (*1*) and those beyond. Transverse section. LM (phase contrast). (Fujita et al. 1978a)

pear to be six to eight cambial cell rows. All evidence currently available indicates that the dividing cambial cells in compression wood are similar in size and shape to those in normal wood, that is, rectangular in outline and always flattened in the radial direction. A tangential displacement of the radial files of the cambial cells in relation to the files of differentiating xylem cells can frequently be observed (Fig. 9.11) (Wardrop and Davies 1964, Kutscha and Gray 1972, Kutscha et al. 1975). This is undoubtedly a sectioning or preparation artifact caused by the fact that the soft and almost semi-fluid active cambial zone is sandwiched between the more solid regions of differentiating xylem and phloem. Electron micrographs reported by Casperson and Zinsser (1965) indicate that the active fusiform cells in compression wood of *Pinus sylvestris* have an angular outline and lack intercellular spaces.

In contrast to the dormant fusiform cambial cells, the active cells are always highly vacuolated and often to such an extent that the cytoplasm forms only a thin layer along the wall. An example of this feature, which is characteristic of

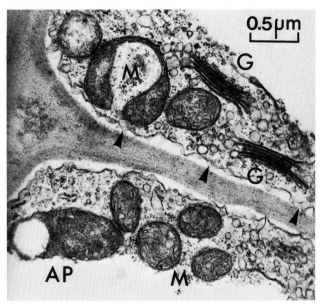

Fig. 9.14. Parietal cytoplasm in two fusiform cells in compression wood of *Abies balsamea* with an amyloplast (*AP*), mitochondria (*M*), and Golgi cisternae (*G*). Vesicles released from the cisternae are filled with a granular material similar to that in the wall. The membrane of some vesicles appears to be fusing with the plasma membrane (*arrowheads*). Transverse section. TEM. (Timell 1979a)

both normal and compression tissues, is shown in Fig. 9.12. The dividing xylem cells in this cambial zone of *Abies balsamea* form a group of Sanio's four, which is followed by a group of expanding four. The thickest tangential wall indicates where the initial had changed from phloem to xylem production. The cambial cells in this figure as well as in Fig. 9.13 are distinctly rectangular in outline and lack intercellular spaces.

Kutscha (1968) and Côté et al. (1968b) claim that the active fusiform cells in compression wood of *Abies balsamea* become vacuolated after a new primary wall has been formed. This could not be confirmed by Timell (1979a), who studied a similar tissue. Actually, most investigators agree that the replacement of the many small vacuoles in the dormant cambial zone with one large, central vacuole occurs at the onset of vernal activity (Srivastava and O'Brien 1966a).

Active cambial cells in compression wood of *Abies balsamea* were found by Kutscha (1968) to contain mitochondria, endoplasmic reticulum, Golgi bodies with associated vesicles, amyloplasts, and lipid droplets. Both Wardrop and Davies (1964) and Timell (1973, 1979a) observed that the endoplasmic reticulum frequently was aligned with the surface of the adjacent cell wall. Golgi bodies were more numerous than in dormant cambial cells, the plasma membrane was often infolded, and ribosomes, frequently arranged in groups as polysomes, were abundant (Timell 1973). In his studies on the formation of the cell plate and the primary wall in fusiform cells of normal and compression

9.4 The Active Cambial Zone

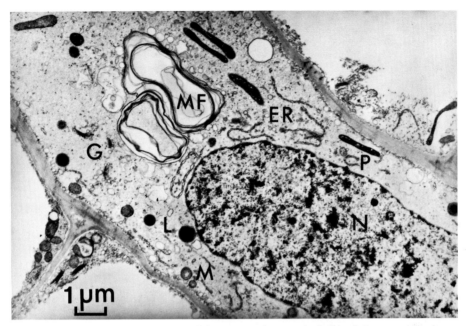

Fig. 9.15. Cambial ray parenchyma cell in compression wood of *Abies balsamea* with a large nucleus (*N*), a few Golgi bodies (*G*), endoplasmic reticulum (*ER*), plastids (*P*), and lipid droplet (*L*). *MF* myelin-like figure. Transverse section. TEM. (Timell 1979a)

woods of *Abies balsamea*, Kutscha (1968) could not observe any differences between the two tissues in these respects. This was later confirmed by Timell (1979a) in his investigation of compression wood of the same species. Golgi cisternae in the parietal cytoplasm (Fig. 9.14) seemed to release vesicles containing a granular material.

The ray cells in the active cambial zone of compression wood in *Abies balsamea* were not always isodiametrical but were sometimes elongated in the radial direction (Timell 1979a), as observed previously in normal wood by Wodzicki and Brown (1973). Ergastic material was more sparse than in the dormant ray cells, and rough endoplasmic reticulum was common, but Golgi bodies were rare (Fig. 9.15). The plasma membrane had many small infoldings, a fact also observed by Kutscha (1968).

Timell (1979a) concluded from his investigation of the active cambial zone in compression wood of *Abies balsamea* that the mode of division, arrangement, and ultrastructural cytology of this zone was the same, regardless of whether normal wood or compression wood was being produced. A comparison with observations reported for active cambial regions in normal wood of various conifer species attests to this (Srivastava and O'Brien 1966a, Murmanis and Sachs 1969b, Murmanis 1970, 1971, Itoh 1971, Barnett 1971a,b, 1973, Tsuda 1975a, Keith and Godkin 1976, Hirakawa et al. 1979, Imagawa 1981, Imagawa and Ishida 1981). It also agrees with the fact that the primary wall has the same

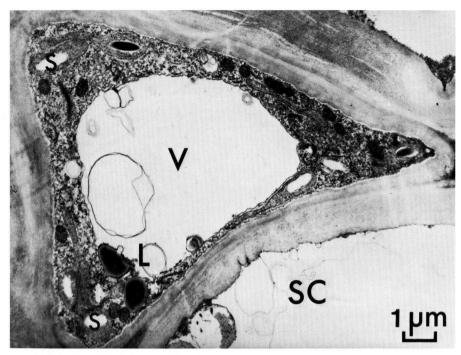

Fig. 9.16. Axial parenchyma cell in differentiating compression phloem of *Abies balsamea* with a large, central vacuole (*V*) and small plastids containing starch (*S*). *SC* sieve cell. Transverse section. TEM. (Timell 1980a)

chemical composition in normal and compression woods (Côté et al. 1968a, Larson 1969b) (Chap. 6.3.3).

9.5 The Differentiating Secondary Phloem

The differentiating secondary phloem associated with formation of compression wood has only been studied by Timell (1980a). As originally observed by Sanio (1873) and later confirmed by Murmanis (1970, 1971), the phloem cells in *Abies balsamea* were arranged in pairs (Fig. 9.12). Differentiation was rapid, with maturation of the sieve cells and formation of tannin in the axial parenchyma cells occurring only six cells from the initials. Except in the region around the nucleus, the cytoplasm was parietal in all developing cells. The mature sieve cells had a thick, lamellar secondary wall, tubular endoplasmic reticulum, and plastids with protein crystals, all features also reported for sieve cells in the hypocotyl of *Pinus resinosa* (Neuberger and Evert 1974, 1975, 1976). Callose was present in the sieve areas.

The axial parenchyma cells in *Abies* are arranged in regular, tangential strands (Srivastava 1963, den Outer 1967, Esau 1969). Kutscha et al. (1975) found that the tannin cells, while continuous in normal phloem of *Abies bal-*

9.5 The Differentiating Secondary Phloem

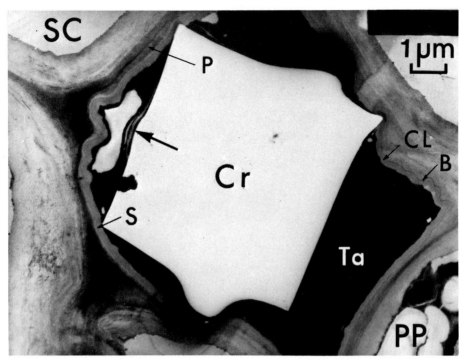

Fig. 9.17. Axial parenchyma cell in differentiating compression phloem of *Abies balsamea* with tannin (*Ta*) and formerly containing a large calcium oxalate crystal (*Cr*) which was coated with a cork layer (*large arrow*). The cell wall consists of primary (*P*) and secondary (*S*) walls, a dark border (*B*), and a cork layer (*CL*). *SC* sieve cell. *PP* phloem parenchyma cell. Transverse section. TEM. (Timell 1980a)

samea, were discontinuous in the compression wood. This, however, is probably not a feature typical of compression wood, for the tangential bands of osmiophilic tannin cells are continuous in compression phloem of *Picea abies* (Timell 1973, Höster 1974). Timell (1980a) found axial parenchyma cells without tannin to be sparse in differentiating compression phloem of *Abies balsamea*. As shown in Fig. 9.16, these cells often contained a large, central vacuole. The dense cytoplasm rapidly acquired amyloplasts and lipid droplets, as in normal phloem (Murmanis and Evert 1967, Barnett 1974b). The content of starch increased with increasing distance from the cambium. Most of the axial parenchyma cells became filled with tannin at an early stage.

In the Pinaceae, some axial parenchyma cells in the phloem acquire large crystals of calcium oxalate, usually embedded in tannin and covered with a cork layer (Barnett 1974b). The ultrastructure and formation of these crystal cells in the phloem of *Larix decidua* have been studied in great detail by Wattendorf (1968, 1969a, b) and by Wattendorf and Schmid (1973). A crystal cell in compression phloem of *Abies balsamea* is shown in Fig. 9.17. The ultrastructure of the wall was essentially the same as that described by Wattendorf (1969a, b). Crystal cells were abundant in the compression phloem and were located sur-

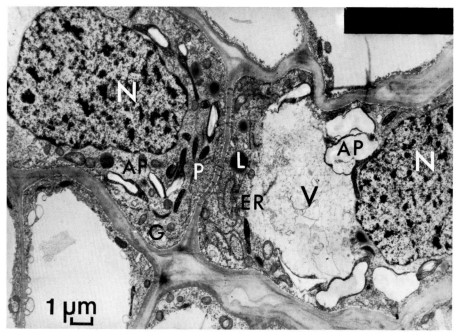

Fig. 9.18. Portions of two differentiating ray parenchyma cells in compression phloem of *Abies balsamea*. The cells contain a nucleus (*N*) and a dense cytoplasm with plastids (*P*), amyloplasts (*AP*), lipid droplets (*L*), endoplasmic reticulum (*ER*), and Golgi cisternae (*G*). *V* a large vacuole. Transverse section. TEM. (Timell 1980a)

prisingly close to the cambium, as noted previously for normal phloem by several investigators (Abbe and Crafts 1939, Grillos and Smith 1959, Srivastava 1963). Similar crystal cells had previously been observed in differentiating compression phloem of *Larix decidua, Pinus resinosa,* and *Tsuga canadensis,* but not in *Picea abies* (Timell, unpublished observations).

The cytoplasm of the ray parenchyma cells increased in density with progressing differentiation, but these cells always had fairly large vacuoles, as also observed by Tsuda (1975b). Figure 9.18 shows that they contained no tannin but lipid droplets, amyloplasts, and plastids. As they matured, the ray parenchyma cells accumulated increasing amounts of starch. Kutscha et al. (1975) noted that collapse of dead sieve cells was more extensive in compression than in normal phloem of *Abies balsamea.* This is undoubtedly a result of the higher radial growth rate of the compression wood. Other differences between the bark on the upper and lower sides of leaning conifer stems are discussed in Chap. 4.13.

Timell (1980a) concluded from his observations that there were few, if any, differences between differentiating normal and compression phloems. The sieve cells, axial parenchyma cells, and ray parenchyma cells in the compression phloem of *Abies balsamea* were similar to those observed in normal phloem of species such as *Larix leptolepis* (Imagawa 1981), *Metasequoia glyptostroboides* (Kollmann and Schumacher 1961, 1962a,b, 1963), *Pinus radiata* (Barnett

9.6.1.1 Cell Enlargement

1974a,b), *P. resinosa* (Neuberger and Evert 1974, 1975, 1976), and *P. strobus* (Evert and Alfieri 1965, Srivastava and O'Brien 1966b, Murmanis and Evert 1966, 1967, Alfieri and Evert 1968a,b, Srivastava 1969, Chafe and Doohan 1972).

9.6 The Differentiating Secondary Xylem

9.6.1 The Differentiating Tracheids

9.6.1.1 Cell Enlargement

During the enlargement phase of cell differentiation, the former cambial cells increase in size both longitudinally and transversely. At this stage they are still enclosed only by the thin and highly extensible primary wall. After the cells have ceased to enlarge, deposition of the secondary wall is initiated. Wardrop and Dadswell (1952) observed that in differentiating compression wood in branches of *Pinus radiata*, the tracheids became rounded in outline, and intercellular spaces appeared before the S_2 layer had been deposited. Their light micrographs reveal that this occurred while the S_1 layer was being formed. Wardrop and Davies (1964) later presented conclusive evidence that the intercellular spaces develop prior to the deposition of both the S_1 and the S_2 layers. They emphasized that the spaces are not produced by forces created in the course of the cell wall thickening but appear as a result of physiological conditions existing during the phase of cell enlargement, for example a high turgor pressure. According to Wardrop and Davies, the intercellular spaces have a schizogenous origin and are formed when the cells assume a rounded outline by separating along the middle lamella. The presence of cytoplasm within the spaces is in agreement with this hypothesis.

Casperson and Zinsser (1965) also observed that the developing xylem cells became rounded and that intercellular spaces appeared while the cells were enlarging and before any secondary wall had developed. Figure 9.19, which probably represents tissue at the end of the growing season, shows that at this time the transition from an angular to a circular cell outline can be very rapid. At earlier stages the change is more gradual, as can be seen from Fig. 9.13. Timell (1979a) noted that the characteristic rounding of the developing compression wood tracheids began immediately after cell division had terminated and while the cells were enclosed only by a primary wall. He also pointed out that the compression wood stimulus probably is received by the enlarging young tracheids. This question is discussed in Chap. 12.5.2. Mio and Matsumoto (1982) recently stated that the intercellular spaces develop at an early stage of the differentiation of the tracheids in *Pinus thunbergii*.

Figure 9.13, reported by Fujita et al. (1978a), gives a good picture of the different stages in the development of the tracheids in compression wood. The fusiform cambial cells are radially flattened and are followed in file E by five enlarging cells. The first two of these (-6 and -5) are still angular, but in the following three the outline of the cells becomes more and more rounded.

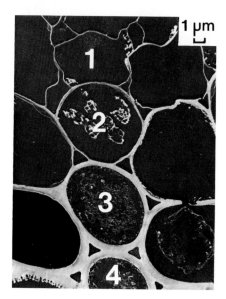

Fig. 9.19. Differentiating compression wood tracheids in *Pinus sylvestris*. The enlarging cells assume a rounded outline very rapidly. Cell *1* possesses only a primary wall, cell *2* has begun to form an S_1 layer, which is complete in cell *3*, and in cell *4* deposition of S_2 has been initiated. Transverse section. TEM. (Casperson and Zinsser 1965)

Bamber (1980) has suggested that tensile forces in the primary wall and the S_1 layer, at a stage of reduced turgor pressure, cause the tracheids to contract, resulting in their rounded outline and creation of intercellular spaces. This contraction is stated to be transverse because of the flat helix of microfibrils in the S_1 layer.

9.6.1.2 Formation of the S_1 Layer

Following cell enlargement, the next phase of differentiation entails the deposition of a secondary wall. It is generally assumed that this thickening of the cell wall is initiated only after the cell has attained its final size and form, and that cells can enlarge only as long as they are enclosed only by the thin, extensible primary wall.

While this is usually the case, it would appear that developing compression wood tracheids are able to change their shape at a stage when the secondary wall is being formed. Results reported by Wardrop and Davies (1964), Casperson and Zinsser (1965) (Fig. 9.19), Fujita et al. (1978a) (Fig. 9.13), and Timell (1979a) all suggest that the differentiating tracheids in compression wood assume their final, uniformly rounded outline while the outer portion of the S_1 layer is being deposited. In Fig. 9.13 the layer S_1 is developing in the cell rows marked (−2) and (−1). All of the former and several of the latter still have the irregular outline characteristic of the earlier, enlarging cells, marked (−6) to (−3). The initiation of the strongly birefringent S_1 layer can easily be observed in polarized light, as shown in Fig. 9.20. Compression wood tracheids at an early stage of S_1 formation are seen in more detail in Fig. 9.21. The uneven cross sectional outline of the younger of these cells is quite obvious. Wardrop and

9.6.1.2 Formation of the S_1 Layer

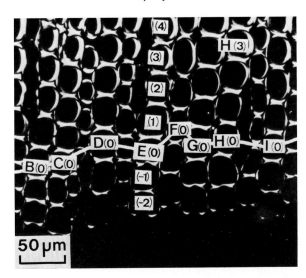

Fig. 9.20. The transverse section in Fig. 9.13 viewed in polarized light. The cambial cells, which are enclosed by only a primary wall, show complete extinction. Birefringence appears first in the cell row marked (–2), indicating the beginning of S_1 deposition. (Fujita et al. 1978a)

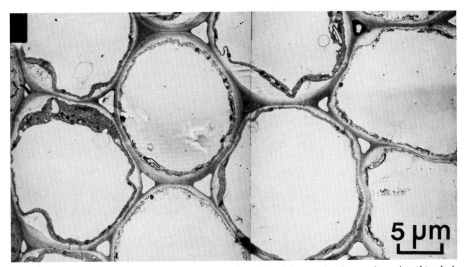

Fig. 9.21. Differentiating compression wood tracheids with a parietal cytoplasm in *Abies balsamea*. The two cells to the *right*, closest to the cambium, which are in the process of depositing an S_1 layer, have not yet assumed their final, rounded outline. Transverse section. TEM

Dadswell (1952) noted that formation of the secondary wall began at the center of compression wood tracheids while there was still only a primary wall at the cell tips. They also observed, as already mentioned, that radial enlargement could occur after deposition of the S_1 layer had been initiated. These observations agree with those described above and suggest that changes in the size and shape of a differentiating tracheid take place while wall thickening is in progress.

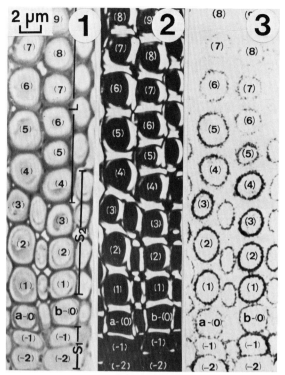

Fig. 9.22. Compression wood tracheids in *Cryptomeria japonica*. **1** Phase contrast. **2** Polarized light. **3** Radioautograph. Tritiated glucose is being deposited in large amounts in cells (−2) and (−1) where the S_1 layer is developing. Formation of S_2 is commencing in cell (0). Oblique (45°) section. (Fujita and Harada 1979)

Fujita et al. (1978a, b) distinguish between three consecutive stages in the differentiation of compression wood tracheids. The initial formation of the S_1 layer is accompanied by lignification of the intercellular region. During the second stage, the S_2 layer is rapidly deposited. This is followed by the third stage when both S_1 and S_2 become lignified. In an autoradiographic study of the formation of the secondary wall in compression wood, Fujita and Harada (1978) administered D-glucose-6-^{3}H to the lower stem side of potted, leaning 3-year-old *Cryptomeria japonica* saplings (Fujita et al. 1981). As can be seen in Fig. 9.22, the S_1 layer was being deposited in the cells marked (−2) and (−1). The autoradiograph suggests, according to the investigators, an abundant incorporation of tritiated glucose in these cells, while incorporation has decreased noticeably in the cells marked (0), where formation of S_2 has commenced. Whether this can be attributed much significance is, however, somewhat doubtful, for the two (0) cells actually seem to have incorporated as much radioactive glucose as some of the cells marked a (−1) or b (−2).

Plants utilize D-glucose for synthesis of both cellulose and various hemicelluloses, such as the galactan, galactoglucomannans, and xylan present in

9.6.1.2 Formation of the S_1 Layer

compression wood. According to Fujita and Harada (1978), cellulose can be assumed to be added to the wall by apposition, whereas the hemicelluloses are incorporated into the wall by intussusception (Ray 1967, Takabe et al. 1981a, 1983, 1984). On the basis of this assumption, Fujita and Harada (1978) concluded from their results that formation of the S_1 layer entails an initial deposition of cellulose which is followed by encrustation with hemicelluloses. It should perhaps be noted that no radioactive xylan can be expected to be synthesized from D-glucose-6-^{3}H, since the D-xylose precursor of this polysaccharide is formed by loss of the carbon at position 6 in the glucose.

As already mentioned in Sect. 9.4, Timell (1979a,b) found that vacuolation of the fusiform cells occurred at the very beginning of the cambial activity. The large, central vacuole persisted throughout the period of formation and enlargement of the primary wall, as can be seen in Fig. 9.12. The tracheids remained highly vacuolated during the early phase of the deposition of the S_1 layer, as shown in Fig. 9.21. All investigators agree that at the time of initiation of the S_1 layer in the tracheids of compression wood, the cytoplasm is present as a thin, parietal layer that closely follows the developing cell wall (Wardrop and Davies 1964, Casperson and Zinsser 1965, Kutscha 1968, Côté et al. 1968b, Fujita et al. 1978b, Timell 1979b).

Cronshaw and Wardrop (1964) and Wardrop (1965) observed that both tracheids and fibers in normal wood were conspicuously vacuolated at the stage when the S_1 layer was being deposited. Wardrop and Davies (1964) found the same to be true of compression wood tracheids. Fujita et al. (1978b) confirmed this and added that the cells remained remarkably vacuolated not only throughout the stages of S_1 formation but also after the deposition of S_2 had begun. In contrast to these results, Timell (1979b) found that, as the formation of the S_1 layer progressed, the central vacuole decreased in size in many of the cells, and that it had disappeared entirely by the time S_1 had been completed, as shown in Fig. 9.23. The reason for this discrepancy is not known. In most of the previous investigations (Wardrop and Davies 1964, Casperson and Zinsser 1965, Côté et al. 1968b) the cytoplasm was poorly preserved, which is, perhaps, not surprising in view of the fact that the differentiating xylem is one of the plant tissues that is most difficult to fix adequately (Srivastava and O'Brien 1966a, Kidwai and Robards 1969, Timell 1973). This explanation cannot, however, be applied to the investigation of Fujita et al. (1978b), where the cytoplasm was well preserved throughout. Further examinations of longitudinal instead of transverse sections may offer new information.

Kutscha (1968) and Côté et al. (1968b) found that the S_1 layer in compression wood tracheids of *Abies balsamea* was deposited in the form of loosely arranged lamellae which later became compacted. Often the recently formed lamellae were convoluted, and more so than in corresponding normal wood (Fig. 9.28), although these folds disappeared on compaction of the wall. Timell (1979b), using similar tissues from the same species, made similar observations, as shown in Figs. 9.23 and 9.24. He also found that folding of the lamellar wall was especially extensive at the cell corners where the S_1 layer has long (Sanio 1873) been known to attain its greatest thickness (Fig. 9.23).

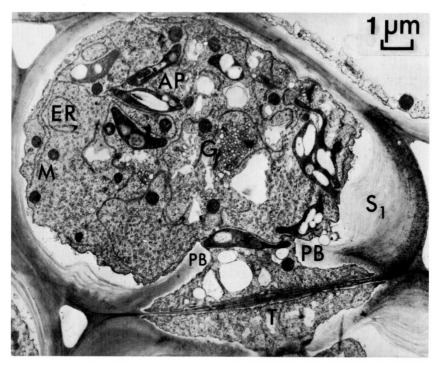

Fig. 9.23. Differentiating tracheid with a bordered pit in compression wood of *Abies balsamea* at the time of formation of the S_1 layer. The dense cytoplasm contains mitochondria (*M*), endoplasmic reticulum (*ER*), Golgi cisternae associated with many Golgi vesicles (*G*), and numerous amyloplasts (*AP*). The plasma membrane is in close contact with the cell wall and the pit borders (*PB*). The torus (*T*) has not been completed. Transverse section. TEM. (Timell 1979b)

Regardless of whether it was strictly parietal or filled the entire wall, the cytoplasm remained the same throughout the development of the S_1 layer in the tissues observed by Timell (1979b). It was always dense, and especially so in the pit regions as shown in Fig. 9.23. The "poor" cytoplasm reported by Fujita et al. (1978b) could not be observed. Throughout the formation of the S_1 layer, the plasma membrane was in close contact with the developing wall, a fact noted by Fujita et al. (1978b) and Timell (1979b). In the material examined by Timell (1979b), Golgi bodies and endoplasmic reticulum were both abundant. The Golgi cisternae were often located close together within certain regions where they were surrounded by a dense mass of derived vesicles which were evidently released from the rims of the cisternae (Figs. 9.23 and 9.24). Many of them contained a granular material of matrix substances, probably hemicelluloses. They could also be seen between the plasma membrane and the wall, evidently discharging their content to the latter (Fig. 9.24). As had been observed earlier by Wardrop and Davies (1964), the endoplasmic reticulum was frequently parallel with the adjacent cell wall. Contrary to the observations of these investigators, Timell (1979b) noticed the presence of many mitochondria,

9.6.1.2 Formation of the S_1 Layer

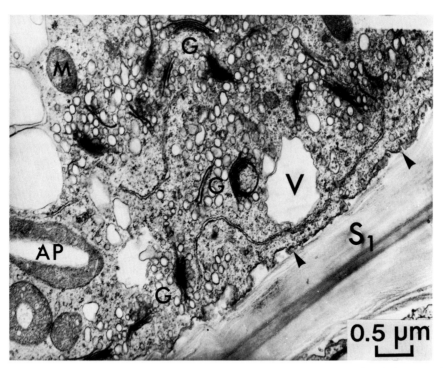

Fig. 9.24. A portion of a differentiating tracheid in compression wood of *Abies balsamea* at the time of completion of the S_1 layer. There are numerous Golgi cisternae (*G*) with vesicles, some of which appear to be discharging their content to the wall (*arrowheads*). *AP* amyloplast; *V* vacuole. Transverse section. TEM. (Timell 1979b)

as can be seen from Fig. 9.23. Both plastids and amyloplasts were common, but lipid material was sparse.

It should perhaps be noted that some of the differentiating tracheids examined by Timell (1979b) remained vacuolated throughout the stage of S_1 formation, as had been observed by Wardrop and Davies (1964) and Fujita et al. (1978b). The reason why some cells became filled with cytoplasm as differentiation proceeded, while others remained vacuolated, is not clear. Unfortunately, only transverse sections were examined in this study.

In most respects the formation of the S_1 layer in compression wood tracheids does not seem to differ from the similar event in normal wood of either softwoods (Cronshaw and Wardrop 1964, Murmanis and Sachs 1969b, 1973, Tsuda 1975b) or hardwoods (Wooding and Northcote 1964, Cronshaw 1965a, b, Wardrop 1965, Robards 1968). In some of these investigations, for example in the thorough study reported by Tsuda (1975b), it is clear that the cytoplasm is parietal at the stage of S_1 deposition. If Timell's (1979b) observations are correct, compression wood differs from normal wood in this respect. As is evident from the next section, this situation is the same at the stage of S_2 formation.

9.6.1.3 Formation of the S_2 Layer

As was described in Chap. 4.4.5, it is in the structure of the S_2 layer of its tracheids that compression wood differs most from normal, coniferous wood. Similarly, the differentiation of compression wood differs most from that of normal wood in the formation of the S_2 layer. The structure of this nonuniform layer is complex. The outer, solid one third has a very low concentration of polysaccharides and is as highly lignified as the compound middle lamella (Wood and Goring 1971). Following my suggestion, this layer is now commonly referred to as the $S_2(L)$-layer (Chap. 4.4.5.3), although Harada and his co-workers prefer to call it "the outer S_2". The remaining, inner part of the S_2 is a thick, moderately lignified wall layer which is smooth in certain primitive gymnosperms and in root compression wood (Timell 1978a,b). In all other cases, this portion of S_2 is penetrated by deep fissures, previously known as spiral striations, but now usually referred to as helical cavities. These cavities seem to branch inward and always terminate on reaching the $S_2(L)$ layer. They are invariably oriented in the same direction as the cellulose microfibrils. The inner portion of the S_2 layer as a result consists of helically arranged ribs on ridges, often rounded on the lumen surface and at times bulging into the lumen (Chap. 4.4.5.4).

The formation of the helical cavities has been a subject of considerable controversy for many years, as already mentioned. According to one view, they are schizogenous in origin and are produced as a result of a tangential contraction of the microfibrils in S_2. According to one hypothesis, this contraction occurs when the wall becomes lignified. Other investigators have come to the conclusion that the helical ribs are preformed and are produced by an irregular deposition of cell wall material. Other aspects of the formation of the S_2 layer in compression wood tracheids have also been controversial, for example the question whether the plasma membrane temporarily disappears during the deposition of S_2.

Few observations have been reported on the formation of the polysaccharide portion of the $S_2(L)$ layer. Figure 9.25 shows a developing compression-wood tracheid in *Abies balsamea* at a time when the cellulose microfibrils of this layer are being deposited. The $S_2(L)$ layer consists of slightly folded lamellae, deposited at a distance from one another in contrast to the more compacted microfibrils of the S_1 layer (Timell 1979b). The dense cytoplasm fills the entire cell at this stage and seems to be deeply folded. As can be seen from Fig. 9.26, the plasma membrane is in close and continuous contact with the cell wall. Mitochondria, Golgi cisternae with their associated vesicles, and plastids, some of them with starch and several amoeboid, are abundant during this phase of the differentiation.

The helical cavities in compression wood were first mentioned by Sanio in 1873. Hartig described and illustrated them with unrivaled accuracy in 1901 (Figs. 2.7 and 2.8). A few years later, Gothan (1905) considered their possible origin. He believed that the helical cavities were created when sapwood was transformed into heartwood, where the dryer conditions caused the cell wall to split. The fallacy of this interpretation was soon demonstrated by Krieg (1907),

9.6.1.3 Formation of the S$_2$ Layer

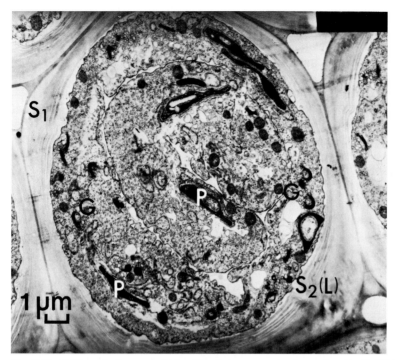

Fig. 9.25. Differentiating tracheid in *Abies balsamea* at the time when the S$_2$ (L) layer was being formed. The inner portion of the cytoplasm is seemingly separated from the outer and is folded. The dense cytoplasm contains many Golgi bodies (*G*), mitochondria (*M*), and plastids (*P*). The plasma membrane is throughout in close contact with the lamellar wall. Transverse section. TEM. (Timell 1979b)

who pointed out that the spiral striations in compression wood can be observed in green wood and not only in heartwood but also in sapwood. He compared the striations with the spiral thickenings present in genera such as *Pseudotsuga* and *Taxus* and expressed as his opinion that both are "products of the living protoplasm". Krieg (1907) emphasized that a clear distinction must be made between the regular spiral striations and the larger, irregular checks now known to be drying artifacts (Chap. 4.4.5.4.8).

Onaka (1949) suggested that the longitudinal expansion that compression wood undergoes in a living tree probably is caused by the swelling that occurs when the tissue becomes lignified. He also proposed that, as the tracheid wall thickens on lignification, the wall substance becomes compressed in the tangential direction. This force, together with the expansion along the grain, could cause cracks to appear in the wall, proceeding in a radial direction along the microfibrils. The helical cavities in compression wood are accordingly of schizogenous origin, the first time this seems to have been suggested.

Fifteen years later, Wardrop and Davies (1964) reported their observations on the differentiation of compression wood in *Actinostrobus pyramidalis* and *Pinus radiata*. They found that the formation of the S$_2$ layer was associated with striking cytological changes. A large number of small vesicles appeared at this stage, many of them evidently derived from the Golgi cisternae, and mitochondria, plastids, and amyloplasts could be observed. The cells were no longer vacuolated. No plasma membrane could be recognized until the very end of differentiation, when it suddenly reappeared and could be seen to penetrate into the helical cavities, which had been formed in the meantime. In some cells, a membrane resembling the plasma membrane was located at some distance from the developing wall. There were numerous vesicles in the region between this membrane and the wall. Wardrop and Davies (1964) found these observations to be quite different from those made with developing normal wood, but they rejected the notion that some of them could have been artifacts resulting from inadequate fixation. It should be noted that they used permanganate as a fixative, having concluded that use of osmium tetroxide offered no advantage. They found that fixation was extremely variable and experienced difficulties with the embedding of the fixed tissues. During most of the period when the helical cavities were being developed, no plasma membrane appeared to be present, and they suggested that the two phenomena were probably related.

As to the origin of the helical cavities, Wardrop and Davies (1964) considered two possibilities, namely either that the helical ribs were produced by the cytoplasm or that fissures developed in the inner portion of the S_2 layer. It is well known how helical thickenings are deposited in protoxylem cells by accumulation of wall material at regular intervals along the primary wall. As pointed out by Wardrop and Davies and demonstrated by Hepton et al. (1956), such thickenings consist of concentrically arranged lamellae, giving a cross section such as that shown in Fig. 9.26.1. Wardrop and Davies observed that in the helical ribs in compression wood, the lamellae were oriented parallel to the surface of the lumen, as indicated in Fig. 9.26.2, suggesting a schizogenous origin of the cavities. Noting that the fissures were always parallel with the cellulose microfibrils, Wardrop and Davies concluded that the cavities must have arisen as a result of a tangential contraction of the originally continuous S_2 lamellae in a direction perpendicular to that of the microfibrils. They added that this process might entail extrusion of water into the cell.

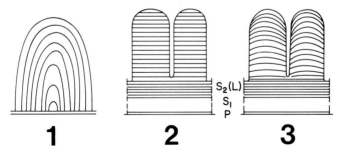

Fig. 9.26. Diagrammatic representation of the pattern of lamellation in helical protoxylem thickenings (**1**) and in the helical ribs of compression wood tracheids (**2** and **3**). In **2** all lamellae are parallel with the lumen surface, while in **3** the later lamellae are curved. (**1** and **2** redrawn from Wardrop and Davies 1964)

9.6.1.3 Formation of the S_2 Layer

In a simultaneous investigation on the formation of compression wood in *Pinus sylvestris*, Casperson and Zinsser (1965) made observations similar to those reported by Wardrop and Davies (1964). The fixative was osmium tetroxide-potassium bichromate, but, as pointed out by the investigators, the protoplast was often poorly fixed. In some of the published electron micrographs the protoplasm is totally absent, but in a few it is at least partly preserved, with the plasma membrane located at a short distance from and parallel to the developing wall. Unlike Wardrop and Davies, Casperson and Zinsser concluded that the helical ribs were formed as a result of regularly interrupted deposition of wall material. The wall accordingly grew in the form of individual ribs. Some of them became thicker and others became thinner toward the lumen, giving an impression of inward branching of the cavities.

The different interpretations offered by Wardrop and Davies (1964) on the one hand and by Casperson and Zinsser (1965) on the other prompted Kutscha (1968) and Côté et al. (1968b) to subject this problem to a renewed investigation. This electron microscopic study was concerned with the differentiation of the xylem in both normal and compression woods of *Abies balsamea*. Similar trees from the same location were later used by Kutscha and Gray (1972), Kutscha et al. (1975), and Timell (1979a, b, 1980) for collection of compression-wood tissues. Kutscha (1968) did not limit himself to fixation with permanganate, as erroneously claimed by Fujita et al. (1973), but also used the superior glutaraldehyde-osmium tetroxide fixative.

In an earlier report (Anonymous 1963) it had been suggested that cavity formation in compression wood might be related to the folding of the cell wall lamellae referred to previously. This folding was initiated in the S_1 layer and later continued throughout S_2. Kutscha (1968) observed that recently deposited lamellae in secondary walls tended to be folded in both normal and compression woods, but more so in the latter tissue. These irregular folds were oriented perpendicularly to the orientation of the microfibrils, whereas the helical cavities were always parallel with the latter. There was accordingly no relation between the folds and the cavities.

Incomplete fixation makes it impossible to draw many conclusions regarding the cytological events in Kutscha's material. Neither of the two fixatives used had served to preserve the tracheid protoplast intact and especially not the permanganate. As usual, better results were obtained with ray cells. The plasma membrane was either withdrawn from the developing wall or could not be observed at all. Golgi bodies producing vesicles were present and also microtubules which were parallel with the microfibrils in the adjacent wall.

The developing S_1 and S_2 layers both consisted of convoluted, loosely arranged lamellae which only gradually became more compacted, as can be seen in Figs. 9.27 and 9.28. Although most of the S_2 layer had been deposited in the two tracheids shown in Fig. 9.28, helical cavities are completely absent in both. Sometimes, at a somewhat later stage in differentiation, one tracheid containing no cavities could be observed next to another with well-developed cavities, as shown in Fig. 9.29. In the cell to the left, helical cavities are fully developed and only lignification of the ribs remains. In the cell to the right, by contrast, the wall still consists of continuous, folded lamellae, although about three quarters

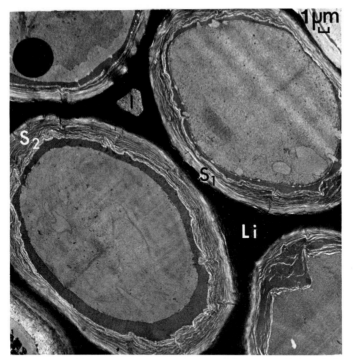

Fig. 9.27. Differentiating tracheids in compression wood of *Abies balsamea* at the time of formation of the outer portion of S_2. The lamellae are more compacted and less folded in S_1 than in the S_2 layer. Lignin (*Li*) has been deposited in the middle lamella and in the outer part of the S_1 layer. *I* intercellular space. Transverse section fixed with potassium permanganate and stained with lead citrate. TEM. (Kutschka 1968, Côté et al. 1968b)

of S_2 had been formed. Since the cavities usually extend over two thirds of the total width of the S_2 layer, they must have originated here by fission to at least one half of their total, final depth.

Imagawa (personal communication, 1974) made similar observations when examining differentiating compression wood of *Larix leptolepis* in ultraviolet light. As can be seen in Fig. 9.30, tracheids with S_2 fissured by helical cavities occurred next to cells with a solid secondary wall of the same width.

Kutscha (1968) (Côté et al. 1968b) could follow the sequence of cavity formation by examining radial and transverse sections of tissues at different stages of differentiation. After many continuous wall lamellae had been deposited, an initial contraction seemed to occur in the plane of the lamellae, perpendicular to the orientation of the microfibrils, as shown in Fig. 9.31. On continued contraction, the cavities progressed further into the wall. During this phase, the plasma membrane was either located at a distance from the developing wall or could not be observed at all, as noted previously by Wardrop and Davies (1964). The finding by these investigators that the plasma membrane penetrated into the cavities at the completion of differentiation could not be confirmed. The initial helical ribs created on contraction of the wall grew cen-

9.6.1.3 Formation of the S_2 Layer

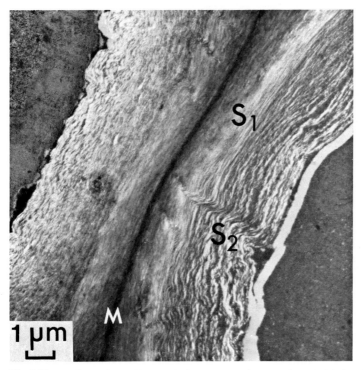

Fig. 9.28. Portions of two differentiating compression-wood tracheids in *Abies balsamea*. Both the S_1 and the S_2 layers are moderately folded. Despite the considerable thickness of the wall, there is no sign of any helical cavities. The lamellae are more compacted in S_1 than in S_2. *M* middle lamella. Transverse section. TEM. (Courtesy of N. P. Kutscha)

tripetally toward the center of the lumen by deposition of additional wall material on their top. The lamellae formed during this final, maturation stage, as a result were no longer parallel with the surface of the lumen but were arranged as indicated in Fig. 9.26.3. Strands of microfibrils were seen to connect the developing ribs throughout (Fig. 9.31), thus offering further evidence for the schizogenous origin of the cavities. It should perhaps be noted here that these connecting strands could possibly have been drying artifacts.

If the helical cavities are a result of a tangential contraction of the lamellae in S_2, as suggested by Wardrop and Davies (1964) and by Kutscha (1968), the cause of this contraction becomes of interest. Wardrop and Davies vaguely mentioned the possibility of a change in physical texture of the secondary wall. One such change has been considered by Boyd (1973b), namely lignification, which he believes causes a contraction of the inner part of S_2. Boyd's treatment is based on his earlier observations (Boyd 1950, 1972, 1973a) on shrinkage, lignification, and stresses in normal and compression woods (Chaps. 7.3.3 and 17.4). Strangely, little or no mention is made of previous investigations on the formation of the helical cavities in compression wood, and available information concerning the distribution of lignin in this wood is not considered.

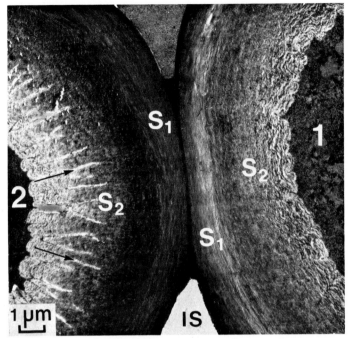

Fig. 9.29. Two adjacent tracheids at different stages of differentiation in compression wood of *Abies balsamea*. The inner portion of the S_2 layer in tracheid *1* is folded and lacks cavities. Tracheid *2* contains fully developed cavities (*arrows*) and ridges. Note that the inner part of the wall in this cell is still unlignified. *IS* intercellular space. Transverse section fixed with permanganate and stained with lead citrate. TEM. (Kutscha 1968)

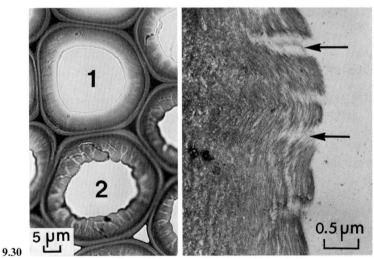

Fig. 9.30. Differentiating compression wood tracheids in *Larix leptolepis*. The younger tracheid *1* has a solid S_2 layer while the more mature tracheid *2* contains helical cavities. Transverse section. UVLM. (Imagawa, personal communication 1974)

Fig. 9.31. Opening of cavities (*arrows*) in a compression wood tracheid in *Abies balsamea*. Radial section fixed with glutaraldehyde-osmium tetroxide. TEM. (Kutscha 1968, Côté et al. 1968b)

9.6.1.3 Formation of the S_2 Layer

The key assumption in Boyd's approach is that lignification causes the same dimensional changes in the different tracheid wall layers as does water. Boyd (1973b) assumes that lignin and water occupy the same sites in the woody cell wall. Wood swells in water, and similarly it has been found that lignification is accompanied by swelling (Frey 1926, Preston and Middlebrook 1949). A complicating factor with lignification is, however, that it proceeds gradually from the primary wall toward the lumen. To obtain a measure of the effect of lignification on the dimensions of the tracheid wall layers, Boyd applied the theory for shrinkage of wood on drying from water developed by Barber and Meylan (1964) (Chap. 7.3.3.1). The later contribution by Barber (1968) was not deemed relevant, notwithstanding the curved wall in compression wood tracheids. The theory requires information concerning the ratios of the modulus of elasticity of the cellulose microfibrils to the modulus of rigidity of the matrix material, that is, hemicelluloses and lignin. With some approximations, Boyd (1973b) arrived at the following moduli ratios for the different layers in compression wood tracheids, namely for the middle lamella-primary wall 5−10, for the S_1 layer 50, for the $S_2(L)$ layer 10−20, and for the inner S_2 again 50. The two values for M + P and $S_2(L)$ do not appear unreasonable considering the similar lignin content of these two regions (Goring and Wood 1971). The value is about 50 for the entire secondary wall in normal wood tracheids. Using these ratios, Boyd could calculate the strain ratios in length, width, and wall thickness of the cells with the aid of the relationships developed by Barber and Meylan. The results obtained indicated that circumferentially there would be a slight expansion on lignification in the M + P and the S_1 layer, a very small contraction in $S_2(L)$, and a substantial contraction in the inner portion of the S_2 layer. In the last two layers the microfibril angle was assumed to be 45°−50°. The transition from a slight expansion to a considerable contraction of the inner S_2 would be sharp and occur at microfibril angles between 45° and 50°. For all angles there should be a significant increase in cell wall thickness on lignification. All wall layers except the inner S_2 would be expected to expand noticeably in the longitudinal direction. The inner part of S_2, on the other hand, would elongate only moderately and only at angles larger than 45°.

As a result of the overall contraction of the inner S_2, strong transverse tensile forces must develop in this layer as lignification proceeds. The development of transverse splits would accordingly be predicted. Obviously, the strong covalent bonds linking the glucose residues together in the cellulose chains are not broken. Instead, the weaker hydrogen bonds binding the cellulose chains together laterally are ruptured, that is, the splits develop at an angle perpendicular to the direction of the microfibrils. Boyd argues that the first fissures would occur in small regions where bonding is weak. These would subsequently tend to grow, and long, continuous checks running in the same direction as the microfibrils would be produced. Not yet lignified, inner portions of S_2 may also rupture as a result of strain induced by lignification of adjacent regions. Boyd suggests that new carbohydrate material would probably be deposited onto the ribs thus formed. When the ribs finally become lignified, one would expect further opening of the already existing fissures. The helical cavities would accordingly tend to become wider toward the lumen. Boyd's (1973b) concept of the origin of the helical cavities in compression wood is based on the assumption that lignification of the inner portion of the S_2 layer precedes or at least coincides with the appearance of the helical cavities. Whether this is really the case will be discussed after the process of lignification has been described in Sect. 9.6.1.4.2.

In the first of a series of investigations by Harada and his co-workers on the formation of compression wood, Fujita et al. (1973) subjected the development of the helical cavities and ridges to a renewed examination. The differentiation of the tracheids in compression wood of *Cryptomeria japonica* was studied by polarizing microscopy and transmission and scanning electron microscopy. In order to obtain truly transverse sections of the helical cavities and ridges, the investigators sectioned their wood specimens at an angle of 45°, a technique originally recommended by Krieg (1907) and later used also by Casperson (1959). The appearance of the orthogonal and parallel sides in a compression-wood tracheid in such a section is shown in Fig. 4.71. According to Fujita et al. (1973) transverse sections of compression wood, as ordinarily used, can easily be misinterpreted because of the inclination of the helical cavities.

Fujita et al. (1973) (Fujita 1981) found that fixation with potassium permanganate caused extensive swelling and convolution of the developing tracheid wall, whereas use of glutaraldehyde followed by osmium tetroxide preserved both the wall and the cytoplasm. They observed that the helical cavities were present at a vary early stage of the deposition of the inner part of the S_2 layer, as shown in Fig. 9.32. The ridges originally formed gradually grew toward the inside. During the final phase some of them increased in size, protruding into the lumen, while others remained small and gave the impression of being suppressed. The entire sequence of events as visualized by Fujita et al. (1973) is illustrated in Fig. 9.33. Contrary to the earlier results of Wardrop and Davies (1964), Fujita et al. could observe the plasma membrane throughout the development of the secondary wall. It followed closely the surface of the wall and penetrated into the cavities. A fission of the wall lamellae as a result of a tangential contraction of the inner part of S_2 could not be observed.

Harada and his co-workers concluded from their findings that the helical cavities in compression wood tracheids are not schizogenous in origin but are formed as a consequence of an interrupted deposition of cell wall substance. The earlier, contrary results obtained by Wardrop and Davies (1964) and by Côté et al. (1968b) were attributed to unsatisfactory fixation and sectioning techniques.

In continuation of their investigations on the formation of the tracheids in compression wood of *Cryptomeria japonica*, Fujita et al. (1978b) noted that the cytoplasm became dense at the onset of the deposition of the S_2 layer. Microtubules were abundant along the plasma membrane, where their orientation was the same as that of the cellulose microfibrils in the adjacent wall. Only a few Golgi bodies could be observed at this stage. The investigators believed that the microtubules were directly involved in the deposition of the microfibrils. At the stage when the S_2 layer rapidly increased in thickness (cells 2, 3, and 4 in Figs. 9.13 and 9.22), Golgi cisternae became numerous, as shown in Fig. 9.34, although their associated vesicles were quite small. The vesicles

9.6.1.3 Formation of the S$_2$ Layer

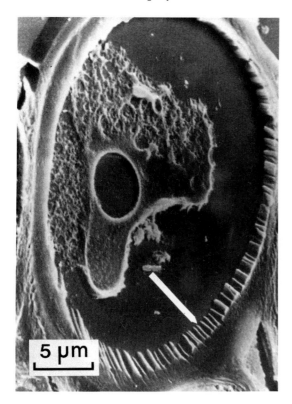

Fig. 9.32. Differentiating compression wood tracheid in *Cryptomeria japonica* at an early stage of the formation of S$_2$, showing the presence of helical cavities and ridges (*arrow*). Oblique (45°) section fixed with glutaraldehyde-osmium tetroxide. SEM. (Fujita et al. 1973)

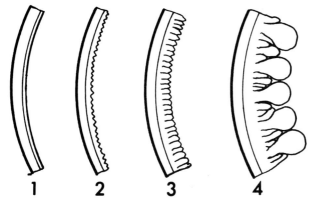

Fig. 9.33. Schematic outline of the formation of the helical cavities and ribs (1 → 4) in a compression wood tracheid. (Fujita et al. 1973)

seemed to increase in size with increasing distance from the cisternae. When the helical ridges were being completed, the cytoplasm contained many Golgi cisternae and large vesicles. The plasma membrane followed closely the outline of the growing ridges and penetrated deeply into the cavities, as can be seen in Fig. 9.35. Microtubules could still be observed, but at this stage they were irregularly distributed, being either clustered together on top of the ridges or lo-

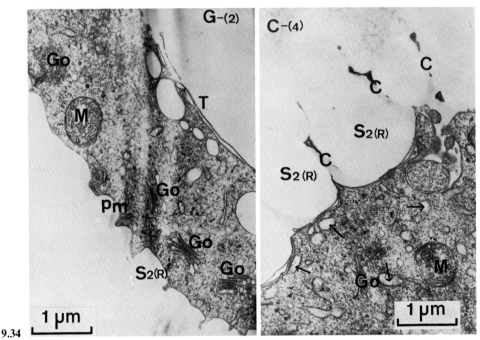

Fig. 9.34. Portion of a differentiating tracheid in *Cryptomeria japonica* at the time of initiation of the helical ribs in S_2 [$S_{2(R)}$]. The plasma membrane (*pm*) adheres closely to the wall. There are numerous Golgi bodies (*Go*) and associated cisternae. *M* mitochondrion; *T* tonoplast. Oblique (45°) section. TEM. (Fujita et al. 1978b)

Fig. 9.35. The same tissue as in Fig. 9.34 at a later stage of differentiation, showing the presence of cytoplasm (*C*) within the cavities and large vesicles (*arrows*). Oblique (45°) section. TEM. (Fujita et al. 1978b)

cated singly within the deep cavities. The electron micrographs published by Fujita et al. (1978b) provide little or no evidence for a possible role of the endoplasmic reticulum, yet the investigators concluded that this organelle was probably involved in the formation of the cell wall hemicelluloses, while the Golgi apparatus was assumed to take part in the subsequent deposition of lignin.

Fujita and Harada (1978), in their autoradiographic study, referred to earlier, found that silver grains were distributed throughout the S_2 at the beginning of the deposition of this layer (cells 1–4 in Fig. 9.22). At a later stage, numerous silver grains were concentrated at the inner surface of the wall, as shown in Fig. 9.36. Still later, when formation of S_2 was almost complete, the grains occurred throughout the tracheid wall. Fujita and Harada concluded from their observations that the S_2 layer first grows by apposition of cellulose microfibrils and that hemicelluloses are later added by intussusception. Inspection of their autoradiograph suggests, however, that the tracheids seen here were all in the process of depositing the $S_2(L)$ layer. If this is correct, not only the last but also

9.6.1.3 Formation of the S_2 Layer

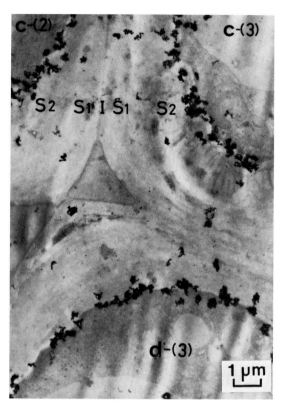

Fig. 9.36. Autoradiograph of differentiating compression wood tracheids in *Cryptomeria japonica* at the time of formation of the inner S_2. Tritiated glucose is being deposited onto the inner surface of this layer. Oblique (45°) section. TEM. (Fujita and Harada 1978)

the first phase in the formation of S_2 must entail deposition of hemicelluloses. The scarcity of microfibrils and the high concentration of galactan in the $S_2(L)$ layer (Chap. 6.3.3) agree with such an interpretation.

Fujita and Harada (1978) noted a lack of radioactivity within the cytoplasm when cellulose was assumed to be added to the wall and concluded that organelles in the cytoplasm were probably not involved in the biosynthesis of the cellulose microfibrils. At the presumed time of formation of the hemicelluloses, on the other hand, radioactive glucose was observed in the cytoplasm, suggesting that these matrix polysaccharides were synthesized in the latter and then transported to the wall. These observations are in agreement with current views of the biosynthesis of cellulose (Brown 1982) and hemicelluloses (Delmer 1977, Ericson and Elbein 1980).

In his study of the formation of compression wood in *Abies balsamea*, Timell (1979b) found that the cytoplasm changed spectacularly at the onset of the formation of the S_2 layer in the tracheids. The cytoplasm, which filled the entire cell, at this stage seemed to be thrown into deep folds, as shown in Fig. 9.25, where lamellae are being deposited to form the outer portion of S_2. Folding of the cytoplasm was seldom observed close to the tracheid tip, as illustrated in Fig. 9.37, obtained at a time when the inner part of S_2 was being formed. Golgi

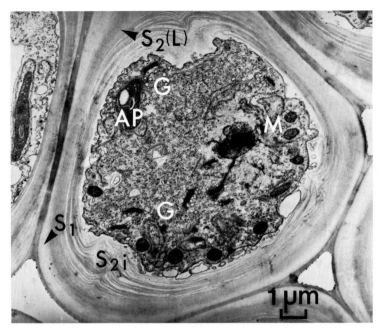

Fig. 9.37. Differentiating compression wood tracheids in *Abies balsamea* at the time of formation of the inner S_2 (S_{2i}). The cytoplasm contains Golgi cisternae (G) and vesicles, mitochondria (M), and amyloplasts (AP). There are no helical cavities in the lamellar, folded cell wall. Transverse section. TEM. (Timell 1979b)

Fig. 9.38. Rupture of the inner S_2 in a tracheid in compression wood of *Abies balsamea* (*arrow*). Transverse section. TEM

bodies and plastids were abundant during this phase. Large vesicles, obviously derived from the Golgi cisternae, were exceedingly numerous. They seemed to be of a uniform size, regardless of their distance from the cisternae. The plasma membrane remained in close contact with the developing wall throughout the deposition of S_2.

As can be seen in Fig. 9.37, the wall portion immediately inside the $S_2(L)$ layer at first consisted of slightly folded, concentric lamellae. No helical cavities

9.6.1.3 Formation of the S₂ Layer

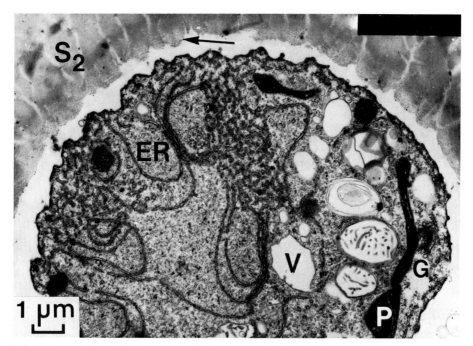

Fig. 9.39. A differentiating tracheid in *Abies balsamea* at the time of completion of S$_2$. The cytoplasm contains an extensive system of endoplasmic reticulum (*ER*), Golgi bodies (*G*), and plastids (*P*). Wall material is being added to the top of the lamellar helical ridges (*arrow*). The separation of plasma membrane and wall is a fixation artifact. *V* vacuole. Transverse section. TEM. (Timell 1979b)

or ridges could be observed at this stage. Soon, however, splits seemed to develop within the lamellae (Fig. 9.38), exactly as described previously by Wardrop and Davies (1964) and by Kutscha (1968). These fissures continued to grow outward until they reached the S$_2$(L) layer. After the helical ridges had been formed, new wall material was added to their top from the cytoplasm which during this phase contained an extensive system of endoplasmic reticulum, as can be seen in Fig. 9.39. Some ridges now grew inward more rapidly than others and many of them protruded into the lumen. The plasma membrane and cytoplasm penetrated partly into the cavities. The protoplasm continued to occupy the entire cell in the maturing or mature tracheids, as shown in Fig. 9.40. During this final phase of the differentiation, Golgi cisternae and mitochondria were abundant.

The folding of the cytoplasm observed by Timell (1979b) has not been reported previously and could possibly have been an artifact. Both Wardrop and Davies (1964) and Timell noted the absence of a large, central vacuole while the S$_2$ layer was being formed. The evidence presented by Fujita et al. (1978a, b) is inconclusive on this point. There may be a difference in this respect between normal and compression woods, since the cytoplasm appears to be parietal during formation of both S$_1$ and S$_2$ in normal tracheids (Cronshaw

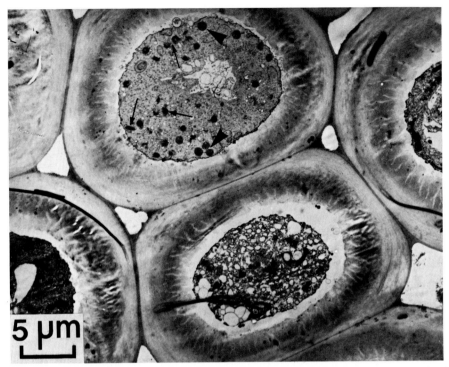

Fig. 9.40. Mature tracheids in compression wood of *Abies balsamea*. The cells are completely filled with necrotic cytoplasm, containing mitochondria (*arrowheads*) and Golgi bodies (*small arrows*). Transverse section. TEM. (Timell 1979b)

and Wardrop 1964, Murmanis and Sachs 1969, Tsuda 1975b). The S_2 layer is lamellar in compression wood tracheids. Kerr and Goring (1975, 1977) and Ruel et al. (1978) have shown that this layer consists of lamellae with an interlamellar distance of 7 – 8 nm in normal wood of spruce and fir.

Timell (1979b) noted that in his tissues the cytoplasm was denser and also filled the cells more completely than in normal wood during the deposition of the S_2 layer. His impression was that of a high cellular activity at the time when these tissues had been collected in late July. This agrees with the findings of Kutscha et al. (1975), discussed in Sect. 9.7.2.

Waterkeyn et al. (1982), who believe that the 1,3-glucan (callose, laricinan) in compression wood is located within the helical ribs in the form of thin, helical bands (Chap. 6.3.3), have recently suggested that the helical cavities are formed by a contraction of these bands of callose at the time of the death of the cell. They do not state how this contraction occurs, except to mention that it is physical in nature. The suggestion that it should occur when the protoplasm ceases to function is difficult to reconcile with the facts that the first signs of openings in the inner S_2 occur while this layer is still being formed and before it has become lignified. It is also clear that cell wall material is added to the top of the helical ribs for some time after the latter have developed (Fig. 9.39).

9.6.1.4 Lignification

9.6.1.4.1 Lignification of Normal Wood. Differentiation of wood cells has traditionally been divided into four phases, namely cell division, cell enlargement, cell wall thickening, and lignification. The fourth of these phases, that involving encrustation of the cell wall with lignin, is in most cases followed by the death of the cell. It has been known ever since the pioneering investigations of Sanio (1873) that the deposition of lignin lags behind the formation of the polysaccharide constituents of the wall.

The biosynthesis of lignin by dehydrogenation polymerization of cinnamyl alcohols was established in a series of brilliant investigations by Freudenberg and his co-workers (Freudenberg 1968). According to Freudenberg, lignin precursors are probably formed in the cambial zone and are transported centripetally from this region into the differentiating xylem where they polymerize. While there is evidence for this view, there is now much more compelling reason to believe that the lignin precursors arise within each differentiating cell, an interpretation that has long been championed by Wardrop (1957, 1964, 1965, 1971, 1976). Many facts argue in favor of an intracellular lignification. The helical thickenings in primary xylem are known to be selectively lignified (Wardrop 1957, Hepler et al. 1970); a guaiacyl-p-hydroxyphenyl lignin is present in the middle lamella but a guaiacyl lignin in the secondary wall of the conifers (Whiting and Goring 1982); a guaiacyl lignin occurs in the primary but a guaiacyl-syringyl lignin in the secondary xylem of arboreal angiosperms (Wardrop and Bland 1959); in hardwoods the middle lamella and secondary wall of the vessels have a guaiacyl lignin, while the wall of the fibers and the ray parenchyma cells contain a syringyl lignin (Fergus and Goring 1970a,b, Musha and Goring 1975a,b, Gromov et al. 1977, Hardell et al. 1980); the S_2 layer in compression-wood tracheids is unevenly lignified, and in tension wood fibers the innermost gelatinous layer is almost invariably unlignified; there is no lignin deposited in the torus of the bordered pits of the Pinaceae (Thomas 1975); in normal wood of some softwoods and hardwoods and in opposite wood the S_3 layer has a higher concentration of lignin than the layers S_1 and S_2; in the sapwood of the diploxylon pines the ray parenchyma cells remain unlignified (Balatinecz and Kennedy 1967); at the end of the growing season, each cell evidently ceases to lignify independently of its neighbors (Wardrop 1957, Wardrop and Bland 1959), and where a lignified cell is next to an unlignified one, the middle lamella between them remains unlignified (Wardrop 1976).

Compared to the enormous amount of research that has been devoted to the structure and biosynthesis of lignin, the process of lignification has received scant attention. Whatever the reason for this situation, it cannot reside in technical difficulties, for of all wood components none is more readily visualized than lignin, as described in Chap. 6.4.1.

Sanio (1873) noted that lignin first appeared within or close to the primary wall at the corners of softwood tracheids, an observation later confirmed by Kerr and Bailey (1934). Wardrop (1957) and Wardrop and Bland (1959) studied the lignification of the xylem in *Pinus radiata*, using ultraviolet and polarized light microscopy. In later investigations, tissues were stained with permanganate and observed in the electron microscope (Wardrop 1965, 1971, 1976). Wardrop found that lignification began at the time when deposition of the secondary wall was commencing. Lignin could first be detected in the primary wall at the cell corners, as had originally been observed by Sanio (1873). Lignification progressed from this region into the middle lamella and through the remainder of the primary wall, first along the tangential and later along the radial walls. By the time the S_1 layer had been completed, lignin deposition was still confined to the cell corners. Lignification subsequently proceeded continuously from the outer toward the inner portions of the secondary wall, always lagging behind the deposition of the wall polysaccharides. Throughout the lignification of the secondary wall, lignin continued to be formed in the intercellular region. Later observations indicated that lignin first developed within the middle lamella and sometimes even in the outer part of the S_1 layer before the primary wall became lignified (Wardrop 1971). Both initiation and termination of the lignification occurred cell by cell, the entire process evidently being controlled by each individual cell.

In an autoradiographic study, Saleh et al. (1967) administered tritiated ferulic acid, a lignin precursor, to rooted branches of *Populus trichocarpa*. In agreement with Wardrop's (1957) earlier observations, lignification was found to begin in the primary wall at the cell corners. In contrast to the results obtained by Wardrop, deposition of lignin first extended into the middle lamella along the radial rather than along the tangential walls. Lignin continued to be formed in the intercellular region until the end of the lignification process, as observed previously by Wardrop and Bland (1959).

Kutscha and Schwarzmann (1975) (Schwarzmann and Kutscha 1973) subjected the lignification of the xylem in *Abies balsamea* to a thorough investigation, using potassium permanganate for visualizing the lignin in the electron microscope. Unlike previous investigators, they found that lignification began in the middle lamella between the pit borders of adjacent tracheids. Shortly afterward, lignin formation commenced in the initial pit borders and, at the same time at the tracheid cell corner farthest from the cambial zone, but it could not be decided whether this occurred in the outer portion of the primary wall or in the intercellular region closest to this wall. As lignification proceeded along the cell corner, lignin was also deposited in the outer part of the S_1 layer, while the primary wall between the middle lamella and the outer S_1 remained unlignified. The entire S_1 layer at the cell corners was fully lignified before the process began to extend along the tracheid walls.

Among the four cell walls, the tangential wall farthest from the cambial zone was the first to lignify, followed by the two radial walls. The tangential wall closest to the cambium began to lignify last. These observations could explain the contradictory results obtained by Wardrop (1957) and by Saleh et al. (1967). This sequence of lignification was modified in tracheids adjacent to a ray parenchyma cell. In such cells the radial wall next to the ray cell was the first to become lignified. The tangential wall most remote from the cambial region lignified next, followed by the other radial wall and, finally, the tangential wall closest to the cambium. Wardrop (1957) (Wardrop and Bland 1959) found in ring-barking experiments that tracheid walls next to a ray parenchyma cell remained unlignified. The observations reported by Kutscha and Schwarzmann (1975) must represent the normal situation, for tracheids are known to survive better the higher their incidence of contact with ray parenchyma cells.

Kutscha and Schwarzmann (1975) noted that lignification of the S_2 layer was initiated before the primary wall was fully lignified. Lignification of the secondary wall progressed gradually toward the lumen, lagging behind formation of the polysaccharide components. When the warty layer was deposited, the S_3 was still not fully lignified. In bordered pits, lignification proceeded toward the apertures. The authors' conclusion that the torus also became lignified has not been supported by other findings (Thomas 1975). The pit membrane remained unlignified.

Some of the results reported by Kutscha and Schwarzmann (1975) were later confirmed by Imagawa et al. (1976), who studied the lignification of the tracheids in *Larix leptolepis*, using ultraviolet microscopy. They found that lignification commenced in the middle lamella and simultaneously at all cell corners, except that cell corners adjacent to a ray parenchyma cell were always the first to lignify. Lignification began in the pit border at the same time. The primary wall at the cell corners remained unlignified for some time.

The process of lignification in normal tracheids of *Pinus thunbergii* was studied by Takabe et al. (1981 b). Lignin deposition was first observed at the cell corners and in the pit regions, after which it began in the intercellular region, where it continued until the time of formation of the S_3 layer. Lignification proceeded from the outside toward the inside of the secondary wall. The rate of encrustation increased during the last phase, when lignin was rapidly deposited within all cell wall layers.

Saka and Thomas (1982) applied the SEM-EDXA technique (Chap. 6.4.1) to a study of the course of lignin deposition in *Pinus taeda* tracheids. Confirming previous observations, they found that lignification began at the cell corners within the middle lamella-primary wall region. Lignification was rapid in this zone. When it was half-way completed, the secondary wall commenced to lignify more slowly, with lignin deposition proceeding gradually through the S_1, S_2, and S_3 layers in this order. It is evident from this brief review that the sequence of lignification in normal conifer wood is now relatively well established, although some details are still uncertain.

9.6.1.4.2 Lignification of Compression Wood. What was probably the first investigation of the lignification of any tree tissue was reported in 1873 by Sanio in his study of compression wood from the lower side of *Pinus sylvestris* branches. Sanio concluded from his careful observations that lignification was initiated at the cell corners of the compression wood tracheids, and that it gradually progressed through the wall toward the lumen. His observations seem to have been overlooked by all later investigators (Timell 1980 c). Wardrop and Bland (1961), in a study of the lignification of compression wood in *Pinus radiata*, noted that formation of lignin began at the end of the period of surface enlargement of the tracheids. Compared to normal wood, lignification of the S_1 layer was considerably reduced.

Wardrop and Davies (1964), in their investigation on the differentiation of compression wood, did not examine the lignification of their tissues. Casperson and Zinsser (1965) mention only that by the time lignin begins to form in the outer portion of the S_2 layer, the helical cavities are already fully developed. In his histochemical study of the differentiation of compression wood in *Pinus radiata*, Scurfield (1967) examined the progress of lignification, using different stains for visualizing the lignin. He found that the first lignin was deposited at the apices of intercellular spaces, where the walls of adjacent tracheids approached one another. The lignin began to penetrate the S_2 layer either in this location or at the cell corners. The protoplasm had disappeared before the entire S_2 layer had become lignified. As indicated by Scurfield (1967) himself, this disappearing of the protoplasm from still differentiating cells could have been a fixation artifact.

Kutscha (1968) briefly studied the lignification of the tracheids in compression wood of *Abies balsamea*. Tissues fixed with potassium permanganate were examined in the electron microscope. As had been noted by Wardrop and Bland (1961), lignin appeared at a stage when cell enlargement had ceased and formation of the S_1 layer had commenced. Lignin was first observed at the cell corners, where three or four of the rounded tracheids were meeting. It seemed to be deposited here in the intercellular region and at the same time in the outermost portion of the S_1 layer. The primary wall remained unlignified, as can be seen in Fig. 9.41. A later stage is shown in Fig. 9.27. The primary wall and most of S_1 are now lignified. The outer part of S_2 is being deposited in the form of loosely compacted, folded lamellae. Figure 9.29 illustrates a still later stage. In the maturing tracheid to the left, the primary wall and the S_1 and the $S_2(L)$ layers had become lignified, and lignification seems to be extending into the inner part of S_2. The helical ridges are still unlignified, as had previously been observed by Casperson and Zinsser (1965). Kutscha (1968) concluded from his observations that lignification proceeded in the same manner in compression wood as in normal wood. A more detailed investigation of lignification in *Abies balsamea* was later carried out by Kutscha and Schwarzmann (1975), as already mentioned, but this did not include observations on compression wood.

The progress of lignification in compression wood can be followed in light micrographs published by Kutscha and Gray (1972), who used some of the specimens of *Abies balsamea* previously collected by Kutscha (1968). Best results were obtained by staining with 2-thiobarbituric acid, phloroglucinol, and safranin-Celestine Blue B, respectively. Careful inspection of these micrographs reveals that the first traces of lignin appear in the second row of the radially expanded tracheids. The lignin seems to be deposited close to the primary wall at the cell corners. From here it spreads into the middle lamella and at the same time across the S_1 layer. Lignin is laid down in the inner part of the S_2 layer beginning with the 6th cell row from the cambium. By the 10th or 11th row, lignification is apparently complete. Lignification of the ray cells lags behind that of the tracheids.

A close examination of Kutscha's original light micrographs indicates that lignification was initiated in this compression wood in the same manner as was later reported by Kutscha and Schwarzmann (1975) for normal wood. The first

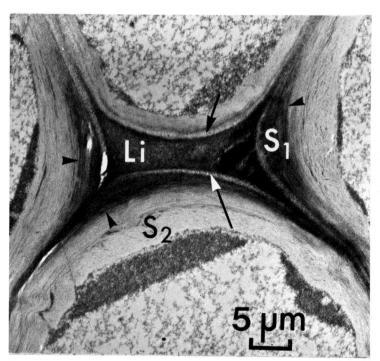

Fig. 9.41. Compression wood tracheid in *Abies balsamea* at the time of initiation of S_2. The middle lamella and the outer portion of S_1 (*arrowheads*) have been lignified (*Li*) at the cell corners but not the primary wall (*arrows*). Transverse section fixed with potassium permanganate and stained with lead citrate. TEM. (Kutscha 1968)

traces of lignin can be seen at the cell corners most distant from the cambium. Lignification then progressed to the other cell corners and subsequently to the tangential wall farthest from the cambial zone. The remaining walls seemed to become lignified more rapidly than in normal tracheids, perhaps because of the rounded outline of the cells.

The most extensive investigations on the process of lignification in compression wood are those by Harada and his co-workers. Fujita et al. (1978a, b) found that in compression wood of *Cryptomeria japonica*, lignification took place in two separate stages. During the first phase, lignin accumulated in the middle lamella and at the cell corners at a time when the S_1 layer was being deposited. The second phase involved lignification of the entire secondary wall and occurred when formation of the S_2 layer had been completed. At this stage, numerous, large vesicles appeared in the cytoplasm, presumably derived from Golgi cisternae and assumed by the investigators to contain lignin precursors. They could be seen to fuse with the plasma membrane (Fig. 9.42). Their content was removed on treatment with acid chlorite, a delignifying agent. The investigators suggested that the small vesicles observed by Wardrop (1965, 1971) to accumulate between the wall and the plasma membrane might be particles of lignin precursors. The fact that these vesicles were stained by potassium per-

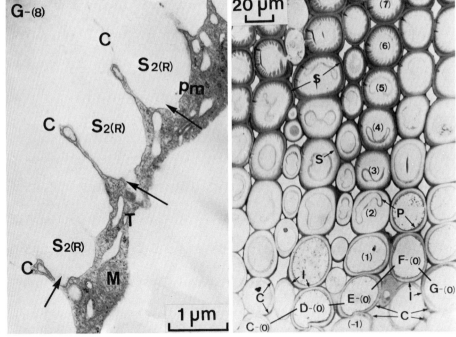

Fig. 9.42. Differentiating compression wood tracheid in *Cryptomeria japonica* at the time when the S_2 layer was being lignified. Large vesicles in the thin, parietal cytoplasm seem to be fusing with the plasma membrane (p_m) (*arrows*). Note the presence of cytoplasm deep within the helical cavities (*C*). $S_2(R)$ helical ribs in S_2. Oblique (45°) section. TEM. (Fujita et al. 1978b)

Fig. 9.43. Differentiating compression wood tracheids in *Cryptomeria japonica* photographed in ultraviolet light at 280 nm. Lignin is deposited at the cell corners in cells (*−1*) to (*2*). Lignification of the secondary wall begins at cell (*3*). Transverse section. (Fujita et al. 1978a)

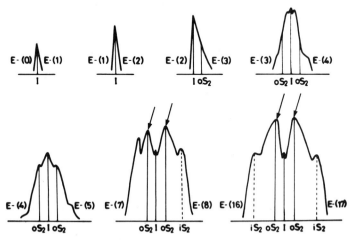

Fig. 9.44. Densitometer tracings across double cell walls in the tracheids shown in Fig. 9.43. The lignification of the middle lamella is completed in cells *E-3* to *E-4*. Note that beginning with cells *E-7* to *E-8* the concentration of lignin in the S_2 (L) layer (S_2) (*arrows*) exceeds that in the intercellular region (*I*). (Fujita et al. 1978a)

9.6.1.4 Lignification

manganate cannot, however, be regarded as very strong evidence in favor of this suggestion, and the presence of the vesicles outside the plasma membrane may have been a fixation artifact.

The course of lignification in compression wood of *Cryptomeria japonica* was studied in ultraviolet light by Fujita et al. (1978a). A transverse section of differentiating tracheids is shown in Fig. 9.43. In file E in this section, lignification of the intercellular space and the cell corner was first observed in the cell marked (-1) and continued until cell (3). The $S_2(L)$ layer became lignified in cells (3) to (7). The investigators designated the former lignin as "primary wall lignin" and the latter as "secondary wall lignin", emphasizing that these two lignins were deposited independently of each other.

The progress of lignification from cell E-0 to E-17 in Fig. 9.43 can be followed with the aid of the densitometer tracings shown in Fig. 9.44. Lignification of the $S_2(L)$ layer was initiated in cell E-3, and shortly thereafter it began in the inner part of S_2 in E-4. The investigators noted that lignin entered the middle lamella regions both before and after the S_2 layer had been formed. Inspection of Fig. 9.44 reveals, however, that deposition of intercellular lignin had been completed in cells E-3 to E-4, since the peak height henceforth remained constant. Encrustation of the $S_2(L)$ region seems to have continued until the end while the inner part of S_2 was fully lignified in cell E-8. It is notable that the concentration of lignin in this case is higher in the $S_2(L)$ layer than in the middle lamella. The somewhat higher lignin concentration in the S_2 area close to the lumen might be an artifact, caused by the presence of the warty layer.

In normal wood of *Pinus radiata*, Wardrop and Bland (1959) (Wardrop 1965, 1971) found that lignification of the intercellular region continued unabated while the secondary wall was being lignified. According to Imagawa et al. (1976) lignification of the secondary wall of normal tracheids in *Larix leptolepis* proceeded continuously and gradually. The results reported by Fujita et al. (1978a) indicate that compression wood may be lignified differently. As repeatedly emphasized by these investigators, the deposition of lignin in the intercellular region in compression wood appears to represent a phase distinctly separate from the later lignification of the secondary wall. It is possible that the presence of intercellular spaces between the tracheids in compression wood could be responsible for this difference. The suggestion by Fujita et al. (1978a) that lignification should proceed in the same manner in normal wood and compression wood is not supported by their own findings and those reported by other investigators with normal wood. The matter has recently been briefly discussed by Yumoto et al. (1982a).

The results obtained by Fujita et al. (1978a) were later confirmed by Fujita and Harada (1979) in an autoradiographic study of the lignification of compression-wood tracheids in *Cryptomeria japonica*. Two lignin precursors were administered to the differentiating cells, namely ^{3}H-L-phenylalanine and ^{3}H-ferulic acid. Radioactivity from these precursors first accumulated within the cytoplasm of the cambial cells, as shown in Fig. 9.45.1. At the stage when S_1 was being deposited, radioactive lignin was localized in the middle lamella and at the cell corners (Fig. 9.45.2). During the formation of the S_2 layer, radioactivity was again concentrated within the cytoplasm. Vesicles were abundant in

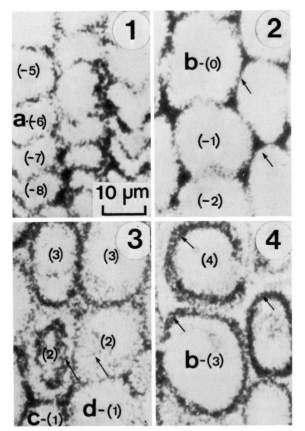

Fig. 9.45. Autoradiograph of differentiating compression wood tracheids in *Cryptomeria japonica*. **1** Cambial cells. **2** Cells depositing S_1 and outer S_2. **3** Cells depositing S_2. **4** Early stage of the lignification of the secondary wall. (Fujita and Harada 1979)

the cytoplasm at this stage, as shown in Fig. 9.45.3. Both the vesicles and the Golgi cisternae seemed to contain ^3H, but it could not be decided whether the tritium was derived from lignin precursors or from protein. Following completion of the S_2 layer, there was a massive incorporation of radioactive lignin into the outer portion of this layer (Fig. 9.45.4).

Kutsuki and Higuchi (1981) determined the activities of five enzymes involved in the biosynthesis of guaiacyl lignin in opposite and compression woods of *Metasequoia glytostroboides* and *Thuja orientalis*, namely phenylalanine ammonia lyase, caffeate 3-*O*-methyltransferase, p-hydroxycinnamate CoA ligase, cinnamyl alcohol dehydrogenase, and peroxidase. The activities of the first four enzymes were 2.6 and 2.8 times higher in compression wood than in opposite wood in these two species. Peroxidase exhibited the same activity in the two types of tissues. Larger amounts of lignin precursors could accordingly be expected to be available in compression wood than in opposite wood, leading to a higher lignin content of the former. Kutsuki and Higuchi (1982) also

9.6.1.4 Lignification

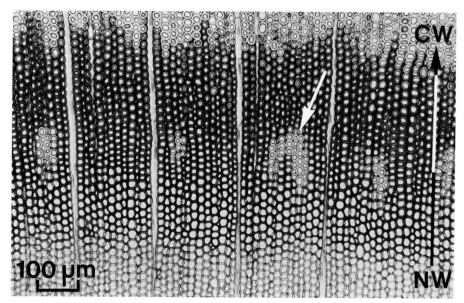

Fig. 9.46. Transition of normal wood to compression wood in *Picea glauca*, showing the presence of several tangentially arranged regions with incompletely lignified tracheids (*arrow*). Transverse section stained with hematoxylin-orange G. LM. (Yumoto et al. 1982a)

compared enzyme activities in the synthesis of p-hydroxyphenyl lignin in opposite and compression woods of the same two species. Phenylalanine ammonia-lyase, p-hydroxycinnamate CoA ligase, cinnamyl alcohol dehydrogenase, peroxidase, and cinnamate 4-hydroxylase all exhibited higher activities in compression wood than in opposite wood. The fact that the activity of caffeate 3-*O*-methyltransferase was much higher in compression wood (Kutsuki and Higuchi 1981) is surprising. The presence of guaiacyl-p-hydroxyphenyl lignin in compression wood must evidently be attributed to an enhanced availability of precursors of the p-hydroxyphenylpropane units. As pointed out by the investigators, the lignin in the compression wood should be of the bulk type (Sarkanen 1971) with a higher than normal proportion of condensed units. This has also been found to be the case (Lee 1968) (Chap. 5.9.2).

In their investigations on the transition from normal wood to compression wood in tilted *Picea glauca* trees, Yumoto et al. (1982a) found that lignification commenced in or near the middle lamella at the cell corners and then proceeded, first along the tangential, and subsequently along the radial walls. In material collected in August, they observed, in a transitional zone between normal and compression woods, tangentially arranged areas with incompletely lignified tracheids, as illustrated in Fig. 9.46. No explanation could be offered for this phenomenon.

9.6.1.5 Formation of the Warty Layer

The nature of the warty layer in compression wood tracheids is discussed in Chap. 4.4.8. Kutscha (1968) found that this layer was formed in the same manner in normal and compression wood tracheids of *Abies balsamea*. The warts in compression wood were apparently formed outside the plasma membrane by dense cytoplasmic material. Mitochondria and Golgi bodies were still present in the cytoplasm at this stage, but the entire protoplasm appeared necrotic.

9.6.1.6 Formation of the Bordered Pits

It is evident from many investigations that the cytoplasm in differentiating xylem is most readily fixed within the pit chamber (Cronshaw and Wardrop 1964, Cronshaw 1965a, b, Wardrop 1965, Murmanis and Sachs 1969b, Timell 1973, 1979b, Tsuda 1975b). The most likely explanation for this is, first, that the cytoplasm probably is protected within the narrow confines of the pit chamber and, second, that it is never vacuolated and therefore readily fixed.

Kutscha (1968) studied the formation of the bordered pits in the tracheids of normal and compression woods of *Abies balsamea*. The pit membrane, the torus and the pit border were formed in the same manner in the two tissues. Lignification began in the intercellular region between two initial pit borders and shortly afterward progressed into the border, as later described for normal fir wood by Kutscha and Schwarzmann (1975).

In his study of the differentiating compression xylem in the same species, Timell (1979b) examined bordered pits at different stages of development. Cross sections of pits at the beginning and at the end of the formation of the S_1 layer in the tracheids are shown in Figs. 9.23 and 9.47, respectively. The dense cytoplasm contained Golgi cisternae with associated vesicles, polysomes, mitochondria, plastids, and amyloplasts. The torus was traversed by plasmodesmata and median cavities. Contrary to the observations of Wardrop (1965), the plasma membrane closely followed the contour of the developing pit border. The initial pit border had been completed before the secondary wall had been initiated, as observed in normal wood by Wardrop and Dadswell (1957) and by Murmanis and Sachs (1969a, b). Formation of the torus commenced at the same time as that of the secondary wall. The general nature of the cytoplasm within the pit chamber suggested an intense cellular activity in this region.

9.6.1.7 Formation of the First-Formed Tracheids

The first five to ten rows of tracheids formed at the beginning of the growing season in compression wood differ in some respects from those formed later, as was first observed by Sanio (1873) and later confirmed by Cieslar (1896) and Hartig (1896). The ultrastructure of these early tracheids is described in Chap. 4.5. According to Côté et al. 1967, they have the angular outline and thin

9.6.1.7 Formation of the First-Formed Tracheids

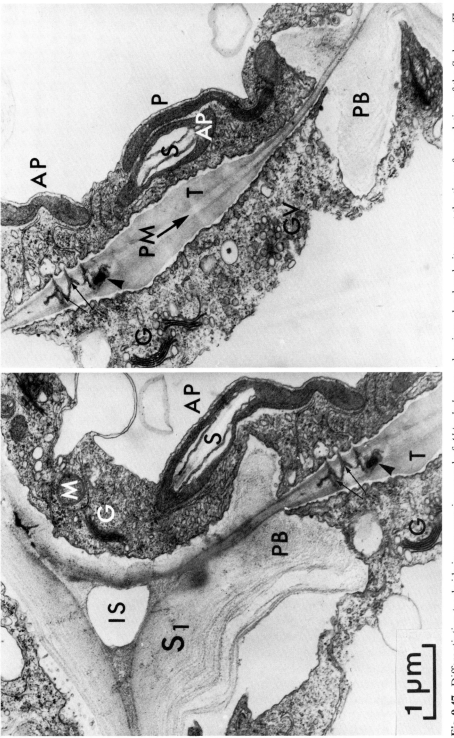

Fig. 9.47. Differentiating tracheids in compression wood of *Abies balsamea*, showing a bordered pit area at the time of completion of the S_1 layer. The cytoplasm contains Golgi bodies (*G*) and vesicles (*GV*), mitochondria (*M*), plastids (*P*), and amyloplasts (*AP*). *PB* pit border; *PM* pit membrane with plasmodesmata (*small arrows*) and a median cavity (*arrowheads*). *IS* intercellular space; *T* torus. Transverse section. TEM. (Timell 1979 b)

wall of normal earlywood tracheids and lack intercellular spaces. In many genera, but apparently not in *Pinus*, helical cavities are usually absent or only poorly developed. In all other respects, however, these cells possess the ultrastructural characteristics of ordinary compression wood tracheids, including a thick S_1 layer, a highly lignified $S_2(L)$ layer, and a complete absence of an S_3 layer. Their chemical composition is that of the tracheids formed later. The unique distribution of lignin typical of ordinary compression wood tracheids is found also in the first-formed cells (Côté et al. 1967, Wood and Goring 1971, Fukazawa 1974, Yoshizawa et al. 1982) (Chap. 6.4.2, Fig. 4.118).

In a study of the formation of the first-formed tracheids in compression wood of *Picea abies*, Höster (1974) attempted to explain why these cells differ from those formed later. The dormant cambial zone consisted of five to six cells which could not be distinguished from one another. When the cambium became active, however, it could be established that the dormant cambial zone contained one overwintering phloem mother cell, one initial cell, one xylem mother cell, two daughter cells and occasionally one xylem tissue cell. Groups of two or four cells were observed in the active cambial zone, as shown in Fig. 9.10. The latter must have consisted of Sanio's four, as pointed out by Timell (1979b).

At the beginning of the growing season, five xylem tissue cells originated from the two daughter cells and four from the mother cell. An additional four cells were formed if a xylem tissue cell was also present in the dormant cambium. The first 9–13 tracheids were evidently derived from overwintering cambial cells which had been deposited at the end of the previous season. According to Höster, these 9–13 tracheids are the so-called first-formed tracheids that lack some of the attributes of ordinary compression wood tracheids. Their angular outline and absence of intercellular spaces and helical cavities may be directly related to their origin, and the fact that the overwintering cambial cells had been formed at a time, late in the growing season, when the availability of indoleacetic acid is known to be low. As a result, these cells are not destined to differentiate into ordinary compression wood tracheids.

The hypothesis developed by Höster (1974) is attractive, but it is clear that some questions still remain to be answered. First, there have been several cases reported, where also the first-formed tracheids in a compression wood growth ring were of the fully developed type (Sanio 1873, Jaccard 1912, Core et al. 1961, Yoshizawa et al. 1982), as mentioned in Chap. 4.5. Second, the number of cell rows of first-formed tracheids can vary over a wider range than is implicit in Höster's hypothesis, namely from only 3–4 to as many as 20. It is not clear why the last-formed tracheids deposited at the same time as the overwintering cambial cells are fully developed compression wood tracheids (Timell 1972b) if a deficiency of auxin were involved.

An alternative explanation is based on the results obtained by Kennedy and Farrar (1965) and by Larson (1969a, b) (Chaps. 6.3.3 and 12.5.2). According to this theory, some factor in the chain of events linking the reception of the gravistimulus to the physiological reaction is missing for a certain, and probably brief, period of time at the initiation of xylem development in the spring. No stimulus is received during cell enlargement, and the cells remain angular in outline. Obviously, the stimulus becomes operative very soon, and the S_1 layer which is now formed is of the compression wood type, as is also the S_2 layer. Other features, such as the lack of an S_3 layer, the chemical composition of the secondary wall, and the distribution of lignin in this wall, are also those of typical compression wood. The fact that the secondary wall is thin could be a consequence of the rapid growth at this early stage and the lack of photosynthate.

One aspect which cannot be explained by either of these two hypotheses is the presence of helical cavities in the first-formed tracheids in *Pinus* while these cavities are often absent in genera such as *Abies, Larix, Picea,* and *Tsuga*. Harris (1976) and Yoshizawa et al. (1981, 1982) have suggested that compression wood develops less readily in earlywood than in latewood, but this, of course, does not explain the marked difference between *Pinus* and the other genera.

As also pointed out by Yoshizawa et al. (1982), different species might possess a different sensitivity to the compression wood stimulus. They have tentatively attributed this to differences in time of bud break, time of initiation of cell division, and availability of growth hormones in the cambial zone. It is tempting to attribute the nature of the first-formed earlywood to a lack of auxin at the beginning of the growing season, since an excess of auxin is known to stimulate the formation of compression wood (Chap. 13.4). According to recent investigations, however, reactivation of the cambium in the early spring does not depend on a supply of indoleacetic acid from active buds, shoots, and needles, and formation of indoleacetic acid in these organs appears to continue during the winter (Savidge et al. 1982, Savidge and Wareing 1982). Obviously, more evidence is needed before the origin of the first-formed tracheids in compression wood is clear. There can be little doubt that this problem is only part of the larger subject of the physiology of compression wood formation.

9.6.2 The Differentiating Ray Cells

There is considerable evidence that the ray parenchyma cells have the same chemical composition (Hoffmann and Timell 1972a, b) (Chap. 6.3.2) and ultrastructure (Timell 1972a) (Chap. 4.7.3) in normal and compression woods. As could be expected, the same applies to their mode of formation. Timell (1979b) found that in differentiating compression wood of *Abies balsamea* the ray parenchyma cells expanded to several times their original length in the radial direction so that the mature ray cell eventually extended along 7–8 tracheids. Both primary and secondary walls were present. Fujikawa and Ishida (1975) and Yamamoto et al. (1977) have shown that in normal wood of *Abies* spp. both S_1 and S_2 layers occur in the secondary wall.

A relatively thin cytoplasm, often with an extended system of endoplasmic reticulum but with few Golgi bodies, filled the entire cell at the early stages (Timell 1979b). Fukazawa and Imagawa (1970) observed numerous amyloplasts in differentiating normal ray parenchyma cells of *Abies sachalinensis*. Lipid droplets were, in contrast, the only ergastic material present in the compression wood ray cells. At the time when the S_2 layer was being completed in the tracheids, the cytoplasm in the ray cells became denser; it contained a few

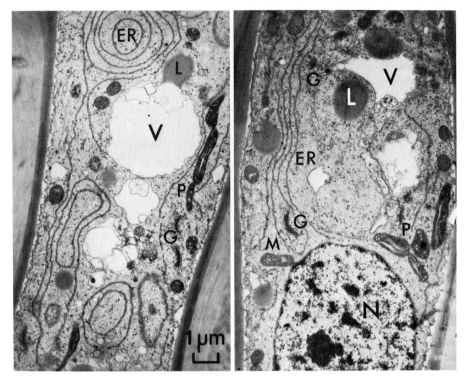

Fig. 9.48. Two transverse sections of a long ray parenchyma cell in differentiating compression wood of *Abies balsamea* at the time of formation of the S_2 layer in the tracheids. The dense cytoplasm contains mitochondria (M), a few Golgi bodies (G), endoplasmic reticulum (ER), numerous plastids (P), and some lipid droplets (L). N nucleus. V vacuole. TEM. (Timell 1979 b)

Golgi bodies, endoplasmic reticulum, plastids, and lipid material, as shown in Fig. 9.48. The nucleus was always elongated and occasionally contained two nucleoli. There were half-bordered pits and intercellular spaces between the ray cells and the tracheids.

9.7 The Rate of Formation of Compression Wood

9.7.1 Introduction

There is surprisingly little information available regarding the time required for a xylem cell in trees to develop from a cambial to a mature cell. The cambial cells complete their division within a certain period of time. The first part of the time needed for differentiation is devoted to enlargement and the second and larger part to cell wall thickening, that is, formation of the secondary wall. Lignification is generally assumed to require the same time as deposition of this wall.

9.7.1 Introduction

The different time periods reported for softwood cambial cells to develop into tracheids are both approximate and variable. This is, of course, to be expected, as the duration of each phase depends on the vigor of the tree, which is determined by the species involved, the age of the tree, its location, the latitude, and the climate. Gregory and Wilson (1968) found, for example, that similar *Picea glauca* trees grew more rapidly in Alaska than in Massachusetts and that within each region the rate of growth was positively correlated with tree vigor. In his study of tracheid development of *Tsuga canadensis* trees growing in Massachusetts, Skene (1972) noted that vigorous trees had a larger number of cambial cells and required a shorter generation time than specimens of less vigor.

As already mentioned, agreement between reported time periods is only moderate. According to Wodzicki and Peda (1963), tracheid enlargement ("radial growth") was completed within 1−2 weeks and cell wall thickening within 2−4 weeks in *Abies alba*. Whitmore and Zahner (1966), by contrast, found that radial growth proceeded for 3−5 weeks and wall thickening for 3−7 weeks in *Pinus resinosa*. It is, however, necessary in studies of this kind to take into consideration the fact that the times needed for both enlargement and secondary wall formation vary with the time of the growing season. All investigators agree that in the differentiation of normal softwood xylem, the enlargement phase requires a longer time at the beginning than at the end of the season, whereas the rate of formation of the secondary wall decreases continuously as the season progresses. In *Pinus radiata* growing in Australia, Skene (1969) found that the time needed for enlargement of the tracheids decreased from 20 to 10 days while the period of cell wall thickening increased from 25 to 70 days. Corresponding values for *Tsuga canadensis* were 18 to 9 days and 10 to 50 days, respectively (Skene 1972). Wodzicki (1971) found that in *Pinus sylvestris* in Poland radial enlargement at first proceeded for 30 days, while the last-formed cells needed only 10 days. The time required for secondary wall formation increased from 14 to 56 days.

In their investigation of the seasonal activity of the cambium and its derivatives in *Abies balsamea* growing in New York, Kutscha et al. (1975) followed the differentiation of the tracheids in normal wood of a single tree, obtaining the results shown in Table 9.1. Again, cell enlargement proceeded more rapidly at the end than at the beginning of the growing season, although in this case the maximum time required for enlargement occurred in the middle of June and the minimum by mid July, only 1 month later. The time needed for formation of the secondary wall increased from 6 days in late May to 27 days in late August. Imagawa and Ishida (1970) and Mahmood (1971) have also reported that the rate of cell enlargement increases, while that of cell wall thickening decreases during the season.

According to Wilson (1964), cell division is completed in the cambial zone of *Pinus strobus* within 2.5 weeks. For *Pinus radiata*, Skene (1969) found this time to be longer, namely 4 weeks. Unlike the times needed for differentiation, it did not vary with the season.

Because tree cells cannot be observed continuously in the course of their development, the numbers of cells in the different growth zones instead have to

Table 9.1. Number of days required for completion of the tracheid wall at different times of the growing season in normal (NW) and compression (CW) woods of *Abies balsamea*. (Modified from Kutscha et al. 1975)

Time of season	Enlargement		Maturation		Total differentiation	
	NW	CW	NW	CW	NW	CW
Late May	8	2	6	17	14	19
Mid June	11	19	13	37	24	56
Mid July	2	7	12	28	14	35
Mid August	3	1	22	8	25	9
Late August	5	3/4	27	8	32	8
Early September						5

be determined at different times of the season. The time required for each phase can be obtained by a method first suggested by Whitmore and Zahner (1966) and later further developed by Skene (1969). The sampling errors entailed in this procedure are evidently appreciable and they are, moreover, correlated in the graphs used for computation. Only conspicuous seasonal trends can therefore be attributed experimental significance (Skene 1969).

9.7.2 The Rate of Formation of Compression Wood

Casperson (1963) bent a 5-year-old *Picea abies* tree into a horizontal position in opposite directions, keeping the tree in each position for 1 week. Provided the cambial cells were able to perceive the gravistimulus within 24 h, Casperson estimated that 1.3 compression wood and 0.7 normal wood cells were produced each day. In their study of wood formation in tilted seedlings of *Larix laricina* and *Pinus banksiana* (Chap. 12.5.2), Kennedy and Farrar (1965) found that one new compression wood tracheid was formed each day and that differentiation was completed within 20 days. Cell division required 10 days, cell enlargement 2 days, wall thickening 4 days, and lignification also 4 days.

As discussed in Chap. 6.3.3, Larson (1969a, b) studied the incorporation of ^{14}C in developing tracheids of *Pinus resinosa* in normal earlywood and latewood, compression wood, and opposite wood. In trees that had been tilted to an angle of 45° before treatment with $^{14}CO_2$, one new tracheid was added each day. The metabolic activity was much lower in the opposite wood, and addition of a new tracheid here required 4 days. Other trees were tilted at the time they were treated with $^{14}CO_2$, which had the effect of doubling the original rate of one tracheid every second day to one tracheid per day. Formation of compression wood was obviously associated with a considerable increase in the rate of cambial division and xylem differentiation, including a rapid deposition of the S_1 layer.

9.7.2 The Rate of Formation of Compression Wood

Nix (1974) induced formation of compression wood by application of indoleacetic acid to *Pinus elliottii* seedlings, as described in Chap. 13.4. He found that, in comparison with normal wood, the formation of compression wood was characterized by an increased rate of cell division, while the time required for cell enlargement and wall deposition remained the same. There was a decrease in the total differentiation time.

In a study of the rate of formation of compression wood, Kutscha et al. (1975) found that in *Abies balsamea* cambial activity was initiated and terminated in compression wood at the same time as in normal wood. The technique used by Skene (1969) for estimating the rate of division in the cambial zone does not take into account the fact that some cambial initials, albeit a minority, divide to produce a phloem and not a xylem mother cell. Kutscha et al. (1975) calculated that the average ratio of xylem to phloem cells was 14:1 in normal wood but 21:1 in compression wood. The same number of phloem cells were formed in the two tissues, but there were 65% more xylem cells produced in the compression wood.

The seasonal changes in the rate of enlargement and wall thickening observed by Kutscha et al. (1975) with *Abies balsamea* compression wood differed decisively from those found for normal wood in this species. As can be seen in Table 9.1, enlargement of the tracheids required 2 days in May but 19 days only a few weeks later. The rate increased rapidly during July and August, and by the end of the latter month the enlargement phase lasted less than 1 day. The time required for deposition of the secondary wall reached a maximum of 37 days by the middle of June. Contrary to the situation in normal wood, the rate increased rapidly thereafter. During the latter half of August the secondary wall was completed within 8 days. By early September, the entire differentiation was concluded within the astonishingly short period of 5 days, compared to over 30 days for normal wood.

Deactivation of the cambial zone began shortly afterward, by the middle of September. It is clear, as pointed out by Timell (1980b), that in this case only a few days separated the period of maximum radial growth and the onset of deactivation, and that the termination of growth in compression wood was more abrupt than in normal wood.

Kutscha et al. (1975) found that in normal wood the cell division cycle required from 7 to 25 days with an average of 13 days. In compression wood this time varied over a considerably wider range, namely from 4 to 52 days, with an average of 20 days. It should perhaps be mentioned at this point that while the specimens used by Kutscha et al. to study the formation of normal wood were all obtained from a single tree, the compression wood samples were collected from different trees and also from different parts of these trees. The tree-to-tree variation was great and might have affected some of the results obtained, as pointed out by the investigators themselves.

In young *Picea glauca* trees tilted at an angle of 45°, Yumoto et al. (1982a) found that 3–4 new compression-wood cells were formed each day in July and 2.0–2.5 in August. The rate of division in the cambial zone was constant throughout the month of July. These are rates two to four times higher than those reported by Kennedy and Farrar (1965) and by Larsson (1969b), who ob-

Table 9.2. Number of compression wood (CW) and normal wood (NW) tracheid rows formed in stems of *Taxus cuspidata* first inclined at an angle of 60° for 5, 10, and 20 days, respectively, and subsequently oriented vertically for 20 days. The number of such treatments was four in experiments 1 and 2 and three in experiment 3. (Yoshizawa et al. 1984b)

Experiment No.	Time of inclination, days	Wood	Number of tracheid rows formed				Total	Average
			Arc No.					
			1	2	3	4		
1	5	CW	5	4	5	4	18	4.5
		NW	4	6	6	6	22	5.5
2	10	CW	8	9	11	9	37	9.3
		NW	6	6	8	7	27	6.8
3	20	CW	18	22	21		61	20.3
		NW	6	8	6		20	6.7

served only one new tracheid per day. The fact that the rate of formation of compression wood was almost twice as high in July as in August was confirmed by the deposition of twice as much compression wood in July. These results are contrary to those obtained by Kutscha et al. (1975), who found the rate to reach a maximum in August. It is possible that Yumoto et al. (1982a) used unusually vigorous spruce trees. It should also be noted here that both the normal wood and the compression wood examined in these investigations were of the juvenile type. In normal wood of *Larix leptolepis*, Imagawa and Ishida (1970) observed only 0.7 cambial divisions per day.

Yoshizawa et al. (1985) have recently reported results of tilting experiments similar to those carried out by Kennedy and Farrar (1965). Young *Taxus cuspidata* trees were fixed at an angle of 60° to the vertical and were kept in this position for 5, 10, and 20 days, respectively. After 20 days in a vertical position, the trees were again tilted for the same periods of time. These treatments, which were repeated from early May until late October, resulted in formation of alternating arcs of compression wood and normal wood on the lower side of the stem. The number of tracheid rows in each arc is shown in Table 9.2. As could be expected, the number of rows of compression wood tracheids increased with increasing time of inclination. The figures show that one new tracheid was added each day, a rate exactly the same as that observed previously by Casperson (1963), Kennedy and Farrar (1965), and Larson (1969b), but significantly lower than that found by Yumoto et al. (1982a). Normal tracheids, by contrast, were produced at the much lower rate of one new cell every 3rd to 4th day, as had also been observed by Larson (1969b).

In summary, it is clear that the longitudinal tracheids are formed at a three to four times higher rate in compression wood than in corresponding normal wood. In four investigations, it has been found that one new tracheid is produced per day in compression wood. The rate obviously varies in the course of the growing season, but exactly how is currently uncertain since available evidence is both scarce and contradictory.

Dadswell and Wardrop (1949) have mentioned that in conifers the vascular cambium becomes active at an earlier date on the lower side of branches and inclined stems than on the upper. According to my own observations, the cambium evidently remains active on the lower side of leaning or bent conifer stems long after it has become dormant on the upper side. Citing this observation, Westing (1968) has suggested that perhaps the growing season is longer in compression wood than in normal wood as a consequence of a lateral redistribution of growth substances. Kutscha et al. (1975), by contrast, found that the cambium became dormant at the same time in normal and compression woods, but this observation was made with trees growing under quite different conditions and cannot therefore be assigned general validity. More recently, Yumoto and Ishida (1982) have confirmed my observations. In their *Picea glauca* trees, tilted at an angle of 45°, cambial activity terminated on the upper side of the stem in the early fall, while actively differentiating xylem occurred on the lower side as late as by mid December. They also found that the cambium became active on the upper side at an earlier date than on the lower, again in disagreement with the results obtained by Kutscha et al. (1975).

9.7.3 Conclusions

In fully developed compression wood, thin-walled tracheids are produced only for a short period in the spring. The results obtained by Kutscha et al. (1975), albeit scanty, suggest that these cells differentiate at the same rate as normal tracheids. Lack of photosynthate is probably responsible for their thin wall. All subsequent cells have the same, thick wall notwithstanding the widely varying rates at which they differentiate. Obviously, wall thickness, at least in this case, is more strongly correlated with availability of photosynthate than with the rate of wall deposition, a fact previously doubted by Skene (1969). Thick-walled tracheids continue to be formed until the very end of the season, at which time they seem to mature within only a few days. This high rate of differentiation, which is in striking contrast to the low rate prevailing in normal wood at this time, cannot be fully explained at the present time. Since compression wood is believed to be formed under conditions of a high concentration of auxin, it would be tempting to attribute its rapid differentiation to an abundant supply of auxin throughout the season. With this assumption, however, it becomes difficult to explain the unusually low rate of differentiation found by Kutscha et al. to prevail earlier in the season.

The findings reported by Yumoto et al. (1982a) for the rate of formation of compression wood in leaning *Picea glauca* trees is more in accord with what is known about the formation of normal conifer wood, for example the more rapid cell division in July than in August. A comparison with the results obtained by Kutscha et al. (1975) is perhaps not justified, for Kutscha collected his compression wood from the lower side of the bend in branches that had developed into new stems following removal of the leader, and his findings might have been atypical. All investigators agree that compression wood is formed at a significantly higher rate than is normal wood (Chap. 10.3.3). This is also obvious from the fact that compression wood generally occurs within unusually wide growth rings. The evidence presented by Kutscha et al. (1975) notwithstanding, it is likely that formation of compression wood is initiated earlier and terminated later than that of normal wood within a growing season.

9.8 The Transition Between Normal Wood and Compression Wood

It has long been known that mild and moderate compression woods lack some of the attributes of the more often examined pronounced type. Kennedy and Farrar (1965) could show that the time during the period of cell division and differentiation when the compression wood stimulus is received had a decisive influence on the properties of the tracheids produced. Typical compression wood tracheids were formed only if the cells were exposed to the stimulus throughout development. If, for example, the stimulus was received at the end of the period of wall thickening, the only effect was an increased deposition of lignin. These and other results obtained in this investigation are discussed in Chap. 12.5.2. No direct attempt was made in this study to examine the transition between normal wood and compression wood.

Wardrop and Davies (1964), as mentioned in Chap. 13.4, induced formation of compression wood in seedlings of *Pinus radiata* by application of indoleacetic acid. They observed that, when deposition of compression wood gradually ceased, the helical cavities were the first to disappear at the same time as an S_3 layer appeared. When Casperson and Hoyme (1965) applied lanoline paste alone or a paste containing a very small amount of indoleacetic acid to *Picea abies* trees (Chap. 13.4), an abnormal wood formed with structural features of both normal and compression woods. The tracheids had a rounded outline and a thick wall and were surrounded by intercellular spaces, but they also possessed an S_3 layer and lacked helical cavities.

Fukazawa (1973), in a study of formation of compression wood and righting in tilted seedlings of *Abies sachalinensis*, observed the appearance of the helical cavities as recovery proceeded. He found that the cavities were originally more widely spaced than later, an observation reported at the same time also by Fujita et al. (1973). Their depth gradually decreased as the compression wood slowly changed into normal wood. In his later report on the distribution of lignin in compression wood of the same species, Fukazawa (1974) gave a detailed account of the transition between compression wood on the lower side and normal wood on the flank of an inclined stem. These results are discussed in Chap. 6.4.2 and illustrated in Figs. 6.20 – 6.23.

Results of a more detailed study of the transition between normal and compression woods were reported 5 years later by Fujita et al. (1979). A 12-year-old *Cryptomeria japonica* tree was tilted at an angle of 45 ° in early May and fixed with a stake in this position for 40 days, following which it was restored to a vertical orientation. The tree was cut down in the fall, and wood specimens were collected at a height of 1 m above the ground where five growth rings had been formed. A somewhat older *Pinus resinosa* tree was treated similarly. Wood sections were examined under the light microscope. Staining with safranin was

9.8 The Transition Between Normal Wood and Compression Wood

used for observing the distribution of lignin in the tracheids. The presence or absence of an S_3 layer was observed under polarized light. Best results were obtained with the *Cryptomeria* tree, and only those will be considered here.

The gradual transition from normal earlywood to compression wood was found to be associated with a well-defined series of changes in the ultrastructure of the tracheids, as shown in Figs. 9.49.1 and 9.49.2. First, the layer S_3 disappeared, which resulted in a temporary reduction in the wall thickness. Next, the heavily lignified $S_2(L)$ layer began to be formed at the same time as the entire S_2 layer increased in thickness, and the microfibril angle became flatter. During the third phase, helical cavities and ridges developed, while the tracheids assumed a rounded outline. The appearance of intercellular spaces constituted the fourth and last phase. The investigators pointed out that lignification of the secondary wall could have been expected to be the first process to occur in the transformation of normal wood into compression wood. They offered as a tentative explanation the suggestion that the compression wood stimulus might have been unable to effect an additional lignification of the secondary wall because of a lack of lignin precursors in the cytoplasm (Chap. 12.5.2).

The transition from compression wood to normal wood is clearly illustrated in Figs. 9.49.3 and 9.49.4. The first compression wood characteristic to disappear was the helical cavities and ridges. Despite the appearance of an S_3 layer, the next step was a reduction in wall thickness. During the third phase, lignification abated, and at the fourth stage, the intercellular spaces disappeared. Although not mentioned by the investigators, the tracheids evidently ceased to have a rounded outline at the time when an S_3 layer became apparent. The continued high level of lignification was attributed by the investigators to the presence in the cytoplasm of precursors formed there at an earlier stage. As pointed out by Fujita et al. (1979), the changes taking place during the two transitions between normal and compression woods are mutually consistent. With the exception of the lignification phase, they are inversely related to the events involved in the differentiation of the tracheids in compression wood.

Similar transitions between normal and compression woods were later observed by Park et al. (1980) in their study of the structure of branch wood in *Pinus densiflora*. Within an annual ring, tracheids with an S_3 layer were first replaced by cells with an S_2 lacking helical cavities. Subsequently weak cavities appeared, followed by fully developed ones. The reverse order was observed when compression wood was gradually transformed into normal wood, the S_3 layer finally being added to tracheids with no helical cavities in the S_2.

The transition between normal wood and compression wood in *Cephalotaxus, Taxus, Torreya,* and *Pseudotsuga,* studied by Yoshizawa et al. (1982) is discussed and illustrated in Chap. 4.4.7. In the first three genera both tissues posses helical thickenings, whereas in *Pseudotsuga* the helical thickenings in the normal wood are replaced by helical cavities and ribs in the compression wood (Chap. 4.4.5.4).

The fact that the helical cavities are the first anatomical feature to disappear when the compression wood stimulus ceases to operate or is diminished has also been reported by Kennedy and Farrar (1965), Fukazawa (1974) and

688

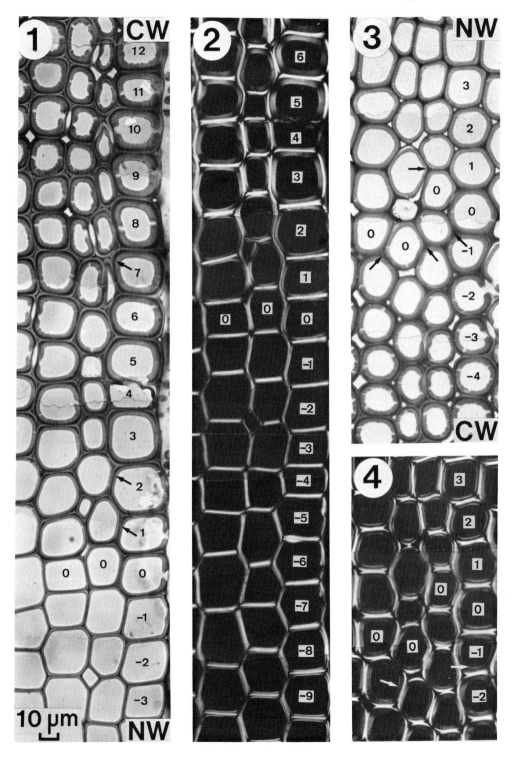

9.8 The Transition Between Normal Wood and Compression Wood

Harris (1976). They are frequently also absent in the first-formed compression wood growth ring as well as in the compression wood formed by some primitive gymnosperms (Timell 1978b) (Chap. 8.3.2). It would appear that the helical cavities and ribs not only constitute a late evolutionary feature of compression wood but also develop only under conditions of optimum stimulus.

Yumoto et al. (1982a,b) (Yumoto and Ishida 1979) recently reported results obtained in a thorough investigation of the transition from normal wood to compression wood in young *Picea glauca* trees inclined at an angle of 45° under such conditions that no external compressive or tensile stresses could have existed within the stems. Those aspects of these experiments that relate to the perception of the gravistimulus are discussed in Chap. 12.5.2. The gradual change from normal to compression wood was observed either by light microscopy or by a combination of scanning electron and ultraviolet light microscopy, the latter a novel technique that gave excellent results. Two experiments were carried out, one in July and another in August. Only the former will be referred to here.

No compression wood features could be observed on the lower side of the stems 2 days after the trees had been inclined (Yumoto et al. 1982a). After 4 days, the differentiating xylem tracheids suddenly assumed a rounded outline with intercellular spaces, and the seemingly normal S_2 layer had become slightly thicker and more lignified than normal. The last two features, which occurred almost simultaneously, were more pronounced after 6 and 8 days. After 15 days the tissue contained 15 rows of mature compression wood tracheids. The change from normal wood to compression wood occurred earlier in the upper than in lower portions of the stem, a fact also observed by Kennedy and Farrar (1965) in their seedlings.

An overall view of the transition from normal wood to compression wood is presented in Fig. 9.50 (Yumoto et al. 1982b). The pattern shown here was observed throughout the growing season and the trees sampled. The S_3 layer was the first anatomical feature of the normal wood to disappear. This is well illustrated in Fig. 9.51. Sometimes S_3 reappeared after having first ceased to be formed and then disappeared again. In one case the transition layer between S_3 and S_2 could also be distinguished in a tracheid between cells with and without S_3, as shown in Fig. 9.52. Following the disappearance of the S_3 layer, there occurred a change in the lignin distribution in the S_2 layer, as evidenced by ultraviolet light absorption. Strangely, this commenced with a higher absorption in the inner region of S_2, as can be seen in the latewood region shown in Fig. 9.53. This unusual pattern of lignification is difficult to explain, as pointed out by the

---

Fig. 9.49. Transition between normal and compression woods in *Cryptomeria japonica*. **1** and **2** Transition from normal wood to compression wood photographed in ordinary and in polarized light. Cell (*0*) is the first to have an S_2 (L) layer (*arrows*). The S_3 layer has disappeared beginning with cell (*–4*). **3** and **4** Transition from compression wood to normal wood photographed in ordinary and in polarized light. Cell (*0*) is the last to have an S_2 (L) layer. *Black arrows* indicate absence of intercellular spaces, clearly a variable trait. The S_3 layer appears first in cell (*–1*) (*white arrows*). Transverse sections. (Fujita et al. 1979)

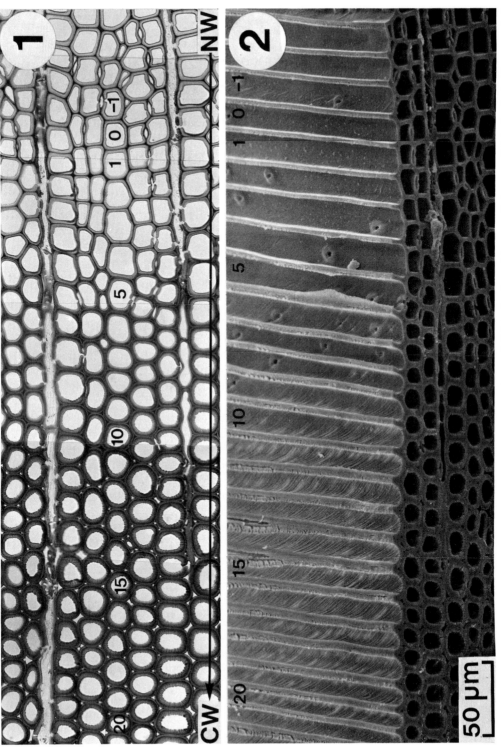

Fig. 9.50. Transition from normal wood to compression wood in *Picea glauca*. Cell 1 is the first without an S₃ layer, cell 5 is the first with a rounded outline and the beginning of an S₂(L) layer and helical cavities, and cell 11 is the first fully developed compression wood tracheid. **1** Transverse sec-

9.8 The Transition Between Normal Wood and Compression Wood

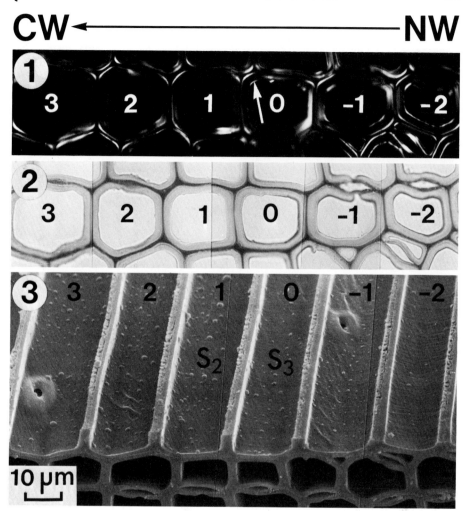

Fig. 9.51. Transition from normal wood (*NW*) to compression wood (*CW*) in *Picea glauca*. The S_3 layer (*arrow*) is still present in cell *0* but has disappeared in cell *1*. This change cannot be detected in the ultraviolet photomicrograph. **1** Transverse section. PLM. **2** Transverse section. UVLM. **3** Radical surface. SEM. (Modified from Yumoto et al. 1982b)

investigators. They suggested tentatively that if an abundant supply of lignin precursors were formed after the outer part of S_2 had been lignified in a normal fashion, only the inner, unlignified part would become highly lignified, the outer portion having become impermeable to the precursors. At a somewhat later stage, lignin began to accumulate in the outer part of S_2, the $S_2(L)$ layer.

After the inner part of S_2 had become lignified, the wall began to increase in thickness, and the tracheids became rounded in outline. The first cell with a high concentration of lignin in $S_2(L)$ was also the first to assume a rounded outline. When this had happened, the helical cavities and ribs began to form

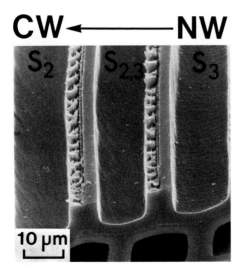

Fig. 9.52. Transition from normal wood (*NW*) to compression wood (*CW*) in *Picea glauca*, showing a transition tracheid with an innermost $S_{2,3}$ layer. Radial section. SEM. (Modified from Yumoto et al. 1982b)

within the inner S_2. Sometimes the helical cavities developed gradually (Fig. 9.51), but in other cases they formed abruptly (Fig. 9.53). The rounding of the tracheids preceded the appearance of intercellular spaces, as can be seen from Fig. 9.51. Epithelial cells in horizontal resin canals changed abruptly from flat and thin-walled to oval and thick-walled (Fig. 4.129).

In their similar study of the transition of normal wood into compression wood, Fujita et al. (1979) observed an initial reduction in the thickness of the tracheid wall, a fact which they attributed to the early disappearance of the S_3 layer. Yumoto et al. (1982b) observed a similar decrease but with the difference that in their case this took place after the S_3 had already disappeared. They concluded that the phenomenon was unrelated to the disappearance of S_3, and that it was the S_2 layer itself that became thinner. Fujita et al. (1979) had found that, following the lignification of $S_2(L)$ and the thickening of S_2, the microfibril angle in this layer became larger. Yumoto et al. (1982b) observed no such increase in microfibril angle because the tissues that they studied consisted of juvenile wood where this angle can be as large in normal wood as in compression wood (Chap. 6.2.1).

In summary, the sequence of changes occurring when normal wood is transformed into compression wood as observed by Fujita et al. (1979) and Yumoto et al. (1982a,b) is the following: (1) disappearance of the S_3 layer, (2) a temporary reduction in thickness of S_2, (3) lignification of the secondary wall with a simultaneous increase in wall thickness, (4) an increased microfibril in S_2, (5) a rounding of the tracheid outline, (6) formation of helical cavities, and (7) appearance of intercellular spaces.

Yoshizawa et al. (1984, 1985) recently studied the transition between normal wood and compression wood in *Taxus cuspidata* and *Torreya nucifera*, both species that form helical thickenings instead of helical cavities and ribs in their compression wood tracheids. The transition from normal wood to compression

9.8 The Transition Between Normal Wood and Compression Wood

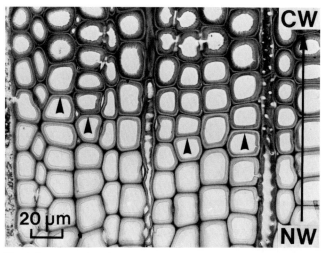

Fig. 9.53. Transition from normal wood (*NW*) to compression wood (*CW*) in *Picea glauca*. Some of the transitional tracheids have a high UV absorbance (high lignin concentration) in the inner part of their S_2 layer (*arrowheads*). Transverse section. UVLM. (Modified from Yumoto et al. 1982b)

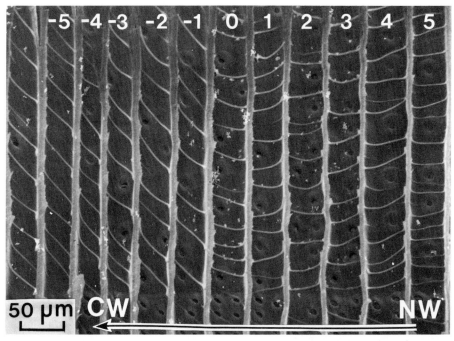

Fig. 9.54. Tracheids in *Taxus cuspidata* during transition from normal wood (*NW*) to compression wood (*CW*). Cell *0* is the first tracheid with no S_3 layer. Note the slow change in the orientation of the thickenings, from a flat S helix of the normal wood to a steeper and less variable Z helix in the compression wood. Radial surface. SEM. (Yoshizawa et al. 1984)

wood occurred in four successive steps: (1) a change in the orientation of the thickenings from an S helix to a Z helix, (2) disappearance of the S_3 layer, (3) increased formation of lignin and thickening of the secondary wall, and (4) appearance of a rounded cell outline. When compression wood was transformed into normal wood, these events took place in the reversed order: (1) the helix of the thickenings changed from a Z to an S type, (2) the S_3 layer appeared, (3) the lignin content decreased, and (4) the tracheids assumed an angular outline. These sequences of events are the same as those observed by Fujita et al. (1979) and by Yumoto et al. (1982a,b), with the difference that these investigators examined species with helical cavities and ribs rather than helical thickenings in their compression wood.

In the transition from normal wood to compression wood, the orientation of the thickenings in *Taxus cuspidata* changed gradually from an S to a Z helix, as illustrated in Fig. 9.54. The actual transition seems to have occurred in the cell marked 0, and this was also the first cell where no S_3 layer could be detected. The almost transverse helical thickenings were attached to S_3 in the normal wood and followed the orientation of the microfibrils in this layer. In the compression wood, they were attached to the lumen surface of the S_2 layer, and both microfibrils and thickenings were oriented at an angle of 45° to the fiber axis. The change in the direction of the thickenings took place either immediately before or at the same time as the disappearance of the S_3 layer.

When normal wood was transformed into compression wood, the microfibril angle in the S_2 layer increased rapidly (Chap. 4.4.5.2). In the tilting experiments with *Taxus cuspidata* described in Chap. 9.7.2, Yoshizawa et al. (1985) found that with each transition between normal and compression woods, the change in the microfibril angle in S_2 was paralleled by a change in the direction of the helical thickenings, as is clearly brought out in Fig. 9.55. A direct causal relationship between these two events, as suggested by the investigators, is, however, not likely, since the thickenings are attached to S_2 in compression wood and to S_3 in normal wood.

It was also observed in these experiments (Yoshizawa et al. 1985) that when normal wood gradually changed into compression wood, the inner region of the S_2 layer became highly lignified prior to the appearance of the $S_2(L)$ layer with its high concentration of lignin. As pointed out by the investigators, this is exactly the same observation as that made previously by Yumoto et al. (1982b) (Fig. 9.53). These investigators had also found that in the transition from normal wood to compression wood there occurred a temporary reduction in the thickness of the tracheid wall that was caused by a decrease in the thickness of the S_2 layer and not as a consequence of the disappearance of S_3, as suggested earlier by Fujita et al. (1979). Yoshizawa et al. (1982b) made the same observation, agreeing with Yumoto et al. (1982b) that the S_2 layer became thinner at a stage when S_3 had already disappeared.

In the transition from compression wood to normal wood, the $S_2(L)$ layer disappeared very gradually and slowly. The reduction in microfibril angle in S_2 was also quite slow. The gradual change in the direction of the helical thickenings occurred before any S_3 layer could be observed, an interesting observation. Cell 0 in Fig. 9.56 has an S_3 layer and its thickenings form an S helix. In cells 1,

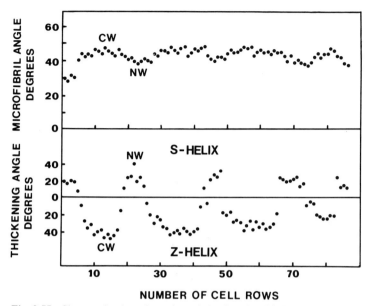

Fig. 9.55. Changes in the orientation of the S_2 microfibrils and the helical thickenings on repeated tilting and righting of *Taxus cuspidata* trees. Each maximum in the microfibril angle in the arcs of compression wood (*CW*) is synchronous with each maximum in the angle of the Z helix of the thickenings. The minima in the microfibril angle in the arcs of normal wood (*NW*) coincide with the maximima in the angle of the S helix of the thickenings. (Modified from Yoshizawa et al. 1985)

3, and 4 there is no S_3 layer, and the Z type helices are parallel with the microfibrils in the S_2 layer. The transitional tracheid 2 is most unusual, for here the thickenings form an S helix that is attached to S_2, crossing the mcirofibrils in this layer. It is evident, as had also been observed previously, that the orientation of helical thickenings and that of the microfibrils in the layer to which they are attached are not always the same (Chap. 4.4.7).

On the basis of these and other results, Yoshizawa et al. (1982a) presented an informative model of how differentiating tracheids respond to the initiation and termination of the gravistimulus. This model and their discussion of its implications are dealt with in Chap. 12.5.2.

9.9 General Conclusions

As could be expected, compression wood is in many respects formed in the same way as normal wood. Like the dormant cambial zone in normal coniferous wood, that in compression wood is organized in groups of Sanio's four and expanding four (Timell 1980b). In the active cambial region the same groups occur on the xylem side of the initial and pairs of phloem cells on the phloem side (Timell 1979a). As far as can be judged from the limited information available, compression phloem seems to differentiate exactly as normal phloem

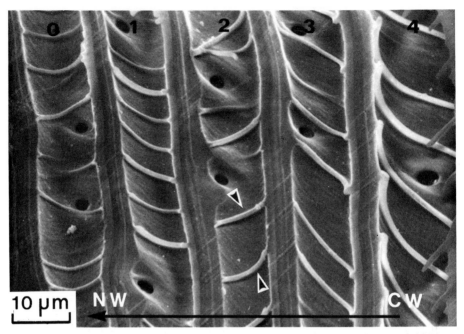

Fig. 9.56. Tracheids in *Taxus cuspidata* during transition from compression wood (*CW*) to normal wood (*NW*). The transitional cell *2*, which still lacks an S_3 layer, has thickenings organized in an S helix which are attached to S_2, crossing the microfibrils in this layer (*arrowheads*). In the subsequent cell *1* the thickenings are organized in a Z helix as in tracheids *3* and *4*. Cell *0*, where the helix is of the S type, is the first tracheid with an S_3 layer. Radial section. SEM. (Yoshizawa et al. 1984)

(Timell 1980a). In the differentiating xylem, the rapid rounding of the originally angular fusiform cells is the most conspicuous feature. This process evidently takes place at the end of the enlargement phase, immediately prior to the initiation of the S_1 layer (Wardrop and Davies 1964, Timell 1979a). The tracheids assume their final outline while this layer is being deposited. The rounding of the cell wall is associated with the appearance of intercellular spaces between the tracheids.

The cytoplasm appears to be denser and less frequently parietal in compression than in normal xylem during cell wall thickening, especially when the S_2 layer is being deposited (Wardrop and Davies 1964). Several reports indicate that the cytoplasm remains parietal in normal wood at this stage (Cronshaw and Wardrop 1964, Murmanis and Sachs 1969b, Tsuda 1975b). The observation by Wardrop and Davies (1964) that the plasma membrane temporarily disappears while the cavities develop in S_2 has not been confirmed by later investigators. Fujita et al. (1978b) have suggested that the hemicelluloses should be derived from vesicles originating in the endoplasmic reticulum. The results reported by Ray et al. (1976) and Robinson et al. (1976) render this unlikely. These matrix substances are undoubtedly synthesized in the Golgi cisternae and transported to the wall in the Golgi vesicles. The developing secondary wall

9.9 General Conclusions

in compression wood tracheids is distinctly lamellar. It has been shown that the S_2 layer in tracheids and fibers of wood consists of lamellae with an interlamellar distance of 7–8 nm (Kerr and Goring 1975, 1977, Ruel et al. 1978).

Wardrop and Davies (1964), Kutscha (1968), Côté et al. (1968b) and Timell (1979b) found that the developing secondary wall was more convoluted in compression wood than in normal tracheids, especially when potassium permanganate was used as a fixative. Fujita et al. (1973) considered this folding to be an artifact, caused by the permanganate. The same folded lamellae have, however, been observed when other fixatives, such as glutaraldehyde followed by osmium tetroxide, have been used (Casperson 1964, Casperson and Zinsser 1965, Kutscha 1968, Fujita and Harada 1978, Timell 1979b). Folds have also been reported to occur in developing walls of both normal (Kutscha 1968) and tension (Wardrop 1964, 1965) woods.

Two different opinions continue to coexist concerning the origin of the helical cavities and ridges in compression wood tracheids. The earlier view is that they result from a tangential contraction of the microfibrils in the S_2 layer. Additional wall material is subsequently added to the top of the newly created helical ridges (Wardrop and Davies 1964, Kutscha 1968, Côté et al. 1968b, Timell 1979b, Waterkeyn 1982). According to Casperson and Zinsser (1965) and Fujita et al. (1973), the cavities and ridges are created by regular interruptions in the deposition of wall material from the cytoplasm. As mentioned by Timell (1979b), this view is without question more attractive than the former, but it fails to explain the fission of a previously solid cell wall repeatedly observed. Meylan (1981) has recently pointed out that if an S_3 layer were present in the tracheids of compression wood, it would probably prevent the cavities from forming, adding that the S_1 layer probably offers some restraint. It is clear that the microfibril angle alone is not decisive, for helical cavities are lacking in severe compression wood of many species and usually in the first-formed earlywood. They are also absent in normal juvenile wood despite the large microfibrillar angle of its tracheids.

Boyd (1973b) has concluded that lignification of the inner portion of the S_2 layer in compression wood tracheids causes this layer to contract circumferentially, thereby creating helical cavities. Such a contraction might well occur but, as pointed out by Timell (1979b) (Meylan 1981), this hypothesis has one serious drawback: it is clear from several studies that the helical cavities appear in this portion of the S_2 before this layer has become lignified (Fig. 9.29) (Casperson and Zinsser 1965, Kutscha 1968, Fujita et al. 1978a).

Fujita et al. (1978a) distinguish between three separate phases in the differentiation of the tracheids in compression wood, namely formation of S_1 and lignification of the primary wall, a rapid deposition of S_2, and lignification of the entire secondary wall. There can be no question that this is the general course of lignin formation, but the sequence probably represents somewhat of an oversimplification. It is clear from the results obtained by Kutscha (1968) that the primary wall is not yet fully lignified by the time lignin has been deposited in the outer part of the S_1 layer, as is also the case in normal wood (Kutscha and Schwarzmann 1975). Wardrop (1965, 1971, 1976) has shown that in normal coniferous wood, lignin continues to be formed in the middle lamella while it is being deposited in the secondary wall. It would be surprising if the same did not occur in compression wood. Although the evidence is meager, it would appear that lignification of both normal and compression wood tracheids is initiated at the cell corners and proceeds from there first into the tangential wall most distant from the cambium. Citing an earlier investigation by Pickett-Heaps (1968) on wall formation in wheat seedlings, Fujita et al. (1978b) have adduced evidence in favor of the view that the Golgi cisternae and vesicles are involved in the process of lignification. More studies using various lignin precursors are, however, necessary before this question can be settled.

The suggestion by Höster (1974) that the first-formed compression wood tracheids should be derived from over-wintering cambial cells is attractive but needs to be tested, since it leaves some facts unexplained. The extremely high rate of differentiation of the tracheids in compression wood at the end of the growing season observed by Kutscha et al. (1975) is entirely unexpected, considering the low rate prevailing in normal wood at this time, and it does not agree with the findings reported later by Yumoto et al. (1982a). This interesting problem deserves more attention. Any further studies should incorporate more frequent and technically improved sampling of trees of comparable vigor, and should include both normal and compression woods of several conifer species.

It is interesting to note that the helical cavities are among the last characteristics to appear when normal wood is gradually transformed into compression wood, and that they are the first to disappear when a transition from compression into normal wood occurs, while the opposite is true with respect to lignin content. These observations, as well as the results obtained by Kennedy and Farrar (1965), suggest that the exceptionally high lignin content of the tracheids in compression wood is perhaps a more characteristic feature of this wood than are the more spectacular cavities and ridges in the inner S_2. The fact that cavities can be absent even in pronounced compression wood agrees with this conclusion.

References

Abbe LB, Crafts AS 1939 Phloem of white pine and other coniferous species. Bot Gaz 100:695–722

Alfieri FJ, Evert RF 1968a Seasonal development of the secondary phloem in Pinus. Am J Bot 55:518–528

Alfieri FJ, Evert RF 1968b Observations on albuminous cells in Pinus. Planta 78:93–97

Anonymous 1963 Structure of plant cells: cell-wall organization and cell-wall formation. CSIRO Aust Div For Prod Rep 1963/1964:8

Balatinecz JJ, Kennedy RW 1967 Maturation of ray parenchyma cells in pine. For Prod J 17(10):57–64

Bamber RK 1980 The origin of growth stresses. IUFRO Conf Laguna, Philippines, 1978, For Comm NSW Repr 715 WT, 7 pp

Bannan MW 1955 The vascular cambium and radial growth in Thuja occidentalis. Can J Bot 33:113–138

Bannan MW 1962 The vascular cambium and tree ring development. In: Kozlowski TT (ed) Tree growth. Ronald, New York, 3–21

Bannan MW 1968 Anticlinal divisions and the organization of conifer cambium. Bot Gaz 129:107–113

Barber NF 1968 A theoretical model of shrinking wood. Holzforschung 22:97–103

Barber NF, Meylan BA 1964 The anisotropic shrinkage of wood, a theoretical model. Holzforschung 18:146–156

Barnett JR 1971 a Winter activity in the cambium of Pinus radiata. NZ J For Sci 1:208–222

Barnett JR 1971 b Electron microscope preparation techniques applied to the light microscopy of the cambium and its derivatives in Pinus radiata. J Microsc 94:175–180

Barnett JR 1973 Seasonal variation in the ultrastructure of the cambium in New Zealand grown Pinus radiata D Don. Ann Bot 37:1005–1011

Barnett JR 1974a Secondary phloem in Pinus radiata D Don. 1. Structure of differentiating sieve cells. NZ J Bot 12:245–260

Barnett JR 1974b Secondary phloem in Pinus radiata D Don. 2. Structure of parenchyma cells. NZ J Bot 12:261–274

Baur PS, Walkinshaw CH 1974 Fine structure of tannin accumulations in callus cultures of Pinus elliottii (slash pine). Can J Bot 52:615–619

Boyd JD 1950 Tree growth stresses. III. The origin of tree growth stresses. Aust J Sci Res B-3:294–309

Boyd JD 1972 Tree growth stresses. V. Evidence of an origin in differentiation and lignification. Wood Sci Technol 6:251–266

Boyd JD 1973a Compression wood force generation and functional significance. NZ J For Sci 3:240–258

Boyd JD 1973b Helical fissures in compression wood cells: causative factors and mechanics of development. Wood Sci Technol 7:92–111

Brown RM, Jr (ed) 1982 Cellulose and other natural polymer systems. Biogenesis, structure, and degradation. Plenum, New York London, 519 pp

Butterfield BG 1975 Terminology used for describing the cambium. IAWA Bull 1975(1):13–14

Casperson G 1959 Mikroskopischer und submikroskopischer Zellwandaufbau beim Druckholz. Faserforsch Textiltech 10:536–541

Casperson G 1963 Über die Bildung der Zellwand beim Reaktionsholz. II. Zur Physiologie des Reaktionsholzes. Holztechnologie 4:33–37

Casperson G 1965 Über die Entstehung des Reaktionsholzes bei Kiefern. In: Aktuelle Probleme der Kiefernwirtschaft, Symp Eberswalde 1964, Deutsch Akad Landwirtsch, Berlin DDR, 523–528

Casperson G, Hoyme E 1965 Über endogene Faktoren der Reaktionsholzbildung. 2. Untersuchungen an Fichte (Picea abies Karst). Faserforsch Textiltech 16:352–358

Casperson G, Zinsser A 1965 Über die Bildung der Zellwand beim Reaktionsholz. 3. Zur Spaltenbildung im Druckholz von Pinus sylvestris L. Holz Roh-Werkst 23:49–55

Catesson AM 1964 Origine, fonctionnement et variations cytologiques saisonières du cambium de l'Acer pseudoplatanus (Acéracées). Ann Sci Nat (Bot) 12 ser 5:229–498

Catesson AM 1974 Cambial cells. In: Robards AW (ed) Dynamic aspects of plant ultrastructure. McGraw-Hill, London, 359–390

Catesson AM 1980 The vascular cambium. In: Little CHA (ed) Control of shoot growth in trees. Proc IUFRO workshop on xylem physiology, Mar For Res Centre, Fredericton NB, 12–40

Catesson AM, Roland 1981 Sequential changes associated with cell wall formation and fusion in the vascular cambium. IAWA Bull (ns) 2:151–162

Chafe SC, Doohan ME 1972 Observations on the ultrastructure of the thickened sieve cell wall in Pinus strobus L. Protoplasma 75:67–78

Chafe SC, Durzan DJ 1973 Tannin inclusions in cell suspension cultures of white spruce. Planta 113:251–262

Cieslar A 1896 Das Rothholz der Fichte. Cbl Ges Forstwes 22:149–165

Core HA, Côté WA, Jr., Day AC 1961 Characteristics of compression wood in some native conifers. For Prod J 11:356–362

Côté WA, Jr., Day AC, Kutscha NP, Timell TE 1967 Studies on compression wood. V. Nature of the compression wood formed in the early springwood of conifers. Holzforschung 21:180–186

Côté WA, Jr., Kutscha NP, Simson BW, Timell TE 1968a Studies on compression wood. VI. Distribution of polysaccharides in the cell walls of tracheids of compression wood. Tappi 51:33–40

Côté WA, Jr, Kutscha NP, Timell TE 1968b Studies on compression wood. VIII. Formation of cavities in compression wood tracheids of Abies balsamea (L) Mill. Holzforschung 22:138–144

Cronshaw J 1965a The organization of cytoplasmic components during the phase of cell wall thickening in differentiating cambial derivatives of Acer rubrum. Can J Bot 43:1401–1407

Cronshaw J 1965b Cytoplasmic fine structure and cell wall development in differentiating xylem elements. In: Côté WA, Jr (ed) Cellulose ultrastructure of woody plants. Syracuse Univ Press, Syracuse, 99–124

Cronshaw J, Wardrop AB 1984 Organization of cytoplasm in differentiating xylem of Pinus radiata. Aust J Bot 12:15–23

Delmer DP 1977 The biosynthesis of cellulose and other cell wall polysaccharides. In: Loewus FA, Runeckles VC (eds) The structure, biosynthesis, and degradation of wood. Plenum, New York London, 45–77

den Outer RW 1967 Histological investigations of the secondary phloem of gymnosperms. Meded Landbouwhogesch Wageningen 67-7:1–119

Ericson MC, Elbein AD 1980 Biosynthesis of cell wall polysaccharides and glycoproteins. In: Press J (ed) The biochemistry of plants. A comprehensive treatment. Vol 3 Carbohydrates: structure and function. Academic Press, New York, 589–616

Esau K 1969 The phloem. Gebrüder Borntraeger, Berlin Stuttgart, 505 pp

Evert RF 1977 Phloem structure and histochemistry. Ann Rev Plant Physiol 28:199–222

Evert RF, Alfieri FJ 1965 Ontogeny and structure of coniferous sieve cells. Am J Bot 52:1058–1066

Evert RF, Deshpande BP 1970 An ultrastructural study of cell division in the cambium. Am J Bot 57:943–961

Fergus BJ, Goring DAI 1970a The location of guaiacyl and syringyl lignins in birch xylem tissue. Holzforschung 24:113–117

Fergus BJ, Goring DAI 1970b The distribution of lignin in birch wood as determined by ultraviolet microscopy. Holzforschung 24:118–124

Freudenberg K 1968 The constitution and biosynthesis of lignin. In: Freudenberg K, Neish AC Constitution and biosynthesis of lignin. Springer, New York, 45–122

Frey A 1926 Submikroskopische Struktur der Zellmembranen. Polarisationsoptische Studie zum Nachweis der Richtigkeit der Mizellartheorie. Jahrb Wiss Bot 56:195–223

Fujikawa S, Ishida S 1975 Ultrastructure of ray parenchyma wall of softwood. Mokuzai Gakkaishi 21:445–456

Fujita M 1981 Deposition of major cell wall components in the differentiating tree xylem cells. Kyoto Univ Dep Wood Sci Technol, 92 pp

Fujita M, Harada H 1978 Autoradiographic investigations of cell wall development. I. Tritiated glucose assimilation in relation to cellulose and hemicellulose deposition. Mokuzai Gakkaishi 24:435–440

Fujita M, Harada H 1979 Autoradiographic investigations of cell wall development. II. Tritiated phenylalanine and ferulic acid assimilation in relation to lignification. Mokuzai Gakkaishi 25:89–94

Fujita M, Saiki H, Harada H 1973 The secondary wall formation of compression wood tracheids. On the helical ridges and cavities. Bull Kyoto Univ For 45:192–203

Fujita M, Saiki H, Harada H 1978a The secondary wall formation in compression wood tracheids. II. Cell wall thickening and lignification. Mokuzai Gakkaishi 24:158–163

Fujita M, Saiki H, Harada H 1978b The secondary wall formation of compression-wood tracheids. III. Cell organelles in relation to cell wall thickening and lignification. Mokuzai Gakkaishi 24:353–361

Fujita M, Saiki H, Sakamoto J, Araki N, Harada H 1979 Cell wall structure of transitional tracheids between normal wood and compression wood. Bull Kyoto Univ For 51:247–256

Fujita M, Takabe K, Harada H 1981 Autoradiographic investigations of cell wall development. III. Method of quantitative and chemical analyses of light microscopic autoradiography adapted to a series of semi-thin section. Mokuzai Gakkaishi 27:337–341

Fukazawa K 1973 Process of righting and xylem development in tilted seedlings of Abies sachalinensis. Res Bull Coll Exp For Hokkaido Univ 30:103–124

Fukazawa K 1974 The distribution of lignin in compression- and lateral-wood of Abies sachalinensis using ultraviolet microscopy. Res Bull Coll Exp For Hokkaido Univ 31:87–114
Fukazawa K, Imagawa H 1970 Some preliminary observations of ray parenchyma cell by the scanning electron microscopy. Res Bull Coll Exp For Hokkaido Univ 27:79–90
Gothan W 1905 Zur Anatomie lebender und fossiler Gymnospermenhölzer. Abhandl Königl Preuss Geol Landesanst 44:1–108
Gregory RA 1971 Cambial activity in Alaskan white spruce. Am J Bot 58:160–171
Gregory RA, Wilson BF 1968 A comparison of cambial activity of white spruce in Alaska and New England. Can J Bot 46:733–734
Grillos SJ, Smith FH 1959 The secondary phloem of Douglas-fir. For Sci 5:377–388
Gromov VS, Evdokimov AM, Abramovich TS, Khrol YS 1977 (Distribution of lignin in wood and topochemistry of its delignification studied by ultraviolet microspectrophotometry. 1. Lignin distribution in the cell walls of birchwood tracheary elements.) Khim Drev 6:73–79
Hardell HL, Leary GJ, Stoll M, Westermark U 1980 Variations in lignin structure in defined morphological parts of spruce. Svensk Papperstidn 83:44–49
Harris RA 1976 Characterization of compression wood severity in Pinus echinata Mill. IAWA Bull 1976(4):47–50
Hartig R 1896 Das Rothholz der Fichte. Forstl-Naturwiss Z 5:96–109, 157–169
Hartig R 1901 Holzuntersuchungen. Altes und Neues. Julius Springer, Berlin, 99 pp
Hepler PK, Fosket DE, Newcomb EH 1970 Lignification during secondary wall formation in Coleus: an electron microscopy study. Am J Bot 57:85–96
Hepton CEL, Preston RD, Ripley GW 1956 Electron microscopic observations on the secondary wall of the protoxylem of Cucurbita. Nature 177:660–661
Hirakawa Y, Ishida S, Ohtani J 1979 A SEM observation of organelles in the cambial and living xylem cells of Todomatsu (Abies sachalinensis). Res Bull Coll Exp For Hokkaido Univ 36:459–468
Hoffmann GC, Timell TE 1972a Polysaccharides in ray cells of normal wood of red pine (Pinus resinosa). Tappi 55:733–736
Hoffmann GC, Timell TE 1972b Polysaccharides in ray cells of compression wood of red pine (Pinus resinosa). Tappi 55:871–873
Holdheide W 1951 Anatomie mitteleuropäischer Gehölzrinden. In: Freund H (ed) Mikroskopie des Holzes und des Papiers V(1). Umschau, Frankfurt, 193–367
Höster HR 1974 On the nature of the first-formed tracheids in compression wood. IAWA Bull 1974(1):3–9
Imagawa H 1981 Study of the seasonal development of the secondary phloem in Larix leptolepis. Res Bull Coll Exp For Hokkaido Univ 38:31–44
Imagawa H, Fukazawa K, Ishida S 1976 Study of the lignification in tracheids of Japanese larch, Larix leptolepis Gord. Res Bull Coll Exp For Hokkaido Univ 33:127–138
Imagawa H, Ishida S 1970 Study of the wood formation in trees. 1. Seasonal development of the xylem ring in Japanese larch stem, Larix leptolepis Gord. Res Bull Coll Exp For Hokkaido Univ 22:373–394
Imagawa H, Ishida S 1981 An observation of the cambial cells in Larix leptolepis by semiultrathin sections. Res Bull Coll Exp For Hokkaido Univ 38:45–54
Itoh T 1971 On the ultrastructure of dormant and active cambium of conifers. Wood Res (Kyoto Univ) 45:23–35
Itoh T, Hayashi TS, Kishima T 1968 Cambial activity and radial growth in Sugi tree (Japanese Cryptomeria). Wood Res (Kyoto Univ) 51:23–45
Jacquiot C 1984 Sur la physiologie de cambium des arbres forestiers. Rev For Fran 36:113–121
Keith CT, Gordon SE 1976 Fixation of juvenile cambium from two coniferous species for ultrastructural study. Wood Fiber 8:177–200
Kennedy RW, Farrar JL 1965 Tracheid development in tilted seedlings. In: Côté WA, Jr (ed) Cellular ultrastructure of woody plants. Syracuse Univ Press, Syracuse, 419–453
Kerr AJ, Goring DAI 1975 The ultrastructural arrangement of the wood cell wall. Cellul Chem Technol 9:563–573

Kerr AJ, Goring DAI 1977 Lamellation of the fiber wall of birch wood. Wood Sci 9:136–139

Kerr T, Bailey IW 1934 The cambium and its derivative tissues. X. Structure, optical properties and chemical composition of the so-called middle lamella. J Arnold Arbor 15:327–349

Kidwai P, Robards AW 1969 The appearance of differentiating vascular cells after fixation in different solutions. J Exp Bot 20:664–670

Kollmann R, Schumacher W 1961 Über die Feinstruktur des Phloems von Metasequoia glyptostroboides und seine jahreszeitlichen Veränderungen. I. Das Ruhephloem. Planta 57:583–607

Kollmann R, Schumacher W 1962a Über die Feinstruktur des Phloems von Metasequoia glyptostroboides und seine jahreszeitlichen Veränderungen. II. Vergleichende Untersuchungen der plasmatischen Verbindungsbrücken in Phloemparenchymzellen und Siebzellen. Planta 58:366–386

Kollmann R, Schumacher W 1962b Über die Feinstruktur des Phloems von Metasequoia glyptostroboides und seine jahreszeitlichen Veränderungen. III. Die Reaktivierung der Phloemzellen im Frühjahr. Planta 59:195–221

Kollmann R, Schumacher W 1963 Über die Feinstruktur des Phloems von Metasequoia glyptostroboides und seine jahreszeitlichen Veränderungen. IV. Weitere Beobachtungen zum Feinbau der Plasmabrücken in den Siebzellen. Planta 60:360–389

Krieg W 1907 Die Streifung der Tracheidenmembran im Koniferenholz. Bot Zbl Beih 21:245–262

Kutscha NP 1968 Cell wall development in normal and compression wood of balsam fir, Abies balsamea (L) Mill. Ph D Thesis State Univ Coll For, Syracuse NY, 231 pp

Kutscha NP, Gray JR 1972 The suitability of certain stains for studying lignification in balsam fir, Abies balsamea (L) Mill. Life Sci Agr Exp Sta Univ ME, Orono, Tech Bull 53, 50 pp

Kutscha NP, Hyland F, Schwarzmann JM 1975 Certain seasonal changes in balsam fir cambium and its derivatives. Wood Sci Technol 9:175–188

Kutscha NP, Schwarzmann JM 1975 The lignification sequence in normal wood of balsam fir (Abies balsamea). Holzforschung 29:79–84

Kutsuki H, Higuchi T 1981 Activities of some enzymes of lignin formation in reaction wood of Thuja orientalis, Metasequoia glyptostroboides and Robinia pseudoacacia. Planta 152:365–368

Kutsuki H, Higuchi T 1982 Activities of some enzymes of lignin formation in reaction wood of Thuja orientalis and Metasequoia glyptostroboides. Wood Sci Technol 15:287–291

Larson PR 1969a Incorporation of ^{14}C in the developing walls of Pinus resinosa tracheids (earlywood and latewood). Holzforschung 23:17–26

Larson PR 1969b Incorporation of ^{14}C in the developing walls of Pinus resinosa tracheids: compression wood and opposite wood. Tappi 52:2170–2177

Lee, VPFF 1968 Structural differences in lignin formation between normal and compression wood of Douglas fir. M S Thesis, Univ WA Seattle, 52 pp

Little CHA 1970a Derivation of the springtime starch increase in balsam fir (Abies balsamea). Can J Bot 48:1995–1999

Little CHA 1970b Seasonal changes in carbohydrate and moisture content in needles of balsam fir (Abies balsamea). Can J Bot 48:2021–2028

Little CHA, Bonga JM 1974 Rest in the cambium of Abies balsamea. Can J Bot 52:1723–1730

Mahmood A 1968 Cell grouping and primary wall generations in the cambial zone, xylem, and phloem in Pinus. Aust J Bot 16:177–195

Mahmood A 1971 Numbers of initial-cell divisions as a measure of activity in the yearly cambial growth pattern in Pinus. Pak J For 21:27–42

Meylan BA 1981 Reaction wood in Pseudowintera colorata – A vessel-less dicotyledon. Wood Sci Technol 15:81–92

Mio M, Matsumoto T 1982 On intercellular spaces in Kuromatsu (Pinus thunbergii Parl). Bull Kyushi Univ For 52:107–114

Mischke K 1890 Beobachtungen über das Dickenwachstum der Coniferen. Bot Zbl 44:39–43, 65–71, 97–102, 137–142, 169–175

Miyakawa MM, Fujita H, Saiki H, Harada H 1973 The cell-wall structure of the secondary phloem elements in Cryptomeria japonica D Don. Bull Kyoto Univ For 45:181–191

Murmanis L 1970 Locating the initial in the vascular cambium of Pinus strobus L by electron microscopy. Wood Sci Technol 4:1–14

Murmanis L 1971 Structural changes in the vascular cambium of Pinus strobus L during an annual cycle. Ann Bot 35:133–141

Murmanis L, Evert RF 1966 Some aspects of sieve cell ultrastructure in Pinus strobus. Am J Bot 53:1065–1078

Murmanis L, Evert RF 1967 Parenchyma cells of secondary phloem in Pinus strobus. Planta 73:301–318

Murmanis L, Sachs IB 1969a Structure of pit border in Pinus strobus L. Wood Fiber 1:7–17

Murmanis L, Sachs IB 1969b Seasonal development of secondary xylem in Pinus strobus L. Wood Sci Technol 3:177–193

Murmanis L, Sachs IB 1973 Cell-wall formation in secondary xylem of Pinus strobus L. Wood Sci Technol 3:173–188

Musha Y, Goring DAI 1975a Distribution of syringyl and guaiacyl moieties in hardwoods as indicated by ultraviolet microscopy. Wood Sci Technol 9:45–58

Musha Y, Goring DAI 1975b Cell dimensions and their relationship to the chemical nature of the lignin from the wood of broad-leaved trees. Can J For Res 5:259–268

Neuberger DS, Evert RF 1974 Structure and development of the sieve-element protoplast in the hypocotyl of Pinus resinosa. Am J Bot 61:360–374

Neuberger DS, Evert RF 1975 Structure and development of sieve areas in the hypocotyl of Pinus resinosa. Protoplasma 84:109–125

Neuberger DS, Evert RF 1976 Structure and development of sieve cells in the primary phloem of Pinus resinosa. Protoplasma 87:27–37

Newman IV 1956 Pattern of meristems of vascular plants. 1. Cell partition in living apices and in the cambial zone and in relation to the concepts of initial cells and apical cells. Phytomorphology 6:1–19

Nix L 1974 The role of growth regulators in tracheid differentiation of southern pines. Ph D Thesis Univ GA, Athens, 69 pp

Onaka F 1949 (Studies on compression and tension wood.) Mokuzai Kenkyo, Wood Res Inst Kyoto Univ 1, 88 pp, Transl For Prod Lab Can 83 (1956), 99 pp

Parham RA, Kaustinen HM 1976 Differential staining of tannin in sections of epoxy-embedded plant cells. Stain Technol 51:237–240

Park S, Saiki H, Harada H 1980 Structure of branch wood in Akamatsu (Pinus densiflora Sieb et Zucc). II. Wall structure of branch wood tracheids. Mem Coll Agr Kyoto Univ 115:33–44

Perry TO 1971 Dormancy of trees in winter. Science 171:29–36

Philipson WR, Ward JM 1965 The ontogeny of the vascular cambium in the stem of seed plants. Biol Rev 40:534–579

Philipson WR, Ward JM, Butterfield BG 1971 The vascular cambium. Its development and activity. Chapman and Hall, London, 182 pp

Pickett-Heaps JD 1968 Xylem wall deposition. Radioautographic investigations using lignin precursors. Protoplasma 65:181–205

Pomeroy MK, Siminovitch D 1971 Seasonal cytological changes in the living bark and needles of red pine (Pinus resinosa) in relation to adaptation to freezing. Can J Bot 48:953–967

Preston RD, Middlebrook M 1949 The fine structure of sisal fibres. J Text Inst 40:715–722

Ray PM 1967 Radioautographic study of cell wall deposition in growing plant cells. J Cell Biol 35:659–674

Ray PM, Eisinger WR, Robinson DG 1976 Organelles involved in cell wall polysaccharide formation and transport in pea cells. Ber Deutsch Bot Ges 89:121–146

Robards AW 1968 On the ultrastructure of differentiating secondary xylem in willow. Protoplasma 65:449–464

Robinson DG, Eisinger WR, Ray PM 1976 Dynamics of the Golgi system in wall matrix polysaccharide synthesis and secretion by pea cells. Ber Deutsch Bot Ges 89:147–161

Romberger JA 1963 Meristems, growth, and development in woody plants. An analytical review of anatomical, physiological, and morphogenic aspects. US Dep Agr Tech Bull 1293, 214 pp

Ruel K, Barnoud F, Goring DAI 1978 Lamellation in the S_2 layer of softwood tracheids as demonstrated by scanning transmission electron microscopy. Wood Sci Technol 12:287–291

Saka S, Thomas RJ 1982 A study of lignification in loblolly pine tracheids by the SEM-EDXA technique. Wood Sci Technol 16:167–179

Saleh TM, Leney L, Sarkanen KV 1967 Radioautographic studies of cottonwood, Douglas fir, and wheat plants. Holzforschung 21:116–120

Sanio K 1873 Anatomie der gemeinen Kiefer (Pinus silvestris L). Jahrb Wiss Bot 9:50–126

Sarkanen KV 1971 Precursors and their polymerization. In: Sarkanen KV, Ludwig CH (eds) Lignins. Occurrence, formation, structure and reactions. Wiley-Interscience, New York, 95–163

Sauter JJ 1966a Untersuchungen zur Physiologie der Pappelholzstrahlen. I. Jahresperiodischer Verlauf der Stärkespeicherung im Holzstrahlparenchym. Z Pflanzenphysiol 55:246–258

Sauter JJ 1966b Untersuchungen zur Physiologie der Pappelholzstrahlen. II. Jahresperiodische Änderungen der Phosphataseaktivität im Holzstrahlparenchym und ihre mögliche Bedeutung für den Kohlenhydratstoffwechsel und den aktiven Assimilattransport. Z Pflanzenphysiol 55:349–362

Savidge RA, Heald JK, Wareing PF 1982 Non-uniform distribution and seasonal variation of endogenous indol-3yl-acetic acid in the cambial region of Pinus contorta Dougl. Planta 155:89–92

Savidge RA, Wareing PF 1982 Apparent auxin production and transport during winter in the nongrowing pine tree. Can J Bot 60:681–691

Schmid R 1976 The elusive cambium – another terminological contribution. IAWA Bull 1976(4):51–59

Schwarzmann JM, Kutscha NP 1973 Preliminary observations on lignification in tracheids of balsam fir. 31st Ann Proc Electron Microsc Soc Am, New Orleans LA, 464–465

Scurfield G 1967 Histochemistry of reaction wood differentiation in Pinus radiata D Don. Aust J Bot 18:377–392

Skene DS 1969 The period of time taken by cambial derivatives to grow and differentiate into tracheids in Pinus radiata D Don. Ann Bot 33:253–262

Skene DS 1972 The kinetics of tracheid development in Tsuga canadensis Carr and its relation to tree vigour. Ann Bot 36:179–187

Srivastava LM 1963 Secondary phloem in the Pinaceae. Univ CA Publ Bot 36(1):1–142

Srivastava LM 1969 On the ultrastructure of cambium and its vascular derivatives. III. The secondary walls of the sieve elements of Pinus strobus. Am J Bot 56:354–361

Srivastava LM, O'Brien TP 1966a On the ultrastructure of cambium and its vascular derivatives. I. Cambium of Pinus strobus L. Protoplasma 61:257–276

Srivastava LM, O'Brian TP 1966b On the ultrastructure of cambium and its vascular derivatives. II. Secondary phloem of Pinus strobus L. Protoplasma 61:277–293

Takabe K, Fujita M, Harada H, Saiki H 1981a The deposition of cell wall components in differentiating tracheids of Sugi. Mokuzai Gakkaishi 27:249–255

Takabe K, Fujita M, Harada H, Saiki H 1981b Lignification process of Japanese black pine (Pinus thunbergii Parl) tracheids. Mokuzai Gakkaishi 27:813–820

Takabe K, Fujita M, Harada H, Saiki H 1983 Changes in the composition and the absolute amount of sugars with the development of Cryptomeria tracheids. Mokuzai Gakkaishi 29:183–189

Takabe K, Fujita M, Harada H, Saiki H 1984 Incorporation of the label from ^{14}C-glucose into cell-wall components during the maturation of Cryptomeria tracheids. Mokuzai Gakkaishi 30:103–109

Thomas RJ 1975 The effect of polyphenol extraction on enzyme degradation of bordered pit tori. Wood Fiber 7:207–215

Timell TE 1972a Beobachtungen an Holzstrahlen im Druckholz. Holz Roh-Werkst 30:267–273

Timell TE 1972b Nature of the last-formed tracheids in compression wood. IAWA Bull 1972(4):10–17

Timell TE 1973 Ultrastructure of the dormant and active cambial zone and the dormant phloem associated with formation of normal and compression woods in Picea abies (L) Karst. SUNY Coll Environ Sci For, Syracuse, Tech Publ 96, 94 pp

Timell TE 1978a Helical thickenings and helical cavities in normal and compression woods of Taxus baccata. Wood Sci Technol 12:1–15

Timell TE 1978b Ultrastructure of compression wood in Ginkgo biloba. Wood Sci Technol 12:89−103

Timell TE 1979a Formation of compression wood in balsam fir (Abies balsamea). I. Ultrastructure of the active cambial zone and its enlarging derivatives. Holzforschung 33:137−143

Timell TE 1979b Formation of compression wood in balsam fir (Abies balsamea). II. Ultrastructure of the differentiating xylem. Holzforschung 33:181−191

Timell TE 1980a Formation of compression wood in balsam fir (Abies balsamea). III. Ultrastructure of the differentiating phloem. Holzforschung 34:5−10

Timell TE 1980b Organization and ultrastructure of the dormant cambial zone in compression wood of Picea abies. Wood Sci Technol 14:161−179

Timell TE 1980c Karl Gustav Sanio and the first scientific description of compression wood. IAWA Bull (ns) 1:147−153

Tsuda M 1975a The ultrastructure of the vascular cambium and its derivatives in coniferous species. I. Cambial cells. Bull Tokyo Univ For 67:158−226

Tsuda M 1975b The ultrastructure of the vascular cambium and its derivatives in coniferous species. II. Xylem cells and phloem parenchyma cells. Bull Tokyo Univ For 68:25−89

Tsuda M, Shimaji K 1971 Seasonal changes of cambial activity and starch content in Pinus densiflora Sieb et Zucc. J Jap For Soc 53:103−107

Wardrop AB 1957 The phase of lignification in the differentiation of wood fibers. Tappi 40:225−243

Wardrop AB 1964 The structure and formation of the cell wall in xylem. In: Zimmermann MH (ed) The formation of wood in forest trees. Academic Press, New York, 87−134

Wardrop AB 1965 Cellular differentiation in xylem. In: Côté WA, Jr (ed) Cellular ultrastructure of woody plants. Syracuse Univ Press, Syracuse, 61−97

Wardrop AB 1971 Occurrence and formation in plants. In: Sarkanen KV, Ludwig CH (eds) Lignins. Occurrence, formation, structure and reactions. Wiley-Interscience, New York, 19−41

Wardrop AB 1976 Lignification of the plant cell wall. Appl Polym Symp 28:1041−1063

Wardrop AB, Bland DE 1959 The process of lignification in woody plants. In: Kratzl K, Billek G (eds) Biochemistry of wood. Pergamon, London, 92−116

Wardrop AB, Bland DE 1961 Lignification in reaction wood. Abstr Pap 140th Meet Am Chem Soc, Chicago IL, 5E

Wardrop AB, Dadswell HE 1952 The nature of reaction wood. III. Cell division and cell wall formation in conifer stems. Aust J Sci Res B-5:385−398

Wardrop AB, Dadswell HE 1957 Variations in the cell wall organization of tracheids and fibres. Holzforschung 11:33−41

Wardrop AB, Davies GW 1964 The nature of reaction wood. VIII. The structure and differentiation of compression wood. Aust J Bot 12:24−38

Waterkeyn L, Caeymaex C, Decamps E 1982 La callose des trachéides du bois de compression chez Pinus silvestris et Larix decidua. Bull Soc R Bot Belg 115:149−155

Wattendorf J 1968 Entstehung der Calciumoxalat-Kristalle im Cytoplasma junger Rindenzellen der Lärche (Larix decidua Mill). Naturwissenschaften 55:186

Wattendorf J 1969a Feinbau und Entwicklung der Calciumoxalat-Kristallzellen in der Rinde von Larix decidua Mill. Ber Deutsch Bot Ges 81:343−346

Wattendorf J 1969b Feinbau und Entwicklung der verkorkten Calciumoxalat-Kristallzellen in der Rinde von Larix decidua Mill. Z Pflanzenphysiol 60:307−347

Wattendorf J, Schmid H 1973 Prüfung und perjodatreaktive Feinstrukturen in den suberinisierten Kristallzell-Wänden der Rinde von Larix und Picea. Z Pflanzenphysiol 68:422−431

Westing AH 1968 Formation and structure of compression wood in gymnosperms II. Bot Rev 34:51−78

Whiting P, Goring DAI 1982 Chemical characterization of tissue fractions from the middle lamella and secondary wall of black spruce tracheids. Wood Sci Technol 16:261−267

Whitmore FW, Zahner R 1966 Development of the xylem ring in stems of young pine trees. For Sci 12:198−210

Wilson BF 1964 A model for cell production by the cambium of conifers. In: Zimmermann MH (ed) The formation of wood in forest trees, Academic Press, New York London, 19–36

Wilson BF, Wodzicki TJ, Zahner R 1966 Differentiation of cambium derivatives: proposed terminology. For Sci 12:438–440

Wodzicki TJ 1971 Mechanism of xylem differentiation in Pinus silvestris L. J Exp Bot 22:670–687

Wodzicki TJ, Brown CL 1973 Cellular differentiation of the cambium in the Pinaceae. Bot Gaz 134:139–146

Wodzicki TJ, Peda T 1963 Investigation on the annual ring of wood formation in European silver fir (Abies pectinata D C). Acta Soc Bot Polon 32:609–618

Wood JR, Goring DAI 1971 The distribution of lignin in stem wood and branch wood of Douglas fir. Pulp Pap Mag Can 72(3):T95–T102

Wooding FBP, Northcote DH 1964 The development of the secondary wall of the xylem in Acer pseudoplatanus. J Cell Biol 23:327–337

Worrall J 1971 Absence of "rest" in the cambium of Douglas-fir. Can J For Res 1:84–89

Yamamoto K, Fukazawa K, Ishida S 1977 Study of the cell-wall development of ray parenchyma in genus Pinus using ultraviolet microscopy. Res Bull Coll Exp For Hokkaido Univ 34:79–96

Yoshizawa N, Idei T, Okamoto K 1981 Structure of inclined grown Japanese black pine (Pinus thunbergii Parl). 1. Distribution of compression wood and cell wall structure of tracheids. Bull Utsunomiya Univ For 17:89–105

Yoshizawa N, Itho T, Shimaji K 1982 Variation in features of compression wood among gymnosperms. Bull Utsunomiya Univ For 18:45–64

Yoshizawa N, Koike S, Idei T 1984 Structural changes of tracheid wall accompanied by compression wood formation in Taxus cuspidata and Torreya nucifera. Bull Utsunomiya Univ For 20:59–76

Yoshizawa N, Koike S, Idei T 1985 Formation and structure of compression-wood tracheids induced by repeated inclination in Taxus cuspidata. Mokuzai Gakkaishi 31:325–333

Yumoto M, Ishida S 1979 (Growth and compression wood formation in Picea glauca bent to several inclinations.) Proc Hokkaido Branch Jap Wood Res Soc 11:29–34

Yumoto M, Ishida S 1982 Studies on the formation and structure of the compression wood cells induced by artificial inclination in young trees of Picea glauca. III. Light microscopic observation on the compression wood cells formed under five different angular displacements. J Fac Agr Hokkaido Univ 60:337–351

Yumoto M, Ishida S, Fukazawa K 1982a Studies on the formation and structure of the compression wood cells induced by artificial inclination in young trees of Picea glauca. I. Time course of the compression wood formation following inclination. Res Bull Coll Exp For Hokkaido Univ 39:137–162

Yumoto M, Ishida S, Fukazawa K 1982b Studies on the formation and structure of the compression wood cells induced by artificial inclination in young trees of Picea glauca. II. Transition from normal to compression wood revealed by a SEM-UVM combination method. J Fac Agr Hokkaido Univ 60:312–335

Yumoto M, Ishida S, Fukazawa K 1983 Studies on the formation and structure of the compression wood cells induced by artificial inclination in young trees of Picea glauca. IV. Gradation of the severity of compression wood tracheids. Res Bull Coll Exp For Hokkaido Univ 40:409–454